Statistical Thermodynamics for Pure and Applied Sciences

Frederick Richard Wayne McCourt

Statistical Thermodynamics for Pure and Applied Sciences

Statistical Thermodynamics

Frederick Richard Wayne McCourt
University of Waterloo
Waterloo, ON, Canada

This work contains media enhancements, which are displayed with a "play" icon. Material in the print book can be viewed on a mobile device by downloading the Springer Nature "More Media" app available in the major app stores. The media enhancements in the online version of the work can be accessed directly by authorized user.

Additional material to this book can be downloaded from http://extras.springer.com.

ISBN 978-3-030-52008-3   ISBN 978-3-030-52006-9   (eBook)
https://doi.org/10.1007/978-3-030-52006-9

This Springer imprint is published by the registered company Springer Nature Switzerland AG
The registered company address is: Gewerbestrasse 11, 6330 Cham, Switzerland

Preface

The concepts of thermodynamics and statistical mechanics are fundamental physical concepts that underlie much of chemical engineering, chemistry, materials science, and physics. Thermodynamics treats matter as a continuum and is therefore not concerned with the atomic or molecular nature of its constituents, while statistical mechanics is concerned with the effective ways to deal with very large numbers of constituents (atoms or molecules) using statistical concepts, such as probability, averages, and fluctuations, coupled with the fundamental laws of physics that govern individual atoms and molecules and their interactions at the microscopic level. Statistical thermodynamics combines these two subdisciplines of science in order to be able to understand and predict macroscopic properties of specific material systems in terms of the attributes of their microscopic constituents.

While the present book does focus mainly upon the employment of statistical mechanical concepts and methodology to evaluate and interpret thermodynamic properties in terms of atomic and/or molecular attributes of the bulk matter constituents, it also examines a number of other topics, such as symmetry effects in molecular spectra (in Chap. 6), an introduction to transport phenomena (in Chap. 7), paramagnetism, magnetic susceptibilities of lanthanides, and an introduction to ferromagnetism (in Chap. 8), and electrons in metals and semiconductors, and the Bose–Einstein condensation (in Chap. 10).

The structure of this book is as follows:

Basic background material, including equations of state for classical and quantum ideal gases and concepts derived from mathematical statistics, such as ensembles, fluctuations, statistical average, variance, and standard deviation, is reviewed in Chap. 1.

Chapter 2 provides a short treatment of macroscopic thermodynamics. For those who have taken a previous course on thermodynamics, it may serve as a reminder of those aspects of thermodynamics (including a discussion of thermodynamic engine cycles that may be more relevant for engineers) that will be needed in later chapters. For those who have not had previous exposure to thermodynamics, it will serve to introduce the relevant thermodynamic concepts.

Chapter 3 introduces and illustrates the main statistical ensembles that are employed in treating thermodynamic systems, while Chap. 4 makes the relevant general connections between statistical values and a number of thermodynamic functions and concepts.

Chapter 5 focuses upon atomic systems and examines the contributions to thermodynamic functions and properties associated with the various degrees of freedom possessed by individual atoms. This chapter also considers an ensemble of simple harmonic oscillators and its application to obtain expressions for the heat capacities of crystalline monatomic solids.

The focus of Chap. 6 is upon molecular systems, and it examines the contributions to thermodynamic functions and properties associated with the various degrees of freedom possessed by individual molecules. This chapter also includes a discussion of intensity alternations in molecular spectra that arise because of nuclear interchange symmetry in highly symmetric molecules.

Chapter 7 covers the employment of classical statistical mechanics to extend the thermodynamics of gaseous systems beyond the ideal gas limit using the virial equation of state and includes discussion of the Liouville and Boltzmann equations that provide a lead-in to nonequilibrium phenomena.

Chapter 8 extends thermodynamics to include electric and magnetic effects in polarizable media. Thorough treatments are given of the thermodynamics of spin systems and magnetic susceptibilities of paramagnetic salts. An introduction to ferromagnetism and the Ising model is also provided.

Chapter 9 briefly discusses chemical equilibrium and the activated complex model for chemical kinetics.

Chapter 10 deals with the quantum statistics associated with the behaviours of systems of fermions (with half-odd-integer spins) and bosons (with zero and integer spins). Specifically, quantum statistics is applied to fermion systems, such as the ^{3}He isotope and electrons in metals or semiconductors, and to boson systems, such as the ^{4}He isotope and photons. The chapter concludes with a brief introduction to the density matrix.

Seven appendices covering the aspects of combinatorial analysis, multivariate calculus and infinite series, the Stirling approximation, atomic and molecular term symbols, spherical and symmetric top molecules, and reviews of relevant solid state and Hamiltonian mechanics that may prove useful to readers have been developed for completeness.

Although this book deals only with a small fraction of equilibrium statistical mechanics, its aim has been to provide a preparation that will suffice for further adventures into this exciting realm of science. The material covered here is fairly traditional, and bits of it can be found in many texts written around the subject. Two aspects of the subject matter covered in this introduction to statistical thermodynamics are (1) to provide 'derivations' of the laws of thermodynamics and (2) to provide an introduction to the important areas of Fermi–Dirac and Bose–Einstein statistics that necessarily play roles at the level of nanoscale systems, where quantum mechanics reigns.

I am grateful to Dr. Kevin Bishop for carrying out at my request the calculations for and preparations of Figs. 1.4, 1.5, and 7.2 for Lennard–Jones argon and to Dr. Lee Huntington for preparing Figs. 10.2 and 10.11. I am also grateful to my colleague, Professor Pierre-Nicholas Roy, for his help and encouragement, as well as his testing of previous versions of this book in his statistical thermodynamics course for nanotechnology engineering students. Finally, I am most grateful to my wife, Janet L. McCourt, for her steadfast encouragement and support throughout the years during which this book was developed.

Waterloo, ON, Canada   Frederick R. W. McCourt

Contents

Chapter 1
Basic Background Material

This chapter introduces a number of basic terms and concepts that are fundamental to a statistical mechanical description of the properties of bulk matter. It begins with some basic definitions, followed by brief considerations of the equation of state for ideal gases whose constituents are either classical (typified by hard spheres) or quantum mechanical (typified by photons) in nature. A number of statistical concepts, such as fluctuation, ensemble, ensemble average, variance, and standard deviation, are defined and illustrated.

1.1 Introduction

There are two general approaches to the description of physical systems. The *macroscopic* approach in many ways represents the world as we see it, with matter considered in 'large' units, while the *microscopic* approach looks at a physical system in terms of the individual constituent atoms and molecules, i.e., in units of matter that are considered 'small'. We shall come to appreciate what is meant by 'large' and 'small' as we proceed. More specifically, our employment of these terms is outlined below.

Macroscopic A physical system is described macroscopically in terms of a set of appropriate variables, many of which will already be familiar. We tend to separate these variables into two distinct classes, referred to as 'mechanical variables', exemplified by volume, V; number, N; pressure, P; internal energy, U; enthalpy, H; rate constant, k; surface tension, γ; and 'non-mechanical variables', exemplified by temperature, T; entropy, S; Helmholtz energy, A; and Gibbs energy, G.

Microscopic As we have earlier indicated, a physical system may be described in terms of its component atoms and/or molecules, so that its behaviour is examined on a fine scale, for which we have developed two means of description. These are:

© Springer Nature Switzerland AG 2021
F. R. W. McCourt, *Statistical Thermodynamics for Pure and Applied Sciences*,
https://doi.org/10.1007/978-3-030-52006-9_1

(a) *Classical mechanics*. If we are to use classical mechanics, we must assume that we are able to state in some manner the values of the position coordinates x, y, z and momentum components p_x, p_y, p_z, plus any other relevant dynamical variables for each atom or molecule in the physical system. Each such set of values then constitutes what we shall refer to as a classical *microstate* of the system under consideration.

(b) *Quantum mechanics*. If we are to use quantum mechanics, we must be able to specify the wave (or state) function $\psi_\alpha(x, y, z, \ldots)$ for each atom or molecule in the system, including the attendant quantum numbers, which we have represented collectively by the subscript α on ψ. Each such set of wave functions we shall refer to as a quantum *microstate* of the system under consideration.

Of course it is appealing to attempt to explain or interpret macroscopic behaviour in terms of laws governing the individual microscopic constituents of the system, provided that we are able in some way to work out the detailed behaviour of so many particles. For a typical macroscopic amount, which may consist of a mole of particles, the fundamental question is, 'How do we deal with 10^{23} or so particles in interaction?'. Indeed, the problem is intractable from such a point of view, and so we are in need of a new branch of physics. We shall still require the laws of mechanics (quantum or classical) and electromagnetism, since they are concerned with motions and properties of atoms and molecules and with radiation. However, when we relinquish the idea of arriving at a detailed description of the particle motions, we need to invent new concepts that are capable of answering the types of questions that we wish to ask, questions such as:

(1) What is entropy, and why does it always increase in spontaneous processes?
(2) What is the absolute entropy of a given chemical system?
(3) Can the equilibrium constant for a chemical reaction be predicted from a knowledge of the properties of the individual reactants and products?
(4) How do biological systems work?
(5) How can we understand liquefaction?

To answer such questions as these, the concepts that we must invent should be extremely powerful. In all likelihood, such powerful concepts will pose a certain intellectual challenge. Fortunately, however, we shall not require much advanced mathematics! Indeed, much of the developmental work was carried out more than a century ago by the founding fathers of the subject, James Clerk Maxwell, Ludwig Boltzmann, and Josiah Willard Gibbs.

The objective of our study is then to describe the properties of matter—gases, liquids, solids, plasmas, and polymers—from a microscopic point of view, i.e., by maximizing the use of the laws of mechanics and electromagnetism. We should compare this aim with that of thermodynamics and fluid mechanics, in which matter is described solely in macroscopic terms (without invoking any knowledge of its molecular structure). In general, such a goal would involve the entire field of *Equilibrium and Nonequilibrium Statistical Mechanics*, and this would clearly be a truly daunting task. We shall therefore reduce our goal to that of learning about and understanding mainly the subfield of *Statistical Thermodynamics*, which has the derivation of the laws of thermodynamics from fundamental principles as one of its main objectives. To see what it is that we are trying to accomplish, we shall first consider a simple example that illustrates both the distinction between the microscopic and macroscopic descriptions of matter and the procedures that we wish to follow in the remainder of this book.

1.2   The Ideal Gas

The concept of an ideal gas is, in principle, an abstraction. It is traditionally represented in terms of a collection of N noninteracting point particles confined to a vessel of volume V, and in thermal equilibrium with the vessel walls (or 'surroundings'). Thermal equilibrium is characterized by a common temperature T. The bulk gas behaviour is then represented by an 'equation of state' that may also be termed an 'ideal gas law' that relates the pressure, P, that the ideal gas particles exert on the vessel walls to the number of particles (or the number of moles), the gas volume, and the temperature of the system. We shall consider two models for the ideal gas, one classical and the other quantum mechanical.

The most commonly encountered ideal gas is that of noninteracting point particles that obey the laws of classical mechanics; we shall see that this model gives rise to the canonical ideal gas law,

$$PV = Nk_BT ,$$

in which $k_B \equiv 1.380649 \times 10^{-23}\,\mathrm{J\,K^{-1}}$ is the Boltzmann constant. Note that this traditional representation of ideal gas particles ignores interparticle and/or particle–wall interactions that must exist in order for the gas to achieve thermal equilibrium with its surroundings, minimally via energy exchange between gas particles and the container walls. This ideal gas model provides an increasingly accurate description of gas behaviour as its (number) density $n \equiv N/V$ decreases. Indeed, this idealized equation of state has become so ubiquitous that it is commonly referred to as *the* ideal gas equation of state. Let us note that the classical ideal gas is described in terms of three independent variables, namely, pressure P, volume V, and temperature T.

Photons (which are quanta of electromagnetic field energy) can be considered to be ideal quantum mechanical particles, if for no other reason than that they cannot be described in terms of classical mechanics. As they have rest mass zero and travel at the speed of light, they are also inherently relativistic. A thought experiment along lines proposed by Leff [1] will allow us a means for envisaging the construction of a photon gas from energy stored in the container walls. Consider a cylindrical container closed at one end and fitted with a frictionless piston, with the entire assembly surrounded by an ultrahigh vacuum chamber and maintained

at temperature T, and with the piston face initially in contact with the back wall
of the cylinder. We shall also stipulate that all interior walls of the container be
ultrasmooth. This initial configuration thus represents the zero-volume state for our
photon gas, with $N = 0$ photons present. Upon moving the piston very slowly away
from the end wall of the cylinder, the container walls (including the piston face)
will emit photons into the space created by the receding piston. Some of these
photons will scatter elastically off the ultrasmooth walls of the container, some
will be absorbed by the walls, and new photons will be emitted by the walls. A
dynamic equilibrium will be established once the average absorption and emission
rates balance.[1]This construct gives us a fully quantum mechanical photon gas having
volume V and temperature T as independent variables. As photons can be emitted
and absorbed, the number of photons is not an independent variable but depends
upon both V and T, i.e., $N \equiv N(T, V)$.

1.2.1 Classical Ideal Gas Equation of State: Microscopic Derivation

As we are all aware, the ideal gas law relates the pressure, volume, and temperature
of an ideal gas (a gas in which interparticle interactions can be ignored). You will
also remember that this law is utilized to calculate one of the three variables in terms
of known values of the other two. Perhaps the most difficult of the three variables
to interpret is the *pressure* of a gas. At a qualitative level, we can view it either
macroscopically as a force per unit area, or microscopically as due to myriads of
molecules impinging upon a surface, thereby causing both the molecules and the
surface to recoil.

Let us introduce a *model* for explaining gas pressure. We shall attempt to carry
out a quantitative description of gas pressure based upon a model (which is idealized
in order to dispense with irrelevant details and to emphasize the essential elements).
Incidentally, this is typical of scientific procedure as a first step in the study of a
phenomenon. For our model, let us consider a frictionless piston at one end of a
rigid cylinder (see Fig. 1.1). Let the area of the piston face be $\mathcal{A}$. We shall fill the
piston between the rigid end wall and the piston with atoms of some gas. Let us
represent these atoms as rigid (hard) spheres in constant motion. The atoms, in
moving around, will collide with one another, the walls of the chamber, and the
piston face. If we have a vacuum to the right of the piston (see Fig. 1.1), then the
piston will be pushed to the right, and to maintain the piston at position x, a force
F will have to be applied from the right. The relevant question at this point is 'How
much force?'. Let us *define* the pressure in terms of force and area as

---

[1]In this sense, it would seem that Aristotle may inadvertently have been correct with his postulate
'Horror vacui' (commonly expressed as 'Nature abhors a vacuum').

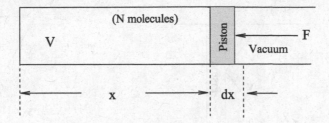

Fig. 1.1 Cylinder-piston schematic for 1-dimensional motion

$$P = \frac{F}{A}. \tag{1.2.1}$$

Now we shall make use of the fundamental microscopic law governing the motion of the atoms, i.e., Newton's Second Law, which states that the force on a particle equals its rate of change of momentum, or in symbols,

$$F = ma = \frac{\mathrm{d}p}{\mathrm{d}t}. \tag{1.2.2}$$

This tells us that to find the force on the piston face we must calculate the net momentum change per second given to it by the impinging atoms, i.e., we must find the momentum accumulated by the piston through gas atom bombardment. We shall carry this task out as a two-step process:

1. calculate the momentum delivered to the piston by a single collision;
2. calculate the number of collisions/second.

The product of these two quantities will give us the total momentum per second received by the piston. For step 1, we must remember that we are dealing with an 'ideal piston', which is thus a 'perfect molecular reflector' (in other words, we speak of elastic collisions only occurring), so that no energy is received by the piston, only momentum. From Fig. 1.2, if **v** is the velocity of an atom of mass m before a collision with the piston face, and $\mathbf{v}'$ is the velocity following such a collision, then the change of momentum $\Delta(mv_x)$ is given by $2mv_x$. This completes the first step of our calculation. For step 2, we recognize that we have N atoms in a volume V, so that the quantity

$$n = \frac{N}{V} \tag{1.2.3}$$

represents the number of molecules per unit volume (it is referred to as the *number density*). Now consider an atom moving along the x-direction with velocity v_x. In time Δt, it will hit the piston face, provided that it is not further away than the distance $v_x \Delta t$; this holds for all molecules, and hence the volume occupied by the atoms that are going to hit the piston face in time Δt is $v_x \Delta t \, A$. The number of

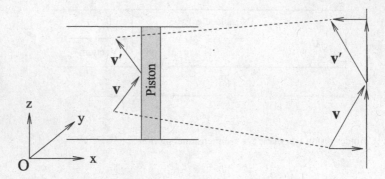

Fig. 1.2 Microscopic dynamics at the piston face

molecules that are going to hit the piston face is thus $nv_x \Delta t \mathcal{A}$, so that the number of hits per second is $nv_x \mathcal{A}$. Hence, the force exerted on the piston is given by

$$F = (nv_x \mathcal{A}) \times (2mv_x),$$

and the pressure is thus given by

$$P = \frac{F}{\mathcal{A}} = 2nmv_x^2.$$

To arrive at a feeling for the maximum total momentum transferrable to a surface by molecules impinging upon it, consider one mole of N_2 molecules all travelling at a thermal speed $v_{th} \simeq (2k_B T/m)^{\frac{1}{2}}$, with the N_2 molecules delivering momentum $p_{th} = \sqrt{2mk_B T}$ (cf. $p = \sqrt{2mE}$) to the surface upon impinging upon it. The thermal speed for a N_2 molecule is given by

$$v_{th}(N_2) \simeq \sqrt{\frac{2 * 1.38 \times 10^{-23} \times 300}{1.0623 \times 28 \times 10^{-27}}} \approx 420\, \mathrm{m\,s^{-1}}.$$

As $1\,\mathrm{m\,s^{-1}}$ is equivalent to $0.36\,\mathrm{km\,hr^{-1}}$, we see that a N_2 molecule will be travelling at approximately $151\,\mathrm{km\,hr^{-1}}$. Each molecule carries a momentum of

$$p_{th}(N_2) \equiv m_{N_2} v_{th}(N_2) \approx 196.2 \times 10^{-25}\,\mathrm{kg\,m\,s^{-1}},$$

so that N_0 nitrogen molecules all striking the surface at once will bring a total momentum of $N_0 p_{th}(N_2) = 6.023 \times 10^{23} \times 196.2 \times 10^{-25}\,\mathrm{kg\,m\,s^{-1}} = 11.8\,\mathrm{kg\,m\,s^{-1}}$. Avogadro's number of N_2 molecules striking a surface simultaneously at $v_{th} \simeq 151\,\mathrm{km\,hr^{-1}}$ is thus equivalent to a one kilogram object travelling at $11.8\,\mathrm{m\,s^{-1}}$.

However, we are not yet done, since we still need some refinements. Let us now examine them. If we wished to apply our expression obtained above directly to calculate the pressure for the gas as a whole, we would have to assume that all the atoms in the cylinder between the piston face and the end wall travel with the same x-component of velocity and that they all move in the same direction. This is, of course, unrealistic, since neither assumption would hold, as we well know. How do we correct for such assumptions? In the first place, since all the atoms have v_x^2-values that are different, we must 'take an average'; we shall denote this average by $\langle v_x^2 \rangle$. Further, we must remember that on average, as many atoms will have x-components of velocity that could be designated by $-v_x$, and that since $(-v_x)^2 = v_x^2$, when we replace v_x^2 by its average, we obtain a value that is twice as large as we should have, so that the pressure should be given now by

$$P = nm\langle v_x^2 \rangle .$$

Thus far, we have only considered motion along the x-axis, but as far as atoms are concerned, there is nothing special about the x-direction: it will thus be true that $\langle v_x^2 \rangle = \langle v_y^2 \rangle = \langle v_z^2 \rangle$. Now, if this is true, we can employ the trick

$$\langle v_x^2 \rangle = \tfrac{1}{3}\langle v_x^2 + v_y^2 + v_z^2 \rangle = \tfrac{1}{3}\langle v^2 \rangle$$

and obtain for the pressure the relation

$$P = nm\tfrac{1}{3}\langle v^2 \rangle = \tfrac{2}{3}n\left(\tfrac{1}{2}mv^2\right) . \tag{1.2.4}$$

Because $n = N/V$, an equivalent form for this equation is

$$PV = \tfrac{2}{3}N\left(\tfrac{1}{2}mv^2\right) ,$$

in which $N\langle \tfrac{1}{2}mv^2 \rangle$ represents the total kinetic energy of the atoms in the gas (for hard spheres, this is also the total energy). Let us call the total energy U, so that we can also write[2]

$$PV = \tfrac{2}{3}U . \tag{1.2.5}$$

We might now ask, 'What about the factor $\tfrac{2}{3}$?'. Is it always that, independent of the gas in the cylinder? In fact, it is found to vary with the kind of gas that one is using in the cylinder, so that to be general enough, we should allow for this factor to vary.

---

[2]In thermodynamics, one refers to U as the *internal* energy, meaning the energy contained within the thermodynamic system at rest, in contrast to the *external* energy associated, for example, with the kinetic energy of translation and rotation of a macroscopic system as a whole. The thermodynamic internal energy thus consists of the total kinetic and potential energies of the constituent particles comprising the macroscopic system.

To conform with tradition, we shall write this factor in terms of the heat capacity ratio γ as $\gamma - 1$, to obtain the expression

$$PV = (\gamma - 1)U.$$ (1.2.6)

For a monatomic gas (which is relatively well represented by our hard-sphere model), $\gamma \equiv C_P/C_V$, with C_P and C_V, the heat capacities at constant pressure P and volume V (defined formally in Sects. 2.2 and 2.3), has the value 5/3. The differential work, dW, done on the gas in compressing it by moving the piston by $-dx$ is

$$dW = F(-dx) = -P\mathcal{A}dx = -P\,dV.$$

Now assume that the compression is adiabatic, i.e., there is no energy added or removed during the process: this implies that work can be converted into internal energy, so that $P\,dV = -dU$. Thus, if we take the differential of expression (1.2.6), assuming that γ is a constant, we obtain

$$P\,dV + V\,dp = (\gamma - 1)\,dU.$$

If we now combine these last two results, we obtain the differential equation

$$\gamma\frac{dV}{V} + \frac{dP}{P} = 0,$$

which has the solution

$$\gamma \ln V + \ln P = \ln C,$$

or

$$PV^\gamma = C.$$ (1.2.7)

Under adiabatic conditions, when the compression of a gas can only increase its temperature, because the work done cannot go anywhere else but into internal energy, we thus predict that the product of pressure and volume to the power γ is a constant. For a monatomic gas (such as He, Ar, etc.), $PV^{5/3} = \text{const.}$, and this works. Of course, you may already have seen this result previously as an application of the First Law of Thermodynamics: however, you should note that we have now obtained this same result from basic microscopic considerations and from Newton's laws of motion.

Our simple model gives Eq. (1.2.5) as the ideal gas equation of state. This is also one of the forms obtained in thermodynamics. For a gas, the thermodynamic internal energy U is the sum of all kinetic and potential energies for the gas particles that, in the present case, are hard-spheres, which possess no energies other than translational. Moreover, because we are considering only an ideal gas, we have just the hard-sphere interaction potential between pairs of gas particles. We should allow for this by replacing U by U_{tr}, the internal energy associated with the translational motion of the atoms/molecules of the gas. Thus, our equation of state should actually read

$$PV = \tfrac{2}{3}U_{tr}. \tag{1.2.8}$$

We shall later (in Chap. 4) see that U_{tr} for an ideal gas depends only upon the equilibrium temperature T of the gas and is given by

$$U_{tr}(T) = \tfrac{3}{2}Nk_BT = \tfrac{3}{2}\left(\frac{N}{N_0}\right)(N_0k_B)T, \tag{1.2.9a}$$

or

$$U_{tr}(T) = \tfrac{3}{2}nRT, \tag{1.2.9b}$$

in which n here represents the *number of moles* of gas, and $R \equiv N_0k_B$ is the universal gas constant, with N_0 the Avogadro number and k_B the Boltzmann constant. Upon combining this result with Eq. (1.2.5), we see that we have, indeed, obtained a version of the well-known ideal gas equation of state $PV = nRT$ from our model.

1.2.2 Quantum Ideal Gas Equation of State

A microscopic derivation of the equation of state relevant to a gas consisting of fully quantum mechanical particles, as exemplified by a photon gas [1], has both similarities to and differences from the corresponding derivation of the equation of state for a gas consisting of typical classical gas particles. For example, the assumption of isotropy of space is common to both derivations. For a photon gas, however, we shall need to assume that there is a well-defined average energy density $\langle h\nu\rangle(T)$ that depends solely upon the temperature T of the walls of the container and that the number of photons emitted into an empty chamber is directly proportional to its volume.[3]

An argument showing that the energy spectrum of isotropic radiation in equilibrium with matter at temperature T, also referred to as 'blackbody radiation', is independent of all other factors has been given by Landsberg [2], who has also shown that the addition of a small quantity of matter, capable of absorbing and

---

[3]Strictly speaking, the photon gas is an example of an ideal gas representing what we shall later refer to as an *open* thermodynamic system.

emitting photons at all frequencies, to radiation in an enclosure with perfectly reflecting walls will essentially catalyse thermal equilibration between the radiation and the walls of the enclosure. This mechanism thus allows arbitrary radiation to be converted into isotropic radiation in thermodynamic equilibrium at temperature T.

If we designate the number of photons per unit volume having velocity components c_x lying in the velocity range c_x to $c_x + dc_x$ by $n(c_x, T)dc_x$, then the isotropy of space implies that the average number of photons lying within the specified velocity range and moving to the right is equal to the average number of photons $n(-c_x, T)dc_x$ moving to the left. We note that photons, unlike classical particles, all have fixed speed c, so that c_x varies only because the direction of motion changes. Moreover, as for classical particles, no special significance can be attached to any of the three Cartesian axis directions. The photon number density, $n(T) = N(T, V)/V$, is then given by

$$n(T) \equiv \int_{-c}^{c} n(c_x, T)dc_x = 2 \int_0^c n(c_x, T)dc_x .$$

Consider photons in a cylindrical container of length L, cross-sectional area $\mathcal{A}$, and volume $V = \mathcal{A}L$ (see Fig. 1.2) that are moving to the right with x-components of velocity c_x to $c_x + dc_x$ and located at a distance $c_x \Delta t$ from the end wall. In a time interval $\Delta t \ll L/c$, the average number of photons within the volume element $\mathcal{A}c_x \Delta t$ that will collide with the end wall is given by $(\mathcal{A}c_x \Delta t)n(c_x, T)dc_x$. The x-component of photon momentum will be $p_x = pc_x/c$, so that the momentum change arising from an elastic photon–wall collision will be $2p_xc_x/c$, thereby giving an average force $2p_xc_x/(c\Delta t)$ exerted on the wall during time Δt.

If we allow for a distribution of photon frequencies ν, then thermodynamically, the average force exerted on the wall by collisions with photons having average energy $\langle h\nu \rangle$ and x-component of velocity lying between c_x and $c_x + dc_x$ will be given by

$$d\langle F \rangle = \frac{2\langle h\nu \rangle c_x}{c^2} \mathcal{A}c_x n(c_x, T)dc_x .$$

The pressure P associated with a photon gas is thus defined as

$$P = \frac{2\langle h\nu \rangle}{c^2} \int_0^c n(c_x, T)c_x^2 dc_x = \frac{N(T, V)\langle h\nu \rangle}{Vc^2} \overline{c_x^2},$$

which, upon taking into account that $\overline{c_x^2} = \overline{c_y^2} = \overline{c_z^2} = \frac{1}{3}c^2$, gives

$$P = \frac{1}{3} \frac{\langle h\nu \rangle N}{V}, \tag{1.2.10}$$

for the equation of state for a photon gas. Upon identifying $N\langle h\nu \rangle$ as the internal energy, $U(T, V)$, of the photon gas, we see that the pressure and the internal energy of the ideal quantum photon gas are related by

$$P = \frac{1}{3} \frac{U(T, V)}{V}.$$ (1.2.11)

We shall see in Chap. 10 that a photon gas has an internal energy $U(T, V)$ given by

$$U(T, V) = \frac{4\sigma V}{c} T^4,$$ (1.2.12)

in which σ is the Stefan-Boltzmann constant

$$\sigma \equiv \frac{2\pi^5 k_B^4}{15c^2 h^3} \simeq 5.67037 \times 10^{-8} \, \text{W} \, \text{K}^{-4} \, \text{m}^{-2}.$$

It will prove useful to write the internal energy expression (1.2.11) for $U(T, V)$ in terms of the internal energy density $u(T) \equiv U(T, V)/V$, i.e.,

$$U(T, V) = Vu(T),$$ (1.2.13)

with the internal energy density $u(T)$ for the photon gas given explicitly as

$$u(T) = \frac{4\sigma}{c} T^4.$$ (1.2.14)

Substitution of Eq. (1.2.12) into Eq. (1.2.11) gives the photon pressure P as

$$P(T) = \frac{4\sigma}{3c} T^4,$$ (1.2.15)

which may be compared with Eq. (1.2.5) for the classical ideal gas, namely,

$$PV = \tfrac{2}{3} U_{\text{trans}}(T) - Nk_B T.$$

1.3 Fluctuations

We begin by considering a dilute gas of N atoms, all separated by distances that are on average much larger than the average thermal de Broglie wavelength (more about this later). Further, we shall assume that the container plus atoms is completely isolated from the rest of the universe and that it has been there for an extremely long time. Following Reif [3], we shall attempt to examine what is happening in the container by examining *instantaneous snapshots* of the myriad of moving particles. For reference purposes, as illustrated in Fig. 1.3, we shall place an imaginary partition that separates the box into two equal-volume parts. For each snapshot,

Fig. 1.3 Isolated container for sharing by N atoms, with n atoms in the left half, and n' atoms in the right half, of the container

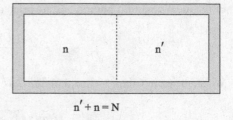

$$n' + n = N$$

we shall count the number of particles in each half of the container, giving numbers n, n'. Obviously, $n+n' = N$ serves as a *constraint* on our system. Moreover, as there is no special reason why we should find more particles in one half of the container than in the other half, we should *ordinarily* expect to find $n \simeq n'$.

If we have two halves of the container, and if the presence of any one particle does not affect the presence of any other particle in the container, then for N particles there will be 2^N configurations, and of these, only one will correspond to all N particles in a particular half of the container. Thus, the probability for finding all N particles in, say, the left half of the container is $1/2^N$. To make this argument, we shall draw upon the frequentist interpretation of probability, which states that the probability of occurrence of an event is the *relative frequency* with which that particular event occurs: for example, if we have two particles in the (equally divided) container, then we will have probabilities $p_0 = \frac{1}{4}$, $p_1 = \frac{1}{2}$, $p_2 = \frac{1}{4}$, while for four particles in the container, we will have probabilities $p_0 = \frac{1}{16}$, $p_1 = \frac{1}{4}$, $p_2 = \frac{3}{8}$, $p_3 = \frac{1}{4}$, $p_4 = \frac{1}{16}$.

In general, if $C(n)$ represents the number of possible ways of distributing atoms in the container so that n of them are found in a specified half of the container, then the probability for finding n atoms in that half is

$$p_n = \frac{C(n)}{2^N}. \tag{1.3.1}$$

We note that $n = 0$ or $n = N$ implies a single configuration, which in turn implies that $C(N) = C(0) = 1$, if N is large and if n is close to N or to 0, $C(n) \ll 2^N$, and $p_n \ll 1$. As such cases are rarely realized in nature, we refer to them as *non-random, orderly*, or *improbable*. $C(n)$ will be maximal when $n \simeq n' \simeq \frac{1}{2}N$, in which case p_n is larger, so that such configurations are more probable: we therefore refer to such situations as *random, disorderly*, or *probable*.

To see what actually happens, we shall consider a series of snapshots, shown in Figs. 1.4 and 1.5, that have been generated from computer calculations[4] by solving Newton's equations of motion for a given set of initial conditions.

Figure 1.4a shows the outcome from a computer-generated series of snapshots for a set of four argon atoms that interact via a Lennard-Jones (12,6) potential

---

[4]Figures 1.4 and 1.5 were inspired by figures appearing in chapters 1 and 2 of Ref. [3].

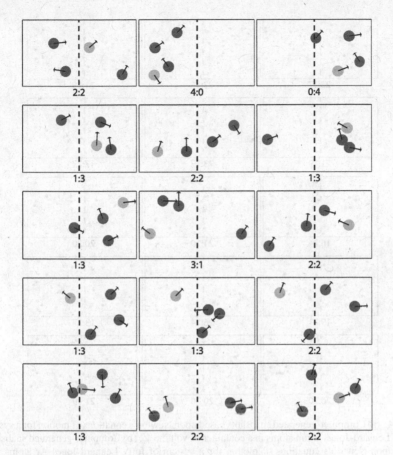

Fig. 1.4 (Continued)

energy function (PEF), starting with a set of assumed initial conditions, then evolving according to Newton's laws of motion. Each atom is represented by a filled circle: the arrows indicate the velocity directions for the Ar atoms at the time that the snapshot was generated. A Lennard-Jones (12,6) PEF has the form $V_{LJ}(R) = \mathcal{D}_e[(R_e/R)^{12} - (R_e/R)^6]$, with R the separation between a pair of interacting atoms, while R_e and $-\mathcal{D}_e$ represent, respectively, the position and depth of the minimum in the interaction energy. This set of snapshots corresponds to a time sufficiently long after the start of the simulation for the 4-atom system to have attained an essentially random, that is, no longer time-dependent, distribution of the atoms. We say that a system that satisfies this condition has attained equilibrium. For such a very small number of atoms, however, we should perhaps not be too surprised to observe that all four atoms are in one half (either left or right) in approximately 2 out of the 15 frames, rather than the 2 out of 16 frames that would correspond to our previous discussion. Consistent with this behaviour, we see that there are also 7 out of 15 frames in which one particle is in one half, three in the other half (differing

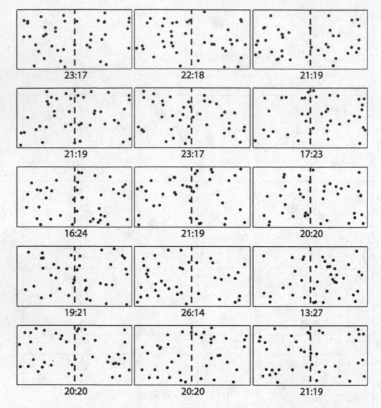

Fig. 1.4 (a) Computer-generated snapshots based upon Newton's equations of motion for a system of four Lennard-Jones argon atoms in a container of volume V. (b) Computer-generated snapshots based upon Newton's equations of motion for a system of forty Lennard-Jones Ar atoms in a container of volume V [this figure has been generated and provided by Dr. Kevin Bishop]

from 8 out of 16), and 5 out of 15 frames (rather than 6 out of 16) in which two particles are in each half. We should also note that the distribution of partitions of the four atoms lacks the symmetry between the 1:3 and 3:1 distributions that we might have anticipated. We should recall, however, that probability theory and most statistical predictions are only valid for large numbers of observations on, or large numbers of participants in, a process.

If we now increase the number of Ar atoms involved by a factor of 10 and examine a similar set of fifteen computer-generated snapshots for 40 atoms, as shown in Fig. 1.4b, we see that the numbers still vary from frame to frame, though not as much as they did for the 4-atom case. Indeed, although the 20:20 partition (split) occurs in 4 out of the 15 frames, the deviations from a 20:20 partition of the atoms between the two halves in all but one case do not exceed 4. This result strongly suggests that the distribution of partitions has become quite sharp even for the small number of atoms being considered. These variations in the distribution of partitions of the N particles into the two equal-volume halves of the container represent what we shall refer to as *local fluctuations* in the number of particles. Such fluctuations occur even though the system under study has attained equilibrium.

1.4   Statistical Ensembles

We wish to employ probability arguments in order to overcome the necessity of following very large numbers of particles in interaction over very long periods of time. We may only use probabilities when the number of experiments or *trials* is extremely large. A nice trick invented by Gibbs to make it easier for us to do this is, rather than fixing our attention on the actual system A, to imagine $\mathcal{N}$ mental copies of it, with $\mathcal{N}$ a very large number. The technical name for a large set of objects under the same conditions is *ensemble*. This procedure has two distinct advantages:

1. There is no worry about establishing identical initial conditions each time a real experiment is to be performed on A.
2. We can imagine the preparation of the members of the ensemble in the same condition and then imagine an infinite number of hands performing at the same time and in the same fashion, the same experiment.

If there are $\mathcal{N}_r$ members of the ensemble that exhibit event r, then the probability for that event is defined to be

$$p_r = \frac{\mathcal{N}_r}{\mathcal{N}}, \qquad (\text{as } \mathcal{N} \to \infty).$$

Let us note that the probability of occurrence of one event depends crucially upon the information that we have about the system A.

To give some idea of the arguments lying behind the use of an ensemble, rather than attempting to carry out a detailed time-dependent study of the behaviours of large collections of particles, we may again turn to computer-generated snapshots of the time evolution of a system of Lennard-Jones (LJ) Ar atoms. Figure 1.5a–c shows a series of snapshots from a set of computations with 40 Ar atoms satisfying Newton's laws of motion, but with different (randomly generated) initial conditions (ICs).

Figure 1.5a shows a series of snapshots generated from the initial time, $t = 0$, and moving forward for a series of four out of, in principle, a very large number of computations: the three frames shown in each row are snapshots representing a series of time steps that still show significant differences in the time evolution associated with four sets of arbitrarily selected initial conditions. Each column represents a probabilistic statement at the time t at which the snapshots were generated: we can see that these probabilities are clearly time-dependent.

The frames shown in Fig. 1.5b, c have all been generated a significant time later in the evolutionary development of the ensemble and show that the various snapshots

System
number

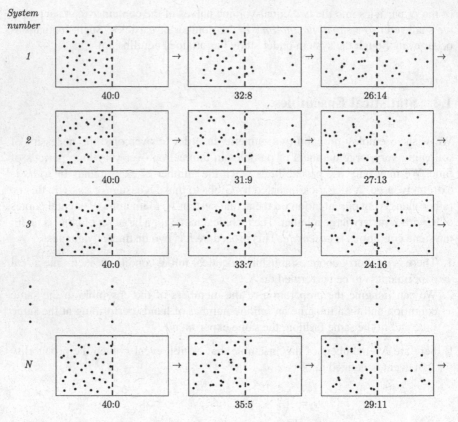

Fig. 1.5 (continued)

are beginning to look more similar both across the rows and down the columns. As a consequence, the probability distributions corresponding to the various columns have become less obviously time-dependent.

Figure 1.5b, c shows only small differences between snapshots obscrved either horizontally (i.e., along a row) or vertically (i.e., down a column): in other words, the system has essentially become time-independent, and equilibrium has been attained. These snapshots aid us in understanding the dynamic nature of an equilibrium. Moreover, the sets of frames shown in Fig. 1.5b, c give us a clear definition of what 'equilibrium' means: *An isolated macroscopic system is said to be in equilibrium if a statistical ensemble of such systems is time-independent.* As we are interested at present only in equilibrium statistical thermodynamics, we are hence interested only in calculating p_r, and *not* in calculating $p_r(t)$.

Although calculations of the type illustrated in Fig. 1.5 may prove to be very convincing to us, they do not constitute a proof that it must always be so. We thus place our belief in the appropriateness of this behaviour in the form of a postulate, sometimes referred to as the 'first postulate of statistical mechanics', namely,

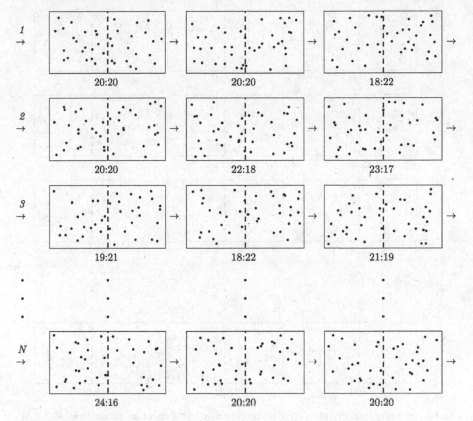

Fig. 1.5 (continued)

- **Postulate**: The long-time average of a mechanical variable in a thermodynamic system is equal to its ensemble average in the limit that the number of members of the ensemble tends to infinity.

Of course, before we can make use of this postulate, we shall need to be able to compute an ensemble average. To accomplish this task, we will require a knowledge of the relative probabilities of occurrence of the various (quantum) states making up the individual members of the ensemble. We shall return to this topic in the following chapter.

Many different ensembles may be constructed. However, because the three most important thermodynamic environments can be described either as isolated systems (with a fixed number of particles, N; fixed volume, V; and fixed total energy, E) or as a closed isothermal system (with fixed N, fixed V, and a fixed temperature, T) or as an open isothermal system (with fixed chemical potential, μ, fixed V, and fixed T), we shall focus mainly upon three representative ensembles, appropriately referred to as the microcanonical ensemble (fixed N, V, E), the canonical ensemble (fixed N, V, T), and the grand ensemble (fixed μ, V, T). In the event that we are

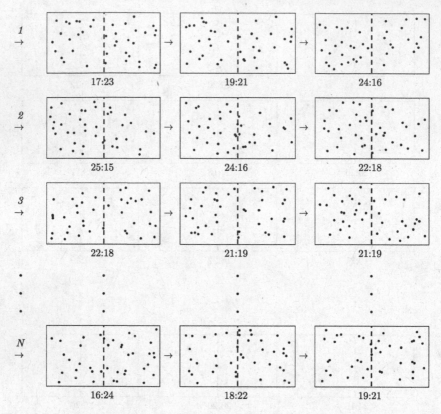

Fig. 1.5 (**a**) Early time evolution of four representative members of an ensemble of 40 LJ Ar atoms initially confined to the left half of a container of volume V. (**b**) Behaviour of the mid-time evolution of the four representative members of an ensemble of 40 LJ Ar atoms initially confined to the left half of a container of volume V. (**c**) Behaviour of the long-time evolution of the four representative members of an ensemble of 40 LJ Ar atoms initially confined to the left half of a container of volume V [this figure (a–c) has been generated and provided by Dr. Kevin Bishop]

considering multicomponent systems, N and μ represent sets of distinct particle numbers $\{N_1, N_2, \dots\}$ and chemical potentials $\{\mu_1, \mu_2, \dots\}$, respectively, making up the total system. Moreover, although we shall focus in the main upon the volume V as the (typical) mechanical variable, we should recognize that each non-pressure-volume thermodynamic work term will be associated with a corresponding mechanical variable, such as area, length, magnetization, for example.

1.4.1 Ensemble Averages

Let us consider a variable v having M possible values, together with the corresponding probabilities for their occurrences: the most complete possible description of this system is given by knowing each value $\{v_1, v_2, \ldots, v_M\}$ of v and the corresponding probabilities of occurrence $\{p_1, p_2, \ldots, p_M\}$. If there are $\mathcal{N}$ similar systems in the ensemble, then v has the value v_r in $\mathcal{N}_r = \mathcal{N}p_r$ of them. With so many possibilities in principle, we should be willing to settle for a little less detail; that is, we might be interested in first instance only in the *average value* of the variable (other names for this quantity are: *mean value, ensemble average*) defined by

$$\langle v \rangle \equiv \frac{1}{\mathcal{N}}(\mathcal{N}_1 v_1 + \mathcal{N}_2 v_2 + \cdots + \mathcal{N}_M v_M) = \sum_{r=1}^{M} p_r v_r. \qquad (1.4.1)$$

We can define the average of a function of the variable v in a similar way:

$$\langle f(v) \rangle \equiv \sum_{r=1}^{M} p_r f(v_r). \qquad (1.4.2)$$

Notice that the averaging procedure is a linear operation, in that

$$\langle cf(v) \rangle = c\langle f(v) \rangle, \qquad (1.4.3)$$

and

$$\langle f(v) \pm g(v) \rangle = \langle f(v) \rangle \pm \langle g(v) \rangle. \qquad (1.4.4)$$

We should, however, be cautious with regard to the average of the product of two such functions of the same variable. Let us begin with the definition of the average of a product function, namely

$$\langle f(u)g(v) \rangle = \sum_{r=1}^{M} \sum_{s=1}^{M} p_{rs} f(u_r)g(v_s), \qquad (1.4.5)$$

in which p_{rs} represents what is known as the joint probability (see the appendix). If we can *assume* (or if we know) that the variables are statistically independent, so that $p_{rs} = p_r p_s$, we can simplify the expression for the average of the product in the following way:

$$\langle f(u)g(v) \rangle = \sum_{r=1}^{M} \sum_{s=1}^{M} p_r p_s f(u_r)g(v_s)$$

$$= \sum_{r=1}^{M} p_r f(u_r) \sum_{s=1}^{M} p_s g(v_s) \qquad (1.4.6)$$

$$= \langle f(u) \rangle \langle g(v) \rangle.$$

All that has been said above is fine, provided that we know all values of p_r, but what do we do to find such averages without knowing all these values? Let us examine a concrete case, that of an ideal gas of N statistically independent atoms enclosed in a container of volume V_T and divided into two parts, with volumes V and V', and let us construct an ensemble of such containers, each with N ideal gas atoms. For each member of this ensemble, if p is the probability that an atom will be found in volume V, then $q \equiv 1 - p$ is the probability that the atom will be found in volume V'. Since at equilibrium we expect the atoms to be uniformly distributed throughout the volume V_T, we may represent the probabilities p and q as

$$p \equiv \frac{V}{V_T}, \qquad q \equiv \frac{V'}{V_T}.$$

We have shown in Appendix A that the probability $p(n, N)$ that n out of the N atoms will be found in volume V (with $N - n$ hence found in V') is given by the binomial distribution

$$p(N, n) = \frac{N!}{n!(N - n)!} p^n q^{N-n}, \tag{1.4.7}$$

in which $N! \equiv N(N - 1)(N - 2) \ldots 2 \cdot 1$ is termed factorial N. The behaviour of $p(n, N)$ for small values of N is illustrated in Fig. 1.6 by a series of bar graphs, from which it may be seen, even for relatively small values of N, that the distribution $p(n, N)$ narrows and grows as N increases. We may therefore anticipate that the

Fig. 1.6 Bar graph representation of the evolution of the binomial distributions for $N = 2, 4, 8,$ and 20

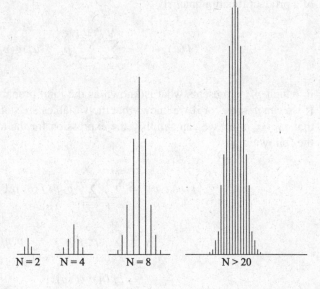

probability distribution will be quite sharply peaked for large values of N and that the value $\bar{n}$ corresponding to the position of the peak is also quite large. For $N = 10^{12}$ say, $\bar{n} \simeq 5 \times 10^{11}$ or so, and changing n by one unit, i.e., by 1 part in 10^{12}, will make little change in $p(N, n)$. This is a way of saying that we have an expression that reminds us of the condition for a differential, namely,

$$|p(N, n+1) - p(N, n)| \ll p(N, n).$$

This also suggests that we may profit from approximating this distribution by an equivalent continuous distribution so that we can utilize the calculus for our analysis of its behaviour. As we are actually only really interested in the region of the maximum in $p(n, N)$, let us henceforth focus upon it. From the calculus, the condition for a maximum is

$$\frac{dp}{dn} = 0,$$

with $p(n) \equiv p(n, N)$ given by Eq. (1.4.7). As this expression has several factors that depend upon n, it suggests that it would be useful to employ the logarithmic derivative here. Thus, let us work with $\ln p$, which is given by

$$\ln p(N, n) = \ln N! - \ln n! - \ln(N - n)! + n \ln p + (N - n) \ln q,$$

rather than with p itself. The appropriate derivative expression is hence

$$-\frac{d}{dn} \ln n! - \frac{d}{dn} \ln(N - n)! + \ln p - \ln q = 0,$$

or, equivalently,

$$\frac{d}{dn} \ln n! + \frac{d}{dn} \ln(N - n)! = \ln \frac{p}{q}. \qquad (1.4.8)$$

If we now employ the Stirling approximation (see Appendix C) for $\ln x!$, i.e., $\ln x! \simeq x \ln x - x$, which implies that $(d/dx) \ln x! \simeq \ln x$, we obtain the result

$$\frac{d}{dn} \ln p(N, n) = 0 \simeq \ln \left(\frac{N - \bar{n}}{\bar{n}} \right) + \ln \frac{p}{q}.$$

We may further simplify this expression into

$$\ln \left\{ \frac{N - \bar{n}}{\bar{n}} \frac{p}{q} \right\} = 0,$$

from which we obtain the result

$$\frac{N - \bar{n}}{\bar{n}} \frac{p}{q} = 1, \tag{1.4.9}$$

or

$$Np = \bar{n}(p + q) \equiv \bar{n}. \tag{1.4.10}$$

From this last result, we see that Np is thus the most probable value for n.

It will now be instructive for us to investigate the behaviour of $\ln p(N, n)$ around its maximum, by using the Taylor series expansion expression:

$$\ln p(N, n) = \ln p(N, \bar{n}) + \frac{d \ln p(N, n)}{dn}\bigg|_{n=\bar{n}} (n - \bar{n})$$

$$+ \frac{1}{2!} \frac{d^2 \ln p(N, n)}{dn^2}\bigg|_{n=\bar{n}} (n - \bar{n})^2 + \dots . \tag{1.4.11}$$

We therefore need to evaluate the second derivative of $\ln p(N, n)$ at $n = \bar{n}$: we obtain in general the result

$$\frac{d^2}{dn^2} \ln p(N, n) = -\frac{N}{n(N - n)}, \tag{1.4.12}$$

from which the specific value at $n = \bar{n}$ is obtained as

$$\frac{d^2}{dn^2} \ln p(N, n)\bigg|_{n=\bar{n}} = -\frac{N}{\bar{n}(N - \bar{n})} = -\frac{1}{Npq}. \tag{1.4.13}$$

If we now return to our Taylor expansion expression for $\ln p(N, n)$, and substitute this result into it, we find that

$$\ln p(N, n) = \ln p(N, \bar{n}) - \frac{(n - \bar{n})^2}{2Npq}, \tag{1.4.14}$$

or equivalently, for $p(N, n)$ we find the approximate expression

$$p(N, n) = p(N, \bar{n}) \exp\left\{-\frac{(n - \bar{n})^2}{2Npq}\right\}. \tag{1.4.15}$$

This expression is referred to as the Gaussian distribution.

We may now determine the constant in Eqs. (1.4.14) and (1.4.15) by utilizing the normalization condition for probabilities, namely

$$\sum_n p(N, n) = 1, \quad \text{or} \quad \int_0^N dn \, p(N, n) = 1, \tag{1.4.16}$$

so that, using Eq. (1.4.14) in the integral form of the normalization condition, we have that

$$p(N, \overline{n}) \int_0^N dn \ e^{-(n-\overline{n})^2/(2Npq)} = 1.$$

Evaluation of the integral gives us a final expression for the Gaussian probability distribution, namely

$$p(N, n) = \frac{1}{\sqrt{2\pi Npq}} \ e^{-(n-\overline{n})^2/(2Npq)} , \qquad (1.4.17)$$

in which N, n are both very large.

We can now use this expression for $p(N, n)$ to evaluate the average value of n itself in the following way:

$$\langle n \rangle \equiv \sum_n np(N, n) \simeq \frac{1}{\sqrt{2\pi Npq}} \int_0^N dn \ ne^{-(n-\overline{n})^2/(2Npq)},$$

from which we obtain the result

$$\langle n \rangle = \overline{n} . \qquad (1.4.18)$$

Hence, for this special distribution, the average value of n is also its most probable value.

1.4.2 Variance and Standard Deviation

Let us consider a set of $\mathcal{N}$ 'measurements' of a quantity v compared with its mean value $\langle v \rangle$. The deviation of an individual 'measurement' from the mean value is expressed by $\Delta v = v - \langle v \rangle$. The set of measurements is, by construction, symmetrically displaced about the mean $\langle v \rangle$ (see Fig. 1.7), so that

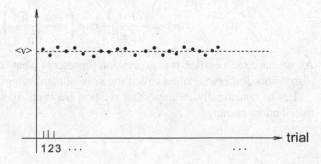

Fig. 1.7 Illustration of the average $\langle v \rangle$ for a set of repeated measurements of v

$$\langle \Delta v \rangle = \langle v - \langle v \rangle \rangle = \langle v \rangle - \langle v \rangle = 0\,.$$

We know that there is scatter of the individual measurements about the mean value, so that $\langle \Delta v \rangle$ does not provide for us a measure in the mean of how much scatter there is in the set of measurements. Obviously, in order to measure the scatter, we shall need a quantity that is always positive; such a quantity is given by $(\Delta v)^2$. Let us now calculate the mean of this quantity:

$$\langle (\Delta v)^2 \rangle = \langle v^2 \rangle - \langle v \rangle^2\,. \tag{1.4.19}$$

This quantity is known as the *dispersion* or *variance* of the set of measurements. If we wish to deal with a quantity that is on the same level as v itself, we can then take the square root of the variance, which gives us a new quantity, known as the *standard deviation*. This new quantity then represents the root-mean-square (RMS) deviation for v.

The simplest examples of variance and standard deviation are those for n itself:

$$\langle (\Delta n)^2 \rangle \equiv \sum_n p(n)[n - \langle n \rangle]^2 = \sum_n p(n)(n - \overline{n})^2$$

$$\simeq \frac{1}{\sqrt{2\pi Npq}} \int_0^N dn\ e^{-(n-\overline{n})^2/(2Npq)}(n - \overline{n})^2 \tag{1.4.20}$$

$$\approx \frac{2}{\sqrt{2\pi Npq}} \int_0^\infty dy\ y^2 e^{-y^2/(2Npq)}$$

$$= Npq\,.$$

From this expression for the variance, we immediately obtain the expression for the standard deviation as

$$\sigma_n \equiv \sqrt{\langle (\Delta n)^2 \rangle} = \sqrt{Npq}\,. \tag{1.4.21}$$

We can now utilize this result to write the probability solely in terms of the mean value $\overline{n}$ and the standard deviation σ_n as parameters characterizing its 'position' on the n-axis and its 'width':

$$p(n) = \frac{1}{\sqrt{2\pi}} \frac{1}{\sigma_n} \exp\left\{ -\frac{(n - \overline{n})^2}{2\sigma_n^2} \right\}\,. \tag{1.4.22}$$

As we have seen earlier, this expression represents what is known as the *Gaussian distribution*, otherwise often called the *normal distribution*.

Let us note finally an important relation between σ_n and $\langle n \rangle$ for the Gaussian distribution, namely,

$$\frac{\sigma_n}{\langle n \rangle} = \frac{(Npq)^{\frac{1}{2}}}{Np} = \left(\frac{q}{p}\right)^{\frac{1}{2}} \frac{1}{\sqrt{N}} , \qquad (1.4.23)$$

which tells us that the scatter in a set of 'measurements' is inversely proportional to $\sqrt{N}$. Thus, for example, to decrease the scatter by a factor 10, we must increase the number of measurements by a factor 100.

1.5 Problems for This Chapter

1. What is the probability of throwing a total of six points or less with two 'honest' dice?
2. If in a factory producing bolts, there is a probability 0.05 that a defective bolt will be produced, what is the average number of defective bolts, $\langle n \rangle$, in a total of 4000 bolts?
3. A hand of 13 cards is dealt at random from a pack of 52 playing cards. Show that the probability that the hand contains all four aces is 11/4165. Calculate the probabilities that the hand will contain 3 aces, 2 aces, 1 ace, and no aces.
4. A robot takes steps along a straight line, moving either forwards or backwards with equal probability. If the length of each step is 1 dm, calculate the probability that after taking N steps, the robot will be found $+n$ dm from its starting point.
5. Four bases (A, C, T, and G) appear in DNA. Assume that the appearance of each base in a DNA sequence is random.

   (a) What is the probability of observing the sequence AAGACATGCA?
   (b) What is the probability of observing the sequence GGGGGAAAAA?
   (c) What are the corresponding probabilities of observing the sequences in parts (a) and (b) above if the probability of observing A is twice the probability of observing G, C, or T?

6. The weight W associated with the occurrence of an event resulting from N trials (in which the outcome of an individual trial is either outcome 1 or outcome 2) that has N_1 occurrences of outcome 1 and N_2 occurrences of outcome 2 (necessarily equal to $N - N_1$) can be defined as

$$W(N, N_1) \equiv \frac{N!}{N_1!(N - N_1)!} .$$

   Show that for $2N$ coin tosses, the most probable event is N heads and N tails, that is, $N_H = N_T = N$.
7. Obtain an explicit expression for $W_{max} \equiv W(2N, N)$ for $2N$ binary outcome trials.

8. Show that the ratio of the weight $W(2N, N_H)$ associated with an event other than the most probable event to the weight of the most probable event for $2N$ binary outcome trials is given by

$$\ln \left(\frac{W}{W_{\text{max}}} \right) = -N_H \ln \left(\frac{N_H}{N} \right) - N_T \ln \left(\frac{N_T}{N} \right)$$

   when $2N$ is a sufficiently large number.

9. We may define a 'deviation index' α that provides a measure of the difference in the numbers of heads and tails obtained upon having completed a sequence of $2N$ coin tosses as $\alpha \equiv (N_H - N_T)/(2N)$. Show that expressions for N_H and N_T can also be obtained in terms of N and α as $N_H = N(1 + \alpha)$ and $N_T = N(1 - \alpha)$.

10. Use the results of Problems 8 and 9 to show that for $|\alpha| \ll 1$, the ratio W/W_{max} may be approximated as

$$\frac{W}{W_{\text{max}}} \simeq e^{-2N\alpha^2} .$$

11. Three tetrahedra are embedded in a host of transparent spheres in such a way that one full face of each tetrahedron is visible. Each tetrahedron has 1 red, 1 blue, 1 white, and 1 green face. How many ways can these tetrahedra be arranged so that two of the visible faces have the same colour?

References

1. H.S. Leff, Amer. J. Phys. **70**, 792 (2002)
2. P.T. Landsberg, *Thermodynamics and Statistical Mechanics*, ch. 13.1 (Oxford University Press, Oxford, 1978)
3. F. Reif, *The Berkeley Physics Course: Statistical Physics*, vol. 5. (McGraw Hill, New York, 1965)

Chapter 2
Macroscopic Thermodynamics

This chapter introduces those aspects of thermodynamics directly of concern to statistical mechanics. The natures of thermodynamic work and 'heat', the roles of thermodynamic internal energy and entropy in formulating the laws of thermodynamics, the concept of heat capacities, definitions of the various thermodynamic state functions and their uses are discussed and illustrated. The concepts of thermodynamic heat engines/refrigerators and their connections to thermodynamic cycles should be of interest to those with engineering backgrounds. Stability of isolated thermodynamic systems is addressed in terms of increases in the system entropy associated with spontaneous irreversible processes. Multicomponent systems and the roles played by the Gibbs energy and the chemical potentials for the various components are investigated. The thermodynamics of real gases is considered briefly, and illustrated in terms of the virial equation of state for model Van der Waals fluids, two-phase regions, and the Joule–Thomson expansion.

Toward the end of Sect. 1.1 we referred to the thermodynamic internal energy U without concerning ourselves with the meaning of the term 'thermodynamic'. The branch of experimental science that involves the study of the properties of bulk continuum matter and the processes associated with changes in these bulk properties is traditionally termed thermodynamics. Several definitive, and reasonably recent, treatments of thermodynamics are available [1–4].

The main thrust of thermodynamics is to identify relationships amongst global properties of macroscopic systems. In particular, the pressure, P, volume, V, temperature, T, and internal energy U are the most commonly encountered such properties: in many cases they are also the most important properties of macroscopic systems. If a closed mathematical relation connecting the relevant global macroscopic properties for a particular thermodynamic system can be obtained, it is referred to as the *equation of state* pertaining to that system. Thermodynamics is, in general, concerned with the utilization and conversion of energy, and provides general restrictions on macroscopic (versus microscopic) energy transfers

F. R. W. McCourt, *Statistical Thermodynamics for Pure and Applied Sciences*,
https://doi.org/10.1007/978-3-030-52006-9_2

based upon a fundamental set of (three) laws, referred to as the three Laws of Thermodynamics, plus an additional postulate, sometimes referred to as the zeroth law of thermodynamics. Before proceeding with a discussion of thermodynamics, however, we shall require definitions of a number of traditional terms that are typically employed in this discipline. More extensive discussions of the fundamental aspects of thermodynamics can be found in Ref. [5–7].

2.1 Basic Thermodynamic Definitions

The term *thermodynamic system*[1] refers to a (macroscopic) portion of the physical world that can be fully described in terms of a set of its physical attributes referred to as thermodynamic variables; the rest of the physical world, with which a system of interest is in contact (and with which it may in general exchange energy and mass) is referred to as the *surroundings*. A gross characterization of such systems may be made on the basis of how they interact with their surroundings. For example, two systems are said to be in *thermal contact* if the only interactions between them involve no exchange of matter and involve energy transfers that do not require work to be done by one system on the other. In general, systems fall into three categories: a system that is incapable of exchanging either energy or mass with its surroundings is said to be an *isolated system*, a system that is only in thermal contact with its surroundings is said to be a *closed system* and, finally, a system that may exchange both energy and mass with its surroundings is said to be an *open system*. We shall have occasion to encounter all three of these categories, although not always necessarily in the context of macroscopic thermodynamics.

A more refined characterization of a particular system may be carried out in terms of the physical properties associated with a set of macroscopic coordinates, by which we mean coordinates that are determined via measurements representing averages over regions of space containing many atoms and/or molecules and over time spans that are many orders of magnitude longer than those associated with particle motions. In addition, we shall be concerned with changes that involve macroscopically large amounts of energy. Physical properties that have been characterized in this manner are then termed *thermodynamic properties*, and are further categorized as *intensive* if they are independent of the mass of the system (examples are pressure, P, and temperature, T, which we know from experiment are independent of the mass/size of the thermodynamic system) and as *extensive* if they do depend upon the mass of the system (such as the volume, V, and number of atoms or molecules, N).[2] This characterization is often referred to loosely as 'size dependence'. However, it is worth noting that the determination of extensivity requires all intensive properties to be held fixed.

---

[1] For simplicity, we shall henceforth often refer to a 'thermodynamic system' simply as a 'system'.
[2] See also the discussion at the end of subsection (vi) of Sect. B.1.2.

To illustrate more precisely what we mean by extensive and intensive thermo-dynamic properties, we shall examine the specific example of a single-component system, made up of a macroscopic sample of a pure chemical substance which, for simplicity, we shall take to be an ideal gas. The relevant thermodynamic equation of state is then the ideal gas law, $PV = Nk_BT$. The mass of the system is given by $M = Nm$, with m the mass of an individual ideal gas particle. Because the number of moles, n, in the system is given by $n = N/N_0$, with N_0 the Avogadro number, it is clear that n is an extensive property of the system, as it depends directly upon the amount of the substance. Recall that the temperature and pressure of a system are known from experiment to be intensive properties. It will then be clear that the volume, V, of this ideal gas system will be proportional to N, so that V is an extensive property of this system. We have seen that the internal energy of an ideal gas is given by $U = \frac{3}{2}Nk_BT$, so that U is also an extensive property of this system. These are special cases of a more general behaviour associated with what are known in mathematics as homogeneous functions (see Sect. B.1.2). It is by now well-established that all extensive thermodynamic state functions are homogeneous functions of first order in the extensive variables and that all intensive thermodynamic state functions are homogeneous functions of zeroth order in the extensive variables.

Intensive properties may be further employed to characterize a system as spatially *homogeneous* if they are continuous functions of position throughout the region of space occupied by the system or as spatially *heterogeneous* if they are not. In particular, a heterogeneous system will typically consist of two or more distinct homogeneous regions separated by surfaces of discontinuity; each distinct homogeneous region in a heterogeneous system is termed a *phase*.

A system that is simultaneously in a state of *mechanical equilibrium* (in which no unbalanced force exists either in the interior of the system or between the system and its surroundings) and a state of *chemical equilibrium* (i.e., suffering no spontaneous change in its internal composition or transfer of matter from one part of the system to another, however, slowly) is said to be in a state of *thermal equilibrium* should there be no spontaneous change in the thermodynamic properties characterizing its mechanical and chemical equilibrium when it is separated from its surroundings and isolated. We note in particular that all parts of a system in thermal equilibrium are said to be at the same temperature.

A system is said to be in *thermodynamic equilibrium* should all requirements for mechanical, chemical, and thermal equilibrium of the system be satisfied simultaneously. Moreover, if two closed equilibrium systems are placed in thermal contact, each system will either remain in equilibrium upon contact or it will not. Temperature is the property of a system that determines whether or not it will remain in equilibrium upon thermal contact with another system. Should a system not remain in equilibrium once thermal contact with another system has been established, we say that the temperatures of the two systems are unequal. When two closed systems at different temperatures, say T_1 and T_2, are brought into thermal contact, both systems undergo changes that lead to final states of thermal equilibrium that correspond to a common temperature that lies intermediate between

T_1 and T_2. A slight variant of this argument is sometimes referred to as the Zeroth Law of Thermodynamics, namely, if two systems are independently in thermal equilibrium with a third system, then they are in thermal equilibrium with one another. This result provides a means for the quantitative definition of temperature through the third system serving as a thermometer.

As a thermodynamic (macroscopic) equilibrium system (similarly, its surroundings) will not tend to change state with time, it may be described in terms of time-independent macroscopic, or thermodynamic, variables. Should any of the conditions required for mechanical, chemical, or thermal equilibrium not be satisfied, the system is said to be in a thermodynamic nonequilibrium state. Nonequilibrium states are thus fundamentally different from equilibrium states in that they cannot be described in terms of thermodynamic variables referring to the system as a whole.

2.2 An Introduction to Thermodynamics

We shall focus our attention firstly upon the thermodynamics of pure substances. We have seen in Sect. 1.2[3] that the equation of state for a macroscopic system consisting of N classical ideal gas particles is

$$PV = Nk_{\mathrm{B}}T ,\tag{2.2.1}$$

relating the global thermodynamic properties P, V, and T. The ideal gas equation of state thus constrains the allowed values of P, V, and T such that once any two of them have been assigned (independent) values, the value of the third property is fixed. An ideal gas consisting of N noninteracting particles can thus be described thermodynamically via its equation of state.

We may rewrite Eq. (1.2.8) for the internal energy $U(T)$ of a classical ideal monatomic gas as

$$U(T) = \tfrac{3}{2}Nk_{\mathrm{B}}T ,\tag{2.2.2a}$$

from which it is clear that U is an extensive thermodynamic variable, as it is proportional to the mass of the system via the number N of ideal gas particles. We can also see from Eq. (2.2.2a) that the internal energy of such an ideal gas depends only upon temperature and, in particular, does not depend upon volume. This means that for an N-particle classical ideal gas, changes in the internal energy, dU, depend only upon changes in temperature, dT, and hence

---

[3]The first two values for a section or equation number identify chapter and section values associated with that item.

$$dU = \left(\frac{dU}{dT} \right) dT \,.$$

We shall write this expression as

$$dU = C_V \, dT \,, \tag{2.2.2b}$$

in which $C_V(T)$ is defined as

$$C_V(T) \equiv \left(\frac{\partial U}{\partial T} \right)_V \,,$$

and is known as the heat capacity at constant volume. This definition arises more generally from the (total) differential in the internal energy dU obtained when U is a function of T and V (such as occurs for a nonideal gas, for example), i.e.,

$$dU = \left(\frac{\partial U}{\partial T} \right)_V dT + \left(\frac{\partial U}{\partial V} \right)_T dV = C_V \, dT + \left(\frac{\partial U}{\partial V} \right)_T dV \,.$$

We have also seen in Sect. 1.2 that for an adiabatic change (in which there is no direct energy exchange between the thermodynamic system and its surroundings) the change in the internal energy of an ideal gas will be given by the work δW done on the ideal gas in compressing it, so that dU is also given by the work δW as

$$dU = -P \, dV \,. \tag{2.2.2c}$$

Upon equating expressions (2.2.2b) and (2.2.2c) for dU, we obtain the constraint

$$C_V \, dT + P \, dV = 0 \,. \tag{2.2.2d}$$

If we now replace the pressure in terms of T and V from the ideal gas law expression, we may rearrange Eq. (2.2.2d) into the form

$$C_V \frac{dT}{T} + N k_B \frac{dV}{V} = 0 \,, \tag{2.2.2e}$$

from which we obtain, upon integration over T and V from initial, i, to final, f, (thermodynamic) states, the condition

$$C_V \ln\left(\frac{T_f}{T_i} \right) + N k_B \ln\left(\frac{V_f}{V_i} \right) = 0 \,. \tag{2.2.3}$$

This condition means that in carrying out an adiabatic process that takes an ideal gas from an arbitrary initial thermodynamic state characterized by (T_i, V_i) to a final thermodynamic state characterized by (T_f, V_f), we are restricted to choosing arbitrarily either the final temperature T_f or the final volume V_f, but not both, as to do so would entail a violation of the relation (2.2.3). Although we have determined this restriction by considering a rather special type of system, it happens that it extends beyond the ideal gas; indeed, it applies to reversible adiabatic expansions of any gas.

As it is known that the number of independent thermodynamic state functions for a system having a fixed mass equals the number of work terms plus one [6], then for a system that possesses only pressure–volume work, there will be two independent state functions, which we would normally have chosen as T and V. However, we have just seen for an adiabatic change that the constraint (2.2.3) only allows us to choose the final value either of T or of V, but not of both T and V: this restriction therefore implies that there must be some other thermodynamic state function for the system that happens not to change value during a reversible adiabatic expansion. We shall label this (at present) unknown state function by S, so that its change, ΔS, for a reversible adiabatic process is given by

$$\Delta S = S_f - S_i \equiv 0 \, .$$

This thermodynamic state function has been given the name *entropy*; as a consequence, adiabatic processes are therefore also referred to as *isentropic*. Entropy is an inherently difficult property to understand and appreciate, especially from a strictly thermodynamic perspective, as there is no known means for actually performing its direct measurement. We shall, however, be able to arrive at a deeper understanding and develop a more comprehensive appreciation of this seemingly mysterious, yet very important, thermodynamic state function once we have had an opportunity to examine it from the perspective of statistical mechanics.

The three laws of thermodynamics may be expressed in a number of different ways. For our present purposes, the most appropriate formulation involves the thermodynamic energy, entropy, and temperature, in terms of which the first law is a statement of the conservation of energy, the second law is a statement of the increase of the total entropy (of the universe), while the third law states that the absolute temperature for a thermodynamic system remains positive. Thermodynamics is thus concerned with the discovery and nature of universal laws that govern the utilization and conservation of energy on a macroscopic level. It will therefore be of no surprise that these laws should be recovered naturally from a microscopic description of natural systems, such as that provided by statistical mechanics.

The thermodynamic formulation of the first law of thermodynamics couples the conservation of energy with the knowledge that for a macroscopic system there are two means available for the *transfer of energy* between subsystems of an isolated system, namely *work done* and *heat exchanged*. If we split the universe into two subsystems, one of which is the thermodynamic system of interest, the other the rest of the universe, and labelled as the 'surroundings', then work provides a means for energy transfer due to unbalanced forces between the thermodynamic system and its surroundings. By convention, the amount of work done *on* a system (by its surroundings) *increases* the internal energy U of the system by an amount ΔU

expressed as the work, W, done on the system. If the forces involved in such a process of energy transfer (between two subsystems) are only infinitesimally unbalanced at any time during the process in which energy is transferred as work, then that process is said to be *reversible*. We should always understand, however, that true reversibility is an idealization that cannot be fully accomplished.

Experimentally, it is known that a change of the thermodynamic state of a system separated from its surroundings by walls that allow energy to pass through them can be effected simply by raising or lowering the temperature of the surroundings, with no work performed either on or by the system. This mode of energy exchange, associated with a temperature difference (gradient) between the system and its surroundings, is called heat transfer. The amount of energy thereby transferred between the system and its surroundings is referred to as the heat, Q, and is taken to be positive if it represents an energy input to the system. Note that because heat and work are defined in terms of specific processes and represent modes of energy transfer, they are not thermodynamic state functions. Moreover, as they both depend, in general, upon the specific process employed or, equivalently, on the path taken, their differentials are designated by δQ and δW, respectively, rather than by dQ and dW. Similarly, their integrals, written

$$\int_i^f \delta Q \equiv \int_C \delta Q = Q \quad \text{and} \quad \int_i^f \delta W \equiv \int_C \delta W = W,$$

cannot be written as differences between values of Q and W for corresponding initial (i) and final (f) thermodynamic states of the system. The differential δQ is referred to as an inexact differential and the integral $\int_C \delta Q$ is referred to as a line integral.

2.2.1 Internal Energy, Entropy: The First Two Laws

The First Law of Thermodynamics is essentially a statement of the conservation of energy in a closed system that is at rest. Such a system has zero kinetic energy and its potential energy can be set to zero: the only type of energy with which we then need to be concerned for such a system is thus the sum of the kinetic and potential energies of the microscopic particles the make up the macroscopic system. The First Law (1st Law) can also be thought of as a statement of the equivalence of heat and work, and is often expressed mathematically in the differential form

$$dU = \delta Q + \delta W. \tag{2.2.4a}$$

An equivalent version of the first law is the integral form, given as

$$\Delta U = Q + W, \tag{2.2.4b}$$

with ΔU the change in the internal energy for the process as a whole, while Q and W are the heat exchanged and the work done during the process as a whole.

Because it is possible both for a thermodynamic system to perform work on its surroundings (which we shall refer to as work done *by* the system) and for its surroundings to do work *on* a system, we shall require a sign convention for W. For the First Law as expressed by Eq. (2.2.4b), the convention is that W represents the work done *on* the system by its surroundings, so that a negative value for W signifies that work has been done *by* the system on its surroundings. Similarly, as energy can be transferred as 'heat' either to the thermodynamic system from its surroundings or from the system to its surroundings, we need to be clear on the sign convention for Q: the traditional convention (herein adopted) is that Q is positive when energy is transferred to the system from its surroundings, corresponding to an increase in the energy of the system.

Dimensionally, it will be clear that the product PV represents pressure–volume work, as pressure is a force per unit area, so that PV has dimensions of force times distance, i.e., work. The incremental work done by (reversibly) changing the volume of a gaseous thermodynamic system by an amount dV will thus be represented by $-P\,dV$ according to the sign convention that has been adopted herein: the minus sign ensures that work is done on the system by the surroundings in the compression of a gas, for which $dV < 0$. The pressure–volume work done in changing the system volume from an initial value V_i to a final value V_f is thus given by

$$W = -\int_{V_i}^{V_f} P\,dV,\tag{2.2.5}$$

with the pressure, P, determined from the equation of state. For example, for an ideal classical gas, the pressure is given by $P = Nk_BT/V$, so that the work done *on* the system in carrying out an isothermal compression from volume V_i to volume V_f $(V_f < V_i)$ is

$$W = -Nk_BT\int_{V_i}^{V_f}\frac{dV}{V} = -Nk_BT\ln\left(\frac{V_f}{V_i}\right) > 0,$$

while the work done *by* the system in an isothermal expansion $(V_f > V_i)$ will be

$$W = -Nk_BT\ln\left(\frac{V_f}{V_i}\right) < 0.$$

As the internal energy of an ideal classical gas depends only upon the temperature, ΔU will be zero for any isothermal change, so that the energy transferred (as heat) between the system surroundings (also referred to as the reservoir) and the thermodynamic system will be the negative of the system work. Hence, compression of an ideal classical gas necessitates, via the First Law, that Q(system) be negative, corresponding to energy transferred (as heat) to the surroundings.

If the thermodynamic system is contained within *adiabatic* walls (which do not permit energy transfer through them), then $Q \equiv 0$, and $\Delta U \equiv U_f - U_i = W$, in which W is the work done on the thermodynamic system. Note that the change ΔU depends only upon the initial and final thermodynamic states of the system and is independent of the path taken for the process. Because ΔU is path-independent, an equivalent reversible path can often be found between the initial and final thermodynamic states should the equation of state for the system be known. This is a rather important observation, as the computation of Q and W can be carried out explicitly for such a reversible path, and their sum provides the value of ΔU for the actual process, whether or not the process is itself reversible.

When we apply the differential expression (2.2.4a) for the first law to a thermodynamic system in which only pressure–volume work $\delta W = -P_{ext}\,dV$ is possible, we obtain dU as

$$dU = \delta Q - P_{ext}\,dV. \tag{2.2.6}$$

A closed simple chemical system (having, in general, a fixed mass and composition for which surfaces and external fields are unimportant) satisfies this requirement. If the process that we wish to consider is adiabatic (i.e., $\delta Q \equiv 0$), then Eq. (2.2.6) gives

$$dU = -P\,dV, \tag{2.2.7a}$$

with the pressure P expressible as a function of V via the equation of state should the process happen also to be reversible. For an irreversible adiabatic process, however, Eq. (2.2.6) gives simply

$$dU = -P_{ext}\,dV, \tag{2.2.7b}$$

for which P_{ext} cannot in general be expressed as a function of V using the equation of state.

If we consider a reversible isothermal expansion of an ideal gas, the work done is given by $\delta W = -P\,dV$, while from Eq. (2.2.2b), $\Delta U = 0$ for an isothermal process ($dT \equiv 0$), so that Eq. (2.2.4a) for the first law gives the heat transferred as

$$\delta Q = P\,dV = Nk_B T\,d\ln V, \tag{2.2.8a}$$

as the pressure P is given as $P = Nk_B T/V$ from the ideal gas equation of state. Integration of Eq. (2.2.8a) from initial state i (with volume V_i) to final state f (with volume V_f) gives

$$\int_i^f \frac{\delta Q}{T} = Nk_B \int_{V_i}^{V_f} d\ln V = Nk_B \ln\left(\frac{V_f}{V_i}\right).$$

Moreover, as T is held fixed during the process, we may write $(\delta Q)/T = \delta\left(\dfrac{Q}{T}\right)$, so that we obtain

$$\frac{Q}{T} = Nk_B \ln\left(\frac{V_f}{V_i}\right). \tag{2.2.8b}$$

Thus, because the temperature is fixed for an isothermal process, the final volume V_f is determined by the amount of energy Q transferred into the system from the surroundings; V_f can thereby be arbitrarily set (despite the fact that both T_i and T_f have been fixed). This result, like the restriction (2.2.3) obtained earlier for an adiabatic process, also extends beyond the ideal gas, thereby implying that in general the entropy must change, with the consequence that $\Delta S \neq 0$ for any reversible isothermal process. Because S changes during an isothermal process, we are thereby free to choose both V_i and V_f arbitrarily.

For a general change of state for an ideal structureless gas, we obtain $(\delta Q)/T$ from the first law as

$$\frac{\delta Q}{T} = C_V\, d \ln T + Nk_B\, d \ln V \tag{2.2.9a}$$

or in an equivalent integral form as

$$\int_i^f \frac{\delta Q}{T} = C_V \int_{T_i}^{T_f} d \ln T + Nk_B \int_{V_i}^{V_f} d \ln V$$

$$= C_V \ln\left(\frac{T_f}{T_i}\right) + Nk_B \ln\left(\frac{V_f}{V_i}\right). \tag{2.2.9b}$$

As the right-hand side of Eq. (2.2.9b) depends only upon the initial and final thermodynamic states characterized by the thermodynamic variables (T, V), the quantity $(\delta Q)/T$ on the left-hand side of Eq. (2.2.9b) must represent an exact (total) differential that is zero for an adiabatic process and gives the result (2.2.8b) for an isothermal process. This behaviour suggests that T^{-1} serves as an integrating factor that then enables us to make the identification

$$dS = \frac{\delta Q}{T}, \tag{2.2.10a}$$

with the consequence that the macroscopic change ΔS is given by

$$\int_i^f \frac{\delta Q}{T} = S_f - S_i \equiv \Delta S. \tag{2.2.10b}$$

The Second Law of Thermodynamics is essentially a statement of the impossibility of attaining 100% efficiency in the conversion between heat and work, and is sometimes more colourfully expressed as the 'principle of no free lunch'. The thermodynamic state function entropy, S, is the major player in this law, and is often expressed in the form of an inequality as

$$dS_{sys} \geq \frac{\delta Q_{sys}}{T}, \tag{2.2.11a}$$

that applies for any process during which there is a change of state of the thermodynamic system. The equality applies for a reversible process, the inequality for an irreversible process. An equivalent integral form for the second law is

$$\Delta S \geq \int_{i}^{f} \frac{\delta Q}{T}, \tag{2.2.11b}$$

with the equality applying for *reversible paths* between points i and f, and the inequality applying for irreversible paths. Because the only means typically available to us for actual evaluation of ΔS in thermodynamics is by means of a reversible path, we must define carefully what we mean in practice by this term, as a reversible path is not truly achievable in principle. We therefore approximate a reversible path in practice by taking the system along the path via a sequence of changes that are carried out sufficiently slowly that for any given step of the process the thermodynamic system is maintained at equilibrium. Such a sequence of changes is referred to as a *quasistatic process*. A quasistatic process is referred to as 'internally reversible' if the interaction between the system and its surroundings is irreversible: this means that the term 'reversible' is typically reserved exclusively for reversible system processes coupled with reversible system-surroundings interactions.

Example 2.1 Isothermal expansion of a thermodynamic system.

When a thermodynamic system undergoes an isothermal expansion at temperature T, heat must be imported from its surroundings (a heat reservoir) in order to maintain the system at a constant temperature during the expansion. Importation of heat Q into the system from its surroundings (at temperature $T_{surr} \geq T$) generates an overall change ΔS in entropy given by

$$\Delta S = (\Delta S)_{system} + (\Delta S)_{surr}$$

$$\geq \frac{Q}{T} - \frac{Q}{T_{surr}} = \frac{T_{surr} - T}{T T_{surr}} Q.$$

As we know from the Second Law of Thermodynamics that $\Delta S \geq 0$ in general, and that $\Delta S = 0$ only for a reversible process, we see from the above expression for ΔS that if $T_{surr} > T$, importation of heat from the heat reservoir is irreversible. Note that even if T_{surr} is only infinitesimally greater than T, the process is irreversible. By the same token, should T_{surr} be even infinitesimally lower than T, heat transfer is disallowed by the Second Law. Thus, the transfer of heat from a heat reservoir to the system may technically take place reversibly only when T_{surr} *precisely* matches T.

Moreover, as the heat transfer process only approaches reversibility for temperature differences that exceed T by infinitesimal amounts, thereby requiring exceedingly long times to effect the transfer of a finite amount of heat energy Q, reversible heat flow processes are necessarily quasistatic.

We may therefore conclude that essentially all isothermal changes are irreversible due to the irreversible nature of the system-surroundings interaction (i.e., thermal contact) that typically enables the transfer of heat between a system and its surroundings. Our expression for ΔS, when coupled with the Second Law requirement that $\Delta S \geq 0$, also clearly establishes that heat flows spontaneously from higher to lower temperatures, and may only be transferred reversibly should the temperature of the surroundings precisely match that of the thermodynamic system.[4]   □

From Eq. (2.2.10b) we see that the reciprocal temperature essentially serves as an integrating factor that enables us to replace the (inexact) differential δQ in terms of the total (exact) differential dS for the entropy, so that the original first law expression (2.2.4a) becomes

$$dU = T \, dS - P \, dV \qquad (2.2.12)$$

which, because of the role played by the differential of the entropy, is commonly referred to as the 'Combined First and Second Laws' expression. As both U and S are path-independent functions, we know that

$$\Delta U = \int_i^f dU = U_f - U_i \, ; \qquad \Delta S = \int_i^f dS = S_f - S_i \, , \qquad (2.2.13)$$

with the values ΔU and ΔS depending only upon the values of U and S at the thermodynamic state end-points for the line integrals. Thermodynamic functions having this property are known as (thermodynamic) *state functions*. This expression for the combined first and second laws serves as the basis for much of the development of chemical thermodynamics for closed thermodynamic systems.

We shall now apply the combined First and second law expression (2.2.12) to obtain an important relation between the partial derivatives $\left(\dfrac{\partial T}{\partial V}\right)_S$ and $\left(\dfrac{\partial P}{\partial S}\right)_V$. We note firstly, that in accordance with expression (2.2.12), the independent variables for the internal energy are S and V. Moreover, if we note that the total differential dU in Eq. (2.2.12) may also be represented mathematically as

$$dU \equiv \left(\frac{\partial U}{\partial S}\right)_V dS + \left(\frac{\partial U}{\partial V}\right)_S dV \, , \qquad (2.2.14a)$$

we see that the internal energy is thus formally a function of the thermodynamic variables S and V; as these variables for U arise through the first and second laws

---

[4] We shall often indicate the completion of either an interlude or an example with a terminal square.

of thermodynamics, they are often referred to as the 'natural' independent variables for U. We see from Eq. (2.2.14a) that the internal energy provides the appropriate thermodynamic energy state function for the description of adiabatic (isentropic) and constant-volume (isochoric) thermodynamic processes.

As the partial derivatives of U with respect to S and V are themselves functions of S and V, this tells us that in Eq. (2.2.12), the temperature T and pressure P are thus (implicitly) functions of S and V, so that we may write Eq. (2.2.12) for present purposes more explicitly as

$$dU = T(S, V)\,dS - P(S, V)\,dV\,. \tag{2.2.14b}$$

Moreover, comparison between Eqs. (2.2.14a), (2.2.14b) then allows us to identify T and P as the first partial derivatives of the internal energy $U(S, V)$, specifically as

$$T = \left(\frac{\partial U}{\partial S}\right)_V \quad \text{and} \quad P = -\left(\frac{\partial U}{\partial V}\right)_S, \tag{2.2.14c}$$

respectively. Equality of the mixed second partial derivatives of U as a function of the independent variables S and V, namely,

$$\frac{\partial^2 U}{\partial V \partial S} = \frac{\partial^2 U}{\partial S \partial V},$$

then gives the equality,

$$\left(\frac{\partial T}{\partial V}\right)_S = -\left(\frac{\partial P}{\partial S}\right)_V. \tag{2.2.14d}$$

From this equality, known as the first Maxwell relation, we see that the rate of change of entropy with respect to pressure in an isochoric process can be obtained from the rate of change of volume with respect to temperature in an adiabatic process.

We may utilize the first law expression (2.2.4a) in the form

$$\delta Q = dU + P_{\text{ext}}\,dV\,,$$

to see that for an isochoric thermodynamic process we obtain

$$(\delta Q)_V = dU\,. \tag{2.2.15a}$$

Hence, the integrated form of the first law, when applied to an (isochoric) process, then gives

$$Q_V = \Delta U \equiv U_{\text{f}} - U_{\text{i}}\,. \tag{2.2.15b}$$

We have earlier employed a quantity that we referred to as the heat capacity at constant volume, C_V, and defined as

$$C_V \equiv \left(\frac{\partial U}{\partial T}\right)_V = \lim_{\Delta T \to 0} \frac{(\delta Q)_V}{\Delta T}, \qquad (2.2.15c)$$

with the final step above representing the experimental operational definition of C_V. The experimental determination of the isochoric heat capacity thus consists of supplying a series of accurately known, but sequentially decreasing, quantities of energy δQ into a system of fixed volume, measuring the attendent temperature increases ΔT, and extrapolating the ratio $\delta Q/\Delta T$ to $\Delta T = 0$ to obtain the experimental value of C_V.

2.3 New Thermodynamic State Functions

As many more chemical processes take place under constant pressure, rather than at constant volume, it would be both convenient and useful to define a thermodynamic state function H that, in analogy with Eq. (2.2.15b), gives

$$Q_P = \Delta H \equiv H_f - H_i \qquad (2.3.1a)$$

directly as the difference between its values in the initial and final thermodynamic states connected by an isobaric process. To determine the form taken by this new thermodynamic state function, we may begin again with the first law expression

$$\delta Q = dU + P\,dV,$$

which becomes

$$\delta Q = d(U + PV) - V\,dP,$$

upon adding and subtracting $V\,dP$ on its right-hand side. It is clear from this expression that the heat added to a thermodynamic system for which the pressure is fixed will be given via expression (2.3.1a) as the difference ΔH in the thermodynamic state function H, given by $U + PV$, between the initial and final thermodynamic states. By analogy with Eq. (2.2.15a) the differential heat supplied in a constant pressure (isobaric) measurement is given as

$$(\delta Q)_P = dH \qquad (2.3.1b)$$

which, by analogy with the heat capacity at constant volume, may be defined via the sequence of equalities

$$C_P = \lim_{\Delta T \to 0} \frac{(\delta Q)_P}{\Delta T} \equiv \left(\frac{\partial H}{\partial T}\right)_P \qquad (2.3.1c)$$

as the heat capacity at constant pressure.

The thermodynamic state function H plays much the same role for isobaric processes that the internal energy U plays for isochoric processes, and is known as the enthalpy. It may also be represented as the Legendre transform (see Sect. B.2)

$$H = U + PV \qquad (2.3.2)$$

of the internal energy $U(S, V)$ that replaces the independent extensive variable V with the independent intensive variable P, so that the enthalpy can be written as $H = H(S, P)$.

If we form the total differential dH using Eq. (2.3.2), we obtain

$$dH = dU + PdV + VdP$$

which, upon employing expression (2.2.12) for dU, becomes

$$dH = TdS + VdP. \qquad (2.3.3)$$

Because this expression for dH also arises from the first and second laws, we therefore say that S and P are the natural thermodynamic variables for enthalpy. It will be clear from Eq. (2.3.3) that the enthalpy provides the appropriate thermodynamic energy state function for the description of isentropic and isobaric thermodynamic processes.

Comparison of Eq. (2.3.3) with the formal mathematical representation,

$$dH \equiv \left(\frac{\partial H}{\partial S}\right)_P dS + \left(\frac{\partial H}{\partial P}\right)_S dP,$$

for the total differential dH shows that the dependent variables temperature and volume are related to enthalpy by

$$T = \left(\frac{\partial H}{\partial S}\right)_P \qquad \text{and} \qquad V = \left(\frac{\partial H}{\partial P}\right)_S, \qquad (2.3.4a)$$

respectively. Equality of the mixed second partial derivatives of H then gives the second Maxwell relation

$$\left(\frac{\partial T}{\partial P}\right)_S = \left(\frac{\partial V}{\partial S}\right)_P, \qquad (2.3.4b)$$

which carries an interpretation similar to that for the first Maxwell relation, but involving isentropic and isobaric processes.

Along the same line of reasoning that we have employed to define the enthalpy thermodynamic state function for use in the description of adiabatic and isobaric processes, it is convenient to define similarly a new thermodynamic state function to be employed in the descriptions of isothermal and isochoric processes and having temperature and volume as its natural-independent variables. This may be accomplished via a Legendre transform of the internal energy U that replaces the extensive thermodynamic natural variable S by the corresponding intensive thermodynamic variable T, namely,

$$A \equiv U - TS, \tag{2.3.5a}$$

whose total differential, dA, can be obtained as

$$dA = -SdT - PdV. \tag{2.3.5b}$$

From Eq. (2.3.5b) we see that the natural-independent thermodynamic variables for A, which is known as the Helmholtz energy, are T and V. Note that this also requires that we treat the entropy and pressure as functions of T and V.

It will now also be clear that thermodynamic analyses of processes that occur either under isothermal or under isobaric conditions will be aided by employing another new thermodynamic state function, G, obtained using the Legendre transform

$$G = H - TS \tag{2.3.6a}$$

of the enthalpy H or via the Legendre transform

$$G = A + PV \tag{2.3.6b}$$

of the Helmholtz energy A or even via the double Legendre transform

$$G = U + PV - TS \tag{2.3.6c}$$

of the internal energy U. This thermodynamic state function is known as the Gibbs energy, and has the total differential

$$dG = -SdT + VdP, \tag{2.3.6d}$$

from which we see that its natural-independent thermodynamic variables are indeed T and P.

We see from expression (2.3.6d) that for an isothermal process, integration over pressure from an initial pressure P_i to a final pressure P_f gives

$$\int_{P=P_i}^{P=P_f} dG_T = \int_{P_i}^{P_f} V\, dP.$$
(2.3.7a)

For an ideal gas we thus obtain

$$(\Delta G)_T \equiv G(T, P_f) - G(T, P_i) = Nk_B T \ln\left(\frac{P_f}{P_i}\right).$$
(2.3.7b)

If we choose the initial pressure to be $P^\circ \equiv 1$ bar, we obtain

$$G(T, P_f) - G(T, P^\circ) = Nk_B T \ln\left(\frac{P}{P^\circ}\right)$$

and, if we define the standard Gibbs energy $G^\circ(T)$ via $G^\circ(T) \equiv G(T, P^\circ)$, we may then express the Gibbs energy in the form

$$G(T, P) = G^\circ(T) + Nk_B T \ln\left(\frac{P}{P^\circ}\right).$$
(2.3.7c)

Example 2.2 Illustration of the usefulness of a Maxwell relation.

We begin with the differential form for the combined first and second laws of thermodynamics as given in Eq. (2.2.12), namely, $dU = T\, dS - P\, dV$. We shall consider the entropy to be a function of T and V, thereby allowing us to replace the total differential dS by

$$dS = \left(\frac{\partial S}{\partial T}\right)_V dT + \left(\frac{\partial S}{\partial V}\right)_T dV,$$

to obtain a modified version of the combined first and second laws, which we may express as

$$dU = T\left(\frac{\partial S}{\partial T}\right)_V dT + \left[T\left(\frac{\partial S}{\partial V}\right)_T - P\right] dV$$

$$\equiv \left(\frac{\partial U}{\partial T}\right)_V dT + \left(\frac{\partial U}{\partial V}\right)_T dV.$$
(2.3.8)

Upon comparing the coefficients of dV in each of these two equations for dU, we see that

$$\left(\frac{\partial U}{\partial V}\right)_T = T\left(\frac{\partial S}{\partial V}\right)_T - P.$$
(2.3.9)

To eliminate the partial derivative involving the entropy S from our expression, we may employ one of the four best-known Maxwell relations, namely the reciprocal of Eq. (2.3.4b),

$$\left(\frac{\partial P}{\partial T}\right)_V = \left(\frac{\partial S}{\partial V}\right)_T,$$

to obtain an expression for the partial derivative of the internal energy U with respect to the volume, V, of the thermodynamic system. We thereby find that the volume dependence of U may be determined from the relation

$$\left(\frac{\partial U}{\partial V}\right)_T = T\left(\frac{\partial P}{\partial T}\right)_V - P \equiv T^2\left[\frac{\partial}{\partial T}\left(\frac{P}{T}\right)\right]_V. \tag{2.3.10}$$

We may determine directly from this expression that for an ideal gas, with equation of state (2.2.1), the internal energy U must be independent of the system volume V, as the partial derivative term on the right-hand side of Eq. (2.3.10) reduces to P. This result is consistent with the expression $U \equiv U(T) = \frac{3}{2}Nk_BT$ obtained in Sect. 1.2 for an ideal gas of structureless classical particles (i.e., particles that possess no form of energy other than translational). We have also seen that in such a case the coefficient of the differential dV in the work term of the combined first and second law expression reduces simply to $-P$; for a nonideal gas for which the internal energy does depend upon volume, it will not generally be possible simply to replace the external pressure P_{ext} by the gas pressure P alone during an isothermal thermodynamic change.   □

Another useful way to look at the entropy is to start with the combined first and second laws expression (2.2.12) in the form

$$T\,dS = dU + P\,dV$$

and, by replacing the total differential dU by its formal mathematical definition, to obtain dS as

$$dS = \frac{1}{T}\left(\frac{\partial U}{\partial T}\right)_V dT + \frac{1}{T}\left(\frac{\partial U}{\partial V}\right)_T dV + \frac{P}{T}dV. \tag{2.3.11}$$

This expression can be further simplified to give

$$dS_{\text{ideal gas}} = \frac{C_V(T)}{T}\,dT + \frac{Nk_B}{V}\,dV, \tag{2.3.12a}$$

since $\left(\frac{\partial U}{\partial T}\right)_V \equiv C_V(T)$, while $\left(\frac{\partial U}{\partial V}\right)_T = 0$, and $P/T = Nk_B/V$ for an ideal gas. We have written $C_V(T)$ in general, as C_V is a function of temperature for molecular ideal gases in particular.

Integration of Eq. (2.3.12a) gives the absolute entropy of an ideal gas as

$$S_{\text{ideal gas}} = S_0(N) + C_V \ln T + Nk_B \ln V, \tag{2.3.12b}$$

with $S_0(N)$ the integration constant. This expression for the absolute entropy, however, is not extensive: for example, a simple calculation for a doubling of both the number of particles in and the volume occupied by an ideal classical gas does not result in a doubling of the absolute entropy of that thermodynamic system.

We shall employ a mathematical artifice in order to arrive at a correct form for the absolute entropy: we note that Eq. (2.3.12a) is the sum of two terms, but only the first term has the normal format in which products of thermodynamic variables occur as products between extensive and intensive variables, as exemplified, for example, by PV, TS, μN. We shall therefore rewrite Eq. (2.3.12a) as

$$dS_{\text{ideal gas}} = \frac{C_V(T)}{T}\, dT + \frac{Nk_B}{(V/N)}\, d\left(\frac{V}{N}\right) \tag{2.3.13a}$$

which, upon integration, gives the absolute entropy as

$$S_{\text{ideal gas}} = S_0 + C_V \ln T + Nk_B \ln\left(\frac{V}{N}\right). \tag{2.3.13b}$$

We shall see later (in Chap. 5) that the integration constant, $S_0(N)$, for an N-particle ideal classical gas is obtained as

$$(S_0)_{\text{ideal gas}} = \left[\frac{5}{2} + \frac{3}{2}\ln\left(\frac{2\pi m k_B}{h^2}\right)\right] Nk_B \equiv Ns_0. \tag{2.3.14}$$

Note that as $S_0(N)$ is proportional to N, it is an extensive quantity. Upon carrying out a doubling of both the number of particles and the volume occupied by an ideal classical gas, it will be clear that the absolute entropy of that gas is also doubled, so that extensivity of the absolute entropy is maintained by Eq. (2.3.13b).

Either Eq. (2.3.12a) or Eq. (2.3.13a) may be utilized to obtain a correct expression for the entropy change ΔS between an initial (reference) state (T_0, V_0) and a final state (T, V) for an ideal gas of classical structureless particles, namely,

$$(\Delta S)_{\text{ideal gas}} = Nk_B \ln\frac{V}{V_0} + C_V \ln\frac{T}{T_0}, \tag{2.3.15}$$

with $C_V = \frac{3}{2}Nk_B$.

Example 2.3 Thermodynamic functions for a classical ideal gas.

We have seen that the internal energy, $U(T, V)$, for a thermodynamic system can be obtained in principle via Eq. (2.3.8), while the enthalpy, H, is given in terms of $U(T, V)$ by Eq. (2.3.2), and the Helmholtz and Gibbs energies may then be determined from Eqs. (2.3.5a) and (2.3.6c), respectively.

We begin by recalling that an ideal gas is defined by its equation of state, $PV = Nk_BT$. Because we see from Eq. (2.3.9) that the internal energy U for an ideal gas does not depend upon volume, we may obtain $U(T)$ for an ideal gas via Eq. (2.3.8) as

$$U_{\text{ideal gas}}(T) = \int_0^T C_V(T)\, dT\,,$$

with the constant of integration equal to zero, as $U(0) = 0$. For an ideal gas of particles that possess only translational motion, C_V is temperature-independent and has the value $C_V = \frac{3}{2}Nk_B$, so that $U_{\text{ideal gas}}(T) = \frac{3}{2}Nk_BT$.

To obtain $H_{\text{ideal gas}}(T)$ from $U_{\text{ideal gas}}(T)$, we have only to employ the equation of state for the ideal gas to replace PV in Eq. (2.3.2) by Nk_BT to give

$$H_{\text{ideal gas}}(T) = \int_0^T C_V(T)\, dT + Nk_BT\,,$$

or, for an ideal gas of structureless particles, $H_{\text{ideal gas}}(T) = \frac{5}{2}Nk_BT$.

We may now employ our expression for $U_{\text{ideal gas}}(T)$, together with expression (2.3.13b), to obtain the Helmholtz energy $A_{\text{ideal gas}}(T, V)$ for an ideal gas of structureless particles as

$$A_{\text{ideal gas}}(T, V) = -Nk_BT\left[\ln\left(\frac{V}{N}\right) + \frac{3}{2}\ln\left(\frac{2\pi mk_BT}{h^2}\right) + 1\right]. \qquad (2.3.16)$$

A similar expression for $G_{\text{ideal gas}}(T, P)$ can be obtained from $H_{\text{ideal gas}}(T)$ and the ideal gas law. $\qquad\qquad\qquad\qquad\qquad\qquad\qquad\qquad\qquad\qquad\qquad\qquad\qquad\quad\square$

Example 2.4 Thermodynamic functions for a quantum ideal gas.

Let us consider specifically a photon gas in thermal equilibrium, contained in a cylinder that is closed at one end, sealed by a movable piston, and maintained at a constant temperature T. According to Eq. (1.2.12), an isothermal quasistatic volume change from V to $V + \Delta V$ gives rise to a change, ΔU, in the internal energy U of

$$\Delta U = \frac{4\sigma}{c}\, T^4 \Delta V\,.$$

Because the pressure, P, of a photon gas is a function of T alone, the work done on the photon gas during this process is given, via $W = -\int P(T)dV$, as

$$W = -\frac{4\sigma}{3c}\, T^4 \Delta V\,.$$

From the First Law of Thermodynamics, Eq. (2.2.4b), the energy transfer into or out of the photon gas will then be given by

$$Q = \frac{4\sigma}{3c}\, T^4 \Delta V\,.$$

The energy thus transferred will be seen either as an increase (for $\Delta V > 0$) or as a decrease (for $\Delta V < 0$) in the number of photons making up the photon gas. This is a consequence of N *not* being an independent variable for a photon gas.

Because a slow (quasistatic) isothermal volume change is a reversible change, the entropy change is given by Q/T, giving ΔS as[5]

$$\Delta S = \frac{4\sigma}{3c} T^3 \Delta V, \qquad \Delta V \equiv V_f - V_i.$$

We may now employ this result, with $V_i = 0$, $N = 0$, corresponding to the internal piston face being flush with the back of the cylinder, to create the photon gas by moving the piston quasistatically to a position at which $V = V_f$, with N photons in the cylinder. As the internal energy U equals zero for $V = 0$, it thus follows from the First Law of Thermodynamics that $Q = 0$, whence $S \equiv Q/T = 0$.

The above result for ΔS implies that the entropy of a gas of N photons occupying a container of volume V at temperature T will be [8, 9]

$$S(T, V) = \frac{4\sigma}{3c} VT^3.$$

The total energy required to carry out such an isothermal expansion of a photon gas from volume 0 to volume V is given by $Q_{exp} = TS$. Because the pressure of a photon gas depends only upon T, the expansion occurs at constant pressure, so that $Q_{exp} \equiv Q_P$ corresponds to the change ΔH_{exp} in the enthalpy $H = U + PV$. We may thus deduce that the enthalpy of a photon gas in equilibrium at temperature T is given by

$$H(T, V) = \frac{4\sigma}{3c} VT^4.$$

This clearly shows that the enthalpy H represents the energy required to form a photon gas at temperature T plus that needed to carry out the pressure–volume work necessary to generate the space that the photon gas occupies.   □

It may also be interesting to see how an adiabatic change is characterized for a quantum ideal gas. We first recall that an adiabatic change is characterized by allowing no energy to be exchanged between the thermodynamic system and its surroundings. In the context of a photon gas, this means explicitly that no photons may be absorbed by or emitted from the container walls during the process: this is equivalent to saying that the container walls must be perfectly reflecting mirrors. Hence, an adiabatic volume change for a photon gas requires that the number of photons in the gas remain constant, so that both the entropy, $S(T, V)$, of the photon gas and the number, $N(T, V)$, of photons in the gas must remain constant. From the

---

[5]Note that this entropy change differs from the logarithmic volume dependence of the entropy change for an isothermal volume change for the classical ideal gas, given by Eq. (2.3.12b).

expression that we have just obtained for the entropy of a photon gas, we can see that constancy of the entropy during an adiabatic change in volume means that the temperature must change as $V^{-\frac{1}{3}}$ in order to compensate the volume change: this result, taken together with the expression for the pressure, $P(T)$, for a photon gas implies that the pressure changes as $V^{-\frac{4}{3}}$ or, equivalently, that pressure and volume are related by

$$PV^{\frac{4}{3}} = \text{const}.$$

Although this result may be reminiscent of the condition $PV^{\gamma} = \text{const.}$, with $\gamma = C_P/C_V = \frac{5}{3}$, that applies to adiabatic changes in a classical ideal monatomic gas, the resemblance is purely superficial, as C_P does not, in fact, exist for a photon gas, since it is not possible to vary the temperature while simultaneously holding the pressure fixed, as would be necessary for a determination of C_P.

2.4   Expression for Heat Capacity Difference

The heat capacity of a thermodynamic system is defined as the rate of change with temperature of the heat taken up by the system in a reversible process that is carried out under specified conditions, such as at constant volume, V, or constant pressure, P. Expressions for the heat capacities at constant volume and at constant pressure have been given in Eqs. (2.2.15c) and (2.3.1c), respectively.

To obtain a general expression for the difference between C_P and C_V, we begin with the thermodynamic defining relations for C_P and C_V. We may utilize the formal definition (2.3.2) of the enthalpy to obtain an equivalent expression,

$$C_P = \left(\frac{\partial H}{\partial T}\right)_P = \left(\frac{\partial U}{\partial T}\right)_P + P\left(\frac{\partial V}{\partial T}\right)_P,$$

that will prove to be helpful in obtaining the desired result for the difference $C_P - C_V$ for a general thermodynamic system. Indeed, this expression for C_P leads to $C_P - C_V$ being given as

$$C_P - C_V = \left(\frac{\partial U}{\partial T}\right)_P - \left(\frac{\partial U}{\partial T}\right)_V + P\left(\frac{\partial V}{\partial T}\right)_P.$$

Now, by starting from the formal mathematical expression for the total differential dU of the internal energy as a function of T and V, then treating V as a function of T and P and forming the total differential dV for $V(T, P)$, dU can be obtained as

$$dU = \left[\left(\frac{\partial U}{\partial T}\right)_V + \left(\frac{\partial U}{\partial V}\right)_T \left(\frac{\partial V}{\partial T}\right)_P\right] dT + \left(\frac{\partial U}{\partial V}\right)_T \left(\frac{\partial V}{\partial P}\right)_T dP.$$

Upon comparing this result with the formal mathematical definition

$$dU \equiv \left(\frac{\partial U}{\partial T}\right)_P dT + \left(\frac{\partial U}{\partial P}\right)_T dP \,,$$

for the internal energy as a function of T and P, we see that for these two expressions for dU to be equivalent, it is necessary that

$$\left(\frac{\partial U}{\partial T}\right)_P = \left(\frac{\partial U}{\partial T}\right)_V + \left(\frac{\partial U}{\partial V}\right)_T \left(\frac{\partial V}{\partial T}\right)_P \,.$$

With this result, the difference $C_P - C_V$ becomes

$$C_P - C_V = \left(\frac{\partial U}{\partial V}\right)_T \left(\frac{\partial V}{\partial T}\right)_P + P \left(\frac{\partial V}{\partial T}\right)_P \,,$$

or

$$C_P - C_V = \left[\left(\frac{\partial U}{\partial V}\right)_T + P\right]\left(\frac{\partial V}{\partial T}\right)_P \,. \qquad (2.4.1a)$$

This expression for $C_P - C_V$ is a general thermodynamic result.

For an ideal gas, for which the equation of state is the ideal gas law (2.2.1) and the internal energy (2.2.2a) depends only upon the temperature T, we have

$$\left(\frac{\partial V}{\partial T}\right)_P = \frac{Nk_B}{P} \,, \qquad \left(\frac{\partial U}{\partial V}\right)_T = 0 \,,$$

so that the difference between C_P and C_V becomes simply

$$C_P - C_V = Nk_B \,. \qquad (2.4.1b)$$

An explicit expression for $U(T, V)$ is often not available for a more general thermodynamic system, so that the first factor on the right-hand side of Eq. (2.4.1a) cannot readily be evaluated. Moreover, it is also not directly accessible to experimental determination. However, if we employ Eq. (2.3.10) to substitute this factor by $T\left(\frac{\partial P}{\partial T}\right)_V$, then employ the permutation rule

$$\left(\frac{\partial P}{\partial T}\right)_V \left(\frac{\partial V}{\partial P}\right)_T \left(\frac{\partial T}{\partial V}\right)_P = -1$$

for the variables (P, T, V) to replace $\left(\frac{\partial P}{\partial T}\right)_V$, we obtain the result

$$C_P - C_V = -T \left(\frac{\partial V}{\partial T}\right)_P^2 \left(\frac{\partial P}{\partial V}\right)_T. \tag{2.4.1c}$$

The partial derivatives appearing in this expression for $C_P - C_V$ can be converted into an expression that involves experimentally accessible quantities via the definitions

$$\kappa_T \equiv -\frac{1}{V}\left(\frac{\partial V}{\partial P}\right)_T, \qquad \alpha \equiv \frac{1}{V}\left(\frac{\partial V}{\partial T}\right)_P, \tag{2.4.2}$$

with κ_T and α called, respectively, the isothermal compressibility and the thermal expansivity. The heat capacity difference is thus given in terms of experimentally determinable quantities as

$$C_P - C_V = \frac{V T \alpha^2}{\kappa_T}. \tag{2.4.3}$$

Note that it is already clear from this expression that C_P can never be less than C_V, as κ_T is always positive.

If, in returning briefly to Eq. (2.3.10) for the partial derivative that governs the manner in which the internal energy changes with volume, we also note that the right-hand side can be simplified, we obtain the expression

$$\left(\frac{\partial U}{\partial V}\right)_T = T^2 \left[\frac{\partial}{\partial T}\left(\frac{P}{T}\right)\right]_V. \tag{2.4.4}$$

This result was first obtained by Helmholtz in the nineteenth century and is now often referred to as the Helmholtz equation. It is worth noting also that as this expression for $\left(\frac{\partial U}{\partial V}\right)_V$ is valid for an arbitrary thermodynamic system, it may be employed to obtain a general expression for the internal energy. To do so, we shall carry out the integration of the total differential

$$dU = \left(\frac{\partial U}{\partial T}\right)_V dT + \left(\frac{\partial U}{\partial V}\right)_T dV$$

$$= C_V(T)\, dT + \left(\frac{\partial U}{\partial V}\right)_T dV$$

for a closed thermodynamic system.

Interlude To accomplish the integration, we shall firstly recall that a line integral of a total differential of a multivariate function, such as $z(x, y)$, is independent of the path taken between the end-points, so that

$$\int_{(x_i, y_i)}^{(x_f, y_f)} dz(x, y) = z(x_f, y_f) - z(x_i, y_i).$$

Because a line integral for such a function is independent of the path taken, we may evaluate it in terms of a path whose component parts involve changes in only one of the independent variables at a time. Thus, for $z(x, y)$ we may evaluate our line integral along a path whose first part involves integration over x from x_i to x_f while y is held fixed at $y = y_i$, while the second part involves integration over y from y_i to y_f as x is held fixed at $x = x_f$. This choice of path results in the integral taking the form

$$\int_{(x_i, y_i)}^{(x_f, y_f)} dz(x, y) = \int_{x_i}^{x_f} \left(\frac{\partial z}{\partial x}\right)_{y=y_i} dx + \int_{y_i}^{y_f} \left(\frac{\partial z}{\partial y}\right)_{x=x_f} dy.$$

Line integrals are often encountered in the context of processes involving the temperature (T), pressure (P), and volume (V), of a (thermodynamic) system. These processes are commonly illustrated via Pressure–volume (PV) diagrams in which a point (V, P) in the plane represents an equilibrium state of the system, while a curve in the (V, P) plane represents a quasistatic process (i.e., a thermodynamic process that evolves through a succession of equilibrium states). Thermodynamic processes represented by such curves are assumed to be free of dissipative effects and are, as such, reversible. Note that, as real processes are not quasistatic processes, they cannot be depicted using such a diagram. □

For the determination of the internal energy from the total differential dU, we must identify an appropriate thermodynamic reference state. We could choose, for example, the state characterized by a fixed volume V_0 and temperature $T = 0$.[6] It will also be convenient to include the number of particles/atoms/molecules, N, as it typically appears explicitly in the equation of state. Although N is thus not a true thermodynamic variable for a closed thermodynamic system, in which mass cannot be exchanged between the system and its surroundings, it does remain a parameter that both characterizes the system and reflects its size. We shall hence distinguish it from the 'true' thermodynamic variables using a semicolon instead of a comma, thereby treating it as a 'parameter' for the thermodynamic system.

---

[6]Note that we could equally well choose a reference temperature other than $T = 0$, as thermodynamics almost always involves changes in thermodynamic properties, such as ΔU, ΔH, ΔS, etc., rather than the corresponding absolute values U, H, S, and so on.

2.5   Thermodynamic Engines

Very early on during the initial development of thermodynamics as a discipline, engineers took an interest in its potential application in the design of more efficient versions of the then recently-invented 'heat engines'. An important contribution to what is now the Second Law of Thermodynamics was made by Carnot [10], who was interested in the efficiency of various types of heat engines, which quite generally work in a cyclic manner by drawing energy from a higher-temperature thermal reservoir, exporting a portion of it in the form of work, and discharging the remainder into a lower-temperature thermal reservoir. This process is typically shown schematically as in Fig. 2.1. Maximal efficiency of such a conversion of thermal energy into work will occur when the cyclic process is carried out reversibly. However, as reversible processes are idealizations, the attainable efficiency of any conversion of thermal energy (more colloquially, 'heat') into work will be less than the maximal, reversible, value.

If we assume that the thermodynamic system plus surroundings forms a closed thermodynamic system, then the Second Law of Thermodynamics tells us that $(\Delta S)_{\text{total}} \geq 0$ in general, with the equality holding when all processes are reversible. Let us write $(\Delta S)_{\text{total}}$ as

$$(\Delta S)_{\text{total}} = (\Delta S)_{\text{sys}} + (\Delta S)_{\text{surr}}. \tag{2.5.1}$$

Taking the thermodynamic system through a closed cycle of reversible processes results in $(\Delta S)_{\text{sys}} = 0$, while $(\Delta S)_{\text{surr}}$ is given in terms of the heat Q_{import} imported

Fig. 2.1 Cartoon illustrating
a typical 'heat engine' cycle

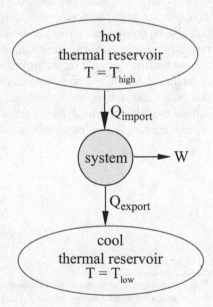

into the system from a heat reservoir at temperature T_{high} and the heat exported from the system into a heat reservoir at temperature T_{low} as

$$(\Delta S)_{surr} = -\frac{Q_{import}}{T_{high}} + \frac{Q_{export}}{T_{low}} . \qquad (2.5.2)$$

We note that $(\Delta S)_{surr}$ includes entropy produced irreversibly during the isothermal heat transfers. If we solve Eq. (2.5.2) for Q_{export}, we obtain

$$Q_{export} = \frac{T_{low}}{T_{high}} Q_{import} + T_{low}(\Delta S)_{surr} . \qquad (2.5.3)$$

According to the First Law of Thermodynamics, the change, ΔU, in the internal energy U when the working thermodynamic substance (referred to more simply as the 'heat engine' or the 'system') is carried through a closed thermodynamic cycle, is given in terms of the work, $W \equiv -W_{export}$, done on the system, Q_{import}, and Q_{export} as

$$(\Delta U)_{cycle} = 0 = -W_{export} + Q_{import} - Q_{export} . \qquad (2.5.4)$$

The work W_{export} is thus given by

$$W_{export} = Q_{import} - Q_{export}$$
$$= Q_{import} \left(1 - \frac{T_{low}}{T_{high}}\right) - T_{low}(\Delta S)_{surr} , \qquad (2.5.5)$$

upon utilizing Eq. (2.5.3) for Q_{export}.

Should the heat engine be reversible, so that $(\Delta S)_{surr} = 0$, then $W_{export} = W_{export}^{(rev)}$ is maximal, namely,

$$W_{export}^{(rev)} = Q_{import} \frac{T_{high} - T_{low}}{T_{high}} = Q_{import} \left(1 - \frac{T_{low}}{T_{high}}\right) , \qquad (2.5.6)$$

and W_{export} may be expressed as

$$W_{export} = W_{export}^{(rev)} - T_{low}(\Delta S)_{surr} . \qquad (2.5.7)$$

Relation (2.5.7) for W_{export} has been termed the work-entropy relation for a heat engine [11], and displays the role of the Second Law of Thermodynamics explicitly. Relation (2.5.7) is obtained upon assuming that all steps in the operating cycle of the heat engine have been carried out reversibly, so that $(\Delta S)_{system} = 0$. Should any of the steps in the operating cycle involve irreversible changes, then $(\Delta S)_{surr}$ in Eq. (2.5.7) must be replaced by $(\Delta S)_{engine} = (\Delta S)_{surr} + (\Delta S)_{cycle}$, with $(\Delta S)_{cycle} > 0$ representing the entropy created by irreversible changes occurring within the system. An increase of entropy is thus always associated with unavailable exportable work. We note also that all entropy generated via irreversible steps in the engine cycle are exported to the low-temperature heat reservoir during the steps involved in returning the system to its initial state.

We shall define the efficiency η of the heat engine as the ratio of the work exported from the system to the heat imported from the hot reservoir, namely, as

$$\eta \equiv \frac{W_{\text{export}}}{Q_{\text{import}}} = \eta_{\text{max}} - \frac{T_{\text{low}}(\Delta S)_{\text{surr}}}{Q_{\text{import}}}, \qquad (2.5.8a)$$

with η_{max} given via Eq. (2.5.6) by

$$\eta_{\text{max}} = 1 - \frac{T_{\text{low}}}{T_{\text{high}}}, \qquad (2.5.8b)$$

and referred to as the maximum efficiency of the heat engine. In particular, because $T_{\text{low}}(\Delta S)_{\text{surr}}$ represents work that is inaccessible by virtue of irreversible processes associated with heat transfers between the system/engine and its surroundings, the maximal work available for export that can be generated by a heat engine thus occurs for a fully reversible engine, for which $(\Delta S)_{\text{surr}} = (\Delta S)_{\text{surr}}^{(\text{rev})} \equiv 0$, in which case, $\eta = \eta_{\text{max}}$. Otherwise, $\eta < \eta_{\text{max}}$.

A perfect engine, i.e., an engine capable of converting Q_{import} fully into exportable work would give (from Eq. (2.5.7))

$$Q_{\text{import}} = W_{\text{export}}^{(\text{perf})} = Q_{\text{import}} \left(1 - \frac{T_{\text{low}}}{T_{\text{high}}} \right) - T_{\text{low}}(\Delta S)_{\text{surr}},$$

from which we would find that $(\Delta S)_{\text{surr}}$ would be given by

$$(\Delta S)_{\text{surr}} = -\frac{Q_{\text{import}}}{T_{\text{high}}} < 0,$$

which violates the Second Law of Thermodynamics. This result is the source of the Kelvin–Planck version of the Second Law, namely, that 'There is no perfect engine: hence, it is impossible to construct a cyclic device which has no effect other than the conversion of heat into an equivalent amount of work'.

2.5.1 The Carnot Engine and Engine Cycle

A Carnot engine [6] is a device that enables a thermodynamic system to exchange mechanical work with its surroundings and to exchange energy reversibly with two thermal reservoirs characterized by temperatures T_{high} and T_{low}. The most common process by which a Carnot engine operates is that devised by Carnot himself, namely, a series of four steps by which the thermodynamic system (or 'working substance') is taken from an initial thermodynamic state to a final thermodynamic state coincident with the initial state. The four steps in the Carnot cycle are thus:

1. a reversible isothermal expansion in thermal contact with the thermal reservoir characterized by temperature $T = T_{high}$;
2. a reversible adiabatic expansion in which the temperature of the system decreases from T_{high} to T_{low};
3. a reversible isothermal compression in thermal contact with the thermal reservoir characterized by temperature $T = T_{low}$;
4. a reversible adiabatic compression in which the system temperature increases from T_{low} to T_{high}, and the system is returned to its initial thermodynamic state.

The importance of the Carnot cycle in applied thermodynamics is reflected in the Carnot Theorem, which states that

> All Carnot cycles operating between the same temperature have the same efficiency.

Note that there is no caveat on this theorem to limit its range of validity by, for example, restricting it to ideal gases or to fluids. Proofs of this theorem may be found in many standard thermodynamics textbooks, such as that by Klotz and Rosenberg [1]. The fact that the Carnot cycle provides a means for measuring the ratio of two temperatures implies that a temperature scale may be established relative to an arbitrary multiplicative constant, so that the assignation of a specific value for the temperature of an arbitrarily chosen standard system then allows the temperatures of all other systems to be determined uniquely in terms of values that are directly proportional to the temperature chosen for the standard system [1, 7]. The absolute (or Kelvin) temperature scale was established in this manner by Lord Kelvin in 1854, by assigning the value 273.16° to the triple point[7]of water.

It is perhaps worth commenting at this point on the continued use of Celsius temperatures in many aspects of engineering research and practice, including the measurement and reporting of thermodynamic properties. The Celsius temperature scale (units °C), which can be related to the absolute thermodynamic (or Kelvin) temperature scale via the prescription $T/°C \simeq T/K - 273.15$, allows negative temperatures, has an incorrect zero and, although it gives temperature differences correctly, it gives incorrect temperature ratios. Thus, as has been emphasized by Callen [7], it really does not qualify as a thermodynamic temperature scale.

The Carnot Cycle for a Classical Ideal Gas

In general, a Carnot cycle consists of a sequence of reversible expansions and compressions beginning with and ending at a particular thermodynamic state,

---

[7]The triple point for a substance is defined as the temperature and pressure at which its solid, liquid, and gas phases are in thermodynamic equilibrium.

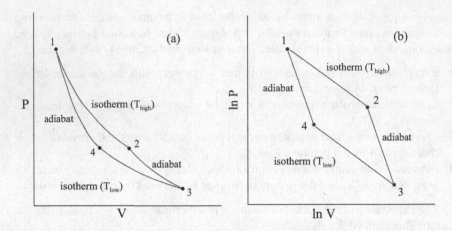

Fig. 2.2 Carnot cycle in the PV-plane for a classical ideal monatomic gas. (**a**) Traditional representation. (**b**) Representation using logarithmic scales on the pressure and volume axes

expressed in terms of a particular pressure–volume point in the PV-plane, as shown in Fig. 2.2. The most common Carnot cycle consists of an isothermal expansion of the working fluid from thermodynamic state 1 to thermodynamic state 2, followed by an adiabatic expansion from state 2 to state 3, an isothermal compression from state 3 to state 4, and finally, an adiabatic compression to return the working fluid from state 4 to the initial state 1. Such a cycle, progressing in a clockwise direction in the PV-plane, is said to move in the forward direction.

We shall consider a classical ideal gas to exemplify this process. As illustrated in Fig. 2.2, the Carnot cycle is associated with the path in the PV-plane taken by a working (ideal) gas in passing from an initial point (V_1, P_1) by a reversible isothermal expansion at temperature T_{high} to the point (V_2, P_2), thence via an adiabatic expansion, during which the temperature changes from T_{high} to T_{low}, from (V_2, P_2) to (V_3, P_3), followed by a reversible isothermal compression at temperature T_{low} from (V_3, P_3) to (V_4, P_4) and, finally, via an adiabatic compression in which the temperature is taken from T_{low} back to T_{high} and returned to the initial thermodynamic state (V_1, P_1).

The first step in the Carnot cycle involves the work

$$W_{1 \to 2}(\text{gas}) = -\int_{V_1}^{V_2} P \, dV = -\int_{V_1}^{V_2} \frac{N k_B T_{\text{high}}}{V} \, dV$$

$$= -N k_B T_{\text{high}} \ln\left(\frac{V_2}{V_1}\right) < 0, \quad V_2 > V_1,$$

done on the gas. Because $U_{\text{system}}(T, V) \equiv U_{\text{ideal gas}}(T)$, $\Delta U_{1 \to 2}(\text{gas}) = 0$ for an isothermal change, and hence $Q_{1 \to 2}(\text{gas}) = -W_{1 \to 2}(\text{gas})$, so that

$$Q_{\text{import}} \equiv Q_{1\to2}(\text{gas}) = Nk_B T_{\text{high}} \ln\left(\frac{V_2}{V_1}\right) > 0 \qquad (2.5.9a)$$

represents the heat imported by the system from its surroundings during the reversible isothermal expansion.

For the adiabatic expansion of the second step of the Carnot cycle, we have $Q_{2\to3}(\text{gas}) \equiv 0$, so that[8]

$$W_{2\to3}(\text{gas}) = (\Delta U)_{2\to3} = \tfrac{3}{2} Nk_B (T_{\text{low}} - T_{\text{high}}). \qquad (2.5.9b)$$

The reversible isothermal compression of the third step in the Carnot cycle involves the work done on the gas, namely,

$$W_{3\to4}(\text{gas}) = -\int_{V_3}^{V_4} P\,dV = -Nk_B T_{\text{low}} \ln\left(\frac{V_4}{V_3}\right) > 0,$$

which implies (because $V_4 < V_3$) that the heat $Q_{3\to4}(\text{gas}) = -W_{3\to4}(\text{gas})$ will be negative. Thus,

$$Q_{\text{export}} \equiv Q_{3\to4}(\text{surr}) = -Nk_B T_{\text{low}} \ln\left(\frac{V_4}{V_3}\right) > 0 \qquad (2.5.9c)$$

represents the amount of heat exported to the surroundings during the compression. Lastly, the adiabatic compression of the final step of the Carnot cycle, with $Q_{4\to1}(\text{gas}) \equiv 0$, leads to

$$W_{4\to1}(\text{gas}) = -\tfrac{3}{2} Nk_B (T_{\text{low}} - T_{\text{high}}). \qquad (2.5.9d)$$

The total work done on the gas through the closed Carnot cycle $1 \to 2 \to 3 \to 4 \to 1$ is thus $W(\text{cycle})$, given by

$$W(\text{cycle}) = W_{1\to2}(\text{gas}) + W_{2\to3}(\text{gas}) + W_{3\to4}(\text{gas}) + W_{4\to1}(\text{gas})$$

$$= -Nk_B T_{\text{high}} \ln\left(\frac{V_2}{V_1}\right) + \tfrac{3}{2} Nk_B (T_{\text{low}} - T_{\text{high}}) - Nk_B T_{\text{low}} \ln\left(\frac{V_4}{V_3}\right)$$

$$- \tfrac{3}{2} Nk_B (T_{\text{low}} - T_{\text{high}})$$

or

$$W(\text{cycle}) = -Nk_B \left[T_{\text{high}} \ln\left(\frac{V_2}{V_1}\right) + T_{\text{low}} \ln\left(\frac{V_4}{V_3}\right) \right]. \qquad (2.5.10)$$

---

[8]For molecular ideal gases, the factor $\tfrac{3}{2}$ is replaced by $(1-\gamma)^{-1}$, with $\gamma \equiv C_P/C_V$.

If we note from Fig. 2.2 that (V_1, P_1) and (V_4, P_4) lie on the same adiabat, and that according to Eq. (2.3.12b) for the entropy of an ideal classical (monatomic) gas, $VT^{\frac{3}{2}}$ is constant for an adiabatic process, then it follows that V_4 and V_1 are related by

$$V_4 = V_1 \left(\frac{T_{\text{high}}}{T_{\text{low}}} \right)^{\frac{3}{2}} .$$

Moreover, as (V_2, P_2) and (V_3, P_3) are also points on a common adiabat, V_3 and V_2 are similarly related by

$$V_3 = V_2 \left(\frac{T_{\text{high}}}{T_{\text{low}}} \right)^{\frac{3}{2}} .$$

From these two conditions, we see that

$$\frac{V_4}{V_3} = \frac{V_1}{V_2} , \tag{2.5.11}$$

so that $W(\text{cycle})$ for an ideal monatomic gas becomes

$$W(\text{cycle}) = Nk_{\text{B}}(T_{\text{low}} - T_{\text{high}}) \ln \left(\frac{V_2}{V_1} \right) \equiv -W_{\text{export}} , \tag{2.5.12}$$

with W_{export} representing the exportable work from the cycle.

The efficiency η, as defined via the first half of Eq. (2.5.8a) can be written as

$$\eta = \frac{Q_{\text{import}} - Q_{\text{export}}}{Q_{\text{import}}} = 1 - \frac{Q_{\text{export}}}{Q_{\text{import}}} . \tag{2.5.13}$$

This expression satisfies the requirements of the First Law of Thermodynamics, but does not yet incorporate restrictions introduced by the Second Law that led to the second half of Eq. (2.5.8a). Specifically, the entropy transferred from the working substance (system) to the cold reservoir during the third step of the Carnot cycle must be greater than (or, for a reversible change, at least equal to) the entropy transferred from the hot reservoir to the working substance in the first step. Thus, we may write

$$\frac{Q_{\text{export}}}{T_{\text{low}}} \geq \frac{Q_{\text{import}}}{T_{\text{high}}} , \quad \text{or} \quad \frac{Q_{\text{export}}}{Q_{\text{import}}} \geq \frac{T_{\text{low}}}{T_{\text{high}}} ,$$

so that

$$\eta \leq 1 - \frac{T_{\text{low}}}{T_{\text{high}}} , \tag{2.5.14}$$

with the equality holding only for a fully reversible cycle (for which $(\Delta S)_{\text{surr}}$ = 0).

We see from the thermodynamics of the Carnot cycle for a classical ideal gas that thermal energy supplied to the ideal gas by the hot thermal reservoir during the isothermal expansion of Step 1 is obtained from the First Law as

$$Q_{\text{import}} \equiv Q_{1\to2}(\text{gas})$$

$$= -W_{1\to2}(\text{gas}) = Nk_B T_{\text{high}} \ln\left(\frac{V_2}{V_1}\right),$$

which is positive, as $V_2 > V_1$. Similarly, the thermal energy exported to the cool reservoir during the isothermal compression of the ideal gas is given by

$$Q_{\text{export}} \equiv Q_{3\to4}(\text{surr})$$

or

$$- Nk_B T_{\text{low}} \ln\left(\frac{V_4}{V_3}\right) = Nk_B T_{\text{low}} \ln\left(\frac{V_3}{V_4}\right),$$

which is also positive, as $V_3 > V_4$.

Thus, for a classical ideal gas, the efficiency, η_{max}, for the Carnot cycle is given explicitly as

$$\eta_{\text{max}} = 1 - \frac{T_{\text{low}}}{T_{\text{high}}} \frac{\ln(V_3/V_4)}{\ln(V_2/V_1)}. \qquad (2.5.15)$$

Employment of Eq. (2.5.11) in expression for η_{max} then gives

$$\eta_{\text{max}} = 1 - \frac{T_{\text{low}}}{T_{\text{high}}}, \qquad (2.5.16)$$

in accordance with the Carnot theorem.

Interlude Figure 2.2b enables a simpler derivation of the final expression for η_{max}. Employment of logarithmic scales for the pressure and volume axes can greatly simplify both the geometric rendering of the Carnot cycle and the simplification of the expression for the maximal efficiency for the Carnot cycle [12]. The condition $PV = K$ (K a constant) that defines an isotherm in the PV-plane is replaced by

$$\ln P = - \ln V + \ln K,$$

which has the form of a straight line with slope -1 and intercept $\ln K$, while the condition $PV^\gamma = K'$ that defines an adiabat in the PV-plane is replaced by

$$\ln P = -\gamma \ln V + \ln K',$$

which similarly defines a straight line, but with slope $-\gamma$ and intercept $\ln K'$. A Carnot cycle is thus represented simply by a parallelogram in the PV-plane (but with logarithmic scales), one pair of sides having slopes -1, the other pair of sides having slopes $-\gamma$, as seen in Fig. 2.2b, with intersection points identified by the integers 1, 2, 3, and 4, as in the more traditional representation of the Carnot cycle illustrated in Fig. 2.2a.

We see directly from this figure that the geometric property of the equality of the lengths of the opposite sides of a parallelogram translates directly into the equality of their projections onto the volume axis: for example, the projection of the upper isotherm onto the $\ln V$-axis is $\ln V_2 - \ln V_1$, while the projection of the lower isotherm onto the $\ln V$- axis is $\ln V_3 - \ln V_4$. The equality of these projections gives directly

$$\ln V_2 - \ln V_1 = \ln V_3 - \ln V_4$$

or

$$\ln \left(\frac{V_2}{V_1} \right) = \ln \left(\frac{V_3}{V_4} \right).$$

Of course, there is nothing special about our selection of the parallelogram sides representing the two isotherms. Precisely the same result is obtained from the equality of the projections onto the $\ln V$-axis of the parallelogram sides representing the two adiabats, namely, $\ln V_4 - \ln V_1 = \ln V_3 - \ln V_2$.   □

The Carnot Cycle for a Quantum Ideal Gas

A specific Carnot cycle for a photon gas (as the primary exemplar of a quantum ideal gas) is illustrated in Fig. 2.3. An isothermal change corresponding to the transition from $(V_1 = 0, P_1)$ to (V_2, P_2) in this Carnot diagram has already been discussed in Example 2.4.

We may utilize Eq. (1.2.12) for the internal energy of the photon gas to obtain the changes ΔU for each of the four segments of the Carnot cycle, namely,

$$\Delta U_{12} = u(T_{\text{expn}})V_2 \,, \quad \Delta U_{23} = -\frac{1}{3} \int_{V_2}^{V_2 + \Delta V} u(T)\, dV \,,$$

$$\Delta U_{34} = -(V_2 + \Delta V)u(T_{\text{comp}}) \,, \quad \Delta U_{41} = 0 \,.$$

Now, as $U(T, V)$ is a function of state, the net change in the internal energy for a closed-loop cycle must be zero: hence, by summing the above four contributions to ΔU for the full Carnot cycle, we obtain

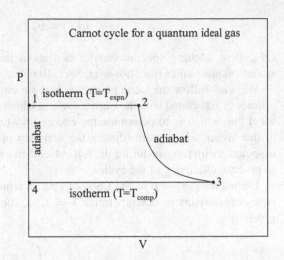

Fig. 2.3 Carnot cycle in the PV-plane for a quantum ideal gas (a gas of photons). Reproduced from [9] with the permission of the American Association of Physics Teachers

$$u(T_{\text{expn}})V_2 - \tfrac{1}{3}u(T_{\text{m}})\Delta V - u(T_{\text{comp}})(V_2 + \Delta V) = 0,$$

in which T_{m} is a temperature such that $T_{\text{comp}} < T_{\text{m}} < T_{\text{expn}}$ and satisfies the intermediate-value theorem. For $T_{\text{expn}} - T_{\text{comp}} \ll T_{\text{comp}}$ and $\Delta V \ll V$, we may approximate ΔV by the differential dV and, upon approximating the difference between the internal energy densities for the isothermal segments of the Carnot cycle by du, we may replace $u(T_{\text{comp}})$ by

$$u(T_{\text{comp}}) = u(T_{\text{m}}) + [u(T_{\text{comp}}) - u(T_{\text{m}})] \le u(T_{\text{m}}) + du,$$

so that to first order in differentials, $\Delta U_{\text{Carnot}} = 0$ is reasonably approximated by the differential equation

$$V\,du - \tfrac{4}{3}u(T)\,dV \simeq 0.$$

Upon utilizing the condition $(\Delta S)_{\text{cycle}} = 0$ for the entropy as well, we may replace $u(T)\,dV$ by

$$u(T)\,dV = V\,du - \frac{uV}{T}\,dT = 0,$$

whence the Carnot cycle for a photon gas requires that the internal energy density $u(T)$ satisfy the differential equation

$$\frac{du}{u} - 4\frac{dT}{T} = 0.$$

Integration of this differential equation gives $u(T)$ as

$$u(T) = bT^4 \, .$$

Of course, nothing specific can be said about the constant b without access to statistical mechanics (see, however, Sect. 10.6.4).

We shall follow the same procedure that we employed in the calculation of the efficiency associated with the Carnot cycle in which the working fluid was a classical ideal gas, which is to determine the energy transferred from the thermal reservoir at the higher temperature (during the first step of the cycle) plus the total work associated with the gas during the full cycle, then use the ratio of these two energies to give the efficiency of the cycle.

During the first step of the Carnot cycle, in which the photon gas undergoes an isothermal expansion at temperature $T = T_{\text{expn}}$, the work done by the photon gas is given by

$$W_{1 \to 2}(\text{gas}) = -\int_{V_1}^{V_2} P \, dV$$

$$= -P_{\text{expn}}(V_2 - V_1) \, ,$$

which is negative, as $V_2 > V_1$, while the change in the internal energy is

$$(\Delta U)_{1 \to 2}(\text{gas}) = u(T_{\text{expn}})(V_2 - V_1) \, .$$

By the First Law, the energy exchanged with the thermal reservoir at temperature T_{expn} will be

$$Q_{1 \to 2}(\text{surr}) = (\Delta U)_{1 \to 2}(\text{gas}) - W_{1 \to 2}(\text{gas})$$

$$= [u(T_{\text{expn}}) + P_1](V_2 - V_1) \, ,$$

which is, as may be expected, positive. Recall that $Q_{1 \to 2}(\text{surr}) > 0$ means that the photon gas has imported energy from its surroundings, so that

$$Q_{1 \to 2}(\text{surr}) = -[u(T_{\text{expn}}) + P_1](V_2 - V_1)$$

$$= -\frac{16\sigma}{3c} T_{\text{expn}}^4 (V_2 - V_1) \, ,$$

using Eqs. (1.2.12). During the second step of the Carnot cycle, the photon gas undergoes an adiabatic expansion from (V_2, P_1) at temperature T_{expn} to (V_3, P_3) at temperature T_3 (which will define the temperature T_{comp} at which the isothermal compression will take place in the third step of the Carnot cycle). Because $PV^{\frac{1}{3}} = K$ for an adiabatic process, we can determine that the work done by the gas is

$$W_{2\to 3}(\text{gas}) = -\int_{V_2}^{V_3} \frac{K}{V^{\frac{1}{3}}} \, dV$$

or

$$W_{2\to 3}(\text{gas}) = 3(P_3 V_3 - P_2 V_2).$$

During the third step of the Carnot cycle, which is a reversible isothermal compression at temperature T_{comp} from (V_3, P_{comp}) to (V_1, P_{comp}), we find that the work done on the photon gas is

$$W_{3\to 4}(\text{gas}) = -\int_{V_3}^{V_4} P_{\text{comp}} \, dV = -P_{\text{comp}}(V_4 - V_3),$$

which is positive, as $V_3 > V_4$.

During the final step of the cycle, which is a reversible adiabatic compression from (V_4, P_{comp}) at T_{comp} to (V_1, P_{comp}) at T_{expn}, the work done by the gas has the same form as that obtained from step 2, namely,

$$W_{4\to 1}(\text{gas}) = 3(P_{\text{expn}} V_1 - P_{\text{comp}} V_4).$$

The total work associated with the photon gas for the Carnot cycle is thus

$$\begin{aligned}
W(\text{cycle}) &= W_{1\to 2}(\text{gas}) + W_{2\to 3}(\text{gas}) + W_{3\to 4}(\text{gas}) + W_{4\to 1}(\text{gas}) \\
&= -P_{\text{expn}}(V_2 - V_1) + 3(P_{\text{comp}} V_3 - P_{\text{expn}} V_2) - P_{\text{comp}}(V_4 - V_3) \\
&\quad + 3(P_{\text{expn}} V_1 - P_{\text{comp}} V_4)
\end{aligned}$$

or

$$W(\text{cycle}) = 4[P_{\text{expn}}(V_1 - V_2) + P_{\text{comp}}(V_3 - V_4)].$$

For the Carnot cycle illustrated in Fig. 2.3, $V_1 = V_4 = 0$ and, as $P_{\text{expn}} = 4\sigma T_{\text{expn}}^4/(3c)$, while $P_{\text{comp}} = 4\sigma T_{\text{comp}}^4/(3c)$, $W(\text{cycle})$ becomes

$$W(\text{cycle}) = \frac{16\sigma}{3c}[T_{\text{comp}}^4 V_3 - T_{\text{expn}}^4 V_2].$$

The efficiency η_{max} for taking a photon gas (specifically, blackbody radiation) through the Carnot cycle is hence given by

$$\eta_{\text{max}} = \frac{W(\text{cycle})}{Q_{1\to 2}(\text{surr})}$$

$$= 1 - \frac{T_{\text{comp}}^4 V_3}{T_{\text{expn}}^4 V_2} .$$

As both V_3 and V_2 lie on the same adiabat, we know that $T_{\text{comp}}^3 V_3 = T_{\text{expn}}^3 V_2$, so that η_{max} simplifies further to

$$\eta_{\text{max}} = 1 - \frac{T_{\text{comp}}}{T_{\text{expn}}},$$

which is precisely the same result obtained for an ideal classical gas.

2.5.2 Reverse Carnot Engine (Refrigeration Cycle)

The Carnot cycle illustrated in Fig. 2.2 for a classical ideal gas proceeds in a clockwise (CW) direction in the PV-plane through the sequence of (PV)-points $1 \rightarrow 2 \rightarrow 3 \rightarrow 4 \rightarrow 1$, and may thus be referred to as a CW Carnot cycle. A (Carnot) heat engine operates between two heat reservoirs, one at a higher temperature, T_{high}, from which it imports heat Q_{import} (in accordance with Fig. 2.1), performs work $(-W)$ on its surroundings, and exports heat $-Q_{\text{export}}$ (with $Q_{\text{export}} > 0$) into the reservoir at temperature T_{low}.

If we examine the outcome of the reverse of this CW Carnot cycle, with the first step starting from point 2 in Fig. 2.2a, then proceeding in the counterclockwise (CCW) direction through the cycle $2 \rightarrow 1 \rightarrow 4 \rightarrow 3 \rightarrow 2$, we see that it corresponds to a reversal of the three directional arrows associated with Q_{import}, W, and Q_{export} in Fig. 2.1. This 'reverse Carnot engine' will thus import heat Q_{import} from the reservoir at temperature T_{low} and, driven by work W done on the working substance, export heat $-Q_{\text{export}}$ to the reservoir at temperature T_{high}. Each CCW cycle will thus transfer heat energy from the lower-temperature heat reservoir to the hotter-temperature heat reservoir. A CCW Carnot cycle is therefore termed a refrigeration cycle.

A device that employs a working fluid to import heat from a low-temperature heat reservoir (which, as implied by the term 'reservoir', may be assumed to have an infinite capacity) and, coupled with work done on the working fluid, export heat to a high-temperature heat reservoir, is termed a *heat pump*. In particular, should the high-temperature heat reservoir be replaced by a finite-mass system (such as a building) as the 'reservoir', this process provides the thermodynamic basis for commercial heat pumps, whereby an electrical device provides the work input needed to export heat from the low-temperature heat reservoir (essentially the surroundings of the building) to the high-temperature 'reservoir'. If, however, the low-temperature reservoir is replaced by a finite-mass system as the 'reservoir' then, following a (sufficiently large) number of completed refrigeration cycles, that finite-mass system will have been cooled to a lower temperature, and will therefore have been refrigerated.

As the internal energy U is a state function, we know that $\Delta U \equiv 0$ for a closed thermodynamic cycle, so that by the First Law (essentially the conservation of energy), we have

$$\Delta U(\text{cycle}) \equiv 0 = W + Q_{\text{import}}^{\text{CCW}} - Q_{\text{export}}^{\text{CCW}},$$

with W, $Q_{\text{import}}^{\text{CCW}}$, $Q_{\text{export}}^{\text{CCW}}$ all positive, or

$$Q_{\text{export}}^{\text{CCW}} = W + Q_{\text{import}}^{\text{CCW}}. \tag{2.5.17}$$

Were we to attempt to evaluate an efficiency η^{CCW} using Eq. (2.5.13) as the ratio of Q_{export}^{CCW} to $Q_{\text{import}}^{\text{CCW}}$, we would obtain a negative value for it, which is meaningless (see also Problem 21). For this reason, we need to identify a meaningful indicator of the effectiveness of a CCW (or refrigeration) cycle. Such an indicator is afforded by the ratio of the heat, $Q_{\text{import}}^{\text{CCW}}$, imported from the lower-temperature reservoir to the external work input that, according to the First Law, is given by the difference between $Q_{\text{export}}^{\text{CCW}}$ and $Q_{\text{import}}^{\text{CCW}}$. This ratio is referred to as the 'coefficient of performance', which we shall designate ε_{COP}. Thus, ε_{COP} is defined in general for a refrigeration cycle as

$$\varepsilon_{\text{COP}} \equiv \frac{Q_{\text{import}}^{\text{CCW}}}{W} = \frac{Q_{\text{import}}^{\text{CCW}}}{Q_{\text{export}}^{\text{CCW}} - Q_{\text{import}}^{\text{CCW}}}. \tag{2.5.18}$$

For an ideal gas working system, we may compute $Q_{\text{import}}^{\text{CCW}}$ from the isothermal expansion step as

$$Q_{\text{import}}^{\text{CCW}} \equiv Q_{4\to 3}(\text{gas}) = Nk_B T_{\text{low}} \ln\left(\frac{V_3}{V_4}\right) \tag{2.5.19a}$$

and $Q_{\text{export}}^{\text{CCW}}$ from the isothermal compression step via

$$Q_{2\to 1}(\text{gas}) = Nk_B T_{\text{high}} \ln\left(\frac{V_1}{V_2}\right) \equiv -Q_{\text{export}}^{\text{CCW}}. \tag{2.5.19b}$$

We have already seen via Eq. (2.5.11) that the two adiabats of the Carnot cycle require the volumes V_i ($i = 1, \ldots, 4$) to be such that the condition $V_2 V_4 = V_1 V_3$ is met.

More specifically, in terms of our expressions for $Q_{\text{import}}^{\text{CCW}}$ and $Q_{\text{export}}^{\text{CCW}}$ for an ideal gas working substance and a reverse Carnot cycle, we see that ε_{COP} is given by

$$\varepsilon_{COP} = \frac{Nk_B T_{low} \ln(V_2/V_1)}{Nk_B(T_{high} - T_{low}) \ln(V_2/V_1)} = \frac{T_{low}}{T_{high} - T_{low}}. \qquad (2.5.20)$$

As may be seen from Eqs. (2.5.16) for $\eta^{CW} \equiv \eta_{max}$ and (2.5.20), the coefficient of performance for a Carnot refrigeration cycle is related to the efficiency of the corresponding Carnot engine cycle by

$$\varepsilon_{COP} = \frac{1}{\eta^{CW}} - 1. \qquad (2.5.21)$$

2.5.3 Curzon–Ahlborn Endoreversible Engine Cycle

The Carnot cycle has been discussed within the context of a reversible engine, which is defined as a working substance that is coupled reversibly to the external world (in the form of a pair of thermal reservoirs characterized by temperatures T_{high} and T_{low}) and which undergoes only reversible thermodynamic transformations during a closed four-step operational cycle. It is well known that the efficiency of an engine is maximized by coupling it to two thermal reservoirs via reversible isothermal processes. However, it will be of no surprise that because a reversible isothermal process must, in principle, be carried out infinitely slowly in order that the working substance remains in thermal equilibrium with the thermal reservoir to which it is coupled, a reversible engine will deliver no power, as an infinite time would be required for completion of a working cycle. Hence, to obtain a finite power output from a thermodynamic cycle, it becomes necessary to affect energy/heat transfers to and from the thermal reservoirs irreversibly. Removal of the condition that the engine be coupled to the external world only via reversible thermodynamic processes provides a broader type of engine, referred to as an *endoreversible* engine [13], and defined as an engine whose working substance undergoes only reversible thermodynamic transformations during each operational cycle. The class of endoreversible engines thus includes engines that are coupled to their surroundings via reversible processes (reversible engines) and engines that are coupled to their surroundings via irreversible processes. Thus, an understanding of endoreversible engines also provides an introduction to the study of finite step-by-step processes [14, 15].

The thermodynamics pertinent to an endoreversible engine may be modelled along the lines that we have utilized previously to analyse the Carnot cycle. To this end, we have a standard ideal heat engine composed of a working substance that is enclosed in a cylinder with walls that can be made conducting (diathermic walls having a constant thermal conductivity), thereby allowing thermal energy to be transferred irreversibly either from the thermal source (reservoir) characterized by temperature T_{high} to the working substance at temperature T_1 or, similarly, from the working substance at temperature T_2 to the thermal sink (reservoir) characterized by temperature T_{low} (with $T_{high} > T_1 > T_2 > T_{low}$), or made insulating for the adiabatic steps in the engine cycle. In addition, we shall require the cylinder to be fitted with a piston that can be employed to perform work on the external world.

Irreversible energy transfers thus occur across finite temperature differences $(\Delta T)_1 \equiv T_{high} - T_1$ and $(\Delta T)_2 \equiv T_2 - T_{low}$. We shall assume that the energy fluxes q_1 and q_2 thereby associated with these irreversible processes are given by

$$q_1 \simeq \frac{Q_{1\to 2}(\text{surr})}{t_1} = \sigma_1(\Delta T)_1, \tag{2.5.22a}$$

and

$$q_2 \simeq \frac{Q_{3\to 4}(\text{system})}{t_2} = \sigma_2(\Delta T)_2, \tag{2.5.22b}$$

in which t_1 and t_2 are the time intervals associated with the two energy transfers, and σ_1, σ_2 are the thermal conductances[9] of the working chamber walls when they are in thermal contact with the reservoirs. Further, if we assume that adiabatic changes occur on time scales much shorter than t_1 and t_2, we may employ Eqs. (2.5.22) to obtain an expression for the total time $t = t_1 + t_2$ associated with a typical endoreversible cycle, namely,

$$t = \frac{1}{\sigma_{high}} \frac{Q_{1\to 2}^{rev}(\text{system})}{T_{high} - T_1} + \frac{1}{\sigma_{low}} \frac{Q_{3\to 4}(\text{surr})}{T_2 - T_{low}}. \tag{2.5.23}$$

The energy input, $Q_{1\to 2}(\text{system})$, from the thermal reservoir at temperature T_{high} and the energy, $Q_{3\to 4}(\text{surr})$, exported to the thermal reservoir at temperature T_{low} can be determined from our analysis of the thermodynamics of the Carnot cycle as

$$Q_{1\to 2}^{rev}(\text{system}) = Nk_B T_1 \ln\left(\frac{V_2}{V_1}\right), \tag{2.5.24a}$$

which is positive, since $V_2 > V_1$. Similarly, the energy to be exported during the isothermal compression step in the Carnot cycle is given by

$$Q_{3\to 4}^{rev}(\text{system}) = Nk_B T_2 \ln\left(\frac{V_4}{V_3}\right),$$

which is negative, as this represents energy to be transferred irreversibly to the surroundings as $Q_{3\to 4}(\text{surr}) = -Q_{3\to 4}^{rev}(\text{system})$, or

---

[9]The thermal conductance for an 'engine' wall is proportional to the thermal conductivity (units: $W\,m^{-1}\,K^{-1}$) of the substance from which it has been constructed, and to the area of the wall that is in thermal contact with the working substance, and is inversely proportional to the thickness of the wall (net units: $W\,K^{-1}$).

$$Q_{3\to 4}(\text{surr}) = Nk_B T_2 \ln\left(\frac{V_3}{V_4}\right). \tag{2.5.24b}$$

Upon substituting expressions (2.5.24) for $Q_{1\to 2}^{\text{rev}}(\text{system})$ and $Q_{3\to 4}(\text{surr})$ for the Carnot cycle into Eq. (2.5.23) for t and utilizing relation (2.5.11) between the volumes of the system during the Carnot cycle, we obtain t as

$$t = \left[\frac{1}{\sigma_1}\frac{T_1}{T_{\text{high}} - T_1} + \frac{1}{\sigma_2}\frac{T_2}{T_2 - T_{\text{low}}}\right] Nk_B \ln\left(\frac{V_2}{V_1}\right). \tag{2.5.25a}$$

From Eq. (2.5.10), we see that the network for a Carnot cycle for a system consisting of an ideal classical monatomic gas and operating reversibly between two thermal reservoirs (one characterized by temperature T_1, the other by temperature T_2, with $T_2 < T_1$) is given by

$$W^{\text{rev}}(\text{system}) = Nk_B(T_2 - T_1)\ln\left(\frac{V_2}{V_1}\right),$$

so that $-W^{\text{rev}}(\text{system}) > 0$ represents the net energy output of the Carnot cycle connected irreversibly to two thermal reservoirs, one characterized by temperature $T_{\text{high}} > T_1$, the other characterized by temperature $T_{\text{low}} < T_2$. The total time taken for the entire process can now be written as

$$t = \left[\frac{1}{\sigma_1}\frac{T_1}{T_{\text{high}} - T_1} + \frac{1}{\sigma_2}\frac{T_2}{T_2 - T_{\text{low}}}\right]\frac{-W^{\text{rev}}(\text{system})}{T_1 - T_2}. \tag{2.5.25b}$$

We may now utilize expression (2.5.25b) to obtain the power exported by this arrangement as

$$P \equiv \frac{-W^{\text{rev}}(\text{system})}{t} = (T_1 - T_2)\left[\frac{1}{\sigma_1}\frac{T_1}{T_{\text{high}} - T_1} + \frac{1}{\sigma_2}\frac{T_2}{T_2 - T_{\text{low}}}\right]^{-1},$$

or

$$P(T_1, T_2) = \frac{\sigma_1\sigma_2(T_1 - T_2)(T_{\text{high}} - T_1)(T_2 - T_{\text{low}})}{T_1\sigma_2(T_2 - T_{\text{low}}) + T_2\sigma_1(T_{\text{high}} - T_1)}. \tag{2.5.26}$$

The power output from an endoreversible engine thus depends upon the operating temperatures T_1 and T_2 of the Carnot cycle, so that we should now determine the temperatures most appropriate for its operation. We therefore wish to determine operating temperatures $T_{1,\text{opt}}$ and $T_{2,\text{opt}}$ that maximize the power output of the Carnot cycle.

By defining two new variables x and y as [13]

$$x \equiv T_{\text{high}} - T_1, \quad y \equiv T_2 - T_{\text{low}}, \tag{2.5.27}$$

we may express Eq. (2.5.26) in a mathematically more tractable form as

$$P(x, y) = \frac{\sigma_1 \sigma_2 xy(T_{high} - T_{low} - x - y)}{\sigma_2 T_{high} y + \sigma_1 T_{low} x + xy(\sigma_1 - \sigma_2)}. \tag{2.5.28}$$

To obtain an expression for the maximum power produced by an endoreversible engine, we must determine optimal values of x and y (in terms of σ_1, σ_2, T_{high}, T_{low}) that maximize expression (2.5.28). To accomplish this task, we require that

$$\left(\frac{\partial P}{\partial x}\right)_y = 0 \quad \text{and} \quad \left(\frac{\partial P}{\partial y}\right)_x = 0 \tag{2.5.29}$$

simultaneously. These two conditions may be employed to show that the optimal values of x and y are related by

$$y_{opt} = \left(\frac{\sigma_1 T_{low}}{\sigma_2 T_{high}}\right)^{\frac{1}{2}} x_{opt} \tag{2.5.30}$$

and ultimately, that x_{opt} obeys the quadratic equation

$$\left(1 - \frac{\sigma_1}{\sigma_2}\right)\left(\frac{x_{opt}}{T_{high}}\right)^2 - 2\left[1 + \left(\frac{\sigma_1 T_{low}}{\sigma_2 T_{high}}\right)^{\frac{1}{2}}\right]\left(\frac{x_{opt}}{T_{high}}\right) + \left(1 - \frac{T_{low}}{T_{high}}\right) = 0. \tag{2.5.31}$$

Optimal values $T_{1,opt}$ and $T_{2,opt}$ for the temperatures of the thermal reservoirs for the Carnot cycle can be obtained by combining the solutions to Eq. (2.5.31) and an equivalent quadratic equation for y_{opt}/T_{low} with the definitions $T_{1,opt} = T_{high} - x_{opt}$, $T_{2,opt} = T_{low} + y_{opt}$, to give

$$T_{1,opt} = g(T_{high}, T_{low}, \sigma_1, \sigma_2)\sqrt{T_{high}}; \quad T_{2,opt} = g(T_{high}, T_{low}, \sigma_1, \sigma_2)\sqrt{T_{low}}, \tag{2.5.32}$$

in which $g(T_{high}, T_{low}, \sigma_1, \sigma_2) \equiv (\sqrt{\sigma_1 T_{high}} + \sqrt{\sigma_2 T_{low}})/(\sqrt{\sigma_1} + \sqrt{\sigma_2})$.

Finally, we note that the efficiency of an endoreversible engine is given by the efficiency η_{opt} of the Carnot cycle operated at temperatures $T_1 = T_{1,opt}$ and $T_2 = T_{2,opt}$ obtained by optimization of the power output of the Carnot cycle, namely,

$$\eta_{opt} = 1 - \frac{T_{2,opt}}{T_{1,opt}}. \tag{2.5.33a}$$

Substitution of the results (2.5.32) for $T_{1,opt}$ and $T_{2,opt}$ then gives η_{opt} as

$$\eta_{opt} = 1 - \frac{\sqrt{T_{low}}}{\sqrt{T_{high}}}. \tag{2.5.33b}$$

Table 2.1 Computed and observed heat engine efficiencies[a]

Power source	T_{low} /K	T_{high} /K	η_{max} /%	η_{opt} /%	η_{obs} /%
Coal-fired steam plant	298	838	64.1	40	36
CANDU nuclear reactor	298	573	48.0	28	30
Geothermal steam plant	353	523	32.3	17.5	16

[a]Data from Table 2.1 of Ref. [13], with the permission of the American Association of Physics Teachers and the authors

As pointed out by Callen [7], that η_{opt} depends only upon the temperatures of the thermal reservoirs and, in particular, is independent of the thermal conductances σ_1 and σ_2, is a remarkable result, indicating that optimal efficiency may well be independent of the mechanism for irreversible heat transfer between an endore-versible engine and its thermal reservoirs. Indeed, Curzon and Ahlborn themselves recognized this possibility, and carried out a comparison between η_{max} (for a Carnot engine), η_{opt} for an endoreversible engine, and η_{obs} determined for three realistic heat engines (power plants). Their comparison is given in Table 2.1.

2.5.4 The Otto and Diesel Engine Cycles

Two thermodynamic cycles that have been proposed as models for first approximations to actual practical engines are the Otto and Diesel cycles. The Otto cycle provides a relatively crude approximation to the operation of a typical four-stroke gasoline engine by first compressing the working fluid adiabatically, then heating it isochorically, followed by an adiabatic expansion (the so-called power stroke), and finally cooling it isochorically to its initial state. However, the initial compression step is not quasi-static (hence not isentropic), and the heating step at constant volume does not account for the internal combustion process. The Diesel cycle first compresses the working fluid adiabatically (as for the Otto cycle), then heats the fluid at constant pressure (the combustion step), expands the fluid adiabatically and finally, cools the fluid at constant volume to its initial state. Generally speaking, both Otto and Diesel cycles involve four different temperatures, only two of which are associated with heat reservoirs. All four temperatures would be required to have associated heat reservoirs were it required to carry out all four steps reversibly.

The Otto Cycle

The Otto cycle is illustrated in Fig. 2.4a by a parallelogram in a $\ln P$ vs. $\ln V$ plot [12]. As the initial compression of the working fluid (which is a hydrocarbon-air mixture in a typical gasoline engine) from (V_1, P_1) to (V_2, P_2) and the expansion (generated in a typical gasoline engine by internal combustion) from (V_3, P_3) to (V_4, P_4) are treated as reversible adiabatic processes, we have $Q_{1 \to 2}^{rev}(\text{fluid}) \equiv 0$ and

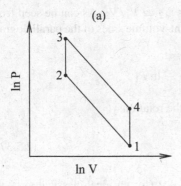

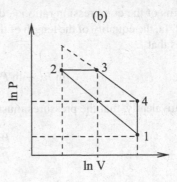

Fig. 2.4 Otto and Diesel thermodynamic cycles using logarithmic scales on the pressure and volume axes: (**a**) Otto cycle, (**b**) Diesel cycle

$Q_{3 \to 4}^{\text{rev}}(\text{fluid}) \equiv 0$, so that the energy imported, namely, $Q_{\text{import}}^{\text{rev}} \equiv Q_{2 \to 3}^{\text{rev}}(\text{fluid})$, with

$$Q_{2 \to 3}^{\text{rev}}(\text{fluid}) = (\Delta U)_{2 \to 3} = C_V(T_{\max} - T_{\text{high}}) > 0, \qquad (2.5.34a)$$

occurs during the isochoric compression step from (V_2, P_2) to (V_3, P_3), while the energy exported, namely, $Q_{\text{export}}^{\text{rev}} \equiv -Q_{4 \to 1}^{\text{rev}}(\text{fluid})$, with

$$Q_{4 \to 1}^{\text{rev}}(\text{fluid}) = (\Delta U)_{4 \to 1} = C_V(T_{\min} - T_{\text{low}}) < 0, \qquad (2.5.34b)$$

occurs during the isochoric expansion step from (V_4, P_4) to (V_1, P_1) that returns the working fluid to its initial state. The total work performed by the working fluid during the Otto cycle is thus $Q_{\text{import}}^{\text{rev}} - Q_{\text{export}}^{\text{rev}}$, or

$$W(\text{cycle}) = C_V[(T_{\max} - T_{\text{high}}) - (T_{\text{low}} - T_{\min})]. \qquad (2.5.35)$$

The efficiency, η_{Otto}, of an Otto engine is thus given by

$$\eta_{\text{Otto}} \equiv 1 - \frac{Q_{\text{export}}^{\text{rev}}}{Q_{\text{import}}^{\text{rev}}}$$

$$= 1 - \frac{T_{\text{low}} - T_{\min}}{T_{\max} - T_{\text{high}}}. \qquad (2.5.36a)$$

The Otto cycle efficiency is traditionally expressed in terms of pressure and volume. If we utilize the ideal (classical) gas law, we see that η_{Otto} may also be written as

$$\eta_{\text{Otto}} = 1 - r_{12} \frac{P_4 - P_1}{P_3 - P_2} \qquad (2.5.36b)$$

in terms of the compression ratio r_{12} defined as $r_{12} \equiv V_1/V_2$. As can be seen from Fig. 2.4a, the equality of the length of the constant-volume sides of the parallelogram means that

$$\ln P_3 - \ln P_2 = \ln P_4 - \ln P_1 \,,$$

or equivalently, that the pressures must satisfy the relation

$$\frac{P_3}{P_2} = \frac{P_4}{P_1} \,. \tag{2.5.37a}$$

Moreover, as the slope of the initial adiabat is $-\gamma$ (for an ideal classical gas), we also have

$$\frac{\ln P_2 - \ln P_1}{\ln V_2 - \ln V_1} = -\gamma \,,$$

from which we obtain the pressure–volume relation

$$\frac{P_2}{P_1} = r_{12}^{\gamma} \,. \tag{2.5.37b}$$

Relations (2.5.37) enable us to re-express the efficiency of the Otto cycle as

$$\eta_{\text{Otto}} = 1 - \frac{1}{r_{12}^{\gamma-1}} \,. \tag{2.5.38}$$

The lowest and highest temperatures, $T_{\min}$ and $T_{\max}$, respectively, are assumed to be known, with $T_{\min}$ fixed by the ambient temperature and $T_{\max}$ determined by the combustion characteristics of the specific gasoline-air mixture being considered. The intermediate temperatures, T_{high} and T_{low}, while not predetermined, may not be varied freely, however, as the total entropy change, $(\Delta S)_{\text{cycle}}$, must be zero. Moreover, as the two adiabatic steps require that $S_2 = S_1$, or $(\Delta S)_{1 \to 2} = 0$, and $S_3 = S_4$, or $(\Delta S)_{3 \to 4} = 0$, the entropy constraint is thereby reduced to

$$(\Delta S)_{\text{cycle}} = (\Delta S)_{2 \to 3} + (\Delta S)_{4 \to 1} = 0 \,,$$

from which we may deduce directly that T_{high} and T_{low} are constrained to take values such that

$$T_{\text{high}} T_{\text{low}} = T_{\min} T_{\max} \,. \tag{2.5.39}$$

We may employ this constraint on the temperatures to obtain a suggestive form for the total work $W(\text{cycle}) \equiv W_{\text{cycle}}(\text{fluid})$ for the fluid, namely,

$$W(\text{cycle}) = C_V[(T_{\min} - 2\sqrt{T_{\min}T_{\max}} + T_{\max}) - (T_{\text{high}} - 2\sqrt{T_{\text{high}}T_{\text{low}}} + T_{\text{low}})]$$

$$= C_V[(\sqrt{T_{\min}} - \sqrt{T_{\max}})^2 - (\sqrt{T_{\text{high}}} - \sqrt{T_{\text{low}}})^2] . \tag{2.5.40a}$$

To optimize the total work associated with the Otto cycle,[10] we first employ the constraint (2.5.39) to express T_{low} in Eq. (2.5.40a) in terms of T_{high}. Then, upon setting the first derivative of the total work, $W'(\text{cycle})$, given by

$$W'(\text{cycle}) \equiv \frac{dW}{dT_{\text{high}}} = -C_V\left[1 - \frac{T_{\min}T_{\max}}{T_{\text{high}}^2}\right],$$

to zero for $T_{\text{high}} = T_{\text{high,opt}}$, we obtain

$$T_{\text{high,opt}} = \sqrt{T_{\min}T_{\max}} \tag{2.5.40b}$$

as the optimum value for T_{high}. The second derivative test gives $W''(T_{\text{high,opt}}) < 0$, so that this optimum value represents a maximum. Thus, from expression (2.5.40a) for the total work associated with the Otto cycle, we find that $W_{\max}(\text{cycle})$, which is attained for $T_{\text{low,opt}} = T_{\text{high,opt}}$, is given by

$$W_{\max}(\text{cycle}) = C_V(\sqrt{T_{\min}} - \sqrt{T_{\max}})^2 . \tag{2.5.40c}$$

Indeed, this result may be deduced simply upon inspection of Eq. (2.5.40a).

We are now also able to determine the optimized efficiency,

$$\eta_{\text{opt}} \equiv 1 - \frac{T_{\text{low,opt}} - T_{\min}}{T_{\max} - T_{\text{high,opt}}} ,$$

for the Otto cycle. From the constraint condition (2.5.39) on T_{high} and T_{low}, we see that $T_{\text{low,opt}} = T_{\text{high,opt}} = \sqrt{T_{\min}T_{\max}}$, so that η_{opt} is given by

$$\eta_{\text{opt}} = 1 - \frac{\sqrt{T_{\min}}}{\sqrt{T_{\max}}} . \tag{2.5.41}$$

This result is the same as that obtained in Eq. (2.5.33b) for the efficiency of an endoreversible engine (i.e., for the Curzon–Ahlborn cycle). However, the present result has been attained through optimization of the work produced during a reversible Otto cycle, rather than via the introduction of irreversible thermodynamic couplings between a (reversible) Carnot engine and the hot and cold thermal reservoirs employed in the two isothermal steps of the Carnot cycle.

---

[10] The optimization discussed here is due to Leff [14].

If, in Eq. (2.5.36a) for η_{Otto}, we employ the adiabaticity constraint (2.5.39) on the non-externally-fixed temperatures T_{high} and T_{low} to eliminate T_{low}, we may also express η_{Otto} in a form analogous to that for the Carnot efficiency, that is,

$$\eta_{\text{Otto}} = 1 - \frac{T_{\text{min}}}{T_{\text{high}}}. \qquad (2.5.42a)$$

Note, however, that as the highest temperature achieved in an Otto cycle is T_{max} and the lowest temperature is the initial temperature T_{min}, both T_{high} and T_{low} (with $T_{\text{high}} > T_{\text{low}}$) lie between these two values. The Otto efficiency given by Eq. (2.5.42a) will thus be smaller than the maximal efficiency η_{max}, given by

$$\eta_{\text{max}} = 1 - \frac{T_{\text{min}}}{T_{\text{max}}}, \qquad (2.5.42b)$$

that would be obtained for a Carnot cycle operating between thermal reservoirs characterized by temperatures T_{max} and T_{min}. We may also express the total work, $W(\text{cycle})$, given by Eq. (2.5.40a) as a function of the efficiency η_{Otto} as

$$W(\text{cycle}) = C_V T_{\text{max}} \eta_{\text{Otto}} \frac{\eta_{\text{max}} - \eta_{\text{Otto}}}{1 - \eta_{\text{Otto}}}, \qquad (2.5.43)$$

from which we also find that η_{Otto} is given by Eq. (2.5.41), as might be anticipated. Moreover, as we require $W(\text{cycle}) \geq 0$, η_{Otto} is restricted to values such that $0 \leq \eta_{\text{Otto}} < 1$.

If we normalize $W(\text{cycle})$ by dividing Eq. (2.5.43) by $W_{\text{max}}(\text{cycle})$, then we obtain

$$\frac{W(\text{cycle})}{W_{\text{max}}(\text{cycle})} = \frac{\eta_{\text{Otto}}(\eta_{\text{max}} - \eta_{\text{Otto}})}{\eta_{\text{opt}}^2(1 - \eta_{\text{Otto}})} \qquad (2.5.44)$$

for the normalized total work for the Otto cycle. From this result, it is clear that $W(\text{cycle})$ goes to zero both for an Otto cycle having zero efficiency ($\eta_{\text{Otto}} = 0$) and for one having the Carnot efficiency ($\eta_{\text{Otto}} = \eta_{\text{max}}$). Also, as the normalized work per cycle has only a single maximum, this means that for each value of $W(\text{cycle})/W_{\text{max}}(\text{cycle})$ other than 0 or 1, an Otto engine can perform work at two different efficiencies, one lying above, the other lying below η_{opt}.

The traditional compression volume ratio r_{12} for the initial step of the Otto cycle may be obtained from Eq. (2.5.36a) as a function of the efficiency as

$$r_{12} \equiv \frac{V_1}{V_2} = (1 - \eta_{\text{Otto}})^{-\frac{1}{\gamma+1}}. \qquad (2.5.45)$$

As r_{12} is an increasing monotonic function of η_{Otto} on the physically relevant interval $0 \leq \eta_{\text{Otto}} \leq \eta_{\text{max}} < 1$, its minimal value $r_{12}^{\text{min}} = 1$ is attained for $\eta_{\text{Otto}} = 0$, and its maximal value

$$r_{12}^{\max} = \left(\frac{T_{\max}}{T_{\min}}\right)^{\frac{\gamma}{\gamma-1}},$$

is attained for $\eta_{\text{Otto}} = \eta^{\max}$: both extreme values for r_{12} correspond to $W(\text{cycle}) = 0$ (as has already been mentioned). We are now in a position to interpret $r_{12} \rightarrow r_{\min} = 1$ as the Otto cycle illustrated in Fig. 2.4a contracting horizontally to a vertical line that encloses zero area and produces zero work, so that although the energy input along path $2 \rightarrow 3$ remains finite, the vanishing of the network leads to an efficiency of zero. Similarly, $r_{12} \rightarrow r_{12}^{\max}$ corresponds to the Otto cycle contracting vertically, so that the two adiabatic steps have paths that approach one another, with the energy imported and work output approaching zero simultaneously at the same rate, so that the efficiency remains finite.

The Diesel Cycle

As has been pointed out by Leff [14], Diesel's original intention in designing the thermodynamic cycle that now bears his name was to use air as the working fluid and, by compression in a cylinder, raise it to a temperature sufficiently high for a hydrocarbon fuel injected into the cylinder to undergo constant-temperature combustion as the resultant gas mixture expanded against the cylinder piston. His ultimate hope was to be able to convert all of the energy released by the combustion process isothermally into work, so that his cycle would simulate the Carnot cycle and thereby achieve close to maximal efficiency.

An idealized version of the Diesel cycle is illustrated in Fig. 2.4b. It begins with an adiabatic heating of the fluid from (V_1, P_1) to (V_2, P_2), followed by an isobaric heating of the fluid by internal combustion from (V_2, P_2) to (V_3, P_2), then the power stroke, represented by an adiabatic expansion from (V_3, P_2) to (V_4, P_4), and finishing with isochoric cooling from pressure P_4 back to the initial pressure, P_1. Thus, for the Diesel cycle, energy imported from the surroundings occurs during the constant pressure second step and, for an ideal gas, it is given via the First Law expression as

$$Q_{2 \rightarrow 3}^{\text{rev}}(\text{fluid}) = (\Delta U)_{2 \rightarrow 3} - W_{2 \rightarrow 3}^{\text{rev}}$$

$$= (C_V + Nk_{\text{B}})(T_{\max} - T_{\text{high}}) > 0,$$

so that Q_{import} is given by

$$Q_{\text{import}} \equiv C_P(T_{\max} - T_{\text{high}}). \tag{2.5.46a}$$

The temperature $T_{\max}$ attained at completion of the combustion step is the maximum temperature for the Diesel cycle.

Similarly, the heat (energy) export occurs during the final, isochoric, step: as $W^{rev}_{4\to1}$(fluid) $\equiv 0$ for an isochoric process, $Q_{export} \equiv -Q^{rev}_{4\to1}$(fluid). Moreover, $Q^{rev}_{4\to1}$(fluid) is given by

$$Q^{rev}_{4\to1}(\text{fluid}) = (\Delta U)_{4\to1}$$
$$= C_V(T_{min} - T_{low}) < 0,$$

since $T_{min} < T_{low}$. Thus, we see that

$$Q_{export} = C_V(T_{low} - T_{min}). \tag{2.5.46b}$$

The efficiency for the Diesel cycle is thus

$$\eta_{Diesel} = 1 - \frac{Q_{export}}{Q_{import}}$$
$$= 1 - \frac{C_V(T_{low} - T_{min})}{C_P(T_{max} - T_{high})}$$

or

$$\eta_{Diesel} = 1 - \frac{T_{low} - T_{min}}{\gamma(T_{max} - T_{high})}. \tag{2.5.47}$$

It is traditional to express the efficiency for the Diesel cycle in terms of pressure and volume as

$$\eta_{Diesel} = 1 - \frac{1}{\gamma}\frac{P_4V_4 - P_1V_1}{P_3V_3 - P_2V_2}$$

or, since $V_4 = V_1$ and $P_3 = P_2$ for a Diesel engine, as

$$\eta_{Diesel} = 1 - \frac{P_1V_1}{\gamma P_2V_2}\frac{\frac{P_4}{P_1} - 1}{\frac{V_3}{V_2} - 1}.$$

From Fig. 2.4b, the trapezoidal geometry of the Diesel cycle gives the slope of the adiabatic power cycle path ($3 \to 4$) as

$$\frac{\ln P_4 - \ln P_1}{\ln V_2 - \ln V_3} = -\gamma, \quad \text{or} \quad \frac{P_4}{P_1} = \left(\frac{V_3}{V_2}\right)^\gamma, \tag{2.5.48a}$$

while the initial adiabatic compression from (V_1, P_1) to (V_2, P_2) similarly gives

$$\frac{\ln P_1 - \ln P_2}{\ln V_1 - \ln V_2} = -\gamma, \quad \text{or} \quad \frac{P_2}{P_1} = \left(\frac{V_1}{V_2}\right)^{\gamma}. \tag{2.5.48b}$$

These two relations allow a further simplification of η_{Diesel} to

$$\eta_{\text{Diesel}} = 1 - \frac{r_{32}^{\gamma} - 1}{\gamma(r_{32} - 1)r_{12}^{\gamma-1}} \tag{2.5.49}$$

in terms of the volume compression and expansion ratios r_{12} and r_{32}, namely,

$$r_{12} \equiv \frac{V_1}{V_2}, \qquad r_{32} \equiv \frac{V_3}{V_2}, \tag{2.5.50}$$

respectively, that represent the volume ratios achieved in the initial compression step and in the power expansion step of the Diesel cycle.

The total work for a Diesel cycle is given by

$$W_{\text{cycle}}(\text{system}) = C_P(T_{\text{max}} - T_{\text{high}}) - C_V(T_{\text{low}} - T_{\text{min}}),$$

so that the net work, $W(\text{cycle})$, available for *export* from a Diesel engine is

$$W(\text{cycle}) = Q_{\text{in}} - Q_{\text{out}}$$
$$= C_P(T_{\text{max}} - T_{\text{high}}) - C_V(T_{\text{low}} - T_{\text{min}}). \tag{2.5.51}$$

As for the Otto cycle, the net entropy change,

$$(\Delta S)_{\text{cycle}} = C_P \ln\left(\frac{T_{\text{max}}}{T_{\text{high}}}\right) + C_V \ln\left(\frac{T_{\text{min}}}{T_{\text{low}}}\right) \equiv 0,$$

imposes the constraint

$$T_{\text{min}} T_{\text{max}}^{\gamma} = T_{\text{low}} T_{\text{high}}^{\gamma} \tag{2.5.52}$$

on the temperatures appearing in the Diesel cycle. We may then employ this constraint to replace the temperature variable T_{low} in Eq. (2.5.51) to obtain

$$W(\text{cycle}) = C_P(T_{\text{max}} - T_{\text{high}}) - C_V T_{\text{min}} \left(\frac{T_{\text{max}}^{\gamma}}{T_{\text{high}}^{\gamma}} - 1\right) \tag{2.5.53}$$

for the available work. Optimization of $W(\text{cycle})$ with respect to T_{high} gives $T_{\text{high,opt}}$ as

$$T_{\text{high,opt}} = (T_{\text{min}} T_{\text{max}}^{\gamma})^f, \qquad f \equiv (\gamma + 1)^{-1}, \tag{2.5.54}$$

while the entropy constraint (2.5.52) gives $T_{\text{low,opt}} = T_{\text{high,opt}}$. The second derivative of $W(\text{cycle})$,

$$\frac{\mathrm{d}^2 W}{\mathrm{d}T_{\text{high}}^2} = -\gamma(\gamma + 1)C_V \frac{T_{\min} T_{\max}^{\gamma}}{T_{\text{high}}^{\gamma+2}} < 0,$$

so that the optimized value for $W(\text{cycle})$ is, as expected, a maximum. The final expression for the optimized available work for a Diesel cycle is somewhat more complicated than that for the Otto cycle, and takes the form

$$W_{\max}(\text{cycle}) = C_V T_{\max}[\gamma(1 - \tau^f) - \tau^f(1 - \tau^{1-f})], \tag{2.5.55}$$

in which τ is the temperature ratio $\tau \equiv T_{\min}/T_{\max}$ between the highest and lowest temperatures in the Diesel cycle. The efficiency of the optimized Diesel cycle is given by Eq. (2.5.47), with $T_{\text{low}} = T_{\text{high}} = T_{\text{high,opt}}$, as

$$\eta_{\text{Diesel}}^{\text{opt}} = 1 - \frac{T_{\text{high,opt}} - T_{\min}}{\gamma(T_{\max} - T_{\text{high,opt}})}$$

or, expressed in terms of the temperature ratio τ, as [14]

$$\eta_{\text{Diesel}}^{\text{opt}} = 1 - \tau^f \frac{1 - \tau^{1-f}}{\gamma(1 - \tau^f)}. \tag{2.5.56}$$

The optimized expressions (2.5.55) and (2.5.56) for $W_{\max}(\text{cycle})$ and $\eta_{\text{Diesel}}^{\text{opt}}$ are more complex than those obtained for the Otto cycle. In particular, the Diesel cycle efficiency depends upon the heat capacity ratio γ for the working fluid, while those for the Carnot, Otto, and Curzon–Ahlborn cycles depend solely upon the ratio of the highest and lowest temperatures for the cycle.

2.5.5 Counterclockwise Otto Cycles

In this subsection, we shall follow fairly closely a recent discussion of counter-clockwise (CCW) pressure–volume working cycles other than Carnot cycles due to Dickerson and Mottmann [16]. A pair of CCW Otto cycles are shown in Fig. 2.5. Panel (a) shows a CCW Otto cycle that functions as a refrigeration cycle that takes the working substance (for convenience, an ideal gas) through a counterclockwise sequence of steps via the closed cycle $1 \rightarrow 2 \rightarrow 3 \rightarrow 4 \rightarrow 1$. The first step of the cycle begins with the working substance at point 1, (V_1, P_1) lying on the isotherm corresponding to the temperature T_{high} of the high-temperature heat reservoir.

Following its isolation from the two heat reservoirs, the system (ideal gas) is expanded adiabatically from point 1 to point 2, (V_2, P_2), at which the system

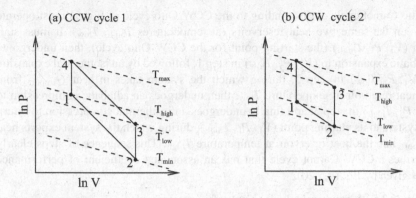

Fig. 2.5 Counterclockwise (CCW) Otto cycles. (**a**) Diagram depicting a CCW Otto cycle that is a refrigeration cycle; (**b**) Diagram depicting a CCW Otto cycle that is *not* a refrigeration cycle

attains its lowest temperature, which we shall call T_{min}. It is then placed in thermal contact with the heat reservoir at temperature T_{low}, and the pressure is increased isochorically to point 3, $(V_3 = V_2, P_3)$ and temperature T_{low}. In accordance with the Second Law, the ideal gas imports heat Q_{import} from the heat reservoir at temperature T_{low} during this step. As $W \equiv 0$ for an isochoric change, Q_{import} is given by

$$Q_{import} = (\Delta U)_{2 \to 3} = \tfrac{3}{2} N k_B (T_{low} - T_{min}). \qquad (2.5.57a)$$

The system is again isolated from the two heat reservoirs and is compressed adiabatically to point 4, $(V_4 = V_1, P_4)$, with the ideal gas achieving its highest temperature, which we shall call T_{max}: this compression requires external work, W, to be done on the system. Finally, the ideal gas working system is placed in thermal contact with the reservoir at temperature T_{high}, and the pressure is then decreased isochorically from P_4 to P_1, thereby returning the system to its thermodynamic starting state at point 1. In accordance with the Second Law, the working system exports heat Q_{export}, given by

$$Q_{export} = -Q_{4 \to 1}(\text{gas}) \equiv -(\Delta U)_{4 \to 1} \qquad (2.5.57b)$$

$$= \tfrac{3}{2} N k_B (T_{max} - T_{high}), \qquad (2.5.57c)$$

to the high-temperature heat reservoir during this final step. If we employ these results for Q_{import} and Q_{export} in Eqs. (2.5.19) defining the coefficient of performance, ε_{COP} for this Otto CCW cycle, we obtain

$$\varepsilon_{COP}^{Otto} \equiv \frac{Q_{import}}{Q_{export} - Q_{import}} = \frac{T_{low} - T_{min}}{T_{max} - T_{high} - (T_{low} - T_{min})}. \qquad (2.5.58)$$

The Carnot cycle corresponding to the CCW Otto cycle of Fig. 5a must operate between the same two heat reservoirs (at temperatures T_{high}, T_{low}). It must start from (V_1, P_1, T_{high}) (the starting point for the CCW Otto cycle), then undergo an adiabatic expansion to (V_2', P_2', T_{low}) in step 1, followed by an isothermal expansion to (V_3', P_3', T_{low}) in step 2, during which the system takes in heat Q_{import} from the heat reservoir at temperature T_{low}, then undergoes an adiabatic compression to (V_4', P_4', T_{high}) in step 3, and finally undergoes an isothermal compression to return the system to its starting point (V_1, P_1, T_{high}), during which the system exports heat Q_{export} into the heat reservoir at temperature T_{high}. This sequence of steps clearly describes a CCW Carnot cycle that has an associated coefficient of performance ε_{COP} given by Eq. (2.5.20).

Example 2.5 A counterclockwise Otto refrigeration cycle.

Let us consider a CCW Otto cycle [16] involving an ideal gas working fluid operating between a pair of heat reservoirs having temperatures T_{high} and T_{low}.

We shall examine a CCW Otto cycle that begins with volume V_1 and pressure P_1 at temperature T_{high} (the hotter heat reservoir), and for which $T_{max} = 2T_{high}$, while $V_2 = 5V_1$. The resultant Otto CCW cycle sequence of four steps is illustrated in Fig. 5a. It may also be represented as the cycle $(V_1, P_1, T_{high}) \to (5V_1, P_2, T_{min}) \to (5V_1, P_3, T_{low}) \to (V_1, P_4, 2T_{high}) \to (V_1, P_1, T_{high})$. This Otto cycle leads to (see Problem 20) pressures $P_2 \simeq 0.06840P_1$, $P_3 \simeq 0.13680P_1$, $P_4 = 2P_1$, and to temperatures $T_{low} \simeq 0.6840T_{high}$ and $T_{min} \simeq 0.3420T_{high} = \frac{1}{2}T_{low}$.

We may utilize these results to evaluate the coefficient of performance, ε_{COP}, for this CCW Otto cycle as

$$\varepsilon_{COP}^{Otto} = \frac{T_{low} - T_{min}}{T_{max} - T_{high} - (T_{low} - T_{min})}$$

$$= 0.5198$$

or $\varepsilon_{COP}^{Otto} \approx 0.52$. The corresponding value for a CCW Carnot cycle operating between the same two heat reservoirs is given via Eq. (2.5.20) as

$$\varepsilon_{COP}^{Carnot} = \frac{T_{low}}{T_{high} - T_{low}} = 2.165.$$

Upon noting that, as has been emphasized by Dickerson and Mottmann [16], it is a corollary of the Second Law that any refrigeration cycle operating between two heat reservoirs must have a coefficient of performance that is smaller than that for a Carnot CCW cycle operating between the same two heat reservoirs, it becomes clear that this CCW Otto cycle does indeed operate as a refrigeration cycle.   □

Let us now consider a CCW Otto cycle that starts from (V_1, P_1) at temperature T_{low}, rather than at T_{high}. By analogy with the CCW Otto cycle considered in Example 2.5, let us fix T_{max} to be $2T_{low}$ but, in this case, undergoes an expansion from volume V_1 to volume $V_2 = 2V_1$ rather than from V_1 to $5V_1$. Corresponding calculations utilizing the ideal gas equation of state give pressures $P_2 \simeq 0.3150 P_1$, $P_3 \simeq 0.6300 P_1$, $P_4 = 2P_1$, and temperatures $T_{\text{high}} \simeq 1.260 T_{\text{low}}$, $T_{\min} \simeq 0.6300 T_{\text{low}}$. The Otto cycle $(V_1, P_1, T_{\text{low}}) \rightarrow (2V_1, P_2, T_{\max}) \rightarrow (2V_1, P_3, T_{\text{high}}) \rightarrow (V_1, P_4, 2T_{\text{low}}) \rightarrow (V_1, P_1, T_{\text{low}})$ thus becomes $(V_1, P_1, T_{\text{low}}) \rightarrow (2V_1, 0.3150 P_1, 0.6300 T_{\text{low}}) \rightarrow (2V_1, 0.6300 P_1, 1.260 T_{\text{low}}) \rightarrow (V_1, 2P_1, 2T_{\text{low}}) \rightarrow (V_1, P_1, T_{\text{low}})$. Notice that in this CCW Otto cycle, heat $Q_{\text{import}} = \frac{3}{2} N k_B (T_{\text{high}} - T_{\text{low}}) = 0.630 \frac{3}{2} N k_B T_{\text{low}}$ is imported from the higher-temperature heat reservoir, while heat $Q_{\text{export}} = \frac{3}{2} N k_B (T_{\max} - T_{\text{low}}) = \frac{3}{2} N k_B T_{\text{low}}$ is exported into the lower-temperature heat reservoir. Not only is this the reverse of the manner in which a refrigeration cycle typically operates, it also requires work to be done on the system in order that $Q_{\text{export}} > Q_{\text{import}}$.

Because this Otto cycle starts at a point (V_1, P_1) on the T_{low} heat reservoir isotherm, the corresponding Carnot cycle must evolve along an adiabat from $(V_1, P_1, T_{\text{low}})$ to a point $(V_2', P_2', T_{\text{high}})$, which requires the working substance to move from right to left along the adiabat. It then imports heat from the T_{high}-reservoir as it expands isothermally to a point $(V_3', P_3', T_{\text{high}})$. Similarly, the system moves from left to right along a second adiabat to arrive at $(V_4', P_4', T_{\text{low}})$ and finally moves from right to left as it compresses to return to the starting point along the T_{low}-isotherm. The corresponding Carnot cycle is thus a CW Carnot cycle, so that this CCW Otto cycle cannot be a refrigeration cycle. Indeed, we see that this CCW Otto cycle not only imports heat from the T_{high}-reservoir but it also requires external work to be done on the system in order to export the heat into the T_{low}-reservoir. It has been aptly, and somewhat sarcastically, referred to [16] as a 'cold pump'!

A calculation of the coefficient of performance for this CCW Otto cycle employing Eq. (2.5.58) gives

$$\varepsilon_{\text{COP}}^{\text{CCW}} = \frac{(1.260 - 0.6300) T_{\text{low}}}{(2 - 1.260 - 1 + 0.6300) T_{\text{low}}} \simeq 1.703 \,,$$

while a formal calculation of the coefficient of performance for the equivalent corresponding Carnot cycle can be carried out based upon Eq. (2.5.19) as

$$\varepsilon_{\text{COP}}^{\text{Carnot}} = \frac{Q_{\text{import}}}{Q_{\text{export}} - Q_{\text{import}}}$$

$$= \frac{1.260 T_{\text{low}}}{(1 - 1.260) T_{\text{low}}} \approx -4.85 \,.$$

That this is a meaningless value provides yet another indication that this CCW Otto cycle cannot be a refrigeration cycle.

We see that four criteria, namely, that

(a) the cycle contains two adiabats in order to enable the working substance to achieve both a maximal temperature $T_{\max}$ that exceeds the temperature T_{high}

of the hotter heat reservoir and a minimal temperature T_{min} that lies below the temperature T_{low} of the cooler heat reservoir;

(b) the range of temperatures ($T_{min} < T < T_{low}$) over which the cycle imports heat lies entirely below the range of temperatures ($T_{max} > T > T_{high}$) over which it exports heat;

(c) the value of ε_{COP} satisfies the Second Law requirement that $\varepsilon_{COP}^{Carnot} > \varepsilon_{COP}$;

(d) the comparison Carnot cycle proceeds in the CCW direction,

must be met in order that a CCW cycle other than a CCW Carnot cycle be a valid refrigeration cycle.

2.6 The Second Law and Stability

We have noted earlier that entropy is a thermodynamic state function, so that the entropy change ΔS that occurs when a thermodynamic system passes from an initial thermodynamic state i to a final thermodynamic state f by *any* valid thermodynamic process depends only upon the values of the entropy for the initial and final thermodynamic states and is independent of the path taken between them. We also have seen (see Eqs. (2.2.9)) that an integrating factor T^{-1} can be introduced for the inexact differential δQ for a reversible process so that $\delta Q/T \equiv dS$. It thus follows that for a reversible isothermal process, the ratio Q_{rev}/T of a reversible heat flow to the surroundings gives the entropy change, ΔS, for that process.

Because entropy is a state function, the net entropy change for a closed cycle in which the system is first taken from state i to state f by a reversible isothermal process, then returned from state f to state i by another reversible isothermal process must be zero, i.e., $(\Delta S)_{cycle}^{rev} = 0$. Consider now what happens when a final system state f is obtained from the initial state i of an isolated system via a spontaneous irreversible process, after which the system is placed in contact with its surroundings and returned to the initial state via a reversible process. However, as this process is a cyclic process between thermodynamic states i and f, i.e., i → f → i, we have $(\Delta S)_{cycle} = 0$. The Second Law of Thermodynamics, in the form (2.2.10b), tells us that the overall entropy change, $(\Delta S)_{cycle}$, is given by

$$(\Delta S)_{cycle} = 0 > \int_i^f \frac{\delta Q_{irr}}{T} + \int_f^i \frac{\delta Q_{rev}}{T} \equiv \oint \frac{\delta Q}{T} \, .$$

This expression, when read in reverse and considered together with $(\Delta S)_{cycle}^{rev} = 0$, is often referred to as the Clausius inequality.

As the first integral vanishes because $\delta Q_{irr} = 0$ for an isolated system (as no energy exchange can occur between an isolated system and its surroundings), while the second integral gives $S_i - S_f$, we thus find that

$$(\Delta S)_i \rightarrow f = S_f - S_i > 0 \, ,$$

so that the entropy of the isolated thermodynamic system has increased in passing from initial state i to final state f via a spontaneous irreversible process. By considering the universe as a closed system and recognizing that naturally-occurring processes are typically spontaneous irreversible processes, this argument provides the source of the statement that the entropy of the universe always increases or, equivalently, tends to a maximum. It will be useful to consider two examples that illustrate the nature of a spontaneous irreversible change.

Example 2.6 Irreversible vs. reversible changes.

We examine in this example the (spontaneous irreversible) expansion of an ideal gas of volume V_i into a connected empty chamber of volume V_c to occupy a final volume $V_f = V_i + V_c$. To obtain an expression for the change $\Delta S = S_f - S_i$ in entropy for this expansion, we shall utilize a reversible path (as the change in any state function for a process connecting two thermodynamic states is independent of the path and depends only upon its values for the two states).

In particular, we shall employ a reversible isothermal expansion of the ideal gas (for which $\Delta U = 0$) into the empty chamber. The First Law then necessitates that $\delta Q_{rev} = -\delta W_{rev}$ and, as $\delta W_{rev} = -P\,dV$ with P given by $P = Nk_B T/V$ for an ideal gas, we see that $(\Delta S)_{sys}^{rev}$ is thus given as

$$(\Delta S)_{sys}^{rev} \equiv \int_{V_i}^{V_f} \frac{\delta Q_{rev}}{T} = -\int_{V_i}^{V_f} \frac{\delta W_{rev}}{T} = Nk_B \int_{V_i}^{V_f} \frac{dV}{V}.$$

Because $(\Delta S)_{sys}^{irr} = (\Delta S)_{sys}^{rev}$ for this expansion, we have

$$(\Delta S)_{sys} \equiv (\Delta S)_{sys}^{irr} = Nk_B \ln \frac{V_f}{V_i} > 0,$$

and the entropy of the system increases due to the spontaneous irreversible expansion of the ideal gas.

To see how to distinguish the way in which an irreversible expansion differs from a reversible expansion of an ideal gas system, we shall examine the corresponding entropy changes for the surroundings.

(a) To evaluate ΔS for a reversible isothermal expansion, we may use the fact that because $\Delta U = 0$ the expanding gas must have acquired heat (energy) $Q_{rev} = -W_{rev} = Nk_B T \ln(V_f/V_i)$ from its surroundings. This, in turn, means that the entropy of the surroundings will have decreased by an amount $Q_{rev.}/T$, i.e.,

$$(\Delta S)_{surr}^{rev} = -\frac{Q_{rev}}{T} = -Nk_B \ln \frac{V_f}{V_i}.$$

However, as the statement $\Delta S \geq 0$ applies only to an isolated system, it will be clear that we must examine $(\Delta S)^{rev}$ for the relevant isolated system, namely, the ideal gas plus its surroundings (i.e., effectively the entire universe). Thus, the appropriate entropy change to be considered is given by

$$(\Delta S)_{\text{total}}^{\text{rev}} = (\Delta S)_{\text{sys}}^{\text{rev}} + (\Delta S)_{\text{surr}}^{\text{rev}}.$$

Because $(\Delta S)_{\text{surr}}^{\text{rev}}$ is the negative of $(\Delta S)_{\text{sys}}^{\text{rev}}$, we see that $(\Delta S)_{\text{total}}^{\text{rev}} = 0$ for the reversible isothermal expansion of an ideal gas.

(b) For the irreversible (isothermal) expansion of the system (the ideal gas), we have $\Delta U \equiv 0$ and $W_{\text{irr}} = 0$, as no work is done during this expansion. By the First Law we therefore have $Q_{\text{irr}} = 0$, i.e., no heat (energy) is delivered to the system from its surroundings. However, we note that as $(\Delta S)_{\text{sys}}^{\text{irr}}$ is still equal to $(\Delta S)_{\text{sys}}^{\text{rev}}$, $(\Delta S)_{\text{surr}}^{\text{irr}} = 0$ for the irreversible expansion of an ideal gas. The total entropy change is therefore

$$(\Delta S)_{\text{total}}^{\text{irr}} = (\Delta S)_{\text{sys}}^{\text{irr}} + (\Delta S)_{\text{surr}}^{\text{irr}} = N k_B \ln \frac{V_f}{V_i},$$

and thus $(\Delta S)_{\text{total}}^{\text{irr}} > 0$.   □

Example 2.7 The mixing of two ideal gases.

We shall now consider two ideal gases A and B each initially in chambers of volume V_A and V_B but separated by a removable impermeable membrane. Each gas is at temperature T and under pressure P. Removal of the impermeable membrane separating the two gases then enables them to mix (by expanding spontaneously and independently of one another), with each gas thereby occupying the total volume $V_f = V_A + V_B$. To examine the entropy change for the resultant (spontaneous irreversible) mixing process, we may once again make use of the fact that S is a thermodynamic state function in order to employ a reversible isothermal expansion to evaluate the entropy change $(\Delta S)^{\text{irr}} \equiv (\Delta S)_{\text{mix}}$ resulting from the mixing process.

For gas A we obtain (from Eq. (2.2.9b), for example)

$$\Delta S_A = N_A k_B \ln \frac{V_A + V_B}{V_A},$$

while for gas B we obtain

$$\Delta S_B = N_B k_B \ln \frac{V_A + V_B}{V_B},$$

with N_A and N_B the numbers of molecules A and B in the two ideal gas subsystems. We may employ the ideal gas equation of state (with $T_A = T_B \equiv T$ and $P_A = P_B \equiv P$) to express ΔS_A and ΔS_B as

$$\Delta S_A = N_A k_B \ln \frac{N}{N_A} \quad \text{and} \quad \Delta S_B = N_B k_B \ln \frac{N}{N_B},$$

with $N = N_A + N_B$ the total number of molecules in the gas mixture. The total entropy change for mixing of the two gases is thus

$$(\Delta S)_{\text{mix}} \equiv \Delta S = \Delta S_{\text{A}} + \Delta S_{\text{B}} \,,$$

or, in terms of the mole fractions $x_{\text{A}} \equiv N_{\text{A}}/N$, $x_{\text{B}} \equiv N_{\text{B}}/N$ of the ideal gases A and B in the final mixture,

$$(\Delta S)_{\text{mix}} = (N_{\text{A}} + N_{\text{B}})k_{\text{B}} \left[x_{\text{A}} \ln \frac{1}{x_{\text{A}}} + x_{\text{B}} \ln \frac{1}{x_{\text{B}}} \right].$$

Equivalently, the molar entropy of mixing is given as

$$(\Delta \overline{S})_{\text{mix}} = -R[x_{\text{A}} \ln x_{\text{A}} + x_{\text{B}} \ln x_{\text{B}}] \,,$$

with $R = N_0 k_{\text{B}}$ the universal gas constant. Note that $(\Delta \overline{S})_{\text{mix}} > 0$, as the mole fractions x_{A}, x_{B} are smaller than 1. This is, of course, consistent with gas mixing being a spontaneous irreversible process. $\square$

We know that the entropy of a system can change either via the transport of entropy to or from its surroundings or via the creation of entropy within the system itself by means of spontaneous irreversible processes [17]. We also know from the Clausius version, $dS \geq (\delta Q)/T$, of the Second Law that $dS = (\delta Q)/T$ for a reversible process, $dS > (\delta Q)/T$ for an irreversible process, and that S may be expressed as $S = S(T, V)$ for a simple closed system or as $S = S(T, V, \xi)$ for a closed system that contains chemically reactive species (see Sect. 2.7). We shall therefore introduce $(\delta Q')/T$ to represent the entropy production by spontaneous irreversible processes within a system or by non-spontaneous irreversible processes arising from interactions between a system and its surroundings, via

$$dS = \frac{\delta Q}{T} + \frac{\delta Q'}{T} \,, \tag{2.6.1a}$$

with $\delta Q' > 0$ for an irreversible process and $\delta Q' = 0$ for a reversible process.[11] Equivalently, we may write δQ as

$$\delta Q = T dS - \delta Q' \,. \tag{2.6.1b}$$

If we substitute expression (2.6.1b) for δQ into the first law expression for dU, we obtain

$$dU = T dS - P dV - \delta Q' \,, \tag{2.6.2a}$$

---

[11] Prigogine and Defay [5] is likely still the most clearly written and definitive text. See also, however, Kirkwood and Oppenheim [6] and the classic reference for nonequilibrium thermodynamics by De Groot and Mazur [17].

from which we may deduce that for any irreversible process that takes place at constant S and V, we have

$$\delta Q' = -dU > 0.$$ (2.6.2b)

This result is more commonly included in a summary statement

$$dU \leq 0,$$ (2.6.2c)

in which the equality holds for a reversible thermodynamic process and the inequality holds for an irreversible thermodynamic process. The result (2.6.2c) tells us that an irreversible change at constant S and V results in a decrease in the internal energy, so that U thus plays the part of an indicator of irreversible processes for changes that occur at constant S and V. The non-occurrence of irreversible process when U remains constant for a closed system and the inevitability of spontaneous irreversible processes occurring when U decreases for a closed system gives rise to the terminology 'thermodynamic potential' for U in association with adiabatic isochoric processes.

If we consider systems for which S and P, rather than S and V, are the relevant system variables, then the appropriate thermodynamic potential will be the enthalpy, H. Hence, irreversible processes at constant S and P will, because dH takes the form

$$dH = TdS + VdP - \delta Q',$$ (2.6.3a)

then lead to

$$\delta Q' = -dH > 0,$$ (2.6.3b)

so that irreversible processes will, in this case, be indicated by a decrease in the enthalpy of the system. Similarly, we may consider systems for which T and V or T and P are the relevant system variables. If the relevant system variables are T and V, the appropriate thermodynamic potential is then the Helmholtz energy, A, from which we find that

$$dA = -SdT - PdV - \delta Q',$$ (2.6.4a)

and hence

$$\delta Q' = -dA > 0,$$ (2.6.4b)

or equivalently,

$$dA \leq 0,$$ (2.6.4c)

provides the relevant condition for irreversible processes occurring under isothermal–isochoric conditions.

A similar line of reasoning for irreversible processes occurring at constant T and P, brings the Gibbs energy, G, into play as the relevant thermodynamic potential. Thus, with dG given by

$$dG = -SdT + VdP - \delta Q', \qquad (2.6.5a)$$

we obtain the condition

$$\delta Q' = -dG > 0, \qquad (2.6.5b)$$

or, equivalently,

$$dG \leq 0, \qquad (2.6.5c)$$

for irreversible processes that take place under isothermal–isobaric conditions.

System stability is thus strongly correlated with entropy creation (or production). The major mechanical means for entropy production is via viscous flow within a system (microscopically associated, for example, with momentum transfer via collisions): no viscous flow occurs when a system is in mechanical equilibrium. Similarly, the major thermal means of entropy production is via energy (heat) transport between two regions of a system that are at different temperatures: no heat transport occurs within a system when it is in thermal equilibrium. For an open system, concentration (number density) differences may create entropy by diffusion within each phase: no diffusion (mass transport) occurs within homogeneous phases of a system.

Consider now a thermodynamic system for which mechanical and thermal equilibrium have already been established. We shall refer to a system that is in mechanical and thermal equilibrium, but which has not yet fully established chemical or phase equilibrium, as being in *partial equilibrium*. The only form of entropy production that can occur is then associated with chemical reactions or with the transport of matter from one phase to another. Moreover, we shall consider here only systems for which T, P, and composition within each phase are uniform.

For a chemically reacting system, we may think of the degree of advancement ξ as the relevant variable to consider, in which case, we may write in analogy with Eq. (2.6.5b)

$$\delta Q' = \mathbb{A} \, d\xi \geq 0, \qquad (2.6.6)$$

with the inequality corresponding to spontaneous reaction and the equality corresponding to equilibrium. The symbol $\mathbb{A} \equiv \mathbb{A}(x, y, \xi)$, with x, y representing a pair of thermodynamic variables such as T, P or T, V, is called the affinity, a thermodynamic function introduced in 1922 by the Belgian chemist de Donder, after whom the inequality (2.6.6) is named. As chemical reactions are most often carried out at constant T and P, we shall choose them to be the appropriate physical variables to describe a chemically reacting system. Entropy production is thus uniquely associated with the chemical reaction, so that $\delta Q'$ will be independent of any changes in T and P that may occur concurrently.

2.7   Extension to Multicomponent Systems

We shall now consider extending our thermodynamic system from a pure substance to a mixture of substances, with corresponding numbers, n_α, of moles of chemical species α. The generalization of the internal energy then becomes $U \equiv U(S, V, \{n_\alpha\})$, rather than simply $U \equiv U(S, V, n)$. Whether we are allowed to describe multicomponent systems in terms of a closed thermodynamic system rather than requiring an open system depends upon whether or not we have mass sinks or sources. If the processes that take place within the system satisfy the conservation of mass principle, then they may be treated within the context of a closed thermodynamic system, just as processes that satisfy the conservation of energy principle maybe treated in the context of an isolated system.

If we return briefly to the differential form for the combined first and second laws for a pure substance, $dU = T dS - P dV$, we note that it is structured as a sum involving the products of intensive thermodynamic variables multiplying differentials of extensive thermodynamic variables. Extension of the differential form for the combined first and second laws of thermodynamics thus logically requires that we add terms with the same structure, i.e., intensive thermodynamic variables multiplying differentials of extensive thermodynamic variables. The obvious extensive thermodynamic variables are the numbers of moles of substances constituting the thermodynamic system. We shall see that the corresponding intensive thermodynamic variables are quantities called *chemical potentials*, which are designated by the symbol μ_α. The combined first and second law expression for a multicomponent system thus becomes

$$dU = T\, dS - P\, dV + \sum_{\alpha=1}^{M} \mu_\alpha \, dn_\alpha \,, \tag{2.7.1}$$

with P the total pressure for the mixture and M the number of distinct chemical components in it. This expression reduces to Eq. (2.2.12), which applies to a single-component closed system, as $dn = 0$ for a closed system.

If we recall that U is a homogeneous function, then we may apply Euler's theorem (see Sect. B.1.2) to $U \equiv U(S, V, \{n_\alpha\})$ to give

$$U = S \left(\frac{\partial U}{\partial S}\right)_{V,\{n_\alpha\}} + V \left(\frac{\partial U}{\partial V}\right)_{S,\{n_\alpha\}} + \sum_{\alpha=1}^{M} n_\alpha \left(\frac{\partial U}{\partial n_\alpha}\right)_{S,V,n_\beta \neq n_\alpha} .$$

However, from expression (2.7.1) we know that

$$\left(\frac{\partial U}{\partial S}\right)_{V,n_{\{\alpha\}}} = T \,; \qquad \left(\frac{\partial U}{\partial V}\right)_{S,n_{\{\alpha\}}} = -P \,; \qquad \left(\frac{\partial U}{\partial n_\alpha}\right)_{S,V,n_\beta \neq n_\alpha} = \mu_\alpha \,,$$

so that U may also be written as

$$U = TS - PV + \sum_{\alpha=1}^{M} \mu_\alpha n_\alpha \qquad (2.7.2)$$

for a multicomponent system. Notice that this also gives us a proper general definition of the chemical potential for chemical species α. Equations (2.7.1) and (2.7.2) are referred to as the Euler forms for dU and U, respectively.

We notice also that if we form the total differential dU from the Euler form for U, we obtain the most general result

$$dU = T\,dS - P\,dV + \sum_\alpha \mu_\alpha\,dn_\alpha + S\,dT - V\,dP + \sum_\alpha n_\alpha\,d\mu_\alpha \,. \qquad (2.7.3)$$

Comparison of this extended result with the combined first and second law differential expression thus requires that

$$0 = S\,dT - V\,dP + \sum_{\alpha=1}^{M} n_\alpha\,d\mu_\alpha \,, \qquad (2.7.4)$$

which is known as the Gibbs–Duhem relation. This is an important relation, as it tells us that not all of the intensive variables $T, P, \{\mu_\alpha\}$ can be independent quantities for an open system.

The Gibbs energy G is defined for an open system in the same manner that it is defined for a closed system, i.e., as the double Legendre transform

$$G \equiv U - TS + PV \,, \qquad (2.7.5)$$

so that the chemical potentials μ_α and numbers of moles n_α are related to G by

$$G = \sum_{\alpha=1}^{M} \mu_\alpha n_\alpha \,. \qquad (2.7.6)$$

For a pure substance, we may reduce this expression to $\mu = G/n$, from which we see that the chemical potential of a pure substance is simply the corresponding molar Gibbs energy.

We may represent a general gas-phase chemical reaction system in terms of a balanced chemical reaction of the type

$$\nu_A A(g) + \nu_B B(g) \rightleftharpoons \nu_C C(g) + \nu_D D(g), \tag{2.7.7}$$

in which A, B, C, D represent chemical species and the ν_α (α = A, B, C, D) are the stoichiometric coefficients dictated by the overall conservation of mass that applies to a chemical reaction. If we consider the reaction to proceed from left to right, we may say, based upon Eq. (2.7.7), that if ν_A, ν_B moles of chemical species A, B (reactants) are consumed in the reaction and ν_C, ν_D moles of chemical species C, D (products) are formed, then one unit of reaction has occurred, so that when the reaction has advanced by ξ units, the corresponding numbers of moles of each chemical species will be

$$n_A = n_A(0) - \nu_A \xi, \qquad n_B = n_B(0) - \nu_B \xi, \tag{2.7.8a}$$

and

$$n_C = n_C(0) + \nu_C \xi, \qquad n_D = n_D(0) + \nu_D \xi, \tag{2.7.8b}$$

with $n_\alpha(0)$ (α = A, B, C, D) the initial number of moles of chemical species α. The quantity ξ is termed the extent of reaction, sometimes referred to equivalently as the de Donder degree of advancement. It provides an unambiguous means for defining the rate of a chemical reaction. Notice that as the stoichiometric coefficients ν_α have no units, ξ has units mol.

As most gas-phase reactions occur under constant pressure and at constant temperature, the relevant thermodynamic state function is the Gibbs energy. From Eq. (2.7.1), we may write the appropriate version of the combined First and Second Laws expression as

$$dU = TdS - PdV + \sum_{\alpha=A}^{D} \mu_\alpha dn_\alpha. \tag{2.7.9}$$

From Eqs. (2.7.8), we may express the incremental changes dn_α in the numbers of moles of the components taking part in the chemical reaction (2.7.7) in terms of the appropriate stoichiometric coefficient and the incremental change, $d\xi$, in the extent of reaction, thereby enabling dU to be obtained as

$$dU = TdS - PdV + [\nu_C \mu_C + \nu_D \mu_D - \nu_A \mu_A - \nu_B \mu_B]d\xi \tag{2.7.10a}$$

or as

$$dU = TdS - PdV - \mathbb{A}d\xi, \tag{2.7.10b}$$

in terms of the thermodynamic property $\mathbb{A}$ defined as

$$\mathbb{A}(S, V, \xi) \equiv -[\nu_C \mu_C + \nu_D \mu_D - \nu_A \mu_A - \nu_B \mu_B] \tag{2.7.11}$$

and called the (chemical) affinity. From Eq. (2.7.10b), it is clear that the affinity is also related to the internal energy via

$$\mathbb{A}(S, V, \xi) = -\left(\frac{\partial U}{\partial \xi}\right)_{S,V},$$

and represents the isentropic-isochoric rate of change of $U(S, V, \xi)$ with respect to the degree of advancement of the chemical reaction.

Based upon expression (2.7.10b) for dU and the Legendre transforms H, A, and G of U given by Eqs. (2.3.2), (2.3.5a), (2.3.6c), the total differentials dH, dA, and dG, are related to the affinity via

$$dH = TdS + VdP - \mathbb{A}d\xi, \tag{2.7.12a}$$

$$dA = -SdT - PdV - \mathbb{A}d\xi, \tag{2.7.12b}$$

$$dG = -SdT + VdP - \mathbb{A}d\xi, \tag{2.7.12c}$$

respectively, so that $\mathbb{A}$ is related to H, A, and G by

$$\mathbb{A} = -\left(\frac{\partial H}{\partial \xi}\right)_{S,P} = -\left(\frac{\partial A}{\partial \xi}\right)_{T,V} = -\left(\frac{\partial G}{\partial \xi}\right)_{T,P} \tag{2.7.13}$$

in terms of the relevant partial derivatives of H, A, and G.

As chemical reactions are most often carried out under conditions of constant temperature and total pressure, Eq. (2.7.12c) will be the relevant expression for discussing the thermodynamics associated with (inherently irreversible) chemical reactions, such as that represented by Eq. (2.7.7). Hence, for reaction (2.7.7) occurring at constant temperature, T, and total pressure, P, the affinity $\mathbb{A}$ will be parameterized by the chosen values of T and P, and will be a function of the extent of reaction, ξ. The change in the Gibbs energy, $(\Delta G)_{\text{react}} \equiv (\Delta G)_{T,P}$, is traditionally given as

$$(\Delta G)_{T,P} = \nu_C \mu_C(T, P_C) + \nu_D \mu_D(T, P_D) - \nu_A \mu_A(T, P_A) - \nu_B \mu_B(T, P_B)$$

or, equivalently, in terms of the de Donder affinity as

$$(\Delta G)_{T,P} = -\mathbb{A}(T, P; \xi), \tag{2.7.14}$$

since the partial pressures P_α are parameterized by P and are functions of ξ.

If reaction (2.7.7) is carried out at constant T, P, and starts out with $n_A(0) = \nu_A 1_m$, with 1_m representing one mole, $n_B(0) = \nu_B$, $n_C(0) = n_D(0) = 0$, we may write the numbers of moles of the chemical species present in the reaction mixture in terms of the extent of reaction ξ as $n_A = \nu_A(1_m - \xi)$, $n_B = \nu_B(1_m - \xi)$, $n_C = \nu_C\xi$, $n_D = \nu_D\xi$, so that the mole fractions, x_α, of the four participating species are given as $x_\alpha = n_\alpha(\xi)/n_{\text{tot}}(\xi)$ or, more explicitly, as

$$x_A = \frac{\nu_A(1_m - \xi)}{n_{\text{tot}}(\xi)}, \quad x_B = \frac{\nu_B(1_m - \xi)}{n_{\text{tot}}(\xi)}, \quad x_C = \frac{\nu_C\xi}{n_{\text{tot}}(\xi)}, \quad x_D = \frac{\nu_D\xi}{n_{\text{tot}}(\xi)},$$

$$\tag{2.7.15a}$$

with $n_{\text{tot}}(\xi)$ given by

$$n_{\text{tot}}(\xi) = (\nu_A + \nu_B)1_m + (\nu_C + \nu_D - \nu_A - \nu_B)\xi. \tag{2.7.15b}$$

The corresponding partial pressures are related to the total pressure via Dalton's law of partial pressures, so that $P_\alpha(\xi) = x_\alpha(\xi)P$.

The Gibbs energy $G_{T,P}(\xi)$ for an ideal gas mixture that may undergo reaction (2.7.7) can now be written in the form

$$
\begin{aligned}
G_{T,P}(\xi) &= \sum_{\alpha=A}^{D} n_\alpha \mu_\alpha(T, P_\alpha) \\
&= \sum_{\alpha=A}^{D} n_\alpha(\xi)[\overline{G}_\alpha^\circ(T) + RT \ln P_\alpha(\xi)]. \tag{2.7.16}
\end{aligned}
$$

Note that in writing this expression, we have made use of the fact that the molar Gibbs energy of a pure substance is the chemical potential for that substance, and subsequently utilized Eq. (2.3.7c) to tease out the pressure dependence of $\mu_\alpha(T, P)$. Substitution of expressions (2.7.15) into Eq. (2.7.16) gives $G_{T,P}(\xi)$ for reaction (2.7.7) as

$$G_{T,P}(\xi) = G_{\text{react}}^\circ + RT\left[\sum_{\alpha=A,B} \nu_\alpha(1_m - \xi)\ln\frac{\nu_\alpha(1_m - \xi)}{n_{\text{tot}}(\xi)} + \sum_{\alpha=C,D} \nu_\alpha\xi\ln\frac{\nu_\alpha\xi}{n_{\text{tot}}(\xi)}\right]$$

$$+ n_{\text{tot}}(\xi)RT\ln P, \tag{2.7.17a}$$

with G_{react}° given by

$$G_{\text{react}}^\circ(T; \xi) = \nu_A 1_m \overline{G}_A^\circ + \nu_B 1_m \overline{G}_B^\circ + (\Delta G^\circ)_{\text{react}}\,\xi, \tag{2.7.17b}$$

in terms of the change $(\Delta G^\circ)_{\text{react}}$ of the Gibbs energy corresponding to completion of one unit of reaction (2.7.7), and defined as

$$(\Delta G^\circ)_{\text{react}}(T) \equiv (\nu_C \overline{G}_C^\circ + \nu_D \overline{G}_D^\circ - \nu_A \overline{G}_A^\circ - \nu_B \overline{G}_B^\circ)1_m, \tag{2.7.17c}$$

in terms of the standard molar Gibbs energies $\overline{G}_\alpha^\circ(T)$. Values for these molar Gibbs energies may normally be obtained from the extensive tabulations of thermochemical data generated by the U. S. National Institute for Standards and Technology (NIST) and known as the JANAF (for Joint Army, Navy, Air Force) tables [18].[12] By convention the Gibbs energy of formation is set to zero for an element in its most stable form at 1 bar pressure and the temperature of interest.

The negative of the partial derivative

$$
\left(\frac{\partial G}{\partial \xi}\right)_{T,P} = (\Delta G^\circ)_{\text{react}} + RT \left[-\sum_{\alpha=A,B} \nu_\alpha 1_{\text{m}} \ln[\nu_\alpha(1-\xi)] - \frac{dn_{\text{tot}}}{dt} \ln n_{\text{tot}}(\xi) \right.
$$

$$
\left. + \sum_{\alpha=C,D} \nu_\alpha 1_{\text{m}} \ln(\nu_\alpha \xi) \right] + \frac{dn_{\text{tot}}}{dt} RT \ln P \qquad (2.7.18)
$$

of the Gibbs energy with respect to the extent of reaction ξ at fixed temperature and total pressure then gives the affinity $\mathbb{A}(T, P; \xi)$.

Typical behaviours of $G_{T,P}(\xi)$ and $\mathbb{A}(T, P; \xi)$ are shown in Fig. 2.6. We see from panel (a) of this figure that G passes through a minimum at $\xi = \xi_{\text{e}}$, which corresponds to equilibrium for the chemical reaction, and as the affinity is proportional to the derivative of G with respect to ξ, it vanishes at $\xi = \xi_{\text{e}}$. When expressed in terms of the partial pressures of the component, Eq. (2.7.18) may be rewritten in the well-known form

$$
\left(\frac{\partial G}{\partial \xi}\right)_{T,P} = (\Delta G)_{\text{react}}(T, \xi) = (\Delta G^\circ)_{\text{react}}(T) + RT \ln \mathbb{Q}(\xi), \qquad (2.7.19)
$$

in which $(\Delta G^\circ)_{\text{react}}$, which depends only upon temperature T, is defined in terms of the chemical potentials of the reaction participants as

$$
(\Delta G^\circ)_{\text{react}}(T) \equiv \nu_C \mu_C^\circ(T) + \nu_D \mu_D^\circ(T) - \nu_A \mu_A^\circ(T) - \nu_B \mu_B^\circ(T), \qquad (2.7.20a)
$$

while $\mathbb{Q}(\xi)$, called the *reaction quotient*, is defined in terms of the partial pressures P_α as

$$
\mathbb{Q}(\xi) \equiv \frac{[P_C(\xi)/P^\circ]^{\nu_C}[P_D(\xi)/P^\circ]^{\nu_D}}{[P_A(\xi)/P^\circ]^{\nu_A}[P_B(\xi)/P^\circ]^{\nu_B}}. \qquad (2.7.20b)
$$

Note that some care must be exercised with the interpretation of $(\Delta G^\circ)_{\text{react}}$ and $\mathbb{Q}$. In particular, we note that $(\Delta G)_{\text{react}}$, which represents the change in the Gibbs energy for $\xi = 1$, has units J mol^{-1}, and has no significance should the coefficients ν_α not be the stoichiometric coefficients for the balanced reaction. Moreover,

---

[12]Note, however, that $\overline{G}^\circ$ and $\overline{H}^\circ$ are given (without overbars) relative to the enthalpy at the fixed reference temperature $T_{\text{f}} = 298.15$ K.

Fig. 2.6 Typical dependence
of (**a**) the Gibbs energy,
$G_{T,P}(\xi)$, and (**b**) the
chemical affinity, $\mathbb{A}(T, P; \xi)$,
on the extent of reaction, ξ,
for a chemical reaction
occurring under constant
temperature and constant total
pressure conditions

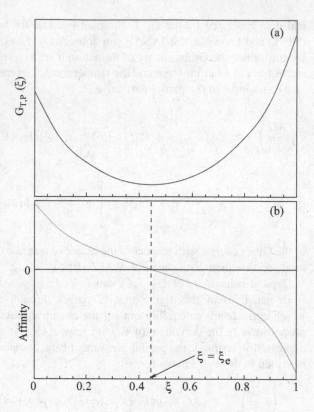

because $(\Delta G°)_{\text{react}}$ is defined in terms of Gibbs energies of the pure components of reaction (2.7.7) in their standard states at temperature T and under pressure $P° = 1$ bar, the partial pressures appearing in the defining expression for the reaction quotient $\mathbb{Q}$ are all relative to $P°$, so that $\mathbb{Q}$ is unitless. Although $P°$ is not traditionally included in the defining relation for the reaction quotient, it has been included in Eq. (2.7.20b) to emphasize this particular point.

Chemical equilibrium corresponds to the vanishing of the affinity or, equivalently, to the Gibbs energy reaching its minimum value, with the consequence that

$$RT \ln \mathbb{Q}(\xi_e) \equiv RT \ln K_P(T) = -(\Delta G°)_{\text{react}}(T) \,, \qquad (2.7.21\text{a})$$

in which K_P is the equilibrium constant, determined by the partial pressures of the equilibrium reaction components, i.e.,

$$K_P(T) = \frac{[P_C(\xi_e)/P°]^{\nu_C}[P_D(\xi_e)/P°]^{\nu_D}}{[P_A(\xi_e)/P°]^{\nu_A}[P_B(\xi_e)/P°]^{\nu_B}} \,. \qquad (2.7.21\text{b})$$

Expression (2.7.21b) clearly emphasizes that $K_P(T)$ applies to the chemical equilibrium mixture of reactants (here A, B) and products (here C, D). Note also that the value obtained for $K_P(T)$ depends upon the particular form given for the balanced chemical reaction to which it applies, and requires specification of the standard states employed for each reactant and product species. The first observation is readily illustrated by rewriting Eq. (2.7.7) as

$$A(g) + \nu'_B B(g) \rightleftharpoons \nu'_C C(g) + \nu'_D D(g).$$

with $\nu'_\alpha \equiv \nu_\alpha/\nu_A$ ($\alpha =$ B, C, D). Upon employing Eq. (2.7.21b) to obtain $K'_P(T)$ for this form for the balanced chemical reaction between A, B, C, and D, we obtain

$$K'_P(T) = \frac{P_C^{\nu'_C} P_D^{\nu'_D}}{P_A P_B^{\nu'_B}} = \left(\frac{P_C^{\nu_C} P_D^{\nu_D}}{P_A^{\nu_A} P_B^{\nu_B}} \right)^{\frac{1}{\nu_A}},$$

from which it is clear that $K_P(T) = [K'_P(T)]^{\nu_A}$.

For chemical reactions for which $\nu_A + \nu_B \neq \nu_C + \nu_D$, the equilibrium constant expression will contain a factor $P^{\nu_C+\nu_D-\nu_A-\nu_B}$ that gives rise to a version of the Le Chatelier Principle of first year general chemistry courses. For example, for a unimolecular dissociation/association reaction of the type

$$A(g) \rightleftharpoons C(g) + D(g),$$

$K_P(T)$ will have the form

$$K_P(T) = \frac{\xi_e^2 P}{(1 - \xi_e)^2}$$

which, when solved to give ξ_e as a function of pressure P, leads to the conclusion that increases in the total pressure in the reaction vessel shifts the equilibrium towards the reactant side of the reaction, in this case giving rise to an increase in the amount of A at equilibrium.

We have seen for the general gas-phase reaction (2.7.7) that the change in the Gibbs energy, $(\Delta G)_{react}(T, \xi)$, associated with one unit of reaction is given by Eq. (2.7.19) in terms of the reaction quotient $\mathbb{Q}$ and the standard Gibbs energy change, $(\Delta G^\circ)_{react}(T)$. If, in Eq. (2.7.19), we replace $(\Delta G^\circ)_{react}(T)$ by $-RT \ln K_P(T)$, we may express $(\Delta G)_{react}$ as

$$(\Delta G)_{react}(T, P, \xi) = RT \ln \frac{\mathbb{Q}(P, \xi)}{K_P(T)}. \qquad (2.7.22)$$

From this form for $(\Delta G)_{react}$, we see that $(\Delta G)_{react} = 0$ for $\mathbb{Q} = K_P$, while for $\mathbb{Q} < K_P$, $(\Delta G)_{react} < 0$, so that its increase as the reaction proceeds toward equilibrium will lead to an increase in the partial pressures of the products and a corresponding decrease in the partial pressures of the reactants that is consistent with the reaction proceeding from reactants to products as written in Eq. (2.7.7). We may thus say, equivalently, that for $(\Delta G)_{\text{react}} < 0$, the reaction proceeds spontaneously from left to right, while for $(\Delta G)_{\text{react}} > 0$ (i.e., $\mathbb{Q} > K_P$), the reaction proceeds spontaneously from right to left, until equilibrium is established.

2.8 Thermodynamics of Real Gases

We may utilize the Helmholtz equation (2.4.4) to examine how the internal energy changes for a real thermodynamic system during an isothermal volume change. Substitution of Eq. (2.4.4) into Eq. (2.3.8) for dU, followed by integration over volume for an isothermal volume change from V_0 to V gives the internal energy $U(T, V; N)$ as

$$U(T, V; N) = U(T, V_0; N) + \int_{V_0}^{V} T^2 \left[\frac{\partial}{\partial T} \left(\frac{P}{T} \right) \right]_V dV \qquad (2.8.1)$$

for a working expression for the computation of $U(T, V; N)$.

In the same spirit that we have considered the internal energy of a general thermodynamic system, we may obtain $C_V(T)$ from Eq. (2.8.1) as

$$C_V(T) \equiv \left(\frac{\partial U}{\partial T} \right)_V = C_{V_0}(T) + \frac{\partial}{\partial T} \int_{V_0}^{V} T^2 \left[\frac{\partial}{\partial T} \left(\frac{P}{T} \right) \right]_V dV,$$

which, upon simplification of the second term, gives

$$C_V(T) = C_{V_0}(T) + \int_{V_0}^{V} T \left(\frac{\partial^2 P}{\partial T^2} \right)_V dV. \qquad (2.8.2a)$$

For a real gas, for example, we may invoke the ideal gas (low number density) model for the reference system, thereby obtaining $C_V(T)$ as

$$[C_V(T)]_{\text{real gas}} = [C_V(T)]_{\text{ideal gas}} + \lim_{V_0 \to \infty} \int_{V_0}^{V} T \left(\frac{\partial^2 P}{\partial T^2} \right)_V dV. \qquad (2.8.2b)$$

Should the equation of state be available in the form of an expression giving P as an explicit function of T, V, and N, then the second partial derivative in Eq. (2.8.2a) may also be obtained explicitly, after which the integration over volume may be performed. Notice, however, that we may already deduce from expression (2.8.2a) that should the pressure P depend linearly upon temperature, the heat capacity at constant volume, C_V, will be equal to that in the reference system. In particular, for a real gas that satisfies this criterion, we see from Eq. (2.8.2b) that C_V will be equal to the heat capacity at constant volume for an ideal gas.

For a closed thermodynamic system (fixed N) we may obtain an explicit expression for the internal energy for a real gas via Eq. (2.8.1) if we utilize the fact that because the number density N/V tends to zero as the volume of the real gas goes to infinity, it tends to behave as an ideal gas, and hence $U_{\text{real gas}}(T, V; N)$ tends to $U_{\text{ideal gas}}(T; N)$. This behaviour thus allows us to write

$$U_{\text{real gas}}(T, V; N) - U_{\text{ideal gas}}(T; N) = \lim_{V_0 \to \infty} \int_{V_0}^{V} T^2 \left[\frac{\partial}{\partial T} \left(\frac{P}{T} \right) \right]_V dV \tag{2.8.3}$$

for the internal energy of a real gas.

To obtain an expression for the Helmholtz energy, $A(T, V; N)$, we begin with the total differential

$$dA = \left(\frac{\partial A}{\partial T} \right)_V dT - P \, dV \tag{2.8.4a}$$

obtained upon utilizing Eqs. (2.3.5a) and (2.3.9) to replace the volume partial derivative of A. As $A(T, V; N)$ is a thermodynamic state function, we may obtain an expression analogous to Eq. (2.8.1) for the internal energy, namely,

$$A(T, V; N) = A(T, V_0; N) - \int_{V_0}^{V} P \, dV , \tag{2.8.4b}$$

for an isothermal expansion of a real gas. From Eq. (2.8.4b), the difference between the Helmholtz energy $A(T, V; N)$ for a real fluid and that for the ideal gas has two terms: the first term gives the difference between $A(T, V_0; N)$ for the real fluid and the ideal gas, and the second term involves an integral of the difference $P_{\text{real fluid}}$ $P_{\text{ideal gas}}$ over volume. The first of these two terms vanishes in the ideal gas limit that V_0 goes to infinity, so that we may write directly the result

$$A_{\text{real fluid}}(T, V; N) - A_{\text{ideal gas}}(T, V; N) = -\lim_{V_0 \to \infty} \int_{V_0}^{V} (P_{\text{real fluid}} - P_{\text{ideal gas}}) \, dV , \tag{2.8.5}$$

for the Helmholtz energy. By the same line of reasoning, the difference in the Gibbs energy between a real fluid and the ideal gas is given as

$$G_{\text{real fluid}}(T, V; N) - G_{\text{ideal gas}}(T, V; N) = -\int_{0}^{P} (V_{\text{real fluid}} - V_{\text{ideal gas}}) \, dP . \tag{2.8.6}$$

Finally, as $A = U - TS$, we may also obtain an expression for the difference between the entropy for a real system and that for an the ideal gas from expressions (2.8.5) and (2.8.3) for $A(T, V; N)$ and $U(T, V; N)$.

The volume of a condensed-phase thermodynamic system (i.e., liquid or solid) is typically dictated by a combination of molecular size and intermolecular forces, and normally depends only weakly upon temperature and pressure. Moreover, as molecular sizes and intermolecular forces vary from molecule to molecule, the equation of state (often expressed in terms of the system volume as a function of temperature and pressure) will be specific to the substance constituting the thermodynamic system.

If we therefore consider the volume, V, to be a function of T and P, then we may express the total differential, dV, for the volume as

$$dV = \left(\frac{\partial V}{\partial T}\right)_P dT + \left(\frac{\partial V}{\partial P}\right)_T dP$$

$$= \frac{1}{V}\left(\frac{\partial V}{\partial T}\right)_P V dT + \frac{1}{V}\left(\frac{\partial V}{\partial P}\right)_T V dP$$

or, upon introducing the thermal expansivity, α, and the isothermal compressibility, κ_T, defined in Eq. (2.4.2), as

$$dV = \alpha V\, dT - \kappa_T V\, dP\,. \tag{2.8.7}$$

Division on both sides of this equation by V, followed by integration, allows us to obtain

$$V(T, P) = V(T_0, P_0)e^{\alpha(T-T_0)-\kappa_T(P-P_0)} \tag{2.8.8a}$$

as the equation of state for a liquid or solid substance.

Typical values of α for liquids are found to lie between 10^{-3} and $10^{-4}\,\mathrm{K}^{-1}$, while values of α for solids are typically between one and two orders of magnitude smaller. Similarly, typical values of κ_T for liquids are of order $10^{-5}\,\mathrm{bar}^{-1}$ and for solids of order $10^{-7}\,\mathrm{bar}^{-1}$. Given the small magnitudes of α and κ_T, a reasonable approximate equation of state for a liquid or solid thermodynamic system is thus

$$V(T, P) \simeq V(T_0, P_0)[1 + \alpha(T - T_0) - \kappa_T(P - P_0)]\,. \tag{2.8.8b}$$

2.8.1 The Van der Waals Model

The most widely-known model for a real thermodynamic system is that proposed for a fluid (that is, applying to both gaseous and liquid fluid phases) by Van der Waals in 1875. The equation of state for a Van der Waals fluid can be written either in the form

$$\left(P + \frac{aN^2}{V^2}\right)(V - Nb) = Nk_{\mathrm{B}}T \qquad (2.8.9a)$$

or, more usefully for present purposes, as

$$P = \frac{Nk_{\mathrm{B}}T}{V - Nb} - \frac{aN^2}{V^2}, \qquad (2.8.9b)$$

which gives the equation of state in terms of the pressure as an explicit function of
V and T. The parameter a in the Van der Waals equation of state is associated with
the existence of attractive forces between pairs of particles, while b represents the
actual, or excluded, volume of the fluid that is occupied by an individual particle
due to its finite size.

A note of caution must be interjected here. Although Eqs. (2.8.9) are forms of
the well-known Van der Waals model equation of state for a fluid, they are not in
the forms most commonly found in typical (macroscopic) thermodynamics texts, as
they explicitly show the role of the number of particles, N, of the fluid. The major
differences lie in the means of representation of the volume variable. We may write
the Van der Waals equation of state (2.8.9a) in a number of equivalent ways, most
commonly in terms of the molar volume, here defined as $\overline{V} \equiv V/n$, with n the
number of moles, or in terms of the volume per particle, $v \equiv V/N$, as

$$\left(P + \frac{\overline{a}}{\overline{V}^2}\right)(\overline{V} - \overline{b}) = RT \qquad (2.8.9c)$$

or

$$\left(P + \frac{a}{v^2}\right)(v - b) = k_{\mathrm{B}}T, \qquad (2.8.9d)$$

respectively, with equivalent versions of Eq. (2.8.9b) for the pressure. Here, we
employ V to symbolize exclusively the macroscopic volume of the thermodynamic
system. However, in some thermodynamics texts, V is employed to designate the
molar volume (rather than using $\overline{V}$ for this purpose), so that the tabulated Van der
Waals constants a and b given in such texts are in fact values for our $\overline{a}$ and $\overline{b}$, and
should therefore be employed in conjunction with Eq. (2.8.9c).

Example 2.8 Internal energy for a Van der Waals fluid.

The equation of state (2.8.9b) for the pressure, $P(T, V)$, of a Van der Waals fluid,
together with Eq. (2.3.10), gives

$$\left(\frac{\partial U}{\partial V}\right)_T = T^2\left[\frac{\partial}{\partial T}\left(\frac{P}{T}\right)\right]_V = \frac{aN^2}{V^2}.$$

The internal energy $U_{\mathrm{VdW}}(T, V; N)$ may therefore be expressed in the spirit of
Eq. (2.8.3) as

$$U_{\text{VdW}}(T, V; N) = U_{\text{ideal gas}}(T; N) - \frac{aN^2}{V}.$$

For an atomic Van der Waals gaś, whose particles possess only translational motion, for example, the internal energy is thus given as

$$U_{\text{VdW}}(T, V; N) = \tfrac{3}{2} N k_{\text{B}} T - \frac{aN^2}{V}.$$

This result is consistent with the association of the second term in expression (2.8.9b) with the existence of an attraction between Van der Waals atoms or molecules. Such an attraction must necessarily depend upon the average separation between the Van der Waals particles. It should thus be no surprise that (isothermal) volume changes will cause changes in the internal energy of the Van der Waals fluid that will be manifested as a volume dependence of the internal energy.   □

Example 2.9 Enthalpy, entropy, and Helmholtz energy for a Van der Waals fluid.

We have already obtained an expression for the internal energy for a Van der Waals fluid of structureless particles in Example 2.8 above. The enthalpy for a Van der Waals fluid is then given as

$$H_{\text{VdW}}(T, V; N) = U_{\text{VdW}}(T, V; N) + (PV)_{\text{VdW}}$$

or, using the expression for $U_{\text{VdW}}(T, V; N)$ obtained in Example 2.8 plus the Van der Waals equation of state, as

$$H_{\text{VdW}}(T, V; N) = H_{\text{ideal gas}}(T; N) + N k_{\text{B}} T \frac{Nb}{V - Nb} - \frac{2aN^2}{V}.$$

The absolute entropy for a Van der Waals fluid may be determined from Eq. (2.3.12a) in the form

$$dS_{\text{VdW}} = \frac{C_V}{T} dT + \frac{N k_{\text{B}}}{V - Nb} dV,$$

as

$$S_{\text{VdW}}(T, V; N) = S_0 + C_V \ln T + N k_{\text{B}} \ln(V - Nb),$$

which has the same form as the absolute entropy for an ideal gas, but with the volume V replaced by the free volume $V - Nb$ for the Van der Waals fluid.

The Helmholtz energy may now be obtained as

$$A_{\text{VdW}}(T, V; N) = U_{\text{VdW}}(T, V; N) - T S_{\text{VdW}}(T, V; N),$$

while the Gibbs energy can be obtained similarly from $A_{\text{VdW}}(T, V; N)$ via Eq. (2.3.6b), for example. Note that it is not convenient to provide the enthalpy and Gibbs energy explicitly in terms of their natural thermodynamic variables T and P, as the allowed values of V for a Van der Waals fluid are given by the roots of a cubic equation in V (see Eq. (2.8.9a)). $\square$

2.8.2   The Virial Equation of State

Accurate equation of state measurements (i.e., experimentally-determined PVT behaviour) for real gases show systematic deviations from the ideal gas law. These deviations may conveniently be represented by the generalization

$$PV = Nk_BT\,\mathcal{Z}(P, T) \qquad (2.8.10a)$$

of the ideal gas law. The factor $\mathcal{Z}(P, T)$, which we may express as the ratio

$$\mathcal{Z}(P, T) \equiv \frac{PV}{(PV)_{\text{ideal}}} = \frac{PV}{Nk_BT}, \qquad (2.8.10b)$$

is called the compressibility factor, and has the value 1 for an ideal gas. Because real gases approach ideal gas behaviour as the pressure tends to zero, we have the limiting behaviour

$$\lim_{P \to 0} \mathcal{Z}(P, T) = 1,$$

so that $\mathcal{Z}(P, T) > 1$ implies that $P_{\text{real gas}} > P_{\text{ideal gas}}$, while $\mathcal{Z}(P, T) < 1$ implies that $P_{\text{real gas}} < P_{\text{ideal gas}}$ for the same molar volume $\overline{V}$. This factor is commonly represented graphically in terms of isotherms on a $\mathcal{Z}$ vs. P plot.

Based upon the observation that high-temperature isotherms for large molar volumes of real gases do not differ significantly from ideal gas isotherms, Kamerlingh–Onnes proposed in 1917 a pressure power series representation of the equation of state, with the leading term given by the ideal gas law. Specifically, he proposed that the product $P\overline{V}$ be represented as a power series in the gas pressure P as

$$P\overline{V} = RT(1 + B_2'P + B_3'P^2 + \cdots), \qquad (2.8.11a)$$

or, perhaps more conveniently, as a power series in the molar (number) density $\overline{\rho}$ (the reciprocal of the molar volume $\overline{V}$), as

$$P\overline{V} = RT(1 + B_2\overline{\rho} + B_3\overline{\rho}^2 + \cdots). \qquad (2.8.11b)$$

Equations of this type are referred to as *virial equations of state*, and the coefficients in the expansion are referred to as the second, third, etc., pressure or density *virial coefficients*. A comparison between Eqs. (2.8.11a) and (2.8.10b), for example, shows that the compressibility factor $\mathcal{Z}(P, T)$ is given by

$$\mathcal{Z}(P, T) = 1 + B_2'(T)P + B_3'(T)P^2 + \cdots . \qquad (2.8.12)$$

Figure 2.7 illustrates the behaviour of isotherms $\mathcal{Z}_T(P)$ for gaseous nitrogen for temperatures $T = 203, 293$, and 673 K. These three curves illustrate three typical behaviours observed for compressibility factor isotherms: for the $T = 203$ K isotherm, we see that the limiting slope, $(\partial \mathcal{Z}/\partial P)_T|_{P=0}$, is negative, so that as a function of pressure the isotherm first falls below 1, reaches a minimum, and then increases thereafter, while for the $T = 673$ K isotherm, $(\partial \mathcal{Z}/\partial P)_T|_{P=0}$ is positive, so that the isotherm is always greater than 1 for all nonzero pressures. For temperature $T = 293$ K, the $\mathcal{Z}_T(P)$ isotherm remains at approximately 1 for pressures up to about 125 bar, which means that N_2 at this temperature behaves as an ideal gas over an extensive pressure range. If we think in terms of the virial expansion (2.8.12), this means that the pressure second virial coefficient, $B_2'(T)$, is very close to zero for nitrogen gas at this temperature. More generally, the

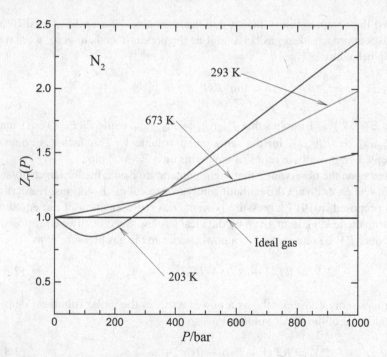

Fig. 2.7 Compressibility factor isotherms illustrating deviations from ideal gas behaviour for molecular nitrogen

temperature T_B for which the second virial coefficient vanishes is known as the Boyle temperature, T_B.

To obtain an equation for the Gibbs energy for a fluid that is described by the pressure virial equation (2.8.11a), we begin with Eq. (2.8.6) with $V_{\text{real fluid}}$ obtained from the virial equation of state (2.8.11a), which becomes, with $P_i = P_{\text{ideal}}$,

$$\int_{P_{\text{ideal}}}^{P} dG_T = Nk_B T \int_{P_{\text{ideal}}}^{P} \left[\frac{1}{P} + B_2'(T) + B_3'(T)P + \cdots \right] dP.$$

The Gibbs energy is then given by the virial series

$$G(T, P) = G(T, P_{\text{ideal}}) + Nk_B T \left[\ln\left(\frac{P}{P_{\text{ideal}}} \right) + B_2'(T)(P - P_{\text{ideal}}) \right.$$

$$\left. + \frac{1}{2} B_3'(T)(P^2 - P_{\text{ideal}}^2) + \cdots \right].$$

The expression for $G(T, P)$ can be simplified by starting from $G(T, P_{\text{ideal}})$, then choosing $P_{\text{ideal}} = 0$ for the lower limit of the integral over pressure, thereby obtaining

$$G(T, P) = G^{\circ}(T) + Nk_B T \left[\ln\left(\frac{P}{P^{\circ}} \right) + B_2'(T)P + \frac{1}{2} B_3'(T)P^2 + \cdots \right]$$
$$(2.8.13)$$

as the expression for the Gibbs energy of a nonideal gas described in terms of the pressure virial equation of state.

Although an expression of this type is rather appealing, we encounter the same problem that we have seen for the equation of state itself, i.e., the Gibbs free energy is no longer universal, as it was for the ideal gas. We see this explicitly from Eq. (2.8.13), as the virial coefficients differ from one nonideal gas to another. As will be apparent in Chap. 9, for example, such an expression would not be particularly convenient for the calculation of equilibrium constants for chemical reactions, so how do we overcome this problem?

2.8.3 Fugacity

To resolve this inconvenience, we shall generalize the concept of pressure *by defining* a new thermodynamic function, $f(T, P)$, called *fugacity*, via

$$G(T, P) \equiv G^{\circ}(T) + Nk_B T \ln\left(\frac{f}{f^{\circ}} \right), \qquad (2.8.14a)$$

so that all gas nonideality is assigned to the fugacity $f(T, P)$. Of course, we must have the condition

$$\lim_{P \to 0} f(T, P) = P \,, \tag{2.8.14b}$$

in order that the generalized expression reduces to the ideal gas expression in the limit of an ideal gas (as all gases behave ideally for pressures sufficiently low).

If we rewrite our virial expression for $G(T, P)$ in the form

$$G(T, P) = G^\circ(T) + N k_B T \ln \left\{ \frac{P}{P^\circ} e^{[B_2'(T)P + \frac{1}{2} B_3'(T) P^2 + \cdots]} \right\} \,,$$

and compare the result with Eq. (2.8.14a), we see that the fugacity must obey the equation

$$\frac{f(T, P)}{f^\circ} = \frac{P}{P^\circ} \exp \left\{ B_2'(T)P + \frac{1}{2} B_3'(T) P^2 + \cdots \right\}.$$

This result suggests that $f^\circ = P^\circ$, and that $f(T, P)$ has the form

$$f(T, P) = P \exp \left\{ B_2'(T)P + \frac{1}{2} B_3'(T) P^2 + \cdots \right\}. \tag{2.8.15}$$

- Notice that $G^\circ(T)$ appearing in Eqs. (2.8.13), (2.8.14a) is the same quantity, namely, the Gibbs energy for the ideal gas at a pressure of 1 bar. This means that the standard state of the nonideal gas is 1 bar *once it has been adjusted* to ideal behaviour, i.e., $f^\circ = P^\circ$.
- Notice also that this choice of standard state both allows all gases to be brought to a single common state (that of a hypothetical ideal gas at $P = P^\circ$) and, as we shall see, provides a means for calculating $f(T, P)$ at any temperature and pressure.

Let us consider the processes illustrated in Fig. 2.8. We wish to obtain an expression for the Gibbs energy change, denoted by ΔG_1, that occurs when we pass from the nonideal gas at temperature T and pressure P to a (hypothetical) ideal gas at temperature T and pressure P.

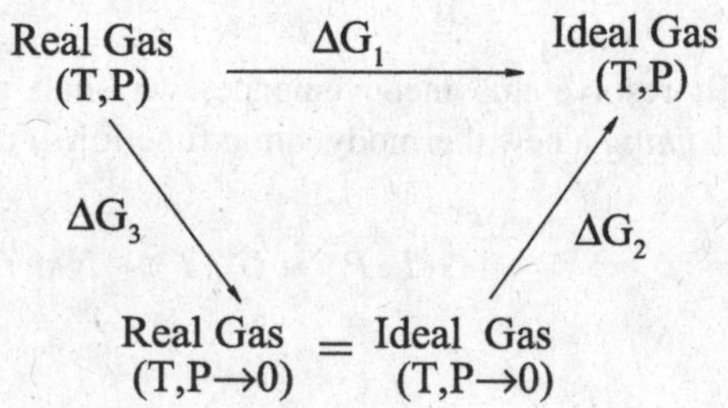

Fig. 2.8 Cartoon to aid in the derivation of a working expression for the fugacity of a real gas

For the direct step we see that ΔG_1 may be written as

$$\Delta G_1 = G_{\text{ideal}}(T, P) - G(T, P)$$

$$= Nk_{\text{B}}T \left[\ln\left(\frac{P}{P^\circ}\right) - \ln\left(\frac{f}{f^\circ}\right) \right]$$

or

$$\Delta G_1 = Nk_{\text{B}}T \ln\left(\frac{P}{f}\right),$$

given that $f^\circ = P^\circ \equiv 1$ bar.

The change ΔG_1 in the Gibbs energy must equal the overall change in the Gibbs energy $\Delta G_2 + \Delta G_3$ that would be obtained by passing through the state corresponding to $T, P \to 0$ at which a nonideal gas becomes an ideal gas. We may utilize the mathematical requirement

$$\left(\frac{\partial G}{\partial P}\right)_T = V$$

of the total differential (2.3.6d) in order to evaluate the Gibbs energy changes ΔG_2 and ΔG_3, namely,

$$\Delta G_2 = \int_P^{P_{\text{ideal}}} \left(\frac{\partial G}{\partial P}\right)_T dP = \int_P^{P_{\text{ideal}}} V \, dP,$$

and

$$\Delta G_3 = \int_{P_{\text{ideal}}}^P V_{\text{ideal}} \, dP = \int_{P_{\text{ideal}}}^P \frac{Nk_{\text{B}}T}{P} \, dP.$$

Addition of these two results gives

$$\Delta G_3 + \Delta G_2 = \int_{P_{\text{ideal}}}^P \left(\frac{Nk_{\text{B}}T}{P} - V\right) dP = Nk_{\text{B}}T \int_{P_{\text{ideal}}}^P \left(\frac{1}{P} - \frac{V}{Nk_{\text{B}}T}\right) dP.$$

The equality $\Delta G_1 = \Delta G_2 + \Delta G_3$ allows us to obtain the basic expression for determining the fugacity for the nonideal gas, namely,

$$\ln\left(\frac{f}{P}\right) = \int_0^P \left(\frac{V}{Nk_{\text{B}}T} - \frac{1}{P}\right) dP. \qquad (2.8.16)$$

Notice that the limit $P_{\text{ideal}} \to 0$ does not pose a problem for this integral since, for sufficiently low pressures, the real gas behaves as an ideal gas, in which case $V/(Nk_{\text{B}}T) = 1/P$, so that the integral vanishes, and $\ln(f/P) = 0$, so that $f = P$.

Because the deviation of the ratio f/P from unity directly indicates the extent of deviation of the nonideal gas from ideality, the ratio f/P is called the *fugacity coefficient*, and has been assigned the symbol γ, i.e.,

$$\gamma \equiv \frac{f}{P}.$$
(2.8.17a)

For an ideal gas, the fugacity coefficient has the value $\gamma = 1$ and, more generally, is given via Eq. (2.8.16) as

$$\ln \gamma = \int_{P_{\text{ideal}}}^{P} \left(\frac{V}{Nk_{\text{B}}T} - \frac{1}{P} \right) dP.$$
(2.8.17b)

We may also express the fugacity coefficient in terms of the compressibility factor $\mathcal{Z}(T, P)$ as

$$\ln \gamma = \int_{0}^{P} \frac{\mathcal{Z}(T, P) - 1}{P} \, dP.$$
(2.8.18)

For the pressure virial equation of state (2.8.12) $\ln \gamma$ is given as

$$\ln \gamma = \int_{0}^{P} \left[B_2'(T) + B_3'(T)P + \cdots \right] dP$$

or

$$\ln \gamma = B_2'(T)P + \tfrac{1}{2}B_3'(T)P^2 + \cdots.$$
(2.8.19)

It is known that the compressibility factor $\mathcal{Z}(T, P)$ satisfies a law of corresponding states, i.e., the compressibility factor values for many gases fall on the same $\mathcal{Z}(T^*, P^*)$ curves. Thus, if we replace $\mathcal{Z}(T, P)$ by $\mathcal{Z}(T^*, P^*)$, and change the integration variable from P to P^* by dividing both dP and P by the critical pressure P_{c} to obtain

$$\ln \gamma = \int_{0}^{P^*} \frac{\mathcal{Z}(T^*, P^*) - 1}{P^*} \, dP^*,$$
(2.8.20)

we see that $\ln \gamma$, and hence γ itself, is also a universal function of T^* and P^*. This gives us a means of calculating $\gamma(T, P)$ for any gas once we know its critical constants.

Chemical reactions in real gas mixtures can be handled in much the same fashion as chemical reactions in ideal gas mixtures by making use of the fugacity as a generalization of pressure. Thus, we write

$$\mu_\alpha(T, P) = \mu_\alpha^\circ(T) + RT \ln \frac{f_\alpha}{f^\circ}$$

rather than employing Eq. (2.3.7c) to determine the chemical potential of chemical species α. Recall also that $f^\circ = P^\circ = 1$ bar. The change in the Gibbs energy for the chemical reaction (2.7.7) for real gases is then obtained as

$$(\Delta G)_{\text{react}}(T, P, \xi) = (\Delta G^\circ)_{\text{react}}(T) + RT \ln \frac{(f_C(\xi)/f^\circ)^{\nu_C}(f_D(\xi)/f^\circ)^{\nu_D}}{(f_A(\xi)/f^\circ)^{\nu_A}(f_B(\xi)/f^\circ)^{\nu_B}},$$

rather than as given by Eq. (2.7.19). At equilibrium, $(\Delta G)_{\text{react}} = 0$, the fugacities achieve their equilibrium values, $f_\alpha = f_\alpha(\xi_e)$, and

$$(\Delta G^\circ)_{\text{react}}(T) = -RT \ln K_f,$$

with K_f the relevant equilibrium constant. Upon employing the defining relation for the activity coefficients γ_α for the various chemical species in the reaction mixture, we may relate K_f to K_P via

$$K_f = K_\gamma K_P,$$

with $K_\gamma \equiv \gamma_C^{\nu_C}\gamma_D^{\nu_D}/(\gamma_A^{\nu_A}\gamma_B^{\nu_B})$. A deviation of K_γ from one thus provides a measure of the nonideality of the gases participating in the chemical reaction.

2.8.4 The Law of Corresponding States

We shall see that the nature of the deviations from ideal gas behaviour is illustrated well by the Van der Waals gas, for which the compressibility factor takes the form

$$\mathcal{Z}_{\text{VdW}}(P, T) = \frac{V}{V - Nb} - \frac{aN}{k_B T V} \tag{2.8.21a}$$

$$= \frac{1}{1 - \dfrac{Nb}{V}} - \frac{aN}{k_B T V}. \tag{2.8.21b}$$

The form for the first term in Eq. (2.8.21b) facilitates its expansion as a power series in $Nb/V \equiv b/v$ for $Nb \ll V$, which will be the case for a gas density not too far removed from that of an ideal gas. The expanded form can be written as

$$\mathcal{Z}_{\text{VdW}}(P, T) \simeq 1 + \left(b - \frac{a}{k_B T}\right)\frac{1}{v} + \frac{b^2}{v^2} + \cdots \tag{2.8.22a}$$

$$\equiv 1 + B_2^{\text{VdW}}(T)\rho + B_3^{\text{VdW}}(T)\rho^2 + \cdots, \qquad (2.8.22\text{b})$$

which gives the (number) density virial expansion for a Van der Waals gas. A different form for the virial equation, referred to as the pressure virial equation, can be obtained by making use of the relation between V^{-1} and P: the result is

$$\mathcal{Z}_{\text{VdW}}(P, T) = 1 + \frac{1}{k_B T}\left(b - \frac{a}{k_B T}\right)P + \frac{a}{(k_B T)^3}\left(2b - \frac{a}{k_B T}\right)P^2 + \cdots$$
$$(2.8.22\text{c})$$

$$\equiv 1 + B_2'(T)P + B_3'(T)P^2 + \cdots. \qquad (2.8.22\text{d})$$

We note that the expansion in powers of v^{-1} gives $\mathcal{Z}_{\text{VdW}}(P, T)$ as an implicit function of P only, while Eqs. (2.8.22c), (2.8.22d) are more useful to us in examining the explicit dependence of $\mathcal{Z}_{\text{VdW}}(P, T)$ on the gas pressure. Notice that for a Van der Waals gas, only the second virial coefficient is a function of temperature, while all higher virial coefficients are constants. This is perhaps not too surprising, as only the contribution to $B_k(T)$ arising from the attractive forces is temperature dependent in the Van der Waals model equation of state. In this sense, the Van der Waals model can certainly be said to be unrealistic.

From Eq. (2.8.22c), we see that the initial slope of a compressibility factor isotherm will be positive if $k_B T b > a$ and negative if $k_B T b < a$. In order to establish the relationship between the Van der Waals a and b parameters and atomic/molecular excluded-volume and attractive-force effects, we shall consider briefly two interaction potential models for which the second virial coefficient can be evaluated exactly (see Eq. (7.2.31)). Specifically, we shall consider the hard-sphere (HS) model, in which the interaction takes the form

$$V_{\text{HS}}(R) = \begin{cases} \infty, & R \leq \sigma \\ 0, & R > \sigma, \end{cases}$$

with σ the HS diameter (also the distance of closest approach of the centres of the hard spheres), and the Sutherland model, in which the interaction energy is given by

$$V_{\text{S}}(R) = \begin{cases} \infty, & R < \sigma \\ -C_n R^{-n}, & R \geq \sigma. \end{cases}$$

The most common value for n is $n = 6$, corresponding to the induced dipole-induced dipole long-range London dispersion interaction: we shall examine only this case.

Evaluation of Eq. (7.2.31) for the HS potential model gives the second virial coefficient $B_2^{\text{HS}}(T)$ for a HS gas as

$$B_2^{HS}(T) = \frac{2\pi\sigma^3}{3} = 4v_{HS}\,,$$

which is independent of temperature and equal to four times the volume v_{HS} of an individual hard sphere. For the Sutherland model, with $n = 6$, we obtain $B_2(T)$ as

$$B_2^{SM}(T) = B_2^{HS} - 2\pi \int_\sigma^\infty \left[e^{\beta C_6/R^6} - 1\right] R^2 \, dR\,,$$

with β equal to $1/(k_B T)$. We cannot in general evaluate this expression further exactly, but for $\beta C_6/R^6 \ll 1$, we may expand the integrand and retain only the first non-vanishing term to give

$$B_2^{SM}(T) = \frac{2\pi\sigma^3}{3} - \frac{2\pi C_6}{3k_B T \sigma^3}\,.$$

If we compare this result with $B_2^{VdW}(T)$ given in Eqs. (2.8.22), we see that the Van der Waals parameters a and b may be obtained as

$$a = \frac{2\pi C_6}{3\sigma^3}\,, \qquad b = \frac{2\pi\sigma^3}{3}\,,$$

so that a is thus associated (through C_6) with the long-range attractive dispersion interaction and b corresponds to four times the volume v_{HS} associated with a hard sphere of diameter σ (often referred to as the 'forbidden' or 'excluded' volume). Given this association of the parameters a and b with attraction and the volume occupied by an individual atom/molecule, a positive *initial* slope for a compressibility factor isotherm thus indicates that molecular size effects dominate, while a negative initial slope indicates that attractive forces dominate.

Clearly, the Van der Waals equation of state gives a qualitative explanation of real gas behaviour through $\mathcal{Z}(P, T)$. Note, however, that as the Van der Waals equation of state is a function of two parameters a and b that depend upon the nature of the specific gas being considered, we have lost the universality that we had with the ideal gas equation of state.

Another way of viewing the Van der Waals equation is in the form of a cubic equation in the volume V, namely,

$$V^3 - N\left(b + \frac{k_B T}{P}\right) V^2 + \frac{aN^2}{P} V - \frac{aN^3 b}{P} = 0\,. \qquad (2.8.23)$$

Figure 2.9 illustrates the behaviour of the Van der Waals pressure isotherms for a and b parameters characterizing CO_2. We see from this figure that for temperatures $T > 304.25$ K, the isotherms are monotone decreasing functions of molar volume, while for temperatures below 304.25 K, the Van der Waals equation for CO_2 gives strongly non-monotonic behaviour over a fairly extensive range of molar volume $\overline{V}$.

Fig. 2.9 Pressure–Volume
isotherms for carbon dioxide
modelled according to the
Van der Waals equation of
state (2.8.9c) using Van der
Waals parameters
$\overline{a} = 3.658$ dm^6 bar mol^{-2},
$\overline{b} = 0.0429$ dm^3 mol^{-1}

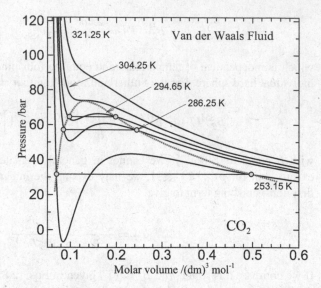

What, if anything, does the non-monotonic behaviour of a Van der Waals $P\overline{V}$-isotherm mean? To answer this question, we need to ask if such behaviour is physically meaningful. We shall see in Sect. 7.3 of Chap. 7 that $\left(\dfrac{\partial P}{\partial \overline{V}}\right)_{T,N} \leq 0$ for a $P\overline{V}$-isotherm, so that if $\left(\dfrac{\partial P}{\partial \overline{V}}\right)_{T,N} > 0$ for a section of a Van der Waals isotherm, that portion of the plot must represent an unphysical behaviour that has been introduced by the mathematical representation of the Van der Waals equation of state $P \equiv P(\overline{V}, T)$. Any looping of the isotherms in Fig. 2.9 for temperatures below 304.25 K must therefore be a manifestation of the Van der Waals model. This unphysical looping of Pv-isotherms for a Van der Waals fluid was recognized by Maxwell quite soon after Van der Waals proposed his model: Maxwell's solution, which is still employed, was to replace the unphysical region of the isotherm by a flat portion based upon what he termed an equal-area (i.e., above and below the horizontal line) construction to determine its end-points.

To understand why the Maxwell construction is appropriate, we need to examine the total differential, dG, for the Gibbs energy of a pure substance in a single phase (such as a gas or liquid), which is given by

$$dG = -S\,dT + V\,dP$$

or, in terms of the specific entropy $s \equiv S/N$ and volume v, as

$$d\mu = -s\,dT + v\,dP\,, \tag{2.8.24}$$

with $\mu \equiv G/N$ being the chemical potential for a pure substance in a specified phase. In the event, for example, that we have both liquid and gaseous phases of the substance present at the same point (v, P) of the Pv-diagram, the total Gibbs energy is given by $G = G_\ell + G_g$ and the total number of atoms/molecules is given by $N = N_\ell + N_g$, in which the subscripts ℓ, g denote the liquid and gas phases, respectively. The total differential of G at the specified point (v, P) in the phase diagram can then be expressed as

$$dG = \left(\frac{\partial G_\ell}{\partial N_\ell}\right)_{N_g} dN_\ell + \left(\frac{\partial G_g}{\partial N_g}\right)_{N_\ell} dN_g. \tag{2.8.25a}$$

Now, as the total number of atoms/molecules of the substance is fixed, $dN \equiv 0$, and $dN_\ell = -dN_g$. Given that the partial derivatives in expression (2.8.25a) for dG define the chemical potentials μ_ℓ and μ_g for the two phases of the substance under consideration, we may express dG as

$$dG = (\mu_g - \mu_\ell)\, dN_g . \tag{2.8.25b}$$

At thermodynamic equilibrium, $dG = 0$ and, as $dN_g \neq 0$, thermodynamic equilibrium between two phases of the same substance then requires that the chemical potentials associated with those phases be equal, i.e., in the present case, $\mu_\ell = \mu_g$.

Returning to Fig. 2.9, let us focus upon the $T = 286.25$ K isotherm and examine Eq. (2.8.24) for that portion lying between the two (yellow) open circle symbols, both of which correspond to a fixed pressure P_0 and which, for convenience, we shall designate as the initial, i, and final, f, values. Let us integrate both sides of Eq. (2.8.24) over this interval, noting that the first integral on the right-hand side vanishes (because $dT \equiv 0$) and carrying out an integration by parts of the second integral, to obtain

$$\int_{\mu_i}^{\mu_f} d\mu = (Pv)\,\Big|_i^f - \int_{v_i}^{v_f} P\, dv$$

or

$$\mu_\ell - \mu_g = P_0(v_f - v_i) - \int_{v_i}^{v_f} P\, dv$$

$$= \int_{v_i}^{v_f} (P_0 - P)\, dv . \tag{2.8.26}$$

Hence, under thermodynamic equilibrium conditions for which $\mu_\ell = \mu_g$, the areas above and below the horizontal line $(P = P_0)$ in Fig. 2.9 must be equal in order for the equilibrium condition $\mu_\ell = \mu_g$ for phase co-existence to be satisfied.

Fig. 2.10 Illustration of the
two-phase region for a Van
der Waals fluid

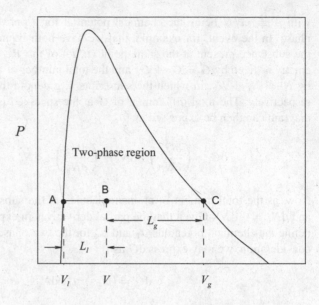

The locus obtained from the end-points of the constant pressure portions of
isotherms for temperatures lying below 304.25 K for CO_2 that thus determines
the two-phase region of the PV plot is illustrated in Fig. 2.10. This curve is
consequently referred to as the co-existence curve. The horizontal line AC in
Fig. 2.10, called a 'tie-line', is associated with a specific isotherm and is obtained
via the Maxwell construction for a two-phase equilibrium region of the fluid. The
volume at the right-hand endpoint, C, of the tie-line represents the volume, V_g, of
the (Van der Waals) fluid occupied by the pure gaseous state at pressure P_0, while
the volume at the left-hand endpoint, A, of the tie-line represents the volume, V_ℓ,
occupied by the pure liquid state at pressure P_0. Volumes V corresponding to points
(V, P_0) on the isotherm (e.g., B in Fig. 2.10), with $V_\ell < V < V_g$, are equilibrium
mixtures of gaseous and liquid phases of the (Van der Waals) fluid in which the mole
fraction $x_g \equiv n_g/n$ of the gaseous phase is given by

$$x_g = \frac{V - V_\ell}{V_g - V_\ell},$$ (2.8.27a)

while the mole fraction of the liquid phase is given as $x_\ell = 1 - x_g$. Equivalently,
the ratio of the numbers of moles of the substance in the liquid and gaseous phases
is given by

$$\frac{n_\ell}{n_g} = \frac{x_\ell}{x_g} = \frac{V_g - V}{V - V_\ell},$$ (2.8.27b)

a result often referred to [19] as the lever rule.[13] For example, the lever rule as applied to the liquid-gaseous mixture at total pressure P_0 and having total volume V
(located at B in Fig. 2.10) may be expressed as $n_\ell L_\ell = n_g L_g$, with $L_\ell = V - V_\ell$
and $L_g = V_g - V$ representing the lengths of the tie-line from the end-points A and
C, respectively.

The isotherm in Fig. 2.10 that has only an inflection point is called the critical
isotherm, its temperature is called the critical temperature, T_c. The inflection point
is called the critical point of the Van der Waals fluid, and is characterized by the
critical volume, V_c, and critical pressure, P_c. The critical values T_c and V_c for a Van
der Waals fluid may be obtained from the conditions

$$\left(\frac{\partial P}{\partial V}\right)_{T_c} = \left(\frac{\partial^2 P}{\partial V^2}\right)_{T_c} = 0$$

at the inflection point, with P_c obtained from the equation of state (2.8.9b) as $P_c = P(T_c, V_c)$. However, a simpler procedure for obtaining the critical values T_c, V_c, and
P_c for the Van der Waals fluid is to recognize that the cubic Van der Waals equation
(2.8.23) always has three roots and that, for temperatures below T_c there will be
three distinct real roots (leading to the loop structures of the isotherms), at $T = T_c$
the three real roots coalesce to give an inflection point, and for $T > T_c$ there will
be only one real root plus two complex conjugate roots (giving rise to the simple
monotonic isotherms). Hence, for $T = T_c$, Eq. (2.8.23) can be written explicitly as
$(V - V_c)^3 = 0$ or, in expanded form, as

$$V^3 - 3V_c V^2 + 3V_c^2 V - V_c^3 = 0. \tag{2.8.28a}$$

By equating the coefficients of the powers of V in Eq. (2.8.23) for $T = T_c$, $P = P_c$
with the coefficients of those same powers of V in Eq. (2.8.28a) gives

$$3V_c = N\left(b + \frac{k_B T_c}{P_c}\right), \qquad 3V_c^2 = \frac{aN^2}{P_c}, \qquad V_c^3 = \frac{abN^3}{P_c}. \tag{2.8.28b}$$

The three expressions (2.8.28b) are readily solved for V_c, P_c, T_c to give

$$V_c = 3Nb, \qquad T_c = \frac{8a}{27k_B b}, \qquad P_c = \frac{a}{27b^2}. \tag{2.8.28c}$$

These expressions are clearly consistent with V being an extensive variable, while
T and P are intensive variables.

If we obtain the Van der Waals constants a and b in terms of V_c and P_c from
expressions (2.8.28c) as

---

[13]For a more complete discussion of two-phase equilibria, phase stability conditions, and the lever
rule for a single-component system, see Ch. 18 (especially § 18.4) of [19].

$$a = \frac{3 P_c V_c^2}{N^2}, \qquad b = \frac{V_c}{3N},$$

and substitute these results into the Van der Waals equation of state (2.8.9a), then divide the left-hand side of Eq. (2.8.9a) by $P_c V_c$ and the right-hand side by $P_c V_c$ in the form $3N k_B T_c / 8$, we obtain

$$\left(\frac{P}{P_c} + \frac{3 V_c^2}{V^2} \right) \left(\frac{V}{V_c} - \frac{1}{3} \right) = \frac{8T}{3 T_c}.$$

Upon defining dimensionless variables P_R, V_R, and T_R via

$$P_R \equiv \frac{P}{P_c}, \qquad V_R \equiv \frac{V}{V_c}, \qquad T_R \equiv \frac{T}{T_c}, \qquad (2.8.29a)$$

which are referred to as reduced variables, the Van der Waals equation takes the form

$$\left(P_R + \frac{3}{V_R^2} \right) (V_R - \tfrac{1}{3}) = \tfrac{8}{3} T_R, \qquad (2.8.29b)$$

which is then referred to as the reduced Van der Waals equation of state. We may also establish that the compressibility factor $\mathcal{Z}(P, T)$ for a Van der Waals fluid can be obtained in reduced form as

$$\mathcal{Z}(V_R, T_R) = \frac{V_R}{V_R - \frac{1}{3}} - \frac{9}{8 V_R T_R}. \qquad (2.8.30)$$

As no reference to the Van der Waals constants a and b for a particular gas appears in the reduced forms (2.8.29b) and (2.8.30), they thus provide laws of corresponding states that apply to all Van der Waals fluids. Of course, the specific P, V, T values that give rise to a particular set of P_R, V_R, T_R values will clearly be different for different Van der Waals fluids, so that two different Van der Waals fluids will have the same properties when they are compared under corresponding conditions, i.e., when they have the same values of reduced pressure, volume, and temperature. It can be shown that any two-parameter equation of state, such as the Van der Waals equation of state examined here, may be obtained in a reduced form, and hence obeys a law of corresponding states.

2.8.5 Joule–Thomson Inversion

Let us now consider a cylinder that contains two pistons, one on each side of a fixed porous plug (see Fig. 2.11). Now consider a thermodynamic process in which

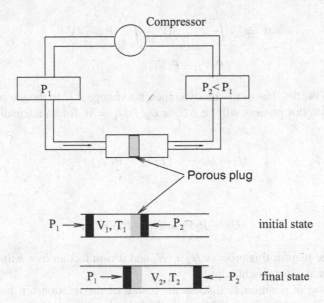

Fig. 2.11 Cartoon illustrating the Joule–Thomson expansion process

we begin with a fixed mass of a gas with volume V_1 at temperature T_1 in front of the first piston face and in front of the porous plug, then subject it to a constant external pressure P_1 so that the gas is forced through the plug into a volume that is created as the second piston, which is subjected to a fixed external pressure $P_2 < P_1$ retreats from the face of the plug. Because P_2 is less than P_1, the entire mass of gas will pass through the porous plus and will occupy a volume V_2 at temperature T_2 under pressure P_2. We shall also carry out this process under adiabatic conditions which, as we have seen, involves no exchange of heat with the surroundings, so that $Q = 0$. Because the external pressure P_2 is less than P_1, the gas will have expanded in passing through the porous plug, and $V_2 > V_1$. In general, we expect that $T_2 \neq T_1$: we have, however, still to determine what conditions determine both if this is so and the nature of the inequality.

For a fixed amount of gas passing through the porous plug under adiabatic conditions, we have a thermodynamic process for which we have:

- initial state: volume V_1 at temperature T_1 under a constant external pressure $P_{\text{ext},1} = P_1$
- final state: volume V_2 at temperature T_2 under a constant external pressure $P_{\text{ext},2} = P_2$

As the differential work is given by $\delta W = -P_{\text{ext}} \, dV$, the work W done in passing from the initial thermodynamic state to the final thermodynamic state will be given by

$$W = -\int_{V_1}^{0} P_{\text{ext},1}\, dV - \int_{0}^{V_2} P_{\text{ext},2}\, dV$$

$$= -(P_2 V_2 - P_1 V_1).$$

According to the first law of thermodynamics, the change ΔU in the internal energy associated with this process will be $\Delta U \equiv U_2 - U_1 = W$ for an adiabatic process, so that

$$U_2 - U_1 = -(P_2 V_2 - P_1 V_1)$$

or

$$U_2 + P_2 V_2 = U_1 + P_1 V_1 .$$

Hence, we see that for this process $H_2 = H_1$ and it is in fact an *isenthalpic* process (i.e., it occurs at constant enthalpy).

Now we are in position to discuss the nature of the relationship between the temperatures T_2 and T_1. To do this, we shall need to know how the temperature and pressure are related for an isenthalpic gas expansion. To accomplish this goal, we shall examine the partial derivative

$$\mu_{\text{JT}} \equiv \left(\frac{\partial T}{\partial P}\right)_H ; \tag{2.8.31}$$

known as the Joule–Thomson coefficient. Note that although we may not strictly think of partial derivatives as ratios of differentials they behave in many ways as if they were, in the sense that

$$\mu_{\text{JT}} > 0 \Rightarrow \begin{cases} T \text{ increases as P increases} \\[1em] T \text{ decreases as P decreases} \end{cases}$$

$$\mu_{\text{JT}} < 0 \Rightarrow \begin{cases} T \text{ decreases as P increases} \\[1em] T \text{ increases as P decreases} \end{cases}$$

which we may think of as equivalent to $\Delta T/\Delta P$ for ($\Delta P > 0, \Delta T > 0; \Delta P < 0, \Delta T < 0$) in the first case or to ($\Delta P > 0, \Delta T < 0; \Delta P < 0, \Delta T > 0$) in the second case.

As a first step towards expressing the Joule–Thomson coefficient in terms of more convenient thermodynamic quantities, we may utilize the permutation rule

$$\left(\frac{\partial T}{\partial P}\right)_H \left(\frac{\partial H}{\partial T}\right)_P \left(\frac{\partial P}{\partial H}\right)_T = -1$$

to write μ_{JT} alternatively as

$$\mu_{JT} = -\frac{\left(\frac{\partial H}{\partial P}\right)_T}{\left(\frac{\partial H}{\partial P}\right)_T} = -\frac{1}{C_P}\left(\frac{\partial H}{\partial P}\right)_T .$$

However, this expression for μ_{JT} is still not in terms of partial derivatives that may readily be accessed. But, upon recalling that the natural thermodynamic variables for the enthalpy are T and P, we can transform the combined first and second laws version of the total differential dH given in Eq. (2.3.3) into a total differential expressed in terms of T and P by considering the entropy S to be a function of T and P, thereby obtaining

$$dH = T\left(\frac{\partial S}{\partial T}\right)_P dT + \left[T\left(\frac{\partial S}{\partial P}\right)_T + V\right]dP ,$$

so that

$$\left(\frac{\partial H}{\partial P}\right)_T = T\left(\frac{\partial S}{\partial P}\right)_T + V .$$

Finally, the fourth Maxwell relation,

$$\left(\frac{\partial S}{\partial P}\right)_T = -\left(\frac{\partial V}{\partial T}\right)_P .$$

may now be employed to give

$$\left(\frac{\partial H}{\partial P}\right)_T = V - T\left(\frac{\partial V}{\partial T}\right)_P ,$$

from which we obtain the Joule–Thomson coefficient μ_{JT} as

$$\mu_{JT} = \frac{1}{C_P}\left[T\left(\frac{\partial V}{\partial T}\right)_P - V\right]. \tag{2.8.32}$$

If we evaluate μ_{JT} for the ideal gas, we find that $\mu_{JT}(\text{ideal gas}) = 0$, which means that the expansion of an ideal gas through a porous plug is an isothermal process. Of course, this should not come as a surprise to us, as we have seen that the enthalpy of an ideal gas depends, like the internal energy, only upon temperature.

We may therefore conclude that $\mu_{JT} \neq 0$ may occur only for nonideal, i.e., real, gases.

As the best-known model for real gas behaviour is the Van der Waals fluid, let us examine the consequences of $\mu_{JT} \neq 0$ for it. From the Van der Waals equation of state (2.8.9b), we may obtain $\left(\dfrac{\partial V}{\partial T}\right)_P$ as

$$\left(\frac{\partial V}{\partial T}\right)_P = \frac{1}{T}\frac{V - Nb}{1 - \dfrac{2Na(V - Nb)^2}{k_B T V^3}},$$

with the consequence that μ_{JT}^{VdW} is given by

$$\mu_{JT}^{VdW} = \frac{1}{C_P}\left[\frac{V - Nb}{1 - \dfrac{2Na(V - Nb)^2}{k_B T V^3}} - V\right]. \tag{2.8.33}$$

Expansion of the reciprocal term in this expression gives the approximate result

$$\mu_{JT}^{VdW} \simeq \frac{1}{C_P}\left\{(V - Nb)\left[1 + \frac{2Na(V - Nb)^2}{k_B T V^3} - \cdots\right] - V\right\}$$

$$= \frac{1}{C_P}\left[-Nb + \frac{2Na(V - Nb)^3}{k_B T V^3} - \cdots\right].$$

Upon multiplying out $(V - Nb)^3$ and retaining only those terms that are linear in the Van der Waals parameters a and b, μ_{JT}^{VdW} is approximated by

$$\mu_{JT}^{VdW} \approx \frac{N}{C_P}\left(\frac{2a}{k_B T} - b\right). \tag{2.8.34}$$

In this approximation, expression (2.8.34) represents a competition between $2a/(k_B T)$ and b: there will clearly be some temperature for which $2a/(k_B T) = b$, so that μ_{JT}^{VdW} vanishes. The temperature at which the Joule–Thomson coefficient vanishes is designated as T_i and is termed the *inversion temperature*, as it represents the temperature at which the Van der Waals fluid ceases to heat upon expansion and instead cools upon expansion.

We can see from Eq. (2.8.34) that in lowest approximation the inversion temperature for a Van der Waals fluid will have the value

$$T_i = \frac{2a}{bk_B}.$$

More generally, however, the vanishing of the Joule–Thomson coefficient for a Van der Waals fluid requires that V_i and the inversion temperature be related by

$$V_i = \frac{Nb}{1 - \sqrt{\dfrac{bk_B T_i}{2a}}} .$$

Substitution of this expression for the inversion volume into the Van der Waals equation of state leads to the relation

$$(b^2 P_i + a + \tfrac{3}{2} b k_B T_i)^2 = 8abk_B T_i \qquad (2.8.35)$$

between the inversion temperature T_i and the inversion pressure P_i.

If we regard Eq. (2.8.35) as a quadratic equation in the inversion pressure P_i, then the coordinates of the apex of this parabolic curve are obtained by setting the first derivative

$$\frac{dP_i}{dT_i} = \frac{1}{2b^2} \left[\sqrt{\frac{8abk_B}{T_i}} - 3bk_B \right]$$

to zero at $T_i = T_{i,max}$ to obtain

$$T_{i,max} = \frac{8a}{9bk_B} .$$

Substitution of this result into the governing equation (2.8.35) determines the corresponding value for $P_{i,max}$ as $P_{i,max} = a/(3b^2)$. Note that as the second derivative,

$$\frac{d^2 P_i}{dT_i^2} = -\frac{1}{4T_i} \sqrt{\frac{8abk_B}{T_i}} < 0 ,$$

is negative, $(T_{i,max}, P_{i,max})$ indeed represents the maximum in the inversion pressure as a function of inversion temperature.

Finally, we note that if we replace the Van der Waals a and b parameters in terms of the critical temperature T_c and pressure P_c, we obtain a corresponding governing equation for the reduced inversion temperature T_i^* and pressure P_i^*, which is

$$(P_i^* + 12T_i^* + 27)^2 = 1728 T_i^* .$$

A plot of the Joule–Thomson inversion curve for Van der Waals fluids is shown in Fig. 2.12. The Joule–Thomson effect provided the means for the first successful liquefaction of air, and still serves today as the basis for the commercial production of liquid nitrogen and other cryogenic liquids and as the basis for the production

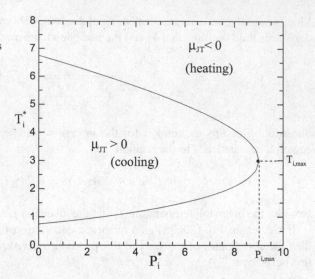

Fig. 2.12 The Joule–Thomson inversion curve for Van der Waals fluids

of liquid natural gases (LNG) employed for transporting methane, ethane, propane, and butane around the globe.

In order to plot the universal $T_i^* - P_i^*$ Van der Waals inversion curve, it is convenient to rewrite our result as a more conventional quadratic equation for T_i^*, namely, as

$$(T_i^*)^2 + \tfrac{1}{6}\,(P_i^* - 45)T_i^* + \tfrac{1}{144}(P_i^* + 27)^2 = 0\,,$$

from which we can obtain values of T_i^* as

$$T_i^* = \tfrac{1}{12}(45 - P_i^*) \pm \sqrt{9 - P_i^*}\,.$$

Although the Van der Waals equation of state provides a qualitative description for the pressure–volume–temperature behaviour of fluids, it is too simple an expression to account quantitatively for the complex PVT properties of general fluids. A large number of more accurate model equations of state have been developed over the previous decades for fluid systems. Many of them, especially the multi-parameter models, may be considered to be essentially sophisticated fitting functions for the representation of experimental data. However, a pair of two-parameter model generalizations of the Van der Waals model that have enjoyed considerable success should perhaps be singled out. The Redlich–Kwong equation,

$$P = \frac{RT}{\overline{V} - B} - \frac{A}{\overline{V}(\overline{V} + B)\sqrt{T}}\,, \qquad (2.8.36a)$$

and the Peng–Robinson equation,

$$P = \frac{RT}{\overline{V} - \beta} - \frac{\alpha}{\overline{V}(\overline{V} + \beta) + \beta(\overline{V} - \beta)}, \qquad (2.8.36b)$$

have proven to be quite successful for simple molecular fluids, especially in the liquid/vapour region. These two equations have been employed extensively by chemical engineers for molar densities ρ ranging from well below $1\ mol\,L^{-1}$ to as much as $15\ mol\,L^{-1}$ over a fairly extensive range of temperatures. Both models represent a considerable improvement over the Van der Waals equation of state for molar densities exceeding about $7\ mol\,L^{-1}$.

2.9 Problems for This Chapter

1. Obtain an expression for the work W done by an ideal classical gas during an isothermal expansion from initial volume V_i to a final volume V_f, and show that the Helmholtz energy A decreases by the same amount.

2. The form for the entropy of an ideal classical gas given in Eq. (2.3.13b) ensures that the entropy is an extensive thermodynamic quantity. Verify the extensivity of S by doubling both the number of ideal gas particles N and the volume V of the container, while holding the pressure and temperature (which are the relevant intensive thermodynamic variables) constant. Determine whether or not the Helmholtz energy A is an extensive quantity.

3. Consider ideal gases A and B, both at temperature T, sharing a container of total volume V, but initially separated into volumes V_A and V_B by an impermeable partition that is then removed to allow the two gases to mix. Show that the pressures P_A and P_B of the two gases in the final mixture satisfy the Dalton law of partial pressures. How are the partial pressures of the mixture components related to their initial pressures?

4. Consider a container of volume V, initially separated by an impenetrable wall of negligible volume into two compartments having unequal volumes V_A and V_B and containing ideal gases A and B, respectively. If the number of molecules are N_A and N_B (in general, $N_A \neq N_B$), and both gases are at temperature T, obtain an expression for the entropy difference, $(\Delta S)_{i\ \rightarrow\ f}$, between the initial state (with the partition) and the final state (without the partition) once equilibrium has been re-established. Show that if $N_A = N_B \equiv N$ and $V_A = V_B \equiv V$, the entropy difference becomes $2Nk_B \ln 2$.

5. Show that if in Problem 4 the ideal gases A and B have both temperature T and pressure P in common, then the entropy difference can be expressed in the form $(\Delta S)_{i \rightarrow f} = -nR(x_A \ln x_A + x_B \ln x_B)$, with $n \equiv n_A + n_B$ the total number of moles in the mixture, and $x_\alpha \equiv n_\alpha/(n_A + n_B)$, $\alpha = $ A, B, the mole fraction of species α in the binary gas mixture.

6. Consider a container of volume V and at temperature T, initially separated by an impenetrable wall of negligible volume into two compartments of volumes V_1 and V_2, each compartment containing an ideal gas A. If the

two compartments contain N_1 and N_2 A molecules, respectively, obtain an expression for the entropy difference, $(\Delta S)_i \rightarrow$ f, between the initial state (with the partition) and the final state (without the partition) once equilibrium has been re-established. Show that if the initial mole fractions $n_1 = N_1/V_1$ and $n_2 = N_2/V_2$ are equal, the entropy of mixing is zero. Why would this not be the case were $n_1 \neq n_2$?

7. Use the fact that the differentials dA and dG of the Helmholtz and Gibbs energies are total differentials to obtain the third and fourth Maxwell relations

$$\left(\frac{\partial S}{\partial V}\right)_T = \left(\frac{\partial P}{\partial T}\right)_V ; \qquad \left(\frac{\partial V}{\partial T}\right)_P = -\left(\frac{\partial S}{\partial P}\right)_T ,$$

8. Show that $V_1 V_3 = V_2 V_4$ for the Carnot cycle discussed in Sect. 5.1.

9. The equation originally proposed by Van der Waals for the thermodynamic equation of state for a nonideal gas was given as $(P + an^2/V^2)(V - nb) = nRT$, with V the total volume occupied by n moles of gas, P and T the gas pressure and the (absolute) temperature, R the universal gas constant, b the volume excluded by virtue of its occupation by the nN_0 finite-sized particles in the container, while an^2/V^2 (proportional to the square of the molar concentration) represents the influence of interparticle attractions between gas particles. Show that Eq. (2.8.9a) is equivalent to the equation proposed by Van der Waals.

10. Show that the Redlich–Kwong equation (2.8.36a) obeys a law of corresponding states.

11. Use the Van der Waals equation (2.8.9c), with $\bar{a} = 3.658\,\mathrm{dm}^6\,\mathrm{bar}\,\mathrm{mol}^{-2}, \bar{b} = 0.0429\,\mathrm{dm}^3\,\mathrm{mol}^{-1}$ to calculate the molar volume for CO_2 at a pressure of $200\,\mathrm{bar}$ and a temperature of $450\,\mathrm{K}$. Express your answer accurate to five significant figures. Hint: Although one may solve the cubic equation for $\overline{V}$ analytically using the Cardan formula, you may instead wish to solve it by using an iterative method that starts from a reasonable zeroth-order guess, say $\overline{V}_c$.

12. One early equation of state for a nonideal gas is the Dieterici equation

$$P = \frac{RT}{\overline{V} - b'}\, e^{-a'/(RT\overline{V})} .$$

Obtain expressions for the Dieterici parameters a', b' in terms of the critical temperature and molar volume, and evaluate them for Xe, for which the critical constants are $T_c = 289.74\,\mathrm{K}, \overline{V}_c = 0.11800\,\mathrm{dm}^3\,\mathrm{mol}^{-1}$, and $P_c = 58.40\,\mathrm{bar}$. Obtain the reduced Dieterici equation of state.

13. Consider a vessel of capacity $1.0\,\mathrm{dm}^3$ that is maintained at a constant temperature of $25\,^\circ\mathrm{C}$. Calculate the pressure that $0.1313\,\mathrm{kg}$ of Xe gas would exert in this vessel if Xe is treated (a) as an ideal gas, (b) as a Van der Waals gas, (c) as a Dieterici gas. The Van der Waals parameters for Xe are $\bar{a} = 4.192\,\mathrm{dm}^6\,\mathrm{mol}^{-2}\,\mathrm{bar}$ and $\bar{b} = 0.0516\,\mathrm{dm}^3\,\mathrm{mol}^{-1}$, while the Dieterici

parameters for Xe are $a' = 5.6850\,\mathrm{dm}^6\,\mathrm{mol}^{-2}$ and $b' = 0.05900\,\mathrm{dm}^3\,\mathrm{mol}^{-1}$. Compare your calculated values with an experimental value deduced from the data provided by Michels et al. [A. Michels, T. Wassenaar, and P. Lowerse, *Physica* **20**, 99–106 (1954)].

14. The expressions obtained in Examples 2.8 and 2.9 for the molar internal energy $\overline{U}(T, \overline{V})$ and entropy $\overline{S}(T, \overline{V})$ for a Van der Waals fluid may be written as $\overline{U}(T, \overline{V}) = \overline{U}_{\text{ideal gas}}(T) - \overline{a}/\overline{V}$ and $\overline{S}(T, \overline{V}) = \overline{S}_{\text{ideal gas}}(T) + R\ln(\overline{V} - \overline{b})$. For fixed temperature and pressure, the Gibbs energy typically exhibits two minima as a function of molar volume, corresponding to values of the molar volumes of the liquid and gaseous phases of the Van der Waals fluid, with the deeper minimum corresponding to the thermodynamically stable phase. Evaluate $\overline{G}$ as a function of $\ln\overline{V}$ from $\overline{V} = 0.02\,\mathrm{dm}^3$ to $\overline{V} = 400\,\mathrm{dm}^3$ for a fixed pressure $P = 1$ bar and for temperatures 354, 374, 394 K, for a Van der Waals fluid having parameters $\overline{a} = 4.843\,\mathrm{dm}^6\,\mathrm{bar}\,\mathrm{mol}^{-2}$, $\overline{b} = 0.016\,\mathrm{dm}^3\,\mathrm{mol}^{-1}$. Plot $\overline{G}$ vs. $\ln\overline{V}$ for each of these three temperatures and discuss the relative stabilities of the liquid and gaseous phases at these temperatures.

15. By employing the criterion that $\Delta\overline{G} = \overline{G}_g - \overline{G}_\ell = 0$ for the process liquid $\Longleftrightarrow$ gas at the normal boiling temperature T_b, determine from the results of Problem 14 the value of T_b for the Van der Waals fluid, as well as the molar volumes of the two co-existing phases at the boiling temperature. Develop expressions for $\Delta\overline{H}_{\text{vap}}(T_b)$ and $\Delta\overline{S}_{\text{vap}}(T_b)$ for this Van der Waals fluid and evaluate them.

16. Use the ideal gas, Van der Waals, and Redlich–Kwong equations of state to compute the pressure of ethane gas at temperature $T = 350$ K and molar density $\overline{\rho} = 14.241\,\mathrm{mol}\,\mathrm{dm}^{-3}$, given that the Van der Waals parameters for ethane are $\overline{a} = 5.56145\,\mathrm{dm}^6\,\mathrm{mol}^{-2}\,\mathrm{bar}$, $\overline{b} = 0.06380\,\mathrm{dm}^3\,\mathrm{mol}^{-1}$, and that the Redlich–Kwong parameters for ethane are $A = 98.831\,\mathrm{dm}^6\,\mathrm{mol}^{-2}\,\mathrm{bar}\,\mathrm{K}^{\frac{1}{2}}$, $B = 0.045153\,\mathrm{dm}^3\,\mathrm{mol}^{-1}$. Compare your computed pressures with the experimental pressure 506.6 bar.

17. The Benedict–Webb–Rubin equation of state, proposed in 1940, is an 8-parameter equation:

$$P = RT\overline{\rho} + (BRT - A - C_0 T^{-2})\overline{\rho}^2 + (bRT - a)\overline{\rho}^3 + a\alpha\overline{\rho}^6$$
$$+ cT^{-2}\overline{\rho}^3(1 + \gamma\overline{\rho}^2)e^{-\gamma\overline{\rho}^2},$$

with $\overline{\rho} \equiv 1/\overline{V}$, and A, B, C_0, a, b, α, c, and γ the eight parameters. For ethane, these parameters have the values $A = 4.25062\,\mathrm{dm}^6\,\mathrm{mol}^{-2}\,\mathrm{bar}$, $B = 0.0627724\,\mathrm{dm}^3\,\mathrm{mol}^{-1}$, $C_0 = 1.81972\times10^5\,\mathrm{dm}^3\,\mathrm{mol}^{-2}\,\mathrm{K}^2\,\mathrm{bar}$, $a = 0.34973\,\mathrm{dm}^9\,\mathrm{mol}^{-3}\,\mathrm{bar}$, $b = 0.011122\,\mathrm{dm}^6\,\mathrm{mol}^{-2}$, $c = 3.3201 \times 10^4\,\mathrm{dm}^9\,\mathrm{mol}^{-3}\,\mathrm{K}^2\,\mathrm{bar}$, $\alpha = 2.43389\times10^{-4}\,\mathrm{dm}^9\,\mathrm{mol}^{-3}$, $\gamma = 0.01118\,\mathrm{dm}^6\,\mathrm{mol}^{-2}$, while the universal gas constant has the value $R = 0.083145\,\mathrm{dm}^3\,\mathrm{mol}^{-1}\,\mathrm{K}^{-1}\,\mathrm{bar}$. Calculate the pressure of an ethane sample with molar volume $\overline{V} = 0.07022\,\mathrm{dm}^3\,\mathrm{mol}^{-1}$ at temperature $T = 350$ K using the Benedict–Webb–

Rubin equation of state, and compare the result obtained with the experimental value $P_{\text{expt}} = 506.6$ bar, and with the pressures predicted by the ideal gas, Van der Waals, and Redlich–Kwong equations of state (as obtained in Problem 16).

18. Show, using Eqs. (2.8.11), that $B_2(T) = RT B_2'(T)$ quite generally, and obtain an expression for $B_3(T)$ in terms of $B_2'(T)$ and $B_3'(T)$.

19. Carry out the calculations for the Carnot cycle using a Van der Waals fluid for the working fluid and show that the final expression for the efficiency η for the Carnot cycle is the same as that for an ideal gas. (Hint: follow the heat trail.)

20. Use the ideal gas equation of state and the adiabaticity condition that PV^γ is constant, with $\gamma \equiv C_P/C_V$ the heat capacity ratio, to verify the values given for P_2, P_3, P_4, $T_{\min}$, and T_{low} for the CCW Otto cycle in Example 2.5.

21. Obtain expressions for $W_{i \to f}$ and $Q_{i \to f}$ for the four steps making up a counterclockwise Carnot cycle using an ideal gas for the working substance. How are $Q_{\text{import}}^{\text{CCW}}$ and $Q_{\text{export}}^{\text{CCW}}$ related to $Q_{\text{import}}^{\text{CW}}$ and $Q_{\text{export}}^{\text{CW}}$? Show explicitly that η^{CCW} defined as $\eta^{\text{CCW}} \equiv W^{\text{CCW}}(\text{cycle})/Q_{\text{import}}^{\text{CCW}}$ is meaningless and that $\varepsilon_{\text{COP}}^{\text{Carnot}}$ must be larger than ε_{COP} for any other refrigeration cycle.

22. We have seen that for a refrigeration cycle in which the goal is to remove heat (by importation) from the low-temperature heat reservoir and to export heat to the high-temperature heat reservoir, the appropriate definition of the coefficient of performance, ε_{COP}, for a refrigeration cycle is $\varepsilon_{\text{COP}} \equiv Q_{\text{import}}/W$, with W the work done on the working substance. If, however, the goal of a refrigeration cycle is the importation of heat into the high-temperature heat reservoir (i.e., of Q_{export} from the working fluid), then the appropriate definition of ε_{COP} for a heat pump must be $\varepsilon_{\text{COP}} \equiv Q_{\text{export}}/W$. Obtain an expression for ε_{COP} for a Carnot heat pump in terms of the temperatures T_{high} and T_{low} of the two heat reservoirs.

23. Obtain the equations for x_{opt} and y_{opt} that arise from the optimization conditions (2.5.29) and use them to obtain the quadratic equation (2.5.31) that determines $x_{\text{opt}}/T_{\text{high}}$.

24. By examining the smaller root of Eq. (2.5.31), show that $x_{\text{opt}}/T_{\text{high}}$ is given by Eq. (2.5.32), and employ your expression for $x_{\text{opt}}/T_{\text{high}}$ together with Eq. (2.5.30) to obtain the expression given in Eq. (2.5.32) for $y_{\text{opt}}/T_{\text{low}}$. Employ these results to obtain expressions for $T_{1,\text{opt}}$ and $T_{2,\text{opt}}$, and show that the efficiency of the Curzon–Ahlborn endoreversible heat engine is indeed given by Eq. (2.5.33b).

25. Utilize the relations obtained in Example 2.4 for the ideal gas volumes and pressures, P_i, V_i ($i = 2, 3, 4$), and temperatures $T_{\min}$, T_{low}, $T_{\max}$ in terms of V_1, P_1, and T_{high} to sketch plausible PV- and TS-diagrams pertaining to the clockwise (CW) idealized Otto (engine) cycle that takes an ideal gas through the closed thermodynamic cycle represented by $(V_1, P_1, T_{\text{high}}) \overset{\text{isochor}}{\longrightarrow} (V_1, P_2, T_{\max}) \overset{\text{adiabat}}{\longrightarrow} (5V_1, P_3, T_{\text{low}}) \overset{\text{isochor}}{\longrightarrow} (5V_1, P_4, T_{\min}) \overset{\text{adiabat}}{\longrightarrow} (V_1, P_1, T_{\text{high}})$. Employ your TS-diagram to argue that the engine efficiency η^{Otto} is less than the efficiency η^{Carnot} of a Carnot engine that operates between heat reservoirs

at temperatures T_{max} and T_{min}. How is your result supported by the outcomes from this particular idealized Otto engine cycle?

26. By carrying out expansions of the type that we used for the Van der Waals equation in Sect. 2.8.2, show that the second and third virial coefficients for a Redlich–Kwong gas are given by

$$B_2(T) = B - \frac{A}{RT^{\frac{3}{2}}} \; ; \quad B_3(T) = B^2 + \frac{AB}{RT^{\frac{3}{2}}} ,$$

in terms of the A and B Redlich–Kwong constants.

27. Calculate the P and T values for which $H_2(g)$ is in a corresponding state with $Xe(g)$ at a temperature of 450.0 K and a pressure of 85.0 bar. You will need the critical temperatures and pressures for these gases: $T_c(Xe) = 289.74$ K, $P_c(Xe) = 58.40$ bar; $T_c(H_2) = 32.94$ K, $P_c(H_2) = 12.93$ bar.

28. The compressibility $\mathcal{Z}$ for a Van der Waals gas has a value $\mathcal{Z} = 1.00084$ at temperature $T = 298.15$ K and pressure $P = 1$ bar. The Boyle temperature for the gas is $T_B = 125$ K. Use $\mathcal{Z} = 1 + B_2^{VdW}(T)/\overline{V}$ and approximate $\overline{V}$ using the ideal gas law (this is reasonable, as $P = 1$ bar is a pressure at which many gases behave essentially as ideal gases). Estimate values for the Van der Waals parameters $\overline{a}$ and $\overline{b}$.

29. Using Eq. (2.8.17b) the fugacity coefficient can be expressed in terms of the molar volume $\overline{V}$ as

$$RT \ln \gamma = \int_{P_{ideal}}^{P} \left(\overline{V} - \frac{RT}{P} \right) dP .$$

Show that $\ln \gamma$ may also be given as

$$RT \ln \gamma = P\overline{V} - RT - \int_{\overline{V}_{ideal}}^{\overline{V}} P d\overline{V} - RT \ln \left(\frac{P}{P_{ideal}} \right) .$$

Show further that for a Van der Waals fluid this expression can be evaluated to give

$$\ln \gamma = \frac{\overline{b}}{\overline{V} - \overline{b}} - \frac{2\overline{a}}{RT\overline{V}} - \ln \left[1 - \frac{\overline{a}(\overline{V} - \overline{b})}{RT\overline{V}^2} \right] ,$$

as both $\overline{a}$ and $\overline{b}$ are small in comparison with $\overline{V}_{ideal}$ for $P \ll 1$ bar.

30. Employ the expressions for the Van der Waals parameters $\overline{a}$ and $\overline{b}$ in terms of the critical constants to obtain the reduced form of the equation for $\ln \gamma$.

31. By noting that for a pure substance the molar Gibbs energy is the chemical potential, i.e., $\overline{G}(T, P) \equiv \mu(T, P)$, show that the fugacity for a Van der Waals fluid may be obtained from

$$\ln|f| = \ln\left|\frac{RT}{\overline{V} - \overline{b}}\right| + \frac{\overline{b}}{\overline{V} - \overline{b}} - \frac{2\overline{a}}{RT\overline{V}}\ .$$

32. For CO_2 considered as a Van der Waals fluid, plot the fugacity as a function of pressure for pressures between 0 and 90 bar at temperature $T = 300$ K. The Van der Waals $\overline{a}$ and $\overline{b}$ parameters for CO_2 are $\overline{a} = 3.6551\,dm^6\,bar\,mol^{-2}$ and $\overline{b} = 0.042816\,dm^3\,mol^{-1}$.

33. Obtain expressions for the thermal expansion coefficient α, the isothermal compressibility coefficient κ_T, and $d\overline{U}$ for a Van der Waals fluid. Show that $d\overline{U}$ for the Van der Waals fluid reduces to the expression for the ideal gas in the ideal gas limit ($\overline{a} = \overline{b} = 0$).

34. The equation

$$(\overline{b}^2 P_i + \overline{a} + \tfrac{3}{2}\overline{b}RT_i)^2 = 8\overline{a}\,\overline{b}RT_i$$

is the molar version of Eq. (2.8.35), and is the form normally encountered in thermodynamics texts. Show that the first derivative of the inversion pressure P_i with respect to the inversion temperature T_i for a Van der Waals fluid is given by

$$\frac{dP_i}{dT_i} = \frac{1}{2\overline{b}^2}\left[\sqrt{\frac{8\overline{a}\,\overline{b}R}{T_i}} - 3\overline{b}R\right].$$

Verify that for $P_{i,\max} = \overline{a}/(3\overline{b}^2)$, the inversion curve equation has only one solution, namely, $T_i = 8\overline{a}/(9R\overline{b})$, and that the approximate result $T_i \simeq 2\overline{a}/(R\overline{b})$ is obtained formally as the *upper* inversion temperature at $P_i = 0$.

35. Employ the inversion equation (2.8.35) to compute the upper and lower inversion temperatures for N_2 treated as a Van der Waals fluid at a pressure of 100 bar, given that $\overline{a} = 1.3661\,dm^6\,bar\,mol^{-2}$ and $\overline{b} = 0.038577\,dm^3\,mol^{-1}$.

36. Employ the result from Problem 29, together with the Van der Waals equation of state in reduced form to plot γ against P_R for $T_R = 1.20$, $T_R = 2.00$, and $T_R = 3.00$.

37. Obtain an expression for $\overline{G}(\xi)$ and the affinity $\mathbb{A}(\xi)$ for the generic chemical reaction $\nu_A A + \nu_B B \rightarrow \nu_C C + \nu_D D$.

38. By considering explicitly the reaction for the formation of water from molecular hydrogen and oxygen, namely,

$$2H_2(g) + O_2(g) \rightleftharpoons 2H_2O(g)\,,$$

obtain expressions for the Gibbs energy, $G(\xi)$, and the affinity, $\mathbb{A}(\xi)$, as functions of the extent of reaction, ξ. Show further, that the equilibrium constant,

$K_P(T)$, is given in terms of the Gibbs energy of formation, $\Delta_f \overline{G}^\circ$, for gaseous water from elemental hydrogen and oxygen as $K_P(T) = \exp\{-2\Delta_f \overline{G}^\circ/(RT)\}$.

39. Use the expression for the affinity as a function of the extent of reaction, ξ, for the formation of water from hydrogen and oxygen (cf. Problem 37) to determine the equilibrium value, ξ_e, for the extent of reaction for temperatures $T = 2000$ and 5000 K for reaction mixtures for which the total pressure is maintained at 1 bar. Employ these equilibrium values of ξ to calculate the partial pressures of the three components, namely, $P_{H_2O}(\xi_e)$, $P_{H_2}(\xi_e)$, $P_{O_2}(\xi_e)$, in the equilibrium mixtures, and confirm that the values thereby obtained give $K_P(T)$ values that are, as they should be, consistent with the values of $K_P(T)$ obtained using the Gibbs energy of formation expression. The Gibbs energy of formation for H_2O has values $\Delta_f \overline{G}^\circ = -135.528$ kJ mol^{-1} at $T = 2000$ K and 40.949 kJ mol^{-1} at 5000 K.

40. At temperature $T = 500$ K both Br_2 and I_2 are gases, and react to form the interhalogen $IBr(g)$ according to the chemical reaction

$$Br_2(g) + I_2(g) \rightleftharpoons 2IBr(g).$$

Obtain an expression for the Gibbs energy and its first derivative as functions of the extent of reaction ξ for a starting mixture of one mole each of $Br_2(g)$ and $I_2(g)$ and no $IBr(g)$. Given that $\overline{G}^\circ_{IBr} = -8.748$ kJ mol^{-1} at 500 K, plot $G(\xi)$ as a function of ξ, obtain ξ_e for this reaction, and evaluate the equilibrium constant K_P. Determine the composition of the equilibrium mixture of $Br_2(g)$, $I_2(g)$, and $IBr(g)$ at 500 K.

41. According to Eq. (2.2.12), which is the expression obtained upon combining the First and Second Laws of thermodynamics, the internal energy U is a function of entropy, S, and volume, V. For an ideal monatomic gas, for which only translational energy states are thermally accessible, use Eqs. (2.3.13b) and (2.3.14) to obtain an expression giving the translational entropy as an explicit function of T, V, and N (a result that was first obtained by purely thermodynamic reasoning by Sackur and Tetrode). Show that $S_{trans}(T, V, N)$ may be rewritten as a function of U_{trans}, V, and N and that, by inverting it, the internal energy U_{trans} can be expressed as an explicit function of S_{trans}, V, and N.

42. Employ the expression obtained in Problem 41 for $U_{trans}(S, V, N)$ for an ideal monatomic gas and the extended First and Second Laws expression $dU(S, V, N) = T\,dS - P\,dV + \mu\,dN$ for a pure substance to obtain expressions for $P(S, V, N)$, $T(S, V, N)$, and the chemical potential $\mu_{trans}(S, V, N)$. Show that the expressions thereby obtained for P, T, and μ are consistent, respectively, with the equation of state, the expression $U_{trans} = \frac{3}{2}Nk_BT$ for an ideal gas, and the chemical potential, μ, as the Gibbs energy per particle for a pure substance.

References

1. I.M. Klotz, R.M. Rosenberg, *Chemical Thermodynamics. Basic Concepts and Methods*, 7th edn. (Wiley, New York, 2008)
2. K.A. Pitzer, *Thermodynamics*, 3rd edn. (McGraw-Hill, New York, 1995)
3. H. Metiu, *Physical Chemistry. Thermodynamics* (Taylor & Francis, New York, 2006)
4. D. Kondepudi, *Introduction to Modern Thermodynamics* (Wiley, Hoboken, 2008)
5. I. Prigogine, R. Defay, *Chemical Thermodynamics* (Longmans, London, 1954)
6. J.G. Kirkwood, I. Oppenheim, *Chemical Thermodynamics* (McGraw-Hill, New York, 1961)
7. H.B. Callen, *Thermodynamics and an Introduction to Thermostatistics*, 2nd edn. (Wiley, New York, 1985)
8. R.E. Kelly, Am. J. Phys. **48**, 714 (1981)
9. H.S. Leff, Am. J. Phys. **70**, 792 (2002)
10. (N.L.) S. Carnot, *Réflexions sur la puissence motrice du feu et sur les machines propres à développer cette puissance* (Pamphlet, Paris, 1824). Translated as *Reflections on the Motive Power of Heat and on Machines Fitted to Develop This Power* by R.H. Thurston (Wiley, New York, 1897)
11. T.V. Marcella, Am. J. Phys. **60**, 888 (1992)
12. L.-Y. Shieh, H.-C. Kan, Am. J. Phys. **82**, 306 (2014)
13. F.L. Curzon, B. Ahlborn, Am. J. Phys. **43**, 22 (1975). See also §4.9 of H.B. Callen, *Thermodynamics and an Introduction to Thermostatistics*, 2nd edn. (Wiley, New York, 1985)
14. H.S. Leff, Am. J. Phys. **55**, 602 (1987)
15. M.H. Rubin, Phys. Rev. A **19**, 1272, 1279 (1979)
16. R.H. Dickerson, J. Mottmann, Am. J. Phys. **84**, 413 (2016)
17. S.R. de Groot, P. Mazur, *Non-equilibrium Thermodynamics* (North-Holland, Amsterdam, 1962)
18. M.W. Chase Jr., *NIST-JANAF Thermochemical Tables*, 4th edn. J. Phys. Chem. Ref. Data, Monograph, vol. 9 (1998), or via the NIST website http://webbook.nist.gov.chemistry
19. J. de Heer, *Phenomenological Thermodynamics with Applications to Chemistry* (Prentice-Hall, Englewood Cliffs, 1986)

Chapter 3
Ensembles: Systems of Particles

This chapter discusses the nature of microscopic configurations and states for
systems of particles, accessibility of microscopic states, introduces the Fundamental
Postulate of statistical mechanics and illustrates it explicitly for small model
systems. The ensemble concept is applied to an isolated system of two subsystems,
one small (but still macroscopic), the other much larger (the reservoir). A canonical
ensemble describes a small system that can exchange energy, but not matter, with
the reservoir, while a grand ensemble describes a small system that can exchange
both energy and matter with the reservoir. The canonical partition function that
determines the probability for the particle to be in a specific energy state is
obtained for a single-particle that can access various types of energy, and is then
generalized to many particles and to mixtures. The grand partition function is
similarly developed for a system that can exchange both energy and matter with
the reservoir.

3.1 Microscopic Configurations

The microscopic arrangements of particles in a container are often called *configurations*: they are also often called *states* (most appropriately when quantum
mechanics is employed). Since most modern students are often more familiar with
quantum mechanical considerations than with classical mechanical ones, it will be
appropriate to utilize the language of quantum mechanics in much of the following.

Each quantum state of an isolated system may be characterized by the energy
associated with that quantum state. Two or more states characterized by the same
energy are said to be *degenerate*. How do we specify the possible quantum states
of a system? In principle we could simply list the values belonging to all relevant
quantum numbers (essentially one for each degree of freedom possessed by the
system as a whole). This would represent a truly enormous number. Indeed, a

© Springer Nature Switzerland AG 2021
F. R. W. McCourt, *Statistical Thermodynamics for Pure and Applied Sciences*,
https://doi.org/10.1007/978-3-030-52006-9_3

completely precise description of an isolated system of particles would need to take into account *all* interactions amongst its particles, and would determine the rigorously exact quantum states of the system. In fact, were the system found in any one of these exact states, it would then remain in that state forever. In practice, no system is ever so isolated from the rest of the environment for these conditions to be obtained, and we *never* know the exact microscopic states of a system of particles in interaction. Thus, we employ *approximate states* determined by taking into account all the important dynamical properties of its particles, while neglecting small residual interactions.

A system initially known to be in one of its approximate quantum states *does not remain* there forever. In the course of time, under the influence of small residual interactions, it will make a transition from that state into another state (provided that there is no violation of one of the laws of mechanics). Such states are called *accessible states* of the system. One of our objectives is to be able to determine those states that are accessible to a given system.

The typical specifications for a system of particles are energy E, pressure P, and so on. Let us call Ω the number of states accessible to the system, and let $\{y_1, y_2, \ldots, y_n\}$ represent the set of all constraints imposed upon the system. Clearly $\Omega = \Omega(y_1, y_2, \ldots, y_n)$, which implies that the number of accessible states is a function of the parameters $y_1, y_2, \ldots, y_n$. The most commonly occurring such parameter is $y = E$. We shall therefore seek an expression for the number of accessible states $\Omega(E)$ with energies lying between E and $E + \delta E$, with $\delta E \ll E$. Of course in addition to $\Omega(E)$, we are also interested in knowing in which one of the accessible states will the system be found. Is this an answerable question? In complete analogy with the tossing of a die, we cannot know in which state our system lies, but we can calculate the probability that the system lies in any one of its accessible states. If, for example, there are Ω accessible states, then it is extremely difficult to see any reason why the system should be preferably in any specific one of these states, and so we guess that the probability will be $p = 1/\Omega$ for any given state. This is known as the *Fundamental Postulate of Statistical Mechanics* or, equivalently, as the *Equal a priori Probability Postulate*.

Fundamental Postulate of Statistical Mechanics. If an isolated system is in equilibrium, it is found with equal probability in each one of its accessible states.

If we combine the Fundamental Postulate with the postulate of long-time averages [stated in the introductory chapter (see Sect. 1.4)], we are led to the conclusion that a single isolated system of a microcanonical ensemble, considered over a very long time period, spends equal amounts of time in each of its accessible quantum states. This result is commonly referred to as the *quantum ergodic hypothesis*. To broaden its application to the canonical and grand ensembles, it suffices to note that these ensembles themselves are isolated systems.

3.1.1 Illustration of the Role of the Basic Postulate

Let us construct an ensemble of $\mathcal{N}$ systems: if $\mathcal{N}_r$ have parameter $y = E_r$, then

$$p_r \equiv \frac{\mathcal{N}_r}{\mathcal{N}} = \frac{\Omega(E_r)}{\Omega(E)}. \tag{3.1.1}$$

To illustrate more concretely how the basic postulate of statistical mechanics enters into these considerations we shall consider as an illustrative example a collection of simple harmonic oscillators (SHOs), each satisfying the same energy quantization rule, namely $\epsilon_v = (v + \frac{1}{2})h\nu_{osc}$, and *localized* in space, so that the oscillators are *identical yet distinguishable*. We shall proceed by examining three examples [1], specifically, a set of three quanta of energy to be shared amongst three oscillators, followed by a set of five quanta of energy to be shared amongst five oscillators, then by a set of five quanta of energy to be shared amongst ten oscillators. Each of these examples will be examined in detail in order to help us to understand the role played by the Fundamental Postulate of Statistical Mechanics enunciated earlier.

(i) Three oscillators sharing three quanta of energy:

There will be three ways in which one of the three oscillators receives all three quanta of energy, while the other two oscillators receive none. Similarly, there are six ways in which one oscillator receives two quanta of energy, a second oscillator receives one quantum of energy, and the third oscillator receives none. Finally, there is only one way in which each of the three oscillators receives one quantum of energy. This method of reasoning gives us two pieces of information about the sharing of three quanta of energy by the three-oscillator system. Firstly, we note that there will be three *generic configurations* (arrangements) of the three oscillators, namely one arrangement in which one oscillator takes up all three quanta of energy, while the other two oscillators receive none; one arrangement in which one oscillator receives two quanta of energy, a second oscillator receives one quantum of energy, and the third oscillator receives none; one arrangement in which each oscillator receives one quantum of energy. We shall call each such generic configuration a *macrostate* of the system, and we shall label it using a Roman numeral. Secondly, we note that for the first macrostate, I, there are three specific arrangements of the oscillators, each of which we shall call a *microstate* of the system. Similarly, we find that there are six microstates corresponding to macrostate II, and one microstate corresponding to macrostate III. In general, the number of microstates corresponding to a given macrostate 'i' is typically referred to as its *degeneracy* Ω_i, $\{i = \text{I, II, III}, \ldots\}$.

A combinatoric expression for the degeneracy of such system macrostates may be expressed in terms of the number of quanta, n_q, available for sharing, the number of oscillators, N, sharing the n_q quanta of energy, and the numbers of oscillators, N_i, that receive i quanta of energy ($i = 0, 1, 2, \ldots, n_q$): this expression for Ω_i is

$$\Omega_i = \frac{N!}{N_0! N_1! \cdots N_q!}, \qquad i = \text{I}, \text{II}, \ldots.$$

Thus, for example, for the sharing of three quanta of energy amongst three oscillators, we obtain

$$\Omega_{\text{I}} = \frac{3!}{2!0!0!1!} = 3, \quad \Omega_{\text{II}} = \frac{3!}{1!1!1!0!} = 6, \quad \Omega_{\text{III}} = \frac{3!}{0!3!0!0!} = 1,$$

to give a total of 10 microstates for the three macrostates. The corresponding probabilities (relative occurrences) for this three-oscillator system to be found in one of the three macrostates are thus $p_{\text{I}} = 0.30$, $p_{\text{II}} = 0.60$, and $p_{\text{III}} = 0.10$.

(ii) Five oscillators sharing five quanta of energy:

For five quanta of energy shared amongst five equivalent yet distinguishable oscillators the same combinatoric procedure leads to seven generic macrostates, corresponding systematically to one oscillator receiving all five energy quanta (macrostate I), one oscillator receiving four energy quanta plus one oscillator receiving one energy quantum (macrostate II), one oscillator receiving three quanta of energy plus one oscillator receiving two quanta of energy (macrostate III), one oscillator receiving three quanta of energy plus two oscillators each receiving one quantum of energy (macrostate IV), two oscillators each receiving two quanta of energy plus one oscillator receiving one quantum of energy (macrostate V), one oscillator receiving two quanta of energy plus three oscillators each receiving one quantum of energy (macrostate VI), and all five oscillators each receiving one quantum of energy (macrostate VII). The corresponding degeneracies, obtained from our combinatoric formula, are

$$\Omega_{\text{I}} = 5, \quad \Omega_{\text{II}} = 20, \quad \Omega_{\text{III}} = 20, \quad \Omega_{\text{IV}} = 30, \quad \Omega_{\text{V}} = 30, \quad \Omega_{\text{VI}} = 20, \quad \Omega_{\text{VII}} = 1,$$

to give a total of 126 microstates and relative probabilities

$$p_{\text{I}} = 0.0397, \quad p_{\text{II}} = p_{\text{III}} = p_{\text{VI}} = 0.1587, \quad p_{\text{IV}} = p_{\text{V}} = 0.2381, \quad p_{\text{VII}} = 0.0077,$$

for the seven macrostates.

(iii) Ten oscillators sharing five energy quanta:

If we carry out the same analysis for the sharing of five quanta of energy amongst a set of ten equivalent yet distinguishable oscillators instead of five, we find that there are still seven macrostates, determined essentially by the manner in which the five quanta of energy are shared amongst the set of ten oscillators. The same systematic determination of the generic macrostates differs only in the numbers of oscillators that receive zero quanta of energy, with $N_0(i) = N_0'(i) + 5$, $i = \text{I}, \text{II}, \ldots, \text{VII}$, and $N_0'(i)$ the number of oscillators receiving zero quanta of energy in the sharing of five quanta of energy amongst five equivalent yet distinguishable oscillators. There are, however, major differences in the corresponding degeneracies associated with the seven resulting macrostates. We obtain degeneracies

$$\Omega_\text{I} = 10, \quad \Omega_\text{II} = \Omega_\text{III} = 90, \quad \Omega_\text{IV} = \Omega_\text{V} = 360, \quad \Omega_\text{VI} = 840, \quad \Omega_\text{VII} = 252,$$

giving a total of 2002 microstates and relative probabilities $p_\text{I} \simeq 0.0050$, $p_\text{II} = p_\text{III} \simeq 0.0450$, $p_\text{IV} = p_\text{V} \simeq 0.1798$, $p_\text{VI} \simeq 0.4196$, $p_\text{VII} \simeq 0.1259$, for the seven macrostates.

A comparison between these last two simple cases, in which five quanta of energy are shared firstly amongst five and then amongst ten oscillators, shows that a doubling of the number of oscillators available to take up the five quanta of energy leads to an approximately 16-fold increase in the total number of microstates available to the system. By continuing with this type of argument we would obtain a result that is characteristic of any assembly with a large number N of units. More specifically, the outcome would be that for very large assemblies by far the largest number of microstates are associated with a relatively small number of macrostates that are not only very similar to the predominant macrostate but also give rise to properties that are essentially indistinguishable on a macroscopic scale.

How does the fundamental postulate enter into this? Well, were it not for this postulate, we would be unable to conclude that the macrostate containing the largest number of microstates is the most probable macrostate: this association of 'largest number' of microstates with the 'most probable' macrostate is directly based upon the inherent assumption that every microstate has precisely the same probability of occurrence. Otherwise, a very large number of microstates occurring in one macrostate with an associated very small probability factor could be equivalent to a much smaller number of microstates in another macrostate, but associated with a proportionally larger probability of occurrence.

Our considerations thus far have been based upon the assumption that the oscillators in a given macrostate do not interact with one another. In reality, this is never the case. Hence, let us examine what effect interactions between the oscillators will have on the potential evolution [2] of a system of particles.[1] To begin with, we shall assume that any interactions amongst oscillators in a given macrostate will require them to be simultaneously located in close proximity to one another, so that two-body interactions will clearly dominate over three-body and other many-body interactions. As a consequence of such an assumption, N-body interactions will be increasingly less probable the higher the value of N: for the following discussion, we shall treat any interactions that require more than two oscillators to be simultaneously in close proximity as improbable. We shall therefore ignore any outcome that would require simultaneous interaction between more than two oscillators in order to occur.

As a concrete illustration of our expectations, let us examine transitions that can take place if we start from macrostate III obtained from sharing five quanta

---

[1] The remainder of this subsection utilizes a line of reasoning introduced in reference [2].

Table 3.1 Transitions between macrostates due to energy-conserving 2-body interactions between the oscillators occupying the system microstates

Old macrostate		New macrostate		# quanta		
Label	Ω	Label	Ω'	exchanged	Weight[a]	Weight[b]
I	10	II	90	1	9	1
I	10	III	90	2	9	1
II	90	III	90	1	1	1
II	90	IV	360	1	8	2
II	90	V	360	2	8	2
III	90	IV	360	1	8	2
III	90	V	360	1	8	2
IV	360	V	360	1	2	2
IV	360	VI	840	1	7	3
V	360	VI	840	1	14	6
VI	840	VII	252	1	6	20

[a]Weight here represents the number of ways that such an exchange can occur for the initial macrostate to convert into the final macrostate via a 2-body energy-conserving interaction between oscillators making up the macrostate
[b]Weight is the weight associated with the reverse transition between the corresponding macrostates

of energy amongst ten equivalent yet distinguishable oscillators. Because the total energy available to the ten oscillators is fixed, any interaction that leads from one macrostate to another must be one that conserves the total energy of the system: this requirement, together with the relative improbability of interactions involving more than two bodies, imposes strong limitations on the accessibility of nearby macrostates. Those transitions between macrostates that are allowed by two-body energy-conserving interactions have been summarized in Table 3.1.

'Weights' have been assigned according to the number of ways that the transition from one given macrostate to another can take place: thus, for example, if we consider macrostates III and V, there are eight ways for an oscillator in the ground level (energy 0) to gain one quantum of energy and there is one way for an oscillator in the 3-quantum energy level to lose one quantum of energy, giving a total of eight ways in which the transition from macrostate III (containing 90 microstates) into macrostate V (containing 360 microstates) can take place. There are, however, only two ways of making the reverse transition from macrostate V to macrostate III, so that the net result over time will be that for these two macrostates, the system will more likely be found in macrostate V than in macrostate III. If we extend our analysis to take in all results displayed in Table 3.1, then we can see that macrostate I, for example, will evolve preferentially into macrostates II and III, and that macrostates II and III, in turn, will evolve preferentially into macrostates IV and V; similarly, macrostates IV, V will both evolve preferentially into macrostate VI (which is the macrostate having the maximal number of microstates for this system). Finally, we note that macrostate VII also evolves into macrostate VI (with a 10:3 weighting).

We have just made a case for the tendency of a system to evolve in time towards the macrostate having the maximal number Ω of microstates. For our small example system, there will likely be only one macrostate having a clearly defined maximal value for Ω (here $\Omega_{VI} = 840$ vs. $\Omega_{IV} = \Omega_V = 360$). Let us note, however, that for a system with a large number of oscillators (10^3, 10^6, 10^9, etc.), there may well be many macrostates that have similarly large numbers Ω of microstates, in which case once the much larger number of macrostates with markedly smaller values of Ω have evolved into the members of this set of maximal macrostates, they will fluctuate amongst one another due to the continuing interactions between the particles making up the system.

The long-time, equilibrium, behaviour of a system of oscillators is described by its occupation of a set of macrostates characterized by similarly large numbers of microstates, so that the ratios of their macrostate degeneracies Ω are close to unity. If, following Schoepf [2], we consider an energy-conserving change from a macrostate with degeneracy Ω to a nearby macrostate with degeneracy Ω' via the interaction between two oscillators in a particular energy microstate such that one oscillator goes into the next-lowest-energy microstate while the other oscillator goes into the next-highest-energy microstate, the equilibrium requirement that $\Omega'/\Omega \approx 1$ can be restated in terms of microstate occupation numbers n_k as

$$\frac{\Omega'}{\Omega} = \frac{n_i(n_i - 1)}{(n_{i+1} + 1)(n_{i-1} + 1)} \approx 1 .$$

For a realistic system of oscillators (even a 'nano-system') the occupation numbers n_i would be sufficiently large that we could meaningfully make the approximation $n_i \pm 1 \approx n_i$, in which case, our equilibrium criterion becomes

$$\frac{\Omega'}{\Omega} \simeq \frac{n_i^2}{n_{i+1} n_{i-1}} \simeq 1 .$$

Upon carrying out a slight rearrangement of this result, this criterion may be re-expressed as

$$\frac{n_i}{n_{i-1}} \simeq \frac{n_{i+1}}{n_i} .$$

Thus, for the special case that we have been considering, which consists of oscillators possessing equally-spaced energy levels, the ratio, r, of the numbers of oscillators in two adjacent energy levels at equilibrium is the same. Mathematically, this means that

$$\frac{n_{i+1}}{n_i} = \frac{n_i}{n_{i-1}} = \cdots = \frac{n_2}{n_1} = \frac{n_1}{n_0} \equiv r ,$$

so that the occupation numbers n_i ($i = 0, 1, 2, \ldots$) form a geometric progression. For the present case, we will have $r < 1$, as the sum over the occupation numbers is constrained to be the total number of oscillators, N, in the system, i.e., $\sum_i n_i = N$. Now, we note that the ratio of the number of oscillators n_i in microstate i to the number of oscillators n_0 in the lowest-lying microstate can be expressed as

$$\frac{n_i}{n_0} = \frac{n_i}{n_{i-1}} \cdot \frac{n_{i-1}}{n_{i-2}} \cdots \frac{n_2}{n_1} \cdot \frac{n_1}{n_0} = r^i . \tag{3.1.2}$$

As suggested by Schoepf [2], one convenient way of expressing r is as $r = e^{-\beta\epsilon}$, with β some positive constant that characterizes the equilibrium condition that we have imposed. The physical behaviour that we would anticipate is clearly satisfied by this expression for r: for closely-spaced energy levels ($\epsilon \approx 0$) r approximates unity, so that all oscillator energy levels are equally populated, while for widely-spaced energy levels ($\epsilon \gg 1$) r approaches zero, so that almost all of the oscillators will be found in the very lowest energy levels.

Substitution of $r = e^{-\beta\epsilon}$ into Eq. (3.1.2) allows us to obtain an expression relating the occupation numbers in our equilibrium system, namely

$$n_i = n_0 e^{-\beta i \epsilon} \tag{3.1.3}$$

which, as we shall see shortly, is just the Boltzmann distribution that holds generally for an equilibrium system of classical particles. We could thus say that having equally-spaced energy levels is a sufficient condition for the establishment of a Boltzmann distribution of particles amongst the available energy states. However, we shall establish more generally that the attainment of a Boltzmann equilibrium distribution does not require equally-spaced energy levels, so that it is not a necessary condition.

We may now ask what happens when a system of oscillators occupying a set of equilibrium-type configurations is allowed to exchange energy with its (external) surroundings. We shall focus here upon the consequences for a system of oscillators that takes up energy from its surroundings, with which it is in (thermal) contact. It will be clear from what has been said thus far that if energy is added to our system of oscillators, the number of microstates Ω associated with a particular macrostate of the system will change. Unlike the changes that occur under energy conservation, where we are restricted to minimally moving two oscillators, one into a higher-energy microstate, the other into a correspondingly lower-energy microstate, an influx of energy can be accommodated by changing the energy state of a single oscillator. If we go from a configuration having macroscopic energy E (and corresponding degeneracy Ω) to a new configuration having macroscopic energy $E' > E$ (and degeneracy Ω'), and if we consider the added energy[2] q to be taken up by a single oscillator, thereby moving it from microstate energy ϵ_j to

---

[2]The amount of energy q being added is considered here to be small enough that the new configuration obtained from its uptake remains a member of the set of already existing equilibrium macrostates.

microstate energy ϵ_k $(k > j)$ such that $q = (k - j)\epsilon$, then Ω' and Ω are related by

$$\frac{\Omega'}{\Omega} = \frac{n_j!(n_k - 1)!}{(n_j - 1)!n_k!} \simeq \frac{n_j}{n_k} ,$$

assuming $n_j, n_k \gg 1$. From this result, we may deduce that the number of microstates in the new macrostate increases when an oscillator is moved from a more populated microstate energy level to a less populated one (i.e., $n_j > n_k$) and decreases in the opposite case.

Also, we note that if the macroscopic states of the system taking up the energy are members of the set of statistical equilibrium states for the system of oscillators, then the occupation numbers are related in each case by Eq. (3.1.3), so that the ratio Ω'/Ω can also be obtained in the form

$$\frac{\Omega'}{\Omega} = \frac{n_j}{n_k} = \frac{n_0 e^{-\beta \epsilon_j}}{n_0 e^{-\beta \epsilon_k}} = e^{\beta(\epsilon_k - \epsilon_j)} = e^{\beta q} , \qquad (3.1.4)$$

in which $q \equiv \epsilon_k - \epsilon_j$ is the energy added to the system. From this result, we see that the change in the degeneracy Ω depends only upon the amount of energy that has been added to the system and is independent of the specific energy levels between which an oscillator has been moved. The same end result is obtained in the event that the added energy is distributed amongst several oscillators, rather than simply taken up by a single oscillator, so that the result in Eq. (3.1.4) also does not depend upon the number of oscillators that are redistributed.

If we take the natural logarithm of both sides of Eq. (3.1.4), we arrive at the result

$$\ln\left(\frac{\Omega'}{\Omega}\right) = \ln \Omega' - \ln \Omega \equiv \Delta \ln \Omega = \beta q , \qquad (3.1.5)$$

which implies that the change in $\ln \Omega$ is proportional to the amount of energy added to ($q > 0$) or removed from ($q < 0$) the system of oscillators.

Now, let us consider two systems, which we shall label as A and B, with system A having N_A oscillators, total energy E_A, and at equilibrium is characterized by β_A; similarly, system B has N_B oscillators, total energy E_B, and at equilibrium is characterized by β_B. Let us place systems A and B in thermal contact, thereby allowing an amount q of energy to be transferred spontaneously from one system to the other, with N_A, N_B, and $E_{tot} = E_A + E_B$ all fixed. Initially, for $\beta_A \neq \beta_B$, each system is in its most likely macrostate, with Ω_A, Ω_B microstates, respectively. Because each of the Ω_A microstates of system A can be combined with each of the Ω_B microstates of system B, the total number of microstates accessible to the composite (combined) system A⊕B is $\Omega_{tot} = \Omega_A \Omega_B$.[3]

---

[3] We shall refer to systems A and B in thermal contact as subsystems of the (composite/combined) system A⊕B.

The number of microstates Ω_{tot} will be maximal if the two subsystems A, B brought into thermal contact have the same values of β, i.e., if $\beta_A = \beta_B$: otherwise, the number of microstates for the composite system will not be at its maximum value, and energy will be redistributed between the two subsystems so as to maximize the product $\Omega_A\Omega_B$. If an amount q of energy is transferred spontaneously from subsystem A into subsystem B, giving new values Ω'_A and Ω'_B for the numbers of microstates in the two subsystems, then the total number of microstates for the composite system A⊕B will change from $\Omega_{tot} = \Omega_A\Omega_B$ to $\Omega'_{tot} = \Omega'_A\Omega'_B$. Now, if we recall that we have seen in Eq. (3.1.5) that the amount, q, of energy taken up by subsystem B (in our case) can be expressed in terms of the natural logarithm of Ω_B as

$$\Delta \ln \Omega_B \equiv \ln \Omega'_B - \ln \Omega_B = \beta_B q_B = \beta_B q \,,$$

and similarly, this same amount of energy q given up by subsystem A can be expressed in terms of $\ln \Omega_A$ as

$$\Delta \ln \Omega_A = \ln \Omega'_A - \ln \Omega_A = \beta_A q_A = -\beta_A q \,,$$

with the negative sign indicating that energy (of amount q) has been transferred out of subsystem A. If we now turn to the composite system A⊕B and ask about the corresponding change for Ω_{tot}, we see that it is given by

$$\Delta \ln \Omega_{tot} = \Delta \ln(\Omega_A\Omega_B) = (\beta_B - \beta_A)q \,. \tag{3.1.6}$$

This result establishes that if a spontaneous energy transfer out of subsystem A and into subsystem B results in an increase in the total number of microstates associated with the composite system A⊕B (so that $\Delta \ln \Omega_{tot} > 0$), then necessarily $\beta_A < \beta_B$: in other words, energy is transferred spontaneously from the subsystem having the smaller value of β to the subsystem having the larger value of β. As has been pointed out by Schoepf, the expression (3.1.6) for $\Delta \ln \Omega_{tot}$ provides an important relationship between the values of β_A and β_B when thermal equilibrium has been achieved for the composite system. Specifically, near equilibrium small energy changes between two subsystems cause only minor changes in the value of Ω_{tot} (and hence in the value of $\ln \Omega_{tot}$) so that the right-hand side of Eq. (3.1.6) must be vanishingly small, in which case $\beta_A = \beta_B$ becomes an expression of thermal equilibrium between two systems in (thermal) contact. We shall say more about this in the next section.

When energy is transferred from one material object to another via thermal contact, we say that the object that is giving up the energy, the donor, is 'hotter', and that the object that is receiving the energy, the acceptor, is 'colder': by convention, we assign a higher temperature to the energy donor, i.e., the 'hotter' object. Thus, when two objects in thermal contact reach a (common) thermal equilibrium, they must have the same temperature. If we now utilize this convention to assign temperatures T_A and T_B to our two objects (systems A and B, respectively), we may say that

(a) if $T_A > T_B$, then the combined (or composite) system is not at equilibrium, and energy therefore transfers spontaneously from system A into system B;
(b) if $T_A = T_B$, then the combined system is at equilibrium.

Comparison between these two statements for spontaneous energy transfer from system A into system B, namely $\beta_A < \beta_B$, or equivalently, $T_A > T_B$, indicates that β and T must be inversely proportional to one another. We cannot at this stage be any more specific about the constant of proportionality other than to say that it must be dimensioned so that β represents a reciprocal energy that characterizes such an (equilibrium) system.

3.2 The Canonical Ensemble: Closed Systems

Let us now consider in greater detail what happens when two systems can exchange energy between them, i.e., they are not individually isolated systems. Can we make any quantitative statements about such systems? The answer to this question turns out to be 'Yes', provided that one of the systems is small in comparison to the other one. For example, if one takes a piece of hot metal and drops it into a lake, the metal can be expected to cool down, and the lake to take up the heat lost by the metal—however, one does not expect the temperature of the lake to be changed by very much.

In our quest let us call the object of interest 'the system A' and its environment 'the heat reservoir A''. Place A in contact with Λ', and wait until equilibrium has been established. We now ask 'What is the probability p_r of finding the system A in any one particular state r of energy E_r?'

We shall denote by $\Omega'(E')$ the number of states accessible to the reservoir A' when its energy is equal to E', i.e., when the energy lies between E' and $E' + \delta E'$, with $\delta E'$ very small compared with the separation between the energy levels of A, but still large enough to contain many possible states of A'. We shall denote by A* the total system, consisting of the system A plus its surroundings A', as illustrated in Fig. 3.1.

We may treat the whole system A* = A $\oplus$ A' as an isolated system, so that E^* must be constant. If we now associate the energy E_r with A as A $\longleftrightarrow E_r$, then the energies E^*, E', E_r are related via

$$E^* = E' + E_r$$
$$E' = E^* - E_r \,, \qquad\qquad (3.2.1)$$
$$E_r = E^* - E' \,.$$

Fig. 3.1 Schematic sketch of systems for establishing the canonical ensemble

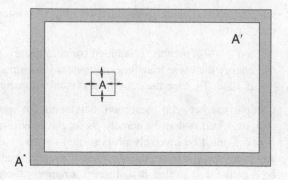

We consider A to be in *one* definite state r, so that the number of accessible states for the combined system A* is proportional to the number of states accessible to A': we shall call this number $\Omega'(E')$,

$$\Omega'(E') = \Omega'(E^* - E_r).\tag{3.2.2}$$

By the fundamental postulate, each of these states is equally likely, so that p_r is proportional to $\Omega'(E^* - E_r)$, and we may therefore write

$$p_r = C'\Omega'(E^* - E_r),\tag{3.2.3}$$

with C' a constant. Thus, we obtain

$$\ln p_r = \ln C' + \ln \Omega'(E^* - E_r).\tag{3.2.4}$$

Now, since $E_r \ll E^*$, we expand in a Taylor series about $E' = E^*$ to obtain

$$\ln \Omega'(E^* - E_r) \equiv \ln \Omega'(E')$$

$$= [\ln \Omega'(E')]_{E'=E^*} + \left[\frac{\partial}{\partial E'} \ln \Omega'(E')\right]_{E'=E^*} (E' - E^*) + \cdots \tag{3.2.5}$$

$$= \ln \Omega'(E^*) - \left[\frac{\partial}{\partial E'} \ln \Omega'(E')\right]_{E'=E^*} E_r + \cdots \tag{3.2.6}$$

Let us designate the first derivative evaluated at E^* by β, i.e., let us set

$$\beta \equiv \left[\frac{\partial}{\partial E'} \ln \Omega'(E')\right]_{E'=E^*},\tag{3.2.7}$$

which is, of course, a *constant* that characterizes the ensemble. If we do this, then the result (3.2.6) can be rewritten as

$$\ln \Omega'(E^* - E_r) \simeq \ln \Omega'(E^*) - \beta E_r, \tag{3.2.8}$$

from which we obtain for $\ln p_r$ the expression

$$\ln p_r = \ln C' + \ln \Omega'(E^*) - \beta E_r, \tag{3.2.9}$$

or

$$p_r = C' \Omega'(E^*) e^{-\beta E_r} \equiv C e^{-\beta E_r}. \tag{3.2.10}$$

This expression for p_r represents the *canonical*, or Maxwell–Boltzmann, distribution: the exponential energy-dependent factor is termed the *Boltzmann factor*, while p_r is called the Boltzmann probability.

We still need to evaluate C: to do this we utilize the normalization condition on p_r, namely $\sum_r p_r = 1$ (the sum runs over all possible states of A, irrespective of energy). Hence, from Eq. (3.2.10) we see that summation over all states requires that $C \sum_r e^{-\beta E_r} = 1$, or

$$C = \frac{1}{\sum_r e^{-\beta E_r}}. \tag{3.2.11}$$

The Boltzmann probability p_r can then be expressed as

$$p_r = \frac{e^{-\beta E_r}}{\sum_r e^{-\beta E_r}} = \frac{e^{-\beta E_r}}{z(\beta)}, \tag{3.2.12a}$$

where we have introduced the definition

$$z(\beta) \equiv \sum_r e^{-\beta E_r}. \tag{3.2.12b}$$

This 'summation over all accessible states of the system' plays a special role in statistical mechanics, and is commonly referred to as the partition function since, as we shall soon see, it serves to determine the partitioning of energy states as a function of the system temperature. Note that the temperature dependence of $z(\beta)$ is explicitly given via β, but that any dependence upon the number of particles, N, making up the system or of the volume, V, occupied by the system comes about implicitly through the dependence of the energies, E_r, of the N-particle system on N and V. The symbol z (or Z) is often used to designate the canonical partition function, as z stands for the German word 'zustandssumme', which translates into English as 'state sum' or 'sum over states'. For a fundamentally quantum mechanical system, the partition function takes the form

$$z(\beta) = \mathrm{Tr}\, e^{-\beta \mathcal{H}}, \tag{3.2.12c}$$

in which $\mathcal{H}$ the Hamiltonian operator and Tr designates a summation over the diagonal elements of $e^{-\beta\mathcal{H}}$: in the energy representation (i.e., the representation consisting of the energy eigenstates of the Hamiltonian $\mathcal{H}$), this sum reduces immediately to Eq. (3.2.12b).

What about the constant β? From its definition (3.2.7) we see that β has dimensions of reciprocal energy. Notice that in our expression for the canonical distribution, there is no reference to the system A′ (bath, reservoir) except that we know that A is in contact with a large system A′ that is characterized by β (through $C = C'\Omega'(E^*)$). In thermodynamics, two systems that are allowed to exchange energy are said to be in *thermal contact*: moreover, if two systems may be brought into thermal contact with *no* exchange of energy taking place, they are said to be at the same temperature. This occurs at equilibrium, which is what we have been describing. Thus, we may associate β with the reciprocal of the energy characteristic of two systems in thermal equilibrium, i.e., $k_B T$, or

$$\beta = \frac{1}{k_B T}. \tag{3.2.13}$$

We obtained $p_r = e^{-\beta E_r}/z$ for a single available state for our subsystem A in contact with reservoir A′. What if the system A has more than one available state at energy E_r? If the degeneracy of the energy level designated by quantum number i is Ω_i, then

$$p_r = \Omega_r e^{-\beta E_r} / \sum_i \Omega_i e^{-\beta E_i}$$

$$\equiv \Omega_r e^{-\beta E_r}/z. \tag{3.2.14}$$

We have seen that for an equilibrium partition function $z(T)$ having the form $z(T) = \sum_i \Omega_i e^{-\beta E_i}$ the relative probability $p_r(T)$ associated with a specific energy level E_r is given by $p_r(T) = \Omega_r e^{-\beta E_r}/z(T)$. These probabilities tell us how a set of particles will be distributed, or partitioned, amongst the available energy levels, so that if we have a set of N such particles, then $N_r = p_r N$ of them will be found to have energy E_r at thermal equilibrium for temperature T. The constraint that the total number of particles is fixed at N, i.e., $\sum_i N_i = N$, then allows us to express p_r in the form $p_r = N_r/N$, so that p_r can equivalently be interpreted as the fractional population associated with energy level E_r. It will often turn out to be convenient to employ this interpretation of p_r.

Example 3.1 Role of (energy) degeneracy.

Let us consider two simple systems, which we shall refer to as systems A and B, both having the same (nondegenerate) ground state energy ϵ_0 and a single excited energy level with energy $\epsilon_1 = \epsilon_0 + 250\,\text{cm}^{-1}$, but with the difference that ϵ_1 for system A is nondegenerate, while ϵ_1 for system B is three-fold degenerate. We have seen that if A and B are both in thermal equilibrium with a common bath (i.e., a much larger system with which they may exchange energy but not mass) at temperature T, then the relative probability for the states of these systems to be occupied will be determined by the Boltzmann distribution via the canonical partition function

$$z(T) = \sum_{\text{levels}} \Omega_{\text{level}} e^{-\beta E_{\text{level}}} ,$$

as

$$p_i(T) = \frac{\Omega_i e^{-\beta E_i}}{z(T)} ,$$

with $\beta = (k_B T)^{-1}$ as usual. Thus, for systems A and B, respectively, we have

$$z_A(T) = e^{-\beta \epsilon_0} + e^{-\beta \epsilon_1}$$
$$= e^{-\beta \epsilon_0}(1 + e^{-359.7/T})$$

and

$$z_B(T) = e^{-\beta \epsilon_0}(1 + 3e^{-359.7/T}) .$$

Note that we have replaced $250\,\text{cm}^{-1}$ by $250\,\text{cm}^{-1}/(0.69503\,\text{cm}^{-1}\,\text{K}^{-1}) = 359.7\,\text{K}$ for calculational convenience: i.e., we have employed the Boltzmann constant in cm^{-1} units as $\overline{k}_B = 0.69504\,\text{cm}^{-1}\,\text{K}^{-1}$, thereby effectively expressing energy in kelvin temperature units.

We shall examine the difference in behaviour between systems A and B due to the presence of the triple degeneracy in ϵ_1 for system B. If we examine the fractional populations of the ground and excited states, given by

$$p_0^A(T) = \frac{e^{-\beta \epsilon_0}}{z_A(T)} = \frac{e^{-\beta \epsilon_0}}{e^{-\beta \epsilon_0}(1 + e^{-359.7/T})}$$
$$= \frac{1}{1 + e^{-359.7/T}} ,$$

$$p_1^A(T) = \frac{e^{-\beta \epsilon_1}}{z_A(T)} = \frac{e^{-\beta \epsilon_0}e^{-359.7/T}}{e^{-\beta \epsilon_0}(1 + e^{-359.7/T})}$$
$$= \frac{e^{-359.7/T}}{1 + e^{-359.7/T}} ,$$

we see that they are independent of the value for ϵ_0. We may thus deduce from this result that the value of ϵ_0 is irrelevant for our purposes, and hence we are free to choose ϵ_0 to be zero. The fractional populations for system B can now simply be written down as

$$p_0^B(T) = \frac{1}{1 + 3e^{-359.7/T}} , \quad p_1^B(T) = \frac{3e^{-359.7/T}}{1 + 3e^{-359.7/T}} .$$

For temperature $T = 450\,°C$, for example, the fractional population of the excited state of system A will be

$$p_1^A = \frac{0.4870}{1.4870} = 0.328 ,$$

while that for system B will be

$$p_1^B = \frac{1.4611}{2.4611} = 0.594 .$$

Thus, we see that the triple degeneracy associated with ϵ_1 in system B has increased the fractional population of the excited state for system B relative to that for system A by just over 80% at a temperature of 450 °C (or 723.15 K).   □

Example 3.2   Silicon and germanium atoms.

To illustrate in a more concrete manner the role played by energy degeneracy, let us compare Si and Ge atoms, whose three lowest energy levels have 1, 3, 5 states, respectively, with the pseudo-atoms Si* and Ge* for which the three lowest energy levels all have 1 state only.

If we represent the (filled shell) electronic configurations of Ne and Ar by [Ne] and [Ar], then the electronic configuration of Si can be written in the form $[Ne](3s)^2(3p)^2$, while that of Ge is given as $[Ar](4s)^2(3d)^{10}(4p)^2$. All atoms possess a hierarchy of energy levels, beginning with the lowest, or ground, energy level, which is equivalent to ϵ_0 from the previous example. Atomic energy levels have been systematically characterized using an agreed-upon convention expressed in the form of atomic term symbols (see Appendix D for an explanation of their origin and meanings). The ground electronic term for both Si and Ge is designated by ^3P, made up of three distinct levels, denoted in turn by $^3P_2$, $^3P_1$, and $^3P_0$. The ground level for the Si atom is the $^3P_0$ level, with the $^3P_1$, $^3P_2$ levels lying [3] 77.115 cm^{-1}, 223.16 cm^{-1}, respectively, above it; the ground level for the Ge atom is also a $^3P_0$ level, with $^3P_1$, $^3P_2$ levels lying [3] 557.13 cm^{-1}, 1409.96 cm^{-1} above it. The next electronic term for each of these atoms is a $^1D_2$ term lying 6298.85 cm^{-1}, respectively, 7125.30 cm^{-1} above the $^3P_0$ levels of the Si and Ge atoms. For atomic levels, the degeneracy is given by $2J + 1$, with J being the subscript on the atomic term symbol: thus, the $^3P_2$, $^3P_1$, $^3P_0$ levels have degeneracies 5, 3, 1, respectively, and each $^1D_2$ term has degeneracy 5.

We shall compare the actual fractional populations for the three electronic levels of the ground ^3P terms of Si and Ge with the fractional populations that would have been obtained were all three ^3P levels nondegenerate.

As in our first example for systems A and B, we shall work with a temperature $T = 723.15$ K (or 450 °C) to keep matters simple. The relevant partition functions are

$$z_{Si}(T) = 1 + 3e^{-111.0/T} + 5e^{-321.1/T} + 5e^{-9063/T} + \cdots,$$

for a Si atom, with the ellipsis ($\cdots$) representing the contributions from all excited terms/levels with energies greater than that of the $^1D_2$ level. The equivalent expression for a Ge atom is

$$z_{Ge}(T) = 1 + 3e^{-801.6/T} + 5e^{-2028.5/T} + 5e^{-10252/T} + \cdots.$$

We shall call our pseudoatom comparator systems Si* and Ge*, each with nondegenerate electronic levels $^3P_0$, $^3P_1^*$, $^3P_2^*$. These pseudoatoms will then have partition functions

$$z_{Si^*}(T) = 1 + e^{-111.0/T} + e^{-321.1/T} + e^{-9063/T} + \cdots,$$

and

$$z_{Ge^*}(T) = 1 + e^{-801.6/T} + e^{-2028.5/T} + e^{-10252/T} + \cdots.$$

For the same system temperature $T = 723.15$ K, the two comparator sets of systems give rise to fractional populations p_r summarized in Table 3.2.

If, on the one hand, we compare the fractional populations obtained for the Si and Si* systems, we see that the fractional population of the $^3P_0$ ground level for Si is about 63% lower, the fractional population for the $^3P_1$ first excited level is about 14% higher, and the fractional population for the $^3P_2$ second excited level is about 84% higher than would have been the case had all three levels of Si been nondegenerate. From a comparison of the Ge and Ge* systems, on the other hand, we see that the fractional population of the $^3P_1$ level is about double, and the fractional population of the $^3P_2$ level is about triple, what they would have been had all three levels of Ge been nondegenerate. Of course, these increased fractional

Table 3.2 Fractional populations for Si and Ge atoms/pseudoatoms

System	Level #	1	2	3
Si	Symbol	$^3P_0$	$^3P_1$	$^3P_2$
	Ω_r	1	3	5
	p_r	0.1475	0.3795	0.4730
Si*	Symbol	$^3P_0$	$^3P_1^*$	$^3P_2$
	Ω_r	1	1	1
	p_r^*	0.4001	0.3432	0.2567
Ge	Symbol	$^3P_0$	$^3P_1$	$^3P_2$
	Ω_r	1	3	5
	p_r	0.4362	0.4319	0.1319
Ge*	Symbol	$^3P_0$	$^3P_1^*$	$^3P_2^*$
	Ω_r	1	1	1
	p_r^*	0.7191	0.2374	0.0435

populations come at the cost of an approximately 40% reduction in the fractional population of the ground level of Ge.

Comparison between the fractional population values for Si* and Ge* illustrates the role of the excitation energies of excited states, with the energy levels for the two excited states about 2.75 times higher on average for Ge* than for Si*, with a consequent relative increase in the fractional population of the ground state of Ge* and decreases in the fractional populations of both excited states. Finally, a comparison between the fractional populations for Si and Ge shows that the degeneracies of the excited levels of Ge more than offset the increased excitation energies to provide overall increases in the fractional populations of the excited electronic levels of the Ge atom.   □

Formula (3.2.14) holds if we are dealing with discrete energy states. What happens if the accessible states are part of a continuum as, e.g., turns out to be an excellent approximation when we consider translational motion? Note that the treatment of translational motion as a special case is particularly justified by the fact that it exclusively gives rise to the volume dependence of the partition function for any ensemble.

3.2.1 Translational States: Continuum Approximation

Let us assume that we are given the volume V and the number of particles N, and that we *measure* the energy E for this system, and find $\epsilon \leq$ energy $\leq \epsilon + \Delta\epsilon$. Let us now define $\Omega(\epsilon, V)\Delta\epsilon$ to be the number of microstates of the system of interest having energies between ϵ and $\epsilon + \Delta\epsilon$. In principle we can always calculate this quantity for any system. Let us do so now for a particle-in-a-box. Consider

$$\Phi(\epsilon, V) \equiv \text{Number of microstates having energy} \leq \epsilon$$

$$\Phi(\epsilon + \Delta\epsilon, V) \equiv \text{Number of microstates having energy} \leq \epsilon + \Delta\epsilon \, .$$

The number of microstates having energy E lying between ϵ and $\epsilon + \Delta\epsilon$ is therefore given by the difference between these two quantities. If we invoke a limiting process whereby $\Delta\epsilon$ is taken to zero, we obtain the partial derivative

$$\lim_{\Delta\epsilon \to 0} \frac{\Phi(\epsilon + \Delta\epsilon, V) - \Phi(\epsilon, V)}{\Delta\epsilon} \equiv \left(\frac{\partial \Phi(\epsilon, N, V)}{\partial \epsilon} \right)_V$$

$$\equiv \Omega(\epsilon, V) \, , \tag{3.2.15}$$

so that $\Omega(\epsilon, V)$ is called the *density of states* for continuum states.

Let us now apply the particle-in-a-box model to obtain an explicit expression for $\Omega(\epsilon, V)$ for translational motion. The expression for the energy for a particle in a 3-D box has the form

Fig. 3.2 Positive octant of a sphere of radius R and a unit cell, both in the reciprocal-lattice space

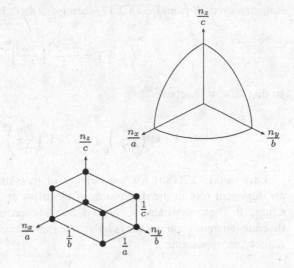

$$\epsilon_{n_x,n_y,n_z} = \frac{h^2}{8m}\left[\frac{n_x^2}{a^2} + \frac{n_y^2}{b^2} + \frac{n_z^2}{c^2}\right]. \tag{3.2.16}$$

It will prove useful to write the factor in square brackets as

$$\frac{n_x^2}{a^2} + \frac{n_y^2}{b^2} + \frac{n_z^2}{c^2} = \frac{8m\epsilon_{n_x,n_y,n_z}}{h^2} \leq \frac{8m\epsilon}{h^2} \equiv R^2, \tag{3.2.17}$$

with R defining the radius of a sphere in the reciprocal-lattice space illustrated in Fig. 3.2.

As each point in the reciprocal-lattice space represents an energy eigenstate $\epsilon_{n_x n_y n_z}$, the number of energy eigenstates having energies less than a given value ϵ is equal to the number of lattice points contained in the positive octant of the sphere defined by Eq. (3.2.17). The number of lattice points is thus obtained by dividing the volume of the octant of the sphere by the volume per point (recall that we are working in the reciprocal-lattice space, rather than in the physical space). Upon examining the rectangular unit cell illustrated in Fig. 3.2, we see that every such unit cell contains one-eighth of each of the eight points located at the corners of the unit cell, and hence each unit cell, of volume $\frac{1}{a}\cdot\frac{1}{b}\cdot\frac{1}{c} = \frac{1}{V}$ (with $V = abc$ being the volume of the physical box), corresponds to one energy eigenstate. Thus, $\Phi(\epsilon, V)$ is given by

$$\Phi(\epsilon, V) = \frac{\frac{1}{8}\left(\frac{4}{3}\pi R^3\right)}{\frac{1}{a}\cdot\frac{1}{b}\cdot\frac{1}{c}} = \frac{1}{6}\pi V R^3.$$

Substitution of R from Eq. (3.2.17) then gives $\Phi(\epsilon, V)$ explicitly as

$$\Phi(\epsilon, V) = \frac{4\pi V}{3} \left(\frac{2m}{h^2}\right)^{\frac{3}{2}} \epsilon^{\frac{3}{2}}, \tag{3.2.18a}$$

so that $\Omega(\epsilon, V)$ becomes

$$\Omega(\epsilon, V) \equiv \left(\frac{\partial \Phi}{\partial \epsilon}\right)_{N,V} = 2\pi V \left(\frac{2m}{h^2}\right)^{\frac{3}{2}} \epsilon^{\frac{1}{2}}. \tag{3.2.18b}$$

Expression (3.2.18b) for the density of translational energy states will play an important role in the discussion of electrons in metals and semiconductors in Chap. 10. These steps also serve to define a procedure for replacing the sum over discrete (quantum) energy states by an integration over a continuum of energy states, namely, the replacement

$$\sum_k \cdots \leftrightarrow \int d\epsilon\, \Omega(\epsilon, V) \cdots. \tag{3.2.19}$$

We may now evaluate the partition function z for translational motion (typified by our consideration of a single particle in a 3-D box). We begin by writing the partition function for translational motion as

$$z(\beta, V) = \int_0^\infty \Omega(\epsilon, V) e^{-\beta \epsilon}\, d\epsilon, \tag{3.2.20}$$

and recognize that z depends explicitly not only upon temperature, via β, but also upon V, via Ω, as given by Eqs. (3.2.18). The partition function for a single particle possessing only translational motion is thus given by

$$z(\beta, V) = 2\pi V \left(\frac{2m}{h^2}\right)^{\frac{3}{2}} \int_0^\infty \epsilon^{\frac{1}{2}} e^{-\beta \epsilon}\, d\epsilon.$$

This expression can be simplified by carrying out the integration, to give the final form

$$z(T, V) = \left(\frac{2\pi m k_B T}{h^2}\right)^{\frac{3}{2}} V \equiv \frac{V}{\Lambda^3(T)}, \tag{3.2.21}$$

in which β has been replaced by $(k_B T)^{-1}$. The quantity $\Lambda(T)$, defined by

$$\Lambda(T) \equiv \frac{h}{\sqrt{2\pi m k_B T}}, \tag{3.2.22}$$

is referred to as the *thermal de Broglie wavelength*. This quantity plays an important role in statistical mechanics, and we shall say more about it later (especially in Chap. 10).

Example 3.3 Translational states of a Ne atom with energy $3k_BT$.

Noble gas atoms, like neon (Ne), have ground electronic states that are separated from their lowest excited electronic states by energies of the order of several electron volts (eV), and hence at room temperature or thereabouts, the only energy states accessible to these atoms are translational states. Let us estimate the number of translational states lying at or below an energy $3k_BT$ that are accessible to a Ne atom in a box of volume $10\,cm^3$ for a temperature of 300 K.

To make our estimate, we shall treat the Ne atom as a particle-in-a-box. As $\Phi(E, V)$, given by

$$\Phi(E) = \frac{4\pi V}{3} \left(\frac{2m}{h^2}\right)^{\frac{3}{2}} E^{\frac{3}{2}},$$

represents the number of translational microstates having energy less than or equal to E, then by setting $E = 3k_BT = 900k_B$, $V = 10\,cm^3$, and using the value 33.210×10^{-27} kg for the mass of a Ne atom, we obtain

$$\Phi(900k_B) = \frac{4\pi \, 10^{-5} m^3}{3} \left(\frac{2 * 33.210 \times 10^{-27}\,kg * 900 * 1.3807 \times 10^{-23}\,J}{(6.6261 \times 10^{-34})^2\,J^2 s^2}\right)^{\frac{3}{2}}$$

$$= \frac{4\pi \, 10^{-5}\,m^3}{3} \left(\frac{8.2535 \times 10^{-46}}{43.905 \times 10^{-68}\,m^2}\right)^{\frac{3}{2}}$$

$$= 3.414 \times 10^{27}.$$

The number of translational states accessible to a Ne atom at temperature 300 K is thus of the order of

$$\Phi(900k_B) = 3 \times 10^{27},$$

which indeed represents a very large number of states. □

Example 3.4 Density of states for an ideal gas in a gravitational field.

This example is based upon a discussion of the entropy of a column of an ideal gas under gravity [4]. Let us consider an isothermal right-circular cylindrical column of an ideal gas of structureless particles of mass m under the influence of a weak gravitational field. To determine the energy density of states $\Omega(V, \epsilon)d\epsilon$ in an energy interval $d\epsilon$, we note that the translational motion of a particle depends quadratically upon the linear momentum **p**, so that the density of translational energy states will be equal to the corresponding density of linear momentum states

in a linear momentum interval d$\mathbf{p}$: that is, $\Omega(V, \epsilon)$dϵ can be obtained by calculating
the number of momentum states associated with d$\mathbf{p}$.

We shall consider translational motion as inherently quantum mechanical, and
thus note that the Heisenberg Uncertainty Principle, which may be expressed in
each dimension ℓ as $\Delta\ell\Delta p_\ell \sim h$, then necessitates the introduction of a factor h^{-3}
into the calculation of the number of momentum states contained in a momentum
interval d$\mathbf{p}$. The expression thereby obtained for the energy density of states is thus

$$\Omega(V, \epsilon)\mathrm{d}\epsilon = \frac{1}{h^3}\left(\int_V \int \mathrm{d}\widehat{\mathbf{p}}\,\mathrm{d}\mathbf{r}\right)\mathrm{d}p$$

$$= \frac{1}{h^3}\left(\int_V \int_0^{2\pi}\int_0^{\pi} p^2\sin\vartheta\,\mathrm{d}\vartheta\,\mathrm{d}\varphi\mathrm{d}\mathbf{r}\right)\mathrm{d}p\,.$$

We shall employ Cartesian coordinates to represent the position, $\mathbf{r}$, of a particle
in the cylindrical column and spherical polar coordinates to represent its momentum
$\mathbf{p}$. As the energy of a particle in a gravitational field is given by

$$\epsilon(\mathbf{r}, \mathbf{p}) = \frac{\mathbf{p}^2}{2m} + mgz\,,$$

with z representing the vertical direction in the cylindrical gas column while g is
the acceleration due to gravity (acting downwards), the integrations over the x and
y coordinates can be carried out trivially, to give $\pi R^2 \equiv \mathcal{A}$, the cross-sectional area
of the cylindrical column of gas, and the integration over the two angles ϑ and φ
associated with the particle momentum gives an additional factor 4π. This gives the
energy density of states as

$$\Omega(V, \epsilon)\mathrm{d}\epsilon = \frac{4\pi\mathcal{A}}{h^3}\left(\int_0^H p^2(z)\,\mathrm{d}z\right)\mathrm{d}p\,,$$

with H the height of the gas column. From the expression for the particle energy,
we obtain the magnitude of the linear momentum and its differential as

$$p = [2m(\epsilon - mgz)]^{\frac{1}{2}}\quad\text{and}\quad \mathrm{d}p = [2m(\epsilon - mgz)]^{-\frac{1}{2}}m\,\mathrm{d}\epsilon$$

so that the energy density of states is thus given as

$$\Omega(\epsilon) = \frac{4\pi\mathcal{A}(2m)^{\frac{1}{2}}m}{h^3}\int_{z_\ell}^{z_u}(\epsilon - mgz)^{\frac{1}{2}}\mathrm{d}z$$

$$= -\frac{4\pi\mathcal{A}(2m)^{\frac{3}{2}}}{3gh^3}\left[(\epsilon - mgz)^{\frac{3}{2}}\right]_\ell^u\,.$$

The lower limit ℓ is always $z = 0$, but the upper limit u depends upon the particle energy: it takes values $z = \epsilon/(mg)$ for $mgH > \epsilon$ and $z = H$ for $\epsilon \geq mgH$. This variable upper limit for the integration over z can be accommodated by utilizing the Heaviside function $\mathsf{H}(mgH/\epsilon)$, defined as

$$\mathsf{H}(x) = \begin{cases} 0, & x > 1 \\ 1, & 0 \leq x \leq 1 \end{cases},$$

to give

$$\Omega(V, \epsilon) = \frac{4\pi \mathcal{A}}{3mg} \left(\frac{2m}{h^2}\right)^{\frac{3}{2}} [\epsilon^{\frac{3}{2}} - (\epsilon - mgH)^{\frac{3}{2}} \mathsf{H}(mgH/\epsilon)].$$

Note that we may also write $\Omega(V, \epsilon)$ as

$$\Omega(V, \epsilon) = \frac{4\pi V}{3mgH} \left(\frac{2m}{h^2}\right)^{\frac{3}{2}} [\epsilon^{\frac{3}{2}} - (\epsilon - mgH)^{\frac{3}{2}} \mathsf{H}(mgH/\epsilon)],$$

with the volume dependence $V \equiv H\mathcal{A}$ displayed explicitly. □

3.2.2 Summary of Forms for the Canonical Partition Function

We have now obtained all relevant versions of the canonical partition function that will be needed for our further employment. We may summarize these three versions of the *single-particle canonical partition function* as

$$z(\beta, V) = \begin{cases} \displaystyle\int_0^\infty \Omega(V, E)e^{-\beta E}\, dE & \text{continuum energy states} \\[2ex] \displaystyle\sum_r e^{-\beta E_r} & \text{discrete energy states} \\[2ex] \displaystyle\sum_i \Omega_i e^{-\beta E_i} & \text{discrete energy levels} \end{cases} \qquad (3.2.23)$$

Thus, the single-particle canonical partition function, or sum over (energy) states, can be represented in two ways for discrete energies: either as a straightforward sum over individual energy states or, equivalently, but often more conveniently, as a sum over distinct energy levels weighted by their respective degeneracies (namely, the number of states having the same energy). This procedure would be, to say the least, tedious in the case of the very closely-spaced energy states associated with the translational motion of any but the smallest-mass particles. For such cases, we typically utilize the continuum approximation in which the summation over energy states is replaced by an integration over energy and the degeneracy factor Ω_i for a given energy state is replaced by the continuous density of states function $\Omega(V, E)$.

3.2.3 Extension to N-Particle Systems

For systems made up of many particles (most often atoms and/or molecules) we could proceed in either of two fashions. Formally, the most obvious means for extending our discussion would be to consider the energy eigenstates entering into the discrete states defining relation (3.2.20) to be the eigenstates of the N-particle system itself, so that the relevant canonical partition function, labelled $Z(\beta, V; N)$, is given by

$$Z(\beta, V; N) \equiv \sum_r e^{-\beta E_r(V;N)}, \tag{3.2.24}$$

in which the energies E_r belong to the (vast number of) individual eigenstates of the N-particle system being considered. While such an approach is certainly correct in principle, it is rather impractical.

A more practical approach is to consider the N-particle system as consisting of many smaller subsystems, reducing it in first instance into subsystems consisting of sets of distinct chemical species A, B, C, ..., made up of N_A, N_B, N_C, ..., respectively, atoms or molecules, with corresponding partition functions $Z_i(\beta, V; N_i)$, $i \in \{A, B, C, \ldots\}$. We shall further reduce the subsystems into smaller subsystems, consisting of individual atoms or molecules, whose partition functions are the single-particle partition functions $z_i(\beta, V)$. To accomplish this goal, we shall employ a two-step procedure. In the first step, we shall consider an N-particle system to be made up of particles that may be considered to be both *statistically independent* and *distinguishable*. The second step will involve taking into account changes that must be made when the system particles are indeed *indistinguishable*.

As usual, we shall treat the energy of any one subsystem as independent of the energy of all other subsystems, which simply means that we shall ignore all interactions between subsystems. This assumption is consistent with our goal of dealing exclusively with equilibrium statistical thermodynamics. To make the concept more concrete, let us consider the simplest case of two subsystems, which we shall refer to as subsystem A and subsystem B. Moreover, we shall label the energy eigenstates of subsystem A by assigning values to a quantum number i, and the energy eigenstates of subsystem B by assigning values to a quantum number j. Then, consistent with the statistical independence and distinguishability of the subsystems, the total system (i.e., $A \oplus B$) is specified by values of the *pair* of quantum numbers ij. The assumption of a weak interaction between A and B means that the total pair energy is given by $\epsilon_{ij} = \epsilon_{Ai} + \epsilon_{Bj}$. Let us note in passing that this is precisely the definition given for an ideal gas in thermodynamics.

To see how the canonical partition function may be evaluated, we begin with its definition as a 'sum over states' of $e^{-\beta \epsilon_{ij}}$ for a system of interest that consists of two distinguishable atoms/molecules, one of chemical species A, the other of chemical species B, in thermal equilibrium with a bath system. The partition function Z_{AB} for this two-particle system is defined by

$$Z_{AB} \equiv \sum_{ij} e^{-\beta \epsilon_{ij}} = \sum_{ij} e^{-\beta \epsilon_{Ai}} e^{-\beta \epsilon_{Bj}}$$

$$= \sum_{i} e^{-\beta \epsilon_{Ai}} \sum_{j} e^{-\beta \epsilon_{Bj}} .$$

If we now recall the definition of the canonical partition function for an individual particle such as A (or B), namely $z_A = \sum_i e^{-\beta \epsilon_{Ai}}$, we may then rewrite our expression for the total partition function z_{AB} as

$$Z_{AB} = z_A z_B . \tag{3.2.25}$$

More generally, we can apply this argument to any set of statistically independent and distinguishable single-particle subsystems to obtain

$$Z = z_A z_B z_C \ldots . \tag{3.2.26}$$

A concrete example of such a splitting of a larger system into smaller independent subsystems is to consider air as a mixture of two ideal gases, oxygen and nitrogen (we shall for the sake of simplicity ignore all the other minor constituents of the air around us). Thus, if we choose to treat air as a mixture of O_2 and N_2 only, nitrogen and oxygen are certainly distinguishable (chemically, if not otherwise) and, in the ideal gas limit, they are statistically independent, so that

$$Z_{air} = Z_{O_2} Z_{N_2} .$$

We can already see from this example chosen to illustrate the concept of distinguishability that were we to extend this argument to Z_{N_2}, we should immediately encounter a problem, as we cannot possibly distinguish one nitrogen molecule from another, no matter how hard we try to do so. We may, however, still consider one nitrogen molecule to be statistically independent of all other nitrogen molecules, as it is equivalent to considering nitrogen to be an ideal gas. Thus, to proceed properly beyond this point, we obviously must now consider the case of statistically independent but *indistinguishable* subsystems.

We shall retain the approximation in which we ignore interactions. Let us examine, as before, two subsystems labelled A and A', and let us compare the result obtained by treating the subsystems as distinguishable with what we obtain by working directly with the definition of the canonical partition function for the total system. Were we simply to ignore the indistinguishability of the A and A' subsystems, we would obtain the result

$$Z_{AA'} = Z_A Z_{A'} \quad \Rightarrow \quad Z_{AA'} = \sum_i e^{-\beta \epsilon_{Ai}} \sum_j e^{-\beta \epsilon_{A'j}} \equiv z^2 \,,$$

because the two summations will be the same. However, from the definition of the canonical partition function we have

$$Z_{AA'} \stackrel{\text{def}}{=} \sum_k e^{-\beta \epsilon_k} \,,$$

in which the values of ϵ_k are given by the possible values of $\epsilon_{Ai} + \epsilon_{A'j}$. Note, however, that because

$$\epsilon_{Ai} + \epsilon_{A'j} = \epsilon_{Aj} + \epsilon_{A'i} \,,$$

we shall have counted this energy *twice* in the product $z_A z_{A'}$ rather than only *once* as required in the definition of $Z_{AA'}$ for the entire system. Thus, we have *overestimated* $Z_{AA'}$ by a factor 2 when we wrote down the result $Z_{AA'} = z^2$. Similarly, if we had N indistinguishable but independent particles (or subsystems), the product $Z = z^N$ similarly *overestimates* the actual value of Z by the number of ways for obtaining the same *total energy*

$$\epsilon = \epsilon_1 + \epsilon_2 + \cdots + \epsilon_N \,,$$

i.e., by $N!$. From this argument we see that we can obtain a good approximation to the canonical partition function for N independent but indistinguishable subsystems by evaluating the canonical partition function for one of these subsystems, raising the result to the power N, and then dividing by $N!$, namely,

$$Z_N(\beta, V) = \frac{z^N(\beta, V)}{N!} \,. \tag{3.2.27}$$

We might well now ask how extend this description to deal with binary mixtures of indistinguishable particles, such as our example of O_2–N_2 mixtures. If we have a mixture of two types of molecules, N_A of type A and N_B of type B, then the partition function Z_{AB} for such a binary mixture will be

$$Z_{AB}(\beta, V; N) = \frac{z_A^{N_A} z_B^{N_B}}{N_A! N_B!} \,; \qquad N = N_A + N_B \,. \tag{3.2.28}$$

This result can be further generalized to multicomponent mixtures should it become necessary.

3.3 The Grand Ensemble: Open Systems

The so-called *grand ensemble* is an ensemble, each of whose members is itself a canonical ensemble with a fixed value for the number of particles, N. In this way we allow the exchange both of energy and mass (via the number of particles), so that the grand ensemble is appropriate for the description of an *open system*. Strictly speaking, we should use the grand ensemble when we wish to describe the thermodynamics of an open system, but as we shall see later on, there is an artifice that will allow us to bypass this necessity so that we may obtain appropriate expressions directly from the canonical ensemble. Let us nonetheless briefly examine the development of the grand ensemble.

Figure 3.3 illustrates figuratively the differences between the three basic ensembles that we have been considering thus far. The microcanonical ensemble has a fixed energy E and is separated from the rest of the universe by thermally-insulating walls (represented here by greyish surrounding walls), the canonical ensemble (referred to as the 'system of interest') is in thermal contact with a second system, normally a very much larger 'heat reservoir' with which it can exchange energy, while the grand ensemble is normally in thermal and material contact with a 'heat' and mass reservoir with which it can exchange both energy and particles of appropriate mass. Couplings between components for the canonical and grand ensembles are indicated, respectively, by the solid and dashed lines that separate the isolated combined systems into two parts. When one component serves as a 'reservoir', it must be very much larger than the 'system of interest'.

As for the canonical ensemble, the number of microstates lying between energy ϵ and $\epsilon + \Delta\epsilon$ is represented by $\Omega(\epsilon, N)\Delta\epsilon$, which implies that the probability for finding the total system (i.e., system of interest plus the reservoir) with energy inside this interval is given by

$$P_{\text{combined system}} = \frac{1}{\Omega(\epsilon, N)\Delta\epsilon}. \tag{3.3.1}$$

Also as for the derivation for the canonical ensemble, we assume that there is only weak contact between the two subsystems, and that the system of interest (A) is small by comparison with the other subsystem (the reservoir). The total energy can be written in the form

$$\epsilon = \epsilon_1 + \epsilon_2 + \epsilon_{12}, \tag{3.3.2}$$

Fig. 3.3 Diagrams illustrating the three most commonly occurring ensembles

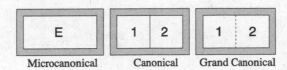

Microcanonical   Canonical   Grand Canonical

in which ϵ_1, ϵ_2 represent the energies of the two subsystems, and ϵ_{12} represents the energy of interaction between them. This energy of interaction will, as usual, be assumed to be so small in comparison with ϵ_1 and ϵ_2 that it may be neglected (weak contact). Further, the total number of particles N is related to the numbers of particles N_1 and N_2 in the two subsystems by

$$N = N_1 + N_2. \tag{3.3.3}$$

If for subsystem 1, which has N_1 particles, we consider the restrictions $\epsilon_1 \rightarrow \epsilon_1 + \mathrm{d}\epsilon_1$, then a certain number of states, which we shall designate by $\Omega_1(\epsilon_1, N_1)\mathrm{d}\epsilon_1$, of the combined system satisfy these restrictions in addition to those for the total energy and number of particles. At the same time we have for subsystem 2 the corresponding energy relation $\epsilon - \epsilon_1 \rightarrow \epsilon - \epsilon_1 + \Delta\epsilon$, in which $\mathrm{d}\epsilon_2 \simeq \Delta\epsilon$, for the $N - N_1$ particles in the subsystem. The corresponding number of states is given by $\Omega_2(\epsilon - \epsilon_1, N - N_1)\Delta\epsilon$. The approximation of $\mathrm{d}\epsilon_2$ by $\Delta\epsilon$ is connected with the assumption that the reservoir (subsystem 2) is much larger than the system of interest, so that the vast majority of $\Delta\epsilon$ resides in the reservoir. The probability $p(\epsilon_1, N_1)\mathrm{d}\epsilon_1$ is obtained by taking the ratio of the total number of states available when the energy and number of particles are partitioned as just described, divided by the total number of states available for all such partitions: hence, we may write

$$p(\epsilon_1, N_1)\mathrm{d}\epsilon_1 = \frac{\Omega_1(\epsilon_1, N_1)\mathrm{d}\epsilon_1 \Omega_2(\epsilon - \epsilon_1, N - N_1)\Delta\epsilon}{\Omega(\epsilon, N)\Delta\epsilon},$$

in which

$$\Omega(\epsilon, N)\Delta\epsilon = \sum_{N_1} \int_0^\epsilon \Omega_1(\epsilon_1, N_1)\Omega_2(\epsilon - \epsilon_1, N - N_1)\mathrm{d}\epsilon_1 \Delta\epsilon. \tag{3.3.4}$$

We can recognize that the total number of microstates available to the entire system at energy ϵ can be obtained by summing over N_1 and integrating over ϵ_1 the expression for the number of microstates for an individual partition. The expression for $p(\epsilon_1, N_1)$ can be simplified to

$$p(\epsilon_1, N_1) = \frac{\Omega_1(\epsilon_1, N_1)\Omega_2(\epsilon - \epsilon_1, N - N_1)}{\Omega(\epsilon, N)}. \tag{3.3.5}$$

To obtain a more useful explicit expression for $p(\epsilon_1, N_1)$ we proceed in much the same way as we did for the canonical ensemble: the most probable energy is found from the condition

$$\frac{\partial p(\epsilon_1)}{\partial \epsilon_1} = 0 \implies \epsilon_{1\mathrm{m}}. \tag{3.3.6}$$

Thermal contact between the two subsystems is again expressed in terms of the parameter β, given in the present case by

$$\beta \equiv \left.\frac{\partial \ln \Omega_1}{\partial \epsilon_1}\right|_{\epsilon_{1m}, N_{1m}} = \left.\frac{\partial \ln \Omega_2(\epsilon_2)}{\partial \epsilon_2}\right|_{\epsilon_{2m}, N_{2m}}. \qquad (3.3.7)$$

Moreover, since N is large (typically of order 10^{23} or so), we may treat N-changes as being continuous, and may then express the material contact between the two subsystems in terms of a parameter γ, defined as

$$\gamma \equiv \left.\frac{\partial \ln \Omega_1}{\partial N_1}\right|_{\epsilon_{1m}, N_{1m}} = \left.\frac{\partial \ln \Omega_2(N_2)}{\partial N_2}\right|_{\epsilon_{2m}, N_{2m}}. \qquad (3.3.8)$$

If we now carry out a double Taylor series expansion of $\ln \Omega_2(\epsilon_2, N_2)$, keeping only the lowest-order non-vanishing terms, we obtain the result

$$\ln \Omega_2(\epsilon_2, N_2) = \ln \Omega_2(\epsilon_{2m}, N_{2m}) + \beta(\epsilon_2 - \epsilon_{2m}) + \gamma(N_2 - N_{2m}) + \cdots,$$

or

$$\Omega_2(\epsilon_2, N_2) \simeq \Omega_2(\epsilon_{2m}, N_{2m})e^{\beta(\epsilon_2 - \epsilon_{2m})}e^{\gamma(N_2 - N_{2m})}. \qquad (3.3.9)$$

We can now utilize the conservation of the number of particles in the form

$$N_1 + N_2 = N_{1m} + N_{2m} \qquad \text{or} \qquad N_2 - N_{2m} = N_{1m} - N_1, \qquad (3.3.10)$$

and the conservation of energy in the form

$$\epsilon_1 + \epsilon_2 = \epsilon_{1m} + \epsilon_{2m} \qquad \text{or} \qquad \epsilon_2 - \epsilon_{2m} = \epsilon_{1m} - \epsilon_1, \qquad (3.3.11)$$

in the exponential factors in Eq. (3.3.9) to obtain the result

$$\Omega_2(\epsilon_2, N_2) = \Omega_2(\epsilon_{2m}, N_{2m})e^{-\beta(\epsilon_1 - \epsilon_{1m})}e^{-\gamma(N_1 - N_{1m})}. \qquad (3.3.12)$$

This result allows us to obtain a useful expression for $\Omega(\epsilon, N)$ from Eq. (3.3.4), namely

$$\Omega(\epsilon, N) = \sum_{N_1} \int_0^\epsilon d\epsilon_1 \, \Omega_1(\epsilon_1, N_1)\Omega_2(\epsilon_2, N_2)$$

$$= \Omega_2(\epsilon_{2m}, N_{2m})e^{\beta\epsilon_{1m}}e^{\gamma N_{1m}} \sum_{N_1=1}^N \int_0^\epsilon \Omega_1(\epsilon_1, N_1)e^{-\beta\epsilon_1}e^{-\gamma N_1}d\epsilon_1.$$

$$(3.3.13)$$

At this point we are unable to proceed further without making some additional approximations. In particular we note that $e^{-\beta\epsilon_1} \rightarrow 0$ as $\epsilon_1 \rightarrow \infty$, and that $e^{-\gamma N_1} \rightarrow 0$ as $N_1 \rightarrow \infty$, so that $\Omega(\epsilon, N)$ can be rather well approximated by

$$\Omega(\epsilon, N) = \Omega_2(\epsilon_{2m}, N_{2m})e^{\beta\epsilon_{1m}}e^{\gamma N_{1m}} \sum_{N_1=1}^{\infty} \int_0^{\infty} \Omega_1(\epsilon_1, N_1)e^{-\beta\epsilon_1}e^{-\gamma N_1}d\epsilon_1$$

$$= \Omega_2(\epsilon_{2m}, N_{2m})e^{\beta\epsilon_{1m}}e^{\gamma N_{1m}}\,\Xi\,, \tag{3.3.14}$$

in which Ξ is the *grand partition function*, and is defined expressly by

$$\Xi \equiv \sum_{N_1=0}^{\infty} \int_0^{\infty} \Omega_1(\epsilon_1, N_1)e^{-\beta\epsilon_1}e^{-\gamma N_1}\,d\epsilon_1\,.$$

As there is no reference to the reservoir in this expression, we may simply drop the system of interest subscripts on ϵ and N, and write instead

$$\Xi(\beta, V, \gamma) \equiv \sum_{N=0}^{\infty} \int_0^{\infty} \Omega(\epsilon, N)\,e^{-\beta\epsilon}\,d\epsilon\,e^{-\gamma N} \tag{3.3.15}$$

for the grand partition function Ξ for continuum energy states, with corresponding expressions being obtained for discrete energy states and for discrete energy levels.

If we now substitute Eqs. (3.3.12) and (3.3.14) into Eq. (3.3.5) for the probability, we find that $p(\epsilon, N)$ takes the form

$$p(\epsilon, N) = \frac{\Omega(\epsilon, N)e^{-\beta\epsilon}e^{-\gamma N}}{\Xi}\,. \tag{3.3.16a}$$

Thermal contact means that the grand (canonical) ensemble is just an ensemble of canonical ensembles [5] (hence the terminology), so that $\beta \equiv (k_B T)^{-1}$, as for the canonical ensemble. By analogy with Eqs. (3.2.12) for the canonical ensemble,[4] which expresses $p(\epsilon)$ as the ratio of the Boltzmann factor $e^{-\beta\epsilon}$ to the canonical partition function $z(\beta, V)$, we may express $p(N)$ for an individual member of the grand ensemble as

$$p(N) = \sum_{r_N} p(\epsilon_{r_N}, N) = \frac{1}{\Xi}\left(\sum_{r_N} \Omega_{r_N}e^{-\beta\epsilon_{r_N}}\right)e^{-\gamma N}\,. \tag{3.3.16b}$$

[4]The term *macroensemble* has been employed for the grand ensemble by Greiner et al. [5] in order to emphasize the sequential relationship between the microcanonical, canonical, and grand (canonical) ensembles.

The quantity $\sum_{r_N} \Omega_{r_N} e^{-\beta \epsilon_{r_N}} e^{-\gamma N}$ is sometimes referred to as the Gibbs factor, in analogy with the term Boltzmann factor for $\Omega_r e^{-\beta \epsilon_r}$ in the discrete levels analogue of Eq. (3.2.10).

We shall see more clearly in Sect. 4.2.1 that $\gamma = -\beta \mu$, where μ is the chemical potential (or Gibbs energy per particle). If we accept this interpretation for the moment, we arrive at formal expressions for the grand partition function analogous to those given in Eq. (3.2.23), namely,

$$\Xi(T, V, \mu) = \begin{cases} \sum_N e^{\beta \mu N} \int_0^\infty \Omega(E, N) e^{-\beta E} \, dE \,, & \text{continuum energy states}\,; \\[2ex] \sum_N e^{\beta \mu N} \sum_{r_N} e^{-\beta E_{r_N}} \,, & \text{discrete energy states}\,; \\[2ex] \sum_N e^{\beta \mu N} \sum_{i_N} \Omega_{i_N} e^{-\beta E_{i_N}} \,, & \text{discrete energy levels}\,. \end{cases}$$

$$(3.3.17)$$

Example 3.5 Single-site model for oxygen uptake by haemoglobin.

The haemoglobin molecule has four adsorption sites, each consisting of an Fe^{2+} ion surrounded by four planar N atoms with another N atom below it, thus enabling an O_2 molecule to 'dock' above (i.e., bind physically to) each site [6]. The simplest, or single-site, model is obtained by treating the binding of a molecule to any given site to be independent of the other three sites. Should O_2 be the only molecule that is capable of occupying a site, then the system has only two possible states, namely, 'occupied' and 'unoccupied': should the 'unoccupied' state of this system be taken as the zero of energy, then the 'occupied' state lies at an energy ϵ of about $-0.70\,\mathrm{eV}$ and the grand partition function for a two-state system is given by

$$\Xi(T, V, \mu) = 1 + e^{-\beta(\epsilon - \mu)} \,.$$

The blood in our lungs is in an approximate thermal/mass equilibrium with the atmosphere, with a partial pressure of O_2 that is approximately 0.2 bar.

The grand partition function is required because there is matter exchange (in the sense that O_2 molecules go from the gas phase into an adsorbed phase on haemoglobin). An equilibrium between O_2 in the air and O_2 molecules adsorbed onto haemoglobin in the blood requires equality between the chemical potentials for these two phases, i.e., $\mu_{ad}(O_2) = \mu_{gas}(O_2)$: this equilibrium will determine the value of μ needed to evaluate Ξ in order to determine the probability (equivalently, fractional occupation of haem sites) $p_{occ} = e^{-\beta(\epsilon - \mu)}/\Xi$. Thus, we see that

$$\mu(T) \equiv \mu_{ad}(O_2) = \mu_{gas}(O_2) = -k_B T \ln \left[\frac{k_B T z_{int}(T)}{P \Lambda^3} \right]$$

or, expressed in terms of molecular parameters,

$$\mu = -k_B T \left[\ln\left(\frac{k_B T}{P \Lambda^3}\right) + \ln\left(\frac{T}{2\Theta_{rot}} \cdot \frac{\omega_{el}}{1 - e^{-\Theta_{vib}/T}}\right) \right].$$

The rotational constant for O_2 is $\overline{B}_0(O_2) = 1.4337\,\text{cm}^{-1}$, and its fundamental vibrational frequency is $\overline{\nu}_{osc}(O_2) = 1556.2\,\text{cm}^{-1}$, equivalent to characteristic rotational and vibrational temperatures $\Theta_{rot} = 2.07\,\text{K}$ and $\Theta_{vib} = 2239\,\text{K}$. Moreover, the electronic term symbol $^3\Sigma_g^-$ for ground state O_2 gives $\omega_{el} = 3$, while the de Broglie wavelength, which is given by $\Lambda(T) = 1.7458 \times 10^{-9}(M^*T)^{-\frac{1}{2}}$, with M^* in amu (32.0 amu for O_2), gives $\Lambda^3 = 5.3859 \times 10^{-33}\,\text{m}^3$ for O_2 in the gas phase at $T \simeq 311\,\text{K}$. Thus, with $k_B T = 4.2938 \times 10^{-21}\,\text{J}$, we obtain

$$\frac{k_B T}{P \Lambda^3} = \frac{4.2938 \times 10^{-21}\,\text{J}}{0.2 \times 10^5\,\text{N m}^{-2}(5.3859 \times 10^{-33})\,\text{m}^3} = 3.9861 \times 10^7.$$

Combining this result with the internal state factor

$$\frac{T}{2\Theta_{rot}} \frac{\omega_{el}}{1 - e^{-\Theta_{vib}/T}} = 225.52$$

gives the chemical potential for O_2 at temperature $T \simeq 311\,\text{K}$ and pressure $P = 0.2$ bar as

$$\mu_{gas}(O_2) = -98.4139 \times 10^{-21}\,\text{J} = -0.6119\,\text{eV}.$$

This value for μ_{gas} then gives $\Xi = 1 + e^{-37.3328(eV)^{-1}(-0.09eV)} = 29.7877$, or $p(\text{occup}) \simeq 96.6\%$ at $T \simeq 311\,\text{K}$ for the single-site haem model. $\square$

3.4   The Isothermal-Isobaric Ensemble

For an ensemble of systems in which the containing walls of the individual systems are both heat conducting and flexible, each system can be characterized by values of temperature T and pressure P. This requires us to place constraints upon the total energy and the total volume for the ensemble: the partition function for this isothermal-isobaric ensemble is traditionally designated by $\Delta(T, P)$.

The isothermal-isobaric ensemble can be developed in much the same way as the grand ensemble by considering the system to consist of two subsystems, one small, the other (typically designated as the reservoir) large. The probability for the small subsystem (labelled by subscript 1) will thus have the form

$$p(\epsilon_1, V_1)\, d\epsilon_1 = \frac{\Omega_1(\epsilon_1, V_1)\Omega_2(\epsilon_2, V_2)}{\Omega(\epsilon, V)}\, d\epsilon_1 , \tag{3.4.1a}$$

with $\Omega(\epsilon, V)$ defined by the requirement that $\int_0^V dV_1 \int_0^\epsilon d\epsilon_1\, p(\epsilon_1, V_1) = 1$ as

$$\Omega(\epsilon, V) \equiv \int_0^V \int_0^\epsilon \Omega_1(\epsilon_1, V_1)\Omega_2(\epsilon - \epsilon_1, V - V_1)\, d\epsilon_1 dV_1 . \tag{3.4.1b}$$

The most probable energy ϵ_{1m} and most probable volume V_{1m} for the small subsystem are determined, respectively, by

$$\left(\frac{\partial \ln p}{\partial \epsilon_1}\right)_{V_1}\Bigg|_{\epsilon_1 = \epsilon_{1m}} = 0 \tag{3.4.2a}$$

and by

$$\left(\frac{\partial \ln p}{\partial V_1}\right)_{\epsilon_1}\Bigg|_{V_1 = V_{1m}} = 0 . \tag{3.4.2b}$$

We shall first examine the conditions that correspond to the system of interest having the most probable energy ϵ_{1m}.

Given expression (3.4.1a) for $p_1 \equiv p(\epsilon_1, V_1)$, we may restate the constraint (3.4.2a) as

$$\frac{\partial \ln p_1}{\partial \epsilon_1}\Bigg|_{\epsilon_1 = \epsilon_{1m}} = \frac{\partial \ln \Omega_1}{\partial \epsilon_1}\Bigg|_{\epsilon_1 = \epsilon_{1m}} + \frac{\partial \ln \Omega_2}{\partial \epsilon_1}\Bigg|_{\epsilon_1 = \epsilon_{1m}} = 0 . \tag{3.4.3}$$

Notice that no term arises from the denominator of expression (3.4.1a) because $\Omega(\epsilon, V)$, given by Eq. (3.4.1b), does not depend upon ϵ_1. Now, as the total energy $\epsilon = \epsilon_1 + \epsilon_2$ is constant, we know that $\dfrac{\partial}{\partial \epsilon_1} = -\dfrac{\partial}{\partial \epsilon_2}$, so that our condition becomes

$$\frac{\partial \ln \Omega_1}{\partial \epsilon_1}\Bigg|_{\epsilon_1 = \epsilon_{1m}} - \frac{\partial \ln \Omega_2}{\partial \epsilon_2}\Bigg|_{\epsilon_2 = \epsilon_{2m}} = 0 ,$$

or

$$\frac{\partial \ln \Omega_1}{\partial \epsilon_1}\Bigg|_{\epsilon_1 = \epsilon_{1m}} = \frac{\partial \ln \Omega_2}{\partial \epsilon_2}\Bigg|_{\epsilon_2 = \epsilon_{2m}} \equiv \beta . \tag{3.4.4a}$$

This condition is precisely the same as that encountered for both the canonical and grand ensembles, hence its identification with the reciprocal thermal energy $(k_B T)^{-1}$.

By applying an equivalent argumentation to the constraint (3.4.2b), we obtain the result

$$\frac{\partial \ln \Omega_1}{\partial V_1}\bigg|_{V_1=V_{1m}} = \frac{\partial \ln \Omega_2}{\partial V_2}\bigg|_{V_2=V_{2m}} \equiv \zeta \,, \qquad (3.4.4b)$$

which introduces a new parameter that we have designated by the Greek letter zeta, ζ.

Along the same lines of reasoning that we applied in our development of the grand ensemble, let us expand $\ln \Omega_2(\epsilon_2, V_2)$ for the bath in terms of a double Taylor series expansion

$$\ln \Omega_2(\epsilon_2, V_2) = \ln \Omega_2(\epsilon_{2m}, V_{2m}) + \frac{\partial \ln \Omega_2}{\partial \epsilon_2}\bigg|_{\epsilon_2=\epsilon_{2m}} (\epsilon_2 - \epsilon_{2m})$$

$$+ \frac{\partial \ln \Omega_2}{\partial V_2}\bigg|_{V_2=V_{2m}} (V_2 - V_{2m}) + \cdots, \qquad (3.4.5)$$

with the ellipsis representing second- and higher-order partial derivative terms in the expansion. Now, if we employ the total energy and volume constraints $\epsilon_1 + \epsilon_2 = \epsilon_{1m} + \epsilon_{2m}$ and $V_1 + V_2 = V_{1m} + V_{2m}$, to replace $\epsilon_2 - \epsilon_{2m}$ and $V_2 - V_{2m}$ by $-(\epsilon_1 - \epsilon_{1m})$ and $-(V_1 - V_{1m})$, respectively, then utilize the definitions of β and ζ from Eqs. (3.4.4) in our expression for the Taylor expansion of $\ln \Omega_2(\epsilon_2, V_2)$ we find that

$$\ln \Omega_2(\epsilon_2, V_2) \simeq \ln \Omega_2(\epsilon_{2m}, V_{2m}) - \beta(\epsilon_1 - \epsilon_{1m}) - \zeta(V_1 - V_{1m}) + \cdots .$$

Exponentiation of both sides of this equation gives

$$\Omega_2(\epsilon_2, V_2) \simeq \Omega_2(\epsilon_{2m}, V_{2m})e^{\beta\epsilon_{2m}+\zeta V_{2m}} e^{-\beta\epsilon_1-\zeta V_1} \,. \qquad (3.4.6)$$

Upon employing Eq. (3.4.6) for $\Omega_2(\epsilon_2, V_2)$ in expression (3.4.1b) for $\Omega(\epsilon, V)$, we obtain

$$\Omega(\epsilon, V) \simeq \int_0^V dV_1 \int_0^\epsilon d\epsilon_1 \Omega_1(\epsilon_1, V_1)\Omega_2(\epsilon_{2m}, V_{2m})e^{\beta\epsilon_{2m}+\zeta V_{2m}} e^{-\beta\epsilon_1-\zeta V_1}$$

$$= \Omega_2(\epsilon_{2m}, V_{2m})e^{\beta\epsilon_{2m}+\zeta V_{2m}} \int_0^V dV_1 \int_0^\epsilon d\epsilon_1 \Omega_1(\epsilon_1, V_1)e^{-\beta\epsilon_1-\zeta V_1} \,.$$

Because of the exponentials in ϵ_1 and V_1, we may extend both upper integration limits to infinity, to give

$$\Omega(\epsilon, V) = \Omega_2(\epsilon_{2m}, V_{2m})e^{\beta\epsilon_{1m}+\zeta V_{1m}} \int_0^\infty \int_0^\infty \Omega_1(\epsilon_1, V_1)e^{-\beta\epsilon_1-\zeta V_1} d\epsilon_1 dV_1 \,,$$

$$(3.4.7)$$

so that our expression for $p_1(\epsilon_1, V_1)d\epsilon_1$ becomes

$$p_1(\epsilon_1, V_1)d\epsilon_1 = \frac{\Omega_1(\epsilon_1, V_1)\Omega_2(\epsilon_{2m}, V_{2m})e^{\beta\epsilon_{1m}+\zeta_{1m}}\, e^{-\beta\epsilon_1-\zeta V_1}}{\Omega_2(\epsilon_{2m}, V_{2m})e^{\beta\epsilon_{1m}+\zeta V_{1m}}\displaystyle\int_0^\infty\int_0^\infty \Omega_1(\epsilon_1, V_1)e^{-\beta\epsilon_1-\zeta V_1}d\epsilon_1 dV_1}\, d\epsilon_1$$

$$= \frac{\Omega_1(\epsilon_1, V_1)e^{-\beta\epsilon_1-\zeta V_1}}{\Delta(\beta, \zeta)}\, d\epsilon_1, \tag{3.4.8}$$

with the partition function $\Delta(\beta, \zeta)$ defined as

$$\Delta(\beta, \zeta) \equiv \int_0^\infty dV \int_0^\infty dE\, \Omega(E, V)\, e^{-\beta E-\zeta V} = \int_0^\infty z(\beta, V)e^{-\zeta V}\, dV. \tag{3.4.9}$$

The final equality in Eq. (3.4.9) relates $\Delta(\beta, \zeta)$ to the single-particle canonical partition function $z(\beta, V)$ of Eq. (3.2.21). We may now examine the connections between these ensembles and thermodynamics.

3.5 Problems for This Chapter

1. Consider two systems of particles, A and B, such that the particles in both systems have the same ground state energy, ϵ_0, and possess a single excited energy level lying $208.51\,\mathrm{cm}^{-1}$ above ϵ_0. The two particle types differ in that the excited energy levels for system A are nondegenerate, while those for system B are doubly degenerate. Assuming that the Boltzmann distribution law applies to each system in turn, calculate the corresponding (absolute) temperature for which the particles in that system have probability 0.15 of occupying the excited state.

2. Consider now two systems of particles, A' and B', having the same ground and first excited states as systems A and B of Problem 1, respectively, but with the addition in each case of a nondegenerate second excited level lying $208.51\,\mathrm{cm}^{-1}$ above the first excited level. Carry out the same calculations as made in Problem 1 for systems A and B. How do your results for the addition of a second excited level differ from those obtained in Problem 1?

3. How many translational states with energies less than $3k_BT$ are available to neon atoms in thermal equilibrium at temperature $T = 300\,\mathrm{K}$ and contained in a box of volume $10\,\mathrm{cm}^3$?

4. Consider particles that possess a ground state together with three excited energy states separated by $100\,\mathrm{cm}^{-1}$ from adjacent states, and consider three sets of 60 particles each populating these four equally-spaced energy levels as shown in the table below. Check that each set corresponds to the same total energy,

ϵ /cm^{-1}	Number of Particles		
	Set A	Set B	Set C
300	8	5	6
200	9	10	10
100	13	20	17
0	30	25	27

determine which set is the most probable and whether it corresponds to a Boltzmann distribution.

5. The ^{14}N nucleus has a nuclear spin I of 1. It is known from quantum mechanics that a spin-1 particle has three m_I states, corresponding to $m_I = -1, 0, +1$, that are degenerate in the absence of an external magnetic field. In the presence of a magnetic field of strength B, however, the m_I-states have energies $E_{m_I} = -\gamma_I \hbar B m_I$, with γ_I the magnetogyric ratio of the nucleus. In a 4.8 T magnetic field, the energy splitting between the state with $m_I = 0$ and either of the $m_I = \pm 1$ states corresponds to a radio frequency of 14.45 MHz. Determine the fractional populations of the three spin states of ^{14}N for temperatures 298 K and 4 K.

6. Evaluate the translational partition function $z_{\mathrm{tr}}(T, V)$ for a particle having the mass of a hydrogen molecule and confined to a volume of $100 \, \mathrm{cm}^3$ at temperature $T = 298$ K. What value does $z_{\mathrm{tr}}(T, V)$ have for a particle with the mass of a nitrogen molecule under the same conditions?

7. At which temperature are there Avogadro's number of translational states available to a particle having the mass of an O_2 molecule and confined to a one litre container?

8. We have seen for particles in a box having macroscopic dimensions (meaning, specifically, dimensions that greatly exceed the thermal de Broglie wavelength associated with such particles) that the summation over energy states appearing in the expression for the single-particle canonical partition function,

$$z_{\mathrm{PiB}}(\beta, V) \equiv \sum_{\{n_x, n_y, n_z\}} \mathrm{e}^{-\beta \epsilon(n_x, n_y, n_z)},$$

may be replaced by an integration over energy ϵ. Show also, starting from expression (3.2.20), that $z(\beta, V)$ is given equivalently as

$$z(\beta, V) = \frac{V}{h^3} \int_0^\infty \mathrm{e}^{-\beta \epsilon(\mathbf{p})} \, \mathrm{d}\mathbf{p},$$

with $\mathbf{p}$ the particle momentum and $\epsilon(\mathbf{p})$ its energy.

9. Show that the density of states $\Omega(V, \epsilon)$ obtained in Example 3.4 for the translational states of a structureless particle in the presence of a gravitational field reduces to the density of states given in Eq. (3.2.18b).

10. Starting from the expression given in Example 3.4 for the density of states for an ideal structureless gas in a gravitational field such that the single-particle energy is given by $\epsilon(\mathbf{p}, z) = \frac{1}{2}\mathbf{p}^2/m + mgz$, with g the acceleration due to the gravitational field that acts in the z-direction, obtain an expression for the corresponding single-particle partition function $z(T, V)$ for an ideal gas of structureless particles. Show that your expression for $z(T, V)$ becomes $z(T, V) = V/\Lambda^3(T)$ for $g = 0$ (i.e., in the absence of a gravitational field).

11. Particles such as photons, neutrinos, and very high-energy electrons or protons (i.e., electrons or protons whose relativistic energy ϵ greatly exceeds their rest-mass energies) are referred to as 'ultrarelativistic'. Show that the relativistic energy for such particles is related to the magnitude of the (relativistic) momentum, p, by $\epsilon \simeq pc$, with c the speed of light, and utilize the result obtained for $z(\beta, V)$ in Problem 8 to obtain an explicit expression for $z(\beta, V)$ for ultrarelativistic particles. [Hint: make use of the relativistic momentum-energy relation $\epsilon^2 = p^2c^2 + \epsilon_0^2$.]

12. Show that the fractional populations $p_v(T)$ of the SHO vibrational states are independent of the vibrational zero-point energy associated with the SHO and evaluate $z_{SHO}(T)$, accurate to four significant figures, for temperatures $T = 100, 250, 500$ K, for a SHO characterized by the oscillator frequency $\nu_{osc} = 1.0418 \times 10^{13}$ s^{-1}. Construct a table of the fractional populations $p_v(T)$ for this set of temperatures, accurate to three significant figures.

13. The $^2\Pi$ ground electronic term of the NO molecule has two doubly degenerate electronic levels that correspond to the term symbols $^2\Pi_{\frac{3}{2}}$ and $^2\Pi_{\frac{1}{2}}$, with the $^2\Pi_{\frac{3}{2}}$ level lying 121.1 cm^{-1} above the $^2\Pi_{\frac{1}{2}}$ level. For the purposes of this exercise, we shall neglect translational, rotational, and vibrational motions associated with the NO molecular framework, and focus upon its electronic nature. In this sense, the ground-term NO molecule thus provides an almost ideal 2-level system with which to work.

   Obtain a value for the characteristic temperature $\Theta \equiv \Theta_{el}$ for the ground electronic term NO molecule, and obtain expressions for the partition function $z_{NO}(T)$ and the fractional populations $p_0(T)$, $p_1(T)$ for the NO($^2\Pi$) molecule. Show that the fractional populations $p_0(T)$ and $p_1(T)$ are the same as those for a nondegenerate 2-level (i.e., 2-state) system. Calculate the fractional populations of the NO ground levels for temperatures such that T/Θ spans the domain [0,10], and plot them versus the reduced variable T/Θ. What can you conclude about the fractional populations for such a 2-level system when the temperature becomes very large (i.e., $T \gg \Theta$)?

14. Evaluate the electronic partition function $z_{el}(T)$ for NO at temperature $T = 298$ K and determine the temperature T for which $z_{el}(T) = 3$.

15. Show that if the degeneracies associated with the ground and excited levels of a 2-level system differ, then a 2-level system with degeneracies is no longer equivalent to a 2-state system. Carry out an explicit calculation for 2-level system in which the energy splitting between the two levels has the same value that is found for the NO molecule, i.e., $\Delta\epsilon = 121.1$ cm^{-1}, the ground level is

nondegenerate and the excited level has degeneracy 3. Discuss any differences between the behaviours of these two systems.

16. For a microcanonical ensemble the number of microstates associated with a given configuration $\{n_0, n_1, n_2, \ldots\}$ with a total of N particles is given by the expression

$$\Omega(n_0, n_1, n_2, \ldots,) = \frac{N!}{n_0! n_1! n_2! \cdots},$$

with it being understood that the total number of particles is N and the total energy for the configuration is E. Provided that the basic postulate of statistical mechanics holds, we may determine the most probable configuration by finding the maximal value for Ω via setting the differential $\mathrm{d}\ln\Omega$ to zero, while recognizing that any changes to the numbers n_i remain subject to the constraints placed upon the total number of particles and the total energy for the configuration. This requirement can be represented as

$$\mathrm{d}\ln\Omega = \sum_i \left(\frac{\partial \ln\Omega}{\partial n_i}\right) \mathrm{d}n_i = 0,$$

taken together with $\sum_i \mathrm{d}n_i = 0$ and $\sum_i \epsilon_i \mathrm{d}n_i = 0$ required in order that N and E have fixed values: these three separate statements can be combined using the method of Lagrange multipliers as

$$\mathrm{d}\ln\Omega = \sum_i \left[\left(\frac{\partial \ln\Omega}{\partial n_i}\right) + \lambda - \beta\epsilon_i\right] \mathrm{d}n_i = 0.$$

This expression involves two Lagrange multipliers, λ and β, one for each constraint that has been imposed. Show that the Boltzmann distribution follows from this expression.

17. Consider a spin-$\frac{1}{2}$ particle, such as an atom with no net electronic angular momentum, but whose nucleus has an intrinsic (nuclear) spin angular momentum $\mathbf{I}$ and an associated magnetic moment $\boldsymbol{\mu} = g_I \mu_\mathrm{N} \mathbf{I}$, with a constant of proportionality that is the product of g_I, known as the nuclear g-factor (specific to the nucleus) and the nuclear magneton μ_N that determines the order of magnitude of nuclear magnetic moments. The interaction between a magnetic moment and an applied (i.e., external) magnetic field is given by the Zeeman Hamiltonian $\mathcal{H} = -\boldsymbol{\mu} \cdot \mathbf{B}$, which determines the energy eigenstates that the spin particle may access when such a field is applied. The direction of the applied magnetic field $\mathbf{B}$ is typically employed to define the z-axis direction. The nuclear angular momentum energy eigenstates are designated (in Dirac notation) $|I\, m_I\rangle$ and have associated (eigen)energies $E_\mathrm{spin} = -M_I g_I \mu_\mathrm{N} B$.

What are the values for the z-component μ_z of the magnetic moment of the spin-$\frac{1}{2}$ nucleus? If we consider our system of interest to consist of a single such spin particle in the presence of the applied magnetic field $\mathbf{B}$ and in thermal equilibrium with an appropriate reservoir at temperature T, determine the probabilities associated with each of its accessible energy states. Obtain an expression for the average magnetic moment $\langle \mu_z \rangle$ for the spin-$\frac{1}{2}$ particle, and determine and discuss its limiting behaviours for $\beta g_I \mu_N B \gg 1$ and $\beta g_I \mu_N B \ll 1$.

References

1. L.K. Nash, *Chemthermo: A Statistical Approach to Classical Chemical Thermodynamics* (Addison-Wesley, Reading MA, 1972)
2. D.C. Schoepf, Am. J. Phys. **70**, 128 (2002)
3. U.S.A. National Institute for Science and Technology (NIST) website: https://physics.nist.gov/PhysRefData/ASD/levels.form.html
4. P.T. Landsberg, J. Dunning-Davies, D. Pollard, Am. J. Phys. **62**, 712 (1994)
5. W. Greiner, L. Neise, H. Stöcker, *Thermodynamics and Statistical Mechanics* (Springer, New York, 1995)
6. A. Ben Naim, *Statistical Thermodynamics for Chemists and Biochemists* (Plenum, New York, 1992)

Chapter 4
Mean Values and Thermodynamics

This chapter develops expressions relating thermodynamic functions to ensemble partition functions. A thermodynamic function that is determined directly from a given ensemble partition function is termed the characteristic function for that ensemble: for the canonical ensemble, it is the Helmholtz energy. Expressions are obtained for each of the traditional thermodynamic functions in terms of the canonical, grand, and isothermal–isobaric ensemble partition functions and their partial derivatives. The canonical ensemble is shown to give simpler formal expressions for U and A, while the isothermal–isobaric ensemble gives simpler formal expressions for H and G: this correlates with what are referred to in thermodynamics as the 'natural' variables for these state functions. It is established that the statistical expression giving the entropy in terms of probabilities is identical for the canonical and grand ensembles.

4.1 Canonical Ensemble: Closed Systems

In order to make the connection between the canonical ensemble and thermodynamics, let us begin with a determination of the mean energy associated with a single-particle canonical distribution. The mean energy, denoted $\langle E \rangle$, is defined as

$$\langle E \rangle \equiv \sum_r p_r E_r = \frac{\sum_r E_r e^{-\beta E_r}}{\sum_r e^{-\beta E_r}}$$

$$= \sum_r E_r e^{-\beta E_r} / z(\beta) , \qquad (4.1.1)$$

in which p_r is the probability that the particle is found in energy state E_r, and the sum is over *all* accessible energy states of the single-particle system. If we now associate the mean value (ensemble average) of E with the per particle

© Springer Nature Switzerland AG 2021

F. R. W. McCourt, *Statistical Thermodynamics for Pure and Applied Sciences*,

https://doi.org/10.1007/978-3-030-52006-9_4

thermodynamic internal energy u, and note that differentiation of $z(\beta, V)$ with respect to the parameter β leads to the relation

$$\sum_r E_r e^{-\beta E_r} = -\frac{1}{\beta} \frac{\partial z}{\partial \beta},$$

then we find that u is given by

$$u \equiv \langle E \rangle = - \left(\frac{\partial \ln z}{\partial \beta} \right)_V \tag{4.1.2a}$$

$$= k_B T^2 \left(\frac{\partial \ln z}{\partial T} \right)_V. \tag{4.1.2b}$$

We thus have a means for calculating the thermodynamic internal energy u per particle from the partition function z without ever performing the actual summation over states that appears in the definition of $\langle E \rangle$.

We have seen in Sect. 3.2.3 that the canonical partition function $Z_N(T, V)$ for a gas consisting of N indistinguishable ideal gas atoms or molecules is given by Eq. (3.2.27). Moreover, it is clear that the average energy $\langle E \rangle_N$ for a system of N such ideal gas atoms or molecules must be N times that for a single ideal gas atom or molecule, and hence we may anticipate that the thermodynamic internal energy U for a canonical ensemble of N such atoms or molecules will be given by

$$U = Nu = Nk_B T^2 \left(\frac{\partial \ln z}{\partial T} \right)_V, \tag{4.1.3a}$$

with u given by Eq. (4.1.2b). Moreover, as we note from Eq. (3.2.27) that

$$N \left(\frac{\partial \ln z}{\partial T} \right)_V = \left(\frac{\partial \ln Z_N}{\partial T} \right)_V,$$

the thermodynamic internal energy U can also be expressed in the form

$$U = k_B T^2 \left(\frac{\partial \ln Z_N}{\partial T} \right)_V. \tag{4.1.3b}$$

Notice that we have not at this point introduced a subscript N for this partial derivative, since technically N is a parameter for a canonical ensemble, rather than an independent variable. There will be occasions when we shall include N in such defining relations, mainly to remind ourselves of the fact that N has a fixed value for any given canonical ensemble, but in effect adjoining N to our canonical ensemble results in the same spirit that in quantum mechanics one adjoins spin to the results appertaining to the non-relativistic Schrödinger equation.

Example 4.1 Internal energy of an ideal gas in a gravitational field [1].

The single-particle canonical partition function for an ideal gas in a (weak) gravitational field can be shown [see Example 7.1, Chap. 7] to be

$$z(T, V) = \frac{V}{\Lambda^3(T)} \frac{1 - e^{-\eta}}{\eta},$$

in which $\Lambda(T)$ is the thermal de Broglie wavelength of Eq. (3.2.22), while $\eta \equiv \beta m g H$, with m the particle mass, $\beta = (k_B T)^{-1}$, g the acceleration due to gravity, H the height of the gas column, parameterizes the gravitational field. The single-particle internal energy $u(T)$ for this ideal gas is given formally as

$$u(T) = -\left(\frac{\partial \ln z}{\partial \beta} \right)_V,$$

from which the internal energy is obtained as

$$u(T) = \frac{3}{2\beta} - \frac{\partial}{\partial \beta} \ln(1 - e^{-\eta}) + \frac{\partial}{\partial \beta} \ln \eta,$$

or, upon simplification,

$$u(T) = \left[\frac{5}{2} - \frac{\eta}{e^\eta - 1} \right] k_B T.$$

This expression for the internal energy of a structureless particle subjected to a weak gravitational field may be rewritten as

$$u(T) = \tfrac{3}{2} k_B T \left[1 + \tfrac{2}{3} \left(1 - \frac{\eta}{e^\eta - 1} \right) \right],$$

so that the second term within the square brackets explicitly represents the contribution due to the presence of the gravitational field.

For a single particle of mass 4 amu and $H = 1$ km, temperature $T = 298$ K, $g = 9.80665$ m s^{-2} (exactly), the gravitational parameter η has a value of approximately 1.6×10^{-2}, which, in turn, leads to a value of approximately 5.1×10^{-4} for the correction factor $\frac{2}{3}[1 - \eta/(e^\eta - 1)]$ in our expression for the internal energy. As $U(T)$ for an ideal gas is given by $U(T) = Nu(T)$, the same correction factor also applies to $U(T)$. This example illustrates why ordinary thermodynamics does not need to take the Earth's gravitational field into account.   □

Now, let us consider the variance, $\langle (\Delta E)^2 \rangle$, and standard deviation, σ_E, of E, namely

$$\langle (\Delta E)^2 \rangle = \langle E^2 \rangle - \langle E \rangle^2 \equiv \sigma_E^2.$$

We have seen how to evaluate $\langle E \rangle$ from z. We shall now see that we may also evaluate $\langle E^2 \rangle$ in much the same way. Let us start with the definition of $\langle E^2 \rangle$, viz.,

$$\langle E^2 \rangle \equiv \sum_r p_r E_r^2,$$

which becomes

$$\langle E^2 \rangle = \frac{1}{z} \left(\frac{\partial^2 z}{\partial \beta^2} \right)_V$$

upon substitution of p_r from Eq. (3.2.12a) into the defining relation, followed by simplification of the resultant expression. If we now introduce this result as well as that for $\langle E \rangle$ into the defining relation for the variance, we find that

$$\langle (\Delta E)^2 \rangle = \frac{1}{z} \left(\frac{\partial^2 z}{\partial \beta^2} \right)_V - \frac{1}{z^2} \left(\frac{\partial z}{\partial \beta} \right)_V^2 \qquad (4.1.4a)$$

$$= \frac{\partial}{\partial \beta} \left(\frac{1}{z} \frac{\partial z}{\partial \beta} \right)_V = \frac{\partial}{\partial \beta} \left(\frac{\partial \ln z}{\partial \beta} \right)_V,$$

which can be further simplified to

$$\langle (\Delta E)^2 \rangle = \left(\frac{\partial^2 \ln z}{\partial \beta^2} \right)_V. \qquad (4.1.4b)$$

As for the transition from canonical ensemble expression (4.1.2b) for the single-particle internal energy u to the N-particle expression (4.1.2c) for U, we see that the single-particle canonical ensemble expression (4.1.4a) passes directly over into the N-particle canonical ensemble expression

$$\langle (\Delta E)^2 \rangle_N = \left(\frac{\partial^2 \ln Z_N}{\partial \beta^2} \right)_V. \qquad (4.1.4c)$$

Let us now examine the statistical mechanical expression for the heat capacity at constant volume, C_V, starting from its definition in terms of the internal energy U. We obtain

$$C_V \equiv \left(\frac{\partial U}{\partial T} \right)_V = \frac{\partial \beta}{\partial T} \left(\frac{\partial U}{\partial \beta} \right)_V$$

$$= -\frac{1}{k_B T^2} \frac{\partial}{\partial \beta} \left(-\frac{\partial \ln Z_N}{\partial \beta} \right)_V,$$

so that

$$C_V = \frac{1}{k_B T^2} \left(\frac{\partial^2 \ln Z_N}{\partial \beta^2} \right)_V .$$

(4.1.5)

Thus, upon comparison of this expression with Eq. (4.1.4c), we see that

$$k_B T^2 C_V = \langle (\Delta E)^2 \rangle ,$$

(4.1.6)

thereby connecting energy fluctuations to the heat capacity at constant volume.

4.1.1 Thermodynamics from the Canonical Ensemble

Let us consider a system of interest A, together with its associated set of energy states E_r, in thermal equilibrium with a bath (or reservoir) at temperature T. We shall characterize the system, for the moment, by a single external parameter x (such as the volume V, for example). Let us now change x slightly, bring it to a new value $x + dx$, and ask what energy changes occur in our system of interest. The energy levels will be functions of the external parameter, so that we should write $E_r = E_r(x)$. We can see that the energy levels do depend upon such parameters in a straightforward way if we consider the example of the particle-in-a-box. The particle-in-a-box has energy given by

$$\epsilon_{n_x n_y n_z} = \frac{h^2}{8m} \left[\frac{n_x^2}{a^2} + \frac{n_y^2}{b^2} + \frac{n_z^2}{c^2} \right]$$

and clearly depends upon the volume V (through a, b, c) and upon the mass m of the particle. If we change x, then there will be a corresponding change $\Delta_x E_r$ in E_r, which can be expressed as follows:

$$\Delta_x E_r = \left(\frac{\partial E_r}{\partial x} \right) dx .$$

(4.1.7)

We should now compare this result with the usual definition of 'work' in mechanics. The basic concept can be expressed as it was by Newton: 'work' = 'force' × 'displacement'. We can also generalize this concept to say: If there is something (call it X) that changes when something else (i.e., x) changes, then the system tends to move away from the state of energy E_r at which it was in equilibrium. However, if the change is produced very slowly, then the system will do work to remain in that state, and

$$\delta w = dE_r = \frac{\partial E_r}{\partial x} dx \equiv -X_r dx ,$$

(4.1.8)

in which we can now identify X_r as a *generalized force*. If we now make the association of 'macroscopic' with 'average behaviour', then we can take the following steps:

$$\delta W = -\langle X \rangle \, \mathrm{d}x; \qquad \langle X \rangle \equiv \left\langle -\frac{\partial E_r}{\partial x} \right\rangle, \qquad (4.1.9)$$

in which the macroscopic generalized force $\langle X \rangle$ is related to the partition function via

$$\langle X \rangle \equiv \sum_r p_r X_r = \sum_r \frac{e^{-\beta E_r}}{Z(\beta)} \left(-\frac{\partial E_r}{\partial x} \right).$$

We may also write this expression in the form

$$\langle X \rangle = \frac{1}{\beta} \frac{\partial \ln Z}{\partial x}, \qquad (4.1.10)$$

so that ultimately the incremental thermodynamic work can be expressed in terms of the partition function and its derivatives as

$$\delta W = -\frac{1}{\beta Z} \frac{\partial Z}{\partial x} \, \mathrm{d}x = -\frac{1}{\beta} \frac{\partial \ln Z}{\partial x} \, \mathrm{d}x. \qquad (4.1.11)$$

For the moment, we shall consider only the case in which the external parameter is the 'volume'. In this case, we obtain explicitly for the incremental work the expression

$$\delta W = -\frac{1}{\beta} \frac{\partial \ln Z}{\partial V} \, \mathrm{d}V. \qquad (4.1.12a)$$

If there are only volume changes in a system such as a gas, then work is related to the average pressure P and the volume change that occurs, namely

$$\delta W = -P \, \mathrm{d}V. \qquad (4.1.12b)$$

From the above two expressions for the incremental work for pressure–volume changes, we may deduce that the equation relating the pressure to the partition function is

$$P = \frac{1}{\beta} \frac{\partial \ln Z(\beta, V)}{\partial V} = k_B T \frac{\partial \ln Z(\beta, V)}{\partial V}. \qquad (4.1.13)$$

In thermodynamics when we have an expression for the *pressure P* in terms of the temperature and volume, we call that expression the *equation of state* for the system.

In order to complete our connection to thermodynamics, it will be necessary to consider the exact, or total, differential for the partition function $Z(\beta, x)$, which is

$$d[\ln Z(\beta, x)] = \frac{\partial \ln Z}{\partial \beta}\, d\beta + \frac{\partial \ln Z}{\partial x}\, dx$$

$$= -\langle E \rangle\, d\beta + \beta \delta W.$$

If we now make the replacement of $\langle E \rangle$ by the thermodynamic internal energy U, followed by a slight algebraic rearrangement of our expression, we obtain

$$d \ln Z = -U\, d\beta + \beta \delta W$$

$$= -d(\beta U) + \beta(dU + \delta W),$$

from which we have the result

$$d[\ln Z + \beta U] = \beta(dU + \delta W). \tag{4.1.14}$$

Again, returning to thermodynamics, we recall that the change in the 'internal energy' plus the work done by the system is the 'heat' δQ absorbed by the system, so that $\beta(dU + \delta W) = \beta \delta Q$. This suggests that we can identify the differential on the left-hand side with the differential of the *entropy* S via

$$\frac{dS}{k_B} \equiv \beta \delta Q, \quad \text{since} \quad dS = \frac{\delta Q}{T}. \tag{4.1.15}$$

Thus, apart from a constant, we can identify the entropy S as

$$S = k_B(\ln Z + \beta U). \tag{4.1.16}$$

When Eq. (4.1.14) is combined with Eqs. (4.1.16) and (4.1.12b), it provides an expression for the combined first and second laws of thermodynamics. Note also that this process has, in addition, given us a microscopic definition of entropy.

Example 4.2 Schottky defects in ionic crystals.

The simplest type of defect in a crystal is a lattice vacancy, in which an atom (or ion) is transferred from a site within the interior of the crystal to its surface. This process requires energy, but it is compensated for by the corresponding increase in the (configurational) entropy of the crystal. Such a simple defect is known as a *Schottky defect*. We shall consider only a simple ionic crystal that initially contains N positive and N negative ions, all with equal absolute charges, and located at the $2N$ crystal lattice sites. In order to retain the net electrical neutrality of the crystal, Schottky defects typically occur in ionic crystals as separated pairs of positive and negative ions.

We shall consider only Schottky defects that appear in the form of n ion pairs at the surface of the crystal. We shall assume that an energy E_f must be expended in the formation of each such ion pair within the lattice and in transporting it to the crystal surface. We shall further assume that the concentration of these defects is sufficiently small that the creation of any one of them may be treated as an independent event.

With these assumptions, the number of ways $\Omega(n)$ of creating n ion-pair defects in the ionic crystal is given by the product of the numbers of ways of independently choosing n positive ions and n negative ions, each from a set of N such ions, i.e.,

$$\Omega(n) = \frac{N!}{n!(N-n)!} \cdot \frac{N!}{n!(N-n)!}.$$

The (configurational) entropy S associated with the formation of these n ion-pair defects will thus be given in terms of $\Omega(n)$ as

$$S(n) = k_B \ln \Omega(n) = k_B \ln \left[\frac{N!}{n!(N-n)!} \right]^2 ,$$

or

$$S(n) = 2k_B \ln \left[\frac{N!}{n!(N-n)!} \right].$$

The Helmholtz energy $A \equiv U - TS$ associated with the formation of n such Schottky defects is then given by $A(n) = U(n) - TS(n)$, in which the internal energy is given in terms of the energy E_f associated with the formation of a single ion-pair defect as $U(n) = nE_f$, and the configurational entropy, $S(n)$, as

$$A(n) = U(n) - TS(n)$$

$$= nE_f - 2k_BT[\ln N! - \ln n! - \ln(N-n)!].$$

Minimization of the expression for $A(n)$ with respect to n will provide an optimized value $\bar{n}$ for the number of Schottky defects. Thus, we must set the derivative of A with respect to n to zero and solve for the value $\bar{n}$ of n that minimizes $A(n)$. From our expression

$$A(n) = nE_f - 2k_BT[\ln N! - \ln n! - \ln(N-n)!]$$

for $A(n)$, we obtain

$$\left. \frac{dA}{dn} \right|_{n=\bar{n}} = E_f - 2k_BT[-\ln\bar{n} + \ln(N-\bar{n})] \equiv 0.$$

We may solve this equation for $(N - \bar{n})/\bar{n}$ as

$$\frac{N - \bar{n}}{\bar{n}} = e^{\frac{1}{2}\beta E_f}.$$

Upon neglecting $\bar{n}$ relative to N, we arrive at the approximation

$$\bar{n} = Ne^{-\frac{1}{2}\beta E_f},$$

for $\bar{n}$.   □

If we now return to expression (4.1.16), we can rewrite it in the form

$$U - TS = -k_B T \ln Z = A. \tag{4.1.17}$$

This relates the Helmholtz energy A to the partition function and the temperature T in a very direct fashion. Equivalently, we may rewrite Eq. (4.1.17) to obtain the partition function Z in terms of the Helmholtz energy A as

$$Z = e^{-\beta A}. \tag{4.1.18}$$

This expression directly relates the canonical partition function to a single thermodynamic state function (in this case, the Helmholtz energy), and possibly the temperature. When we are able to find a thermodynamic state function that is related to the partition function for a particular ensemble in such a simple manner, we say that that state function is the 'characteristic function' for the particular ensemble. The Helmholtz energy function is thus known as the characteristic state function for the canonical ensemble.

Let us now reconsider the entropy and its statistical mechanical interpretation. We might well ask, for example, 'What happens to S when $T \rightarrow 0$?'. To see what does happen, we note that $T \rightarrow 0$ is the same thing as $\beta \rightarrow \infty$, which means that summations of the type $\sum_r e^{-\beta E_r}$ become simply one term, namely $e^{-\beta E_0}$, so that the partition function reduces also to a single term, the internal energy U becomes E_0, and the entropy becomes $k_B \ln \Omega_0$, or

$$Z \underset{T \rightarrow 0}{\rightarrow} \Omega_0 e^{-\beta E_0}, \tag{4.1.19}$$

$$U \underset{T \rightarrow 0}{\rightarrow} U_0 = E_0, \tag{4.1.20}$$

$$S \underset{T \rightarrow 0}{\rightarrow} k_B \ln \Omega_0. \tag{4.1.21}$$

Note that the final of these three expressions tells us that at a temperature of absolute zero the entropy is a constant, independent of all parameters characterizing the system. This is nothing other than a version of the *Third Law of Thermodynamics*.

Now it is useful to re-examine the relationship between S, U, and $\ln Z$:

$$\frac{U}{T} + k_B \ln Z = S = k_B \left(\ln Z + \beta \sum_r p_r E_r \right)$$

$$= k_B \left[\ln Z + \sum_r p_r (\beta E_r) \right],$$

which can be reduced to

$$S = -k_B \sum_r p_r \ln p_r . \tag{4.1.22}$$

We shall see shortly that this result is *one of the most important fundamental results obtained from statistical mechanics*.

To complete the connection between statistical mechanics and thermodynamics, we begin with the observation that in thermodynamics, if any pair of thermodynamic state functions is given, expressions for all other thermodynamic state functions can be obtained from them by various manipulations of partial derivatives. Because the Helmholtz energy A is the characteristic thermodynamic function for the canonical ensemble, it behooves us to see how to obtain all other state functions from A plus one additional state function, which we shall choose to be the internal energy U. This will allow us to make maximal use of Eq. (4.1.17), thereby giving us canonical ensemble expressions for the various thermodynamic functions.

The differential form of the combined first and second laws of thermodynamics gives the total differential dU of the internal energy $U(S, V)$ for a closed thermodynamic system as

$$dU = T dS - P dV ,$$

from which we see that the intensive thermodynamic variables T and P can be defined in principle in terms of partial derivatives of the internal energy with respect to the extensive thermodynamic variables S and V as

$$T = \left(\frac{\partial U}{\partial S} \right)_V \quad \text{and} \quad P = -\left(\frac{\partial U}{\partial V} \right)_S .$$

However, these definitions for T and P are not particularly useful because they both involve the entropy S as an independent variable, and it, unlike volume, is not a quantity for which we have an intuitive feeling. We shall see shortly that this need not cause us any serious concern.

To work with an open thermodynamic system, we need to extend the combined first and second laws expression for dU to allow for mass exchange between different phases of the substance that we are describing thermodynamically. This is achieved by including for a system having two phases in thermodynamic equilibrium, for example, a term $\bar{\mu} dn$, in which $\bar{\mu}$ is termed the (molar) *chemical potential*, and dn is the number of moles of the system substance exchanged between its two phases.[1] As we are working in terms of the number N of atoms or molecules and the chemical potential μ associated with an individual atom or molecule, we shall add μdN to the above expression for dU, to obtain

$$dU = TdS - PdV + \mu dN. \tag{4.1.23}$$

From the thermodynamic definition $A \equiv U - TS$ of the Helmholtz energy A, we obtain the total differential dA of A as

$$dA = dU - TdS - SdT,$$

which, upon utilizing the form (4.1.23) for dU, gives

$$dA = -SdT - PdV + \mu dN. \tag{4.1.24}$$

This expression for the total differential of the characteristic thermodynamic function for the canonical ensemble identifies the independent variables for the thermodynamic system as T, V, and N and leads directly to the connecting relations between S, P, μ, and the canonical partition function $Z(N, T, V)$, namely,

$$S(N, T, V) = -\left(\frac{\partial A}{\partial T}\right)_{N,V} = k_B \ln Z(N, T, V) + k_B T \left(\frac{\partial \ln Z}{\partial T}\right)_{N,V},$$
$$\tag{4.1.25a}$$

$$P = -\left(\frac{\partial A}{\partial V}\right)_{N,T} = k_B T \left(\frac{\partial \ln Z}{\partial V}\right)_{N,T}, \tag{4.1.25b}$$

and

$$\mu(N, T, V) = \left(\frac{\partial A}{\partial N}\right)_{V,T} = -k_B T \left(\frac{\partial \ln Z}{\partial N}\right)_{T,V}. \tag{4.1.25c}$$

Now if we also recall that the internal energy U is given by

$$U = k_B T^2 \left(\frac{\partial \ln Z}{\partial T}\right)_{N,V} \tag{4.1.26}$$

and that the enthalpy H and Gibbs energy G are related to U, PV, and TS by

$$H = U + PV, \qquad G = H - TS, \tag{4.1.27}$$

we see how to proceed in principle to obtain all of thermodynamics from a knowledge of the canonical partition function Z.

---

[1]For a more extensive discussion of the role of the chemical potential, see Chap. 9 of *Molecular Thermodynamics*, D. A. McQuarrie and J. D. Simon [University Science Books, Sausalito, CA, 1999].

4.1.2   Canonical Partition Function for a Mixture

We have seen in Sect. 3.2.3 that the canonical partition function $Z_{AB}(N, T, V)$ for a binary ideal gas mixture of atoms A and B is given via Eq. (3.2.25) in terms of the canonical partition functions $z_A(N_A, T, V)$ and $z_B(N_B, T, V)$ for atoms A and B as

$$Z_{AB}(N, T, V) = \frac{z_A^{N_A}(T, V) z_B^{N_B}(T, V)}{N_A! N_B!}; \quad N = N_A + N_B. \tag{4.1.28}$$

If we now utilize the Stirling approximation (see Appendix C) to write $\ln Z_{AB}$ as

$$\ln Z(N, T, V) = N_A \ln z_A + N_B \ln z_B - N_A \ln N_A + N_A - N_B \ln N_B + N_B,$$

we may obtain expressions for the chemical potential $\mu_A(N_A, T, V)$, for example, as

$$\mu_A \equiv -k_B T \left(\frac{\partial \ln Z}{\partial N_A} \right)_{V, T, N_B}$$

$$= -k_B T \ln \left(\frac{z_A}{N_A} \right). \tag{4.1.29}$$

A completely analogous expression can also be obtained for the chemical potential $\mu_B(N_B, T, V)$ for constituent B.

More generally, the canonical partition function Z_{mix} for a mixture of M chemically distinct species is given by

$$Z_{mix}(T, V_{mix}, N_1, \cdots, N_M) = \prod_{\alpha=1}^{M} Z_\alpha(T, V_{mix}, N_\alpha), \tag{4.1.30}$$

with V_{mix} the volume of the mixture, and the individual canonical partition functions $Z_\alpha(T, V_{mix}, N_\alpha)$ given by Eq. (3.2.27) in terms of the partition function $z_\alpha(T, V_{mix})$ for chemical species α as

$$Z_\alpha(T, V_{mix}, N_\alpha) \equiv \frac{z_\alpha^{N_\alpha}(T, V_{mix})}{N_\alpha!}. \tag{4.1.31}$$

Upon utilizing Eq. (4.1.3b) to obtain the internal energy U_{mix} in terms of the temperature derivative of $\ln Z_{mix}$, we obtain the expression

$$U_{mix}(T, V_{mix}, N_1, \cdots, N_M) = k_B T^2 \sum_{\alpha=1}^{M} \left(\frac{\partial \ln Z_{mix}}{\partial T} \right)_{T, V_{mix}, N_\beta \neq N_\alpha}$$

$$= \sum_{\alpha=1}^{M} U_\alpha(T, V_{mix}, N_\alpha). \tag{4.1.32}$$

We can see from this expression that the internal energy for a multicomponent mixture is thus given as the sum of the internal energies of the component species. As $C_{V,\text{mix}}(T)$ is simply the temperature derivative (at constant volume) of U_{mix}, the same property clearly holds for it.

A similar argument applies to the pressure P_{mix} in a multicomponent mixture, as it is defined through an extension of Eq. (4.1.13) as

$$P_{\text{mix}} = k_B T \left(\frac{\partial Z_{\text{mix}}}{\partial V_{\text{mix}}}\right)_{T,N_{N_\alpha}}$$

$$= k_B T \sum_{\alpha=1}^{M} \left(\frac{\partial \ln Z_\alpha}{\partial V_{\text{mix}}}\right)_{T,N_\alpha} \tag{4.1.33a}$$

or

$$P_{\text{mix}} = \sum_{\alpha=1}^{M} P_\alpha . \tag{4.1.33b}$$

For a mixture of M ideal gases, expression (4.1.33b) will be recognized as a statement of Dalton's law of partial pressures.

If we examine the Helmholtz energy $A_{\text{mix}}(T, V_{\text{mix}}, N_1, \cdots, N_M)$ for a multi-component mixture beginning with the generalization of Eq. (4.1.17) to

$$A_{\text{mix}}(T, V_{\text{mix}}, N_1, \cdots, N_M) = -k_B T \ln Z_{\text{mix}}(T, V_{\text{mix}}, N_1, \cdots, N_M), \tag{4.1.34a}$$

we see that the properties of the natural logarithm give directly

$$A_{\text{mix}}(T, V_{\text{mix}}, N_1, \cdots, N_M) = \sum_{\alpha=1}^{M} A_\alpha(T, V_{\text{mix}}, N_\alpha). \tag{4.1.34b}$$

Multicomponent mixture expressions for the other thermodynamic state functions may be obtained in a similar fashion.

The chemical potential $\mu_\alpha(T, V_{\text{mix}}, N_\alpha)$ is defined by Eq. (4.1.29) for $\alpha = A$, and more generally as

$$\mu_\alpha(T, V_{\text{mix}}, N_\alpha) = -k_B T \left(\frac{\partial \ln Z_{\text{mix}}}{\partial N_\alpha}\right)_{T,V_{\text{mix}},N_\beta \neq N_\alpha}$$

$$= -k_B T \left(\frac{\partial \ln Z_\alpha}{\partial N_\alpha}\right)_{T,V_{\text{mix}}}. \tag{4.1.35}$$

It is clear from this expression for $\mu_\alpha(T, V_{\text{mix}}, N_\alpha)$ that it is not possible to define a chemical potential μ_{mix} for a mixture as a whole and that the individual chemical potentials for the components of the mixture, unlike the chemical potential for a pure substance, do not simply represent the Gibbs energies per component particle.

4.1.3   Microcanonical Ensemble: Isolated System

The microcanonical partition function is $\Omega(E, N, V)$. The microcanonical ensemble is concerned with an isolated system having fixed E, N, V. It can also be obtained as a limiting case of the canonical ensemble. In fact, it is actually a degenerate canonical ensemble in which all systems have the same energy, i.e., we pick out just those systems in a canonical ensemble whose energy is E, remove them from contact with systems having energies different from E, and place them in thermal isolation. In the ensemble so constructed, the probability p_j is given as $e^{-E_j/k_B T}$, and since all E_j are the same for all $\Omega(E, N, V)$ states, and since $\sum_j p_j = 1$, we have

$$p_j = \frac{1}{\Omega} \quad \forall\ j. \tag{4.1.36}$$

Also, because $E_j \equiv E \ \forall\ j$, we have that

$$\langle E \rangle = U = E. \tag{4.1.37}$$

Then, because S is related to the microcanonical ensemble probabilities p_j by $S = -k_B \sum_j p_j \ln p_j$, we have

$$S(U, N, V) = -k_B \sum_j p_j \ln p_j = -k_B \sum_j p_j \ln \frac{1}{\Omega},$$

and hence

$$S = k_B \ln \Omega(U, N, V), \tag{4.1.38}$$

which tells us that the characteristic thermodynamic function for the microcanonical ensemble is the entropy S. Boltzmann wrote $S = k_B \ln W$ and called W the 'number of complexions'.

The result (4.1.38) gives the relation between the quantum mechanical Ω and the entropy S. We note that this expression applies to an *isolated* system. Moreover, we see that should the ground state of the system be nondegenerate, then $\Omega \to 1$ when $T \to 0$, with the consequence that $S \to 0$. For any isolated system, the greater the number of quantum states available to it, the higher will be the entropy: this line of reasoning is the origin of qualitative statements that correlate entropy with 'probability', 'randomness', and 'disorder'. Let us also note that the order of magnitude of $\Omega(E, N, V)$ can be determined from the inverse of Eq. (4.1.38), namely $\Omega = e^{S/k_B}$, by substituting in a typical experimental value for the entropy. A reasonable experimental value for the entropy is $S \simeq N k_B$, so that for finite temperatures, we see that

$$\ln \Omega = N \quad \Rightarrow \quad \Omega \simeq e^N \sim e^{10^{23}},$$

which is indeed, as we might have suspected, a rather large number.

From our statistical mechanical relation,

$$k_B \, d \ln \Omega = dS = \frac{1}{T} \, dU + \frac{P}{T} \, dV \,, \tag{4.1.39}$$

we see that

$$\frac{1}{k_B T} = \left(\frac{\partial \ln \Omega}{\partial U} \right)_V \,, \qquad \bullet \tag{4.1.40a}$$

$$P = k_B T \left(\frac{\partial \ln \Omega}{\partial V} \right)_U \,, \tag{4.1.40b}$$

as the relationships between the microcanonical partition function $\Omega(U, N, V)$ and β and the pressure P.

Example 4.3 Entropy of Mixing.

In Example 2.6, Chap. 2, we examined the entropy difference, $(\Delta S)_{i \to f}$, between an initial thermodynamic state of an isolated system in which two ideal gases A and B are separated by an impermeable membrane into compartments of volumes V_1 (with N_A particles A) and V_2 (with N_B particles B) and a final thermodynamic state in which the membrane has been removed, with each gas accessing the total system volume $V_1 + V_2$. From Eq. (4.1.38), we have $S = k_B \ln \Omega$, so that the entropy for the initial thermodynamic state may be expressed as

$$S_i \equiv S_A + S_B = k_B \ln \Omega(U, V_1, N_A) + k_B \Omega(U, V_2, N_B) \,,$$

while the entropy for the final thermodynamic state may be written as

$$S_f = S_{A \oplus B} = k_B \ln \Omega(U, V_1 + V_2, N_A + N_B) \,,$$

and hence $(\Delta S)_{i \to f}$ becomes

$$(\Delta S)_{i \to f} = k_B [\ln \Omega(U, V_1 + V_2, N_A + N_B) - \ln \Omega(U, V_1, N_A)$$
$$- \ln \Omega(U, V_2, N_B)]$$
$$= k_B \ln \left[\frac{\Omega(U, V_1 + V_2, N_A + N_B)}{\Omega(U, V_1, N_A)\Omega(U, V_2, N_B)} \right] \,.$$

The number of microstates $\Omega(U, V_1 + V_2, N_A, N_B)$ available to the $N_A + N_B$ ideal gas particles in the final volume $V_1 + V_2$ will be the product of the numbers of microstates available to the N_A ideal gas particles in volume V_1 and N_B ideal gas particles in volume V_2 times the number of additional microstates associated with the distinguishability of the A and B particles. We may therefore write

$$\Omega(U, V_1 + V_2, N_A + N_B) = \Omega(U, V_1, N_A)\Omega(U, V_2, N_B) \frac{(N_A + N_B)!}{N_A! N_B!},$$

which gives $(\Delta S)_{i \to f}$ as

$$(\Delta S)_{i \to f} = k_B \ln \left[\frac{(N_A + N_B)!}{N_A! N_B!} \right]$$

$$= k_B [\ln((N_A + N_B)! - \ln N_A! - \ln N_B!].$$

For N_A and N_B sufficiently large, we may utilize the Stirling approximation for $\ln N!$ to obtain

$$(\Delta S)_{i \to f} = (N_A + N_B) k_B [\ln(N_A + N_B) - x_A \ln N_A - x_B \ln N_B],$$

with x_A and x_B the mole fractions of gases A and B in the gas mixture.

If we now identify $(\Delta S)_{i \to f}$ as the entropy of mixing, $(\Delta S)_{\text{mix}}$, we obtain

$$(\Delta S)_{\text{mix}} = (N_A + N_B) k_B \ln \left[\frac{(N_A + N_B)^{x_A + x_B}}{N_A^{x_A} N_B^{x_B}} \right]$$

$$= (N_A + N_B) k_B \ln \left[\left(\frac{N_A + N_B}{N_A} \right)^{x_A} \left(\frac{N_A + N_B}{N_B} \right)^{x_B} \right],$$

which is the same result,

$$(\Delta S)_{\text{mix}} = -(N_A + N_B) k_B [x_A \ln x_A + x_B \ln x_B],$$

obtained via purely thermodynamic arguments. Thus, we may conclude that the entropy of mixing can be associated with the additional microstates arising from the distinguishability of the A and B ideal gas particles. □

4.2 Grand Ensemble: Open Systems

If we examine the form of the grand partition function Ξ,

$$\Xi = \sum_{N=0}^{\infty} e^{-\gamma N} \int_0^{\infty} \Omega(\epsilon_N) e^{-\beta \epsilon_N} \, d\epsilon_N, \tag{4.2.1}$$

we see that we may also write it in the form

$$\Xi = \sum_{N=0}^{\infty} e^{-\gamma N} \, Z(\beta, N) \tag{4.2.2a}$$

$$\equiv \sum_{N} \lambda^N Z_N(\beta), \tag{4.2.2b}$$

in which $Z(\beta, N) \equiv Z_N(\beta)$ is the canonical partition function for N structureless particles, and the quantity $e^{-\gamma} \equiv \lambda$ is called the *absolute activity* (we shall see shortly that $\lambda = e^{\beta\mu}$). This form is in keeping with our interpretation of the grand ensemble as an ensemble of canonical ensembles. We shall therefore also write Ξ more explicitly as

$$\Xi(\lambda, T, V) = \sum_{N} \lambda^N Z_N(T, V). \tag{4.2.3}$$

Note that if we combine Eq. (4.2.3) for $\Xi(T, V, \lambda)$ with Eq. (3.2.27) giving the canonical partition function $Z_N(\beta, V)$ for N indistinguishable atoms/molecules in terms of the corresponding single atom/molecule canonical partition function $z(T, V)$, we obtain the relation

$$\Xi(T, V, \lambda) = \sum_{N=0}^{\infty} \frac{\lambda^N}{N!} z^N(T, V) \tag{4.2.4}$$

connecting $\Xi(T, V, \lambda)$ and $z(T, V)$. As the summation over N on the right-hand side of this expression is precisely the infinite series representation for the exponential of $\lambda z(T, V)$, we see that $\Xi(T, V, \lambda)$ and $z(T, V)$ are related succinctly by

$$\Xi(T, V, \lambda) = e^{\lambda z(T, V)}. \tag{4.2.5}$$

The probability p_{N, r_N} of occurrence of a particular energy state ϵ_{r_N} in the grand ensemble will thus be expressible as the ratio of the corresponding term in the fully discretized form of Eqs. (3.3.17) to Ξ as

$$p_{N, r_N} = \frac{\lambda^N e^{-\beta \epsilon_{r_N}}}{\Xi} = \frac{e^{-\beta \epsilon_{r_N} - \gamma N}}{\Xi}. \tag{4.2.6}$$

We note that the probability p_N associated with a member canonical ensemble of the grand ensemble is given by

$$p_N \equiv \sum_{r_N} p_{N, r_N} = \frac{Z_N(\beta) \lambda^N}{\Xi} \tag{4.2.7}$$

and that the macroscopic thermodynamic (or ensemble average) number $\overline{N}$ is defined as

$$\overline{N} \equiv \sum_N p_N \, N \,. \tag{4.2.8}$$

4.2.1   Thermodynamics from the Grand Ensemble

The average energy $\langle E \rangle$ in the grand ensemble is defined as

$$\langle E \rangle = \sum_{N,r_N} p_{N,r_N} E_{r_N} = \frac{1}{\Xi} \sum_{N,r_N} \epsilon_{r_N} e^{-\beta \epsilon_{r_N}} e^{-\gamma N} \,. \tag{4.2.9}$$

Identification of the ensemble average energy $\langle E \rangle$ with the thermodynamic internal energy U then leads, in much the same manner as Eq. (4.1.2a) was obtained from Eq. (4.1.1), to

$$U = -\left(\frac{\partial \ln \Xi}{\partial \beta} \right)_{\lambda, V} \,. \tag{4.2.10}$$

From the statistical mechanical defining relation $U = \sum_{N,r_N} p_{N,r_N} E_{r_N}$ for the internal energy, we see that the differential of U is given in general by

$$dU = \sum_{N,r_N} p_{N,r_N} \, d\epsilon_{r_N} + \sum_{N,r_N} \epsilon_{r_N} \, dp_{N,r_N} \,. \tag{4.2.11}$$

We shall now examine the second term in this expression in greater detail by making use of the fact that, for the grand ensemble, the expression $p_{N,r_N} = e^{-\beta \epsilon_{r_N}} \lambda^N / \Xi$ implies that

$$\ln p_{N,r_N} = -\beta \epsilon_{r_N} + N \ln \lambda - \ln \Xi \,, \tag{4.2.12}$$

from which we can express the energies ϵ_{r_N} as

$$\epsilon_{r_N} = \frac{1}{\beta} N \ln \lambda - \frac{1}{\beta} \ln \Xi - \frac{1}{\beta} \ln p_{N,r_N} \,.$$

Using this result in the second term of Eq. (4.2.11) gives

$$\sum_{N,r_N} \epsilon_{r_N} \, dp_{N,r_N} = -\frac{1}{\beta} \sum_{N,r_N} [\ln p_{N,r_N} + \ln \Xi - N \ln \lambda] \, dp_{N,r_N}$$

$$= -\frac{1}{\beta} \sum_{N,r_N} \ln p_{N,r_N} \, dp_{N,r_N} - \frac{\gamma}{\beta} \sum_{N,r_N} N \, dp_{N,r_N} , \qquad (4.2.13)$$

as the $\ln \Xi$ term vanishes because $\sum_{N,r_N} dp_{N,r_N} = 0$, while λ and γ are related by $\lambda = e^{-\gamma}$.

Now, since N is always integer, and the average number of particles $\overline{N}$ is given by

$$\overline{N} \equiv \sum_{N} p_N N , \qquad (4.2.14)$$

any infinitesimal change in $\overline{N}$ must come as a result of changes in the values of p_N, so that[2]

$$d\overline{N} = \sum_{N} N \, dp_N . \qquad (4.2.15)$$

As a consequence of this result, Eq. (4.2.13) becomes

$$\sum_{N,r_N} \epsilon_{r_N} \, dp_{N,r_N} = -\frac{1}{\beta} d \left(\sum_{N,r_N} p_{N,r_N} \ln p_{N,r_N} \right) - \frac{\gamma}{\beta} d\overline{N} . \qquad (4.2.16)$$

From Eqs. (4.2.16) and (4.2.11), we see that

$$dU \stackrel{\text{stats}}{=} -\frac{1}{\beta} d \left(\sum_{N,r_N} p_{N,r_N} \ln p_{N,r_N} \right) + \delta W - \frac{\gamma}{\beta} d\overline{N} \qquad (4.2.17)$$

$$\stackrel{\text{thermo}}{=} \delta Q_{\text{rev}} + \delta W + \mu \, dN , \qquad (4.2.18)$$

where we have now associated the thermodynamic variable N and its differential dN with the statistical mechanical quantities $\overline{N}$, $d\overline{N}$, respectively. From the comparison between Eqs. (4.2.17) and (4.2.18), we obtain

$$\delta Q_{\text{rev}} = T \, dS = -\frac{1}{\beta} d \left(\sum_{N,r_N} p_{N,r_N} \ln p_{N,r_N} \right) = -k_B T \, d \left(\sum_{N,r_N} p_{N,r_N} \ln p_{N,r_N} \right) ,$$

or

---

[2]Note also that even were N not an integer, its value is fixed for the individual canonical ensembles that make up the grand ensemble.

$$dS = -k_B\, d\left(\sum_{N,r_N} p_{N,r_N} \ln p_{N,r_N}\right) \quad \Rightarrow \quad S = -k_B \sum_{N,r_N} p_{N,r_N} \ln p_{N,r_N}\,.$$

$$(4.2.19)$$

Notice that *this statistical expression for the entropy has the same form as that obtained earlier* for the canonical ensemble. In fact, *this remains the case for every ensemble*, including those describing nonequilibrium situations.

We also see from Eqs. (4.2.17) and (4.2.18) that the thermodynamic chemical potential μ is related to γ by

$$\mu = -\frac{\gamma}{\beta} = -k_B T \gamma \qquad \text{or} \qquad \gamma = -\beta\mu \equiv -\mu(k_B T)^{-1}\,, \qquad (4.2.20)$$

so that the absolute activity λ becomes[3]

$$\lambda = e^{-\gamma} = e^{\beta\mu}\,. \qquad (4.2.21)$$

Now, from expression (4.2.19) for the entropy, together with Eq. (4.2.12), we find that

$$S = -k_B \sum_{N,r_N} p_{N,r_N}[-\beta\epsilon_{r_N} - \gamma N - \ln \Xi]$$

$$= \frac{U}{T} + k_B \gamma \overline{N} + k_B \ln \Xi\,,$$

or

$$TS = U - \mu\overline{N} + k_B T \ln \Xi\,. \qquad (4.2.22)$$

Thus, we may express $k_B T \ln \Xi$ as

$$k_B T \ln \Xi = TS - U + \mu\overline{N} = \mu\overline{N} - A\,. \qquad (4.2.23)$$

If we now consider the Gibbs energy G, defined in thermodynamics by

$$G = A + PV\,, \qquad (4.2.24)$$

then its differential is

$$dG = dA + P\, dV + V\, dP\,,$$

---

[3]It is a common practice in quantum mechanical treatments to define a dimensionless parameter $\alpha \equiv \beta\mu = -\gamma$ so that the absolute activity is written as $\lambda = e^{\alpha}$.

and, since $dA = -S\,dT - P\,dV + \mu\,d\overline{N}$, dG can be rewritten as

$$dG = -S\,dT + V\,dP + \mu\,d\overline{N},$$

from which we see that the chemical potential for a pure system is given by

$$\left(\frac{\partial G}{\partial \overline{N}}\right)_{T,P} = \mu. \qquad (4.2.25)$$

Now, G is an extensive thermodynamic variable, as is $\overline{N}$, while T and P are both intensive thermodynamic variables, so that if we write $G(\overline{N}, P, T)$ as

$$G(\overline{N}, P, T) = \overline{N}g(P, T),$$

we obtain the result that

$$\left(\frac{\partial G}{\partial \overline{N}}\right)_{T,P} = g(P, T) = \mu,$$

and hence

$$G = \overline{N}\mu. \qquad (4.2.26)$$

We thus see that the chemical potential is simply the Gibbs energy per particle. This result, together with Eq. (4.2.23), gives us the relation

$$k_B T \ln \Xi = \overline{N}\mu - A = G - A = PV, \qquad (4.2.27)$$

from which we see that PV is the characteristic function for the grand ensemble.
  From expression (4.2.27), we see that the differential of PV takes the form

$$d(PV) = dG - dA = \overline{N}\,d\mu + S\,dT + P\,dV, \qquad (4.2.28)$$

from which we may conclude that

$$\Xi = \Xi(T, V, \lambda) = \Xi(T, V, \mu), \qquad (4.2.29)$$

and that $\overline{N}$, S, P, and U are given by

$$\overline{N} = \left(\frac{\partial(PV)}{\partial \mu}\right)_{T,V} = k_B T \left(\frac{\partial \ln \Xi}{\partial \mu}\right)_{T,V}, \qquad (4.2.30a)$$

$$S = \left(\frac{\partial (PV)}{\partial T}\right)_{\mu,V} = k_B \ln \Xi + k_B T \left(\frac{\partial \ln \Xi}{\partial T}\right)_{\mu,V}, \tag{4.2.30b}$$

$$P = \left(\frac{\partial (PV)}{\partial V}\right)_{T,\mu} = k_B T \left(\frac{\partial \ln \Xi}{\partial V}\right)_{T,\mu}, \tag{4.2.30c}$$

$$U = -\left(\frac{\partial \ln \Xi}{\partial \beta}\right)_{\lambda,V} = k_B T^2 \left(\frac{\partial \ln \Xi}{\partial T}\right)_{\lambda,V}. \tag{4.2.30d}$$

Some texts choose to define a 'grand potential' $\Phi(T, V, \mu)$ via

$$\Phi(T, V, \mu) \equiv -k_B T \ln \Xi(T, V, \mu), \tag{4.2.31}$$

so that $PV = -\Phi$, and $\overline{N}$, S, and P are obtained as

$$\overline{N} = -\left(\frac{\partial \Phi}{\partial \mu}\right)_{T,V}, \qquad S = -\left(\frac{\partial \Phi}{\partial T}\right)_{V,\mu}, \qquad P = -\left(\frac{\partial \Phi}{\partial V}\right)_{T,\mu}. \tag{4.2.32}$$

This grand potential Φ thus plays the same role for the grand ensemble that the Helmholtz energy A plays for the canonical ensemble. The internal energy, U, is then given in terms of Φ by

$$U = \Phi - T \left(\frac{\partial \Phi}{\partial T}\right)_{V,\mu}. \tag{4.2.33}$$

Finally, from the association $N \leftrightarrow \overline{N} = \sum_{N,r_N} p_{N,r_N} N$, we have

$$\overline{N} = \frac{\sum_{N,r_N} N e^{-\beta \epsilon_{r_N}} \lambda^N}{\Xi} = \frac{\lambda \sum_{N,r_N} N e^{-\beta \epsilon_{r_N}} \lambda^{N-1}}{\Xi}.$$

This expression relates $\overline{N}$ to Ξ via

$$\overline{N} = \frac{1}{\Xi} \lambda \frac{\partial}{\partial \lambda} \left(\sum_{N,r_N} e^{-\beta \epsilon_{r_N}} \lambda^N\right) = \frac{\lambda}{\Xi} \left(\frac{\partial \Xi}{\partial \lambda}\right)_{T,V},$$

thereby allowing us to associate the thermodynamic N (which is equivalent to our ensemble average number $\overline{N}$) to the grand partition function Ξ through the relation

$$N_{\text{thermo}} = \lambda \left(\frac{\partial \ln \Xi}{\partial \lambda}\right)_{V,T}. \tag{4.2.34}$$

Example 4.4 The Saha equation for a weak plasma.

In this example, we shall utilize the smallest grand ensemble, that of a single electron with access to two 'states', one of which corresponds to $N = 0$ and is represented by the singly charged positive ion A^+, the other to $N = 1$ and is represented by the neutral atom A. The reservoir for this ensemble consists of N_A atoms A, N_{A^+} ions A^+, and the remaining $N_{e^-} - 1 \approx N_{e^-}$ electrons, all of which are contained in the common volume V. Because the electrons and the two chemical species A and A^+ satisfy the chemical (ionization) reaction

$$A^+ + e^- \rightleftharpoons A,$$

their chemical potentials, μ_{e^-}, μ_{A^+}, and μ_A, must be related by $\mu_{e^-} + \mu_{A^+} = \mu_A$, as required by the chemical equilibrium that it represents.[4]

We shall employ the internal state grand partition function in terms of discrete levels ϵ_{i_N} (with associated degeneracies Ω_{i_N}) [see Eq. (3.3.17)] as

$$\Xi(T, V, \mu) = \sum_N \left(\sum_{i_N} \Omega_{i_N} e^{-\beta \epsilon_{i_N}} \right) e^{\beta \mu N}.$$

We choose the energy of the unoccupied state to be the origin for the internal state energy, that is, we shall set $\epsilon_{i_0} = 0$, $\Omega_{i_0} = 1$, $N = 0$, with a corresponding Gibbs factor of 1. With this convention, the occupied state corresponds to energy $\epsilon_i = -I_p$, $\Omega_{i_1} = 1$, $N = 1$, and has a corresponding Gibbs factor $e^{\beta(I_p + \mu)}$, with μ the chemical potential for the electron. We have neglected the double degeneracy associated with electron spin, as it does not affect our final result. Accordingly, the grand partition function is given by the sum of the two Gibbs factors, namely,[5]

$$\Xi(T, \mu) = 1 + e^{\beta(I_p + \mu)}.$$

According to Eq. (3.3.16a), the relevant probabilities are expressed as ratios of the individual Gibbs factors to the partition function as

$$p_{\text{unoccupied}} \equiv p_u = \frac{1}{\Xi}, \qquad p_{\text{occupied}} \equiv p_o = \frac{e^{\beta(I_p + \mu)}}{\Xi}.$$

The ratio of these two probabilities is thus given by

---

[4]This depiction provides a reasonable model for a gas of ground electronic ($^2S_{\frac{1}{2}}$) atomic hydrogen, with its single $1s$-electron, or a gas of alkali atoms, such as Cs, which also have $^2S_{\frac{1}{2}}$ ground electronic terms (due to their lone ns electrons, $n = 2, \cdots, 5$).

[5]Notice that because the energies ϵ_{i_N} for this ensemble are intrinsic properties of the atoms and ions of the chemical species A, A^+, they do not depend upon the container volume, so that Ξ does not depend upon V.

$$\frac{p_u}{p_o} = \frac{1}{e^{\beta(I_p+\mu)}} = \frac{e^{-\beta I_p}}{e^{\beta\mu}}.$$

Because we have chosen to set up our reservoir essentially as a canonical ensemble representing a mixture of N_A atoms A, N_{A^+} ions A^+, and $N_{e^-} - 1$ electrons, the total pressure P for the mixture will be given by Dalton's law of partial pressures, i.e., $P = P_A + P_{A^+} + P_{e^-}$. Moreover, the ratio of the probabilities p_u and p_o may equivalently be represented as the ratio of the reservoir partial pressures P_{A^+} and P_A, so that we obtain the result

$$\frac{P_{A^+}}{P_A} = \frac{e^{-\beta I_p}}{e^{\beta\mu}}.$$

If we now treat the bath electrons as a canonical ensemble of N_{e^-} electrons (considered as an ideal gas, with electron spin neglected), we may determine the chemical potential appearing in this expression via Eqs. (4.1.25c) and (3.2.21, 27) as

$$\mu \equiv \mu_{e^-} = -k_B T \ln\left(\frac{V}{N_{e^-}\Lambda_{e^-}^3}\right) = -k_B T \ln\left(\frac{k_B T}{P_{e^-}\Lambda_{e^-}^3}\right),$$

with $N_{e^-} \simeq N_{A^+} \ll N_A$, so that the electrons can be treated as an ideal gas, with $P_{e^-}V_{e^-} = N_{e^-}k_B T$. In this approximation, $e^{\beta\mu}$ is given by

$$e^{\beta\mu} = \frac{P_{e^-}\Lambda_{e^-}^3}{k_B T}.$$

With this result, the ratio of the partial pressures is given by

$$\frac{P_{A^+}P_{e^-}}{P_A} = \frac{k_B T}{\Lambda_{e^-}^3} e^{-I_p/(k_B T)},$$

which is one form for the Saha equation employed in plasma physics to compute the degree of ionization of atoms in a weak plasma (see also Example 9.3 of Chap. 9, Sect. 9.3).   □

Example 4.5 Effect of an impurity atom in a semiconductor crystal.

For an impurity atom/ion in a semiconductor crystal, including electron spin degeneracy, we have three possible states: one ionized state (with no electron), plus two un-ionized states (with one electron, either spin-up or spin-down) present. Similar to what we have seen in the previous example, the ionized state (here the bottom of the conduction band) is assigned energy $\epsilon = 0$ and number $N = 0$, and the un-ionized doubly degenerate state is assigned energy $\epsilon = -I_p$ and number $N = 1$, giving Gibbs factors $e^0 = 1$ for the first state, and $e^{\beta(I_p+\mu)}$ for the second state.

The grand partition function in this case will be given by

$$\Xi(T, \mu) = 1 + 2e^{\beta(I_p+\mu)} ,$$

and the probability that a donor will ionize is thus

$$p_i = \frac{1}{1 + 2e^{\beta(I_p+\mu)}} .$$

The chemical potential for an ideal gas of conduction electrons is

$$\mu = -k_B T \ln\left(\frac{V z_{int}^{e^-}}{\overline{N}_{ce} \Lambda_{ce}}\right) ,$$

with $z_{int}^{e^-} = 2$ due to the electron spin degeneracy for the spin-$\frac{1}{2}$ electron, while Λ_{ce} is the thermal de Broglie wavelength corresponding to the effective mass m_{ce}^* of a conduction electron.

The probability for the ionization of a donor atom is given by the ratio of the number of conduction electrons, $\overline{N}_{ce}$, to the number of donor atoms, N_d, i.e.,

$$p_i = \frac{\overline{N}_{ce}}{N_d} = \frac{1}{1 + 2e^{\beta I_p}(\overline{N}_{ce}\Lambda_{ce}^3/2V)} = \frac{1}{1 + \overline{N}_{ce}\Lambda_{ce}^3 e^{I_p/(k_B T)}/V} ,$$

from which $\overline{N}_{ce}$ is obtained as

$$\overline{N}_{ce} = \frac{N_c e^{-I_p/(k_B T)}}{4}\left[-1 + \sqrt{1 + \frac{8N_d}{N_c}e^{I_p/(k_B T)}}\right] .$$

We shall plot $\overline{N}_{ce}/N_d$ vs. T/Θ_I, for T ranging from 0 to Θ_I, with Θ_I is given by $\Theta_I \equiv I_p/k_B T$. Thus $\overline{N}_{ce}/N_d$ takes the form

$$\frac{\overline{N}_{ce}}{N_d} = \frac{N_c}{N_d}\frac{e^{-\Theta_I/T}}{4}\left[-1 + \sqrt{1 + \frac{8N_d}{N_c}e^{\Theta_I/T}}\right] .$$

For $N_d = 10^{23}\,\text{m}^{-3}$ and $m_{ce}^* = 0.45m_e$, we may express N_c as

$$N_c = 2\left(\frac{2\pi m_{ce}^* k_B T}{h^2}\right)^{\frac{3}{2}} = 4.613 \times 10^{22}T^{\frac{3}{2}}\,\text{m}^{-3} ,$$

from which we find that $N_c/(4N_d) = 0.1153T^{\frac{3}{2}}$. Moreover, using $k_B = 8.618 \times 10^{-5}\,\text{eV K}^{-1}$, the characteristic temperature for this doped semiconductor has the value $\Theta_I = 510.56$ K. Putting all of this information together, we obtain the specific result

$$\frac{\overline{N}_{ce}}{N_d} = 0.1153T^{\frac{3}{2}}e^{-510.56/T}\left[-1 + \sqrt{1 + 17.342T^{-\frac{3}{2}}e^{510.56/T}}\right]$$

for the ratio of the number of donor-generated conduction electrons to the number of dopant donor atoms in the sample.[6] A plot of $\overline{N}_{ce}/N_d$ vs. T/Θ_I is illustrated in Fig. 4.1.   □

4.2.2 Fluctuations in the Number of Particles

As we have seen in Eq. (4.2.30a), $\overline{N}$ is given in the grand ensemble by

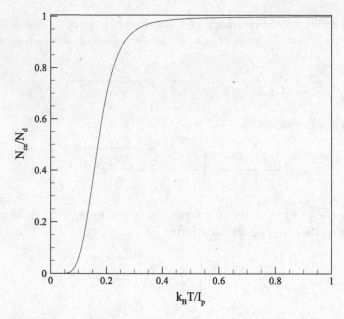

Fig. 4.1 Plot showing the dependence of the ratio of the number of donor electrons released into the conduction band of a Si crystal from dopant P atoms on the temperature of the sample

---

[6]From this expression, we may directly determine the fraction of the dopant P atoms that are ionized at room temperature (say, $T = 294$ K) to be 0.9949. That is, approximately 99.5% of the P atoms are already in their ionized form at room temperature.

$$\overline{N} = k_{\mathrm{B}} T \left(\frac{\partial \ln \Xi}{\partial \mu} \right)_{T,V}, \tag{4.2.35}$$

while according to the definition of $\overline{N}$, we have

$$\overline{N} \, \Xi = \sum_N N Z_N \lambda^N. \tag{4.2.36}$$

From this latter expression, we obtain the result

$$\left(\frac{\partial \overline{N}}{\partial \mu} \right)_{T,V} \Xi + \overline{N} \left(\frac{\partial \Xi}{\partial \mu} \right)_{T,V} = \sum_N N Z_N N \lambda^{N-1} \left(\frac{\partial \lambda}{\partial \mu} \right)_{T,V}. \tag{4.2.37}$$

But, we know that

$$\left(\frac{\partial \lambda}{\partial \mu} \right)_{T,V} = e^{\mu / k_{\mathrm{B}} T} \frac{1}{k_{\mathrm{B}} T}, \tag{4.2.38}$$

so that

$$\left(\frac{\partial \overline{N}}{\partial \mu} \right)_{T,V} \Xi + \overline{N} \left(\frac{\partial \Xi}{\partial \mu} \right)_{T,V} = \sum_N \frac{N^2 Z_N \lambda^N}{k_{\mathrm{B}} T} = \frac{\overline{N^2} \Xi}{k_{\mathrm{B}} T},$$

and

$$\left(\frac{\partial \overline{N}}{\partial \mu} \right)_{T,V} + \overline{N} \left(\frac{\partial \ln \Xi}{\partial \mu} \right)_{T,V} = \frac{\overline{N^2}}{k_{\mathrm{B}} T}. \tag{4.2.39}$$

From Eqs. (4.2.35) and (4.2.39), we see directly that the variance of N is given by

$$\sigma_N^2 = \overline{N^2} - \overline{N}^2 = k_{\mathrm{B}} T \left(\frac{\partial \overline{N}}{\partial \mu} \right)_{T,V}. \tag{4.2.40}$$

Expression (4.2.40) for σ_N^2 is still not in a form that can readily be evaluated in order to provide us with an estimate of the order of magnitude to be expected for σ_N for a typical ensemble, as we do not yet have a means of determining the order of magnitude of the partial derivative of $\overline{N}$ with respect to the chemical potential. To obtain an expression for σ_N^2 that involves experimentally accessible quantities, we must carry out a few standard thermodynamic manipulations. To accomplish this goal, we shall begin with Eq. (4.2.28) for the differential $d(PV)$, use the product rule for differentials, and then simplify to obtain the result

$$\overline{N} \, d\mu - V \, dP + S \, dT = 0, \tag{4.2.41a}$$

which is known in thermodynamics as the Gibbs–Duhem equation. This equation is considered to be one of the fundamental equations of chemical thermodynamics: for a pure substance, it tells us simply that not all of the three intensive thermodynamic variables μ, P, and T can be independent. As the traditional means for expressing this result is to treat the chemical potential as a function of P and T, it will also become convenient for us to rewrite the Gibbs–Duhem equation in the form

$$d\mu = v\,dP - s\,dT\,, \qquad (4.2.41b)$$

with $v \equiv V/\overline{N}$ and $s \equiv S/\overline{N}$ the specific volume and specific entropy.[7]

If we now transform our description of the chemical potential from its natural variables P and T to the new variables v and T by expressing the pressure as a function of v and T, we must replace the differential dP by

$$dP = \left(\frac{\partial P}{\partial V}\right)_T dv + \left(\frac{\partial P}{\partial T}\right)_v dT$$

to obtain $d\mu$ as

$$d\mu = v\left(\frac{\partial P}{\partial v}\right)_T dv + \left[v\left(\frac{\partial P}{\partial T}\right)_v - s\right] dT\,. \qquad (4.2.41c)$$

From this form for the Gibbs–Duhem equation, we see that

$$\left(\frac{\partial \mu}{\partial v}\right)_T = v\left(\frac{\partial P}{\partial v}\right)_T. \qquad (4.2.42a)$$

We shall convert the partial derivative on the left-hand side of Eq. (4.2.42a) into a partial derivative with respect to $\overline{N}$ at constant V and T, and the partial derivative on the right-hand side into a partial derivative with respect to V at constant $\overline{N}$ and T. This conversion leads to the expression

$$\left(\frac{\partial \mu}{\partial \overline{N}}\right)_{T,V} = -\frac{V^2}{\overline{N}^2}\left(\frac{\partial P}{\partial V}\right)_{T,\overline{N}},$$

which, upon reciprocation, gives

$$\left(\frac{\partial \overline{N}}{\partial \mu}\right)_{T,V} = -\frac{\overline{N}^2}{V^2}\left(\frac{\partial V}{\partial P}\right)_{T,\overline{N}}. \qquad (4.2.42b)$$

---

[7]The term *specific* denotes either per particle (as employed herein) or per unit mass. Recall also that the thermodynamics literature employs N, rather than $\overline{N}$ for the (variable) number of particles in a macroscopic (thermodynamic) system.

This result can now be written in terms of the experimental isothermal compressibility coefficient κ_T, defined as

$$\kappa_T \equiv -\frac{1}{V}\left(\frac{\partial V}{\partial P}\right)_{T,\overline{N}}, \tag{4.2.42c}$$

to obtain

$$\left(\frac{\partial \overline{N}}{\partial \mu}\right)_{T,V} = \frac{\overline{N}^2 \kappa_T}{V}. \tag{4.2.42d}$$

Substitution of this result into Eq. (4.2.40) for σ_N^2 gives

$$\left(\frac{\sigma_N}{\overline{N}}\right)^2 = \frac{k_B T \kappa_T}{V}. \tag{4.2.43}$$

As the isothermal compressibility is typically of the order of $V/(\overline{N}k_B T)$, we see finally that the relative standard deviation $\sigma_N/\overline{N}$ is of order $\overline{N}^{-\frac{1}{2}}$. Since for a macroscopic system $\overline{N}$ is typically of the order of the Avogadro number N_0, we see that $\sigma_N/\overline{N} \sim 10^{-12}$, so that the distribution is indeed very narrow. However, this also means that for small systems like nanosystems, the distributions will be significantly broader, with $\sigma_N/\overline{N}$ of order 10^{-4}–10^{-6}.

Let us also examine the behaviour of the grand (canonical) distribution in the neighbourhood of its maximum. From the defining relation for p_N, we may write

$$p_N \equiv \sum_{r_N} p_{r_N} = \sum_{r_N} \frac{\lambda^N e^{-\beta E_{r_N}}}{\Xi} \quad \Rightarrow \quad \Xi\, p_N = Z_N \lambda^N.$$

Differentiation of $\Xi\, p_N$ with respect to N gives the relation

$$\Xi\, \frac{\partial p_N}{\partial N} = \left(\frac{\partial Z_N}{\partial N}\right)_{T,V} \lambda^N + Z_N N \lambda^{N-1} = 0,$$

as the condition for a maximum. At the maximum in p_N, let us set $N = N^*$. From the above condition, we can express N^* as

$$N^* = -\lambda \left(\frac{\partial \ln Z_N}{\partial N}\right)_{T,V}. \tag{4.2.44}$$

Consider now the behaviour of $t_N \equiv \Xi p_N$ as a function of N:

$$t_N = Z_N \lambda^N, \tag{4.2.45a}$$

and

$$t_{N^*} = Z_{N^*} \lambda^{N^*} .$$

The form of the distribution near the maximum is obtained from considering

$$\ln t_N = \ln Z_N + N \ln \lambda$$

$$= \ln t_{N^*} + \left(\frac{\partial \ln t_N}{\partial N} \right)_{N=N^*} (N - N^*) + \left(\frac{\partial^2 \ln t_N}{\partial N^2} \right)_{N=N^*} \frac{(N - N^*)^2}{2} + \cdots ,$$

so that t_N can be expressed near the maximum as

$$t_N = t_{N^*} \exp \left\{ -\frac{(N - N^*)^2}{2\sigma_N^2} \right\} , \qquad (4.2.45b)$$

in which we have set

$$\left(\frac{\partial^2 \ln t_N}{\partial N^2} \right) \Bigg|_{N=N^*} \equiv -\frac{1}{\sigma_N^2} . \qquad (4.2.46)$$

We may now represent Ξ in terms of this distribution as

$$\Xi = \Xi \sum_N p_N = \sum_N t_N = \int_0^\infty t_{N^*} e^{-(N-N^*)^2/2\sigma_N^2} \, dN \qquad (4.2.47)$$

$$= t_{N^*} \sqrt{2} \sigma_N \int_{-N^*/\sqrt{2}\sigma_N}^\infty e^{-x^2} \, dx$$

$$\simeq t_{N^*} \sqrt{2} \sigma_N \int_{-\infty}^\infty e^{-x^2} \, dx .$$

We can now see how to calculate $\overline{N}$ from this distribution. We start by writing $\overline{N} = \sum_N N p_N$, which we then rewrite in the form

$$\overline{N} = \frac{1}{\Xi} \sum_N N t_N = \frac{t_{N^*}}{\Xi} \int_0^\infty e^{-(N-N^*)^2/2\sigma_N^2} N \, dN$$

$$= \frac{t_{N^*}}{\Xi} \left[\int_0^\infty e^{-(N-N^*)^2/2\sigma_N^2} (N - N^*) \, dN + \int_0^\infty N^* e^{-(N-N^*)^2/\sigma_N^2} \, dN \right]$$

$$= \frac{t_{N^*}}{\Xi} \left[\left(\sqrt{2}\sigma_N^2 \right) \int_{-\infty}^\infty x e^{-x^2} \, dx + N^* \sqrt{2} \sigma_N \int_{-\infty}^\infty e^{-x^2} \, dx \right] ,$$

so that we have as a final result

$$\overline{N} = \frac{N^* t_{N^*} \sqrt{2} \sigma_N \int_{-\infty}^{\infty} e^{-x^2} dx}{\Xi} = N^*, \tag{4.2.48}$$

from which we may conclude that the average number equals the most probable number.

4.3 The Isothermal–Isobaric Ensemble

Let us examine briefly discrete versions of the isothermal–isobaric ensemble starting from the fully discretized version of $\Delta(T, \zeta)$, which may be written in the form

$$\Delta(T, \zeta) = \sum_{i, r_i} e^{-\beta E_{r_i} - \zeta V_i}, \tag{4.3.1}$$

in which $E_{r_i}(V)$ is the energy associated with an energy microstate of a member system of volume V_i in the ensemble, and β is, as usual, $(k_B T)^{-1}$.

The probability p_{i, r_i} for finding the member system with volume V_i and energy E_{r_i} is defined as the ratio of an individual term in the summation that defines the partition function to the full sum, i.e., in the present case,

$$p_{i, r_i} = \frac{e^{-\beta E_{r_i} - \zeta V_i}}{\Delta(T, \zeta)}. \tag{4.3.2}$$

We note that by performing the energy summation over r_i for a given value V_i of the volume, we obtain the canonical partition function $z(T, V_i)$ for the member system of the isobaric–isothermal ensemble having volume V_i, i.e.,

$$z(T, V_i) = \sum_{r_i} e^{-\beta E_{r_i}}. \tag{4.3.3}$$

We can employ this result for $z(T, V_i)$ to rewrite the isobaric–isothermal partition function $\Delta(T, \zeta)$ as

$$\Delta(T, \zeta) = \sum_{i} z(T, V_i) e^{-\zeta V_i}. \tag{4.3.4a}$$

Either from this expression for Δ, or from the relation $p_i = \sum_{r_i} p_{i, r_i}$, with p_{i, r_i} given in expression (4.3.2), the probability p_i for finding the system with volume V_i is obtained as

$$p_i = \frac{z(T, V_i) e^{-\zeta V_i}}{\Delta(T, \zeta)}. \tag{4.3.4b}$$

We may also extend this ensemble straightforwardly to handle N-particle systems by replacing the single-particle eigenenergies $E_{r_i}(V_i)$ appearing in Eq. (4.3.3) by the N-particle eigenenergies $E_{r_i}(N, V_i)$ to give $Z(N, T, V_i)$, which can then replace $z(T, V_i)$ in Eqs. (4.3.4) to give $\Delta(N, T, \zeta)$ as

$$\Delta(N, T, \zeta) = \sum_i Z(N, T, V_i) \, e^{-\zeta V_i}, \tag{4.3.5a}$$

with the corresponding probabilities

$$p_i = \frac{Z(N, T, V_i) \, e^{-\zeta V_i}}{\Delta(N, T, \zeta)}. \tag{4.3.5b}$$

Before proceeding further, however, let us establish the association between ζ and thermodynamic variables. To accomplish this task, we begin with the defining expression

$$U \equiv \sum_{i,r_i} p_{i,r_i} E_{r_i} \tag{4.3.6}$$

for the thermodynamic internal energy and from it form the differential dU as

$$dU = \sum_{i,r_i} p_{i,r_i} dE_{r_i} + \sum_{i,r_i} E_{r_i} dp_{i,r_i}. \tag{4.3.7}$$

Let us focus initially upon the second term in this expression. By making use of the defining relation $p_{i,r_i} = e^{-\beta E_{r_i} - \zeta V_i}/\Delta$, we may express E_{r_i} in the form

$$E_{r_i} = -\frac{1}{\beta}[\ln p_{i,r_i} + \zeta V_i + \ln \Delta],$$

thereby allowing us to rewrite the second term of Eq. (4.3.7) as

$$\sum_{i,r_i} E_{r_i} dp_{i,r_i} = -\frac{1}{\beta} \sum_{i,r_i} [\ln p_{i,r_i} + \zeta V_i + \ln \Delta] dp_{i,r_i}. \tag{4.3.8}$$

Only the first term on the right-hand side of Eq. (4.3.8) gives a non-vanishing result, namely,

$$\sum_{i,r_i} \ln p_{i,r_i} dp_{i,r_i} = d\left(\sum_{i,r_i} p_{i,r_i} \ln p_{i,r_i}\right) = -\frac{1}{k_{\mathrm{B}}} dS, \tag{4.3.9a}$$

while the second and third terms vanish, the third term for the same reasons that the term involving $\ln \Xi$ in Eq. (4.2.13) vanished for the grand ensemble. The first term in expression (4.3.7) for dU becomes

$$\sum_{i,r_i} p_{i,r_i} dE_{r_i} = -\frac{\zeta}{\beta} d\overline{V} \qquad (4.3.9b)$$

via reasoning akin to that employed earlier in deducing Eq. (4.2.18) for the grand ensemble. Thus, expression (4.3.7) for dU can be expressed in the form

$$dU = T dS - \frac{\zeta}{\beta} d\overline{V} \qquad (4.3.10a)$$

$$\overset{\text{thermo}}{=} \delta Q_{\text{rev}} + \delta W_{\text{rev}} . \qquad (4.3.10b)$$

Now, as the reversible thermodynamic differential work is given by

$$\delta W_{\text{rev}} \overset{\text{thermo}}{=} -P d\overline{V} ,$$

we see that the thermodynamic pressure P and ζ are related by

$$\zeta = \beta P . \qquad (4.3.11)$$

Substitution of Eq. (4.3.11) for ζ in terms of β and P, together with the identification of β with $(k_B T)^{-1}$, now allows us to express the partition function $\Delta(N, \beta, \zeta)$ of Eq. (3.4.9) in terms of the thermodynamic variables N, T, and P as

$$\Delta(N, T, P) = \sum_i \left[\sum_{r_i} e^{-\beta E_{r_i}(N, V_i)} \right] e^{-\beta P V_i} . \qquad (4.3.12)$$

To determine the characteristic function for this ensemble, we shall begin with expression (4.3.12) for $\Delta(N, T, P)$, in which the summation over r_i is over individual energy microstates associated with a canonical ensemble having volume V_i. The natural logarithm of the corresponding probability $p_{r_i}(V_i)$ is given by

$$\ln p_{r_i}(N, V_i) = -\beta E_{r_i}(N, V_i) - \beta P V_i - \ln \Delta . \qquad (4.3.13)$$

Upon employing this expression in the fundamental form (4.1.22) for the entropy [see also expression (4.2.19)], we obtain

$$S = -k_B \sum_{i,r_i} [p_{r_i}(N, V_i) \ln p_{r_i}(N, V_i)] ,$$

which, upon substitution of $\ln p_{r_i}(N, V_i)$ from expression (4.3.13), gives S as

$$S = \frac{1}{T} \sum_{i,r_i} p_{r_i}(N, V_i) E_{r_i}(N, V_i) + \frac{P}{T} \sum_{i,r_i} p_{r_i}(N, V_i) V_i + k_B \ln \Delta \sum_{i,r_i} p_{r_i}(N, V_i).$$

The first sum/integral in this expression represents the internal energy, the second sum/integral the ensemble average volume $\overline{V}$ [note: the overline does not represent molar volume in this case], and the final sums simply give 1. Thus, we obtain the basic relation

$$TS = U + P\overline{V} + k_B T \ln \Delta \qquad (4.3.14)$$

for the isothermal–isobaric ensemble. Equivalently, we may write

$$-k_B T \ln \Delta = U - TS + P\overline{V} \equiv G, \qquad (4.3.15)$$

from which we see that the Gibbs energy, G, is related to the isothermal–isobaric partition function $\Delta(N, T, P)$ by

$$G(N, T, P) = -k_B T \ln \Delta(N, T, P). \qquad (4.3.16)$$

This relation thus identifies the Gibbs energy as the characteristic thermodynamic function for the isothermal isobaric ensemble.

The important thermodynamic relations

$$S = -\left(\frac{\partial G}{\partial T}\right)_{N,P}, \qquad \overline{V} = \left(\frac{\partial G}{\partial P}\right)_{N,T} \qquad (4.3.17)$$

arise from combining the basic thermodynamic expression

$$dG = -S\,dT + \overline{V}\,dP + \mu\,dN \qquad (4.3.18)$$

for the total differential dG with the mathematical defining expression for the total differential for a function $G(N, T, P)$. Employment of Eq. (4.3.16) in the relations (4.3.17) provides us with expressions giving the entropy $S(N, T, P)$ and ensemble averaged volume $\overline{V}(N, T, P)$ in terms of the isothermal–isobaric partition function, namely, as

$$S = k_B \ln \Delta + k_B T \left(\frac{\partial \ln \Delta}{\partial T}\right)_{N,P}, \qquad (4.3.19a)$$

and

$$\overline{V} = -k_B T \left(\frac{\partial \ln \Delta}{\partial P} \right)_{N,T}. \tag{4.3.19b}$$

The enthalpy H is given by $H = TS + G$, so that

$$H = k_B T^2 \left(\frac{\partial \ln \Delta}{\partial T} \right)_{N,P}. \tag{4.3.20}$$

Finally, we obtain a formal expression for the Helmholtz energy A in the isothermal–isobaric ensemble via the relation $A = G - P\overline{V}$, so that

$$A = G - P\overline{V} = -k_B T \ln \Delta + k_B T P \left(\frac{\partial \ln \Delta}{\partial T} \right)_{N,T}. \tag{4.3.21}$$

Note that a formal expression for the internal energy U for this ensemble may be obtained from the relation $U = H - P\overline{V}$ as

$$U = k_B T^2 \left(\frac{\partial \ln \Delta}{\partial T} \right)_{N,P} + P k_B T \left(\frac{\partial \ln \Delta}{\partial P} \right)_{N,T}. \tag{4.3.22}$$

We thus see that the constant-volume canonical ensemble has simpler formal expressions for U and A (corresponding to the 'natural' thermodynamic variables for these two thermodynamic state functions), while the constant-pressure ensemble has simpler expressions for H and G (corresponding to the 'natural' thermodynamic variables for these two thermodynamic state functions).

4.4   Interconnections Between Ensembles

The symbol Ω was chosen to designate the microcanonical partition function specifically because the degeneracy factor Ω_i for discrete energy levels or, for continuous energies, the density of (energy) states $\Omega(E, V)$ represents the number of microstates accessible to a system having fixed energy E. For a continuous distribution of energy states, we found that the canonical partition function $z(\beta, V)$ for a single structureless particle is related to the microcanonical partition function $\Omega(E, V)$ via Eq. (3.2.20), namely,

$$z(\beta, V) = \int_0^\infty \Omega(E, V) e^{-\beta E} \, dE, \tag{4.4.1}$$

from which we may conclude that the canonical partition function $z(\beta, V)$ is a Laplace transform of the microcanonical partition function $\Omega(E, V)$. The microcanonical partition function for a system of N structureless particles is given [2] by

$$\Omega(E, V, N) = \left(\frac{2\pi m}{h^2}\right)^{\frac{3N}{2}} \frac{V^N}{N!} \frac{E^{\frac{3N}{2}-1}}{\Gamma(3N/2)},$$

so that the canonical partition function for an N-particle system is likewise obtained as a Laplace transform,

$$Z(\beta, V, N) = \int_0^\infty \Omega(E, V, N)e^{-\beta E}\, dE, \tag{4.4.2}$$

of $\Omega(E, V, N)$. An appropriate definition of the inverse Laplace transform correctly gives $\Omega(E, V, N)$ from $Z(\beta, V, N)$. Moreover, this interrelationship between $\Omega(E, V, N)$ and $Z(\beta, V, N)$ explains why the microcanonical and canonical ensembles yield identical final results for the thermodynamic functions of state, as they ultimately contain the same information about the system being treated.

If, in expression (4.2.2a) for the grand partition function (with $\gamma = \beta\mu$), we approximate the sum over N by an integration over N, we see that

$$\Xi(\beta, V, \mu) = \int_0^\infty Z(\beta, V, N)e^{-\beta\mu N}\, dN \tag{4.4.3}$$

gives the grand partition function as a Laplace transform of the canonical partition function $Z(\beta, V, N)$. In a similar fashion, we see from expression (4.3.5a) that by approximating the summation over the volumes V_i of the ensemble members by an integration over continuous volume V, the isobaric–isothermal partition function $\Delta(T, P, N)$ can be obtained as a different Laplace transform of the canonical partition function $Z(\beta, V, N)$, namely,

$$\Delta(T, P, N) = \int_0^\infty Z(T, V, N)e^{-\beta PV}\, dV. \tag{4.4.4}$$

4.5   Problems for this Chapter

1. Show that the expression for the single-particle internal energy given in Example 4.1 for an ideal gas in a (weak) gravitational field reduces to $u(T) = \frac{3}{2}k_B T$ in the zero gravity $g \to 0$ limit. How does $u(T)$ behave for very low temperatures or in a strong gravitational field?

2. Obtain an expression giving the standard deviation σ_E in terms of the heat capacity per particle at constant volume, $c_V = C_V/N$, and show that σ_E/U, with $U \equiv \overline{E}$ the thermodynamic internal energy, behaves as $N^{-\frac{1}{2}}$. What conclusions can thus be drawn from this result with respect to the equivalence or nonequivalence of results obtained for a macroscopic (thermodynamic) system using the microcanonical and canonical ensembles?

3. Determine the relative energy fluctuation $\sigma_E/\overline{E}$ in terms of the number, N, of particles in an ideal classical hard-sphere gas.

4. By establishing that the internal energy U can also be expressed as $\overline{E} = -Z'(\beta)/Z(\beta)$, with the prime denoting differentiation with respect to β, show that $\overline{E^2} = Z''(\beta)/Z(\beta)$, and thereby obtain Eq. (4.1.6). Evaluate $\overline{(E - \overline{E})^2}/\overline{E}^2$ for an ideal gas of N closed-shell atoms, such as Kr.

5. Show that

$$\overline{(E - \overline{E})^3} = k_B^2 T^3 \left[T \left(\frac{\partial C_V}{\partial T} \right)_V + 2C_V \right] ,$$

and obtain an expression for $\overline{(E - \overline{E})^3}/\overline{E}^3$ for an ideal gas of N closed-shell atoms, such as Kr.

6. A proton in a magnetic field can access only two energy states, whose values are given by $E_\pm = \mp\frac{1}{2}\hbar\gamma_p B$, in which γ_p is the magnetogyric ratio ($\gamma_p = 2.67522 \times 10^8$ s^{-1}T^{-1}) and B is the strength of the magnetic field (induction) in Tesla (T). Show that the canonical partition function for a proton can be expressed as $z_p(T, B) = 2\cosh(\frac{1}{2}\beta\hbar\gamma_p B)$, with $\beta \equiv (k_B T)^{-1}$, and obtain an expression for the internal energy $U(T, B)$ for such a proton. Show that $U(T, B) \simeq -\frac{1}{2}\hbar\gamma_p B(1 - 2e^{-\beta\gamma_p B} + \cdots)$ for temperatures such that $\frac{1}{2}\hbar\beta\gamma_p B \gg 1$, and that $U(T, B) \simeq -\frac{1}{4}\beta(\hbar\gamma_p B)^2$ for temperatures such that $\frac{1}{2}\beta\hbar\gamma_p B \ll 1$. What values would $U(0, B)$ and $U(\infty, B)$ have were such temperatures attainable for this system?

7. Let N_+ and N_- (with $N_+ + N_- = N$) designate the numbers of protons in a sample that are aligned, respectively, anti-aligned with an applied magnetic field of strength B. From the fractional populations N_+/N and N_-/N, show that

$$\frac{N_+}{N_-} = e^{\hbar\gamma_p B/(k_B T)} .$$

Show also that the fraction of protons aligned with the field is given by

$$\frac{N_+}{N} = \frac{e^{\hbar\gamma_p B/(k_B T)}}{1 + e^{\hbar\gamma_p B/(k_B T)}} .$$

Is there a temperature for which $N_+ < N_-$? How does your result for N_+/N_- correlate with the statement that one cannot construct a laser based upon a 2-level system?

8. Evaluate the internal energy $u(T)$ per particle, given by Eq. (4.1.2b), for particles possessing only linear momentum $\mathbf{p}$ and mass m and having translational energy $\epsilon(p) = p^2/(2m)$. Use the form for the single-particle partition function $z(\beta, V)$ obtained in Problem 8 of Chap. 3. Obtain expressions for the internal energy $U(T)$, the heat capacity at constant volume C_V, and the pressure P for

an ideal gas of N such particles. Show how the pressure and internal energy are related.

9. Obtain expressions for the internal energy, U_{ur}, the heat capacity at constant volume, C_V^{ur}, and the pressure, P_{ur}, for a gas of N noninteracting ultra-relativistic (ur) particles, and obtain the relation between the pressure and the internal energy for such an ultra-relativistic ideal gas. How does the internal energy, heat capacity, and pressure for an ultra-relativistic gas compare with those for a classical (non-relativistic) gas?

10. Starting from the defining relation (4.2.8) for $\overline{N}$ and (4.2.7) for the probabilities p_N, with $\gamma \equiv -\beta\mu$, show that $\overline{N}$ is obtained from the grand potential $\Phi(T, V, \mu)$ of Eq. (4.2.31) as

$$\overline{N} = -\left(\frac{\partial \Phi}{\partial \mu}\right)_{T,V}.$$

11. Starting from the fundamental expression

$$S(T, V, \mu) \equiv -k_B \sum_{N, r_N} p_{N, r_N} \ln p_{N, r_N}$$

for the entropy in the grand ensemble, show that the grand potential $\Phi(T, V, \mu)$ is given in terms of thermodynamic quantities as

$$\Phi(T, V, \mu) = U(T, V, \mu) - T S(T, V, \mu) - \mu \overline{N}(T, V, \mu).$$

12. Consider the lowest twelve energy *levels* for a particle of mass m in a box of volume V_{box}, namely,

$$\epsilon_{n_x n_y n_z} = \frac{h^2}{8m V_{box}^{\frac{2}{3}}} (n_x^2 + n_y^2 + n_z^2),$$

with n_x, n_y, n_z each taking nonzero positive integer values such that $n_x^2 + n_y^2 + n_z^2 \leq 27$. The ground level for this model thus has energy $\epsilon_0 \equiv 3h^2/(8m V_{box}^{\frac{2}{3}})$: given that the mass m and volume V_{box} are such that $\epsilon_0/k_B = 100\,\text{K}$, find the relative energies $\epsilon_{n_x n_y n_z}/\epsilon_0$ corresponding to $n_x^2 + n_y^2 + n_z^2 \leq 27$, evaluate the partition function $z(T)$ for $T = 200\,\text{K}$, and compute the fractional populations for the 12 energy levels. Carry out the same calculations, still for temperature $T = 200\,\text{K}$, but with the box volume decreased so that $(V_{box}')^{\frac{2}{3}} = \frac{1}{2}(V_{box})^{\frac{2}{3}}$, obtained by isothermally compressing the system. What can you say about how system populations behave when the system volume is changed isothermally?

13. Show that the defining relation $U \equiv N\langle E \rangle = N \sum_r p_r E_r$ and relation (1.2) both give the same result for the contribution $U_{int}(T)$ to the total (thermodynamic) internal energy $U(T)$ arising from the internal states of the

system members. For a 2-state system, in which the system members have internal energies ϵ_0 and $\epsilon_1 = \epsilon_0 + \Delta\epsilon$, express your results in terms of the ratio of the system temperature T to the characteristic temperature Θ_{int}, defined as $\Theta_{\text{int}} \equiv \Delta\epsilon/k_B$. Construct a plot of $U_{\text{int}}(T)/(Nk_B\Theta_{\text{int}})$ vs. T/Θ_{int} for T/Θ_{int} varying between 0 and 10.

14. Obtain an expression for the entropy $S_{\text{int}}(T)$ associated with a 2-state system and determine expression for the behaviour of $S_{\text{int}}(T)$ for $T \ll \Theta_{\text{int}}$ and $T \gg \Theta_{\text{int}}$, with Θ_{int} defined as in Problem 13. Plot the ratio $S_{\text{int}}(T)/(Nk_B)$ as a function of T/Θ_{int} and comment upon the manner in which S_{int} approaches its limiting high- and low-temperature values.

15. Consider a sample of a molecular solid containing N molecules in thermal equilibrium at temperature T. If each individual molecule in this sample may be in only one of two energy eigenstates, so that its energy is either $+\epsilon_0$ or $-\epsilon_0$, and if the solid sample at equilibrium contains N_+ excited molecules having energy ϵ_0, obtain an expression for the configurational entropy $S(N, N_+)$ (i.e., the entropy associated with the number of ways in which the excitation can be distributed amongst the N molecules in the sample).

16. For a molecular crystal of the type considered in Problem 15, in which individual molecules can be found in one of two internal states having energies $-\epsilon_0$ or $+\epsilon_0$, obtain expressions for the probabilities p_+ and p_- that an individual molecule may be found in the states at energies ϵ_0, $-\epsilon_0$, and obtain expressions for the internal state contribution, $U_{\text{int}}(T)$, to the thermodynamic internal energy $U(T)$ and for the contribution $C_V^{\text{int}}(T)$ to the heat capacity $C_V(T)$ for such a solid. Express the heat capacity contribution in terms of the separation $\Delta\epsilon$ between the molecular energy states.

17. Consider N atoms arranged in a regular three-dimensional lattice so as to form a perfect crystal. The displacement of n of these atoms ($1 \ll n \ll N$) from their lattice sites to interstices of the crystal lattice creates an imperfect crystal, with n crystal defects said to be of the Frenkel type. The energy of such an imperfect crystal is higher than that of the corresponding perfect crystal by an amount $U(n) \equiv E(n) = nw$, with w the energy required to move one atom from a crystal site to an interstitial site. By assuming that the number M of interstitial sites into which an atom can enter is similar in magnitude to N and that the concentration of such defects is sufficiently small that the creation of any one defect can be treated as an independent event, argue that the number of configurations associated with n Frenkel defects in a crystal having N lattice sites and M defect sites is given by the number of ways $\Omega(N, M, n)$ of removing n atoms from the N lattice sites and then independently redistributing them among the M interstitial sites. From your expression for $\Omega(M, N, n)$, obtain an expression for the configurational entropy $S(n)$ associated with n Frenkel defects. Note that S depends *parametrically* upon the fixed values N and M.

18. Employ your expression for $S(n)$ from Problem 17 to obtain an expression for the configurational contribution $A(n)$ to the Helmholtz free energy of an imperfect crystal caused by Frenkel defects. As the configurational contribution

to the Helmholtz free energy of an imperfect crystal in thermal equilibrium at temperature T will attain its minimum for n equal to $\bar{n}$ (the equilibrium number of Frenkel defects), minimize $A(n)$ with respect to n, and show that w is given in terms of N, M, and $\bar{n}$ by

$$w = k_{\mathrm{B}}T \ln\left[\frac{(N-\bar{n})(M-\bar{n})}{\bar{n}^2}\right],$$

and show that for a given value of w the equilibrium number of Frenkel defects can be approximated as

$$\bar{n} \simeq \sqrt{MN}\, e^{-w/(2k_{\mathrm{B}}T)}.$$

19. By utilizing expressions (4.1.32–34) for the thermodynamic functions U_{mix}, P_{mix}, and A_{mix}, respectively, together with the thermodynamic defining relations for the enthalpy, H, entropy, S, and Gibbs energy, G, show that H_{mix}, S_{mix}, and G_{mix} for a multicomponent mixture occupying the common volume V_{mix} at the same temperature T are additive.

20. Consider two ideal monatomic gases A and B and show that if the $N!$ factor is not present in the definition of $Z(T, V, N)$, i.e., if $Z(T, V, N)$ for a pure monatomic gas is defined simply as $Z'(T, V, N) \equiv z^N(T, V)$, the Helmholtz energy A' defined as $A' = -k_{\mathrm{B}}T \ln Z'(T, V, N)$ is not an *extensive* thermodynamic quantity, and hence also the entropy S' will not be extensive. We know from experiment that if we double the number of atoms in an ideal gas and at the same time also double the volume occupied by the gas, the entropy doubles. By how much is S' in error when we change N to $2N$ and V to $2V$? Show explicitly that the incorporation of $N!$ into the definition of $Z(T, V, N)$ fixes this problem for the Helmholtz energy, and hence also for the entropy. Why does the internal energy not show a similar problem?

21. The molar entropy of Ar at temperature $T = 298$ K under a pressure $P = 1$ bar is $154.72\,\mathrm{J\,K^{-1}}$. Estimate the number of states Ω available to Ar atoms under these conditions.

22. From the total differential for the characteristic function for the grand ensemble, Eq. (4.2.28), show that the differential of the chemical potential μ can be obtained as

$$d\mu = v\,dP - s\,dT,$$

with v and s the specific (i.e., per particle) volume and entropy, respectively.

23. By considering μ to be a function of V and T, rather than of P and T, show that

(a) $\left(\dfrac{\partial\mu}{\partial v}\right)_T = v\left(\dfrac{\partial P}{\partial v}\right)_T$;   (b) $\left(\dfrac{\partial\mu}{\partial T}\right)_v = v\left(\dfrac{\partial P}{\partial T}\right)_v - s$.

24. Consider the expression obtained in Problem 23(a) for the partial derivative of the chemical potential μ with respect to the specific volume v at constant temperature T. Employ the definition $v \equiv V/N$ to rewrite it in terms of the partial derivative of μ with respect to N (with V held fixed); similarly, rewrite the partial derivative of P with respect to v (with T held fixed) in terms of the partial derivative of P with respect to V (with N held fixed). Employ these results to show that

$$\left(\frac{\partial N}{\partial \mu} \right)_{T,v} = \frac{\kappa_T N^2}{V},$$

in which κ_T the isothermal compressibility of Eq. (2.5.2).

25. Determine the values for the ratio $\bar{n}/N$ of Schottky defects to be expected for samples of crystalline NaF, NaCl, NaBr, and NaI at temperatures $T = 300\,\text{K}$, $500\,\text{K}$, and $800\,\text{K}$ if the E_f values for these substances are $E_f = 2.888, 2., 2.452$, and $2.050\,\text{eV}$. What can you say about the relative abilities of these halides to form Schottky defects?

26. If we assume that the number of interstices M in a crystal is roughly equal to the number of lattice sites N in the crystal, we may compare the equilibrium number n of Frenkel defects with the equilibrium number

$$\frac{\bar{n}}{N} \simeq e^{-\frac{1}{2}\beta E_f},$$

of Schottky defects obtained in Problem 25. Use the values for E_f and w given in the table below to calculate the equilibrium fraction of defects in KCl, KBr, AgCl, and AgBr crystals at room temperature (use $T = 300$ K) to see which type of defect predominates in the various crystals.

Crystal	E_f /eV	w /eV	Crystal	E_f /eV	w /eV
KCl	2.530	3.461	AgCl	1.713	1.318
KBr	2.331	3.158	AgBr	1.499	0.945

27. Show that the term $\sum_{N,r_N} \ln \Xi \, dp_{N,r_N}$ appearing in Eq. (4.2.13) vanishes and, by a similar argument, show that the term $\sum_{i,r_i} V_i \, dp_{i,r_i}$ vanishes in expression (4.3.8) for the internal energy in the isothermal–isobaric ensemble.

28. Use Eq. (4.2.5) for the grand partition function $\Xi(\lambda, T, V)$ to show that the ensemble average number, $\bar{N}$, of particles is given by $\bar{N} = \ln \Xi$, and show that the ideal gas equation of state follows from this expression, with $\bar{N}$ playing the role of the thermodynamic N. Show further, starting from Eqs. (4.2.30c) giving the pressure P and the internal energy U in terms of the grand partition function, that both $P(\bar{N}, T, V)$ and $U(\bar{N}, T, V)$ have the same forms as those obtained from the canonical partition function of Eq. (3.2.27), but with N replaced by $\bar{N}$.

29. Show that the Saha equation obtained in Example 4.4 of Sect. 4.2.1 is unaffected by including both the double degeneracy of the single level of the A atom and the double spin degeneracy of the electron.

30. Consider closed-shell atoms A adsorbed onto adsorption sites of an adsorbent surface with M such sites. The adsorbed atoms are in equilibrium with gas phase atoms A at a pressure such that they can be treated as an ideal gas, the chemical potential for A is μ, and the adsorption energy per atom is $-\varepsilon$. Determine the configurational degeneracy $\Omega(N, M)$ associated with $N \leq M$ atoms adsorbed onto the surface (with a corresponding adsorption energy $-N\varepsilon$). Obtain the canonical partition function for the adsorption of N atoms onto the surface, and use it to obtain the grand partition function for this system. Show that the grand partition function can be written as $\Xi(T, N, \mu) = [1 + e^{\beta N(\varepsilon+\mu)}]^M$. Obtain an expression for $\overline{N}$ for this ensemble, and show that the fractional coverage, $\theta \equiv \overline{N}/M$ for this ensemble results in the Langmuir isotherm

$$\theta = \left[\frac{k_\mathrm{B} T}{P \Lambda^3} e^{-\beta \varepsilon} + 1\right]^{-1} = \frac{P}{P^* + P},$$

with $P^* \equiv k_\mathrm{B} T e^{-\beta \varepsilon}/\Lambda^3(T)$.

31. Suppose now that carbon monoxide, CO, is present in the air that is in contact with the model haemoglobin site discussed in Example 3.5, Chap. 3. CO can also adsorb on a haemoglobin site, so that the single-site model system will now have three states available to it: 'unoccupied', 'occupied by O_2', and 'occupied by CO'. Obtain the grand partition function for this system in terms of the chemical potentials and binding energies for O_2 and CO, and the temperature T. If the chemical potential for CO at a partial pressure that is 1% of the O_2 partial pressure lies approximately 0.12 eV below the chemical potential for O_2, and the binding energy for CO is given as -0.85 eV, determine the probability for the haem site to be occupied by an O_2 molecule.

32. In the vicinity of a cell (where O_2 is to be released by the haemoglobin), the chemical potential of an O_2 molecule in blood is lower than it is for blood in the lungs. Use the model of Example 3.5, Chap. 3, for O_2 in air to calculate and plot the fraction of occupied sites (i.e., the probability for O_2 to be bound to the haem site) as a function of the partial pressure of O_2 by assuming the bound O_2 molecules to be in equilibrium with a hypothetical ideal gas of O_2 molecules at partial pressure P.

33. It happens that the four sites in haemoglobin do not actually behave completely independently, and the tendency of O_2 to bind to a site increases when other haem sites are already occupied. Consider now an improvement to the model of haemoglobin, in which each haemoglobin molecule possesses two sites, rather than only one (or, in reality, four) that can be occupied. This model for the haemoglobin molecule has four possible states involving O_2: one state with energy 0 and no O_2 molecule bound to it ($N = 0$), two different states, each having one site occupied by an O_2 molecule ($N = 1$), and one state with both

sites occupied by O_2 molecules ($N = 2$). Let us choose the energy associated with the unoccupied state as our zero of energy, and let the energy associated with single occupancy of either site to be -0.55 eV, and let the energy for a doubly occupied site be -1.3 eV (which means that the change in energy upon binding the second O_2 molecule is -0.75 eV, rather than simply -0.55 eV). To facilitate your calculations, write the energy ϵ_2 for double occupation as $\epsilon_2 = 2\epsilon_1 + \delta$, with $\delta = -0.2$. Calculate and plot the fraction of occupied sites as a function of the effective partial pressure of O_2 in blood that is in contact with a cell. Why might this behaviour for the fraction of sites occupied by O_2 be preferable (i.e., provide a more realistic representation of how haemoglobin actually functions in our bodies) to the behaviour obtained using the single-site model for haemoglobin?

34. The internal energy U is given in terms of the canonical partition function $Z(T, V, N)$ by Eq. (4.1.3b) and in terms of the isobaric–isothermal partition function by Eq. (4.3.22), while the enthalpy H is given similarly in terms of the canonical partition function by

$$H(T, V, N) = k_B T \left[T \left(\frac{\partial \ln Z}{\partial T} \right)_{V,N} + V \left(\frac{\partial \ln Z}{\partial V} \right)_{T,N} \right]$$

and in terms of the isobaric–isothermal partition function by Eq. (4.3.20). Of course, the final expressions for U and H in terms of T, V, N or T, P, N must be equivalent. Show the equality/equivalence of the expressions for U, H obtained from the two partition functions in the case of an ideal gas.

References

1. P.T. Landsberg, J. Dunning-Davies, D. Pollard, Am. J. Phys. **62**, 712 (1994)
2. W. Greiner, L. Neise, H. Stöcker, *Thermodynamics and Statistical Mechanics* (Springer, New York, 1995), p. 138 (see also pp. 186–190)

Chapter 5
Atomic Systems

This chapter begins with the development of canonical and isothermal–isobaric partition functions for large systems of atoms that have nonreactive ground electronic states and may only access translational states. Expressions are obtained for the internal energy, entropy, pressure, constant-volume heat capacity, Gibbs energy, and the chemical potential associated with translational motion. None of the final thermodynamic expressions for these quantities obtained using the two partition functions differ in the thermodynamic limit. Expressions are also developed for nonzero nuclear spins and excited electronic atomic state contributions to the thermodynamic functions. Vibrational contributions to the thermodynamic internal energy, heat capacity, entropy, and Helmholtz energy are obtained for an ensemble of simple harmonic oscillators. Simple harmonic oscillator concepts are then utilized to develop the Einstein and Debye models for the constant-volume heat capacity of monatomic crystalline solids.

5.1 Ground Electronic Term Atoms

We have at this point covered the most difficult aspects of equilibrium statistical thermodynamics, namely the concepts of 'ensemble', 'partition function', and their connections to thermodynamics itself. What remains to be achieved in order to use the subject matter is to be able to perform an actual evaluation of a partition function and of the quantities that we have derived from it in a formal fashion. In order to accomplish this goal, we have to be able to reduce the problem of evaluating the statistical thermodynamic expressions to their simplest terms. We shall accomplish this goal by reducing the problem ultimately to the evaluation of these quantities in terms of the properties of individual atoms and molecules.

© Springer Nature Switzerland AG 2021
F. R. W. McCourt, *Statistical Thermodynamics for Pure and Applied Sciences*,
https://doi.org/10.1007/978-3-030-52006-9_5

We may employ the thermal de Broglie wavelength $\Lambda(T)$ defined in Chap. 3 to place a condition upon the energy states of a system that is required before the system can be described using continuum arguments, i.e., upon the validity of Eq. (3.2.18b) for $\Omega(\epsilon, V)$. We must have the number of accessible states with energy less than or equal to $k_B T$ very much larger than the number of particles N being considered. As we have seen in Sect. 3.2 that the number of accessible states having energies less than or equal to ϵ is given by Eq. (3.2.18a) for $\Phi(\epsilon)$ with $\epsilon = k_B T$, this condition becomes

$$\Phi(k_B T) \simeq \left(\frac{2\pi m k_B T}{h^2} \right)^{\frac{3}{2}} V \gg N .$$

This result may be expressed equivalently by the condition

$$\frac{\Lambda^3 N}{V} \ll 1 ,$$

which is favoured by low (number) density, N/V, high temperature, T, and large particle mass, m.

5.1.1   The Canonical Partition Function

Let us express the canonical partition function for a gas made up of N independent and indistinguishable particles, each of which possesses only translational motion, in terms of the partition functions for the individual particles in the form

$$Z_N^{tr} \equiv Z_{tr}(\beta, V; N) = \frac{z_{tr}^N(T, V)}{N!} = \frac{V^N}{N! \Lambda^{3N}(T)} , \qquad (5.1.1)$$

in which we have identified $Z_{tr}(\beta, V; 1)$ with $z_{tr}(T, V)$. Note that this formula applies only to particles possessing no motions other than *translational*. We have employed the notation $Z_{tr}(\beta, V; N)$ to emphasize that the number of particles in a canonical ensemble is a parameter that characterizes the ensemble, rather than a true thermodynamic variable like T or V.

Let us now obtain explicit expressions for the internal energy U_{tr}, entropy S_{tr}, and pressure P for translational motion for N particles:

$$U_{tr} = k_B T^2 \left(\frac{\partial \ln Z_N^{tr}}{\partial T} \right)_{N,V} = k_B T^2 \frac{\partial}{\partial T} \ln \left\{ \frac{V^N}{N! \Lambda^{3N}} \right\}$$

$$= k_B T^2 \frac{\partial}{\partial T} \{ N \ln V - \ln N! - 3N \ln \Lambda(T) \} .$$

Upon utilizing Eq. (3.2.22) for the de Broglie wavelength $\Lambda(T)$, $U_{\text{tr}}(T)$ is obtained as

$$U_{\text{tr}}(T) = \tfrac{3}{2} N k_{\text{B}} T . \tag{5.1.2}$$

From this result, we can immediately obtain the expression

$$C_{V,\text{tr}}(T) \equiv \left(\frac{\partial U_{\text{tr}}}{\partial T}\right)_{N,V} = \tfrac{3}{2} N k_{\text{B}} \tag{5.1.3a}$$

for the translational heat capacity $C_{V,\text{tr}}$ for N atoms at constant volume, or equivalently, the expression

$$c_{V,\text{tr}} \equiv \frac{C_{V,\text{tr}}}{N} = \tfrac{3}{2} k_{\text{B}} \tag{5.1.3b}$$

for the translational heat capacity $c_{V,\text{tr}}$ per atom. Similarly, we obtain the pressure P from the defining relation as

$$P = k_{\text{B}} T \left(\frac{\partial \ln Z_N^{\text{tr}}}{\partial V}\right)_{N,T} = \frac{N k_{\text{B}} T}{V} ,$$

or equivalently,

$$PV = N k_{\text{B}} T = \tfrac{2}{3} U_{\text{tr}} \tag{5.1.4}$$

for the equation of state, which we recognize as the ideal gas law. Finally, we can employ the expression

$$S_{\text{tr}} = k_{\text{B}} \ln Z_N^{\text{tr}} + \frac{U_{\text{tr}}}{T} , \tag{5.1.5}$$

to obtain an explicit expression for the entropy associated with purely translational motion. The expression so obtained is known in thermodynamics as the Sackur–Tetrode equation and can be obtained from Eq. (5.1.1) by employing Stirling's approximation for the factorial of N. Remember that Stirling's approximation for $N!$ is obtained from the Stirling approximation for the natural logarithm of $N!$, i.e., $\ln N! \simeq N \ln N - N \ln e = \ln(N/e)^N$. With this approximation, we may write the partition function Z_N^{tr} as

$$Z_N^{\text{tr}} = Z_{\text{tr}}(\beta, V; N) = \frac{V^N}{N! \Lambda^{3N}} \simeq \left(\frac{eV}{N \Lambda^3}\right)^N . \tag{5.1.6}$$

If we now substitute Eq. (5.1.2) for U_{tr} and Eq. (5.1.6) for Z_N^{tr} into Eq. (5.1.5) for S_{tr}, we obtain the Sackur–Tetrode equation,

$$S_{tr}(T, V; N) = Nk_B \ln \left\{ \frac{e^{\frac{5}{2}} V}{N \Lambda^3} \right\}, \tag{5.1.7}$$

for the entropy S_{tr} of an ideal monatomic gas.[1] Sackur and Tetrode established that their expression for the entropy also gave the translational contribution for a molecular gas.

It is important to realize that the Sackur–Tetrode equation applies, strictly speaking, to an ideal gas. As by definition, an ideal gas is made up of point particles of mass m that do not interact, an ideal gas will remain a gas even for $T = 0\,K$. This clearly creates a problem, as any real gas (except ^{3}He) will ultimately condense into a crystalline solid at some temperature $T > 0$. This lack of an interparticle interaction between ideal gas particles is reflected in the low-temperature behaviour of the Sackur–Tetrode equation. In particular, the ideal gas entropy can become (algebraically) less than $-\frac{5}{2}$ as T approaches zero, thereby making the translational entropy negative. As we know from the Third Law of Thermodynamics, entropy cannot be negative. This problem may be avoided by restricting the Sackur–Tetrode equation to the classical regime, that is to values of T, V, N such that $V/[N\Lambda^3(T)] \gg 1$ and $S_{tr}(T, V; N) \geq 0$.

Example 5.1 Entropy of an ideal gas in a gravitational field [1].

The canonical partition function for N indistinguishable particles in a gravitational field is given in terms of the single-particle partition function $z(T, V)$ as

$$Z_N(T, V) = \frac{1}{N!} z^N(T, V),$$

with $z(T, V)$ given in terms of the thermal de Broglie wavelength $\Lambda(T)$ and the gravitational field parameter $\eta = \beta m g H$ (Example 3.4, Chap. 3) by

$$z(T, V) = \frac{V}{\Lambda^3(T)} \frac{1 - e^{-\eta}}{\eta}.$$

As the entropy $S(T, V; N)$ of an ideal gas is given by Eq. (4.1.16) as

$$S(T, V; N) = k_B(\ln Z_N + \beta U),$$

with the internal energy $U(T, V; N)$ given in Eq. (4.1.3a) as

$$U(T, V; N) = Nu(T, V),$$

---

[1]This equation was named in recognition of the two individuals who independently obtained this result: Otto Sackur [*Ann. Physik* **36**, 958 (1911)] and Hugo Tetrode, (aged 17!) [*Ann. Physik* **38**, 434 (1912); Erratum, *ibid.* **39**, 255 (1912)].

we obtain $S(T, V; N)$ as

$$S(T, V; N) = k_{\mathrm{B}} \left[\ln \left(\frac{z^N}{N!} \right) + \beta N u(T, V) \right].$$

Upon employing the Stirling approximation to $\ln N!$ and the expression for $u(T, V)$ obtained in Example 4.1 of Chap. 4, we see that $S(T, V; N)$ is given by

$$S(T, V; N) = N k_{\mathrm{B}} \left[\ln \left(\frac{V}{N\Lambda^3} \right) + \ln \left(\frac{1 - e^{-\eta}}{\eta} \right) + \frac{7}{2} - \frac{\eta}{e^\eta - 1} \right],$$

which we may also write in the form

$$S(T, V; N) = N k_{\mathrm{B}} \ln \left(\frac{V e^{\frac{7}{2}}}{N\Lambda^3} \right) + N k_{\mathrm{B}} \left[\ln \left(\frac{1 - e^{-\eta}}{\eta} \right) - \frac{\eta}{e^\eta - 1} \right].$$

That this expression for the entropy of an ideal gas of structureless classical particles predicts that the entropy can become negative for sufficiently large values of the parameter η, see also Problem 10, indicates that there are problems with this classical result in the presence of a sufficiently strong gravitational field.   □

To obtain expressions for the translational Gibbs energy and chemical potential, we also use this approximation for Z_N^{tr}, together with the defining relation for G in terms of A and PV. We thus obtain

$$G_{\mathrm{tr}}(T, V; N) = A_{\mathrm{tr}} + PV = -k_{\mathrm{B}} T \ln Z_N^{\mathrm{tr}} + N k_{\mathrm{B}} T$$

$$= -N k_{\mathrm{B}} T \ln \frac{eV}{N\Lambda^3} + N k_{\mathrm{B}} T \ln e$$

$$= -N k_{\mathrm{B}} T \ln \left(\frac{V}{N\Lambda^3} \right). \tag{5.1.8}$$

As $V = N k_{\mathrm{B}} T / P$ for an ideal gas, we may also write the final result as a function of temperature and pressure as

$$G_{\mathrm{tr}}(T, P; N) = -N k_{\mathrm{B}} T \ln \left(\frac{k_{\mathrm{B}} T}{P\Lambda^3} \right). \tag{5.1.9}$$

Because the chemical potential μ for a pure substance is merely the Gibbs energy per particle, we see that the translational chemical potential is given by

$$\mu_{\mathrm{tr}}(T, P) = \frac{G_{\mathrm{tr}}(T, P; N)}{N} = -k_{\mathrm{B}} T \ln \frac{k_{\mathrm{B}} T}{P\Lambda^3}. \tag{5.1.10}$$

If we choose to separate out the pressure dependence, we may write $\mu_{tr}(P, T)$ in the conventional manner of chemical thermodynamics as

$$\mu_{tr}(T, P) = \mu_{tr}^{\circ}(T) + k_B T \ln P, \tag{5.1.11}$$

wherein $\mu_{tr}^{\circ}(T)$ is given explicitly by

$$\mu_{tr}^{\circ}(T) = -k_B T \ln\left(\frac{k_B T}{\Lambda^3}\right). \tag{5.1.12}$$

Note that not only have we arrived at the thermodynamic form taken by the chemical potential for a pure substance in the gaseous state, but we have also obtained a formula that allows us to calculate the actual value of the translational contribution to $\mu^{\circ}(T)$ for particles of a given mass m at a specified temperature T.

5.1.2   The Isothermal–Isobaric Partition Function

Expression (5.1.1) for the canonical partition function $Z_N(T, V)$ enables us both to extend the defining relation (3.4.9) for $\Delta(T, \zeta)$ to an isothermal–isobaric ensemble of N structureless atoms possessing only translational states as

$$\Delta(T, P; N) = \int_0^{\infty} Z_N(T, V)e^{-\beta PV}\, dV, \tag{5.1.13}$$

with the translational states treated classically as continuum states and to obtain an explicit closed-form expression for it. The closed-form expression is obtained upon substituting Eq. (5.1.1) into the extended definition (5.1.13) to give

$$\Delta(T, P; N) = \frac{1}{N!\Lambda^{3N}(T)}\int_0^{\infty} V^N e^{-\beta PV}\, dV, \tag{5.1.14}$$

followed by evaluation of the definite integral over volume. For $N \gg 1$, we thereby obtain $\Delta(T, P; N)$ as

$$\Delta(T, P; N) = \left(\frac{k_B T}{\Lambda^3 P}\right). \tag{5.1.15}$$

By substituting this result for $\Delta(T, P; N)$ into Eq. (4.3.19a) for the entropy, we obtain the expression

$$S_{tr}(T, P; N) = N k_B \ln\left(\frac{k_B T e^{\frac{5}{2}}}{\Lambda^3 P}\right), \tag{5.1.16a}$$

which is the Sackur–Tetrode equation for the translational entropy of an ideal gas expressed in terms of N, T, and P, rather than in terms of N, T, and V. Similarly, if we employ Eq. (4.3.22) for the internal energy U, we find that

$$U_{tr}(T; N) = \tfrac{5}{2}Nk_BT - Nk_BT = \tfrac{3}{2}Nk_BT , \qquad (5.1.16b)$$

which is, as we should anticipate, the same result for the translational internal energy that we obtained using the canonical ensemble. Finally, if we employ either Eq. (4.3.21) for $A(T, P; N)$ in terms of $\Delta(T, P; N)$ or the thermodynamic defining relation $A \equiv U - TS$ coupled with expressions (5.1.16a, 5.1.16b), we obtain

$$A_{tr}(T, P; N) = -Nk_BT \ln \left(\frac{k_B T e}{\Lambda^3 P} \right) \qquad (5.1.16c)$$

for the translational Helmholtz energy expressed in terms of N, T, and P, rather than in terms of N, T, and V. As should be expected, none of the final thermodynamic expressions obtained using the isothermal–isobaric ensemble differ from the (equivalent) results obtained from the canonical ensemble, simply because the final macroscopic result should not depend upon the particular ensemble utilized to describe the microscopic behaviour.[2]

5.1.3 Translational Versus Internal State Contributions

We may accommodate situations in which a portion of the N atoms in a canonical ensemble can access internal atomic electronic states at a given temperature T by noting that the total energy, ϵ, of an atom in such an excited electronic state is given by the sum,

$$\epsilon = \epsilon_{tr} + \epsilon_{el} , \qquad (5.1.17)$$

of its translational energy, ϵ_{tr}, and the energy, ϵ_{el}, associated with its (excited) electronic state. We should also note, in passing, that we have typically employed the convention that the energy associated with the electronic ground state of such atoms may be set to zero, thereby serving as the energy origin. We shall retain this convention, so that the energy ϵ_{el} designates the energy of an excited electronic state of an atom relative to the energy ($\epsilon_0 \equiv 0$) of its ground electronic state.

As the translational motion of an atom is rigorously decoupled from its electronic state, we may always write the single-atom canonical partition function as the product

---

[2]This conclusion obviously holds for truly macroscopic systems (i.e., in the thermodynamic limit that both N and V go to infinity, but their ratio N/V, which gives the number density, remains finite). However, it does not necessarily hold for small systems like nanosystems.

$$z(T, V) = z_{\text{tr}}(T, V) z_{\text{el}}(T) \,, \tag{5.1.18}$$

in which the volume dependence of z is exclusively associated with the translational partition function z_{tr}, and the electronic partition function z_{el} is a function of temperature alone. We may now write the N-particle canonical partition function $Z(T, V; N)$, which is defined for indistinguishable classical particles by

$$Z(T, V; N) \equiv \frac{z^N(T, V)}{N!} \,, \tag{5.1.19a}$$

as the product

$$Z(T, V; N) = Z_{\text{tr}}(T, V; N) Z_{\text{el}}(T) \,, \tag{5.1.19b}$$

with the corresponding N-particle canonical partition functions $Z_{\text{tr}}(T, V; N)$ and $Z_{\text{el}}(T; N)$ given by

$$Z_{\text{tr}}(T, V; N) \equiv \frac{z_{\text{tr}}^N(T, V)}{N!} \,, \qquad Z_{\text{el}}(T; N) \equiv z_{\text{el}}^N(T). \tag{5.1.19c}$$

Because we are able to express the N-particle partition function as the product of $Z_{\text{tr}}(T, V; N)$ and $Z_{\text{el}}(T; N)$, every thermodynamic state function that depends upon the logarithm of $Z(T, V; N)$ is thus additive, as can readily be seen for the Helmholtz energy $A(T, V; N) = -k_{\text{B}} T \ln Z(T, V; N)$, which gives directly

$$A(T, V; N) = A_{\text{tr}}(T, V; N) + A_{\text{el}}(T; N) \,, \tag{5.1.20a}$$

with

$$A_{\text{tr}}(T, V; N) = -k_{\text{B}} T \ln Z_{\text{tr}}(T, V; N) \,,$$
$$A_{\text{el}}(T; N) = -k_{\text{B}} T \ln Z_{\text{el}}(T; N) \,. \tag{5.1.20b}$$

Similarly, from Eqs. (4.1.3b) and (5.1.19b), the internal energy $U(T, V; N)$ may be split into two components as

$$U(T, V; N) = U_{\text{tr}}(T, V; N) + U_{\text{el}}(T; N) \,, \tag{5.1.21a}$$

with $U_{\text{tr}}(T, V; N)$ and $U_{\text{el}}(T; N)$ given by

$$U_{\text{tr}}(T, V; N) = k_{\text{B}} T^2 \left(\frac{\partial \ln Z_{\text{tr}}}{\partial T} \right) ; \qquad U_{\text{el}}(T; N) = k_{\text{B}} T^2 \left(\frac{\partial \ln Z_{\text{el}}}{\partial T} \right) . \tag{5.1.21b}$$

It is worth noting that the volume dependence of both the internal energy and the Helmholtz energy resides entirely with their respective translational contributions.

5.2 Why Is the Chemical Potential Negative?

To understand why the chemical potential μ for an ideal classical gas is negative, we shall consider explicitly what happens for a monatomic gas, where we can make use of the Sackur–Tetrode equation for the entropy. We shall also utilize the thermodynamic relation

$$\mu = \left(\frac{\partial U}{\partial N}\right)_{S,V} \tag{5.2.1}$$

obtained from the extension of the combined first and second laws (of thermodynamics) expression $dU = TdS - PdV$ for closed systems to open systems namely,

$$dU = TdS - PdV + \mu dN.$$

For convenience, we shall also write the Sackur–Tetrode equation for the entropy in the (equivalent) form [2]

$$S \equiv S(T, V; N) = Nk_B \left[\ln\left(\frac{V}{N}\right) + \frac{3}{2}\ln\left(\frac{4\pi mU}{3Nh^2}\right) + \frac{5}{2}\right], \tag{5.2.2}$$

with $U = \frac{3}{2}Nk_BT$. The key to understanding why the chemical potential for an ideal classical gas is inherently negative lies in asking how the internal energy U changes with particle number *while holding the volume and entropy of the system constant.*

Let us add a particle to the system, so that N becomes $N + 1$, and let the change in the internal energy associated with the addition of this particle be μ, so that the internal energy U becomes $U + \mu$. To see what value μ must take, let us calculate the entropy $S(N + 1)$ corresponding to $N + 1$ particles by employing the Sackur–Tetrode equation (5.1.7): we obtain

$$S(N + 1) = (N + 1)k_B \left\{\ln\left(\frac{V}{N+1}\right) + \frac{3}{2}\ln\left[\frac{4\pi m(U + \mu)}{3(N+1)h^2}\right] + \frac{5}{2}\right\}.$$

However, our thermodynamic definition of μ requires that $S(N + 1)$ has the same (i.e., unchanged) value of $S(N)$: this requirement constrains μ. To see how the constraint works, let us write

$$S(N+1) = (N+1)k_{\text{B}} \left\{ \ln \left[\frac{V}{N(1+\frac{1}{N})} \right] + \frac{3}{2} \ln \left[\frac{4\pi mU(1+\frac{\mu}{U})}{3Nh^2(1+\frac{1}{N})} \right] + \frac{5}{2} \right\}$$

$$= (N+1)k_{\text{B}} \left\{ \ln \left(\frac{V}{N} \right) - \ln \left(1+\frac{1}{N} \right) + \frac{3}{2} \ln \left(\frac{4\pi mU}{3Nh^2} \right) \right.$$

$$\left. + \frac{3}{2} \ln \left(1+\frac{\mu}{U} \right) - \frac{3}{2} \ln \left(1+\frac{1}{N} \right) + \frac{5}{2} \right\}.$$

If we now subtract $S(N)$ from the above expression for $S(N+1)$, use the approximation $\ln(1+x) \simeq x$ for $x \ll 1$ (recalling that N is very large and that μ will be much smaller than U), then upon neglecting terms of order N^{-1} and μ/U in comparison to 1, we find that

$$S(N+1) - S(N) \approx k_{\text{B}} \left[\ln \left(\frac{V}{N} \right) + \frac{3}{2} \ln \left(\frac{4\pi mU}{3nh^2} \right) \right] + \frac{3}{2} N k_{\text{B}} \frac{\mu}{U}.$$

The requirement that the entropy be held fixed during this process then reduces to requiring that μ be determined by the constraint

$$\frac{3}{2} N k_{\text{B}} \frac{\mu}{U} = -k_{\text{B}} \left[\ln \left(\frac{V}{N} \right) + \frac{3}{2} \ln \left(\frac{4\pi mU}{3Nh^2} \right) \right],$$

or

$$\mu = -\frac{2U}{3Nk_{\text{B}}} k_{\text{B}} \left[\ln \left(\frac{V}{N} \right) + \frac{3}{2} \ln \left(\frac{2\pi mk_{\text{B}}T}{h^2} \right) \right] = -k_{\text{B}}T \ln \left(\frac{V}{N\Lambda^3} \right),$$

which is inherently a negative quantity for $V/(N\Lambda^3) > 1$. Of course, this expression is precisely the expression that we have obtained earlier for μ, as it must be.

Thus, we see that by requiring that the internal energy change μ associated with the addition of particles to an open system occurs in such a way that the entropy and volume of the system remain fixed requires the change in the internal energy to be negative: physically, this means that the addition of particles is made in such a way that the internal energy of the gas because the two p-electrons are indistinguishable cooling the system. From the viewpoint of entropy, we have seen earlier that the addition of particles to an ideal gas causes the number of microstates, and hence the entropy, to increase unless at the same time U is required to decrease sufficiently to suppress any increase in the number of microstates.

To examine the behaviour of the chemical potential μ in a way that will facilitate its comparison with the behaviours of the chemical potential for quantum systems (i.e., systems for which the present description proves to be inadequate, as will be discussed in detail in Chap. 10), we begin with expression (5.1.8) for the Gibbs energy, G, for a pure system in terms of the number density $n \equiv N/V$ and the thermal de Broglie wavelength $\Lambda(T)$ as

$$G = -Nk_BT \ln \left(\frac{1}{n\Lambda^3} \right), \tag{5.2.3}$$

from which we may obtain the chemical potential (Gibbs energy per particle) μ as

$$\mu = k_BT \ln(n\Lambda^3). \tag{5.2.4}$$

From expression (5.2.4), it is clear that μ will be positive, zero, or negative depending upon whether $n\Lambda^3$ is greater than, equal to, or less than 1, respectively. For a fixed value n_0 of the number density, we can obtain a value T_0 for the temperature at which $n_0\Lambda^3$ will have the value unity: this temperature then serves as a characteristic temperature and provides a characteristic thermal energy k_BT_0, which we may use to define a dimensionless chemical potential $\mu/(k_BT_0)$ and a dimensionless temperature T/T_0.

To proceed further, we need to specify a number density. It will be convenient to choose $n_0 \equiv 5 \times 10^{25}\,\mathrm{m}^{-3}$, which is a value close to the number density for an ideal gas at STP (i.e., $T = 298.15$ K, $P = 1$ bar), and then examine the behaviour of $\mu(T)$ for this fixed value of n_0. From the defining relation (3.2.22) for $\Lambda(T)$, we may write $\Lambda^3(T)$ as

$$\Lambda^3(T) \simeq 5.3215 \times 10^{-27}(mT)^{-\frac{3}{2}}\,\mathrm{m}^3, \tag{5.2.5}$$

with mass m expressed in atomic mass units (amu) and temperature T expressed in K. For $n = n_0$, we can write the chemical potential as

$$\frac{\mu(T)}{k_BT_0} = \frac{T}{T_0} \ln \left[\frac{5.3215 \times 10^{-27}n_0}{(mT)^{\frac{3}{2}}} \right]. \tag{5.2.6}$$

To proceed further, it will also be necessary to choose a specific mass.

Let us choose Ar, with mass $m = 39.95$ amu, as a substance with a typical mass. For Ar, we find that $n_0\Lambda^3 = 1$ occurs for temperature $T_0 \simeq 0.0104$ K, a very low temperature indeed, and roughly three orders of magnitude smaller than the freezing temperature ($T_f = 83.8$ K) for Ar. The condition for $\mu_{Ar}(T)$ to be a maximum is given by $(d\mu/dT)_{T=T_{max}} = 0$ or, equivalently, $\ln[n_0\Lambda^3(T = T_{max})] = \frac{3}{2}$. We thus find that $T_{max} \simeq 3.8$ mK, so that $T_{max}/T_0 \simeq 0.3654$. The characteristic behaviour of the chemical potential for a classical gas is shown in Fig. 5.1.

Even for a gas made up of H atoms, the corresponding values for T_0 and T_{max} are $T_0 \simeq 0.414$ K and $T_{max} \simeq 0.152$ K for $n_0 = 5 \times 10^{25}\,\mathrm{m}^{-3}$. It is clear from our discussion above that $n\Lambda^3$ will be less than unity for all physically realizable situations for a classical gas, and hence $\mu(T)$ is, for all practical purposes, always negative.

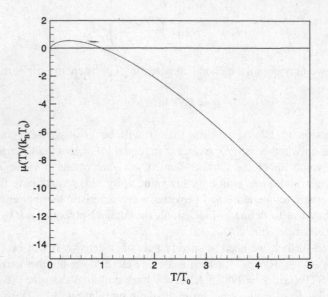

Fig. 5.1 Temperature dependence of the reduced chemical potential for a gas obeying classical (Maxwell–Boltzmann) statistics

5.3   Electronic and Nuclear Spin States

Because excited nuclear spin states lie at energies that exceed thermal energies by many orders of magnitude, they are effectively inaccessible. Consequently, only the ground nuclear spin state must be considered. As will be seen in Sect. 5.3.3, this leads to a considerable simplification of the nuclear spin contribution to the partition function for an individual atom. However, although excited electronic states of many atoms remain inaccessible at typical thermal energies, there are atoms (similarly, ions in crystal lattices) that possess relatively low-lying excited electronic states that are accessible at typical thermal energies, so that they must be considered.

5.3.1   Setting the Stage: Excited Electronic States

We shall consider a specific example in order to develop an expression for the partition function, $z_{el}(T)$, associated with atoms that possess excited electronic states that are accessible at a given temperature. An important atom in the electronics industry and for nanotechnology is the Si atom, which is the third-row equivalent of the C atom in the second row of the periodic table of the elements. From first year general chemistry, we know that its ground (or lowest energy) electronic configuration is $(1s)^2(2s)^2(2p)^6(3s)^2(3p)^2$, which is sometimes also written as $[Ne](3s)^2(3p)^2$, with [Ne] representing the closed-shell electronic configuration of the Ne atom. We also know from introductory quantum mechanics that electronic configurations can be associated with what is called the independent-electron model and described in terms of a central-field Hamiltonian in which an individual electron moves under the influence of a central potential energy function that is due to the net motion of all other electrons about the positively charged nucleus. At this stage, all inter-electron repulsions are ignored.

We shall consider the $(3s)^2$ component of the $[Ne](3s)^2(3p)^2$ electronic configuration of the Si atom first. According to the Pauli (Exclusion) Principle for electrons, no two electrons in a given atom may have the same set of values for all four of the quantum numbers, n, l, m_l, and m_s. Both $(3s)^2$ electrons have principal quantum numbers $n = 3$, azimuthal quantum numbers $l = 0$, and magnetic quantum numbers $m_l = 0$, with the consequence that the Pauli Principle therefore now requires that they have different values for the spin quantum number m_s. We also know that electrons have 'spin' one-half, so that the quantum number m_s may have only one of the two values $\pm\frac{1}{2}$. The two $3s$ electrons are thus said to be 'paired'. This pair of electrons can be represented as a pair-state (or outer product state) for which the total spin S is zero: the associated magnetic spin quantum number is designated as M_S and also necessarily has the value zero. Let us now consider the two $3p$ electrons: we have $n_1 = n_2 = 3$ and $l_1 = l_2 = 1$, but in this case, we have $m_l = 0, \pm 1$ and $m_s = \pm\frac{1}{2}$ available to us as choices to be made. Thus, each of the two $3p$-electrons in principle has six possible pairs of values (m_l, m_s) available to it; however, the Pauli Principle in this case reduces this choice from six to five for the second $3p$-electron in order that not all four of the quantum numbers associated with an individual electron be the same. This means that instead of 36 possible quantum pair-states for the two $3p$ electrons, there will only be 30; however, we are not finished yet, because the two p-electrons are indistinguishable so that we must now halve the number of possible pair-states from 30 to 15. All 15 of these pair-states will have the same (configurational) energy at this level of description. In other words, the ground electronic configuration of the Si atom is 15-fold degenerate in the central-field approximation.

At the next stage of description, we shall consider the effect of including the inter-electron repulsion terms in the Hamiltonian. The most important consequence arising from taking these repulsions into account is that the orbital angular momenta of the valence electrons are coupled, so that if l_i represents the orbital angular momentum (vector) for electron i, then the only orbital angular momentum that is valid for a multi-electron atom at this level of description is the total orbital angular momentum given by the vector sum $L = \sum_i l_i$ for all electrons in the atom. We have saved some effort, however, in that the contribution from the electrons of the Ne 'core' for the configuration gives a net orbital angular momentum of zero, as do the two electrons in the $3s$ sub-shell, with the consequence that we have to consider in detail only those contributions arising from the $3p$ valence electrons of the Si atom. Two vectors l_1 and l_2, each of 'length' 1, can be added vectorially to give resultant vectors L of 'length' 0, 1, and 2, corresponding to addition of the vectors anticollinearly, at an angle of 60°, and collinearly.

We also need to think about what happens with the electron spins at this stage of development. This is a little more complicated to deal with unless we utilize another, more generalized, version of the Pauli Principle that is based on symmetry arguments. Our discussion begins by noting that all fundamental particles, such as electrons, protons, and neutrons, are of two basic types, known as bosons and fermions. These two types of fundamental particle are distinguished by the values of the intrinsic angular momentum $\mathbf{I}$ with which they are associated. The angular momentum $\mathbf{I}$ is referred to as the 'nuclear spin' (with magnitude designated by I) for want of a better terminology. Fundamental bosons and fermions can also combine to form what are referred to as composite bosons and fermions, with net values of I determined by vector addition of the fundamental nuclear spin vectors. A system of many indistinguishable bosons or fermions can be described by a multiproduct wavefunction that must be either symmetric or antisymmetric to the interchange of any pair of particles making up the system. Those particles that are symmetric to this interchange are called bosons and those that are antisymmetric to the interchange are called fermions. It turns out that fermions are fundamental or composite particles that have half-odd-integer values for the magnitudes of their intrinsic angular momenta (normally, nuclear or electronic spins), while bosons are fundamental or composite particles that have zero or integer values for the magnitudes of their intrinsic angular momenta. The best known examples of such particles are electrons, protons, and neutrons, all of which are fermions with 'spin' $\frac{1}{2}$, and neutrinos and photons, both of which are bosons, with 'spins' 0 and 1, respectively.

Let us now return to our specific example of the Si atom, and consider how to represent the wavefunction for the pair-state in which both m_{l_1} and m_{l_2} have value $+1$, so that the total value for the pair-state will be $M_L = m_{l_1} + m_{l_2} = +2$. We have already said that we may represent a pair-state wavefunction as the product function for the states making up the pair, so that the wavefunction $\Psi_{M_L=2}(1, 2)$ for the pair-state will have the form $\Psi_{M_L=2}(1, 2) = \psi_{m_{l_1}=1}(1)\psi_{m_{l_2}=1}(2)$, which is symmetric to the interchange of the electron labels 1 and 2. It can also be established, using the quantum mechanical total electronic orbital angular momentum lowering operator, that all five M_L-states for $L = 2$ are symmetric to this $1 \leftrightarrow 2$ interchange (more generally, all $2L + 1$ M_L-states for total angular momentum L will have the same particle interchange symmetry). We can also establish that the three M_L pair-states for $L = 1$ are antisymmetric to the 1,2 interchange, while the single $M_L = 0$ pair-state corresponding to $L = 0$ is symmetric to the 1,2 interchange.

In Schrödinger quantum mechanics, the spin dependence is adjoined to the orbital angular momentum dependence, so that the total wavefunction for the pair of $3p$ electrons in the valence shell of the Si atom may be represented by

$$\Psi_{\text{total}}(1, 2) = \psi_{\text{orbital}}(1, 2)\psi_{\text{spin}}(1, 2),$$

with both $\psi_{\text{orbital}}(1, 2)$ and $\psi_{\text{spin}}(1, 2)$ having the form of outer product functions. We know from our construction that the orbital product functions all have specified interchange symmetries (symmetric if $L = 0, 2$, and antisymmetric if $L = 1$), but at this point, we do not necessarily know what interchange symmetries, if any, are possessed by the electron spin product functions. We do know, however, that the Pauli Principle for a many-electron system requires that the total wavefunction be antisymmetric to the interchange of any pair (here specifically, the $3p$ pair) of electrons. Hence, for $\Psi_{\text{total}}(1,2)$ to be antisymmetric when the orbital angular momentum pair-state corresponds to $L = 2$, it is necessary that the spin angular momentum pair-state be antisymmetric to the interchange of the two $3p$ electrons. Similarly, for $L = 1$, the spin angular momentum pair-state must be symmetric to this interchange, and for $L = 0$, it must again be antisymmetric. Let us now examine how this can happen.

Two electron spins (each of magnitude $\frac{1}{2}$) can be coupled vectorially to give total electron spins $\mathbf{S} = \mathbf{s}_1 + \mathbf{s}_2$ of magnitudes 1 and 0 (corresponding to $\mathbf{s}_1$ and $\mathbf{s}_2$ combined collinearly or anticollinearly, respectively). These total electron spins have corresponding M_S values 1, 0, −1 for $S = 1$, and $M_S = 0$ for $S = 0$. The same arguments that we used above for the $M_L = +2$ pair-state for $L = 2$ hold for the $M_S = +1$ pair-state for $S = 1$, so that the three total spin $S = 1$ pair-states with $M_S = +1, 0, -1$ are all symmetric to the interchange of the two $3p$ electrons, and the single pair-state for $S = 0$ (with $M_S = 0$) is antisymmetric to this interchange. Thus, the Pauli Principle requires in the case of the two equivalent (indistinguishable) $3p$ valence electrons of Si that the (symmetric) $L = 2$ total orbital angular momentum states must combine with the antisymmetric $S = 0$ total spin angular momentum state, and similarly, the (antisymmetric) $L = 1$ orbital states combine with the symmetric $S = 1$ spin states and the (symmetric) $L = 0$ orbital state with the antisymmetric $S = 0$ spin state. These collections of states are designated by *term symbols* ^1D, ^3P, and ^1S, respectively, and are called the *atomic terms* of the ground electronic configuration of the Si atom.[3] At this stage of development, all states associated with a particular atomic term will be degenerate in energy.

It happens that the terms themselves are split further by the interaction between the spin and orbital angular momenta (a process referred to as spin–orbit coupling). For the Si atom, this coupling only affects the ^3P term (because $S = 0$ for the D and S terms, so that it vanishes for them). The ^3P term is, however, affected and splits into three components (referred to as *levels*), corresponding to magnitudes of the total electronic angular momentum, traditionally written as $\mathbf{J} = \mathbf{L} + \mathbf{S}$, having values $J = 2, 1, 0$. A level with total electronic angular momentum of magnitude J has $2J + 1$ degenerate states associated with it; note that in the present case, the three levels obtained from the ^3P term correspond to a total of $1 + 3 + 5 = 9$ states.

A determination of the relative energies for the three terms (^1S, ^1D, ^3P) would require us to delve too deeply into quantum mechanics. Those who would like to pursue this topic in greater detail should consult a modern quantum chemistry text, such as those by Levine [3] and Levin [4]. However, perhaps the most readable yet fairly complete explanations of the term energies remain those given by Pauling and

---

[3] For additional information on these atomic terms and the term symbols, see Appendix D.

Wilson [5] and Bethe and Jackiw [6] in their now-classic texts. For our purposes, it will suffice to employ the empirical rules set out in 1927 by Hund [7], namely:

1. the term with the highest multiplicity (i.e., value of $2S + 1$) lies lowest in energy;
2. of terms having the same multiplicity, the one having the greatest value of L lies lowest in energy;
3. for ground configurations corresponding to electronic shells that are no more than half-filled, the level for which $J = |L - S|$ lies lowest in energy, while for ground configurations corresponding to electronic shells that are more than half-filled, the level for which $J = L + S$ lies lowest in energy.

Although Hund's observations were based upon spectroscopic data obtained for ground configurations and hold strictly only for those configurations, there are cases of excited configurations for which these same rules still give meaningful results. Note, in particular, however, that Hund's rules say nothing about the relative energies of other than ground-configuration terms.

5.3.2 Excited Electronic State Contributions

Let us now formulate an expression for the electronic contribution to the canonical partition function $z_{el}(T)$ in light of our knowledge of the electronic structure of the Si atom. Quite generally, we have seen that the partition function for a set of states lying at various energies can be written down as

$$z_{el}(T) = \sum_{i=0}^{\infty} \omega_i^e e^{-\beta \epsilon_i^e} = \omega_0^e e^{-\beta \epsilon_0^e} + \omega_1^e e^{-\beta \epsilon_1^e} + \omega_2^e e^{-\beta \epsilon_2^e} + \cdots , \qquad (5.3.1)$$

in which ω_i^e designates the degeneracy of the energy level having energy ϵ_i^e. It is convenient to factor out $e^{-\beta \epsilon_0^e}$ from this expression, to obtain

$$z_{el}(T) = e^{-\beta \epsilon_0^e} \left[\omega_0^e + \omega_1^e e^{-\beta \Delta \epsilon_1^e} + \omega_2^e e^{-\beta \Delta \epsilon_2^e} + \cdots \right], \qquad (5.3.2a)$$

in which $\Delta \epsilon_i^e \equiv \epsilon_i^e - \epsilon_0^e$ gives the energy of level i relative to the ground electronic energy ϵ_0^e. It is often convenient to write $z_{el}(T)$ as

$$z_{el}(T) = e^{-\beta \epsilon_0^e} \widetilde{z}_{el}(T), \qquad (5.3.2b)$$

in which $\widetilde{z}_{el}(T)$ is the electronic partition function relative to the ground electronic energy. When the ground electronic energy is taken to be the zero of energy for the atom, $\widetilde{z}_{el}(T)$ becomes the relevant electronic partition function.

We can make a reasonable estimate of the temperatures for which an excited electronic state of an atom will be important by defining a temperature that characterizes the thermal accessibility of that excited electronic state to the atom. We mean by this that in an equilibrium ensemble of a gas of such atoms, collisions for which the relative kinetic energy of the colliding atoms is at least equal to the energy difference $\Delta\epsilon_i^e$ may excite one of the atoms into the excited electronic state, leaving both atoms travelling more slowly. An appropriate definition of what we shall call a *characteristic temperature*, denoted as T_i^e, is given by

$$T_i^e \equiv \frac{\Delta\epsilon_i^e}{k_B}. \tag{5.3.3}$$

We may employ this definition to rewrite Eq. (5.3.2a) in terms of characteristic temperatures as

$$z_{el}(T) = e^{-\beta\epsilon_0^e}\left[\omega_0^e + \omega_1^e e^{-T_1^e/T} + \omega_2^e e^{-T_2^e/T} + \cdots\right]. \tag{5.3.4}$$

Thus, if $T_i^e \gg T$, so that $T_i^e/T \gg 1$, all terms other than ω_0^e may be neglected; in such a case, $z_{el}(T)$ can be represented very well by

$$z_{el}(T) \simeq \omega_0^e e^{-\beta\epsilon_0^e}. \tag{5.3.5}$$

If we are further able to choose $\epsilon_0^e = 0$ for our ground electronic energy, then $z_{el}(T) = \omega_0^e$ is also independent of temperature. A temperature-independent partition function makes no contribution to the thermodynamic internal energy U or to the heat capacities C_V and C_P (because they are obtained from temperature derivatives of the partition function), but it does contribute to the thermodynamic entropy S and to the thermodynamic Helmholtz and Gibbs energies A and G, respectively. Note, however, that such additive contributions to S, A, and G will cancel out when ΔS, ΔA, and ΔG are determined, so that it is normally not relevant for chemical processes or thermodynamic changes.

Example 5.2 Electronic partition function for the Si atom.

We have seen that the ground term of the Si atom is a ^3P term, and that it is split by spin–orbit coupling into three components, called *levels*, identified as $^3P_2$, $^3P_1$, and $^3P_0$. We also know (see Problem 8) that the $^3P_0$ level (which is nondegenerate, i.e., consists of a single atomic state) lies lower in energy than the other two levels, by 77.12 cm^{-1} for the $^3P_1$ level (with three atomic states) and by 223.16 cm^{-1} for the $^3P_2$ level (with five atomic states). In addition, there is a $^1D_2$ level (with five atomic states) lying 6298.85 cm^{-1} above the ground level and a (nondegenerate) $^1S_0$ level lying 15,394.4 cm^{-1} above the ground level.

Let us utilize Eq. (5.3.4) to obtain an expression for $z_{el}(T)$ for the Si atom: we first determine the characteristic electronic temperatures for the various atomic levels of Si as $T_1^e = \Delta\epsilon_1^e/k_B = 110.0\,\text{K}$, $T_2^e = 321.31\,\text{K}$, $T_3^e = 9063\,\text{K}$, and $T_4^e = 22,150\,\text{K}$, so that $z_{el}(T)$ becomes

$$z_{el}(T) = 1 + 3e^{-110.0/T} + 5e^{-321.3/T} + 5e^{-9063/T} + e^{-22150/T} + \cdots,$$

with the $\cdots$ allowing for electronic terms associated with excited electronic configurations (all lying at considerably higher energies than the $^3P_0$ term of the ground configuration that we have considered thus far). Note that we have also set $\epsilon_0^e = 0$ for the ground energy because an arbitrary shift in the zero of energy affects neither the probabilities for occupation of an energy level nor any of the thermodynamic quantities with which we shall be concerned.

In addition, we note that even the $^1D_2$ term associated with the ground electronic configuration contributes very little to $z_{el}(T)$ for temperatures below 2000 K. It contributes only 0.66% for $T = 2000$ K, while the $^3P_0$, $^3P_1$, and $^3P_2$ levels contribute 12.33%, 35.01%, and 51.94%, respectively, to $z_{el}(T = 2000$ K).   □

Note that for monatomic species the main concern is with low-lying excited electronic states, such as those found in the $^3P_1$ and $^3P_2$ levels of the ground-configuration Si atom. When an atom is in one of these excited electronic states, the energy of the atom is given as

$$\epsilon = \epsilon_{tr} + \epsilon_{el}, \tag{5.3.6}$$

and the corresponding form for the atomic partition function is given by the product

$$z(T, V) = z_{tr}(T, V)z_{el}(T). \tag{5.3.7}$$

5.3.3 Nuclear Spin Contributions to the Canonical Partition Function

The excitation energies associated with excited nuclear states of an atom are very large indeed (typically of order MeV), which means that the nuclear degrees of freedom can be considered to be unexcited, and play a role in the determination of the partition function only in the case that the ground state of a nucleus is degenerate (which typically means that the nucleus has a nonzero nuclear spin I). The nuclear partition function has precisely the same form as that developed above for the electronic partition function, i.e.,

$$z_{nuc}(T) = e^{-\beta\epsilon_0^n}\left[\omega_0^n + \omega_1^n e^{-\beta\Delta\epsilon_1^n} + \omega_2^n e^{-\beta\Delta\epsilon_2^n} + \cdots\right], \tag{5.3.8}$$

with the difference being that $\Delta\epsilon_i^n$ is typically many orders of magnitude greater than the thermal energy $k_B T$, so that $z_{nuc}(T)$ reduces to

$$z_{nuc}(T) = \omega_0^n e^{-\beta\epsilon_0^n}, \tag{5.3.9}$$

which, with $\epsilon_0^n \equiv 0$ chosen for the nuclear ground state energy, becomes simply

$$z_{\text{nuc}} = \omega_0^{\text{n}} = 2I + 1, \tag{5.3.10}$$

with I the nuclear spin angular momentum quantum number associated with the ground state of the atomic nucleus.

As for the electronic contribution to the partition function, the nuclear spin contribution does not affect U or C_V, but does contribute to S, A, and G, though not to ΔS, ΔA, or ΔG.

5.3.4 The Full Canonical Atomic Partition Function

The full partition function for an atom, including translational, electronic, and nuclear contributions, can now be expressed as

$$z_{\text{atom}}(T, V) = e^{-\beta(\epsilon_0^{\text{e}} + \epsilon_0^{\text{n}})} \, \omega_0^{\text{n}} \widetilde{z}_{\text{el}}(T) \, \frac{V}{\Lambda^3(T)},$$

in which $\widetilde{z}$ simply indicates that the ground electronic energy Boltzmann factor has been extracted from $z_{\text{el}}(T)$. It is clear from this expression that it makes sense to set the zero of energy for the calculation of $z_{\text{atom}}(T, V)$ at $\epsilon_0^{\text{e}} + \epsilon_0^{\text{n}}$, so that our final expression for the canonical partition function for an individual atom is

$$z_{\text{atom}}(T, V) = \omega_0^{\text{n}} \widetilde{z}_{\text{el}}(T) \, \frac{V}{\Lambda^3(T)}. \tag{5.3.11}$$

5.4 Atomic Gas Thermodynamic Functions

5.4.1 Atoms Having No Thermally Accessible Excited Electronic States

If the atoms making up the gas have no low-lying (thermally accessible) electronic states, then $z_{\text{el}}(T)$ is independent of temperature and is simply given by $z_{\text{el}} = \omega_0^{\text{e}}$, the degeneracy of the ground electronic state of the atom. The value of ω_0^{e} is given by $2J + 1$, with J the value for the total electronic angular momentum for the multi-electron atom, and provided by the subscript on the atomic term symbol for the ground state. In this case, the electronic contribution to the Helmholtz energy A for an ideal gas of N atoms is simply

$$A_{\text{el}}(T; N) = -Nk_{\text{B}}T \ln \omega_0^{\text{e}}, \tag{5.4.1}$$

and the nuclear contribution is similarly given by

$$A_{\text{nuc}}(T; N) = -Nk_{\text{B}}T \ln \omega_0^{\text{n}}. \qquad (5.4.2)$$

We may now combine these two contributions with that obtained earlier for the translational degrees of freedom to obtain a final expression for the Helmholtz energy for an ideal gas of N atoms:

$$A(T, V; N) = -Nk_{\text{B}}T \ln \left\{ \frac{V e \omega_0^{\text{e}} \omega_0^{\text{n}}}{N \Lambda^3(T)} \right\}. \qquad (5.4.3)$$

There are two possible routes for obtaining an expression for the contributions to the entropy S from the electronic and nuclear degrees of freedom. One is to utilize the defining relation for A in terms of U and TS (also called a Legendre transformation, see Appendix B.2), namely $A = U - TS$, from which we obtain S in terms of A and U as $S = (U - A)/T$. This relation, which we have utilized previously to obtain $S_{\text{trans}}(T, V; N)$, would require us to determine the electronic and nuclear contributions to the internal energy U next. However, another means for obtaining an expression for S utilizes the expression for the total differential dA of $A(T, V; N)$, which is

$$\mathrm{d}A = -S\,\mathrm{d}T - P\,\mathrm{d}V + \mu\,\mathrm{d}N. \qquad (5.4.4)$$

This relation gives us the entropy in terms of A as

$$S = -\left(\frac{\partial A}{\partial T} \right)_{N,V}. \qquad (5.4.5)$$

Using $A_{\text{el}} + A_{\text{nuc}} = -Nk_{\text{B}}T \ln(\omega_o^{\text{e}} \omega_0^{\text{n}})$, we see that $S_{\text{el}} + S_{\text{nuc}}$ is given directly by

$$S_{\text{el}} + S_{\text{nuc}} = Nk_{\text{B}} \ln(\omega_0^{\text{e}} \omega_0^{\text{n}}). \qquad (5.4.6)$$

Adding this result to the translational entropy (expressed as the Sackur–Tetrode equation) gives us the total entropy as

$$S(T, V; N) = Nk_{\text{B}} \ln \left\{ \frac{e^{\frac{5}{2}} V \omega_0^{\text{e}} \omega_0^{\text{n}}}{N \Lambda^3(T)} \right\}. \qquad (5.4.7)$$

To obtain expressions for U_{el} and U_{nuc}, we may either use our statistical mechanical defining relation $U = k_{\text{B}} T^2 \left(\dfrac{\partial \ln Z}{\partial T} \right)_{N,V}$ directly or obtain a thermodynamic relation by employing $A = -k_{\text{B}} T \ln Z$ to eliminate $\ln Z$ from the defining relation, thereby giving

$$U = -T^2 \left(\frac{\partial (A/T)}{\partial T} \right)_{N,V}. \qquad (5.4.8)$$

As both A/T and $\ln Z$ are independent of T and V for the electronic and nuclear degrees of freedom, both U and P have no contributions arising from them. Thus, U and P are determined solely by the translational degrees of freedom for the atoms as

$$U = \tfrac{3}{2} N k_B T, \qquad P = N k_B T / V.$$

Similarly, as the heat capacities at constant volume and pressure, C_V and C_P, are given by the temperature partial derivatives of U and H with the volume (and number of atoms) held fixed, they too will be determined only by the translational contributions, namely, $C_V = \tfrac{3}{2} N k_B$ and $C_P = \tfrac{5}{2} N k_B$, so that $\gamma \equiv C_P/C_V$ has the value 5/3. Finally, from the total differential (5.4.4) for dA, we may also obtain an expression for the chemical potential μ as

$$\mu(T, V) = \left(\frac{\partial A}{\partial N} \right)_{T,V}$$

$$= \mu_{\mathrm{tr}}(T, V) - k_B T \ln(\omega_0^e \omega_0^n)$$

$$= -k_B T \ln \left\{ \frac{V \omega_0^e \omega_0^n}{N \Lambda^3(T)} \right\}. \tag{5.4.9}$$

Moreover, as the chemical potential for a pure substance is the Gibbs energy per particle (here atom), we have $G(T, V; N) = N\mu(T, V)$.

5.4.2   Influence of Excited Electronic States

Some atoms have excited electronic states that are thermally accessible at normal temperatures, meaning temperatures lying roughly in the temperature range of 250 K to 1000 K. Many of the elements in Groups 4A through 7A of the periodic table possess low-lying excited atomic states: of special note are C, O, and F in the second row, Si, S, and Cl in the third row, as well as many of the transition and lanthanide metals and their ions. We shall see in Chap. 8, for example, that such low-lying atomic electronic states play an important role in understanding the magnetic properties of a number of europium and samarium salts.

We shall write the electronic partition function for such an atom as

$$z_{\mathrm{el}}(T) = \sum_{i=0}^{\infty} \omega_i^e \, e^{-\beta \Delta \epsilon_i^e}, \tag{5.4.10}$$

with $\Delta \epsilon_i^e = \epsilon_i^e - \epsilon_0^e$. To obtain an expression for $U_{\mathrm{el}}(T, N)$, we shall employ the fundamental statistical mechanical definition of U to obtain

$$U_{el}(T; N) = N \sum_{i=0}^{\infty} \Delta\epsilon_i^e \omega_i^e \frac{e^{-\beta\Delta\epsilon_i^e}}{z_{el}(T)}. \tag{5.4.11a}$$

Note that if we recognize that $p_i^e = \omega_i^e e^{-\beta\Delta\epsilon_i^e}/z_{el}(T)$ represents the probability that the excited atomic state at energy $\epsilon_i^{e'} \equiv \epsilon_i^e - \epsilon_0^e$ above the ground electronic state is occupied at temperature T, we obtain $U_{el}(T)$ in the form

$$U_{el}(T; N) = N \sum_{i=0}^{\infty} p_i^e \epsilon_i^{e'}$$

or, equivalently, as

$$U_{el}(T; N) = N \sum_{i=0}^{\infty} p_i^e \epsilon_i^e, \tag{5.4.11b}$$

provided that $\epsilon_0^e \equiv 0$ has been chosen as the energy origin for the atomic electronic states. As for the case in which there are no thermally accessible excited electronic states, the pressure, P, is still given by the ideal gas law $P = Nk_BT/V$, as $z_{el}(T)$ is independent of the volume V.

The electronic contribution to the Helmholtz energy A is given by

$$A_{el}(T; N) = -Nk_bT \ln\left(\sum_{i=0}^{\infty} \omega_i^e e^{-\beta\Delta\epsilon_i^e}\right), \tag{5.4.12}$$

from which the electronic contribution to the entropy is obtained as

$$S_{el}(T; N) = Nk_B \frac{\partial}{\partial T} \left\{ T \ln\left(\sum_{i=0}^{\infty} \omega_i^e e^{-\beta\Delta\epsilon_i^e}\right) \right\}$$

$$= Nk_B \ln\left(\sum_{i=0}^{\infty} \omega_i^e e^{-\beta\Delta\epsilon_i^e}\right) + Nk_B \sum_{i=0}^{\infty} \beta\omega_i^e \Delta\epsilon_i^e \frac{e^{-\beta\Delta\epsilon_i^e}}{z_{el}(T)}. \tag{5.4.13}$$

We see that our expression for $U_{el}(T; N)$ now has a nonzero electronic contribution that can be associated with the thermal accessibility at temperature T of excited electronic states. In particular, this means that we will have a nonzero electronic contribution, $C_{V,el}(T; N)$, to the heat capacity at constant volume given by

$$C_{V,el}(T; N) = Nk_B \left[\sum_{i=0}^{\infty} p_i^e (\beta\Delta\epsilon_i^e)^2 - \left(\sum_{i=0}^{\infty} \beta p_i^e \beta\Delta\epsilon_i^e\right)^2 \right], \tag{5.4.14}$$

which has the form of Eq. (4.1.6) relating the heat capacity to the energy variance. The heat capacity at constant pressure, P, is simply related to C_V in the usual way (for an ideal gas) as $C_P = C_V + Nk_B$. The normal complement of thermodynamic functions is rounded out by the electronic contribution, $\mu_{el}(T)$, to the chemical potential, which is given by

$$\mu_{el}(T) = -k_B T \ln z_{el}(T). \tag{5.4.15}$$

The above formulae for the various thermodynamic functions all contain the relatively formidable-looking infinite summation over the electronic states of the atoms making up the ideal gas. If we look back to our example for the Si atom, where we identified explicitly the first few of the (in principle infinite) set of excited states (terms, levels, and configurations) for such an atom, we saw that rather few of the states lay at energies that are thermally accessible, even for temperatures as high as a few thousand degrees. What this means in practice, of course, is that these summations truncate rather quickly, and we have really only a few terms that need to be retained to obtain all electronic contributions that lie within the accuracy required for comparison with experiment.

5.5 Simple Harmonic Oscillator Ensembles

5.5.1 The Simple Harmonic Oscillator

The energy levels for an individual (stationary) simple harmonic oscillator (SHO) are given by

$$\epsilon_v^{SHO} = (v + \tfrac{1}{2})h\nu_{osc}, \qquad v = 0, 1, 2, \cdots,$$

with ν_{osc} the fundamental oscillator frequency. Upon incorporating this expression for the SHO energy into the basic definition of the canonical partition function, we obtain

$$z_{vib}^{SHO}(T) = \sum_{v=0}^{\infty} e^{-\beta \epsilon_v^{SHO}} = \sum_{v=0}^{\infty} e^{-\beta(v+\frac{1}{2})h\nu_{osc}}. \tag{5.5.1}$$

To evaluate this sum over states, we set $e^{-\beta h\nu_{osc}} \equiv y$, so that $z_{vib}^{SHO}(T)$ is given as

$$z_{vib}^{SHO}(T) = y^{\frac{1}{2}} \sum_{v=0}^{\infty} y^v = y^{\frac{1}{2}}\{1 + y + y^2 + \cdots\}. \tag{5.5.2}$$

The sum is simply the geometric series, equal to $(1 - y)^{-1}$ for $y < 1$, so that $z_{\text{vib}}^{\text{SHO}}(T)$ becomes

$$z_{\text{vib}}^{\text{SHO}}(T) = \frac{y^{\frac{1}{2}}}{1 - y} = \frac{e^{-\frac{1}{2}\beta h \nu_{\text{osc}}}}{1 - e^{-\beta h \nu_{\text{osc}}}}. \tag{5.5.3}$$

It is fairly common to define a *characteristic temperature* Θ through the dimensional relation $k_B \Theta =$ energy, or

$$\Theta \equiv \frac{\text{Energy}}{k_B} = \frac{h \nu_{\text{osc}}}{k_B}, \tag{5.5.4}$$

in the present case. By employing this convention, we can write $z_{\text{vib}}^{\text{SHO}}(T)$ as

$$z_{\text{vib}}^{\text{SHO}}(T) = \frac{e^{-\Theta/(2T)}}{1 - e^{-\Theta/T}} = e^{-\Theta/(2T)} \widetilde{z}_{\text{vib}}(T), \tag{5.5.5}$$

in which we have defined a new partition function $\widetilde{z}_{\text{vib}}(T)$, given as

$$\widetilde{z}_{\text{vib}}(T) = \frac{1}{1 - e^{-\Theta/T}}, \tag{5.5.6}$$

which represents the vibrational partition function exclusive of the zero-point energy contribution. This expression thus represents the contribution to the partition function $z_{\text{vib}}^{\text{SHO}}(T)$ that is associated with its excited vibrational states. Moreover, we shall find this form for z_{vib} useful both in Chap. 6 and in Chap. 9.

For low temperatures, by which we mean temperatures such that $T \ll \Theta$, we see that the exponential in the denominator is much smaller than 1, so that it can be neglected, while for high temperatures, we have $\Theta/T \ll 1$, so that we may expand the $e^{-\Theta/T}$ term in the denominator of Eq. (5.5.5) as $e^{-\Theta/T} \simeq 1 - \frac{\Theta}{T} + \cdots$, again giving a simplified expression for $z_{\text{vib}}^{\text{SHO}}(T)$. Note that in the low-temperature case, we cannot say anything about the exponential factor in the numerator, and it remains $e^{-\Theta/(2T)}$. We may summarize the above observations by saying that the vibrational canonical partition function has asymptotic behaviour given by

$$z_{\text{vib}}^{\text{SHO}}(T) = \begin{cases} e^{-\Theta/(2T)}, & \text{(low temperatures)} \\ \frac{T}{\Theta}, & \text{(high temperatures)}. \end{cases} \tag{5.5.7}$$

Finally, we note that an alternative, and relatively compact, representation for $z_{\text{vib}}^{\text{SHO}}(T)$ is

$$z_{\text{vib}}^{\text{SHO}}(T) = \tfrac{1}{2} \operatorname{csch}\left(\frac{\Theta}{2T}\right);$$

this expression can be obtained from the first equality in expression (5.5.5) by multiplying the numerator and denominator by $e^{\Theta/(2T)}$ and recalling the mathematical expressions defining the hyperbolic functions. This form is often utilized in the statistical physics of two-level systems and in solid state physics.

Let us examine the SHO problem from a different perspective. Thus far, we have been primarily interested in the partition function for a single SHO in thermal equilibrium at temperature T with an unspecified reservoir (the surroundings). As a SHO (located at a fixed lattice site) could exchange energy with its surroundings, we obtained an expression for the canonical partition function $z_{\text{int}} \equiv z_{\text{int}}(T; N=1)$. For a canonical ensemble of N such SHOs, the internal state partition function $Z_{\text{int}}^{\text{SHO}}(T; N)$ is then simply given by

$$Z_{\text{int}}^{\text{SHO}}(T; N) = z_{\text{int}}^{N}(T).$$

If, however, we focus on the means by which energy may be transferred via 'thermal contact' between the SHO and its surroundings, we may visualize that a SHO can change the number of units of excitation energy, $h\nu_{\text{osc}}$, either via vibrationally inelastic collisions with individual members of its surroundings (such as an ideal gas of such SHOs) or by exchanging photons carrying energy $mh\nu_{\text{osc}}$ (m, an integer) with a thermal radiation reservoir. From the latter point of view, it makes sense to focus on the degree of excitation of the SHO by expressing the energy of the SHO as

$$\epsilon_n = h\nu_{\text{osc}}\left(n + \tfrac{1}{2}\right), \quad n = 0, 1, 2, \cdots,$$

with n representing the number of units of energy $h\nu_{\text{osc}}$ acquired from the surroundings. In this context, we should like to consider the average energy of the ensemble of eigenstates of an isolated SHO, as given, for example, by

$$\langle \epsilon \rangle(T) = \tfrac{1}{2}h\nu_{\text{osc}} + \frac{h\nu_{\text{osc}}}{e^{\beta h\nu_{\text{osc}}} - 1},$$

in terms of the average degree of excitation $\langle n \rangle(T)$ of the SHO.

To see the connection between these two interpretations, it suffices to consider the ensemble average energy $\langle \epsilon \rangle(T)$ as

$$\langle \epsilon \rangle(T) = \langle (n + \tfrac{1}{2})h\nu_{\text{osc}}\rangle(T)$$

$$= \tfrac{1}{2}h\nu_{\text{osc}} + h\nu_{\text{osc}}\langle n\rangle(T).$$

Comparison of the two expressions for $\langle \epsilon \rangle(T)$ then shows that the average degree of excitation for a SHO in thermal equilibrium at temperature T is thus

$$\langle n \rangle(T) = \frac{1}{e^{\beta h\nu_{\text{osc}}} - 1}.$$

If we examine the low-temperature behaviour of $\langle n \rangle (T)$, we see that it behaves as

$$\langle n \rangle (T) \sim e^{-\beta h \nu_{osc}}, \quad \beta h \nu_{osc} \gg 1,$$

which tells us that the average number of photons of energy $h\nu_{osc}$ decreases exponentially as T decreases for temperatures $T \ll \Theta_{osc}$, with $\Theta_{osc} \equiv h\nu_{osc}/k_B$ the characteristic temperature associated with such an oscillator. For $T \gg \Theta_{osc}$, we see that

$$\langle n \rangle (T) \sim \frac{T}{\Theta_{osc}},$$

which we may interpret as saying that for 'high' temperatures, the number of photons of energy $h\nu_{osc}$ available to the SHO system grows linearly with temperature. An equivalent statement for an ensemble of N SHOs may be deduced from the relation between $\langle \epsilon \rangle (T)$ and $\langle n \rangle (T)$; for temperatures T below Θ_{osc}, the average energy, $\langle \epsilon \rangle (T)$, approaches the zero-point energy, $\epsilon_0 = \frac{1}{2}h\nu_{osc}$, as $e^{-\beta h \nu_{osc}}$ tends to zero.

Once we have an expression for the canonical partition function for a stationary individual SHO, we may readily construct the partition function for a gas of N such oscillators as

$$\begin{aligned} Z(T, V; N) &= \frac{z_{trans}^N(T, V; N)}{N!} z_{SHO}^N(T) \\ &= Z_{trans}(T, V; N) Z_{vib}^{SHO}(T; N), \end{aligned} \tag{5.5.8}$$

in which $z_{trans}(T, V; N)$ is given by Eq. (3.2.21). However, as we are not interested here in the translational motion of a gas of simple harmonic oscillators, but only in the contributions to the thermodynamic state functions arising from their vibrational motions, we may simply omit the translational component of the full N-particle canonical partition function in the following.

Let us designate the vibrational contribution of the simple harmonic oscillators to the internal energy $U(T; N)$ as $U_{vib}^{SHO}(T)$. This vibrational contribution will then be given via the general statistical mechanical defining relation for the internal energy as

$$\begin{aligned} U_{vib}^{SHO}(T; N) &= k_B T^2 \left(\frac{\partial \ln Z_{vib}^{SHO}(T; N)}{\partial T} \right)_N \\ &= N k_B T^2 \frac{d \ln z_{SHO}}{dT}. \end{aligned} \tag{5.5.9}$$

We shall employ $z_{SHO}(T)$ in the form given in Eq. (5.5.5) to obtain

$$U_{\text{vib}}^{\text{SHO}}(T; N) = \tfrac{1}{2} N k_B \Theta + \frac{N k_B \Theta}{e^{\Theta/T} - 1}. \tag{5.5.10}$$

If we note that the first term on the right-hand side of this equation represents the vibrational contribution to the internal energy at temperature $T = 0$, then we may also write

$$U_{\text{vib}}^{\text{SHO}}(T; N) - U_{\text{vib}}^{\text{SHO}}(N, 0) = \frac{N k_B \Theta}{e^{\Theta/T} - 1}, \tag{5.5.11}$$

from which we see that $N k_B \Theta (e^{\Theta/T} - 1)^{-1}$ represents the contribution to the internal energy associated with all thermally accessible excited oscillator states at temperature T. From expression (5.5.11), we see that for low temperatures, i.e., $T \ll \Theta$, $U_{\text{vib}}^{\text{SHO}}(T; N) - U_{\text{vib}}^{\text{SHO}}(N, 0)$ approaches zero exponentially as

$$U_{\text{vib}}^{\text{SHO}}(T; N) - U_{\text{vib}}^{\text{SHO}}(0; N) \sim N k_B \Theta e^{-\Theta/T},$$

while for high temperatures ($T \gg \Theta$), $U_{\text{vib}}^{\text{SHO}}(T; N) - U_{\text{vib}}^{\text{SHO}}(0; N)$ increases linearly with temperature, i.e.,

$$U_{\text{vib}}^{\text{SHO}}(T; N) - U_{\text{vib}}^{\text{SHO}}(0; N) \sim N k_B T.$$

Because $U(0; N)$ is temperature-independent, we may also obtain the vibrational contribution to the heat capacity $C_{V,\text{vib}}^{\text{SHO}}(T; N)$ as

$$
\begin{aligned}
C_{V,\text{vib}}^{\text{SHO}}(T; N) &= \frac{d}{dT} \left[U_{\text{vib}}^{\text{SHO}}(T; N) - U_{\text{vib}}^{\text{SHO}}(0; N) \right] \\
&= \frac{d}{dT} \left[\frac{N k_B \Theta}{e^{\Theta/T} - 1} \right] \\
&= N k_B \left(\frac{\Theta}{T} \right)^2 \frac{e^{\Theta/T}}{(e^{\Theta/T} - 1)^2}.
\end{aligned}
\tag{5.5.12}
$$

We may obtain the low- and high-temperature behaviours of the heat capacity either directly from expression (5.5.12) or from the temperature derivatives of the low- and high-temperature behaviours of $U_{\text{vib}}^{\text{SHO}}(T; N) - U_{\text{vib}}^{\text{SHO}}(0; N)$ as

$$C_{V,\text{vib}}^{\text{SHO}}(T; N) \sim \left(\frac{\Theta}{T} \right)^2 e^{-\Theta/T}, \qquad (T \ll \Theta),$$

and

$$C_{V,\text{vib}}^{\text{SHO}}(T; N) \sim N k_B, \qquad (T \gg \Theta).$$

To complete the picture for a gas of simple harmonic oscillators, we may obtain the vibrational contribution, $A_{\text{vib}}^{\text{SHO}}(T; N)$, to the Helmholtz energy as

$$A_{\text{vib}}^{\text{SHO}}(T; N) = -k_{\text{B}} T \ln z_{\text{SHO}}^{N}(T)$$
$$= \tfrac{1}{2} N k_{\text{B}} \Theta + N k_{\text{B}} T \ln(1 - e^{-\Theta/T}), \tag{5.5.13}$$

which we may rewrite in the form

$$A_{\text{vib}}^{\text{SHO}}(T; N) - U_{\text{vib}}^{\text{SHO}}(N, 0) = N k_{\text{B}} T \ln(1 - e^{-\Theta/T}), \tag{5.5.14}$$

and the entropy associated with the vibrational motion of a gas of simple harmonic oscillators, obtained from the thermodynamic relation

$$S_{\text{vib}}^{\text{SHO}}(T; N) = \frac{1}{T} \left[U_{\text{vib}}^{\text{SHO}}(T; N) - A_{\text{vib}}^{\text{SHO}}(T; N) \right]$$

as

$$S_{\text{vib}}^{\text{SHO}}(T; N) = N k_{\text{B}} \left[\frac{\Theta}{T} \frac{1}{e^{\Theta/T} - 1} - \ln(1 - e^{-\Theta/T}) \right]. \tag{5.5.15}$$

5.5.2 Interlude: Degrees of Freedom

It is traditional in the physical sciences to speak of degrees of freedom possessed by physical systems: the more complex the physical system is, the greater will be the number of degrees of freedom. The idea behind the degree of freedom concept arose from the kinematic description of the various motions that a physical body may undergo, taken together with the equipartition theorem of classical thermodynamics, which states that each quadratic term in the expression for the total energy leads to a contribution of $\tfrac{1}{2} k_{\text{B}} T$ to the thermodynamic internal energy per particle.

Let us consider firstly the simplest type of body, namely a point particle (or, equivalently, a rigid sphere) of mass m, for which the only accessible motion is translation through space. We shall employ, as is common, a Cartesian representation of space, in which there are three mutually perpendicular and constant axes. Arbitrary motion of a point particle through space can then be represented in terms of its (independent) motions along each of these three axis directions. Because a point particle is limited only to translational motions, it is said to possess (only) the three translational degrees of freedom corresponding to its motion along the three Cartesian axis directions, traditionally designated as the x-, y-, and z-directions. These three translational degrees of freedom can also be characterized by the kinetic energies associated with their (translational) motions along the three Cartesian component axes and corresponding to a total (kinetic) energy $E = \tfrac{1}{2} m \mathbf{v}^2 \equiv$

$\frac{1}{2}mv_x^2 + \frac{1}{2}mv_y^2 + \frac{1}{2}mv_z^2$. We note that the total energy E in this case contains three terms, each of which is quadratic in a Cartesian velocity component.

The second simplest type of particle is a one-dimensional simple harmonic oscillator of mass m, which may be represented by a point particle constrained to move symmetrically along a Cartesian direction (say the x-direction) about a fixed position (or origin) due to a constraining (or restoring) force described by Hooke's law $F = -kx$, with a corresponding potential energy $V(x) = \frac{1}{2}kx^2$. Such an oscillator will clearly possess a (translational) kinetic energy $T = \frac{1}{2}mv_x^2$ and an associated Newtonian force $F_T = ma = m\ddot{x}$, with $\ddot{x}$ being the second time derivative of x. The total energy E for a SHO is then the sum of its kinetic and potential energies, namely, $E = \frac{1}{2}mv_x^2 + \frac{1}{2}kx^2$. The total energy E of a one-dimensional SHO is constant and is determined by the initial condition (set at time $t = 0$) imposed either by releasing the particle at a distance $x_{\max}$ from its origin or by impulsively giving the point particle a speed $v_{\max}$ at the origin. The total energy E for each vibrational degree of freedom thus has two quadratic components (i.e., kinetic energy and potential energy), rather than one quadratic component, as for translational motion.

We may determine when the classical equipartition theorem is applicable by examining our statistical mechanical expression for the thermodynamic internal energy per particle, $u(T, V)$, for these two types of motion. For one-dimensional (1D) translational motion modelled as a particle in a 1D box, we may write the 1D analogue of Eq. (3.2.21) for the canonical partition function for three-dimensional (3D) translational motion as

$$z(T, L) = \frac{L}{\Lambda(T)},$$

in which L is the length of the 1D box and $\Lambda(T)$ is the usual thermal de Broglie wavelength. The thermodynamic internal energy per particle, $u(T, L)$, is then given by

$$u(T, L) = k_B T^2 \left(\frac{\partial \ln z}{\partial T}\right)_L = \frac{1}{2}k_B T.$$

We could also have written expression (3.2.21) for 3D translational motion as

$$z(T, L_x, L_y, L_z) = \frac{L_x L_y L_z}{\Lambda^3(T)},$$

with $V = L_x L_y L_z$ the volume of the (rectangular) box, from which it will be clear that the thermodynamic internal energy per particle is given by

$$u(T, L_x, L_y, L_z) = \frac{3}{2}k_B T.$$

This result is consistent with the classical equipartition theorem, with a contribution of $\frac{1}{2}k_B T$ per particle to the internal energy for each degree of translational freedom. If we now examine the internal energy for a classical SHO (i.e., no zero-point energy) in the same way, we obtain

$$u^{SHO}(T) = \frac{d \ln \widetilde{z}_{vib}}{dT} = \frac{k_B \Theta}{e^{\Theta/T} - 1},$$

which, as it stands, does not satisfy the classical equipartition theorem. If, however, we ask what happens when the temperature is such that $T \gg \Theta$, we find that the internal energy is then given by $u_{vib}(T) \equiv u^{SHO}(T) = k_B T$ per oscillator, which indeed satisfies the classical equipartition theorem. We also note, in passing, that $u_{vib}(T) = k_B T$ is consistent with there being two quadratic terms in the expression for the vibrational energy, one kinetic, and the other potential. $\square$

5.6   Monatomic Solids

Consider a crystal made up of atoms that, since they are fixed in space at specific locations, are distinguishable subsystems. For a crystal consisting of N atoms, there will be $3N$ coordinates required for a description of the motion of all N atoms; however, the motion of the crystal as a whole through space takes up three coordinates, and three more coordinates are required in order to describe the rotation of the crystal as a whole in space, leaving $3N-6$ coordinates (which must represent the vibrational degrees of freedom of the crystal). Since for a macroscopic sample, N is of the order of 10^{18}–10^{23}, we shall ignore the 6 relative to $3N$.

In 1907, Albert Einstein introduced a simple model for the description of the vibrational motions of closed-shell atoms in a solid lattice. His model considers each atom as vibrating independently of all other atoms in a spherically symmetric potential field $\varphi(r)$ and examines the case of *small* vibrations of the atoms about their equilibrium positions. For such a model, we can expand $\varphi(r)$ about the equilibrium position as

$$\varphi(r) = \varphi(r_e) + \frac{d\varphi}{dr}\bigg|_{r=r_e} (r - r_e) + \frac{d^2\varphi}{dr^2}\bigg|_{r=r_e} \frac{(r - r_e)^2}{2} + \cdots,$$

or

$$\varphi(r) = \varphi(r_e) + \frac{1}{2}f(r - r_e)^2 + \cdots. \tag{5.6.1}$$

This is the harmonic oscillator approximation, which has ν_{osc} given by

$$v_{\text{osc}} = \frac{1}{2\pi} \sqrt{\frac{f}{m}} = v_{\text{osc}} \left(\frac{V}{N} \right), \qquad (5.6.2)$$

which is therefore a function of V/N (density), because f will be a function of the density.

Atoms interact in pairs via $\varphi(r)$, so that in this case, we cannot ignore the interaction energy of the system; after all, we are dealing with a solid! The energy is given by

$$E = \frac{N\varphi(r_e)}{2} + \epsilon_1 + \epsilon_2 + \cdots + \epsilon_{3N} \equiv \frac{N\varphi(0)}{2} + \epsilon_1 + \epsilon_2 + \cdots + \epsilon_{3N}, \qquad (5.6.3)$$

in which $r_e = 0$ has been chosen as the origin for the pair interaction between an arbitrary atom in the lattice and its nearest neighbours. We shall, for convenience, designate $N\varphi(0)/2$ by W, so that the canonical partition function for the N-atom lattice can be written as

$$Z = e^{-\beta W} z_1 z_2 \cdots z_{3N}. \qquad (5.6.4)$$

Einstein proposed that, in the simplest approximation, each of the $3N$ modes of vibration of the crystal lattice should be assumed to have the same fundamental oscillator frequency, now referred to as v_E, in which case, the canonical partition function, z_{vib}, for each of the $3N$ modes is then precisely the same: we thus obtain Z as

$$Z = e^{-\beta W} z_{\text{vib}}^{3N}. \qquad (5.6.5)$$

An expression for the Helmholtz energy is directly obtained from

$$A = -k_B T \ln Z$$

as

$$A = W - 3N k_B T \ln z_{\text{vib}}.$$

We may write A equivalently as

$$A = W - 3N k_B T \ln \left\{ \frac{e^{-\Theta_E/2T}}{1 - e^{-\Theta_E/T}} \right\},$$

in terms of a characteristic (Einstein) temperature $\Theta_E \equiv h v_{\text{osc}}/k_B$, or as

$$A = W + \tfrac{3}{2} N k_B \Theta_E + 3N k_B T \ln(1 - e^{-\Theta_E/T}). \qquad (5.6.6)$$

Similarly, we obtain the expression

$$U = W + \tfrac{3}{2}Nk_{\mathrm{B}}\Theta_{\mathrm{E}} + 3Nk_{\mathrm{B}}\Theta_{\mathrm{E}}\frac{e^{-\Theta_{\mathrm{E}}/T}}{1 - e^{-\Theta_{\mathrm{E}}/T}} \tag{5.6.7}$$

for the internal energy U, from which the heat capacity, C_V, at constant volume is given as

$$C_V = 3Nk_{\mathrm{B}}\left(\frac{\Theta_{\mathrm{E}}}{T}\right)^2 \frac{e^{-\Theta_{\mathrm{E}}/T}}{(1 - e^{-\Theta_{\mathrm{E}}/T})^2}, \tag{5.6.8a}$$

or, equivalently, as

$$C_V = 3Nk_{\mathrm{B}}\left(\frac{\Theta_{\mathrm{E}}/2T}{\sinh\{\Theta_{\mathrm{E}}/2T\}}\right)^2. \tag{5.6.8b}$$

For T large, we see from Eqs. (5.6.8a) and (5.6.8b) that $C_V \sim 3Nk_{\mathrm{B}}$, which gives for the classical limit, the Dulong and Petit 'law'. For T very small, we see that $C_V \sim 3Nk_{\mathrm{B}}(\Theta_{\mathrm{E}}/T)^2 e^{-\Theta_{\mathrm{E}}/T}$, which explains (at least qualitatively) the low-temperature behaviour of the heat capacity of monatomic solids (often metals) observed by the experimentalists at the beginning of the twentieth century.

Notice that the final result for C_V is of the form

$$\frac{C_V}{Nk_{\mathrm{B}}} = f\left(\frac{\Theta_{\mathrm{E}}}{T}\right), \tag{5.6.9}$$

and hence if we had two different crystals having Einstein temperatures $\Theta_{\mathrm{E}a}$ and $\Theta_{\mathrm{E}b}$, whenever the temperatures for which C_V in the two crystals is to be determined are such that

$$\frac{\Theta_{\mathrm{E}a}}{T_a} = \frac{\Theta_{\mathrm{E}b}}{T_b},$$

then the heat capacities for the a and b crystals are the same. This is an example of the *law of corresponding states* for crystals. We can also readily obtain an expression for the entropy of a monatomic crystal from $S = (U - A)/T$ and our results for U and A. Thus,

$$S = 3Nk_{\mathrm{B}}\left[\frac{\Theta_{\mathrm{E}}/T}{e^{\Theta_{\mathrm{E}}/T} - 1} - \ln(1 - e^{-\Theta_{\mathrm{E}}/T})\right]. \tag{5.6.10}$$

In principle, we could also try to apply the defining relation

$$P = -\left(\frac{\partial A}{\partial V}\right)_{T,N} = k_{\mathrm{B}}T\left(\frac{\partial \ln Z}{\partial V}\right)_{T,N}$$

for the thermodynamic pressure, but we cannot readily obtain an expression for P, or for the equation of state for the monatomic crystal in this direct fashion, because both W and ν_E are functions of the particle density N/V, and we would therefore need to know explicitly how these quantities depend upon the volume V. This difficulty in obtaining an appropriate expression for P means that it will be equally difficult in principle to obtain proper expressions for the thermodynamic state functions $G = A + PV$ and $H = U + PV$. However, as the vapour pressure for crystals is normally extremely small, so that $PV \ll A, U$, for example, we can approximate G by A, and H by U. These approximations are normally very good for crystals. The chemical potential (for a pure substance!) is related to G by $G = N\mu$, so that for the Einstein model the chemical potential μ is well approximated by

$$\mu \simeq \frac{A}{N} = \frac{W}{N} + \tfrac{3}{2}k_B\Theta_E + 3k_B\Theta_E \ln(1 - e^{-\Theta_E/T}). \qquad (5.6.11)$$

It is clear from the left panel of Fig. 5.2 that although the Einstein model accounts for the fact that the heat capacity decreases to zero at 0 K, it is apparent from the comparison given in the right panel that the Einstein model tends to overestimate considerably the manner in which C_V decreases for temperatures close to 0 K. We might have expected that such a simple model should not be able to account for all the features of the temperature dependence of the heat capacity, since it is highly unlikely that all possible normal modes of a crystal consisting of N atoms will be identical. In fact, it can be expected that there should be a nearly continuous distribution of frequencies: having said this much, it must still be admitted that the detailed structure of the spectrum of vibrational frequencies (the so-called phonon spectrum) could be extremely complicated. The only thing that we can say for certain is that there will be a distribution of vibrational frequencies rather than just one frequency, as in the Einstein model.

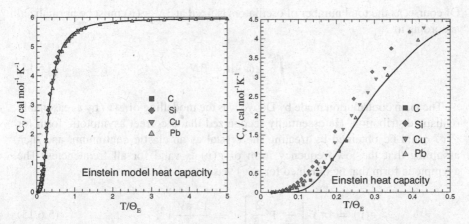

Fig. 5.2 Comparison between experiment and Einstein model calculations for the heat capacity at constant volume for four monatomic solids

In 1913, Peter Debye introduced a meaningful generalization of the Einstein model for monatomic crystals by arguing that if one atom in a crystal is vibrated, it will necessarily affect some of those nearby, and hence it is unrealistic to assume that all normal modes of the crystal have precisely the same frequency. In such a case, the canonical partition function will have the form

$$Z = \mathrm{e}^{-W/k_{\mathrm{B}}T} \prod_{i=1}^{3N} z_i(v_i), \qquad (5.6.12a)$$

in which $z_i(v_i)$ is the canonical partition function for a simple harmonic oscillator with frequency v_i. The logarithm of Z thus becomes

$$\ln Z = -\frac{W}{k_{\mathrm{B}}T} + \sum_{i=1}^{3N} \ln z_i(v_i). \qquad (5.6.12b)$$

There are many normal modes for a typical N of interest (perhaps 10^{20} or so), with the consequence that (as alluded to above) the frequencies will lie very nearly in some sort of continuous distribution. Let us designate by $g(v)$ the function that gives the number of vibrations with frequency v, so that $g(v)\mathrm{d}v$ gives the number of vibrations with frequencies lying between v and $v + \mathrm{d}v$. With the help of this function, we may write $\ln Z$ as

$$\begin{aligned}
\ln Z &= -\frac{W}{k_{\mathrm{B}}T} + \int_0^\infty g(v) \ln z(v)\, \mathrm{d}v \\
&= -\frac{W}{k_{\mathrm{B}}T} + \int_0^\infty g(v) \left[-\frac{hv}{2k_{\mathrm{B}}T} - \ln(1 - \mathrm{e}^{-\beta hv}) \right] \mathrm{d}v.
\end{aligned} \qquad (5.6.13)$$

Of course, as the total number of oscillators is fixed at $3N$, $g(v)$ must be normalized according to

$$\int_0^\infty g(v)\mathrm{d}v = 3N. \qquad (5.6.14)$$

The main contribution made by Debye was the modelling of $g(v)$ by a reasonably realistic distribution. He essentially recognized that a correct asymptotic form for $g(v)$ could be obtained by treating the crystal as an elastic continuum, and then assuming that the low-frequency form of $g(v)$ is valid for all frequencies. The asymptotic form that he obtained for $g(v)$ was

$$g(v) = 4\pi V \left[\frac{1}{c_l^3} + \frac{2}{c_t^3} \right] v^2 \equiv \frac{12\pi V}{c^3} v^2, \qquad (5.6.15)$$

in which c_l and c_t are the longitudinal and transverse phase velocities of sound waves travelling through the crystal, and c is defined by

$$\frac{3}{c^3} \equiv \frac{1}{c_l^3} + \frac{2}{c_t^3}.$$

Notice the similarity between this result and the Rayleigh–Jeans law for low-frequency black-body radiation, which is discussed in most introductory quantum mechanics texts.

Since the low-frequency form for $g(\nu)$, if extended to infinite frequencies, would predict an infinite number of oscillators, and since we know that this number *cannot exceed* $3N$, Debye introduced a cut-off frequency ν_D, beyond which waves cannot propagate in the crystal. This cut-off frequency is determined by the finiteness of the number of normal modes, $3N$, through the condition

$$\int_0^{\nu_D} g(\nu)d\nu = \frac{12\pi V}{c^3} \frac{\nu_D^3}{3} = 3N, \tag{5.6.16a}$$

or

$$\frac{12\pi V}{c^3} = \frac{9N}{\nu_D^3}, \tag{5.6.16b}$$

so that

$$g(\nu) = \begin{cases} 9N\nu^2/\nu_D^3, & 0 \le \nu \le \nu_D, \\ 0, & \nu > \nu_D. \end{cases} \tag{5.6.16c}$$

Now that we know the functional form for $g(\nu)$ for the Debye model, we can proceed to obtain expressions for the various thermodynamic quantities of interest. We shall begin with the internal energy U, which is given by

$$U = k_B T^2 \left(\frac{\partial \ln Z}{\partial T}\right)_{N,V}$$

$$= W + \int_0^\infty \left[\frac{h\nu}{e^{\beta h\nu} - 1} + \frac{1}{2}h\nu\right] g(\nu)\,d\nu$$

$$= W + \frac{9Nh\nu_D}{8} + \frac{9N}{\nu_D^3} \int_0^{\nu_D} \frac{h\nu^3}{e^{\beta h\nu} - 1}\,d\nu. \tag{5.6.17}$$

Another representation of this result may be obtained by setting $x = \Theta_D/T$, with Θ_D the characteristic Debye temperature defined as $\Theta_D \equiv h\nu_D/k_B$, to give

$$U = W + \tfrac{9}{8}Nk_B\Theta_D + 3Nk_BT D(u). \tag{5.6.18}$$

in which $u \equiv \Theta_{\mathrm{D}}/T$, and $D(u)$ is the Debye function given by

$$D(u) \equiv \frac{3}{u^3} \int_0^u \frac{x^3}{e^x - 1}\, dx. \tag{5.6.19}$$

The heat capacity at constant volume for a Debye solid is given by

$$C_V(T) = \frac{\partial}{\partial T}\left[3Nk_{\mathrm{B}}TD\left(\frac{\Theta_{\mathrm{D}}}{T}\right)\right] \tag{5.6.20a}$$

$$= 3Nk_{\mathrm{B}}D\left(\frac{\Theta_{\mathrm{D}}}{T}\right) + 3Nk_{\mathrm{B}}T\left(\frac{\partial D}{\partial T}\right)_{N,V}. \tag{5.6.20b}$$

Note that we stress that either both N and V are explicitly held fixed or the number density $n \equiv N/V$ must be held fixed, as Θ_{D}, like Θ_{E}, is an unknown function of the density. From the defining relations $u \equiv \Theta_{\mathrm{D}}/T$ and (5.6.19) for $D(u)$, we can evaluate the partial derivative of D in Eq. (5.6.20b) as

$$\frac{\partial D}{\partial T} = \frac{3}{T}D(u) - \frac{1}{T}\frac{3u}{e^u - 1},$$

so that we obtain $C_V(T)$ as

$$C_V = 3Nk_{\mathrm{B}}\left\{4D(u) - \frac{3u}{e^u - 1}\right\}. \tag{5.6.21}$$

To examine the high-temperature behaviour of $C_V(T)$ for a Debye solid, we need to examine the behaviour of $D(u)$ and $u(e^u - 1)^{-1}$ as u tends to zero. As both expressions are indeterminate forms for $u \to 0$, we can write $D(u)$ as the ratio of the integral $3\int_0^u \frac{x^3}{e^x - 1}\, dx$ to u^3 and then apply the L'Hôpital rule for limits twice. By doing this, we find that

$$\lim_{u\to 0} D(u) = \lim_{u\to 0} \frac{u}{e^u - 1} = 1,$$

so that the high-temperature limit of $C_V(T)$ for a Debye solid is given as

$$C_V(T) \sim 3Nk_{\mathrm{B}}, \tag{5.6.22a}$$

which, again, is the well-known classical limit, or Dulong-Petit 'Law', for monatomic solids. For low temperatures, $C_V(T)$ can be approximated as

$$C_V(T) \simeq \frac{36Nk_{\mathrm{B}}}{u^3} \int_0^u \frac{x^3}{e^x - 1}\, dx$$

$$\approx \frac{36Nk_B}{(\Theta_D/T)^3} \int_0^\infty \frac{x^3}{e^x - 1} \, \mathrm{d}x.$$

The definite integral in this expression has the value $\pi^4/15$, so that the Debye heat capacity behaves as

$$C_V(T) \sim \frac{12Nk_B\pi^4}{5} \left(\frac{T}{\Theta_D}\right)^3 \tag{5.6.22b}$$

for temperature T such that $T \ll \Theta_D$. This latter result is known as *the Debye T^3 law* for the heat capacity.

Notice also that, just as we have seen for the heat capacity of an Einstein solid, the expression for the heat capacity for a Debye solid can be written as

$$\frac{C_V}{Nk_B} = f(\Theta_D/T), \tag{5.6.23}$$

which expresses the law of corresponding states for a monatomic Debye crystal.

We see from Fig. 5.3 that the Debye model for monatomic crystals gives not only good agreement with the heat capacity data at high temperatures but also excellent agreement with low-temperature heat capacity data, describing the approach to zero Kelvin via a T^3 dependence. Even though the Debye model is still a relatively simplistic model, we see that by allowing the vibrational spectrum to differ from the single frequency that was assumed for the Einstein model, it already provides the correct approach of $C_V(T)$ to zero for monatomic solids at very low temperatures.

As we did in the case of the Einstein model for a monatomic crystal, we shall assume for the Debye model that $PV \ll A$, so that the chemical potential μ will be

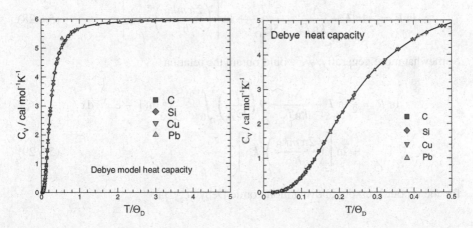

Fig. 5.3 Comparison between experiment and Debye model calculations for the heat capacity at constant volume for four monatomic solids

given at low temperatures by

$$\mu = \frac{A}{N} = \frac{W}{N} + \frac{9h\nu_{\mathrm{D}}}{8} - \frac{\pi^4 k_{\mathrm{B}} T^4}{5\Theta_{\mathrm{D}}^3}. \tag{5.6.24}$$

More generally, μ can be written as

$$\mu = -\Lambda_0 + \frac{9k_{\mathrm{B}} T^4}{\Theta_{\mathrm{D}}^3} \int_0^{\Theta_{\mathrm{D}}/T} \ln(1 - \mathrm{e}^{-x}) x^2 \, \mathrm{d}x, \tag{5.6.25}$$

with Λ_0 representing the *heat of sublimation* per molecule at $0\,\mathrm{K}$, defined by

$$\Lambda_0 \equiv -\frac{W}{N} - \frac{9h\nu_{\mathrm{D}}}{8}. \tag{5.6.26}$$

Such a procedure may be regarded as exact, provided that we use the experimental value for Λ_0 in expression (5.6.25) rather than the model expression given in Eq. (5.6.26).

Now, to obtain an expression for the vapour pressure P of a crystal, we may use the thermodynamic criterion for phase equilibrium to set μ_{crystal} equal to μ_{gas} and then treat the vapour as an ideal gas. Hence, for sufficiently low temperatures, we have

$$\mu_{\mathrm{crystal}} = -\Lambda_0 - \frac{\pi^4 k_{\mathrm{B}} T^4}{5\Theta_{\mathrm{D}}^3} = -k_{\mathrm{B}} T \ln \left\{ \left(\frac{2\pi m k_{\mathrm{B}} T}{h^2}\right)^{\frac{3}{2}} \frac{k_{\mathrm{B}} T}{p} \right\} = \mu_{\mathrm{gas}}. \tag{5.6.27}$$

This equation gives for P the expression

$$\ln P = \tfrac{5}{2} \ln T - \frac{\Lambda_0}{k_{\mathrm{B}} T} - \frac{\pi^4 T^3}{5\Theta_{\mathrm{D}}^3} + \ln \left[\left(\frac{2\pi m k_{\mathrm{B}}}{h^2}\right)^{\frac{3}{2}} k_{\mathrm{B}} \right]. \tag{5.6.28}$$

Somewhat more generally, we would obtain the relation

$$\ln P = \tfrac{5}{2} \ln T - \frac{\Lambda_0}{k_{\mathrm{B}} T} + 9 \left(\frac{T}{\Theta_D}\right)^3 \int_0^{\Theta_{\mathrm{D}}/T} x^2 \ln(1 - \mathrm{e}^{-x}) \, \mathrm{d}x$$

$$+ \ln \left[\left(\frac{2\pi m k_{\mathrm{B}}}{h^3}\right)^{\frac{3}{2}} k_{\mathrm{B}} \right], \tag{5.6.29}$$

for the vapour pressure above a monatomic Debye crystal.

5.7 Problems for This Chapter

1. Consider an ensemble of N atoms in equilibrium at temperature T. Each atom has two energy levels, namely, a ground electronic level with degeneracy ω_1 and energy $\epsilon_{e1} = 0$ and an excited level with degeneracy ω_2 and energy ϵ_{e2}. Obtain an expression for the electronic partition function $z_{el}(T)$ for an individual atom, and use it to obtain an expression for the electronic component of the internal energy U for the N-atom ensemble. Obtain expressions showing the temperature behaviours of U_{el} when T becomes very large and when T becomes very small.

2. A mixture of gaseous krypton and xenon, with the xenon at a concentration of 0.01% that of the krypton, is sprayed onto a metallic surface that is maintained at liquid nitrogen temperature, and a film of the Kr/Xe solid is carefully removed, maintaining it at liquid nitrogen temperature. By treating this film as N xenon atoms isolated in a Kr crystal, with the Xe atoms simply replacing Kr atoms at their crystalline lattice sites, we may consider each Xe atom to be an isolated simple harmonic oscillator, with its SHO energy levels separated by ϵ_{SHO}. If we ignore the zero-point energy associated with the simple harmonic motion of a Xe atom trapped in the Kr lattice, we may write its energy as $E_v = v\epsilon_{SHO}$, with the energy level separation ϵ_{SHO} given in terms of the fundamental oscillator (radial) frequency, ω_{SHO}, as $\epsilon_{SHO} \equiv \hbar\omega_{SHO}$. The partition function for such a SHO is then given by

$$z_{SHO}(T, V) = (1 - e^{-\beta\epsilon_{SHO}})^{-1},$$

with $\beta \equiv (k_B T)^{-1}$, as usual. Obtain an expression for the internal energy $U_{int}(T, V)$ for the N xenon atoms. Show that if the mean internal energy $u(T, V)$ for a Xe atom is given by $u(T, V) = \tilde{u}_{SHO}(T, V)\epsilon_{SHO} \equiv \tilde{u}\,\epsilon_{SHO}$, then β can be obtained as $\ln[(1 + \tilde{u})/\tilde{u}]\,\epsilon_{SHO}$. Express the canonical partition function $z_{SHO}(T, V)$ in terms of $\tilde{u}$, and show that the entropy $s_{SHO}(T, V)$ per Xe atom is given by $s_{SHO}/k_B = S/(Nk_B) = (1 + \tilde{u})\ln(1 + \tilde{u}) - \tilde{u}\ln\tilde{u}$.

3. Determine the value of $\tilde{u}_{SHO}$ for a Xe atom isolated in a Kr matrix prepared at the temperature of liquid argon (boiling temperature 119.93 K) if the oscillator frequency for the Xe atoms is determined to be $\nu_{SHO} = 75.635$ GHz. Compute the corresponding expected value for the entropy s_{SHO}.

4. Consider the Kr/Xe system described in Problem 2 above, and show that a proper quantum mechanical treatment of the SHO model for the vibrational motion of the Xe atoms in the mixed crystal leads to the same expressions if the zero-point motion contribution to the internal energy $U_{SHO}(T, V, N)$ is identified as $U_{SHO}(0, V, N)$ and is subtracted from $U_{SHO}(T, V, N)$.

5. Determine the high-temperature limits of the internal energy, U, and the heat capacity at constant volume, C_V, for a system of N particles whose energy states are given by $\epsilon_m = \alpha m^2$, with α a constant having units of energy, and m having values $0, 1, 2, \cdots$.

6. Rearrange the Sackur–Tetrode equation for the translational entropy of a gas of N particles possessing only translational motion to obtain S_{tr} in terms of the translational internal energy, $U_{tr}(T)$, and the volume, V, of the gas. Invert your result to obtain U_{tr} as a function of S_{tr} and V. Obtain an expression for the total differential dU_{tr} of $U_{tr}(S_{tr}, V)$.

7. The Sackur–Tetrode expression for the molar entropy $\overline{S}$ can be obtained from Eq. (5.1.7) in the form

$$\overline{S} = R\left\{\frac{5}{2} + \ln\left[\left(\frac{2\pi mk_BT}{h^2}\right)^{\frac{3}{2}}\frac{V}{N_0}\right]\right\},$$

with $R \equiv N_0 k_B$ the universal gas constant, N_0 the Avogadro number, m the mass of an individual molecule, and V the volume.

(a) Show that this expression can also be written as

$$\overline{S} = R\ln V + \frac{3}{2}R\ln T + \frac{3}{2}R\ln\overline{M} - 11.073 \text{ cal/mol deg},$$

with $\overline{M} \equiv N_0 m$ the gram molecular weight of the substance and V measured in cm^3. This expression is often found in pre-SI convention literature. [*Warning*: this question is not as simple as it may appear.] What value does the final constant have when V is measured in m^3?

(b) Show that the expression from part (a) becomes

$$\overline{S} = \frac{5}{2}R\ln T + \frac{3}{2}R\ln\overline{M} - R\ln P - aR,$$

and evaluate a for the two cases in which (i) the pressure is in atm, $\overline{M}$ is in $g\,mol^{-1}$, and V is in cm^3 and (ii) the pressure is in bar, $\overline{M}$ is in $kg\,mol^{-1}$, and V is in m^3.

8. Use the Sackur–Tetrode equation to obtain an expression for the ratio of the entropy change at constant pressure to that at constant volume for an ideal monatomic gas.

9. Show that the expression obtained in Example 5.1 for the entropy of an ideal structureless gas in a gravitational field (with acceleration g) reduces to the Sackur–Tetrode equation for the entropy in the absence of gravitation.

10. Show that the internal energy and the entropy of an ideal gas in a gravitational field (with acceleration g) are extensive if the volume of the cylindrical column of gas employed for their derivation in Example 5.1 is doubled via a doubling of the cross-sectional area of the column. What happens if the volume of the cylindrical column is doubled by doubling the height of the column? How might this behaviour be understood?

11. An expression for the entropy $S_\eta(T, V; N)$ for a classical ideal gas in the presence of a gravitational field (characterized by the parameter $\eta \equiv \beta mgH$), with $\beta \equiv (k_BT)^{-1}$, m the mass of an ideal gas particle, g the acceleration due to gravity, and H the height of a cylindrical vessel of volume V, has been given in Example 5.1. Show that $\lim_{\eta \to 0} S_\eta(T, V; N)$ gives the Sackur–Tetrode equation for the entropy of a classical ideal gas (whose particles undergo only translational motions) in the absence of a gravitational field. Show also that for a sufficiently strong gravitational field, the classical expression for $S_\eta(T, V; N)$ can become negative, thereby violating the Third Law of Thermodynamics. Discuss the validity of the classical Sackur–Tetrode equation for this case.

12. The ground electronic level of the F atom has the term symbol $^2P_{\frac{3}{2}}$, and it has a very low-lying first excited electronic level (term symbol $^2P_{\frac{1}{2}}$) lying 404.1 cm^{-1} above the ground electronic level. What is the probability of finding a $F(^2P_{\frac{1}{2}})$ atom for temperatures of 100 K, 500 K, and 2000 K?

13. In Problem 12, we noted that the F atom has both a ground electronic level and a close-lying excited electronic level lying 404.0 cm^{-1} above the ground level. Evaluate the electronic partition function, z_{el}, for the F atom at 298 K, and determine the temperature for which $z_{el}(T) = 1$.

14. Evaluate the electronic partition function for atomic Si at 298 K, given that Si has a ground electronic term ^3P, with the $^3P_0$ level lying lowest in energy (i.e., it is the ground level), and the $^3P_1$ and $^3P_2$ levels lying 77.15 cm^{-1} and 223.31 cm^{-1}, respectively, above the ground level. The next highest electronic term, $^1D_2$, lies 6298.0 cm^{-1} above the ground level. At what temperature will the $^1D_2$ term contribute 0.1 to $z_{el}(T)$?

15. Consider a system made up of one mole of identical, noninteracting, nonlocalized atoms. Each atom can access only three energy levels, with energies and degeneracies: $\epsilon_1 = 0, \omega_1 = 1$; $\epsilon_2/k_B = 100 \text{ K}, \omega_2 = 3$; and $\epsilon_3/k_B = 300 \text{ K}, \omega_3 = 5$. Calculate z plus the average number of atoms in each level at temperature 200 K. Determine the average number of atoms in each level in the high-temperature (saturation) limit.

16. Show that Eq. (5.4.15) for the electronic contribution to the chemical potential for a pure gas indeed provides a logical wrap-up to the normal set of relevant thermodynamic functions for a thermodynamic system.

17. Suppose that a system consists of N noninteracting, nonlocalized, identical atoms A and that each atom can access only two (internal) quantum states (i.e., translational motion is not being considered here) having energies $\epsilon_1 = 0$ and $\epsilon_2 = \Delta\epsilon$. Obtain expressions for z_{int}, Z_{int}, U_{int}, $C_{V,int}$, and S_{int}, and evaluate these quantities for one mole of these atoms at 400 K, if $\Delta\epsilon = 1.0 \times 10^{-20}$ J. Moreover, determine the high-temperature limiting values for $\overline{U}_{int}$, $\overline{C}_{V,int}$, and $\overline{S}_{int}$. Finally, provide physical arguments, in terms of level populations, to support your results.

18. Obtain an expression for a modified vibrational partition function $\widetilde{z}_{\mathrm{vib}}(T)$, defined as $\widetilde{z}_{\mathrm{vib}}(T) \equiv \sum\limits_{v=0}^{\infty} e^{-v\Theta_{\mathrm{vib}}/T}$, in the classical limit that the discrete summation can be replaced by integration over the vibrational quantum number v. When is this result appropriate?

19. Show that the difference between the (classical) limiting value Nk_{B} of $C_V(T)$ and $C_V(T)$ integrated over the temperature for an ensemble of N quantum mechanical simple harmonic oscillators (SHOs) precisely represents the heat capacity contribution associated with the excluded zero-point energy of the N quantum mechanical SHOs.

20. An approximate partition function for a gas of hard spheres can be deduced from the monatomic gas expression simply by replacing the system volume V by the volume available to a gas, each of whose atoms occupies a finite volume typically denoted in thermodynamics by b, namely, replacing V by $V - Nb$ in the expressions for the partition functions for the gas. In this way, obtain expressions for the internal energy U and the pressure P for a system of N hard spheres.

21. If N atoms (or molecules) of a gas adsorbed onto M adsorption sites on a surface are bound sufficiently strongly that they cannot readily leave the surface, but not so strongly that they cannot move from an occupied site to an unoccupied site, then such an adsorbed gas can at low coverage (i.e., $N \ll M$) of the surface be treated as a two-dimensional gas for which the canonical partition function is

$$ Z(T, \mathcal{A}, N) = \frac{1}{N!} \left(\frac{2\pi m k_{\mathrm{B}} T}{h^2} \right)^N \mathcal{A}^N, $$

in which m is the mass of an atom (molecule) and $\mathcal{A}$ is the area of the surface upon which the atoms (molecules) are adsorbed. Obtain an expression for the internal energy, U, and the heat capacity, $C_{\mathcal{A}}$, for such a two-dimensional gas.

22. Obtain an approximation to the molar Einstein heat capacity $\overline{C}_V(T)$ that may be employed at temperatures T such that $T \ll \Theta_{\mathrm{E}}$; similarly, obtain an approximation to $\overline{C}_V(T)$ that applies at temperatures T such that $T \gg \Theta_{\mathrm{E}}$.

23. Show that in the limit that u goes to zero, the Debye function $D(u)$ defined by Eq. (5.6.19) goes to unity.

24. Show that for temperature T such that $T \ll \Theta_{\mathrm{D}}$, expression (5.6.21) for the molar Debye heat capacity $\overline{C}_V(T)$ can be approximated as

$$ \overline{C}_V(T) \simeq \frac{12R\pi^4}{5} \left(\frac{T}{\Theta_{\mathrm{D}}} \right)^3, $$

while for temperature T such that $T \gg \Theta_{\mathrm{D}}$, $\overline{C}_V(T)$ has the value $\overline{C}_V(T) \simeq 3R$.

25. Show, by substituting $\ln \Delta(T, P; N)$ from expression (5.1.15) for the isothermal–isobaric partition function for a gas of N structureless atoms into the formal expression (4.3.19a) for the entropy, that the result is indeed the Sackur–Tetrode equation (5.1.16a).

26. Employ Eq. (4.2.5) for the grand partition function $\Xi(\lambda, T, V)$ together with Eq. (4.2.30b) for the entropy S to show that it is given explicitly in terms of the single-particle canonical partition function z as

$$S = \overline{N} k_B \left[1 + T \left(\frac{\partial \ln z}{\partial T} \right)_V + \ln \left(\frac{z}{\overline{N}} \right) \right],$$

and that it reduces to the Sackur–Tetrode equation for the translational entropy for an (ideal) gas of structureless atoms.

27. Consider an ideal gaseous binary mixture consisting of N_1 A atoms and N_2 B atoms in thermal equilibrium at a temperature T in a container of volume V. Employ Eq. (4.1.28) for the canonical partition function for this binary mixture together with Eq. (5.1.1) to show that the internal energy, U, and the Helmholtz energy, A, are extensive functions. Obtain an expression for the total pressure, P_{mix}, of the binary AB mixture, and show that P_{mix} follows Dalton's law of partial pressures for an ideal gas mixture. Given the extensivity of U and A, what can you say about the entropy, S, the enthalpy, H, and the Gibbs energy, G?

28. An approximate canonical partition function for the Dieterici nonideal gas has the form

$$Z(N, V, T) = \frac{1}{N!} \exp \left\{ N \int_0^V \frac{e^{-aN/(N_0^2 k_B T y)}}{y - NB} \, dy \right\},$$

in which a and B are constants, and N_0 is the Avogadro number. Obtain expressions for the Dieterici equation of state and the internal energy U for this model nonideal gas. Your final expression for U will contain an integral for which there is no closed-form solution.

29. Assume that a monatomic Einstein solid, in which each atom behaves as a simple three-dimensional harmonic oscillator with the same fundamental vibrational frequency, ν_E, is in thermal equilibrium with its vapour, and that an energy ϵ is required to take an atom from the solid phase into the vapour phase. The canonical partition function for the N_s atoms making up the solid phase is, according to the Einstein model, given by

$$Z_s(T, V_s; N_s) = e^{\beta N_s \epsilon} \left[2 \sinh \left(\frac{\Theta_E}{2T} \right) \right]^{-3N_s},$$

while the canonical partition function for the N_g atoms making up the vapour phase is given by

$$Z_g(T, V_g; N_g) = \frac{z_{\text{trans}}^{N_g}(T, V_g)}{N_g!} \, z_{\text{int}}^{N_g}(T).$$

The total number of atoms, $N = N_s + N_g$, in the combined vapour–solid equilibrium system is constant. If the Helmholtz energy for the combined system is given by $A(T, V; N, N_g) = A_s(T; N_s) + A_g(T, V_g; N_g)$, minimize $A(T, V; N, N_g)$ with respect to N_g, and thereby show that N_g is given by

$$N_g = \frac{V}{\Lambda^3} \left[2 \sinh \left(\frac{\Theta_E}{2T} \right) \right]^3 e^{-\beta \epsilon},$$

in which $\Lambda(T)$ is the thermal de Broglie wavelength and $\beta = (k_B T)^{-1}$.

30. By treating the vapour in the previous problem as an ideal gas, show that the vapour pressure P associated with a monatomic crystal is given by

$$P = \frac{k_B T}{\Lambda^3(T)} \left[2 \sinh \left(\frac{\Theta_E}{2T} \right) \right]^3 e^{-\epsilon/(k_B T)}.$$

31. Traditionally, the condition for equilibrium to be established between two phases of the same substance is equality of the chemical potentials for the two phases. Show that minimization of the Helmholtz energy with respect to the number of particles in one of the phases, as carried out in Problem 29, leads to equality of the chemical potentials for the two phases.

References

1. P.T. Landsberg, J. Dunning-Davies, D. Pollard, Amer. J. Phys. **62**, 712 (1994)
2. G. Cook, R.H. Dickerson, Amer. J. Phys. **63**, 737 (1995)
3. I.N. Levine, *Quantum Chemistry*, 7th edn. (Pearson, New Jersey, 2014)
4. F.S. Levin, *An Introduction to Quantum Theory* (Cambridge University Press, Cambridge, 2002)
5. L. Pauling, E.B. Wilson, Jr., *Introduction to Quantum Mechanics* (McGraw-Hill, New York, 1935)
6. H.A. Bethe, R. Jackiw, *Intermediate Quantum Mechanics* (Benjamin, New York, 1968)
7. F. Hund, *Linienspektren und periodisches System der Elemente* (Springer-Verlag OHG, Berlin, 1927)

Chapter 6
Molecular Systems

This chapter determines the contributions of molecular internal states associated with molecular rotations, vibrations, and electronic excitations to thermodynamic functions. Diatomic molecules are treated using rigid-rotor and simple harmonic oscillator approximations, then corrected for small effects, including rotational quantum effects. Special attention is paid to homonuclear diatomic molecules, as nuclear interchange symmetry plays an important role in determining the nature of rotational contributions to thermodynamic properties and affects intensity ratios in molecular spectra. Summary expressions are given for the canonical partition function and for the translational, rotational, vibrational, and electronic contributions to the thermodynamic properties of diatomic and of both linear and nonlinear polyatomic molecules. Third-law and residual entropies are explained in terms of 'frozen in' states. Finally, an examination of the effect of hindered rotational motions is given, using ethane as a typical example.

6.1 Introduction

When we deal with polyatomic molecules we have to deal with translational motion of the polyatomic molecule, and in addition to that, also with the internal rotational, vibrational, and electronic motions associated with the molecule. This makes the derivation of expressions for the thermodynamic functions considerably more difficult than it is for monatomic species. It is nonetheless straightforward, provided that we make use of one or two simple, but remarkably accurate, approximations.

As a first step, we write the energy for a polyatomic molecule in the form

$$\epsilon = \epsilon_{tr} + \epsilon_{int} , \qquad (6.1.1)$$

© Springer Nature Switzerland AG 2021

F. R. W. McCourt, *Statistical Thermodynamics for Pure and Applied Sciences*,
https://doi.org/10.1007/978-3-030-52006-9_6

in which the subscript 'int' stands for 'internal'. We note that the translational and internal motions are rigorously uncoupled, so that we can always write the partition function in the form

$$z = z_{\text{tr}} z_{\text{int}} .$$

(6.1.2)

This result implies, in turn, that the full N-particle canonical partition function Z can also be written as a product, namely

$$Z = Z_{\text{tr}} Z_{\text{int}} ,$$

(6.1.3)

in which the N-particle translational and internal partition functions are defined as

$$Z_{\text{tr}} \equiv \frac{z_{\text{tr}}^N}{N!} , \qquad Z_{\text{int}} \equiv z_{\text{int}}^N ,$$

(6.1.4)

respectively. As a consequence of the partition function being split into the product of two factors, all thermodynamic state functions that depend upon the logarithm of the partition function become additive. A prime example is the Helmholtz energy A:

$$A = -k_{\text{B}} T \ln Z = -k_{\text{B}} T \ln Z_{\text{tr}} - k_{\text{B}} T \ln Z_{\text{int}}$$

$$= A_{\text{tr}} + A_{\text{int}} .$$

(6.1.5)

As a second example we should consider the Gibbs energy, which is given by

$$G = A + PV = \underbrace{A_{\text{tr}} + PV} + A_{\text{int}}$$

$$= \quad G_{\text{tr}} \quad + G_{\text{int}} .$$

(6.1.6)

This basic result of the additivity of the translational and internal state contributions to the thermodynamic state functions means that we do not need to reëvaluate the translational contributions, as we may employ the expressions that we have already obtained upon taking care that we put in the correct molecular mass. From this point onwards we may therefore focus our attention upon the nature of the internal state contributions to the partition function and, through it, to the thermodynamic state functions themselves.

We have established that each of the Helmholtz and Gibbs energies splits naturally into the sum of contributions from the translational and internal degrees of freedom. For the same reason, this is also the case for the thermodynamic internal energy, U, the entropy, S, and the enthalpy, H. In the preceding chapter, we have examined the temperature and volume dependence of the various thermodynamic functions for systems in which the constituents possess only translational degrees of freedom. In this chapter, we shall focus upon the contributions to the thermodynamic properties of substances that arise from the various types of internal states that molecules possess.

For example, we may express the internal state contributions to the internal energy $U(T, V)$ as

$$U_{\text{int}}(T) - U_{\text{int}}(0) = -\frac{N}{z_{\text{int}}}\left(\frac{\partial z_{\text{int}}}{\partial \beta}\right)_V \tag{6.1.7}$$

or, with $z_{\text{int}}(T)$ defined as the sum over internal energy states ϵ_r, with degeneracies ω_r, namely,

$$z_{\text{int}}(T) = \sum_r \omega_r e^{-\beta \epsilon_r}, \tag{6.1.8}$$

as

$$U_{\text{int}}(T) - U_{\text{int}}(0) = \frac{N}{z_{\text{int}}}\sum_r \omega_r \epsilon_r e^{-\beta \epsilon_r}. \tag{6.1.9}$$

The contribution from molecular internal states to the heat capacity at constant volume is then obtained from $U_{\text{int}}(T)$ as

$$k_B T^2 C_{V,\text{int}}(T) - N\left[\frac{1}{z_{\text{int}}}\sum_r \omega_r \epsilon_r^2 e^{-\beta \epsilon_r} \frac{1}{z_{\text{int}}^2}\sum_{r,r'} \omega_r \omega_{r'} \epsilon_r \epsilon_{r'} e^{-\beta \epsilon_r} e^{-\beta \epsilon_{r'}}\right] \tag{6.1.10a}$$

or in the form of the internal energy variance, as

$$k_B T^2 C_{V,\text{int}}(T) = N[\langle \epsilon_{\text{int}}^2 \rangle - \langle \epsilon_{\text{int}}\rangle^2]. \tag{6.1.10b}$$

A slightly different interpretation of the internal state contribution to the heat capacity can be obtained by symmetrizing the first term in expression (6.1.10a) to give

$$\frac{C_{V,\text{int}}(T)}{Nk_B} = \frac{1}{2}\sum_{r,r'} p_r p_{r'}\left(\frac{\Delta \epsilon_{rr'}}{k_B T}\right)^2 \tag{6.1.11a}$$

in terms of the energy level Boltzmann probabilities $p_r \equiv \omega_r e^{-\beta \epsilon_r}/z$. We shall find it helpful later to rearrange this result into a slightly different form, namely,

$$\frac{C_{V,\text{int}}(T)}{Nk_B} = \sum_{r=0}^{\infty} p_r \sum_{r'=r+1}^{\infty} p_{r'}\left(\frac{\Delta \epsilon_{rr'}}{k_B T}\right)^2, \tag{6.1.11b}$$

that emphasizes the connections between heat capacity and inelastic collisional transitions between molecular internal energy levels characterized by quantum numbers r and r'. Such an interpretation arises from our knowledge, for an ideal gas, for example, that the mechanism through which a new thermal equilibrium state is established following a change of system temperature (via energy exchange with the 'surroundings') is that of binary collisions between molecules or of collisions between molecules and container walls. In particular, we shall see that this provides an interesting and useful interpretation of the temperature dependence of $C_{V,\text{int}}(T)$.

The normal procedure in dealing with internal molecular states is to treat diatomic molecules first, both because they represent the simplest molecules, and because they represent an important special class of molecules with which we deal on a frequent basis. Only when the diatomic molecular expressions are well understood are the more general polyatomic molecules considered. We shall follow tradition here, and do the same thing.

6.2   Diatomic Molecules

When we consider diatomic molecules we have to visualize the various degrees of freedom possessed by such molecules. We can split the types of degrees of freedom into four separate classes. In the first place we have three translational degrees of freedom, as for the monatomic case, then we have molecular rotational degrees of freedom (essentially the two end-over-end tumbling motions of the molecules), one vibrational degree of freedom, and electronic degrees of freedom (similar to those of monatomics, but often lying at lower energies). The latter three types of degree of freedom have so far been lumped together as the 'internal' degrees of freedom, as opposed to the 'translational' degrees of freedom.

6.2.1   Setting the Stage

In order to treat the internal degrees of freedom we shall make some simplifying approximations, but ones that are nonetheless very good approximations from the point of view of thermodynamic calculations. Some of these approximations would not be considered good ones to make from the point of view of spectroscopy, however: we must always bear in mind the applications that we wish to consider prior to introducing simplifying approximations.

The first approximation that we shall make is to assume that the electronic and nuclear motions are decoupled. In its simplest terms this assumption means that the motions of the electrons in the molecule are unaffected by the motions of the much heavier, and slower, nuclei, so that we can in principle solve the Schrödinger equation for the electronic motion in the Coulomb field of fixed nuclei. This approximation is based upon the fact that electrons are about 2000 times less

Fig. 6.1 Illustration of potential energy curves for a diatomic molecule as a function of internuclear separation R. $D_e \equiv \mathfrak{D}_e$ is the depth, D_0 the dissociation energy, for the AB pair in the ground electronic state

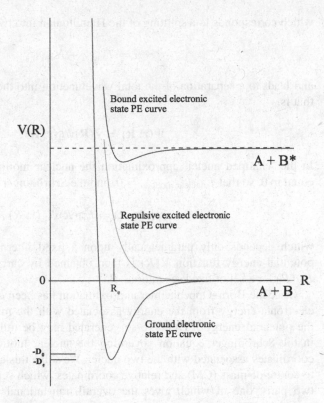

massive than typical nuclei in the molecule, so that the corresponding velocities differ by approximately two orders of magnitude. This approximation is sometimes known as the 'clamped nuclei' approximation, but is more commonly known as the *Born–Oppenheimer approximation*. It is only within this Born–Oppenheimer (BO) approximation that the concept of a potential function holds. Three types of potential energy curve are illustrated in Fig. 6.1: an interaction with an attractive well, typified by a diatomic molecule AB formed from two ground state neutral atoms A and B, a purely repulsive interaction, and an upper (attractive) curve corresponding to a diatomic molecule AB* formed from a ground state atom A and an excited state atom B*, as indicated by the higher-energy asymptote.

The Schrödinger equation for a diatomic molecule can be expressed in the generic form

$$\mathcal{H}\Psi(\mathbf{r}, \mathbf{R}) = E\Psi(\mathbf{r}, \mathbf{R}), \qquad (6.2.1)$$

with $\mathbf{r}$ referring (collectively) to electron coordinates and $\mathbf{R}$ to nuclear coordinates. Now if the total energy E for the molecule can be split into two parts, one corresponding to nuclei, the other to electrons, we may write E as

$$E = E_{\text{nuclear motions}} + E_{\text{electron motions}}, \qquad (6.2.2)$$

which corresponds to a splitting of the Hamiltonian into two parts, namely,

$$\mathcal{H} = \widehat{T}_{\text{nuclei}} + \mathcal{H}_{\text{el}}, \tag{6.2.3}$$

and leads to a separation of the total wavefunction into the product of two factors, that is,

$$\Psi(\mathbf{r}; \mathbf{R}) = \chi(\mathbf{R})\psi_{\text{el}}(\mathbf{r}; \mathbf{R}). \tag{6.2.4}$$

In the 'clamped nuclei' approximation the nuclear motion operator $\widehat{T}_{\text{nuclei}}$ is set equal to 0, so that $E_{\text{nuclear motions}} = 0$, and the Schrödinger equation reduces to

$$\mathcal{H}_{\text{el}}\psi_{\text{el}}(\mathbf{r}; R) = E_{\text{el}}(R)\psi_{\text{el}}(\mathbf{r}; R), \tag{6.2.5}$$

which depends only parametrically upon a fixed internuclear separation R. A potential energy function $V(R)$ is then obtained by carrying out calculations of $E_{\text{el}}(R)$ for a set of fixed separations R.

Once the Born–Oppenheimer approximation has been employed to separate the electronic energy from the energy associated with the motion of the two nuclei, the potential energy (PE) curve so generated may be utilized as the PE function in the Schrödinger equation governing the nuclear motion—in terms of the six coordinates associated with the two nuclei. We solve this equation by transforming to centre-of-mass (CM) and relative coordinates, which separates the equation into two parts, one of which gives the overall translational motion of the diatomic molecule. This equation is solved trivially. The second equation is concerned with the relative motion of the two nuclei, and is more difficult to solve because, for example, the energies are not simply additive. Were we wishing to compute positions of spectroscopic lines using the Schrödinger equation, we would not be able to proceed further except by carrying out accurate numerical solutions of the full Schrödinger equation. However, since we do not require such extreme accuracy for the calculation of thermodynamic quantities (actually, we require about two to three orders less accuracy for the calculation of thermodynamic quantities than we do for the calculation of spectroscopic quantities), we can make two further simplifying approximations at this stage:

- replace $V(R)$ by a parabola which fits near the minimum;
- separate vibrational and rotational motions from one another by assuming that as far as rotations are concerned, the molecule has a fixed internuclear distance, which we shall designate by R_{e}: this approximation is commonly known as the 'rigid-rotor-SHO approximation'.

We can express the end result of these approximations mathematically by writing down the energy for the diatomic molecule in the form

$$\epsilon = \epsilon_{\text{tr}} + \epsilon_{\text{rot}} + \epsilon_{\text{vib}} + \epsilon_{\text{el}}, \tag{6.2.6}$$

which gives a corresponding factorization of the molecular partition function:

$$Z = \frac{z^N}{N!}; \qquad z(V, T) = z_{tr}(V, T)z_{rot}(T)z_{vib}(T)z_{el}(T). \qquad (6.2.7)$$

The translational and electronic factors can be written down immediately from our knowledge of their forms for the monatomic species, as

$$z_{tr}(V, T) = \left(\frac{2\pi M k_B T}{h^2}\right)^{\frac{3}{2}} V, \qquad M = m_1 + m_2, \qquad (6.2.8)$$

and

$$z_{el}(T) = \omega_{el} e^{-\beta \epsilon_{el}}. \qquad (6.2.9)$$

For the separated atoms giving the zero of internal energy, the energy levels of the SHO approximation to the vibrational motion are simply given as $-\mathfrak{D}_e + \frac{1}{2}h\nu$, $-\mathfrak{D}_e + \frac{3}{2}h\nu$, and so on ($\mathfrak{D}_e > 0$), so that

$$z_{el}z_{vib} = \omega_{el}\sum_v e^{-\beta(-\mathfrak{D}_e + \epsilon_v)}$$

$$= \underbrace{\omega_{el}e^{\beta\mathfrak{D}_e}}_{z_{el}(T)} \underbrace{\sum_v e^{-\beta\epsilon_v}}_{z_{vib}(T)}. \qquad (6.2.10)$$

If low-lying electronic states are also present, then a further approximation can be made, namely, that rotational and vibrational states are the same in the different electronic states. The goodness of this approximation can often be tested directly, since a great deal of spectroscopic information is currently available for a very large range of diatomic species.

These approximations give us a simple additivity of most of the thermodynamic functions. Thus, for example,

$$A = -k_B T \ln Z_N = A_{tr} + A_{rot} + A_{vib} + A_{el}, \qquad (6.2.11)$$

in which the various components are related to the internal state partition functions by

$$A_{tr} = -Nk_B T \ln\left(\frac{z_{tr}e}{N}\right); \qquad A_{rot} = -Nk_B T \ln z_{rot}; \qquad (6.2.12)$$

$$A_{vib} = -Nk_B T \ln z_{vib}; \qquad A_{el} = -Nk_B T \ln z_{el}. \qquad (6.2.13)$$

Notice that due to the structures assumed by z_{tr}, z_{rot}, z_{vib}, and z_{el}, the equation of state for the diatomic gas, given by

$$P = k_B T \left(\frac{\partial \ln Z}{\partial V} \right)_{N,T} = N k_B T \left(\frac{\partial \ln z_{tr}}{\partial V} \right)_{N,T} = \frac{N k_B T}{V}, \qquad (6.2.14)$$

does not depend upon z_{rot}, z_{vib}, or z_{el}.

6.2.2 The Diatomic Vibrational Partition Function

We have seen that a diatomic molecule has a total of six degrees of freedom: five of them are taken up by the three translational and two rotational degrees of freedom associated with the molecule as a rigid entity, leaving only a single vibrational degree of freedom. The only type of vibrational motion that a diatomic molecule can undergo is the stretching and compression of the chemical bond holding the diatomic molecule together: the energy associated with this vibrational motion thus depends upon the displacement between the two nuclei at any given time relative to their (equilibrium) displacement determined by the position of the minimum in the potential energy function.

The Simple Harmonic Oscillator Approximation

We shall approximate the potential energy for the vibrational motion of a diatomic molecule as that of a simple harmonic oscillator (SHO) with a potential energy minimum of $-\mathfrak{D}_e$ located at an internuclear separation R_e (see Fig. 6.1), and given by

$$V(R) = -\mathfrak{D}_e + \tfrac{1}{2} k (R - R_e)^2, \qquad (6.2.15)$$

with the zero of energy the separated stationary atoms (SSA) zero, and R_e the equilibrium separation of the two nuclei. When describing the SHO vibrational contribution to the partition function, it will prove more convenient to use the bottom of the potential well as the zero for vibrational energy than to use the SSA zero. Accordingly, we shall drop $\mathfrak{D}_e$ from $V(R)$, and use the form $\omega_{el} e^{\beta \mathfrak{D}_e}$ for z_{el}, as in Eq. (6.2.10).

In order to quantify our description of the vibrational motion of a diatomic molecule, we shall choose a Cartesian axis system, with the z-axis coincident with the internuclear axis. The nuclei, with masses m_1 and m_2, are then located on the z-axis with $z_1 < z_2$ (without loss of generality), as shown in Fig. 6.2. The internuclear distance is then given as $R = z_2 - z_1$, and the SHO potential energy function (6.2.15) becomes

Fig. 6.2 Coordinate system for describing vibrational motion of a diatomic molecule whose constituent atoms, of masses m_1 and m_2, lie along the z-axis

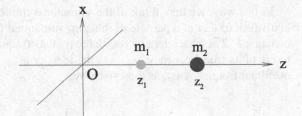

$$V(R) = \tfrac{1}{2} k(z_2 - z_1 - R_e)^2. \tag{6.2.16}$$

If we now utilize Newton's Second law in the forms

$$F_i = m_i a_i = m_i \frac{\mathrm{d}^2 R_i}{\mathrm{d}t^2}; \qquad F_i = -\left(\frac{\partial V}{\partial R_i}\right)_{R_j \neq R_i},$$

and apply them to our diatomic system, we obtain the two equations

$$m_1 \frac{\mathrm{d}^2 z_1}{\mathrm{d}t^2} = -\frac{\partial V(z_1, z_2)}{\partial z_1} = k(z_2 - z_1 - R_e),$$

$$m_2 \frac{\mathrm{d}^2 z_2}{\mathrm{d}t^2} = -\frac{\partial V(z_1, z_2)}{\partial z_2} = -k(z_2 - z_1 - R_e).$$

Upon multiplying the first of these two equations by m_2, the second by m_1, then subtracting the first result from the second, we see that

$$m_1 m_2 \frac{\mathrm{d}^2 (z_2 - z_1 - R_e)}{\mathrm{d}t^2} = -(m_1 + m_2)k(z_2 - z_1 - R_e)$$

or

$$\frac{\mathrm{d}^2 (z_2 - z_1 - R_e)}{\mathrm{d}t^2} = -\left(\frac{1}{m_1} + \frac{1}{m_2}\right) k(z_2 - z_1 - R_e).$$

Upon introducing the reduced mass m_r through the defining relation

$$\frac{1}{m_r} \equiv \frac{1}{m_1} + \frac{1}{m_2}, \tag{6.2.17}$$

and defining the displacement from equilibrium, ξ, via $\xi \equiv z_2 - z_1 - R_e$, we see that our result can be rewritten in the traditional manner in the form of a simple second-order ordinary differential equation, namely,

$$\frac{\mathrm{d}^2 \xi}{\mathrm{d}t^2} = -\frac{k}{m_r} \xi \equiv -\omega_{osc}^2 \xi. \tag{6.2.18}$$

In this way, we may think of the vibrational motion of a diatomic molecule as equivalent to that of a particle of mass m_r suspended from a spring that has a force constant k. The characteristic oscillator (radial) frequency ω_{osc} can be replaced in terms of its true frequency, ν_{osc}, via the identity $\omega_{osc} \equiv 2\pi \nu_{osc}$. The characteristic oscillator frequency ν_{osc} is thus given by

$$\nu_{osc} = \frac{1}{2\pi}\sqrt{\frac{k}{m_r}} . \tag{6.2.19}$$

For a diatomic molecule, the force constant k can be directly related to the strength of the chemical bond between the constituent atoms.

In quantum mechanics, the SHO energy eigenstates are characterized by the (eigen)energies ϵ_v given as

$$\epsilon_v = (v + \tfrac{1}{2})h\nu_{osc} , \tag{6.2.20}$$

in terms of the vibrational quantum number v, which has values $0, 1, 2, \ldots$. The vibrational properties of a diatomic molecule can therefore be represented in lowest order in terms of SHO eigenstates, provided that ν_{osc} is given by Eq. (6.2.19).

The SHO Model Partition Function: Vibrational Contribution to the Thermodynamic State Functions

If we define a characteristic vibrational temperature Θ_{vib} for a diatomic molecule via

$$\Theta_{vib} \equiv \frac{h\nu_{osc}}{k_B} , \tag{6.2.21}$$

then we may write the relevant vibrational partition function directly as

$$z_{vib}(T) = \frac{e^{-\Theta_{vib}/(2T)}}{1 - e^{-\Theta_{vib}/T}} . \tag{6.2.22}$$

The vibrational contributions to the thermodynamic state functions A, U, C_V, and S are then those from a collection of N simple uncoupled one-dimensional harmonic oscillators, much as we have considered in Chap. 4. They thus will have the same functional form, and are given by

$$A_{vib}(T) - \tfrac{1}{2}Nk_B\Theta_{vib} = Nk_BT \ln(1 - e^{-\Theta_{vib}/T}), \tag{6.2.23}$$

$$U_{vib}(T) - \tfrac{1}{2}Nk_B\Theta_{vib} = Nk_B\frac{\Theta_{vib}}{e^{\Theta_{vib}/T} - 1} , \tag{6.2.24}$$

$$C_{V,\text{vib}}(T) = Nk_{\text{B}} \left(\frac{\Theta_{\text{vib}}}{T}\right)^2 \frac{e^{-\Theta_{\text{vib}}/T}}{(1 - e^{-\Theta_{\text{vib}}/T})^2}, \tag{6.2.25}$$

$$S_{\text{vib}}(T) = Nk_{\text{B}} \left[\frac{\Theta_{\text{vib}}/T}{e^{\Theta_{\text{vib}}/T} - 1} - \ln(1 - e^{-\Theta_{\text{vib}}/T})\right]. \tag{6.2.26}$$

These expressions represent the contribution of the excited vibrational levels to the thermodynamic properties of a diatomic gas.

We have isolated the ground vibrational contribution of $\frac{1}{2}Nk_{\text{B}}\Theta_{\text{vib}}$ to the Helmholtz energy $A(T)$ and the thermodynamic internal energy $U(T)$ in anticipation of its important role in the formation of the diatomic molecule. Note that there is no need to present explicit expressions for the vibrational contributions $H_{\text{vib}}(T)$ to the enthalpy and $G_{\text{vib}}(T)$ to the Gibbs energy, as they are equal to $U_{\text{vib}}(T)$ and $A_{\text{vib}}(T)$, respectively.

Beyond the SHO Approximation: Anharmonicity Effects

Upon examination of Fig. 6.3, it can be seen that the SHO approximation will break down for sufficiently large values of the vibrational quantum number v. For values of v for which the quadratic dependence of the SHO approximation for $V(R)$ provides a reasonably accurate representation of the true potential energy function, a quadratic anharmonicity term in the energy expression for E_{vib} may suffice. The vibrational energy is then written traditionally as [1]

$$\overline{E}_{\text{vib}} \equiv \overline{\epsilon}_v = \overline{\omega}_e \left(v + \tfrac{1}{2}\right) - \overline{\omega}_e x_e \left(v + \tfrac{1}{2}\right)^2 + \cdots, \quad v = 0, 1, 2, \ldots, \tag{6.2.27}$$

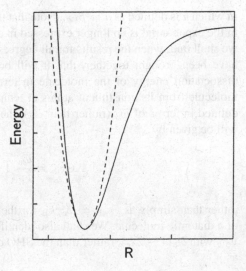

Fig. 6.3 Simple harmonic oscillator approximation to a realistic (anharmonic) diatomic molecule potential energy function

in which $\overline{\omega}_e \equiv \overline{\nu}_{osc}$ is the fundamental oscillator frequency in cm^{-1} units and x_e, called the anharmonicity coefficient, is a small number.

The energy of the ground vibrational level is thus given to a good approximation by

$$\overline{\epsilon}_0 \simeq \tfrac{1}{2}\overline{\omega}_e \left(1 - \tfrac{1}{2}x_e\right) . \qquad (6.2.28)$$

Energies of the excited vibrational levels may now be expressed relative to the ground vibrational level as

$$\overline{\epsilon}_v - \overline{\epsilon}_0 \simeq \overline{\omega}_e v - \overline{\omega}_e x_e(v^2 + v) + \cdots = \overline{\omega}_0 v - \overline{\omega}_0 x_e v(v-1) + \cdots , \qquad (6.2.29)$$

with $\overline{\epsilon}_0 \simeq \tfrac{1}{2}\omega_0$ and $\omega_0 \equiv \omega_e - 2\omega_e x_e$. This final form for the energy expression has been chosen because $\overline{\epsilon}_1 - \overline{\epsilon}_0$ gives the first vibrational spacing, which can be obtained directly from vibrational spectra. Moreover, as pointed out by Mayer and Mayer [2], this form will also lead to more accurate SHO model contributions to the thermodynamic properties as well as provide correction terms associated with the anharmonicity that will be smaller for any given temperature. We shall follow Mayer and Mayer in choosing this form for the vibrational energy levels.

The vibrational partition function is then defined as

$$z_{vib}(T) = \sum_{v=0}^{\infty} e^{-\beta E_{vib}}$$

$$= e^{-\beta \epsilon_0} \sum_{v=0}^{\infty} e^{-[v - x_e v(v-1)]u} , \qquad (6.2.30)$$

in which u is defined as $u \equiv \beta \overline{\omega}_0$. Note that the ground vibrational energy appearing in the exponential is no longer expressed in cm^{-1} units: this has been done because we shall find, when the results for all degrees of freedom for our diatomic molecule have been brought together, that it will be convenient to re-express ϵ_0 and the dissociation energy of the molecule in terms of the energy of formation of the molecule from its constituent atoms at temperature $0\,K$. We note also that with u defined in terms of $\overline{\omega}_0$ (rather than $\overline{\omega}_e$), the characteristic vibrational temperature will be given by

$$\Theta_{vib} \equiv \frac{\overline{\omega}_e}{\overline{k}_B}(1 - 2x_e) , \qquad (6.2.31)$$

rather than simply as $\Theta_{vib} = \overline{\omega}_e/\overline{k}_B$ for the SHO model of the vibrational motion of a diatomic molecule. We shall also identify the fundamental oscillator frequency $\overline{\nu}_{osc}$ with $\overline{\omega}_e(1 - 2x_e)$, rather than the SHO quantity $\overline{\omega}_e$: thus, we shall express $\overline{\nu}_{osc}$ as

$$\overline{\nu}_{osc} \equiv \overline{\omega}_e(1 - 2x_e) , \qquad (6.2.32)$$

as this is the value associated with the $v = 0 \to 1$ transition in the diatomic molecule treated as a SHO. The result (6.2.30) may also be written in the form

$$z_{\mathrm{vib}}(T) = \mathrm{e}^{-\beta\epsilon_0} \sum_{v=0}^{\infty} \mathrm{e}^{-uv} \mathrm{e}^{ux_e v(v-v)} \,. \tag{6.2.33}$$

For $x_e u \ll 1$, we may expand the second exponential in the summand of Eq. (6.2.33), to obtain

$$z_{\mathrm{vib}}(T) \simeq \mathrm{e}^{-\beta\epsilon_0} \sum_{v=0}^{\infty} (\mathrm{e}^{-u})^v [1 + x_e u(v^2 - v) + \cdots]$$

$$= \mathrm{e}^{-\beta\epsilon_0} \left[\frac{1}{1 - \mathrm{e}^{-u}} + x_e u \sum_{v=0}^{\infty} \mathrm{e}^{-uv}(v^2 - v) + \cdots \right]. \tag{6.2.34}$$

This expression can be further simplified by employing parameter differentiation in the second term to obtain

$$z_{\mathrm{vib}}(T) = \mathrm{e}^{-\beta\epsilon_0} \left\{ \frac{1}{1 - \mathrm{e}^{-u}} + x_e u \left[\frac{\mathrm{d}^2}{\mathrm{d}u^2} + \frac{\mathrm{d}}{\mathrm{d}u} \right] \frac{1}{1 - \mathrm{e}^{-u}} + \cdots \right\}$$

$$= \mathrm{e}^{-\beta\epsilon_0} \widetilde{z}_{\mathrm{vib}}(T) \left[1 + \frac{2x_e u}{(\mathrm{e}^u - 1)^2} + \cdots \right], \tag{6.2.35}$$

with $\widetilde{z}_{\mathrm{vib}}(T)$ as defined previously in Sect. 5.5.1. We shall rewrite this result as

$$z_{\mathrm{vib}}(T) = \mathrm{e}^{-\beta\epsilon_0} \widetilde{z}_{\mathrm{vib}}(T) f_{\mathrm{anh}}(T) \,, \tag{6.2.36}$$

in terms of an anharmonic correction factor $f_{\mathrm{anh}}(T)$ given by

$$f_{\mathrm{anh}}(T) = 1 + \frac{2x_e u}{(\mathrm{e}^u - 1)^2} \,. \tag{6.2.37}$$

The result (6.2.37) thus provides a working expression for the lowest-order vibrational anharmonicity correction to the SHO vibrational partition function $z_{\mathrm{vib}}(T)$.

In order to determine the relative importance of anharmonicity terms for the evaluation of thermodynamic state functions, we shall examine its contributions to the vibrational components of the thermodynamic internal energy $U(T)$ and entropy $S(T)$. We have seen that the expression for $U_{\mathrm{vib}}(T)$ for a system of N molecules is defined by the relation

$$U_{\mathrm{vib}}(T) = Nk_{\mathrm{B}}T^2 \frac{\mathrm{d}\ln z_{\mathrm{vib}}(T)}{\mathrm{d}T} \,,$$

and we have shown that for $x_e u \ll 1$, $z_{vib}(T)$ can be approximated by Eq. (6.2.36). We may therefore obtain $\ln z_{vib}(T)$ as

$$\ln z_{vib}(T) = -\beta\epsilon_0 - \ln(1 - e^{-u}) + \ln f_{anh}$$

$$\simeq -\beta\epsilon_0 - \ln(1 - e^{-u}) + \frac{2ux_e}{(e^u - 1)^2} + \cdots . \qquad (6.2.38)$$

Within this approximation the anharmonicity contributions to the thermodynamic state functions are given by

$$A_{vib}^{anh}(T) = -Nk_B \left[\frac{2x_e \Theta_{vib}}{(e^u - 1)^2} + \cdots \right], \qquad (6.2.39)$$

$$U_{vib}^{anh}(T) = Nk_B \frac{2x_e \Theta_{vib}}{(e^u - 1)^3} (2ue^u - e^u + 1), \qquad (6.2.40)$$

$$C_{V,vib}^{anh}(T) = Nk_B \frac{2x_e u^2 e^u}{(e^u - 1)^4} \left(4ue^u - 4e^u + 2u + 4 \right), \qquad (6.2.41)$$

$$S_{vib}^{anh}(T) = Nk_B \frac{4x_e u^2 e^u}{(e^u - 1)^3} , \qquad (6.2.42)$$

in which u is given by $u = \Theta_{vib}/T$, with Θ_{vib} defined via Eq. (6.2.31).

To develop a feeling for how significant the vibrational anharmonicity corrections to thermodynamic state functions will be, let us examine values of anharmonicity corrections for the molar vibrational internal energy and entropy for some typical diatomic gases. As a first example, we shall consider a relatively light molecule, CO, and as a second example we shall consider a relatively heavy molecule, IBr.

Example 6.1 Carbon Monoxide, CO.

The relevant spectroscopic constants may be obtained from the Huber and Herzberg [3] compilation of diatomic molecular spectroscopic constants as $\overline{\omega}_e(CO) = 2169.81358\,cm^{-1}$ and $\overline{\omega}_e x_e(CO) = 13.2881\,cm^{-1}$. From the values for $\overline{\omega}_e$ and $\overline{\omega}_e x_e$, we obtain $x_e(CO) = 0.00612$ and we find that the characteristic vibrational temperature for CO is given as $\Theta_{vib}(CO) = 3083.64\,K$. We shall evaluate the contributions $\overline{U}_{vib}(T)$ and $\overline{S}_{vib}(T)$ for one mole of CO gas at temperatures $T = 300\,K$ (representing room temperature) and $T = 1000\,K$ using the anharmonic correction formulae that we have obtained earlier.

The anharmonicity contribution to the molar vibrational internal energy $\overline{U}_{vib}(T) - N_0\epsilon_0$ at temperature $T = 300\,K$ is found to be about 10^7 times smaller than the SHO model contribution, and is hence negligible for temperatures near room temperature. At $T = 1000\,K$, we find that the anharmonicity contribution of $0.4698\,R$ is 0.32% of the SHO contribution of $147.94\,R$. Similarly, the anharmonicity contribution to the vibrational molar entropy $\overline{S}_{vib}(T)$ is about 10^8 times smaller than the SHO model contribution for $T = 300\,K$, while for $T = 1000\,K$, the anharmonicity contribution of $5.5 \times 10^{-4}\,R$ is about 0.3% of the SHO contribution of $0.19R$. The anharmonic contribution to the thermodynamic properties of CO is thus quite small even for temperatures of the order of $1000\,$K. $\qquad\square$

In hindsight, however, this should not be unexpected, as a temperature of $1000\,$K is still slightly less than one-third the value of the characteristic vibrational temperature, so that many excited vibrational states are still thermally inaccessible to CO molecules even at that temperature.

Example 6.2 Iodine Bromide, IBr.

The spectroscopic constants for IBr are given by Huber and Herzberg [3] as $\overline{\omega}_e(\text{IBr}) = 258.640\,\text{cm}^{-1}$ and $\overline{\omega}_e x_e(\text{IBr}) = 0.8140\,\text{cm}^{-1}$, so that $x_e(\text{IBr}) = 0.00303$ and $\Theta_{\text{vib}}(\text{IBr}) = 381.7\,$K. Here, too, we shall examine the contributions to $U_{\text{vib}}(T) - N_0\epsilon_0$ and $S_{\text{vib}}(T)$ arising from the leading anharmonicity correction to the vibrational energy of a diatomic molecule for the two representative temperatures $300\,$K and $1000\,$K.

For IBr at temperature $300\,$K, the anharmonicity contribution to the molar vibrational internal energy $\overline{U}_{\text{vib}}(T) - N_0\epsilon_0$ is obtained from Eq. (6.2.40) as $0.89R$, which is 0.60% of the value $148.57R$ contributed by the SHO model expression. For temperature $T = 1000\,$K, the anharmonic contribution of $0.15R$ is 1.83% of the SHO model contribution of $821.27R$. The SHO model contribution to the molar entropy of IBr for $T = 300\,$K is found to be $0.824R$, while the anharmonicity contribution is obtained as $0.004R$, and represents 0.5% of the contribution at the level of the SHO model. The corresponding values for the entropy contributions for $T = 1000\,$K are $1.969R$ from the SHO model expression and $0.026R$ from the anharmonicty correction, which is 1.3% of the SHO model contribution. $\qquad\square$

As for the CO example previously considered, the final outcome for IBr should not be too surprising, as $T = 300\,$K is already very close to the characteristic vibrational temperature for IBr, and $T = 1000\,$K is nearly three times as large, so that many IBr vibrational levels will be thermally accessible at both temperatures.

6.2.3 The Diatomic Rotational Partition Function

The Rigid-Rotor Approximation

The concept of angular momentum is quite important in dealing with electronic motion in atoms and with both electronic and nuclear motions in molecules. Its description is simplest for a particle of mass m moving in circular motion about a fixed point. With a small additional step we may also consider the case of the rotational motion of two atoms in a diatomic molecule about their common centre-of-mass. To see how this comes about, let us consider a rigid rotor (RR) consisting of two atoms, with masses m_1 and m_2 separated by a fixed distance $r = r_1 + r_2$ and rotating about their centre-of-mass, as shown in Fig. 6.4.

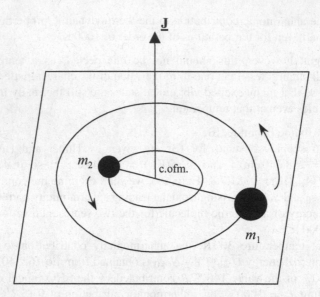

Fig. 6.4 Rotational motion of a pair of bonded atoms about their centre-of-mass

There are two ways in which we can think of the motion of such a diatom. One is to describe the motion in terms of the individual atoms, and obtain an expression for the total kinetic energy of the two atoms in terms of time derivatives of the Cartesian coordinates associated with each atom. Another way of thinking about the motion of the two atoms is to recognize that their motions are coupled by the fixed distance between them, thereby constraining them to move in such a way that their centre-of-mass moves rectilinearly at the same time as the atoms rotate about an axis perpendicular to the plane of their rotation and passing through the centre-of-mass. The first way of thinking about the overall motion leads to mathematically coupled equations that have to be solved simultaneously, which is itself a difficult task. The second way of thinking about the overall motion leads to a decoupling of the equations of motion for the centre-of-mass and relative (rotational) motions. This is the procedure that we shall employ here.

If we locate atoms 1 and 2 at positions $\mathbf{r}_1$ and $\mathbf{r}_2$ relative to a fixed origin O, as shown in Fig. 6.5, then we may describe the position $\mathbf{R} = (X, Y, Z)$ of the centre-of-mass for the diatom via

$$MX = m_1x_1 + m_2x_2; \qquad MY = m_1y_1 + m_2y_2; \qquad MZ = m_1z_1 + m_2z_2,$$

in component form, or by

$$M\mathbf{R} = m_1\mathbf{r}_1 + m_2\mathbf{r}_2, \tag{6.2.43}$$

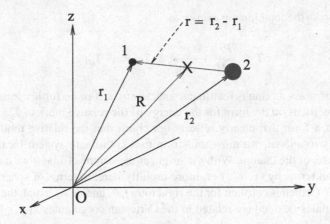

Fig. 6.5 Centre-of-mass position vector **R** and relative position vector **r** in terms of the position vectors r_2 and r_1 for the atoms making up a diatomic molecule

in vector format, in which $M = m_1 + m_2$ is the total mass of the diatomic molecule. The position **R** may be thought of as the location of a 'fictitious particle' of mass M. This defining relation follows from the fundamental physical law of conservation of total linear momentum, $\mathbf{P} = \mathbf{p}_1 + \mathbf{p}_2$, for the motion of the two atoms. At the same time, we know that the position of atom 1 relative to atom 2 is given (see Fig. 6.5) by

$$\mathbf{r} = \mathbf{r}_2 - \mathbf{r}_1 . \tag{6.2.44}$$

We may invert Eqs. (6.2.43) and (6.2.44) to obtain $\mathbf{r}_1$ and $\mathbf{r}_2$ in terms of **R** and **r** as

$$\mathbf{r}_1 = \mathbf{R} - \frac{m_2}{M}\mathbf{r}, \qquad \mathbf{r}_2 = \mathbf{R} + \frac{m_1}{M}\mathbf{r}. \tag{6.2.45}$$

It is now relatively straightforward to establish that the total kinetic energy, given by the sum of the kinetic energies of the two atoms , can be re-expressed as the sum of a term for the centre-of-mass motion plus a term for the relative motion of the first nucleus relative to the second one: thus

$$T = \tfrac{1}{2}m_1\dot{r}_1^2 + \tfrac{1}{2}m_2\dot{r}_2^2$$
$$= \tfrac{1}{2}M\dot{R}^2 + \tfrac{1}{2}m_r\dot{r}^2 , \tag{6.2.46}$$

in which the overdot represents the derivative with respect to time and m_r is the reduced mass of the diatomic molecule.

If we now define the centre-of-mass momentum **P** and the relative momentum **p** in terms of M and m_r by

$$\mathbf{P} \equiv M\dot{\mathbf{R}}; \qquad \mathbf{p} \equiv m_r\dot{\mathbf{r}} , \tag{6.2.47}$$

we may express the total kinetic energy as

$$T = \frac{P^2}{2M} + \frac{p^2}{2m_r} \equiv T_{CM} + T_{rel} \,. \tag{6.2.48}$$

The centre-of-mass motion is rectilinear and, as such, is of no further interest to us. We shall now focus on the form for the energy of the relative motion, T_{rel}.

To obtain a form that clearly reflects our claim that the relative motion of the two atoms is rotational, we must transform to a coordinate system located at the centre-of-mass of the diatom. With the origin at the centre-of-mass, we can express **r** in Cartesian terms by (x, y, z) or, more usefully here, in terms of spherical polar coordinates r (which is constant for the rigid rotor), θ, and ϕ. As usual, the spherical polar coordinates (r, θ, ϕ) are related to the Cartesian coordinates x, y, z by

$$x = r \sin\theta \cos\phi \,, \quad y = r \sin\theta \sin\phi \,, \quad z = r \cos\theta \,. \tag{6.2.49}$$

The kinetic energy of relative motion in Eq. (6.2.48) is given by $\frac{1}{2}m_r \dot{\mathbf{r}}^2 = \frac{1}{2}m_r(\dot{x}^2 + \dot{y}^2 + \dot{z}^2)$. We can obtain time derivatives of the Cartesian coordinates from Eqs. (6.2.49) as

$$\dot{x} = r \cos\theta \cos\phi \, \dot{\theta} - r \sin\theta \sin\phi \, \dot{\phi} \,,$$

$$\dot{y} = r \cos\theta \sin\phi \, \dot{\theta} + r \sin\theta \cos\phi \, \dot{\phi} \,, \tag{6.2.50}$$

$$\dot{z} = -r \sin\theta \, \dot{\theta} \,,$$

where we have used the fact that r is constant (because of the fixed distance between the two atoms), together with the chain and product rules. Using these expressions, $\dot{\mathbf{r}}^2$ can be reduced (by squaring, expanding, and employing trigonometric identities) to

$$\dot{\mathbf{r}}^2 = r^2(\dot{\theta}^2 + \sin^2\theta \, \dot{\phi}^2) \,. \tag{6.2.51}$$

Without loss of generality we may choose $\theta = \pi/2$ and $\dot{\theta} = 0$ by choosing the plane of rotation to be perpendicular to the Cartesian z-axis (equivalently, we could require that the axis of rotation of the two atoms coincides with the z-axis). Thus, we see that $\dot{\mathbf{r}}^2$ is given by

$$\dot{\mathbf{r}}^2 = r^2 \dot{\phi}^2 \,. \tag{6.2.52}$$

The time-rate-of-change of ϕ defines the angular velocity, so that the kinetic energy for the relative motion of the two atoms is given by

$$T_{rel} = = \tfrac{1}{2}m_r \dot{\mathbf{r}}^2 = \tfrac{1}{2}m_r r^2 \omega^2 = \tfrac{1}{2} I \omega^2 \,, \tag{6.2.53}$$

in which $I = m_r r^2$ is the moment-of-inertia of the rotating diatomic molecule. Finally, if we recognize that in classical mechanics the angular momentum J associated with circular (rotational) motion is $J \equiv I\omega$, we see that the kinetic energy for the relative rotational motion of the two atoms is given by

$$T_{rel} = \frac{J^2}{2I}. \tag{6.2.54}$$

Quantum mechanically, the Hamiltonian for the motion of a rigid rotor can be obtained from Eq. (6.2.54) as

$$\mathcal{H}_{rot} = \frac{\hbar^2 \widehat{J^2}}{2I}, \tag{6.2.55}$$

with $\widehat{J^2}$ a (mathematical) self-adjoint operator that does not carry physical units, and has eigenvalues that are real numbers. As the orientation of a rigid rotor is completely specified by two angles, θ and ϕ, we may designate the rigid-rotor eigenfunctions by $\psi_{rot}(\theta, \phi)$, so that the Schrödinger equation for rigid-rotor motion can be written as

$$\mathcal{H}_{rot}\psi_{rot}(\theta, \phi) = F_{rot}\psi_{rot}(\theta, \phi),$$

or, equivalently,

$$\frac{\hbar^2}{2I}\widehat{J^2}\psi_{rot}(\theta, \phi) = E_{rot}\psi_{rot}(\theta, \phi). \tag{6.2.56}$$

We shall designate the eigenvalues of $\widehat{J^2}$ by $j(j+1)$, $j = 0, 1, 2, \ldots$, so that the corresponding (rotational) energies E_{rot} are

$$E_{rot} \equiv \epsilon_j = \frac{\hbar^2}{2I} j(j+1). \tag{6.2.57}$$

Heteronuclear Diatomic Molecules

We shall find that there is a significant difference between the expressions for the rotational partition functions of heteronuclear and homonuclear diatomic molecules. We shall also find that this important difference can ultimately be associated with the interchange symmetries of indistinguishable nuclei.

What we have said in the previous subsection provides us with sufficient information to proceed with an evaluation of the partition function for the rotational motion of heteronuclear diatomic molecules, for which by definition the nuclei are distinguishable by virtue of their different masses. From the basic definition of a partition function in terms of the sum over a set of appropriate energy states, we obtain

$$z_{rot}(T) = \sum_{states} e^{-\beta\epsilon_{states}} = \sum_{j,m} e^{-\beta\epsilon_{jm}} = \sum_{levels} \omega_{levels} e^{-\beta\epsilon_{levels}}$$

$$= \sum_{j=0}^{\infty} (2j+1)e^{-\beta\epsilon_j} , \tag{6.2.58}$$

for the partition function associated with rotational states of a heteronuclear diatomic molecule.

We shall prefer to express z_{rot} in the form

$$z_{rot}(T) = \sum_{j=0}^{\infty} (2j+1)e^{-j(j+1)\Theta_{rot}/T} , \tag{6.2.59}$$

with Θ_{rot}, called 'the characteristic rotational temperature', defined as

$$\Theta_{rot} \equiv \frac{\hbar^2}{2Ik_B} . \tag{6.2.60}$$

At temperatures T such that $T \ll \Theta_{rot}$, the first few terms of the summation will suffice, and we may approximate z_{rot} by

$$z_{rot}(T) \simeq 1 + 3e^{-2\Theta_{rot}/T} + 5e^{-6\Theta_{rot}/T} + \cdots .$$

For high temperatures (i.e., for $T \gg \Theta_{rot}$), the quantum mechanical expression (6.2.59) can be approximated by its classical limit when typical rotational energy level separations are small in comparison with $k_B T$, so that many rotational levels are thermally accessible. The classical limit is obtained by replacing the quantum mechanical sum over the rotational quantum number j by an integral over the classical rotational angular momentum j, namely,

$$z_{rot}^{cl}(T) \simeq \int_0^\infty (2j+1)e^{-j(j+1)\Theta_{rot}/T} \, dj$$

$$= \int_0^\infty e^{-j(j+1)\Theta_{rot}/T} \, d[j(j+1)] .$$

From this last expression we can evaluate $z_{rot}^{cl}(T)$ readily as

$$z_{rot}^{cl}(T) \simeq \left[\frac{e^{-j(j+1)\Theta_{rot}/T}}{-\Theta_{rot}/T} \right]_0^\infty = \frac{T}{\Theta_{rot}} , \tag{6.2.61a}$$

or, equivalently, as

$$z_{\text{rot}}^{\text{cl}}(T) \simeq \frac{2I k_{\text{B}} T}{\hbar^2}, \tag{6.2.61b}$$

in terms of the moment-of-inertia of the molecule.

It is often stated that the classical formula (6.2.61a) may be employed safely for temperatures greater than approximately ten times the characteristic rotational temperature. To test this statement, we shall examine $z_{\text{rot}}(10\Theta_{\text{rot}})$ to see how well the classical result fares. Evaluation of $z_{\text{rot}}(T)$ using the first eleven terms (i.e., terms for which $0 \leq j \leq 10$) of the quantum mechanical sum in Eq. (6.2.59) gives $z_{\text{rot}}(10\Theta_{\text{rot}}) = 10.3410$, accurate to four decimal places. This result may be compared with $z_{\text{rot}}^{\text{cl}}(10\Theta_{\text{rot}}) = 10.0000$ from the classical limit expression (6.2.61a), which thus gives a 3.4% underestimation of $z_{\text{rot}}(T)$ for $T = 10\Theta_{\text{rot}}$. Similarly, if we evaluate the rotational partition function for a temperature $T = 5\Theta_{\text{rot}}$, an 8-term ($0 \leq j \leq 7$) evaluation from the quantum expression gives $z_{\text{rot}} = 5.3472$, again accurate to four decimal places: however, in this case, the classical limit of 5.0000 for $z_{\text{rot}}(T)$ underestimates the true value by 6.5%.

Examination of Table 6.1 shows that the only diatomic molecules that remain in the gaseous state at temperatures T that are at or below their characteristic rotational temperatures are the hydrogen isotopologues. Measurements of the heat capacity $C_{V,\text{rot}}(T)$ for HD, for example, show a rapid rise from $C_{V,\text{rot}}(0) = 0$, reaching a small maximum above the high-temperature classical value Nk_{B} in the vicinity of $T = 50$ K, then falling to Nk_{B} by $T = 130$ K. As we shall see, this behaviour is actually consistent with the traditional expression given by Eq. (6.1.8)′, but it is only observable for molecules that have very small masses.

Table 6.1 Rotational and vibrational characteristic temperatures and dissociation energies for a selection of diatomic molecules[a]

Molecule	Θ_{vib} /K	Θ_{rot} /K	D_0 /eV	Molecule	Θ_{vib} /K	Θ_{rot} /K	D_0 /eV
$H_2(^1\Sigma_g^+)$	5983.2	85.35	4.478	$NO(^2\Pi_{\frac{1}{2}})$	2699.2	2.393	6.497
$HD(^1\Sigma^+)$	5222.5	64.26	4.514	$HF(^1\Sigma^+)$	5695.4	29.58	5.869
$D_2(^1\Sigma_g^+)$	4304.6	43.03	4.556	$HCl(^1\Sigma^+)$	4151.3	15.02	4.433
$N_2(^1\Sigma_g^+)$	3352.2	2.86	9.759	$HBr(^1\Sigma^+)$	3681.1	12.01	3.758
$O_2(^3\Sigma_g^-)$	2239.1	2.07	5.116	$HI(^1\Sigma^+)$	3208.1	9.12	3.054
$F_2(^1\Sigma_g^+)$	1286.5	1.27	1.602	$ICl(^1\Sigma^+)$	548.59	0.164	2.153
$Cl_2(^1\Sigma_g^+)$	797.61	0.35	2.479	$IBr(^1\Sigma^+)$	384.17	0.082	1.818
$Br_2(^1\Sigma_g^+)$	464.96	0.12	1.971	$FCl(^1\Sigma^+)$	1113.4	0.74	2.617
$Na_2(^1\Sigma_g^+)$	226.86	0.22	0.720	$BrCl(^1\Sigma^+)$	633.91	0.22	2.233
$Cs_2(^1\Sigma_g^+)$	60.22	0.02	0.394	$CO(^1\Sigma^+)$	3083.6	2.77	11.092

[a]Values for Θ_{rot}, Θ_{vib}, and D_0 obtained from spectroscopic constants [3] via the relations $\Theta_{\text{vib}} \equiv (\omega_e - 2\omega_e x_e)/k_{\text{B}}$ and $\Theta_{\text{rot}} \equiv (B_e - \frac{1}{2}\alpha_e)/k_{\text{B}}$

To understand the origin of the small maximum in $C_{V,\text{rot}}(T)$, we may utilize an alternative, but equivalent, form for $C_{V,\text{rot}}(T)$ given in Eq. (6.1.11a), namely,

$$\frac{C_{V,\text{rot}}(T)}{Nk_B} = \frac{1}{2} \sum_{j,j'} p_j p_{j'} \left(\frac{\Delta\epsilon_{jj'}}{k_B T}\right)^2, \tag{6.2.62a}$$

in terms of the rotational angular momentum quantum numbers j, j'. The rotational fractional population expression is given by

$$p_j(T) = (2j+1)e^{-j(j+1)\Theta_{\text{rot}}/T}/z_{\text{rot}}, \tag{6.2.62b}$$

with $z_{\text{rot}}(T)$ given by Eq. (6.2.59). The quantity $\Delta\epsilon_{jj'}/k_B$ is given in terms of the characteristic temperature Θ_{rot} and the difference $\Delta_{jj'} \equiv |j' - j|$ between the rotational quantum numbers as

$$\Delta\epsilon_{jj'}/k_B = (2j+1+\Delta j)\Theta_{\text{rot}}\Delta j. \tag{6.2.62c}$$

According to Eqs. (6.2.62), the rotational heat capacity obeys a law of corresponding states, as $C_{V,\text{rot}}(T)$ is solely a function of the ratio T/Θ_{rot}, and hence all diatomic ideal gases will behave in the same fashion. The behaviours of expression (2.5.20a) and its components as functions of T/Θ_{rot} for $T/\Theta_{\text{rot}} \leq 5$ are illustrated in Fig. 6.6.

The corresponding states behaviour of the normalized rotational heat capacity $C_{V,\text{rot}}/(Nk_B)$ as a function of T/Θ_{rot} indeed shows a small maximum for $T/\Theta_{\text{rot}} \simeq 0.75$ and decays to the classical high-temperature value unity by $T/\Theta_{\text{rot}} \simeq 2$. For HD, for example, the maximum should occur at about $T \approx 50\,\text{K}$, and $C_{V,\text{rot}}(T)$ should assume the classical value 1 by $T = 130\,\text{K}$. For comparison, for N_2 the maximum in $C_{V\text{rot}}/(Nk_B)$ would occur for $T \simeq 2\,\text{K}$, and $C_{V,\text{rot}}/(Nk_B)$ would attain its classical value by $T = 6\,\text{K}$. It is also clear from this figure that the maximum in the temperature dependence of $C_{V,\text{rot}}/(Nk_B)$ is almost entirely due to the change in the occupation of the first excited rotational state as the temperature is increased.

Corrections to the Classical Expression

A better approximation to the exact value for $z_{\text{rot}}(T)$ for a diatomic molecule may be obtained from the Euler–Maclaurin summation formula obtained in Section 5 of Appendix B.1. The Euler–Maclaurin summation formula for a sum of the type $\sum_{j=0}^{n} f(j)$, in which $f(j)$ is a function defined on the integers and continuous for all noninteger numbers, is

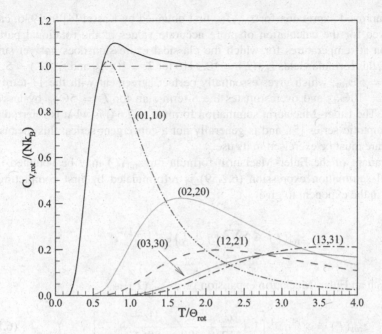

Fig. 6.6 Characteristic temperature dependence of $C_{V,\mathrm{rot}}$ for an ideal diatomic gas as a function of T/Θ_{rot} for $T/\Theta_{\mathrm{rot}} \leq 5$. Contributions associated with changes in the thermal occupations of rotational levels are indicated by pairs of values $(jj', j'j)$ appearing in Eq. (6.2.62a)

$$\sum_{j=0}^{n} f(j) = \int_0^n f(t)\,\mathrm{d}t + \frac{1}{2}[f(0) + f(n)] - \sum_{k=1}^{\infty} \frac{B_{2k}}{(2k)!}[f^{(2k-1)}(0) - f^{(2k-1)}(n)],$$

$$(6.2.63)$$

in which $f^{(k)}(j)$ is the kth derivative of f, evaluated at j, and the B_{2k} are Bernoulli numbers (see Appendix B.1, §5).

For a (rigid) diatomic molecule, the rotational partition function is given by Eq. (6.2.59), so that its Euler–Maclaurin expansion will be that for the function $f(j) = (2j + 1)e^{-\alpha j(j+1)}$, with α defined as $\alpha \equiv \Theta_{\mathrm{rot}}/T$. The leading (integral) term in Eq. (6.2.62) is given by Eq. (6.2.61a) as α^{-1}. Moreover, all contributions from the upper limit infinity to Eq. (6.2.63) vanish due to the exponential in $f(j)$, and we find that $f(0) = 1$, $f^{(1)}(0) = 2 - \alpha$, $f^{(3)}(0) = -\alpha^3 + 12\alpha^2 - 12\alpha$, and $f^{(5)}(0) = \alpha^2[-\alpha^3 + 30\alpha^2 - 180\alpha + 120]$, will suffice for obtaining all contributions up to and including order α^3. The resultant Euler–Maclaurin series expression for $z_{\mathrm{rot}}(T)$ is

$$z_{\mathrm{rot}} = \frac{T}{\Theta_{\mathrm{rot}}}\left[1 + \frac{1}{3}\left(\frac{\Theta_{\mathrm{rot}}}{T}\right) + \frac{1}{15}\left(\frac{\Theta_{\mathrm{rot}}}{T}\right)^2 + \frac{4}{315}\left(\frac{\Theta_{\mathrm{rot}}}{T}\right)^3 + \cdots\right]. \quad (6.2.64)$$

This improved expression for $z_{rot}(T)$, first obtained by Kassel [4] in 1936, can be employed for the calculation of more accurate values of the rotational partition function at temperatures for which the classical expression does not yet suffice. Indeed, this formula yields $z_{rot}(T) = 10.3410$ for $T = 10\Theta_{rot}$ and $z_{rot}(T) = 5.3475$ for $T = 5\Theta_{rot}$, which gives essentially perfect agreement with the 11-term sum for $T = 10\Theta_{rot}$ and overestimates the 8-term sum for $T = 5\Theta_{rot}$ by less than 0.01%. The Euler–Maclaurin summation formula (6.2.64) is what is referred to as an asymptotic series [5], and is generally not a convergent series: this means that some care must be exercised in its use.

A variant of the Euler–Maclaurin formula for $z_{rot}(T)$ may be obtained if the original summation expression (6.2.59) is reformulated by first completing the square in the exponent to give

$$z_{rot}(T) = 2 \sum_{j=0}^{\infty} \left(j + \tfrac{1}{2}\right) e^{-\alpha(j+\frac{1}{2})^2} e^{\alpha/4} \,,$$

for which an Euler–Maclaurin expansion gives $z_{rot}(T)$ as

$$z_{rot}(T) = e^{\alpha/4} \alpha^{-1} \left[1 + \frac{\alpha}{12} + \frac{7\alpha^2}{480} + \frac{31\alpha^3}{8064} + \cdots \right] , \qquad (6.2.65)$$

a result now known as the Mulholland formula [6].

Finally, an asymptotic series that requires fewer terms to yield more accurate values for $z_{rot}(T)$ up to quite high temperatures (of the order of several thousand Kelvin for many diatomic molecules) has been obtained by McDowell [7], who noted that by treating the first two terms within the square brackets of Eq. (6.2.64) as the start of the series expansion for $e^{\alpha/3}$, writing $z_{rot}(T) = e^{\alpha/3} \alpha^{-1} e^{-\alpha/3}[1 + \tfrac{1}{3}\alpha + \tfrac{1}{15}\alpha^2 + \tfrac{4}{315}\alpha^3 + \cdots]$, expanding $e^{-\alpha/3}$, multiplying the two series and, finally, collecting like powers of α, $z_{rot}(T)$ is given as

$$z_{rot}(T) = e^{\alpha/3} \alpha^{-1} \left[1 + \frac{\alpha^2}{90} + \frac{8\alpha^3}{2835} + \cdots \right] . \qquad (6.2.66)$$

Expression (6.2.66) has the double advantage that not only is there no linear correction term, but also the coefficients of all higher-order powers are smaller than those of the corresponding powers in both the Euler–Maclaurin and Mulholland expressions: this means that in practice the series can be terminated sooner to attain a specified accuracy. Indeed, McDowell found that values for the prefactor $\alpha^{-1} e^{\alpha/3}$ alone underestimate the exact values for $z_{rot}(T)$ by less than 0.05% for temperatures of the order of $5\Theta_{rot}$ and by less than 0.005% for temperatures of the order of $20\Theta_{rot}$. An elegant operator treatment of these expansions has been given by Fernández and Tipping [8].

Centrifugal Distortion and Vibration–Rotation Interaction Corrections

The Euler–Maclaurin corrections (typically for temperatures $T < 10\Theta_{\text{rot}}$) discussed in the previous subsection address the differences between the quantum mechanical sum in Eq. (6.2.59) and the classical approximation of Eq. (6.2.61a) for symmetry number $\sigma = 1$. They still assume the rigid-rotor model. However, we also know that because the chemical bonds in diatomic molecules are not truly rigid, they will elongate slightly for rapidly rotating molecules simply due to the centrifugal forces associated with acceleration in the circular orbits of the two atoms about their (common) centre-of-mass. The centrifugal distortion energy can be shown via perturbation theory to have the functional form

$$\epsilon_{\text{cd}}(j) = -D_e[j(j+1)]^2 + H_e[j(j+1)]^3 + \cdots , \qquad (6.2.67)$$

in which D_e and H_e, both positive, are called the quartic and sextic centrifugal distortion constants, respectively. Sextic and higher-order centrifugal distortion constants may play significant roles in the accurate determination of spectral transitions, but for the present discussion it will suffice simply to examine the role of the quartic distortion constant in greater detail.

At the same time that centrifugal distortion contributions to the rotational energy are being considered, we should also consider that rotational energies may depend upon the specific vibrational state of the rotating molecule: this is referred to as the vibration–rotation interaction energy, and is represented empirically in molecular spectroscopy by the introduction of an energy term having the form $\overline{\alpha}(v+\frac{1}{2})j(j+1)$.

The total vibration–rotation internal energy $\overline{\epsilon}_{\text{int}}$ for a nonrigid diatomic molecule can be written as

$$\overline{\epsilon}_{\text{int}} = \overline{\epsilon}_{\text{vib}} + \overline{\epsilon}_{\text{rot}} , \qquad (6.2.68)$$

in which $\overline{\epsilon}_{\text{vib}}$ is the anharmonic vibrational energy, as given, for example, by Eq. (6.2.29), and $\overline{\epsilon}_{\text{rot}}$ is the rotational energy for a nonrigid diatomic molecule, given by

$$\overline{\epsilon}_{\text{rot}} = (\overline{B}_e - \tfrac{1}{2}\delta)j(j+1)[1 - \gamma j(j+1) - v\delta]. \qquad (6.2.69)$$

In this expression, γ and δ are, respectively, the ratios of the quartic centrifugal distortion constant and the vibration–rotation interaction constant to the rotational constant, i.e., $\gamma \equiv \overline{D}_e/\overline{B}_e$ and $\delta \equiv \overline{\alpha}/\overline{B}_e$. Expression (6.2.69) for the rotational energy neglects second-order terms in γ, δ, while the vibrational energy expression (6.2.29) neglects the second-order term $x_e^2\overline{\omega}_e v(v - 1)$. The characteristic rotational temperature can now be introduced into the argument $\beta\epsilon_{\text{rot}}$ via

$$\beta\epsilon_{\text{rot}} = j(j+1)[1 - \gamma j(j+1) - v\delta]\frac{\Theta_{\text{rot}}}{T} , \qquad (6.2.70)$$

with $\Theta_{\rm rot}$ defined as

$$\Theta_{\rm rot} \equiv \frac{(1 - \frac{1}{2}\delta)\overline{B}_{\rm e}}{\overline{k}_{\rm B}}. \tag{6.2.71}$$

To proceed further, we need to recognize that inclusion of the vibration–rotation interaction energy means that we cannot simply treat the rotational and vibrational sums independently when we are dealing with it. Thus, we begin with $z_{\rm vib-rot}(T)$ defined as

$$z_{\rm vib-rot}(T) = {\rm e}^{-\beta\epsilon_0} \sum_{v=0}^{\infty} {\rm e}^{-uv}{\rm e}^{ux_e v(v-1)} \sum_{j=0}^{\infty}(2j+1){\rm e}^{-j(j+1)[1-\gamma j(j+1)-v\delta]\alpha},$$

$$\tag{6.2.72}$$

which we shall approximate as

$$z_{\rm vib-rot}(T) \simeq {\rm e}^{-\beta\epsilon_0} \sum_{v=0}^{\infty} {\rm e}^{-uv}{\rm e}^{ux_e v(v-1)} \sum_{j=0}^{\infty}(2j+1){\rm e}^{-j(j+1)[1-v\delta]\alpha}[1+\gamma j^2(j+1)^2\alpha],$$

$$\tag{6.2.73}$$

with $\alpha \equiv \Theta_{\rm rot}/T$. We shall follow Mayer and Mayer [2] in replacing the rotational summation using the first three terms of an Euler–Maclaurin expansion, namely,

$$\sum_{j=0}^{\infty} f(j) \simeq \int_0^{\infty} f(j)\,{\rm d}j + \tfrac{1}{2}f(0) - \tfrac{1}{12}f'(0), \tag{6.2.74}$$

in which $f(j)$ is the function $f(j) = (2j+1){\rm e}^{-j(j+1)[1-v\delta-\gamma j(j+1)]\alpha}$.

We see that $f(0) = 1$ and $f'(0) = 2 - (1 - v\delta)\alpha$. As the second term in $f'(0)$ will contribute to the Euler–Maclaurin expansion expression in order α^2 (see the previous subsection), we shall ignore it here. By approximating ${\rm e}^{\gamma\alpha j^2(j+1)^2\alpha}$ as $1 + \gamma\alpha j^2(j+1)^2$ and making the substitution $y = \alpha(1-v\delta)j(j+1)$ in the leading (integral) term of Eq. (6.2.74), we obtain the approximate expression

$$\frac{1}{\alpha(1-v\delta)}\int_0^{\infty} {\rm e}^{-y}\left(1 + \frac{\gamma}{\alpha}y^2\right)\,{\rm d}y = \frac{1}{\alpha(1-v\delta)}\left(1 + \frac{2\gamma}{\alpha}\right)$$

$$\simeq \frac{1}{\alpha}\left(1 + \frac{2\gamma}{\alpha} + v\delta\right).$$

We thus obtain

$$\sum_{j=0}^{\infty}(2j+1){\rm e}^{-\alpha j(j+1)[1-\gamma j(j+1)-v\delta]} \simeq \frac{1}{\alpha}\left(1 + \frac{\alpha}{3} + \frac{2\gamma}{\alpha} + v\delta\right) \tag{6.2.75}$$

for the rotational sum.

When this result for the rotational sum is substituted into Eq. (6.2.72) for $z_{vib-rot}(T)$, the summation over the vibrational levels for those terms not containing δ gives $z_{vib}(T)$ [see Eq. (6.2.36)] as a multiplicative factor, while the δ-term gives a multiplicative factor $e^{-u}/(1-e^{-u})^2$, so that $z_{vib-rot}(T)$ can be expressed as

$$z_{vib-rot}(T) \simeq \frac{1}{\alpha(1-e^{-u})}\left[1 + \frac{\alpha}{3} + \frac{\gamma}{\alpha} + \frac{\delta}{e^u - 1} + \frac{2x_e u}{(e^u - 1)^2} + \cdots\right].$$
(6.2.76)

As the vibration–rotation contributions to the thermodynamic functions are all derived from $\ln z_{vib-rot}(T)$ rather than directly from $z_{vib-rot}(T)$ itself, we write

$$\ln z_{vib-rot}(T) \simeq -\ln\alpha + \frac{\alpha}{3} - \ln(1-e^{-u}) + \frac{2\gamma}{\alpha} + \frac{\delta}{e^u - 1} + \frac{2x_e u}{(e^u - 1)^2} + \cdots.$$
(6.2.77)

The first three terms of this expression represent the RR-SHO model approximation, including the leading Euler–Maclaurin correction for rigid rotation, and the final three terms provide the corrections associated, respectively, with (rotational) centrifugal distortion, the vibration–rotation interaction, and (vibrational) anharmonicity.

All three corrections appearing in our expression for $z_{vib-rot}(T)$ will, as can be deduced from Eq. (6.2.72), cause $z_{vib-rot}(T)$ for a fixed temperature to diverge for sufficiently large values of v and/or j. This behaviour is not, however, manifested by our result (6.2.76). Accordingly, we should consider more carefully the behaviour of our quantum mechanical sums. Let us illustrate this specifically, for example, in terms of the role of the centrifugal distortion correction to the rotational energy, and let us focus on the classical limit $z_{rot}^{cl}(T)$, defined via

$$z_{rot}^{cl}(T) \equiv \int_0^\infty e^{-\beta(Bx - D_e x^2)}\, dx$$
(6.2.78)

when the quartic centrifugal distortion contribution to the rotational energy of the diatomic molecule is included. As the centrifugal distortion constant, D_e, entering into Eq. (6.2.78) is positive, the final exponential in $z_{rot}(T)$ will, mathematically, cause $z_{rot}^{cl}(T)$ to diverge: the integrand in Eq. (6.2.78) has value unity for $x = 0$, then decreases until $x = x_m = B/(2D_e)$, at which point it is very nearly zero, then increases without bound. For typical heteronuclear diatomic molecules like HCl or CO in their ground vibrational states, for example, the minimum in the integrand corresponds to rotational quantum numbers j_m of approximately 114 or 397, respectively, and rigid-rotor energies of approximately 136,278 cm^{-1} for an HCl molecule, and 305,162 cm^{-1} for a CO molecule. If these energies are compared with the spectroscopic dissociation energies 35,730 cm^{-1} for HCl and 89,463 cm^{-1} for CO, it is clear that any typical diatomic molecule will have dissociated long before the minimum in the integrand of the classical partition function integral could be attained. Indeed, $Bj(j+1) - D_e[j(j+1)]^2$ already exceeds the CO dissociation energy for $j = 238$, which is well before the value $j_m = 397$ that corresponds to the integrand in Eq. (6.2.78) attaining its minimum value. Moreover, as the sextic centrifugal distortion constant typically has values such that $|H_e/D_e|$ is of order 10^{-4} or smaller, it will not have a large influence on the behaviour of $z_{rot}(T)$.

Upon including the quartic centrifugal distortion contribution to the rotational energy, expanding $e^{\beta D_e j^2(j+1)^2}$ and retaining only the two leading terms gives a modified classical limit for $z_{rot}(T)$ of

$$z_{rot}^{cl}(T) = \frac{T}{\Theta_{rot}}\left[1 + \frac{2D_e}{B}\left(\frac{T}{\Theta_{rot}}\right)\right] \equiv \frac{T}{\Theta_{rot}}f_{cd}, \qquad (6.2.79)$$

which we have written for convenience as a multiplicative correction factor f_{cd} to the rigid-rotor value. Because D_e/B is typically or order 10^{-4} to 10^{-6}, higher-order centrifugal distortion correction contributions, which will contribute even smaller corrections than that for the quartic contribution, will not play significant roles unless the temperature is very high. Nonetheless, an expression for $z_{rot}^{cl}(T)$ that includes both Euler–Maclaurin corrections to the classical limit expression and the first two centrifugal distortion terms in the rotational energy expression has been obtained by McDowell [9], and has the form of Eq. (6.2.66) multiplied by a correction factor, f_{cd}, given as

$$f_{cd} \simeq 1 + \frac{2D_e}{B}\frac{T}{\Theta_{rot}} + 6\frac{6D_e}{B}\left(\frac{2D_e}{B} - \frac{H_e}{D_e}\right)\left(\frac{T}{\Theta_{rot}}\right)^2 + \frac{120D_e^2}{B^2}\left(\frac{D_e}{B} - \frac{H_e}{D_e}\right)\left(\frac{T}{\Theta_{rot}}\right)^3 . \qquad (6.2.80)$$

As pointed out by McDowell [9], terms linear in H_e contribute at about the same level as do terms quadratic in D_e.

Homonuclear Diatomic Molecules

Wolfgang Pauli showed [10] (using quite general group theoretical arguments) that the wavefunction for a set of equivalent fermions must have a specified symmetry (i.e., it must be either symmetric or antisymmetric) to the interchange of any two of them, while the wavefunction for a set of equivalent bosons must have the opposite symmetry under such an interexchange. Which specific symmetry property is to go with fermions, and which with bosons cannot be decided from group theoretical arguments, and an appeal to experiment must be made to see which is which. Pauli was able to establish from experiment that the wavefunction for equivalent fermions must be antisymmetric to the interchange of any two of them, so that the wavefunction for equivalent bosons must therefore be symmetric to their interchange.

As the nuclei in a homonuclear diatomic molecule are indistinguishable, the total wavefunction, Ψ_{total}, for a homonuclear diatomic molecule (which has electronic, vibrational, rotational, translational, and nuclear spin degrees of freedom) must be either symmetric or antisymmetric under the interchange of the two identical nuclei. We find that Ψ_{total} is:

- symmetric if the nuclei have integer (or zero) nuclear spin (bosons);
- antisymmetric if the nuclei have half-odd-integer nuclear spin (fermions).

We shall write the total wavefunction for a diatomic molecule symbolically as the product

$$\Psi_{\text{total}}(1, 2) = \psi'_{\text{total}}(1, 2)\psi_{\text{nuclear-spin}}(1, 2) , \qquad (6.2.81)$$

with 1 and 2 referring to the two nuclei, $\psi'_{\text{total}}(1, 2)$ to the combined translational, rotational, vibrational, and electronic degrees of freedom for the molecule, and $\psi_{\text{nuclear-spin}}(1, 2)$ to the nuclear spin pair-states for the molecule. The overall nuclear interchange symmetry

$$\Psi_{\text{total}}(2, 1) = \pm\Psi_{\text{total}}(1, 2)$$

required by the Pauli Principle will be determined by the product of the interchange symmetries of $\psi'_{\text{total}}(1, 2)$ and $\psi_{\text{nuclear-spin}}(1, 2)$.

It will also be useful for us to write the wavefunction $\psi'_{\text{total}}(1, 2)$ as the product of wavefunctions for the molecular degrees of freedom as

$$\psi'_{\text{total}}(1, 2) = \psi_{\text{trans}}(1, 2)\psi_{\text{rot}}(1, 2)\psi_{\text{vib}}(1, 2)\psi_{\text{el}}(1, 2) ,$$

and then consider the effect of the nuclear interchange on each factor separately. The translational wavefunction, $\psi_{\text{trans}}(1, 2)$, depends only upon the coordinates of the centre-of-mass of the diatomic molecule, and is hence not affected by the interchange of the two nuclei. Similarly, the vibrational wavefunction, $\psi_{\text{vib}}(1, 2)$, depends only upon the magnitude $|\mathbf{R} - \mathbf{R}_e|$ of the separation between the two nuclei, and is thus unaffected by the interchange process. The behaviour of the electronic wavefunction, $\psi_{\text{el}}(1, 2)$, under the nuclear interchange is more complicated, and depends upon the nature of the electronic state in which the molecule is found: we shall see shortly how the nuclear interchange symmetry can be obtained from the spectroscopic term symbol assigned to the electronic state. As most molecules in thermal equilibrium are found in their ground electronic states at normal temperatures, our main concern will be with molecules in their ground electronic states.

Interlude: Electronic State Nuclear Interchange Symmetry

Let us review briefly the symmetry operations relevant to a homonuclear diatomic molecule. Molecular symmetry operations in general consist of rotations about axes, reflections across planes, and inversion, all relative to the centre of the molecule, plus any other distinct combined operations. These operations are represented mathematically by a set of symmetry operators that, together with the identity operator (which leaves all atoms in a molecule unchanged), form a mathematical structure called a group. Molecular symmetry groups are referred to using a generally agreed-upon notation that was introduced by chemists in the early part of the twentieth century.

An n-fold rotation axis passing through the centre of the molecule and corresponding to a rotation about that axis by an angle $2\pi/n$ is denoted by the symmetry operation c_n; the axis having the largest value of n is called the principal rotation axis. The symmetry operations σ_v and σ_h are reflection planes, called, respectively, vertical and horizontal reflection planes, depending upon whether they contain the principal rotation axis (σ_v) or are perpendicular to it (σ_h). The symmetry elements for $D_{\infty h}$, which is of specific interest here, are illustrated in Fig. 6.7, and the placement of a homonuclear diatomic molecule is shown in Fig. 6.8.

The specific symmetry groups (i.e., the set of symmetry operations that leave the molecule indistinguishable following their application) that apply to diatomic molecules (actually, more generally to linear molecules) are designated as $C_{\infty v}$ and $D_{\infty h}$: $C_{\infty v}$ applies to heteronuclear and $D_{\infty h}$ to homonuclear diatomic molecules (more generally, to noncentrosymmetric and centrosymmetric linear molecules, respectively).

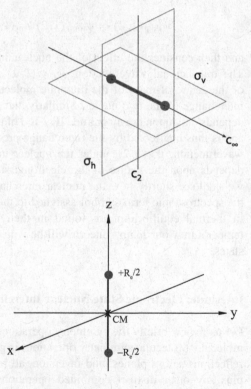

Fig. 6.7 Symmetry elements in the $D_{\infty h}$ point-symmetry group for a homonuclear diatomic molecule. Note that the principal rotation axis is assumed to define the 'vertical direction'

Fig. 6.8 Molecule-fixed Cartesian axis system for the description of the symmetry operations associated with a homonuclear diatomic molecule

We shall treat the interchange of the two identical nuclei as a 2-step process:

(i) rotation of the molecule by an angle π about an axis perpendicular to the bond-axis of the molecule, followed by

(ii) return of the electrons to their original positions.

As in Fig. 6.8, we place the diatomic molecule with its axis lying along the z-axis of the coordinate system, and we place the origin at the centre-of-mass of the molecule. A rotation by an angle π about the x-axis, for example, permutes the nuclei, and at the same time changes the coordinates of the electrons i via

$$(x_i, y_i, z_i) \quad \overset{c_2(x)}{\rightarrow} \quad (x_i, -y_i, -z_i).$$

We now return the *electrons only* to their original coordinates by means of two transformations (which both leave the nuclei unchanged), namely inversion through the centre of the molecule,

$$(x_i, -y_i, -z_i) \quad \overset{i}{\rightarrow} \quad (-x_i, y_i, z_i),$$

followed by reflection across a vertical mirror plane (the yz-plane specifically), to give

$$(-x_i, y_i, z_i) \quad \overset{\sigma_v(yz)}{\rightarrow} \quad (x_i, y_i, z_i).$$

Let us summarize this information in the following way. Electronic functions f that are symmetric/antisymmetric under inversion of their coordinates will be designated as gerade/ungerade (German for even/odd) via a subscript g/u under the inversion operation i, while electronic functions that are even/odd under reflection across a vertical mirror plane σ_v will be designated $+/-$ as follows:

$$\left\{ \begin{array}{l} f_u \overset{i}{\longrightarrow} -f_u \\ f_g \overset{i}{\longrightarrow} +f_g \end{array} \right\} \quad \oplus \quad \left\{ \begin{array}{l} f_+ \overset{\sigma_v}{\longrightarrow} +f_+ \\ f_- \overset{\sigma_v}{\longrightarrow} -f_- \end{array} \right\}$$

Fortunately, the relevant information that we require for the determination of the nuclear interchange symmetry of ψ_{el} is associated with the term symbol assigned to the states. A typical term symbol for a homonuclear diatomic (centrosymmetric linear) molecule has a subscript u or g specifying the symmetry of the electronic wavefunction(s) under inversion of the coordinates and a superscript $+$ or $-$ specifying their symmetry under reflection through a vertical mirror plane. The overall symmetry with which we are concerned is determined by the product of the two symmetries, so that the relevant electronic symmetry can be read directly from the term symbol, as illustrated in Fig. 6.9.

We may summarize the determination of the overall interchange symmetry of the electronic wavefunction as follows: terms for which the (inversion, reflection)

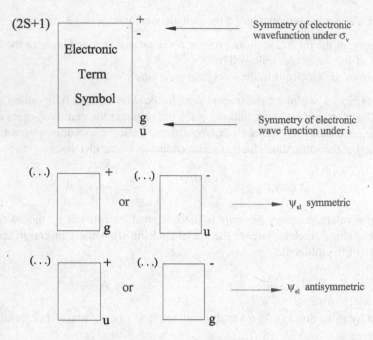

Fig. 6.9 Use of molecular electronic term symbols to determine the symmetry of the electronic wavefunction to nuclear interchange

symmetries are $(g, +)$ or $(u, -)$ are symmetric, while those for which these symmetries are $(g, -)$ or $(u, +)$ are antisymmetric to the nuclear interchange.

The ground electronic state for the vast majority of molecules is symmetric under both operations (i) and (ii) described above: for homonuclear diatomic molecules, the spectroscopic term symbol for such a ground state is Σ_g^+. Thus, for molecules having a Σ_g^+ electronic state, it will be ψ_{rot} that controls the symmetry of ψ'_{total}. □

Symmetries of the Rotational Wavefunctions

Let us focus now upon the behaviour of the rotational wavefunction. To do this, however, requires us to review firstly the way in which inversion works in the spherical polar coordinate system. We accomplish this by beginning with our knowledge of how the inversion operator works on Cartesian coordinates. For the Cartesian coordinates we have the domains $-\infty < x < y$; $-\infty < y < \infty$; $-\infty < z < \infty$, while for spherical polar coordinates we have the domains $0 \leq \rho < \infty$; $0 \leq \theta \leq \pi$; $0 \leq \phi \leq 2\pi$.

From Fig. 6.10 we see that the relation between Cartesian coordinates, (x, y, z), and spherical polar coordinates, (ρ, θ, ϕ), is

$$x = \rho \sin\theta \cos\phi, \qquad y = \rho \sin\theta \sin\phi, \qquad z = \rho \cos\theta.$$

Fig. 6.10 Cartesian and spherical polar coordinate systems

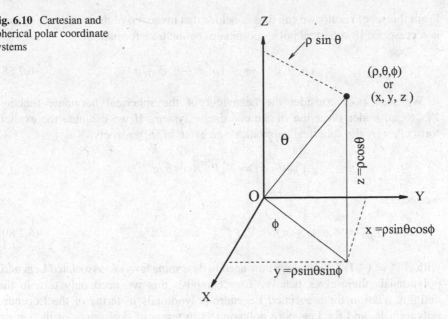

In addition it is easy to see that ρ is given in terms of x, y, z by $\rho = (x^2 + y^2 + z^2)^{\frac{1}{2}}$. The operation of inversion, $\imath$, can be represented in Cartesian coordinates by

$$(x, y, z) \quad \xrightarrow{\imath} \quad (x', y', z') = (-x, -y, -z), \tag{6.2.82}$$

and in spherical polar coordinates by

$$(\rho, \theta, \phi) \quad \xrightarrow{\imath} \quad (\rho', \theta', \phi'), \tag{6.2.83}$$

in which ρ', θ', ϕ' are as yet unknown. From the definition of ρ' in terms of x', y', z', viz. $\rho' = [x'^2 + y'^2 + z'^2]^{\frac{1}{2}}$, we see immediately that

$$\rho' = [(-x)^2 + (-y)^2 + (-z)^2]^{\frac{1}{2}} = \rho. \tag{6.2.84}$$

The equations relating Cartesian and spherical polar coordinates allow us to make the following observations regarding the effect of inversion of the coordinate system:

$$z \to -z \quad \Rightarrow \quad \cos\theta' = -\cos\theta \quad \Rightarrow \quad \theta' = \pi - \theta, \tag{6.2.85}$$

$$x \to -x \quad \Rightarrow \quad \cos\phi' = -\cos\phi \quad \Rightarrow \quad \phi' = \phi \pm \pi, \tag{6.2.86}$$

$$y \to -y \quad \Rightarrow \quad \sin\phi' = -\sin\phi \quad \Rightarrow \quad \phi' = \phi + \pi. \tag{6.2.87}$$

From this set of results we can thus conclude that inversion of the coordinate system is represented in spherical polar coordinates by the transformation

$$(\rho, \theta, \phi) \;\overset{i}{\to}\; (\rho, \pi - \theta, \phi + \pi).\tag{6.2.88}$$

We shall now consider the behaviour of the spherical harmonic function $Y_{j,m}(\theta, \phi)$ under inversion of the coordinate system. If we examine the explicit form taken by the spherical harmonics (see §16.6 of [5]), namely,

$$Y_{jm}(\theta, \phi) = N_j P_j^{|m|}(\cos\theta)e^{im\phi}\,,$$

we see that

$$Y_{jm}(\pi - \theta, \phi + \pi) = N_j P_j^{|m|}(-\cos\theta)e^{im\phi}e^{i\pi m}\,,\tag{6.2.89}$$

with $e^{i\pi m} = (-1)^m$, so that we now need to determine how the associated Legendre polynomials themselves behave. To accomplish this we need only turn to the defining relation for associated Legendre polynomials in terms of the Legendre polynomials, and for Legendre polynomials in terms of derivatives of the square of the sine function. Thus, we have

$$P_j^{|m|}(\cos\theta) \equiv (1 - \cos^2\theta)^{\frac{|m|}{2}} \frac{d^{|m|}}{d\cos\theta^{|m|}} P_j(\cos\theta)\,,$$

and

$$P_j(\cos\theta) \equiv \;\;\mathrm{const} \times \frac{d^j}{d\cos\theta^j}(\cos^2\theta - 1)^j\,.$$

From these two equations we can see immediately that $\cos\theta \;\to\; -\cos\theta \;\Rightarrow$ $P_j(-\cos\theta) = (-1)^j P_j(\cos\theta)$, and therefore that

$$P_j^{|m|}(-\cos\theta) = (-1)^{j+|m|} P_j^{|m|}(\cos\theta)\,,\tag{6.2.90}$$

which has the consequence that

$$Y_{jm}(\pi - \theta, \phi + \pi) = (-1)^{j+|m|} N_j P_j^{|m|}(\cos\theta)e^{im\phi}(-1)^m$$

$$= (-1)^j Y_{jm}(\theta, \phi)\,.\tag{6.2.91}$$

From this last result we may conclude that $Y_{jm}(\theta, \phi)$ is an even (or symmetric) function if the rotational quantum number j is an even integer (or zero), and that $Y_{jm}(\theta, \phi)$ is an odd (or antisymmetric) function if the rotational quantum number j is an odd integer.

The result obtained above allows us to conclude, finally, that ψ'_{total} for a diatomic molecule in a $^1\Sigma_g^+$ electronic state is unchanged by (symmetric to) inversion of the coordinates for j even, and changes sign (is antisymmetric) upon inversion of the coordinates for j odd. This conclusion will be reversed if ψ_{el} is antisymmetric under inversion of the electronic coordinates. All of what we have just said is, of course, exclusive of nuclear spin.

Molecular Hydrogen: The Role of Nuclear Spin Symmetry

Let us now consider the H_2 molecule as a concrete example of how to introduce the effect of the nuclear spin symmetry for homonuclear diatomic molecules. The H atom has 1 proton (spin-$\frac{1}{2}$) which has two magnetic states, which we shall loosely refer to as 'spin-up' and 'spin-down', and symbolize by an 'up-arrow' $\uparrow$, respectively, by a 'down-arrow' $\downarrow$. The hydrogen molecule has two protons, whose 'pair-states' can be symbolized by the combinations

$$\left.\begin{array}{c} \uparrow\uparrow \\ \frac{1}{\sqrt{2}}(\uparrow\downarrow + \downarrow\uparrow) \\ \downarrow\downarrow \end{array}\right\} \quad \text{triplet},$$

$$\frac{1}{\sqrt{2}}(\uparrow\downarrow - \downarrow\uparrow) \quad \text{singlet},$$

corresponding to the total nuclear spins $I = 1$ and $I = 0$, respectively, for H_2.

As for the electronic states of He, for which the (total electron spin) triplet ($^3S_1$) atomic term is called *ortho*-helium, while the singlet $^1S_0$ atomic term is called *para*-helium, we shall refer to the triplet nuclear ($I = 1$) spin case of H_2 as *ortho*-hydrogen (oH_2) and to the singlet ($I = 0$) nuclear spin case as *para*-hydrogen (pH_2). Additional commentary on ortho- and para-helium can be found in the worked example provided in Appendix D.1.

According to the Pauli Principle, the total wavefunction

$$\Psi_{\text{total}} = \psi'_{\text{total}}\psi_{\text{nuclear spin}} \tag{6.2.92}$$

for a homonuclear diatomic molecule, such as H_2, whose nuclei are fermions, must be antisymmetric to the interchange of the two nuclei. Since the ground electronic term for H_2 is $^1\Sigma_g^+$, the interchange symmetry of $\psi'_{\text{total}}(1, 2)$ is determined by that of $\psi_{\text{rot}}(j)$ to be $\psi'_{j,\text{total}}(2, 1) = (-1)^j \psi_{j,\text{total}}(1, 2)$, so that those H_2 molecules in states with j odd must combine with the symmetric (triplet of) nuclear spin states corresponding to a total nuclear spin $I = 1$, while those H_2 molecules in states with j even must combine with the antisymmetric (singlet) nuclear spin state corresponding to a total nuclear spin $I = 0$. In the absence of an externally applied magnetic field the nuclear spin states are strictly degenerate, so that the odd rotational states of H_2 will have a statistical weighting which is three times that of the even rotational states. This weighting has a profound effect both on the spectroscopy and low temperature thermodynamics of bulk H_2 gas.

We have seen from our discussion following Eq. (6.2.81) that ψ'_{total} has four contributing factors, namely, ψ_{tr}, ψ_{vib}, ψ_{el}, and ψ_{rot}, of which ψ_{tr} and ψ_{vib} are always symmetric to the nuclear interchange (as they are unaffected by the interchange process), so that the interchange symmetry of ψ'_{total} depends entirely upon the interchange symmetries of ψ_{el} and ψ_{rot}. Moreover, we have seen that the nuclear interchange symmetry of ψ_{el} can be determined from the electronic term symbol associated with the wavefunction for that electronic state. Electronic wavefunctions ψ_{el} that are symmetric (even) or antisymmetric (odd) under both inversion, ι, of the electronic coordinate system and reflection across a vertical mirror plane, σ_{v}, are symmetric to the nuclear interchange, while electronic wavefunctions that have opposite symmetry under these two operations are antisymmetric to the nuclear interchange. We have also determined that the nuclear interchange symmetry of the rotational wavefunction, $\psi_{\text{rot}} \equiv \psi_{jm}$, ($j$, m are rotational angular momentum quantum numbers) is given by $(-1)^j$.

A key point arising from our consideration of the effect of symmetry upon the behaviour of homonuclear diatomic molecules in a $(g, +)$ electronic term is that the rotational and nuclear spin partition functions cannot be trivially separated into the product of two independent partition functions $z_{\text{rot}}(T)$ and $z_{\text{nuc}}(T)$, but must in general be considered together as a single partition function $z_{\text{rot}-\text{nuc}}(T)$.

For H_2 we thus see that the combined nuclear spin-rotational partition function $z_{\text{rot}-\text{nuc}}$ has in general the structure

$$z_{\text{rot}-\text{nuc}}(T) = z_{\text{ortho}}(T) + z_{\text{para}}(T), \tag{6.2.93}$$

in which the individual partition functions for the ortho and para modifications are given by

$$z_{\text{ortho}}(T) = 3 \sum_{j=1,3,5,\ldots} (2j+1)e^{-j(j+1)\Theta_{\text{rot}}/T}, \tag{6.2.94}$$

and

$$z_{\text{para}}(T) = \sum_{j=0,2,4,\ldots} (2j+1)e^{-j(j+1)\Theta_{\text{rot}}/T}. \tag{6.2.95}$$

We thus see that the restrictions imposed by the Pauli Principle cause $z_{\text{rot}-\text{nuc}}$ for hydrogen not to factor in general into the product of rotational and nuclear spin partition functions.

Although the discussion thus far has been focussed upon molecular hydrogen, it will be clear from what has been said that much the same result, i.e., an inability to represent the rotational and nuclear spin contributions to the partition function for a homonuclear hydrogen molecule in general as the product of rotational and nuclear spin partition functions would also occur for any homonuclear diatomic molecule. Moreover, as it is a manifestation of quantum mechanics, we should expect it to play a more important role for essentially microscopic phenomena. Indeed, as we shall see shortly, there is a profound effect on the rotational spectra of homonuclear diatomic molecules in general, while only the bulk properties, especially the heat capacity, of the homonuclear hydrogen isotopologues H_2, D_2, and T_2 are profoundly affected by this nonseparability.

If we examine the experimental temperature dependence of the molar heat capacity at constant volume for a series of homonuclear diatomic gases, we find that for gaseous N_2, O_2, and Cl_2, for example, $\overline{C}_{V,rot}(T)$ has the value $\frac{5}{2}R$ at all temperatures T for which these substances are gases. For the three homonuclear hydrogen isotopologues, $\overline{C}_{V,rot}(T)$ varies with T for low temperatures but for temperatures above about 350 K, $\overline{C}_{V,rot}(T)$ for these gases has also attained the value $\frac{5}{2}R$. This tells us, therefore, that for sufficiently 'high' temperatures, the bulk (thermodynamic) properties of any homonuclear diatomic gas behaves as does a heteronuclear diatomic gas, for which the factorization $z_{rot-nuc}(T) = z_{rot}(T)z_{nuc}$ is rigorous. As we have already specified for heteronuclear diatomic molecules that in this context 'high' temperature means temperatures T such that $T \gg \Theta_{rot}$, let us examine what happens for a homonuclear diatomic molecule when $T \gg \Theta_{rot}$.

When $T \gg \Theta_{rot}$, we may approximate sums over j by integrals over j, in which case, the sums over j even and j odd are each given by

$$\frac{1}{2}\sum_{\text{all } j}(\cdots) \sim \frac{1}{2}\int_0^\infty (2j+1)e^{-j(j+1)\Theta_{rot}/T}\,dj\,, \tag{6.2.96}$$

or, equivalently,

$$\frac{1}{2}\sum_{\text{all } j}(\cdots) \equiv \frac{T}{2\Theta_{rot}}\,. \tag{6.2.97}$$

When this procedure is applied for bulk hydrogen gas, we therefore obtain the corresponding classical limits for $z_{ortho}(T)$ and $z_{para}(T)$ as

$$z_{ortho}^{cl}(T) = \frac{3T}{2\Theta_{rot}}\,; \qquad z_{para}^{cl}(T) = \frac{T}{2\Theta_{rot}}\,, \tag{6.2.98}$$

with the consequence that the classical limit of Eq. (6.2.93) will be

$$z_{rot-nuc}^{cl}(T) = \frac{3T}{2\Theta_{rot}} + \frac{T}{2\Theta_{rot}} = 4\frac{T}{2\Theta_{rot}}\,. \tag{6.2.99}$$

If we recall that for a proton I_a has the value $\frac{1}{2}$, then the factor 4 in this result simply represents the number of nuclear spin pair-states, i.e., $z_{nuc} \equiv (2I_a + 1)^2$, so that we have

$$z_{\text{rot-nuc}}^{\text{cl}}(T) = (2I_a + 1)^2 \frac{T}{2\Theta_{\text{rot}}} = z_{\text{nuc}} z_{\text{rot}}^{\text{cl}}(T), \qquad (6.2.100)$$

which is the product of the rotational and nuclear spin partition functions, provided that the rotational partition function is $T/(2\Theta_{\text{rot}})$ for a homonuclear diatomic molecule rather than T/Θ_{rot} as for a heteronuclear diatomic molecule. This expression is typically valid to better than 1% for $\Theta_{\text{rot}}/T \le 0.05$.

Let us now examine the percentage of H_2 molecules that belong to the *para* modification as a function of temperature. We begin with the expression for the combined partition function,

$$z_{\text{rot-nuc}}(T) = 3 \sum_{j=\text{odd}} (2j + 1)e^{-j(j+1)\Theta_{\text{rot}}/T} + \sum_{j=\text{even}} (2j + 1)e^{-j(j+1)\Theta_{\text{rot}}/T},$$

$$(6.2.101)$$

with the first and second terms corresponding, respectively, to the contributions from the ortho and para modifications. The percentage of H_2 molecules that belong to the para modification will therefore be given by the ratio

$$\% pH_2 = \frac{z_{\text{para}}(T)}{z_{\text{rot-nuc}}} \times 100\%$$

$$= \frac{\displaystyle\sum_{j \text{ even}} (2j + 1)e^{-j(j+1)\Theta_{\text{rot}}/T}}{3 \displaystyle\sum_{j \text{ odd}} (2j + 1)e^{-j(j+1)\Theta_{\text{rot}}/T} + \sum_{j \text{ even}} (2j + 1)e^{-j(j+1)\Theta_{\text{rot}}/T}} \times 100\%.$$

$$(6.2.102)$$

We shall begin by examining the limiting behaviour of this ratio for $T = 0$ and $T \to \infty$. For very low temperatures, only the terms from the lowest rotational state in each sum will survive, so that the ortho sum reduces to the term for $j = 1$, i.e., $9e^{-2\Theta_{\text{rot}}/T}$, while the para sum reduces to the $j = 0$ term, i.e., 1: the numerator and denominator for our expression thus both become 1 at $T = 0$, and the H_2 is 100% para-H_2. For the high-temperature limit, we can replace the discrete sums over j by integrals, with the consequence that $z_{\text{para}}(T) = T/(2\Theta_{\text{rot}})$ and $z_{\text{ortho}}(T) = 3T/(2\Theta_{\text{rot}})$, giving $z_{\text{rot-nuc}}(T) = T/\Theta_{\text{rot}}$, so that H_2 is only 25% pH_2. Thus, for temperatures T such that $T \gg \Theta_{\text{rot}}$, we obtain what is called 'normal hydrogen', which is simply the high-temperature equilibrium mixture of pH_2 and oH_2 (i.e., $\frac{3}{4}$ ortho plus $\frac{1}{4}$ para). The temperature dependence of the percentage pH_2 in equilibrium hydrogen is displayed in Fig. 6.11.

We see from Fig. 6.11 that at thermal equilibrium, molecular hydrogen consists entirely of pH_2 at $T = 0\,K$, with the percentage pH_2 dropping fairly rapidly with increasing temperature, attaining a 1:1 ratio of pH_2 to oH_2 at about 80 K, and dropping to the 1:3 nuclear spin degeneracy ratio by about $T = 200\,K$, and remaining at the 25% pH_2 mixture ratio thereafter. This high-temperature equilibrium mixture is referred to as normal hydrogen, and designated as nH_2. Because pH_2 and oH_2 have their origin in the coupling of the nuclear spins $I_a = \frac{1}{2}$ of their

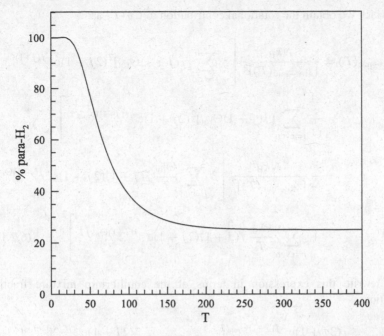

Fig. 6.11 Temperature dependence of the percentage $p\mathrm{H}_2$ in an equilibrium mixture of $o\mathrm{H}_2$ and $p\mathrm{H}_2$

constituent H atoms, their interconversion requires either interaction between an H_2 molecule and a species possessing a magnetic moment to enable the decoupling of the atomic nuclear spin angular momenta of one spin modification followed by a subsequent recoupling to give the other spin modification, or dissociation (typically on a surface) of an H_2 molecule having one value of the net nuclear spin I followed by recombination of the two H atoms to form an H_2 molecule having the other value of I.

Because their characteristic rotational temperatures are so large and their boiling temperatures are so low, the hydrogen isotopologues are the only molecules for which the high-temperature limit does not apply over the entire temperature range for which they are gases. We shall examine in detail only the case of H_2. The rotational contribution to the thermodynamic internal energy, $U_{\mathrm{rot}}(T)$, for equilibrium H_2 is obtained from $z_{\mathrm{rot-nuc}}(T)$, given in Eq. (6.2.101) as

$$U_{\mathrm{rot-nuc}}(T) = \frac{Nk_{\mathrm{B}}\Theta_{\mathrm{rot}}}{z_{\mathrm{rot-nuc}}(T)}\left[3\sum_{j=\mathrm{odd}} j(j+1)(2j+1)e^{-j(j+1)\Theta_{\mathrm{rot}}/T} \right.$$

$$\left. + \sum_{j=\mathrm{even}} j(j+1)(2j+1)e^{j(j+1)\Theta_{\mathrm{rot}}/T} \right], \qquad (6.2.103)$$

from which we obtain the rotational contribution to $C_V(T)$ as

$$
C_{V,\text{rot-nuc}}(T) = \frac{Nk_\text{B}}{[z_{\text{rot-nuc}}(T)]^2}\left[3\sum_{j=\text{odd}}[j(j+1)\Theta_\text{rot}]^2(2j+1)e^{-j(j+1)\Theta_\text{rot}/t} \right.
$$

$$
\left. + \sum_{j=\text{even}}[j(j+1)\Theta_\text{rot}]^2(2j+1)e^{-j(j+1)\Theta_\text{rot}/T} \right]
$$

$$
- \frac{Nk_\text{B}}{[z_{\text{rot-nuc}}(T)]^2}\left[3\sum_{j=\text{odd}}\frac{\Theta_\text{rot}}{T}j(j+1)(2j+1)e^{-j(j+1)\Theta_\text{rot}/T} \right.
$$

$$
\left. + \sum_{j=\text{even}}\frac{\Theta_\text{rot}}{T}j(j+1)(2j+1)e^{-j(j+1)\Theta_\text{rot}/T} \right]^2 . \qquad (6.2.104\text{a})
$$

If we rewrite this expression in terms of the equilibrium mixture fractional populations

$$
p_j^\text{e}(T) \equiv \frac{(2j+1)e^{-j(j+1)\Theta_\text{rot}/T}}{z_{\text{rot-nuc}}(T)}, \quad p_j^\text{o}(T) \equiv \frac{3(2j+1)e^{-j(j+1)\Theta_\text{rot}/T}}{z_{\text{rot-nuc}}(T)},
$$
$$(6.2.104\text{b})$$

then we may express $C_{V,\text{rot-nuc}}(T)$ in the form

$$
C_{V,\text{rot-nuc}}(T) = Nk_\text{B}\left\{ \sum_{j=\text{even}} p_j^\text{e}\left[\frac{j(j+1)\Theta_\text{rot}}{T}\right]^2 + \sum_{j=\text{odd}} p_j^\text{o}\left[\frac{j(j+1)\Theta_\text{rot}}{T}\right]^2 \right.
$$

$$
\left. - \left[\sum_{j=\text{even}} p_j^\text{e}\frac{j(j+1)\Theta_\text{rot}}{T} + \sum_{j=\text{odd}} p_j^\text{o}\frac{j(j+1)\Theta_\text{rot}}{T}\right]^2 \right\}.
$$
$$(6.2.104\text{c})$$

Similar, though simpler, expressions are obtained in the same manner for pure $p\text{H}_2$ and pure $o\text{H}_2$, namely,

$$
C_{V,\text{rot}}^{p\text{H}_2}(T) = Nk_\text{B}\left[\sum_{j=\text{even}} p_j^\text{p}\left(\frac{j(j+1)\Theta_\text{rot}}{T}\right)^2 - \left(\sum_{j=\text{even}} p_j^\text{p}\frac{j(j+1)\Theta_\text{rot}}{T}\right)^2 \right],
$$
$$(6.2.105\text{a})$$

with fractional populations p_j^p given by

$$p_j^{\text{p}}(T) = \frac{(2j+1)e^{-j(j+1)\Theta_{\text{rot}}/T}}{z_{p\text{H}_2}(T)} , \quad z_{p\text{H}_2}(T) \equiv \sum_{j=\text{even}} (2j+1)e^{-j(j+1)\Theta_{\text{rot}}/T} ,$$

(6.2.105b)

and

$$C_{V,\text{rot}}^{o\text{H}_2}(T) = Nk_{\text{B}} \left[\sum_{j=\text{odd}} p_j^{\text{o}} \left(\frac{j(j+1)\Theta_{\text{rot}}}{T} \right)^2 - \left(\sum_{j=\text{odd}} p_j^{\text{o}} \frac{j(j+1)\Theta_{\text{rot}}}{T} \right)^2 \right] ,$$

(6.2.106a)

with fractional populations p_j^{p} given by

$$p_j^{\text{o}}(T) = \frac{(2j+1)e^{-j(j+1)\Theta_{\text{rot}}/T}}{z_{o\text{H}_2}(T)} , \quad z_{o\text{H}_2}(T) \equiv \sum_{j=\text{odd}} (2j+1)e^{-j(j+1)\Theta_{\text{rot}}/T} .$$

(6.2.106b)

Notice that the nuclear spin degeneracy of 3 does not appear in the fractional populations $p_j^{\text{o}}(T)$ of $o\text{H}_2$, as it cancels out between the numerator and denominator. Finally, we may obtain $C_{V,\text{rot}-\text{nuc}}(T)$ for normal hydrogen as

$$C_{V,\text{rot}-\text{nuc}}^{\text{n}}(T) = \tfrac{1}{4} C_{V,\text{rot}-\text{nuc}}^{\text{p}}(T) + \tfrac{3}{4} C_{V,\text{rot}-\text{nuc}}^{\text{o}}(T) .$$

(6.2.107)

Comparisons between calculated and experimental heat capacities of the various spin isotopologues of molecular hydrogen are shown in Fig. 6.12.

Based both upon the difficulty of interconversion between $p\text{H}_2$ and $o\text{H}_2$ and upon the temperature dependence of the thermal equilibrium fractional percentage of $p\text{H}_2$, it becomes clear why it is possible to prepare essentially pure $p\text{H}_2$ but is extremely difficult to prepare $o\text{H}_2$ much in excess of 75% pure. Even though it is possible to separate the $p\text{H}_2$ and $o\text{H}_2$ components of $n\text{H}_2$ using gas chromatography, the half-life (for return to the $n\text{H}_2$ composition) of the $o\text{H}_2$ so separated is only of the order of a few days, while that of pure (or nearly pure) $p\text{H}_2$ is of the order of 3 years. It is also for precisely these same reasons that early experimentalists obtained measurements for $n\text{H}_2$, rather than for equilibrium H_2, which is the origin of the term 'frozen-out' molecular hydrogen mixture for $n\text{H}_2$.

The temperature dependence of the heat capacities $C_{V,\text{rot}}(T)$ as calculated for $p\text{H}_2$, $o\text{H}_2$, and equilibrium H_2 are indicated in Fig. 6.12, together with experimental values obtained by several experimentalists over the period 1912–1925. There was considerable consternation amongst chemists and physicists prior to 1927, until Dennison [17] recognized that the equilibration process between *ortho* and *para* hydrogen might be quite slow, so that what was being measured was simply the high-temperature equilibrium mixture of $p\text{H}_2$ and $o\text{H}_2$, i.e., $n\text{H}_2$, for which the ratio of $p\text{H}_2$ to $o\text{H}_2$ is 1:3. In fact, it is now well-established that in the absence of a suitable catalyst, the conversion process between the two nuclear spin modifications of H_2 has a half-life of approximately 3 years at NTP. Also included in Fig. 6.12 are experimental values obtained in 1929 by Clusius and Hiller [16] for a 95% $p\text{H}_2$, 5%

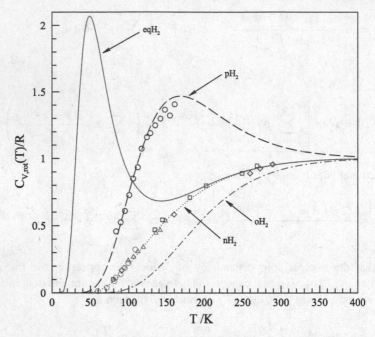

Fig. 6.12 Calculated rotational/nuclear spin contributions to $C_V(T)$ for equilibrium H_2, normal H_2, and the *para* and *ortho* modifications of H_2 as functions of temperature, and their comparison with experiment. Experimental data for nH_2 [11–15]. Data for the 95% pH_2 mixture, Clusius and Hiller [16]

oH_2 mixture: notice that their results are very similar to the computed curve for pure pH_2. The experimental results for nH_2 were summarized by Bonhöffer and Harteck [18] in 1929.

The General Homonuclear Diatomic Molecule X_2

More generally, for a homonuclear diatomic molecule whose nuclei have spin I_a, there are $(2I_a + 1)^2$ total spin pair-states, of which $I_a(2I_a + 1)$ are antisymmetric, and $(I_a + 1)(2I_a + 1)$ are symmetric to the interchange of the two nuclei. Thus, for a homonuclear diatomic molecule whose nuclei have spins I_a, application of the Pauli Principle to Ψ_{total} requires that it be antisymmetric to the interchange of the two nuclei for fermions, i.e., nuclei with spin $I_a = \frac{1}{2}, \frac{3}{2}, \frac{5}{2}, \cdots$, and that it be symmetric to the interchange for bosons, i.e., nuclei with spin $I_a = 0, 1, 2, \ldots$. Because Ψ_{total} has the form of a product, this overall requirement means, in turn, that if the nuclei in a homonuclear diatomic molecule are fermions, then wavefunctions ψ'_{total} that are symmetric to the interchange of the two nuclei must be combined with nuclear spin wavefunctions $\psi_{nuclear\ spin}$ that are antisymmetric under the interchange, and vice versa, while if the nuclei in a homonuclear molecule are bosons, then the nuclear interchange symmetries of ψ'_{total} and $\psi_{\text{nuclear spin}}$ must be the same.

As it happens that a preponderance of homonuclear diatomic molecules have $^1\Sigma_g^+$ ground electronic states, for which we have seen that ψ_{el} is even, the nuclear interchange symmetry of ψ'_{total} is determined solely by the symmetry of the rotational wavefunction. Hence, for a diatomic molecule $X_2(^1\Sigma_g^+)$ whose nuclei are fermions, the combined rotational-nuclear spin partition function takes the form

$$z_{\text{rot-nuc}}(T) = I_a(2I_a + 1) \sum_{j \text{ even}} (2j + 1)e^{-j(j+1)\Theta_{\text{rot}}/T}$$

$$+ (I_a + 1)(2I_a + 1) \sum_{j \text{ odd}} (2j + 1)e^{-j(j+1)\Theta_{\text{rot}}/T} . \qquad (6.2.108)$$

Prominent examples of such diatomic molecules are H_2, F_2, with nuclear spins $I_a = \frac{1}{2}$, Cl_2 (both ^{35}Cl and ^{37}Cl have nuclear spin $I_a = \frac{3}{2}$), and I_2 (with $I_a = \frac{7}{2}$).

Similarly, for $X_2(^1\Sigma_g^+)$ molecules whose nuclei are bosons, we have

$$z_{\text{rot-nuc}}(T) = (I_a + 1)(2I_a + 1) \sum_{j \text{ even}} (2j + 1)e^{-j(j+1)\Theta_{\text{rot}}/T}$$

$$+ I_a(2I_a + 1) \sum_{j \text{ odd}} (2j + 1)e^{-j(j+1)\Theta_{\text{rot}}/T} . \qquad (6.2.109)$$

Prominent examples of such diatomic molecules are D_2, $^{14}N_2$, $^6Li_2$ (these are the only nuclei with $I_a = 1$), $^{12}C_2$, and $^{28}Si_2$ (both the ^{12}C and ^{28}Si nuclei have nuclear spin $I_a = 0$).

Example 6.3 Molecular oxygen.

Molecular oxygen is a special case, as it has a $^3\Sigma_g^-$ ground electronic term. From Fig. 6.9, we see that the electronic wavefunction for this electronic term is antisymmetric to the interchange of the two indistinguishable nuclei of a homonuclear diatomic molecule. This means that $\psi'_{\text{total}}(1, 2)$ for an O_2 molecule will therefore be symmetric to the nuclear interchange for O_2 molecules that are in rotational states for which j is odd, and antisymmetric for O_2 molecules that are in rotational states for which j is even. For an O_2 molecule whose nuclei are equivalent bosons the requirement that the total wavefunction $\Psi_{\text{total}}(1, 2)$ be symmetric to the nuclear interchange results in the odd-j rotational states being coupled with the symmetric nuclear spin pair-states, and vice versa. The opposite outcome will be obtained for an O_2 molecule whose nuclei are equivalent fermions. The most common oxygen isotope by far is ^{16}O, with nuclear spin $I_a = 0$. In this case there is only a single, symmetric, nuclear spin pair-state, and it must therefore be coupled with the odd rotational levels of $^{16}O_2$. Consequently, $^{16}O_2$ molecules possess only odd rotational levels, and the combined rotational-nuclear partition function $z_{\text{rot-nuc}}(T)$ for $^{16}O_2$ is therefore given by

$$z_{\text{rot-nuc}}(T) = \sum_{j=\text{odd}} (2j + 1)e^{-j(j+1)\Theta_{\text{rot}}/T} . \qquad (6.2.110)$$

The same argument holds for the $^{18}O_2$ isotopologue, as the ^{18}O isotope also has nuclear spin $I_a = 0$. The $^{17}O_2$ isotopologue, however, has contributions from both even and odd values of j, as the nuclear spin $I_a = \frac{5}{2}$. As the mixed oxygen molecular isotopologues are all heteronuclear diatomic molecules, they possess no unusual symmetry properties. □

Let us now examine the nature of corrections to rotational partition functions for homonuclear diatomic molecules. We shall proceed by converting the combined rotational-nuclear spin partition function for these molecules into an expression that reduces to the classical value $z_{\text{rot-nuc}}(T) = z_{\text{nuc}}z_{\text{rot}}^{\text{cl}}(T)$, with $z_{\text{nuc}} = (2I_a + 1)^2$ and $z_{\text{rot}}^{\text{cl}}(T) = T/(2\Theta_{\text{rot}})$, plus a difference term traditionally referred to as the exchange term.

If we begin with Eq. (6.2.108) for $z_{\text{rot-nuc}}(T)$ for a homonuclear diatomic molecule whose atomic nuclei are fermions, and if we recognize that the nuclear prefactors for the terms in which the sums are over even, respectively, odd rotational quantum numbers can be written as

$$I_a(2I_a + 1) = \tfrac{1}{2}(2I_a + 1)^2 - \tfrac{1}{2}(I_a + 1)(2I_a + 1) + \tfrac{1}{2}I_a(2I_a + 1)$$

for j even, and as

$$(I_a + 1)(2I_a + 1) = \tfrac{1}{2}(2I_a + 1)^2 + \tfrac{1}{2}(I_a + 1)(2I_a + 1) - \tfrac{1}{2}I_a(2I_a + 1)$$

for j odd, then we see that $z_{\text{rot-nuc}}(T)$ may be rewritten as

$$z_{\text{rot-nuc}}(T) = \tfrac{1}{2}(2I_a + 1)^2 \sum_{j=0}^{\infty}(2j + 1)e^{-j(j+1)\Theta_{\text{rot}}/T}$$

$$-\tfrac{1}{2}(2I_a + 1)\left[\sum_{j=\text{even}}^{\infty}(2j + 1)e^{-j(j+1)\Theta_{\text{rot}}/T}\right.$$

$$\left.- \sum_{j=\text{odd}}^{\infty}(2j + 1)e^{-j(j+1)\Theta_{\text{rot}}/T}\right]. \qquad (6.2.111)$$

This expression may be simplified upon combining the latter two terms of Eq. (6.2.111) into the single sum

$$z_{\text{rot}}^{\text{exc}}(T) = \sum_{j=0}^{\infty}(-1)^j(2j + 1)e^{-j(j+1)\Theta_{\text{rot}}/T} . \qquad (6.2.112)$$

The resultant expression then represents the 'exchange' contribution to the rotational-nuclear spin partition function. A final expression for $z_{\text{rot-nuc}}(T)$ is thus

$$z_{\text{rot-nuc}}(T) = \tfrac{1}{2}(2I_a + 1)^2 \sum_{j=0}^{\infty}(2j + 1)e^{-j(j+1)\Theta_{\text{rot}}/T} + \tfrac{1}{2}(2I_a + 1)z_{\text{rot}}^{\text{exc}}(T),$$

(6.2.113)

when the two nuclei are equivalent fermions.

If we now note that the only difference between $z_{\text{rot-nuc}}(T)$ for homonuclear diatomic molecules whose nuclei are fermions or bosons is that the roles of the nuclear spin factors $I_a(2I_a + 1)$ and $(I_a + 1)(2I_a + 1)$ are interchanged, we see that the only change to Eq. (6.2.113) will be a change of sign in the second term, so that $z_{\text{rot-nuc}}(T)$ for homonuclear diatomic molecules with equivalent boson nuclei is given by

$$z_{\text{rot-nuc}}(T) = \tfrac{1}{2}(2I_a + 1)^2 \sum_{j=0}^{\infty}(2j + 1)e^{-j(j+1)\Theta_{\text{rot}}/T} - \tfrac{1}{2}(2I_a + 1)z_{\text{rot}}^{\text{exc}}(T).$$

(6.2.114)

The first term in each of these equations can be treated in precisely the same way as the discussion in subsection 'Heteronuclear Diatomic Molecules', with the exception that the leading factor will be $T/(2\Theta_{\text{rot}})$ rather than T/Θ_{rot}. This is the source of the well-known symmetry factor σ for homonuclear diatomic molecules. Any difference other than the factor $\tfrac{1}{2}$ arising from the symmetry factor $\sigma = 2$ must arise from the behaviour of the exchange contribution $z_{\text{rot}}^{\text{exc}}(T)$. It has been established that this sum does not have an Euler–Maclaurin expansion [19], so that another method must be found for its evaluation. Indeed, an expression for the exchange sum was first obtained for H_2 in 1933 by Gordon and Barnes [20] using theta functions and their properties. A very similar derivation for the general case was obtained by Kilpatrick et al. [19], who have shown that $z_{\text{rot}}^{\text{exc}}(T)$ is given by

$$z_{\text{rot}}^{\text{exc}}(T) = \left(\frac{\pi}{\alpha}\right)^{\frac{3}{2}} e^{\alpha/4}e^{-\pi^2/(4\alpha)} \sum_{j=0}^{\infty}(-1)^j(2j + 1)e^{-j(j+1)\pi^2/\alpha}, \qquad (6.2.115)$$

with $\alpha \equiv \Theta_{\text{rot}}/T$ as before. For $\alpha = 1$, the summation in Eq. (6.2.115) has the value 1, with an error of less than 1 part in 10^8, and is hence for all intents and purposes identically given by 1. With this result for $z_{\text{rot}}^{\text{exc}}(T)$, McDowell [7] has shown that the optimal classical expression for $z_{\text{rot-nuc}}(T)$ is

$$z_{\text{rot-nuc}}(T) = \tfrac{1}{2}(2I_a+1)^2 e^{\alpha/3}\alpha^{-1}\left[1 + \frac{\alpha^2}{90} + \cdots \pm \frac{\pi^{\frac{3}{2}}}{2I_a + 1}e^{-\alpha/12}e^{-\pi^2/(4\alpha)}\alpha^{-\frac{1}{2}}\right],$$

(6.2.116)

with the upper sign applying when the X nuclei are bosons, the lower sign applying when they are fermions.

Example 6.4 Oxygen, O_2.

It is instructive to see how well the classical formulae work for a homonuclear diatomic molecule. For molecular oxygen, $^{16}O_2$, for example, the full quantum mechanical expression obtained for $z_{rot-nuc}(T)$ is, from Eq. (6.2.109), given by

$$z_{rot-nuc}(T) = \sum_{j=odd}^{\infty} (2j+1)e^{-j(j+1)\Theta_{rot}/T} ,$$

while the classical expression is given by Eq. (6.2.66) to a good approximation as

$$z_{rot-nuc}(T) \simeq \tfrac{1}{2}e^{\Theta_{rot}/(3T)} \frac{T}{\Theta_{rot}} \left[1 + \frac{1}{90}\left(\frac{\Theta_{rot}}{T}\right)^2 \right] , \qquad (6.2.117)$$

and the classical limit is $z_{rot-nuc}(T) = T/(2\Theta_{rot})$. The results obtained for temperatures $T = 5\Theta_{rot}$, $10\Theta_{rot}$, $20\Theta_{rot}$, and 90.5 K, the boiling point for liquid oxygen, are summarized in Table 6.2.

Note that the full quantum mechanical and classical results agree to four decimal places even for $T = 5\Theta_{rot}$, while the classical limit gives values that vary between 6.5% low for $T = 5\Theta_{rot}$ to 0.5% low for $T = 90.5$ K, the boiling temperature for liquid oxygen. □

Summary for Rotational Partition Functions for Diatomic Molecules

If we examine the values assigned to Θ_{rot} for common diatomic molecules, such as those listed in Table 6.1, for example, we see that $\Theta_{rot} \ll T$ for most of them at ambient temperatures, so that we may replace the quantum mechanical sum by an integral. Thus, for sufficiently high temperatures we can write quite generally

$$z_{rot}(T) = \frac{T}{\sigma\Theta_{rot}} , \qquad (6.2.118)$$

in which, σ, referred to as the symmetry number, is assigned the values

$$\sigma = \begin{cases} 1 & \text{for AB molecules} \\ 2 & \text{for } A_2 \text{ molecules}. \end{cases} \qquad (6.2.119)$$

Table 6.2 Comparison between quantum and classical mechanical results for $z_{rot-nuc}(T)$

T	Eq. (6.2.114)	Eq. (6.2.117)	$T/(2\Theta_{rot})$	Error
$5\,\Theta_{rot}$	2.673457	2.673535	2.50000	−6.5%
$10\,\Theta_{rot}$	5.170065	5.170049	5.00000	−3.3%
$20\,\Theta_{rot}$	10.168348	10.168346	10.00000	−1.7%
90.5 K	21.95473	21.95488	21.85990	−0.5%

For temperatures T such that $2\Theta_{rot} < T < 20\Theta_{rot}$, the classical expression for z_{rot} can also be written quite generally as

$$z_{rot}(T) = \frac{T}{\sigma\Theta_{rot}}e^{\Theta_{rot}/(3T)}\left[1 + \frac{1}{90}\left(\frac{\Theta_{rot}}{T}\right)\right]f_{cd}. \qquad (6.2.120)$$

Generally speaking, the rotational partition function can be evaluated in the classical limit for many molecules at room temperature and above. If we wish to obtain values for thermodynamic state functions, the corresponding rotational contributions to the Helmholtz free energy, A, the internal energy, U, the heat capacity at constant volume, C_V, and the entropy, S, for a gas of rigid diatomic molecules are given by

$$A_{rot}(T) = -k_BT \ln Z_{rot} = -Nk_BT \ln\left(\frac{T}{\sigma\Theta_{rot}}\right), \qquad (6.2.121)$$

$$U_{rot}(T) = k_BT^2\frac{d \ln Z_{rot}}{dT} = Nk_BT, \qquad (6.2.122)$$

$$C_{V,rot}(T) = Nk_B, \qquad (6.2.123)$$

$$S_{rot}(T) = \frac{U_{rot} - A_{rot}}{T} = Nk_B \ln\left(\frac{Te}{\sigma\Theta_{rot}}\right). \qquad (6.2.124)$$

6.2.4 Electronic Degree of Freedom

Based upon the separated stationary atoms (SSA) zero of energy for a diatomic molecule and the simple harmonic oscillator (SHO) approximation for the vibrational motion in a diatomic molecule, we see that the product of the electronic and vibrational molecular partition functions can be expressed via Eq. (6.2.10) as

$$z_{el}(T)z_{vib}(T) = \omega_{el}e^{\mathfrak{D}_e/(k_BT)}\frac{e^{-\Theta_{vib}/(2T)}}{1 - e^{-\Theta_{vib}/T}}$$

$$= \omega_{el}e^{D_0/(k_BT)}\tilde{z}_{vib}(T) \qquad (6.2.125)$$

in terms of the experimentally accessible dissociation energy D_0 rather than $\mathfrak{D}_e$. We recall that D_0 is related to $\mathfrak{D}_e$ and the characteristic vibrational temperature Θ_{vib} by $D_0 = \mathfrak{D}_e - \frac{1}{2}k_B\Theta_{vib}$. The modified vibrational partition function $\tilde{z}_{vib}(T)$ is defined as

$$\tilde{z}_{vib}(T) \equiv (1 - e^{-\Theta_{vib}/T})^{-1}. \qquad (6.2.126)$$

By employing these redefined factors, the electronic contributions to the internal energy $U(T, V)$ and the Helmholtz energy $A(T, V)$, for example, are expressed directly as

$$U_{el}(T) = -ND_0, \qquad A_{el}(T) = ND_0 + Nk_BT \ln \omega_{el} \qquad (6.2.127)$$

in terms of experimentally accessible quantities.

The dissociation energy D_0, which represents the amount of energy that has to be added to an individual diatomic molecule to cause it to be dissociated into its component noninteracting atoms (equivalently, at infinite separation) can also be interpreted as the negative of the (internal) energy, $\epsilon_0^\circ = \overline{U}_0^\circ/N_0$, per molecule at $T = 0\,K$ (at which the translational component vanishes). This latter quantity can also be looked upon as the heat of formation of the molecule from its constituent atoms at $0\,K$, i.e., we may write

$$\overline{U}_0^\circ = \Delta \overline{U}_{f0}^\circ = \Delta \overline{H}_{f0}^\circ \,,$$

with the latter equality coming from the ideal gas law $PV = Nk_BT$. This result allows us to express the molar internal energy in the form

$$\overline{U} - \Delta \overline{H}_{f0}^\circ = \tfrac{5}{2} N_0 k_B T + \frac{N_0 k_B \Theta_{vib}}{e^{\Theta_{vib}/T} - 1} \,. \qquad (6.2.128)$$

We note that this form is particularly convenient for comparisons between our calculated contributions from the various internal degrees of freedom, as the right-hand side of this expression no longer contains D_0, which is the least accessible molecular parameter. Determination of the chemical dissociation energy is not as great a problem for diatomic molecules as it is for polyatomic molecules, as we shall see below. Nonetheless, the determination of accurate values for heats of formation has been an item of concern to chemists for more than a century, and they have devised a number of means for determining them for a very large number of substances. Similar expressions to Eq. (2.7.4) for $\overline{U}$ may be obtained for A, H, G and μ°. These are the quantities whose values are listed in the NIST-JANAF[1] thermochemical tables [21].

Some caution must be applied here, however, as our statistical mechanical expressions are based upon the SSA zero of energy, so that $\Delta \overline{H}_{f0}^\circ$ for molecules like N_2 will be nonzero, while the heats of formation of atoms like N will be zero by definition. This must be compared with the more practical convention chosen in thermodynamics, which sets $\Delta \overline{H}_f^\circ$ to zero for the elements in their naturally occurring forms (at earth surface temperatures!). Thus $\Delta \overline{H}_{f0}^\circ$(thermo) for N_2 is zero, while $\Delta \overline{H}_{f0}^\circ$(thermo) for N is nonzero. Let us illustrate the calculation of the heat of

---

[1]JANAF is an acronym standing for Joint Army Navy Air Force: the funding for the creation of these tabulations of thermodynamic data was provided by these three agencies.

Scheme 6.1 Hess law cycle for the chemical reaction between H_2 and Cl_2

Scheme 6.2 Hess law cycle for the reaction of graphite and oxygen

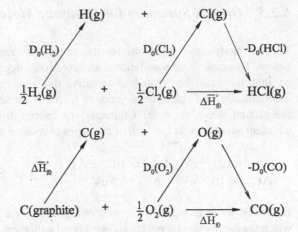

formation briefly, using two examples, namely the formation of HCl from H_2 and Cl_2, and the formation of CO from elemental carbon and O_2.

Example 6.5 The H_2–Cl_2 reaction. The Hess law cycle for this reaction is shown in Scheme 6.1.

This Hess law cycle relates the heat of formation, $\Delta\overline{H}_{f0}^{\circ}$, of HCl at $0\,K$, 1 bar pressure, to the spectroscopic dissociation energies D_0 of H_2, Cl_2, and HCl. From Table 6.1 we have the values $D_0(H_2) = 4.478\,eV$, $D_0(Cl_2) = 2.479\,eV$, and $D_0(HCl) = 4.433\,eV$. The Hess law cycle thus gives us the relation

$$\tfrac{1}{2}D_0(H_2) + \tfrac{1}{2}D_0(Cl_2) - D_0(HCl) = \Delta\overline{H}_{f0}^{\circ}(HCl)\,,$$

so that substitution of the D_0 values gives for $\Delta H_{f0}^{\circ}(HCl)$ the value $-0.955\,eV$ or $-92.111\,kJ\,mol^{-1}$. This value differs by less than 0.1% from the accepted literature value of $-92.127\,kJ\,mol^{-1}$ given in the JANAF tables. $\square$

Example 6.6 Formation of CO from graphite and O_2. The Hess law cycle appropriate to this reaction is shown in Scheme 6.2.

This Hess law cycle requires the heat of formation of gaseous atomic carbon from graphite at $0\,K$, which is $\Delta\overline{H}_{f0}^{\circ} = 711.0\,kJ\,mol^{-1} \simeq 7.370\,eV$, while $D_0(O_2) = 5.116\,eV$, and $D_0(CO) = 11.092\,eV$. Upon employing these values in the Hess law cycle, we obtain $\Delta\overline{H}_{f0}^{\circ}(CO) = -1.164\,eV = -112.32\,kJ\,mol^{-1}$: this value is within 1.2% of the accepted literature value $-113.813\,kJ\,mol^{-1}$ for $\Delta\overline{H}_{f0}^{\circ}(CO)$. The agreement in this case is not quite as good as that obtained in Example 6.5, largely due to the less accurate data for graphite and CO. Note, however, that a difference of only $0.015\,eV$ in the value of $D_0(CO)$ (corresponding to an experimental error of about 0.14%) would already account for the difference between our calculated value of $\Delta\overline{H}_{f0}^{\circ}(CO)$ and the accepted literature value given in the JANAF tables. $\square$

6.2.5 Overall Summary for Diatomic Molecules

The thermodynamic functions for the most commonly occurring case are given below. This case is for translational and rotational degrees of freedom fully excited, so that they can be treated classically, the vibrational mode treated quantum mechanically, and electronic degrees of freedom unexcited (i.e., all molecules are in the ground electronic state). Characteristic energy differences associated with two adjacent energy levels for each of the types of motion mentioned above are given by

$$\left.\begin{array}{ll}\Delta\epsilon_{tr} \sim 10^{-18}\,eV, & \Delta\epsilon_{rot} \sim 10^{-4}\,eV \\ \Delta\epsilon_{vib} \sim 10^{-1}\,eV, & \Delta\epsilon_{el} \sim 5\,eV\end{array}\right\} \quad cf.\ k_B T \sim 10^{-2}\,eV\ for\ T \sim 300\,K.$$

Basically, we may treat the degrees of freedom for a particular motion classically when the energy level separations associated with them are small in comparison with $k_B T$, i.e., $\Delta\epsilon \ll k_B T$. Unexcited implies that $\Delta\epsilon \gg k_B T$; otherwise, the degrees of freedom must be treated quantum mechanically. For the set of characteristic energy values listed above this means that we can treat the translational and rotational degrees of freedom classically, the vibrational degree of freedom quantum mechanically, and the electronic degrees of freedom as unexcited. Such a procedure then gives us the following expressions for the usual thermodynamic functions for a gas made up of diatomic molecules:

$$\overline{A}(T,V) - \Delta\overline{H}_{f0}^\circ = -N_0 k_B T \left\{ \ln\left[\left(\frac{2\pi(m_1+m_2)k_B T}{h^2}\right)^{\frac{3}{2}} \frac{\overline{V}^\circ e}{N}\right] \right.$$

$$\left. + \ln\left(\frac{T}{\sigma\Theta_{rot}}\right) - \ln(1 - e^{-\Theta_{vib}/T}) + \ln\omega_{el}\right\},$$

$$(6.2.129)$$

$$\overline{U}(T,V) - \Delta\overline{H}_{f0}^\circ = N_0 k_B T \left\{\frac{3}{2} + \frac{2}{2} + \frac{\Theta_{vib}}{T}\frac{1}{e^{\Theta_{vib}/T}-1}\right\}, \qquad (6.2.130)$$

$$\overline{C}_V(T) = N_0 k_B \left\{\frac{5}{2} + \left(\frac{\Theta_{vib}}{T}\right)^2 \frac{e^{\Theta_{vib}/T}}{(e^{\Theta_{vib}/T}-1)^2}\right\}, \qquad (6.2.131)$$

$$\overline{S}(T,V) = N_0 k_B \left\{ \ln\left[\left(\frac{2\pi(m_1+m_2)k_B T}{h^2}\right)^{\frac{3}{2}} \frac{\overline{V}^\circ e^{\frac{5}{2}}}{N}\right] + \ln\left(\frac{Te}{\sigma\Theta_{rot}}\right)\right.$$

$$\left. + \frac{\Theta_{vib}}{T}\frac{1}{e^{\Theta_{vib}/T}-1} - \ln(1-e^{-\Theta_{vib}/T}) + \ln\omega_{el}\right\}. \qquad (6.2.132)$$

In addition we have the ideal gas equation of state,

$$PV = Nk_BT,$$

and we can express the chemical potential $\mu(T, P)$ as

$$\mu(T, P) = \frac{G(T, P)}{N} = \mu^\circ(T) + k_BT \ln P, \tag{6.2.133}$$

with the standard chemical potential $\mu^\circ(T)$ given by

$$\mu^\circ(T) = \Delta\overline{H}_{f0}^\circ - k_BT\left\{ \ln\left[\left(\frac{2\pi(m_1 + m_2)k_BT}{h^2}\right)^{\frac{3}{2}} k_BT \right]\right.$$

$$\left. + \ln\left(\frac{T}{\sigma\Theta_{rot}}\right) - \ln(1 - e^{-\Theta_{vib}/T}) + \ln\omega_{el}\right\}. \tag{6.2.134}$$

6.2.6 Direct Evaluation of Internal State Contributions

Not only is it possible to determine the potential energy function for a typical linear molecule, especially a diatomic molecule, extremely accurately from direct fits of an appropriate functional form to the large set of spectroscopic transition frequencies that can be measured with a high degree of precision, but there also exist very accurate computational procedures capable of providing all bound (and quasi-bound) energy levels for the molecule. These energy level values can be utilized directly to compute both the relevant partition function for the molecule and the thermodynamic state functions pertaining to this species. We shall briefly examine the expressions required for the determination of the temperature dependence of the internal state contributions to the thermodynamic state functions in this subsection.

We begin with the fundamental relation (4.1.3b) between the thermodynamic internal energy U and the canonical partition function $Z(T, V; N)$, together with expression (6.1.3) resulting from the rigorous decoupling between translational and internal state motions. We have also seen in Sect. 6.1 that, because the thermodynamic state functions depend in general only upon $\ln Z$ and/or its partial derivatives, the translational and internal state contributions to these functions will be additive. The internal energy $U(T, V; N)$ may then be written in the form

$$U(V; N) = U_{trans}(T, V; N) + U_{int}(T; N), \tag{6.2.135}$$

with all volume dependence contained, and the factor $N!$ included, in the translational component, and U_{int} expressed as a function only of the canonical ensemble thermodynamic *variable*[2] T.

We have argued earlier that if the interaction energies ϵ_{ij} between pairs of the constituent atoms/molecules that make up a canonical ensemble can be neglected relative to the energies ϵ_i and ϵ_j associated with the individual constituents, then the partition function $Z_N(T, V)$ for an ensemble of N indistinguishable atoms/molecules can be represented [see Eq. (3.2.27)] as

$$Z_N(T, V) = \frac{z^N(T, V)}{N!},$$

in terms of the partition function $z(T, V)$ for an individual constituent of the ensemble. Moreover, we have also argued in Sect. 4.1 that the thermodynamic internal energy U for a canonical ensemble of N atoms/molecules will be given by $U(T, V; N) = Nu(T, V)$, with $u(T, V)$ the thermodynamic internal energy per particle which is, in turn, given via Eq. (4.1.1) as the ensemble average energy, $\langle \epsilon \rangle \equiv \sum_i p_i \epsilon_i$, for a single constituent atom/molecule. Due to the separability of the translational and internal state contributions to the partition function, we may apply the same reasoning to the internal state contribution to the thermodynamic internal energy, $U_{int}(T; N)$.

We may express $U_{int}(T; N)$ either via Eq. (4.1.3b) in terms of the temperature partial derivative of $\ln Z_{int}(T; N)$ or, alternatively, in terms of the 'more natural' canonical ensemble variable $\beta \equiv (k_B T)^{-1}$, the so-called dimensionless reciprocal temperature. Thus, in terms of the derivative with respect to β, we write

$$U_{int}(T; N) = -\left(\frac{\partial \ln Z_{int}}{\partial \beta}\right)_N. \qquad (6.2.136a)$$

Upon recalling Eq. (6.1.4) relating $Z_{int}(\beta; N)$ to $z_{int}(\beta)$ by $Z_{int}(T; N) = z_{int}^N(\beta)$, we obtain $U_{int}(T; N)$ as

$$U_{int}(T; N) = -N \frac{d \ln z_{int}}{d\beta}$$

$$= -\frac{N}{z_{int}(\beta)} \frac{dz_{int}}{d\beta}, \qquad (6.2.136b)$$

with $z_{int}(\beta)$ given explicitly by

---

[2] Although we have written $U_{int}(T; N)$, within the context of the canonical ensemble, the only truly independent variable is the temperature T. The number of particles N has, strictly speaking, a fixed value and should be considered as a parameter characterizing particular canonical ensembles: it thus plays a slightly different role in our description of statistical thermodynamics based upon the canonical ensemble than do the truly independent variables T and V.

$$z_{int}(\beta) = \sum_i \Omega_i e^{-\beta \epsilon_i}. \tag{6.2.137}$$

This result can be employed directly to determine the contribution to the thermodynamic internal energy of an atomic gas associated with thermally excited atoms, as the ground atomic term energy may always be set to zero. For molecular gases, however, all excited state energies are referenced to the energy of the lowest energy eigenstate for the molecule which, because of quantum mechanical zero-point motion, will have a nonzero value. We shall designate this zero-point energy as ϵ_0, so that the internal state contribution to the thermodynamic internal energy $U_{int}(T; N)$ can be expressed as

$$U_{int}(T; N) = N[u_{int}(T) + \epsilon_0]. \tag{6.2.138}$$

A similar expression for the internal state contribution to the heat capacity (at constant volume) is, in the same format,

$$C_{V int}(T; N) = \frac{1}{k_B T^2}\left(\frac{\partial^2 \ln Z_{int}}{\partial \beta^2}\right)_N. \tag{6.2.139a}$$

This expression may also be given in terms of $z_{int}(\beta)$ as

$$
\begin{aligned}
C_{V,int}(T; N) &= \frac{N}{k_B T^2} \frac{d^2 \ln z_{int}}{d\beta^2} \\
&= \frac{N}{k_B T^2}\left[\frac{1}{z_{int}}\frac{d^2 z_{int}}{d\beta^2} - \left(\frac{1}{z_{int}}\frac{d z_{int}}{d\beta}\right)^2\right].
\end{aligned}
\tag{6.2.139b}
$$

This last expression is clearly reminiscent of expression (4.1.4a) that we obtained for the variance of the system energy. From Eq. (6.2.137) for $z_{int}(\beta)$, we see that

$$\frac{d^2 z_{int}}{d\beta^2} = (k_B T)^2 \sum_i (\beta \epsilon_i)^2 \Omega_i e^{-\beta \epsilon_i}. \tag{6.2.140}$$

Notice now that if we define a set of dimensionless functions $\{f_n(T)\,|\,n = 0, 1, 2, \cdots\}$ via

$$f_n(T) \equiv \sum_i (\beta \epsilon_i)^n \Omega_i e^{-\beta \epsilon_i}, \qquad n = 0, 1, 2, \ldots, \tag{6.2.141}$$

then $U_{int}(T; N)$ is given in terms of these functions as

$$U_{int}(T; N) - N\epsilon_0 = N k_B T \frac{f_1}{f_0}, \tag{6.2.142a}$$

in which ϵ_0 is the energy of an individual atom/molecule at 0 K. For a diatomic molecule, ϵ_0 can be identified with the spectroscopically determinable dissociation energy, D_0. The heat capacity at constant volume, $C_{V,\text{int}}(T; N)$, can similarly be evaluated in terms of these functions via the expression

$$C_{V,\text{int}}(T; N) = N k_B \left[\frac{f_2}{f_0} - \left(\frac{f_1}{f_0} \right)^2 \right]. \tag{6.2.142b}$$

Because the volume dependence of the thermodynamic state functions all originates from the translational states, the translational contribution to the enthalpy takes the form $H_{\text{trans}}(T, V; N) = U_{\text{trans}}(T, V; N) - PV$, and hence $H_{\text{int}}(T; N) - N\epsilon_0 = U_{\text{int}}(T; N) - N\epsilon_0$. For precisely the same reason, the internal state contributions to the Gibbs and Helmholtz energies are also equal, i.e., $G_{\text{int}}(T; N) - N\epsilon_0 = A_{\text{int}}(T; N) - N\epsilon_0$. The relevant expression for $A_{\text{int}}(T; N)$ is

$$A_{\text{int}}(T; N) - N\epsilon_0 = -N k_B T \ln f_0 \tag{6.2.142c}$$

and, as the entropy $S_{\text{int}}(T; N)$ can be defined in terms of the Helmholtz and internal energies, the appropriate expression for $S_{\text{int}}(T; N)$ is

$$S_{\text{int}}(T; N) = N k_B \left[\frac{f_1}{f_0} + \ln f_0 \right]. \tag{6.2.142d}$$

Expressions (6.2.142) are thus naturally structured to facilitate direct evaluation of the internal state contributions using spectroscopic data.

6.3 Extension to the Ideal Polyatomic Gas

Because we have treated the ideal diatomic molecular gas in full detail in the previous sections, we need not do more in this section than examine the few essential differences between diatomic and polyatomic molecules. Let us consider a polyatomic molecule with N_a constituent atoms. Such a molecule will have a total of $3N_a$ independent degrees of freedom. As for the diatomic molecule that we have already considered in detail, any polyatomic molecule will have 3 translational degrees of freedom associated with its centre-of-mass motion. Significant differences do occur, however, between the partition functions associated with the rotational and vibrational degrees of freedom for diatomic and polyatomic molecules. In particular, there will be 3 rotational degrees of freedom for a nonlinear polyatomic molecule (but still only 2 for a linear polyatomic molecule), leaving $3N_a - 6$ vibrational degrees of freedom for a nonlinear polyatomic molecule (vs. $3N_a - 5$ vibrational degrees of freedom for a linear polyatomic molecule). These vibrational degrees of freedom generally correspond to the IR and Raman spectral transitions in the molecule. Note, however, that in some relatively rare cases, there are vibrational degrees of freedom that are neither IR nor Raman active. The additional vibrational degrees of freedom for polyatomic molecules result in the vibrational partition function being represented by a product function of vibrational partition functions corresponding to the individual independent vibrational degrees of freedom of the polyatomic molecule.

6.3.1 The Rigid-Rotor Simple Harmonic Oscillator Approximation

If we make the SHO approximation for each independent vibrational mode, then we have the product of $3N_a - 6$ (or of $3N_a - 5$) SHO partition functions, namely,

$$z_{\text{vib}}(T) = \prod_{i=1}^{n} \frac{e^{-\Theta_{\text{vib},i}/2T}}{1 - e^{-\Theta_{\text{vib},i}/T}}, \tag{6.3.1}$$

in terms of the characteristic temperatures $\Theta_{\text{vib},i}$ associated with the various vibrational modes of a molecule. The upper limit n in Eq. (6.3.1) will be either $3N_a - 6$ (nonlinear molecule) or $3N_a - 5$ (linear molecule). This change in the form for $z_{\text{vib}}(T)$ gives rise to corresponding changes in U_{vib} and $C_{V,\text{vib}}$ for N molecules [cf. Eqs. (6.2.23 and 6.2.24)]: explicitly,

$$U_{\text{vib}}(T) - \sum_{i=1}^{n} \tfrac{1}{2} k_B \Theta_{\text{vib},i} = N \sum_{i=1}^{n} \frac{k_B \Theta_{\text{vib},i}}{e^{\Theta_{\text{vib},i}/T} - 1} \tag{6.3.2}$$

and

$$C_{V,\text{vib}}(T) = N k_B \sum_{i=1}^{n} \left(\frac{\Theta_{\text{vib},i}}{T} \right)^2 \frac{e^{\Theta_{\text{vib},i}/T}}{(e^{\Theta_{\text{vib},i}/T} - 1)^2}, \tag{6.3.3}$$

in which $\Theta_{\text{vib},i} \equiv h\nu_i/k_B$ is the temperature characteristic of the ith normal mode of vibration. Note that the only difference between these expressions for $U_{\text{vib}}(T)$, $C_{V,\text{vib}}(T)$ and the corresponding expressions for diatomic molecules is the replacement of the single vibrational term by a sum of terms, one term for each of the $3N_a - 5$ or $3N_a - 6$ independent vibrational degrees of freedom possessed by an individual polyatomic molecule. Expressions for $A_{\text{vib}}(T)$ and $S_{\text{vib}}(T)$ can likewise be obtained from the expressions obtained for diatomic molecules in section 'The SHO Model Partition Function: Vibrational Contribution to the Thermodynamic State Functions.'

Rotations are also slightly more complicated for the general polyatomic molecule: for a (rigid) linear molecule, there will be two equivalent end-over-end rotational degrees of freedom which have their axes of rotation perpendicular to the molecular figure axis, and one rotational degree of freedom which has the molecular figure axis as its axis of rotation (and a moment-of-inertia that is essentially zero). As a consequence, the expressions for z_{rot}, U_{rot}, and $C_{V,rot}$ are precisely the same as those for a diatomic molecule. For a rigid nonlinear polyatomic molecule, there are in principle three distinct moments-of-inertia, designated by convention as I_A, I_B, and I_C, so that the high-temperature classical limit for $z_{rot}(T)$ becomes (see Appendix E)

$$z_{rot}(T) = \frac{\sqrt{\pi}}{\sigma} \left(\frac{T^3}{\Theta_A \Theta_B \Theta_C} \right)^{\frac{1}{2}} , \tag{6.3.4}$$

in which Θ_A, Θ_B, and Θ_C are defined in the same manner in terms of the moments-of-inertia I_A, I_B, and I_C as was Θ_{rot} for a diatomic molecule in terms of its single moment-of-inertia I. The quantity σ is the symmetry number for the molecule concerned, and is determined by the order of the symmetry group of all pure rotations which leave the molecule indistinguishable. Thus, for example, $\sigma = 2$ for H_2O, 3 for NH_3, 12 for CH_4 and C_6H_6, and so on.

The rotational contributions to the thermodynamic functions for a sample containing N molecules will now be

$$A_{rot}(T) = -Nk_B T \ln \left[\frac{\sqrt{\pi}}{\sigma} \left(\frac{T^3}{\Theta_A \Theta_B \Theta_C} \right)^{\frac{1}{2}} \right] , \tag{6.3.5}$$

$$U_{rot}(T) = \tfrac{3}{2} Nk_B T , \tag{6.3.6}$$

$$C_{V,rot} = \tfrac{3}{2} Nk_B , \tag{6.3.7}$$

$$S_{rot}(T) = Nk_B \ln \left[\frac{\sqrt{\pi}}{\sigma} \left(\frac{T^3 e^3}{\Theta_A \Theta_B \Theta_C} \right)^{\frac{1}{2}} \right] . \tag{6.3.8}$$

6.3.2   Summary of the RR-SHO Approximation for Polyatomic Molecules

We are essentially done once we have extended the description of the vibrational and rotational contributions from diatomic molecules to polyatomic molecules, as the translational motion is strictly the same (associated, as it is, with the centre-of-mass motion of the molecule) and the same methodology for treating the electronic terms applies. Of course, other complications may ensue, such as how to deal with vibrational modes that are neither infrared nor Raman active (relatively rare occurrences for simple polyatomic molecules, but increasingly more frequent occurrences for large molecules, such as biomolecules and polymers) and anomalous behaviour of the entropy at very low temperature (leading to apparent violations of the usual statements of the third law of thermodynamics). We shall briefly deal with both of these special cases in later subsections.

Because polyatomic molecules are readily divided into the two distinct categories of linear and nonlinear molecules, with each category requiring a slight variant for the treatment of its vibrational and rotational degrees of freedom, we shall summarize the expressions for the thermodynamic state functions separately, with one set of expressions for linear, the other for nonlinear, molecules.

As for diatomic molecules, we shall treat the degrees of freedom for a particular motion classically when the energy level separations associated with them are small in comparison with k_BT. Here, too, with the exception of simple hydrides, the rotational degrees of freedom can be treated classically for all but very low temperatures, and the vibrational degrees of freedom must always be treated quantum mechanically. Although the occurrence of relatively low-lying electronic states is more common for polyatomic molecules than it is for diatomic molecules, we shall not consider such cases here.

The quantity $\mathfrak{D}_e$ may be thought of as the global minimum in the (multidimensional) potential energy surface for the polyatomic molecule. A value of D_0 for a polyatomic molecule may be thought of as the energy required to atomize the molecule into its constituent separated stationary atoms at infinity. Both D_0 and $\mathfrak{D}_e$ may be inferred from spectroscopic measurements, but apart from the special case of diatomic molecules, it is rather difficult to do so. However, as for diatomic molecules, a value of D_0 can be obtained directly from the thermodynamic molar heat of formation, $\Delta \overline{H}_{f0}^{\circ}$, for a molecule, so that the relevant listings in the JANAF Tables [21] are for $\overline{H} - \Delta \overline{H}_{f0}^{\circ}$ and $\overline{A} - \Delta \overline{H}_{f0}^{\circ}$, as for diatomic molecules.

Linear Molecule Expressions

Within the simple harmonic oscillator (SHO), rigid-rotor (RR) approximation, the molecular partition function $z(T, V)$ for a linear molecule is given by

$$z(T, V) = \left(\frac{2\pi M k_B T}{h^2}\right)^{\frac{3}{2}} V \frac{T}{\sigma \Theta_{\text{rot}}} \left[\prod_{i=1}^{3N_a-5} \frac{e^{-\Theta_{\text{vib},i}/2T}}{(1 - e^{-\Theta_{\text{vib},j}/T})}\right] \omega_{e1} e^{\beta \mathfrak{D}_e}.$$

(6.3.9)

The equation of state remains the ideal gas law, $PV = N k_B T$, and the molar thermodynamic state functions $\overline{A}(T, V) - \Delta \overline{H}_{f0}^{\circ}$, $\overline{U}(T, V) - \Delta \overline{H}_{f0}^{\circ}$, $\overline{C}_V(T)$, and $\overline{S}(T, V)$ are given by

$$\overline{A}(T,V) - \Delta\overline{H}_{f0}^{\circ} = -RT \left\{ \ln\left[\left(\frac{2\pi M k_{\mathrm{B}} T}{h^2} \right)^{\frac{3}{2}} \frac{\overline{V}^{\circ} \mathrm{e}}{N} \right] + \ln\left(\frac{T}{\sigma \Theta_{\mathrm{rot}}} \right) \right.$$

$$\left. - \sum_{i=1}^{3N_a - 5} \ln(1 - \mathrm{e}^{-\Theta_{\mathrm{vib},i}/T}) + \ln \omega_{\mathrm{e}1} \right\} , \tag{6.3.10}$$

$$\overline{U}(T,V) - \Delta\overline{H}_{f0}^{\circ} = RT \left\{ \frac{3}{2} + \frac{2}{2} + \sum_{i=1}^{3N_a - 5} \frac{\Theta_{\mathrm{vib},i}/T}{\mathrm{e}^{\Theta_{\mathrm{vib},i}/T} - 1} \right\} , \tag{6.3.11}$$

$$\overline{C}_V(T) = R \left\{ \frac{3}{2} + \frac{2}{2} + \sum_{i=1}^{3N_a - 5} \left(\frac{\Theta_{\mathrm{vib},i}}{T} \right)^2 \frac{\mathrm{e}^{\Theta_{\mathrm{vib},i}/T}}{(\mathrm{e}^{\Theta_{\mathrm{vib},i}/T} - 1)^2} \right\} , \tag{6.3.12}$$

$$\overline{S}(T,V) = R \left\{ \ln\left[\left(\frac{2\pi M k_{\mathrm{B}} T}{h^2} \right)^{\frac{3}{2}} \frac{\overline{V}^{\circ} \mathrm{e}^{\frac{5}{2}}}{N} \right] + \ln\left(\frac{T \mathrm{e}}{\sigma \Theta_{\mathrm{rot}}} \right) \right.$$

$$\left. + \sum_{i=1}^{3N_a - 5} \left[\frac{\Theta_{\mathrm{vib},i}/T}{\mathrm{e}^{\Theta_{\mathrm{vib},i}/T} - 1} - \ln(1 - \mathrm{e}^{-\Theta_{\mathrm{vib},i}/T}) \right] + \ln \omega_{\mathrm{e}1} \right\} . \tag{6.3.13}$$

These are the expressions whose values may be employed for comparison with the JANAF tabulations for linear molecules [21].

Nonlinear Molecule Expressions

Within the simple harmonic oscillator (SHO), rigid-rotor (RR) approximation, the molecular partition function $z(T, V)$ for a nonlinear molecule is given by

$$z(T,V) = \left(\frac{2\pi M k_{\mathrm{B}} T}{h^2} \right)^{\frac{3}{2}} V \frac{\sqrt{\pi}}{\sigma} \left(\frac{T^3}{\Theta_{\mathrm{A}} \Theta_{\mathrm{B}} \Theta_{\mathrm{C}}} \right)^{\frac{1}{2}} \left[\prod_{i=1}^{3N_a - 6} \frac{\mathrm{e}^{-\Theta_{\mathrm{vib},i}/2T}}{(1 - \mathrm{e}^{-\Theta_{\mathrm{vib},i}/T})} \right] \omega_{\mathrm{e}1} \mathrm{e}^{\beta \mathfrak{D}_{\mathrm{e}}} . \tag{6.3.14}$$

As for linear molecules, the equation of state is the ideal gas law, and the molar thermodynamic state functions $\overline{A}(T,V) - \Delta\overline{H}_{f0}^{\circ}$, $\overline{U}(T,V) - \Delta\overline{H}_{f0}^{\circ}$, $\overline{C}_V(T)$, and $\overline{S}(T,V)$ are given by

$$\overline{A}(T,V) - \Delta\overline{H}_{f0}^{\circ} = -RT \left\{ \ln\left[\left(\frac{2\pi M k_{\mathrm{B}} T}{h^2} \right)^{\frac{3}{2}} \frac{\overline{V}^{\circ} \mathrm{e}}{N} \right] + \ln\left[\frac{\sqrt{\pi}}{\sigma} \left(\frac{T^3}{\Theta_{\mathrm{A}} \Theta_{\mathrm{B}} \Theta_{\mathrm{C}}} \right)^{\frac{1}{2}} \right] \right.$$

$$- \sum_{i=1}^{3N_a-6} \ln(1 - e^{-\Theta_{\text{vib},i}/T}) + \ln\omega_{\text{el}} \Bigg\}, \qquad (6.3.15)$$

$$\overline{U}(T,V) - \Delta\overline{H}_{f0}^{\circ} = RT \left\{ \frac{3}{2} + \frac{3}{2} + \sum_{i=1}^{3N_a-6} \frac{\Theta_{\text{vib},i}/T}{e^{\Theta_{\text{vib},i}/T} - 1} \right\}, \qquad (6.3.16)$$

$$\overline{C}_V(T) = R \left\{ \frac{3}{2} + \frac{3}{2} + \sum_{j=i}^{3N_a-6} \left(\frac{\Theta_{\text{vib},i}}{T}\right)^2 \frac{e^{\Theta_{\text{vib},i}/T}}{(e^{\Theta_{\text{vib},i}/T} - 1)^2} \right\}, \qquad (6.3.17)$$

$$\overline{S}(T,V) = R \left\{ \ln\left[\left(\frac{2\pi M k_B T}{h^2}\right)^{\frac{3}{2}} \frac{\overline{V}^{\circ} e^{\frac{5}{2}}}{N} \right] + \ln\left[\frac{\sqrt{\pi} e^{\frac{3}{2}}}{\sigma} \left(\frac{T^3}{\Theta_A \Theta_B \Theta_C}\right)^{\frac{1}{2}} \right] \right.$$

$$\left. + \sum_{i=1}^{3N_a-6} \left[\frac{\Theta_{\text{vib},i}/T}{e^{\Theta_{\text{vib},i}/T} - 1} - \ln(1 - e^{-\Theta_{\text{vib},i}/T}) \right] + \ln\omega_{\text{el}} \right\}. \qquad (6.3.18)$$

These, then, are the expressions whose values may be employed for comparison with the JANAF tabulations for nonlinear molecules [21].

6.3.3 Beyond the RR-SHO Approximation

In the previous subsection we have considered expressions for the partition function and thermodynamic properties of polyatomic gases using the RR-SHO approximation with the rigid-rotor contribution evaluated in the classical limit. We shall refer to this level of description as being the simplest RR-SHO approximation. Corrections to the expressions obtained at this level of description will be of two types. The first type of correction consists in obtaining relevant Euler–Maclaurin type expansions in terms of the variable $\alpha = \Theta_{\text{rot}}/T$ to give more accurate values for the rigid-rotor contributions to the partition function and the thermodynamic functions in the temperature range $5\Theta_{\text{rot}} < T < 200\Theta_{\text{rot}}$. For many non-hydride polyatomic molecules these corrections will be irrelevant because Θ_{rot} will typically be only a small fraction of 1 kelvin. The second type of correction deals with deviations of molecules from RR-SHO behaviour. There are three types of correction that fall into this category, namely, vibrational anharmonicity associated with deviations of the vibrational motions from simple harmonic motion, centrifugal distortion associated with deviations of the rotational motion from rigid-rotor motion, and the vibration–rotation interactions associated with the (weak) dependence of the rotational constant on the vibrational state in which the rotational motion is taking place.

We shall consider only those molecules for which the simplest RR-SHO approximation is sufficient to provide near-quantitative agreement with experimental thermodynamic data. This means that we shall restrict ourselves to the temperature range $10\,\mathrm{K} < T < 3000\,\mathrm{K}$ for many molecular gases, so that the lowest-order corrections to the thermodynamic function values generally do not exceed a few percent of the corresponding RR-SHO contributions. Vibrational anharmonicity, centrifugal distortion, and vibration–rotation interaction corrections may be considered simultaneously for both linear and nonlinear molecules. As Euler–Maclaurin corrections are sensitive to indistinguishable nucleus interchange symmetries, we shall consider corrections to the simplest RR-SHO approximation in three steps.

Vibrational Anharmonicity Corrections

There will be an anharmonicity correction arising from each independent vibrational degree of freedom for a polyatomic molecule. Each such correction will have basically the same form as the anharmonicity correction for a diatomic molecule (see section 'Beyond the SHO Approximation: Anharmonicity Effects'), so that the vibrational partition function, $z_{\mathrm{vib}}(T)$, for an individual polyatomic molecule can be written as

$$z_{\mathrm{vib}}(T) = \mathrm{e}^{-\beta\epsilon_0} \prod_{i=1}^{d} \widetilde{z}_{\mathrm{vib},i}(T)\, f_{\mathrm{anh}}\,, \qquad (6.3.19)$$

with $\widetilde{z}_{\mathrm{vib},i}(T) \equiv (1 - \mathrm{e}^{-u_i})^{-1}$, $d = 3N_a - 5, 3N_a - 6$ for linear, respectively, nonlinear molecules, and f_{anh} is the anharmonicity correction factor. For polyatomic molecules, this factor can be obtained as

$$f_{\mathrm{anh}} = 1 + \sum_{i=1}^{d} \frac{2u_i x_{\mathrm{e},i}}{(\mathrm{e}^{u_i} - 1)^2}\,, \qquad (6.3.20)$$

given that $u_i \equiv \Theta_{\mathrm{vib},i}/T$ and $x_{\mathrm{e},i}$ represents the leading anharmonicity contribution to the vibrational energy associated with the ith vibrational degree of freedom of the molecule.

Corrections to the RR-SHO Model

Corrections associated with the molecular rotational energy, namely, centrifugal distortion, vibration–rotation coupling effects are not quite as straightforward to deal with as are the vibrational anharmonicity corrections. Euler–Maclaurin corrections to the classical limit, in particular, can be quite different for centrosymmetric and noncentrosymmetric polyatomic molecules, as has already been noted for diatomic molecules. We shall consider all three types of correction to the rotational contributions obtained using the classical limit $T \gg \Theta_{\text{rot}}$ separately for linear and nonlinear molecules.

Linear Molecules

Heteronuclear diatomic molecules are the simplest members of the class of noncentrosymmetric linear molecules and homonuclear diatomic molecules are likewise the simplest centrosymmetric linear molecules. Nuclear spin and rotational degrees of freedom may always be decoupled for heteronuclear diatomic molecules, as there are no nuclear interchange effects. Thus, $z_{\text{rot-nuc}}(T)$ may always be written as $z_{\text{rot}}(T)z_{\text{nuc}}$, with $z_{\text{nuc}} = (2I_a + 1)(2I_b + 1)$ and I_a, I_b the nuclear spins of the distinguishable nuclei. For homonuclear diatomic molecules, however, $z_{\text{rot-nuc}}(T)$ cannot, in general, be so decoupled due to the definite nuclear interchange symmetry imposed upon the molecular wavefunctions by the Pauli Principle for indistinguishable particles. We also found [see Eq. (6.2.100)] that in the high-temperature limit $T \gg \Theta_{\text{rot}}$, for which the partition function sums over even and odd rotational quantum numbers are each equal to one-half the partition function sum over all rotational quantum numbers, $z_{\text{rot-nuc}}(T)$ then becomes $z_{\text{rot}}(T)z_{\text{nuc}}$, with $z_{\text{nuc}} = (2I_a + 1)^2$. Precisely this same behaviour will be found for general noncentrosymmetric and centrosymmetric linear molecules, except that z_{nuc} will be given by

$$z_{\text{nuc}} = \prod_{i=1}^{N_a}(2I_i + 1),$$

with I_i the nuclear spin for nucleus i and N_a the number of atoms in the molecule.

Because the moments-of-inertia associated with the end-over-end rotational motions for such molecules will be large, the corresponding characteristic rotational temperatures will typically be fractions of a kelvin. Consequently, any temperature at which the substance is still in gaseous form will correspond to the high-temperature limit, so that there is no longer a need to consider symmetry restrictions explicitly for the purposes of calculating thermodynamic function values for such molecules. Note, however, that we must still consider indistinguishable particle interchange symmetry requirements when dealing with rotational spectroscopy.

Euler–Maclaurin corrections to the classical limit expression for the rotational partition function, as well as centrifugal distortion and vibration–rotation interaction corrections can be expressed in the same way as those for diatomic molecules. In particular, the Euler–Maclaurin corrections to the classical limit for $z_{\text{rot}}(T)$ are given as [7]

$$z_{rot}(T) = \frac{1}{\sigma} e^{\alpha/3} \alpha^{-1} \left[1 + \frac{\alpha^2}{90} + \cdots \right], \tag{6.3.21}$$

in which $\alpha \equiv \Theta_{rot}/T$, as for diatomic molecules, and σ is the symmetry number (having values 1 for noncentrosymmetric, 2 for centrosymmetric linear molecules). Here, the characteristic rotational temperature is defined as

$$k_B \Theta_{rot} \equiv \left(\overline{B}_e - \frac{1}{2} \sum_{i=1}^{d} \overline{\alpha}_i \right).$$

Inclusion of the quartic centrifugal distortion and vibration–rotation interaction corrections further modify $z_{rot}(T)$ to

$$z_{rot}(T) = \frac{1}{\sigma} e^{\alpha/3} \alpha^{-1} \left[1 + \frac{\alpha^2}{90} + \sum_{i=1}^{d} \frac{2x_{e,i} u_i}{(e^{u_i} - 1)^2} + \frac{2\gamma}{\alpha} + \sum_{i=1}^{d} \frac{\delta_i}{e^{u_i} - 1} + \cdots \right];$$
$$\tag{6.3.22}$$

the vibration–rotation interaction parameters are given by $\delta_i = \overline{\alpha}_i / \overline{B}_e$, and the very weak vibrational dependence [22] of the centrifugal distortion constant $\overline{D}_e$ has been ignored.

Nonlinear Molecules

Nonlinear molecules may be divided into three distinct classes according to the relationship amongst the three moments-of-inertia associated with the overall rotational motions of the rigid molecular frames. In general, as has been seen briefly in Sect. 6.3.1, the three principal moments-of-inertia can all be different, in which case we refer to the molecule as an asymmetric top molecule, two of the three principal moments-of-inertia can be equal and different from the third moment-of-inertia (symmetric top molecule) or, finally, all three principal moments-of-inertia can be equal (spherical top molecule). The resultant RR rotational constants are designated A, B, C, with $A > B > C$ by convention. For a symmetric top molecule, two of the three rotational constants will be equal: they are designated B, and the rotational constant associated with rotation about the molecular figure axis is designated A for prolate (rod-shaped) symmetric top molecules and C for oblate (disc-shaped) symmetric top molecules.

Spherical top molecules are the simplest of the nonlinear molecules: they typically have tetrahedral (T_d), octahedral (O_h), or icosahedral (I_h) symmetry. For this reason they differ from centrosymmetric linear molecules, which all have $D_{\infty h}$ symmetry, and three different expressions, one for each symmetry type, will be needed to describe the nuclear spin statistics for them. However, molecules other than hydrides for which only the H atoms contribute to the moment-of-inertia will give rise to characteristic rotational temperatures that are too small for Euler–Maclaurin corrections to make a significant contribution to the rotational partition function at temperatures for which the substances remain gaseous. Indeed, all molecules satisfying this requirement have tetrahedral symmetry: the best known examples are methane (CH_4), fully deuterated methane (CD_4), silane (SiH_4), and germane (GeH_4), with moments-of-inertia having values [23] 5.34117×10^{-47} $kg\,m^2$, $10.633 \times 10^{-47}\,kg\,m^2$, $9.79106 \times 10^{-47}\,kg\,m^2$, and $10.380 \times 10^{-47}\,kg\,m^2$, respectively, and with corresponding characteristic rotational temperatures 7.541, 3.788, 4.11, and 3.88 K.

Corrections to the rigid-rotor classical limit (6.3.4) for the rotational partition function for spherical top molecules have been available in one form or another since Kassel's review [4] in 1936. These corrections have been considered in detail by Fox [24] and by McDowell [25]. The usual Euler–Maclaurin expansion procedure fails for spherical top molecules, and it is necessary to use an alternative procedure involving what are known as (Jacobi) theta functions to obtain appropriate corrections [4, 23]. However, these corrections turn out to be negligible in comparison both to low-temperature quantum effects and to centrifugal distortion corrections. Rotation-vibration interaction corrections remain largely unknown for most polyatomic molecules, including spherical top molecules.

The corrected combined rotational-nuclear spin partition function for the tetrahedral hydrides can be expressed as [4]

$$z_{rot-nuc}(T) = \frac{(2I_a + 1)^4}{\sigma} \sqrt{\pi} e^{\alpha/4} \alpha^{-\frac{3}{2}} \left(1 + \delta' + \frac{15\gamma}{4\alpha}\right), \tag{6.3.23}$$

in which I_a is the nuclear spin of the central nucleus, $\sigma = 12$ is the symmetry number for a tetrahedral molecule, $\alpha \equiv \Theta_{rot}/T$ as usual, $\gamma \equiv D_e/B_e$ as for linear molecules, and δ' is given by

$$\delta' = \frac{16\pi}{3\sqrt{3}} e^{-\pi^2/(9\alpha)} \frac{1}{(2I_a + 1)^2}. \tag{6.3.24}$$

McDowell [25] has shown that this form for $z_{rot-nuc}(T)$ gives rotational partition function values for CH_4 and CD_4 that agree to better than 0.01% with the exact values obtained by Robiette and Dang-Nhu [26] via direct summation for temperatures in the range $10\,K < T < 500\,K$.

A detailed derivation of the various corrections to the rotational partition function for symmetric top molecules, including effects associated with the indistinguishability of nuclei, has been given by McDowell [27]. As many of the most commonly encountered symmetric top molecules belong to the three-fold symmetry group C_{3v}, we shall simply quote McDowell's result for this class of molecules and refer further to his paper for the details and expressions that apply to other classes of symmetric top molecules.

For C_{3v} molecules having structures XY_3 and WXY_3, McDowell's expression for $z_{rot-nuc}(T)$, including the first rigid-rotor correction and the effects of nuclear spin symmetry can be represented by [27]

$$z_{\text{rot-nuc}}(T) = \frac{(2I_a + 1)^2}{\sigma} \sqrt{\frac{\pi B}{A}} \ e^{\alpha(4-m)/12} \alpha^{-\frac{3}{2}} \left[1 + \frac{\alpha^2}{90}(1 - m)^2 + \cdots \right] (1 + \delta'),$$

$$(6.3.25)$$

in which $\sigma = 3$ is the symmetry number for $C_{3\text{v}}$ molecules, $\alpha \equiv \Theta_{\text{rot}}/T$, $m \equiv B/A$ for prolate symmetric top molecules (B/C for oblate tops), and δ' is the nuclear spin symmetry correction factor given by

$$\delta' \simeq \frac{2}{(2I_a + 1)^2} \ e^{\pi^2 m(\alpha m - 6)/(54\alpha)}.$$

$$(6.3.26)$$

Centrifugal distortion effects for symmetric top molecules are significantly more complicated than those for linear and spherical top molecules. We recall that the energy levels for a rigid-rotor symmetric top are given by

$$E(j, k) = Bj(j + 1) + (A - B)k^2,$$

for prolate symmetric tops (A is replaced by C for oblate symmetric tops), with j the usual rotational quantum number and k the quantum number associated with the projection of the rotational angular momentum onto the body-fixed z-axis (which is taken to lie along the molecular symmetry axis). When centrifugal distortion effects are considered, the rather simple energy expression for a prolate symmetric top molecule is replaced at the level of quartic distortions by [28]

$$E(j, k) = Bj(j + 1) + (A - B)k^2 - D_j[j(j + 1)]^2 - D_{jk}j(j + 1)k^2 - D_k k^4 + \cdots$$

$$(6.3.27)$$

with three quartic centrifugal distortion terms, given in terms of three constants D_j, D_k, D_{jk}, rather than only one as found for linear and spherical top molecules. At the next level of approximation, there will be four additional (sextic) centrifugal distortion terms, rather than only the single sextic term found for linear and spherical top molecules. The quartic centrifugal distortion correction terms have been available for a long time [29], and can be represented in terms of a centrifugal distortion correction factor to lowest order in the quartic distortion constants. In its simplest representation, $z_{\text{rot-nuc}}(T)$ has been given by McDowell [25] as

$$z_{\text{rot-nuc}}(T) = \frac{(2I_a + 1)^3}{\sigma} \sqrt{\frac{\pi B}{A}} \ e^{\alpha(4-m)/12} \alpha^{-\frac{3}{2}} \left[1 + \frac{\alpha^2}{90}(1 - m)^2 + \cdots \right] (1 + \delta') f_{\text{cd}}.$$

$$(6.3.28)$$

The centrifugal distortion correction factor, f_{cd}, is given by

$$f_{\text{cd}} = 1 + \rho_0 + \rho_1 \alpha^{-1} + \cdots,$$

$$(6.3.29)$$

in which ρ_0 and ρ_1 are given in terms of the quartic centrifugal distortion constants [26]

$$\rho_0 = -\frac{1}{12B}[(8 + 2m - 4m^2 + 3m^3)D_j + m(2 - 2m + 3m^2)D_{jk} + 3m^3 D_k],$$

$$(6.3.30)$$

and

$$\rho_1 = \frac{1}{4B}[(8 + 4m + 3m^2)D_j + m(2 + 3m)D_{jk} + 3m^2 D_k]. \qquad (6.3.31)$$

McDowell [25] has also given correction terms that include the sextic distortion constants, but they are not needed unless an accuracy much better than 0.01% is required. He has shown, for example, that expression (6.3.28) gives excellent agreement with the direct summation results obtained by Robiette and Dang-Nhu [26] for the substituted methanes CH_3D and CHD_3 at temperatures ranging from 10 to 2000 K. It may well be asked why we should be interested in obtaining closed-form expressions for the rotational partition function when they are as involved as those that we have just seen, especially as powerful computers can generate 'exact' results from the defining summation with relatively little ease once the various spectroscopic constants have been experimentally determined. A very cogent argument for why such expressions should still be of interest to us has been given by McDowell [25], who argues that although direct summation indeed appears to be the simplest route for obtaining accurate values for $z_{rot-nuc}(T)$, especially for low temperatures, it is in fact less straightforward than employing an analytic formula, and more importantly, that values of the partition function thereby obtained are not amenable to simple interpolation, even at lower temperatures, such tabulations are not readily revised, and finally, that any direct summation, no matter how accurately calculated with the rotational constants available at the time that the calculation was carried out, must be completely redone if those constants are refined by later measurements.

It will likely not be too surprising to learn that the various corrections to the rotational partition function for asymmetric top molecules, both to the classical limit for the RR description and to allow for departures from the RR description are even more complicated than those for symmetric top molecules. For these molecular gases, we therefore simply refer the reader to Watson [30] who has given expressions for the Euler–Maclaurin type corrections to the RR classical limit expression, and has shown that they reduce correctly to the symmetric top expressions [27].

6.4 Molecular Spectra: Nuclear Spin Effects

6.4.1 Diatomic Rotational Spectra

We have seen that, apart from the molecular hydrogen isotopologues at temperatures well below room temperature, it suffices to employ the classical limit (or, for temperatures less than approximately $10\Theta_{\text{rot}}$, perhaps an Euler–Maclaurin-type expression) for calculating values of the usual thermodynamic state functions. This procedure is appropriate largely because the thermodynamic state functions, which can be referred to as 'bulk properties' of matter because they are associated with macroscopic amounts of matter, are represented in terms of macroscopic thermal averages over the equilibrium energy distributions of the microscopic molecular states, they will hence be insensitive to the behaviours of individual molecules that occupy the microscopic energy states.

Molecular spectra, however, provide direct probes of the behaviours of individual molecules, and will consequently be highly sensitive to actual equilibrium distributions. The intensity of individual electronic or vibronic spectral transitions involves two factors that depend upon aspects of the initial and/or final microscopic molecular states. The first of these factors, known as the Franck–Condon factor, is determined by the overlap of the molecular wavefunctions for the initial and final electronic states of a spectral transition. The second factor is the population of the initial molecular state. Moreover, because such spectroscopic transitions occur on a very short time scale, typically fractions of a femtosecond, spectra can be obtained both for equilibrium and nonequilibrium systems. We shall, however, only be concerned here with spectra obtained under equilibrium conditions.

Because Franck–Condon factors vary widely from one vibronic transition to another, the effect of initial-state populations is very difficult to disentangle for such transitions: even for ultra-high-resolution spectra at elevated temperatures (of the order of a few thousand kelvin), the role of the initial-state populations is effectively masked. Fortunately, however, the Franck–Condon factors are almost constant for a set of rotational fine-structure transitions within a specific vibronic transition. This means that the population factors play a significant role in determining the intensity patterns of the rotational fine-structure lines. For heteronuclear diatomic molecules, we shall often focus upon infrared rotational fine-structure transitions associated with the vibrational fundamental transition (i.e., the transition $v'' = 0 \longrightarrow v' = 1$) for molecules in their ground electronic states, while for homonuclear diatomic molecules, which do not undergo infrared transitions, we shall instead focus upon Raman transitions for molecules lying in their ground electronic states.

Heteronuclear Diatomic Molecular Spectra

We have seen that the rotational partition function for a heteronuclear diatomic molecule is given by Eq. (6.2.59), viz.,

$$z_{\text{rot}}(T) = \sum_{j=0}^{\infty}(2j+1)e^{-j(j+1)\Theta_{\text{rot}}/T} ,$$

from which the probability for a given molecule to be found in the rotational state characterized by the rotational quantum number j is simply given by

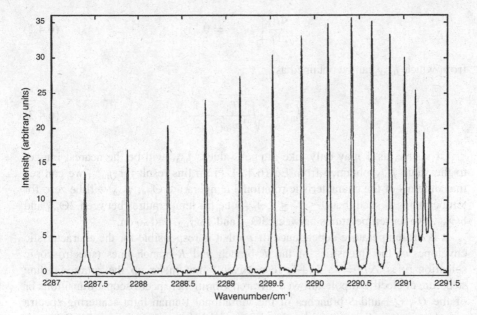

Fig. 6.13 Rotational fine-structure lines corresponding to initial rotational quantum numbers $j'' = 1$ to $j'' = 15$ (from right to left) of the Q-branch of the vibration–rotation Raman spectrum of the $^{14}N^{15}N$ heteronuclear isotopologue of nitrogen. From Bendtsen [32]. Reproduced by permission of John Wiley & Sons, Ltd.

$$p_j(T) = \frac{(2j+1)e^{-j(j+1)\Theta_{rot}/T}}{z_{rot}(T)}. \tag{6.4.1}$$

If we express the number of molecules in each rotational level as $N_j = p_j N$, with N the total number of molecules in the sample, then the relative intensities for rotational spectroscopic transitions will be given directly, all else being equal, by the fractional populations $N_{j''}(T)/N = p_{j''}(T)$ of the initial rotational states. This means, in particular, that the intensity envelope for a specific set of P-branch or R-branch transitions (lines) in an infrared spectrum, similarly, of O-, Q-, or S-branch transitions in a Raman spectrum, will have a maximum that coincides, as can be seen in Fig. 6.13, with the most populous initial rotational state corresponding to the temperature of the sample.[3]

From Eq. (6.2.62b) we see that as a function of rotational quantum number j, the fractional population $p_j(T)$ will have a maximum value for $j = j_{max}$ determined by the condition

---

[3]This is, however, no longer the case for a pure rotational transition, as has been pointed out by Le Roy [31].

$$\frac{\mathrm{d}p_j}{\mathrm{d}j}\bigg|_{j=j_{max}} = 0 \,, \tag{6.4.2}$$

from which j_{max} can be obtained as

$$j_{max} = \sqrt{\frac{T}{2\Theta_{rot}} - \frac{1}{2}} \,. \tag{6.4.3}$$

Of course, as j may only take integer values, j_{max} will be the nearest integer to the value j_{max} obtained from Eq. (6.4.3). From this result for j_{max}, we can see that in terms of the characteristic rotational temperature Θ_{rot}, j_{max} will be zero for temperatures T such that $0 \leq T \leq 2\Theta_{rot}$, one for temperatures between $2\Theta_{rot}$ and $8\Theta_{rot}$, two for temperatures between $8\Theta_{rot}$ and $18\Theta_{rot}$, and so on.

It is the temperature dependence of p_j that is responsible for the characteristic envelopes of the intensities of the P-branch and R-branch lines (spectroscopic selection rules $\Delta j = -1, +1$, respectively) that make up the rotational fine structure of electric-dipole-allowed infrared vibrational spectroscopic transitions or of the O-, Q-, and S-branches of the vibrational Raman light scattering spectra of heteronuclear diatomic molecules. Figure 6.13 shows a portion of the Q-branch ($\Delta v = 1$, $\Delta j = 0$) of the fundamental ($v = 0 \rightarrow v = 1$) vibrational transition in the Raman light scattering spectrum of the $^{14}N^{15}N$ heteronuclear isotopologue of N_2. This spectrum was obtained for a gas sample at temperature $T = 295$ K.

Note that as the transitions illustrated in Fig. 6.13 are from a Q-branch (i.e., transitions obeying the selection rule $\Delta j \equiv j' - j'' = 0$) of a vibration–rotation spectrum, the line positions relative to the pure vibrational transition are given by the product of the difference between the rotational constants B_v in the upper and lower vibrational levels and $j(j+1)$. Because $B_1 < B_0$, the transition frequencies for the sequence in a Q-branch decrease with increasing rotational quantum number j, as is seen in this figure.

Homonuclear Diatomic Molecular Spectra

Because a homonuclear diatomic molecule has neither a permanent electric dipole moment nor a nonzero transition electric dipole moment, it does not support infrared spectral transitions. These molecules do, however, possess non-vanishing electric polarizability tensors, and hence undergo Raman scattering, for which the corresponding (spectroscopic) selection rule is $\Delta j = 0, \pm 2$. For pure rotational Raman scattering, which corresponds to transitions for which both the upper and lower spectroscopic states belong to the same vibrational level (normally the ground, $v = 0$, level), only $\Delta j = +2$ transitions are allowed, so that pure rotational Raman spectra consist soley of an S-branch with both Stokes and anti-Stokes components.

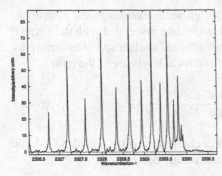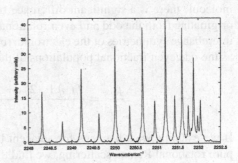

Fig. 6.14 Rotational Q-branch fine-structure transitions in the fundamental $v = 0 \rightarrow v = 1$ vibrational Raman spectrum of the $^{14}N_2$ (on the left) and $^{15}N_2$ (on the right) molecular nitrogen isotopologues. The $^{14}N_2$ spectrum is from Bendtsen and Rasmussen [33], while the $^{15}N_2$ spectrum is from Bendtsen [32]. Reproduced by permission of John Wiley & Sons, Ltd.

Vibrational Raman scattering primarily corresponds to the vibrational selection rule $\Delta v = 1$: for sufficiently high resolution, the vibrational transition can be resolved into fine-structure lines corresponding to rotational transitions satisfying all three rotational selection rules.

The Q-branches for the homonuclear $^{14}N_2$ and $^{15}N_2$ isotopologues are shown in Fig. 6.14. Both isotopologues exhibit clear alternations of intensity between adjacent spectral lines.

We recall that for homonuclear diatomic molecules, identical particle symmetry requirements do not allow the rotational and nuclear spin degrees of freedom to be decoupled. Let us consider specifically a homonuclear diatomic molecule X_2 whose nuclei are fermions (i.e., have half-odd-integer nuclear spin I_a) and whose ground electronic state has a term symbol $^1\Sigma_g^+$. We have obtained the combined rotational-nuclear spin partition function $z_{\text{rot-nuc}}(T)$ for such molecules in section 'Homonuclear Diatomic Molecules' given explicitly as Eq. (6.2.108) for fermion X-nuclei and as Eq. (6.2.109) for boson X-nuclei.

From Eq. (6.2.108), we obtain fractional populations, p_{j_0}, given for odd rotational levels j_0 by

$$p_{j_0}(T) = \frac{(I_a + 1)(2I_a + 1)(2j_0 + 1)e^{-j_0(j_0+1)\Theta_{\text{rot}}/T}}{z_{\text{rot-nuc}}(T)}. \qquad (6.4.4)$$

Similarly, for even rotational levels j_e we obtain fractional populations p_{j_e} given by

$$p_{j_e}(T) = \frac{I_a(2I_a + 1)(2j_e + 1)e^{-j_e(j_e+1)\Theta_{\text{rot}}/T}}{z_{\text{rot-nuc}}(T)}. \qquad (6.4.5)$$

The shape of an envelope for a set of rotational fine-structure transitions (lines) within a given vibrational manifold will be dictated by the fractional populations of the rotational levels in that manifold. However, for a homonuclear diatomic molecule there is a significant difference between the intensities of the transitions originating from the odd and even rotational levels, and associated with the nuclear interchange symmetries of the electronic, rotational, and nuclear spin wavefunctions as the odd/even fractional populations of the rotational levels are in the ratio

$$\frac{p_{j_o}}{p_{j_e}} = \frac{I_a + 1}{I_a} \frac{(2j_o + 1)e^{-j_o(j_o+1)\Theta_{rot}/T}}{(2j_e + 1)e^{-j_e(j_e+1)\Theta_{rot}/T}} . \tag{6.4.6}$$

If we examine what happens for adjacent lines characterized by j_o and $j_o - 1$ for pure rotational Raman scattering, we find that this ratio becomes

$$\frac{p_{j_o}}{p_{j_o-1}} = \frac{I_a + 1}{I_a} \frac{2j_o + 1}{2j_o - 1} e^{-2j_o\Theta_{rot}/T} , \qquad j_o \geq 1 . \tag{6.4.7}$$

Thus, we see that adjacent lines in the Raman scattering spectrum for a homonuclear diatomic molecule have an intensity ratio given by the product of the ratio of the nuclear spin degeneracies associated with the two nuclear spin modifications and a factor that depends upon the rotational quantum number of the odd-j level and the ratio of the characteristic rotational temperature and the temperature of the sample. The second (temperature dependent) factor plays a larger role for small values of the rotational quantum number, but otherwise provides only a slight skewing effect on the intensity alternation across a sequence of rotational fine-structure lines.

As the ^{14}N nucleus has nuclear spin $I_a = 0$ and the ^{15}N nucleus has nuclear spin $I_a = \frac{1}{2}$, the $^{14}N_2$ and $^{15}N_2$ nitrogen isotopologues are homonuclear diatomic molecules whose nuclei are bosons, respectively, fermions. The appropriate forms for the combined rotational-nuclear spin partition function $z_{rot-nuc}(T)$ are thus given by Eq. (6.2.108) and Eq. (6.2.109), with the consequence that the intensities for adjacent j_o'' and $j_e'' = j_o'' + 1$ lines should be 1:2 and 3:1, respectively. While the $j = 0$ transition for $^{15}N_2$ only appears as a small shoulder on the $j = 1$ transition, this transition is clearly distinguishable for $^{14}N_2$ due to the preferential nuclear spin weighting of its even rotational levels.

Example 6.7 Raman Q-branch fine-structure lines for T_2.

The rotational fine structure of the Q-branch of the fundamental Raman vibrational scattering spectrum for molecular tritium, $^3H_2$, is shown in Fig. 6.15.

The ^3H isotope of hydrogen has nuclear spin $I_a = \frac{1}{2}$, so that the intensity alternation for rotational fine-structure lines in Raman scattering spectra of T_2 should be 3:1 for transitions associated with odd versus even j-values were it determined solely by the nuclear spin degeneracy factors. However, because the characteristic rotational temperature for $^3H_2$ is of the order of 30 K, the temperature-dependent factor in Eq. (6.4.7) plays a greater role than it does for many other Raman scattering spectra. We may therefore determine from Eq. (6.4.7) that the intensity ratios I_{j_o}/I_{j_o-1} for temperature 300 K will be approximately 8.1, 3.1, and 2.2 for $j_o = 1, 3, 5$, respectively, which is roughly what is observed in Fig. 6.15. Note also that the wavenumber axis increases from right to left in this figure.   □

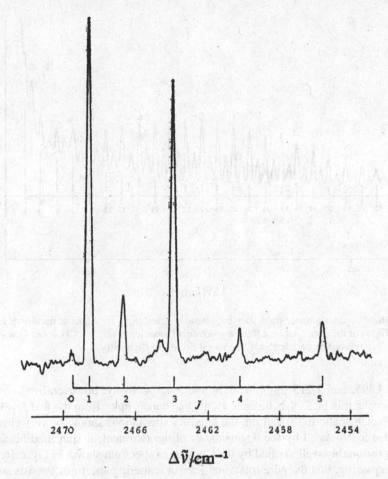

Fig. 6.15 Microdensitometer trace showing intensity alternations for adjacent even–odd j-values for the spectral lines making up the rotational structure of the Q-branch of the $v'' = 0$ to $v' = 1$ vibration–rotation Raman spectrum of room temperature molecular tritium, T_2. From Edwards et al. [34]. Reproduced by permission of The Royal Society of Chemistry

Example 6.8 The pure rotational Raman spectrum of $^{35}Cl_2$.

The pure rotational Raman scattering spectrum for the chlorine $^{35}Cl_2$ isotopologue is shown in Fig. 6.16.

As the ^{35}Cl isotope has nuclear spin $I_a = \frac{3}{2}$, the Raman scattering spectrum for the homonuclear $^{35}Cl_2$ isotopologue shows a characteristic 5:3 odd:even intensity alternation pattern for the transitions characterized by initial rotational quantum numbers j'' associated with the symmetric to antisymmetric ratio of the nuclear spin pair-state degeneracies. Molecular chlorine has a characteristic rotational temperature $\Theta_{rot} = 0.35\,K$, so that for a sample temperature of approximately $300\,K$, the ratio 5:3 of the nuclear spin degeneracy factors is multiplied by factors

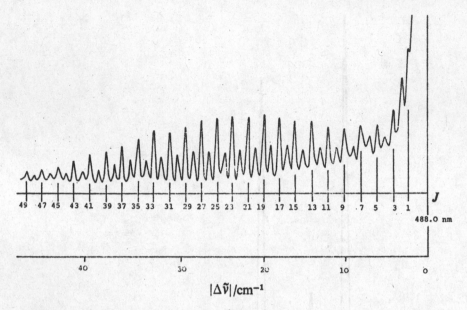

Fig. 6.16 Microdensitometer trace showing intensity alternations for adjacent transitions in the Stokes region of the pure rotational Raman spectrum of room temperature $^{35}Cl_2$. From Edwards et al. [35]. Reproduced by permission of The Royal Society of Chemistry

2.996, 1.395, and 1.215 for $j_o = 1, 3$, and 5, respectively: however, by $j_o = 21$, for example, this factor has become 1.046, and hence, apart from the first few lines in its Raman scattering spectrum, the intensity alternations should be very close to the value 5:3 dictated by the degeneracies of the two nuclear spin modifications. This is reasonably well verified by the Cl_2 Raman spectrum shown in Fig. 6.16. We note, in passing, that the pure rotational Raman scattering spectrum consists solely of an S-branch (for the same reason that the pure rotational infrared spectrum for a heteronuclear diatomic molecule consists solely of an R-branch). Note also that in Fig. 6.16 the wavenumber scale increases from right to left, so that we are indeed looking at an S-branch spectrum.   □

Example 6.9 Rotational Raman spectra of molecular oxygen isotopologues.

Raman scattering by the O_2 isotopologues provides a good illustration of the effects of identical particle symmetry. Oxygen has three stable isotopes, ^{16}O, ^{17}O, and ^{18}O, with natural abundances of 99.757%, 0.038%, and 0.205%, respectively. Features of the various oxygen isotopologues may be illustrated by comparisons amongst the three sets of Raman lines obtained from Raman scattering in a mixture of them, as shown in Fig. 6.17.

Both ^{16}O and ^{18}O have nuclear spins $I_a = 0$, and are hence bosons, while ^{17}O has nuclear spin $I_a = \frac{5}{2}$, and is a fermion. For present purposes, we shall focus upon the O_2 isotopologues formed from ^{16}O and ^{18}O, as they nicely illustrate the consequences for rotational spectroscopy of identical particle symmetry for nuclei with $I_a = 0$. We have seen earlier that $^{16}O_2$ has no $j =$ even rotational levels because $I_a = 0$ for the ^{16}O nucleus. This is also true for the $^{18}O_2$ isotopologue. A consequence of the absence of even j-levels for these two isotopologues is that their rotational Raman spectra show no apparent intensity alternations and will have adjacent lines separated by approximately $8\overline{B}_0$, rather than approximately $4\overline{B}_0$, as found for a normal rotational Raman spectrum. The $^{16}O^{18}O$ isotopologue, however, is a heteronuclear diatomic molecule, so that its Raman spectra will have adjacent lines separated by approximately $4\overline{B}_0$ and will show no intensity alternations. A comparison between the separations of the lines for the $^{18}O_2$ and $^{16}O^{18}O$ isotopologues, for example, clearly illustrates the factor two difference between them.   □

Finally, it is perhaps interesting to note that the presence of intensity alternations in the Raman scattering spectra of homonuclear diatomic molecules played a major role in the discovery of nuclear spin angular momentum itself, as well as in the determination of specific nuclear spin values for many nuclei.

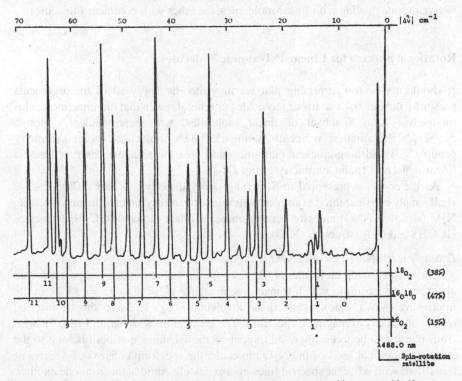

Fig. 6.17 Pure rotational Raman light scattering spectrum of a 15% $^{16}O_2$, 47% $^{16}O^{18}O$, 38% $^{18}O_2$ mixture of oxygen isotopologues. From Edwards et al. [36]. Reproduced by permission of The Royal Society of Chemistry

6.4.2   Polyatomic Molecular Spectra

We have seen in Sect. 6.4.1 that the full quantum mechanical expression for the rotational partition function is needed to explain the intensity patterns both of pure rotational spectra and of the rotational fine structure (typically in the form of P- and R-branches) observed in high-resolution vibrational spectra of diatomic molecules. Moreover, we have seen that there are quite dramatic effects in the form of alternating intensities for adjacent rotational lines and, in some cases, even missing lines in the spectra of homonuclear diatomic molecules. These effects could all be explained via nuclear spin symmetry arguments arising from application of the Pauli Principle for indistinguishable particles.

The arguments utilized in our discussion of rotational spectral intensity patterns for diatomic molecules translate straightforwardly to rotational spectroscopic transitions in linear polyatomic molecules. However, for nonlinear polyatomic molecules, the application of symmetry arguments is considerably more complicated than it is for linear polyatomic molecules. For this reason, we shall split our discussion into two parts, one dealing with linear molecules, the other with nonlinear molecules.

Rotational Spectra for Linear Polyatomic Molecules

It should not be too surprising that we may literally apply all of the arguments presented in Sect. 6.4.1 to linear molecules in general, given that diatomic molecules themselves form a subset of linear molecules, with heteronuclear diatomic molecules the smallest noncentrosymmetric linear molecules (point symmetry group $C_{\infty v}$) and homonuclear diatomic molecules the smallest centrosymmetric linear molecules (point symmetry group $D_{\infty h}$).

As the concepts presented in Sect. 6.4.1 apply directly to linear molecules, we shall simply consider three examples: a simple noncentrosymmetric linear molecule, N_2O (structure NNO) and two centrosymmetric linear molecules, C_2H_2 (structure HCCH), and CO_2 (structure OCO).

Example 6.10 Nitrous oxide, N_2O.

The nitrous oxide molecule, N_2O, has a pure rotational electric-dipole-allowed absorption spectrum in which transitions involving $j'' = 0$ to $j'' = 11$ lie in the microwave (MW) region, while those involving $j'' \geq 12$ lie in the far infrared region of the electromagnetic spectrum. The spectral lines shown in Fig. 6.18 arise from transitions between lower and upper rotational states that both belong to the ground vibrational level of the N_2O molecule: the spectrum consists of a series of transitions with adjacent spectral lines approximately equidistant from one another. Transitions for which both j'' and j' lie in the same vibrational level are necessarily such that $j' - j'' > 0$, so that pure rotational spectra always consist solely of an

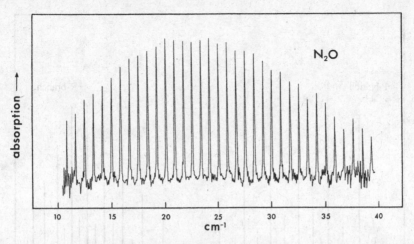

Fig. 6.18 Pure rotational infrared spectrum for the linear triatomic molecule N_2O. Reprinted from Fleming and Cole [37], with permission of Elsevier

R-branch spectrum.[4] Figure 6.18 shows the $j'' = 12$ to $j'' = 46$ portion of the R-branch obtained using a far infrared spectrometer.

The most common nitrous oxide isotopologue by far is $^{14}N_2^{16}O$. Because N_2O has the structure NNO, the two ^{14}N atoms in N_2O are distinguishable (i.e., nonequivalent). The point symmetry group for $^{14}N_2^{16}O$ is thus $C_{\infty v}$, and the combined rotational-nuclear spin partition function takes the form $z_{rot-nuc}(T) = z_{rot}(T)z_{nuc}$, with $z_{nuc} = (2I_N+1)^2(2I_O+1) = 9$ (because $I_a = 1$ for ^{14}N and $I_a = 0$ for ^{16}O). The rotational partition function $z_{rot}(T)$ is given by

$$z_{rot}(T) = \sum_{j=0}^{\infty}(2j+1)e^{-j(j+1)\Theta_{rot}/T}.$$

The rotational constant $\overline{B}_0$ for N_2O has the value [22] 0.4190 cm^{-1}, from which the characteristic rotational temperature is $\Theta_{rot} \simeq 0.603$ K. Figure 6.18 provides a clear example that the maximum intensity line(s) in a pure rotational spectrum do not correspond to the value j_{max} obtained from Eq. (6.5.3). For $T = 298$ K, which was the temperature of the N_2O in the spectrometer cell, Eq. (6.5.3) gives $j_{max} = 15$, while the maximal spectral intensities appearing in Fig. 6.18 corresponds to $23 \leq j'' \leq 28$. □

---

[4]We recall that a sequence of lines in a vibration–rotation spectrum for which $\Delta j \equiv j' - j'' = +1$, with primes and double-primes denoting upper, respectively, lower rotational states for a spectral transition, is referred to as an R-branch, and that a sequence of lines for which $\Delta j = -1$ is called a P-branch.

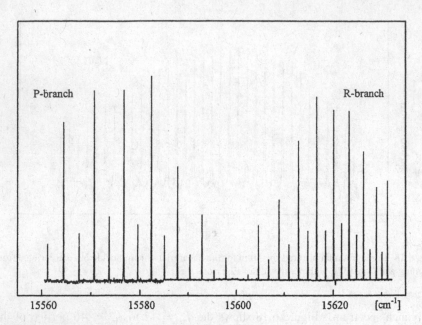

Fig. 6.19 Optoacoustic overtone absorption spectrum of acetylene centred on the fifth vibrational overtone of a local mode. Reprinted from Demtröder [38]

Example 6.11 Acetylene, C_2H_2.

The most common isotopologue of the acetylene molecule, H–C≡C–H is H–^{12}C≡^{12}C–H. As the centre of this molecule lies midway between the two ^{12}C nuclei, acetylene is a centrosymmetric linear molecule with the generic formula X_2Y_2 in which both the ^{12}C nuclei and the H nuclei form pairs of indistinguishable nuclei. However, because the ^{12}C nucleus has spin $I_a = 0$, ^{12}C plays no role in the determination of the combined rotational-nuclear spin partition function. The H nucleus (a proton) has $I_a = \frac{1}{2}$, and hence the nuclear spin symmetry argument for C_2H_2 is precisely the same as that for the H_2 molecule (as can be seen in Fig. 6.19). Thus, the combined rotational-nuclear spin partition function $z_{\text{rot-nuc}}(T)$ is given by

$$z_{\text{rot-nuc}}(T) = \sum_{j=\text{even}} (2j+1)e^{-j(j+1)\Theta_{\text{rot}}/T} + 3 \sum_{j=\text{odd}} (2j+1)e^{-j(j+1)\Theta_{\text{rot}}/T} .$$

We shall examine the rotational fine structure of the vibration–rotation band of C_2H_2 identified simply as a fifth overtone of a local mode. Because this is a vibration–rotation band, we see both a P-branch lying at lower frequencies and an R-branch lying at higher frequencies relative to the band centre corresponding to the pure vibrational transition. The structure of $z_{\text{rot-nuc}}(T)$ tells us that we may anticipate an intensity ratio $I_{j_o}:I_{j_e}$ of 3:1 for adjacent lines in both the P- and R-branches of this vibration–rotation spectrum. The nuclear spin statistical weightings of 3 and 1 play no role in determining the values for $j_{o,\text{max}}$ and $j_{e,\text{max}}$ for the two series of spin-weighted lines in each branch of the rotational fine-structure transitions. C_2H_2 has a rotational constant [22] $\overline{B}_0 = 1.17692\,cm^{-1}$, giving a characteristic rotational temperature $\Theta_{rot} \simeq 1.69\,K$, and a j_{max} value of approximately 9 for temperature $T = 298\,K$. The P- and R-branches of this overtone band in C_2H_2 at temperature $T = 298\,K$ shown in Fig. 6.19 both clearly display 3:1 intensity ratios. ☐

Example 6.12 Carbon dioxide, CO_2.

It is instructive in the case of CO_2 to compare the expected forms for the rotational fine structures of the vibration–rotation bands for three different CO_2 isotopologues. We shall examine the nature of the fine-structure bands for the most abundant isotopologue, $^{12}C^{16}O_2$, and compare our results for it with the structures of the corresponding vibration–rotation bands for two less common isotopologues, specifically, $^{12}C^{17}O_2$ and $^{18}O^{12}C^{16}O$. As we have seen in previous examples, both the ^{12}C and ^{16}O nuclei are bosons with spin $I_a = 0$: the ^{18}O nucleus likewise has spin $I_a = 0$, but the ^{17}O nucleus is a fermion with spin $I_a = \frac{5}{2}$.

Other than producing an overall factor $2I_a + 1 = 1$ in the combined rotational-nuclear spin partition function $z_{rot-nuc}(T)$, the nuclear spin of ^{12}C is irrelevant to our considerations, as the C atom is located at the centre of the CO_2 molecule. Of course, any nucleus located at the centre of a molecule will play no role in a spin symmetry argument.

The $^{12}C^{16}O_2$ isotopologue of CO_2 is a centrosymmetric linear molecule with two equivalent bosonic nuclei, so that if we are considering the fundamental vibrational transition of the ν_1 (symmetric stretch) vibrational mode of CO_2 (which carries σ_g^+ symmetry [22]) for which ψ_{vib} is symmetric to the interchange of the two O nuclei, we may employ Eq. (6.2.109) directly to obtain the relevant form for $z_{rot-nuc}(T)$, namely,

$$z_{rot-nuc}(T) = \sum_{j=even} (2j + 1)e^{-j(j+1)\Theta_{rot}/T}.$$

We thus see from this expression that the rotational fine-structure lines of the ν_1 vibrational transitions in this isotopologue of CO_2 will be associated only with rotational levels for which the quantum number j is an even integer: this means that adjacent lines in all ν_1 infrared spectra will be separated by $4\overline{B}_0$ rather than by $2\overline{B}_0$ because of missing lines that correspond to odd values of j. There will thus be no alternation of intensities in this spectrum, as can be seen from the pure rotational Raman spectrum of the predominant carbon dioxide isotopologue $^{12}C^{16}O_2$ shown in Fig. 6.20.

However, were we dealing with the rotational fine structure of the CO_2 asymmetric stretch, which has (group) symmetry σ_u^+, and is hence characterized by a vibrational wavefunction that is antisymmetric to the interchange of the two O nuclei, just the opposite conclusion would be reached. Thus, for example, rotational fine-structure lines in the ν_3 fundamental vibrational spectrum will involve only rotational levels for which j is odd. This behaviour has indeed been confirmed experimentally [22] long ago.

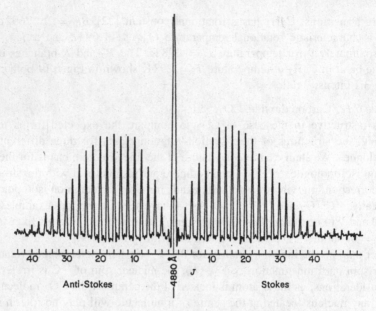

Fig. 6.20 Microdensitometer trace from a photographic plate recording the pure rotational Raman spectrum of $^{12}C^{16}O_2$. This spectrum shows both anti-Stokes and Stokes components of the O-branch and a strong Q-branch. From Edwards [39], with permission by Springer

Although the $^{12}C^{17}O_2$ isotopologue of CO_2 is also a centrosymmetric linear molecule, the equivalent ^{17}O nuclei, with nuclear spin $I_a = \frac{5}{2}$, are fermions, so that $z_{rot-nuc}(T)$ is given by

$$z_{rot-nuc}(T) = 15 \sum_{j=\text{even}} (2j+1)e^{-j(j+1)\Theta_{rot}/T} + 21 \sum_{j=\text{odd}} (2j+1)e^{-j(j+1)\Theta_{rot}/T}$$

according to Eq. (6.2.108). The rotational fine-structure P- and R-branches for this isotopologue of CO_2 will thus have alternating intensities in the ratio 5:7 for $j_e:j_o$ and adjacent lines in each branch will be separated by $2\overline{B}_0$, rather than by $4\overline{B}_0$.

Finally, the $^{18}O^{12}C^{16}O$ isotopologue of CO_2 is a noncentrosymmetric linear molecule, so that the combined rotational-nuclear spin partition function is given by $z_{rot}(T)z_{nuc}$, with $z_{nuc} = 1$ and $z_{rot}(T)$ given by Eq. (6.2.59), as

$$z_{rot-nuc}(T) = \sum_{j=0}^{\infty} (2j+1)e^{-j(j+1)\Theta_{rot}/T} .$$

There will thus be no intensity alternations in the rotational fine-structure infrared spectra of this CO_2 isotopologue, and the separation between adjacent lines in each fine-structure branch will be $2\overline{B}_0$.

We note also that the rotational constant $\overline{B}_0$ will differ slightly for these three isotopologues, so that the characteristic temperatures will also be slightly different (by less than 1%, however): for the present discussion we may hence employ a common value of Θ_{rot} for all three isotopologues.

This set of three carbon dioxide isotopologues illustrates quite dramatically the differences in behaviour that are associated with quantum- statistical effects.   □

Rotational Spectra for Nonlinear Molecules

In general for molecular spectroscopy, the full quantum mechanical form for the rotational partition function is required for any interpretation of the intensity patterns in the rotational fine-structure transitions in infrared spectra. For linear molecules, we were fortunate enough that they all fell into two categories, described by the operations of either the $C_{\infty v}$ point symmetry group (noncentrosymmetric linear molecules) or the $D_{\infty h}$ point symmetry group (centrosymmetric linear molecules). Nonlinear molecules belong to a whole assortment of point symmetry groups, ranging from groups corresponding to a very low degree of symmetry in the molecule to groups corresponding to a very high degree of molecular symmetry. At the low symmetry end are molecules from possessing no symmetry at all (point symmetry group C_1) through molecules possessing a single rotation axis (point symmetry group C_n) or a small number of symmetry elements (such as point symmetry group C_{2v}) to molecules possessing tetrahedral (point symmetry group T_d), octahedral (point symmetry group O_h) or icosahedral (point symmetry group I_h) symmetry. Molecules having low degrees of symmetry (such as S_2, C_{2v}, D_{2h}) are often asymmetric top molecules, those with higher degrees of symmetry (such as C_{3v}, D_2, D_{2d}) are often symmetric top molecules, and those with the highest degrees of symmetry (T_d, O_h, I_h) are spherical top molecules. Both because much molecular spectroscopy is more concerned with spectral frequency patterns than with intensity patterns and because it is often very difficult to measure absolute intensities, the role of the rotational partition function in determining line intensity patterns is not always of great interest except for centrosymmetric linear molecules where the intensity patterns are so comparatively dramatic.

Most polyatomic molecules have large moments-of-inertia and correspondingly small values for their characteristic rotational temperatures. The nuclear spin contributions to the combined rotational-nuclear spin partition function are largest for temperatures that do not exceed $10\Theta_{rot}$ to $20\Theta_{rot}$. For this reason, the only nonlinear molecules for which nuclear spin symmetry effects may be expected to play a role will be spherical top or light symmetric top hydrides, like CH_4, CD_4, SiH_4, CDH_3, or CHD_3. We shall therefore comment briefly only upon the behaviour of this set of molecules.

The relevant expression for $z_{rot-nuc}(T)$ for a tetrahedral molecule like CH_4 has been given explicitly [25] as

$$z_{\text{rot-nuc}}(T) = \frac{(2I_a + 1)^4}{12} \sum_{j=0}^{\infty} (2j + 1)^2 e^{-\alpha j(j+1)}$$

$$+ \frac{(2I_a + 1)^2}{4} \sum_{j=0}^{\infty} (-1)^j (2j + 1) e^{-\alpha j(j+1)}$$

$$+ \frac{4(2I_a + 1)^2}{3\sqrt{3}} \sum_{j=0}^{\infty} (2j + 1) \sin\left[\frac{(2j + 1)\pi}{3}\right] e^{-\alpha j(j+1)}. \qquad (6.4.8)$$

It will be clear from this expression for $z_{\text{rot-nuc}}(T)$ that it will not lead to a simple intensity alternation pattern of the type that we obtained for centrosymmetric linear molecules. The intensity pattern will be quite subtle, and it will likely be difficult to obtain spectra with sufficiently well-determined absolute intensities for much to be learned from them. In any case, such deliberations lie beyond the scope of the present discussion.

6.5　Third-Law Entropy and Residual Entropy

In general the agreement between the values of thermodynamic functions for polyatomic gases calculated from the above formulae and experimental values is excellent. For entropy the calculations are often more precise than experiment: as is illustrated by the values found in Table 6.3.

From the expression

$$dS = \frac{dq_{\text{rev}}}{T},$$

for the exact differential of the entropy, we can calculate $S(T)$ as

Table 6.3 Calculated vs. experimental entropies at $T = 298.16$ K

Gas	S_{calc}[a] /J mol^{-1}K^{-1}	S_{expt}[b] /J mol^{-1}K^{-1}	Gas	S_{calc}[a] /J mol^{-1}K^{-1}	S_{expt}[b] /J mol^{-1}K^{-1}
CO_2	213.8	213.8	CH_4	186.3	186.3
NH_3	192.8	192.8	CH_3Cl	234.4	234.6
NO_2	240.0	240.1	CCl_4	309.2	308.4[c]
ClO_2	249.4	256.8	C_6H_6	270.0[a]	269.2

[a]M.W. Chase, Jr., J. Phys. Chem. Ref. Data, Monograph 9 (1998)
[b]Handbook of Chemistry and Physics, 82nd Edition, D.R. Lide, Editor-in-Chief (CRC Press, Boca Raton, 2001)
[c]E.A. Moelwyn-Hughes, *Physical Chemistry* (Pergamon, London, 1957)

$$S(T) = S(0) + \int_0^T \frac{dq_{rev}}{T} .$$

The second term can be calculated from a knowledge of $c_p(T)$ for an ideal gas, since

$$dq_{rev} = dU + P\,dV$$

and, since for an ideal gas $U = U(T)$ only,

$$C_V = \left(\frac{\partial U}{\partial T}\right)_V = \frac{dU}{dT},$$

and

$$dq_{rev} = C_V\,dT + P\,dV .$$

Now, as all equilibrium states of an ideal gas are represented by

$$PV = Nk_B T ,$$

we have

$$P\,dV + V\,dP = Nk_B\,dT$$

so that

$$dq_{rev} = (C_V + Nk_B)\,dT - V\,dP$$
$$= C_P\,dT - V\,dP .$$

Hence, we see that $S(T)$ can be obtained from

$$S(T) = S(0) + \int_0^T \frac{C_P(T)}{T}\,dT ,$$

in which $C_P(T)$ is obtained from calorimetric data.

By comparing values of $S(T)$ calculated from the statistical mechanical formula and spectroscopic data with those calculated calorimetrically, $S(0)$ is found to be zero for most gases, a result that agrees with Planck's version of the third Law of Thermodynamics: '$S(0)$ for any pure material is zero'. However, some notable exceptions do occur. We might well ask why nonzero values are found for $\acute{S}(0)$. The answer is almost invariably that it is a consequence of unattainability of the lowest state of the system. Let us consider H_2O, CO, N_2O, and CH_3D, all having

$S(0) \neq 0$. For water, $S(0)$ has the value $3.39 \, \text{J} \, \text{mol}^{-1} \text{K}^{-1}$, while $S(0) = 4.64, 4.77, 11.72 \, \text{J} \, \text{mol}^{-1} \text{K}^{-1}$, respectively, for CO, N_2O, and CH_3D.

The following analysis may be carried out for the entropy of water. Let us examine the entropy as calculated from our formulae for water at a temperature of 298.16 K and at a pressure of 1 bar. For the calculated entropy we obtain

S_{calc}	S_{trans}	Sackur–Tetrode	$144.80 \, \text{J} \, \text{mol}^{-1} \text{K}^{-1}$
	S_{rot}	Classical	$43.85 \, \text{J} \, \text{mol}^{-1} \text{K}^{-1}$
	S_{vib}	$\nu_i = 3652, 1592, 3756 \, \text{cm}^{-1}$	$0.00 \, \text{J} \, \text{mol}^{-1} \text{K}^{-1}$
		$S_{\text{calc}} =$	$188.66 \, \text{J} \, \text{mol}^{-1} \text{K}^{-1},$

while Giauque and Stout [41] obtained the following calorimetric results:

S_{exptl}	0–10 K	Debye	$0.092 \, \text{J} \, \text{mol}^{-1} \text{K}^{-1}$
	10–273.16 K	C_p ice	$37.99 \, \text{J} \, \text{mol}^{-1} \text{K}^{-1}$
	273.16 K	melting	$21.99 \, \text{J} \, \text{mol}^{-1} \text{K}^{-1}$
	273.16–298.16 K	C_p water	$6.61 \, \text{J} \, \text{mol}^{-1} \text{K}^{-1}$
	298.16 K	vaporization	$149.36 \, \text{J} \, \text{mol}^{-1} \text{K}^{-1}$
	Compression to 1 bar		$-6.79 \, \text{J} \, \text{mol}^{-1} \text{K}^{-1}$
	gas imperfections		$0.01 \, \text{J} \, \text{mol}^{-1} \text{K}^{-1}$
		$S_{\text{expt}} =$	$185.13 \, \text{J} \, \text{mol}^{-1} \text{K}^{-1},$

with estimated experimental uncertainty $\pm 0.08 \, \text{J} \, \text{mol}^{-1} \text{K}^{-1}$. The quantity

$$S_{\text{calc}}(T) \, - \, S_{\text{expt}}(T) \, \geq \, 0,$$

is referred to as the 'residual entropy' of the substance. From the above results, we see that water has a residual entropy $S(0) = 3.39 \, \text{J} \, \text{mol}^{-1} \text{K}^{-1}$. In many cases the residual entropy can be explained in terms of a partial randomness in the crystalline form at very low temperatures. Another possibility is to consider the role of 'frozen in' rotational states, especially for centrosymmetric molecules.

Let us first examine the case of the simplest centrosymmetric molecule, H_2. For H_2 there can be no partial randomness of the crystalline solid, and we are left with the nuclear spin symmetry considerations associated with *ortho* and *para* hydrogen: as you may recall, nH_2 is a 3:1 metastable (or nonequilibrium) mixture of oH_2 and pH_2. At very low temperatures all pH_2 molecules lie in their ground $j = 0$ rotational states, while all oH_2 molecules lie in their 'ground' $j = 1$ rotational states. We thus have an additional contribution to the spectroscopic entropy of

$$\overline{S}_{\mathrm{spec}}(n\mathrm{H}_2;0) = \tfrac{3}{4}\,\overline{S}_{\mathrm{spec}}(o\mathrm{H}_2;0) + \tfrac{1}{4}\,\overline{S}_{\mathrm{spec}}(p\mathrm{H}_2;0)\,.$$

The entropy at absolute zero for $o\mathrm{H}_2$ will be given by $\overline{S}_{\mathrm{spec}}(o\mathrm{H}_2;0) = R\ln\Omega_{j=1} = R\ln 3$, while the entropy for $p\mathrm{H}_2$ will be given by $\overline{S}_{\mathrm{spec}}(p\mathrm{H}_2;0) = R\ln\Omega_{j=0} = 0$. This gives

$$\overline{S}_{\mathrm{spec}}(n\mathrm{H}_2;0) = \tfrac{3}{4}R\ln 3 \simeq 0.824R \approx 6.86\,\mathrm{J\,mol^{-1}K^{-1}}\,,$$

which should be compared with $\overline{S}_{\mathrm{spec}}(n\mathrm{H}_2) - \overline{S}_{\mathrm{cal}}(n\mathrm{H}_2) = 6.40\,\mathrm{J\,mol^{-1}K^{-1}}$. A similar calculation for $n\mathrm{D}_2$ gives an additional contribution of

$$\overline{S}_{\mathrm{spec}}(n\mathrm{D}_2;0) = \tfrac{1}{3}R\ln 3 \simeq 3.054\,\mathrm{J\,mol^{-1}K^{-1}}\,,$$

to be compared with $\overline{S}_{\mathrm{spec}}(n\mathrm{D}_2) - \overline{S}_{\mathrm{cal}}(n\mathrm{D}_2) = 3.096\,\mathrm{J\,mol^{-1}K^{-1}}$. Note that we need not consider the entropy of mixing, ΔS_{mix}, because it is the same for all temperatures and, as $\overline{S}_{\mathrm{cal}}(T)$ represents the difference between the entropy measured at temperature T and the entropy at absolute zero, the entropy of mixing cancels, and it is $\overline{S}_{\mathrm{spec}}(T)$ *without* the entropy of mixing that must be compared with $\overline{S}_{\mathrm{cal}}(T)$. It is important to note that such a contribution to the spectroscopic entropy for the hydrogen isotopologues gives different values of the residual entropy for H_2 and D_2. This difference is observed.

The case of water ice, for which detailed calculations have been given, is an especially interesting one. In the 1920s and early 1930s the residual entropy of ice was known to be about $4.184\,\mathrm{J\,mol^{-1}K^{-1}}$. An initial explanation of this value was given by Giauque and Ashley [41] along the lines of the successful explanation of the residual entropy of H_2, as being associated with the presence of the metastable ground state of *ortho*-water. However, when improved calorimetric measurements of Giauque and Stout [40] became available, the difference between $\overline{S}_{\mathrm{cal}}$ and $\overline{S}_{\mathrm{spec}}$ was found to be only $3.431\,\mathrm{J\,mol^{-1}K^{-1}}$. Pauling [42] provided an explanation based upon partial randomness in crystalline ice, in which the ordered arrangement corresponding to zero entropy is not attained. From X-ray work it was known that each oxygen atom in ice is tetrahedrally surrounded by four other oxygen atoms at $2.76\,\text{Å}$ distance, bonded by hydrogen bonds. Each H atom is about $0.95\,\text{Å}$ from one oxygen atom (the same as in an isolated water molecule) and $1.81\,\text{Å}$ from another: moreover, they assume positions such that each O atom will have two H atoms attached to it. Pauling made the following assumptions:

(i) In ice each O atom has two H atoms attached to it at distances of about $0.95\,\text{Å}$, forming a water molecule, with the HOH angle being about $105°$;
(ii) each $\mathrm{H}_2\mathrm{O}$ molecule is oriented so that its two H atoms are directed approximately toward two of the four O atoms which surround it tetrahedrally, forming hydrogen bonds;
(iii) the orientation of adjacent water molecules are such that only one H atom lies approximately along each O–O-axis;

(iv) under ordinary conditions the interaction of non-adjacent molecules is not such
      as to appreciably stabilize any one of the many configurations satisfying the
      first three conditions with reference to the others.

An ice crystal can change from one configuration to another by rotation of some
of the molecules or by the motion of some of the hydrogen nuclei, each moving
a distance of about $0.86\,\text{Å}$ from a potential minimum $0.95\,\text{Å}$ from one O atom
to another potential minimum situated a distance $0.95\,\text{Å}$ from another adjacent
O atom. It is likely that both processes occur. We shall designate the number of
configurations available to the crystal by Ω. Near $0\,\text{K}$ the entropy of one mole of
such crystalline ice will be given by $\overline{S} = k_B \ln \Omega$. What will be the value of Ω? We
note that a given molecule can orient itself in one of the six ways while satisfying
condition (ii) above: the probability that the adjacent molecules will permit a given
orientation is $\frac{1}{4}$, so that for one mole of H_2O, the number of configurations will be

$$\Omega = \left(\frac{6}{4}\right)^{N_0} = \left(\frac{3}{2}\right)^{N_0},$$

from which we obtain

$$\overline{S}_0 = k_B \ln \Omega = N_0 k_B \ln \left(\frac{3}{2}\right) \simeq 3.389\,\text{J}\,\text{mol}^{-1}\text{K}^{-1},$$

which should be compared with the value $(3.43 \pm 0.21\,)\text{J}\,\text{mol}^{-1}K^{-1}$ [40].

Another case in which partial randomness in the crystalline form at very low
temperatures offers a credible explanation is that of CO. As CO molecules have
very small dipole moments (≈ 0.1D), they do not have a strong tendency to align
in an energetically favourable manner when CO crystallizes, and the net result
is a random mixture of the two possible orientations CO and OC. As a crystal
of CO is cooled down toward $0\,\text{K}$, each molecule becomes 'frozen in' so that
the number of configurations Ω of the crystal is 2^N, giving an entropy at $0\,\text{K}$ of
$\overline{S}(0) = k_B \ln \Omega = N_0 k_B \ln 2$, rather than $\overline{S}(0) = N_0 k_B \ln 1 = 0$. Perfect randomness
(which would occur were CO to have no dipole moment) would contribute a residual
entropy of $N_0 k_B \ln 2$ to $\overline{S}(0)$ or $5.858\,\text{J}\,\text{mol}^{-1}\text{K}^{-1}$. However, because CO has a
very small dipole moment, perfect randomness will not be achieved, and hence
$\overline{S}(0) \approx 4.602\,\text{J}\,\text{mol}^{-1}\text{K}^{-1}$, rather than $5.858\,\text{J}\,\text{mol}^{-1}\text{K}^{-1}$. The agreement is, in
any case, sufficiently good that we may consider this explanation of the 'frozen
in' configuration to be the correct one. N_2O is similar in its behaviour to CO.
As each CH_3D molecule can assume four different orientations in the crystal at
low temperatures, $\overline{S}(0) \approx N_0 k_B \ln 4 = 11.30\,\text{J}\,\text{mol}^{-1}\text{K}^{-1}$ (cf. $11.71\,\text{J}\,\text{mol}^{-1}\text{K}^{-1}$
observed).

6.6 Effect of Hindered Rotational Motions

Complex molecules often contain groups of atoms connected to one another or to a rigid molecular framework by single-electron chemical bonds. These groups can exercise rotations relative to one another or to the rigid framework in addition to participating in the overall rotational motion of the molecule. Such relative rotational motions are referred to as 'internal' rotations: if two rotating groups are sufficiently close to one another for their electronic charge clouds to interact, then the rotational motion of one group relative to another may generate a potential energy that opposes, or hinders, free rotational motion about the common bond. It seems reasonable to expect that the greater the separation between the two groups of atoms, the freer will the internal rotation be.

For molecules in thermal equilibrium, the magnitude of the potential energy barrier to free rotation relative to thermal energy, $k_B T$, may be employed to delineate three categories of the resultant internal motion of a group of atoms. On the one hand, if the potential energy barrier is large relative to $k_B T$, we say that there is a high barrier to rotation, so that the internal motion will essentially be torsional, and may be treated in the same fashion as any other vibrational motion in the molecule. Typical of this category would be twisting motions about a common multiple bond: perhaps the simplest example of this category of internal motion is the twisting motion of the CH_2 groups in ethene, C_2H_4, about the carbon–carbon double bond. On the other hand, if the potential energy barrier is very small relative to $k_B T$, it may simply be ignored to a good approximation, and the internal motion may be treated as free rotational motion. Typical of this category would be the relative rotational motion of two chemical groups separated by more than one bond length: the simplest example of this category of internal motion is dimethyl acetylene, $H_3C-C{\equiv}C-CH_3$, whose barrier to rotation is essentially zero. The most common category of internal motion is that for which the potential energy barrier to rotation is neither much larger nor much smaller than $k_B T$, in which case all three regimes of motion may be accessed as the temperature is varied, thereby leading to a motion that is quite complex, often requiring numerical evaluation to be carried out. Such calculations often necessitate the use of an appropriate molecular model.

Because we have rotational motion about a bond, the potential energy associated with it must be periodic with period $2\pi/\sigma$, where σ is the symmetry number associated with the group of rotating atoms. The hindering potential $V(\phi)$ must therefore satisfy the condition

$$V(\phi + \tfrac{2\pi}{\sigma}) = V(\phi). \qquad (6.6.1)$$

We shall only have occasion to consider periodic potential energy functions of the form[5]

---

[5]This form for $V(\phi)$ represents the leading term of a Fourier series expansion of the periodic interaction. Although the inclusion of additional terms of the expansion would give a more accurate representation of the interaction, Eq. (6.6.2) normally suffices for comparison with existing experimental data associated with this phenomenon. For C_2H_6 and C_2Cl_6, for example,

$$V(\phi) = \tfrac{1}{2} V_\sigma (1 - \cos \sigma \phi) \,, \tag{6.6.2}$$

which suffices for most cases. Perhaps the simplest and most widely discussed examples of molecules falling into this category are ethane, C_2H_6, and methyl chloroform, CH_3CCl_3.

6.6.1 Setting the Stage

We shall focus here only upon the simplest type of internal rotational motion, specifically, the rotational motion of two symmetric top groups about a common rotation axis. For ethane, for example, this means treating each methyl group as a rigid symmetric top. In classical mechanical terms, we have a rotational kinetic energy, T_{rot}, given as

$$T_{rot} = \tfrac{1}{2} I_1 \omega_1^2 + \tfrac{1}{2} I_2 \omega_2^2 \,, \tag{6.6.3a}$$

with I_1, I_2 the moments-of-inertia of the two tops, and ω_1 and ω_2 their angular velocities. By analogy with the transformation from space-fixed coordinates and momenta (see Sect. 6.2.3), we can transform from space-fixed (rotational) angular velocities ω_1 and ω_2 to centre-of-mass angular velocity Ω and relative angular velocity ω by employing the total angular momentum $J_{tot} = I_{tot}\Omega = I_1\omega_1 + I_2\omega_2$, in which $I_{tot} = I_1 + I_2$ is the total moment-of-inertia. The relative angular velocity is defined as $\omega \equiv \omega_2 - \omega_1$, and is associated with the relative rotational motion of the two tops about the common axis of rotation. We thus have two equations,

$$\Omega = \frac{I_1}{I_{tot}}\omega_1 + \frac{I_2}{I_{tot}}\omega_2 \,, \tag{6.6.3b}$$

$$\omega = \omega_2 - \omega_1 \,, \tag{6.6.3c}$$

to be inverted to give ω_1 and ω_2 as

$$\omega_1 = \Omega - \frac{I_2}{I_{tot}}\omega \,, \tag{6.6.4a}$$

$$\omega_2 = \Omega + \frac{I_1}{I_{tot}}\omega \,. \tag{6.6.4b}$$

---

the second term in the Fourier series expansion is less than 0.005, respectively, 0.02, times the leading term.

Substitution of these results into Eq. (6.6.3a), followed by simplification then gives the total rotational kinetic energy in terms of the centre-of-mass and relative rotational motions as[6]

$$T_{\text{tot}} = \tfrac{1}{2} I_{\text{tot}} \Omega^2 + \tfrac{1}{2} I_r \omega^2 , \tag{6.6.4c}$$

with I_r the reduced moment-of-inertia for the relative motion, i.e.,

$$I_r = \frac{I_1 I_2}{I_1 + I_2} . \tag{6.6.4d}$$

If we view the relative rotational motion of two rigid symmetric tops along their common rotation axis, then the relative angular speed for this motion is given by the time derivative of the angle ϕ between the two groups. The potential energy, $V(\phi)$, describing this relative rotational motion must be periodic: should neither group have symmetry, then $V(\phi)$ will have period 2π; if one of the groups has symmetry number σ_1 and the other symmetry number σ_2, the period will be $2\pi/\sigma$, with σ containing both σ_1 and σ_2 symmetries. The relevant Schrödinger equation governing such one-dimensional internal rotational motion will be

$$\frac{\mathrm{d}^2 \psi_k}{\mathrm{d}t^2} + \frac{2I_r}{\hbar^2}[E - V(\phi)]\psi_k = 0, \tag{6.6.5}$$

with $\psi_k(\phi)$ the relevant hindered rotation wavefunction.

6.6.2 Internal Rotation in Ethane

Ethane can be thought of as two methyl groups joined by a carbon–carbon bond. As mentioned earlier, it provides the prototypical example for hindered internal rotation. Of the $3(8) - 6 = 18$ vibrational degrees of freedom for C_2H_6, one corresponds to the relative rotational motion of the two methyl groups. As this motion turns out to be neither IR nor Raman active,[7] no direct spectroscopic information is available for it. However, if the heat capacity $C_V(T)$ is determined

---

[6]In general, there will be a weak coupling between the end-over-end overall rotational angular momentum of the molecule and the internal rotational angular momenta of its subunits. For a reasonably comprehensive discussion of these aspects, see §9h and Appendices 16, 18 of Pitzer [44].

[7]The reason that this internal motion is neither IR nor Raman active is connected to the nature of the symmetry group for the ethane molecule. The specific motion involved does not belong to an irreducible representation of the symmetry group of ethane that corresponds to any of the irreducible representations for which Cartesian components of the electric dipole moment vector $\mu^{(e)}$ or the polarizability tensor α form a basis.

experimentally, and the contributions from the translational and rotational motions, together with those contributions from the 17 vibrational modes of ethane for which spectroscopic data are available are subtracted from it, the remainder must be the contribution to $C_V(T)$ associated with the hindered rotational motion.

From an examination of the Newman projection of the two CH_3 groups, it is clear that there is maximal repulsion between them when they are eclipsed, and minimal repulsion between them when they are staggered, so that the internal interaction potential has $\cos 3\phi$ symmetry for $0 \leq \phi \leq 2\pi$. This case has been examined in detail by Ercolani [44]. As the eclipsed and staggered forms of ethane have different interaction energies, rotation about the C–C bond will be nonuniform, and the rotational motion is thereby hindered. On the one hand, for energies small enough that the barrier, V_3, to free internal rotation is very large, the result will be (torsional) vibrations of the CH_3 groups, while on the other hand, for very high energies, V_3 will be comparatively small, and essentially free rotational motion will take place. For energies lying between these extremes, the motion will be much more complicated: it is this more complicated motion that is referred to as hindered rotational motion.

We shall briefly examine the two limiting cases discussed earlier, for which the relative energy of rotation is either much smaller or much greater than the barrier height, V_3, associated with Eq. (6.6.5). In the former case, the angle ϕ will be confined near the minimum of one of the three equivalent potential wells of $V(\phi)$: without loss of generality, we may choose the centre of this well as the origin for ϕ. With this choice of origin ϕ will necessarily be small, and we may expand $\cos 3\phi$, retaining only the first two terms, to obtain

$$V(\phi) = \tfrac{1}{2}k_\phi \phi^2 ,$$

which is a SHO potential energy with force constant $k_\phi = \tfrac{9}{2}(\tfrac{1}{2}V_3/I_r)^{\frac{1}{2}}$. The accepted barrier height for C_2H_6 is $V_3 = 12.35\,\text{kJ mol}^{-1}$ or $V_3 = 2.0505 \times 10^{-20}\text{J}$ in SI units. For C_2H_6, I_r is given simply by $\tfrac{1}{2}I_{CH_3}$, with I_{CH_3} the moment-of-inertia of a rigid methyl group undergoing rotation about the ethane C–C bond.

For rotational energies much greater than V_3, i.e., lying well above the top of the barrier, we may replace $V(\phi)$ by its angular average $\tfrac{1}{2}V_3$, to approximate Eq. (6.6.5) by the Schrödinger equation for a free rotor, but with a shift of the zero of the energy scale upwards by $\tfrac{1}{2}V_3$. The energy levels will thus be

$$E_k = \frac{\hbar^2 k^2}{2\pi} + \tfrac{1}{2}V_3 , \qquad k = 0, \pm 1, \pm 2, \ldots .$$

We should hence expect the energy eigensolutions of Eq. (6.6.5) to be similar to those of a triply degenerate SHO for energies near the bottom of the wells and to be similar to those of a shifted free rotor for energies well above the top of the potential energy barriers. Such a comparison has been given by Pitzer [43] and by Ercolani [44], and is also shown in Fig. 6.21.

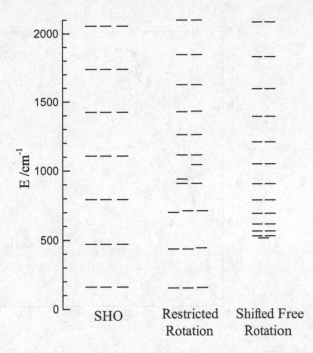

Fig. 6.21 Comparison of energy levels for simple harmonic motion, shifted free rotational motion, and restricted rotational motion. Energy levels have been obtained from numerical solution of the Schrödinger equation (6.6.5) and reported in Table 1 of Ref. [44]. Reproduced with permission of the American Chemical Society

Accurate evaluation of the canonical partition function for a molecule like C_2H_6 can also be made by taking advantage of the merging of the hindered rotational energy levels with those of the corresponding shifted free rotor at sufficiently high rotor energies, so that the hindered rotor canonical partition function, $z_{hr}(T)$, for an ethane molecule can be approximated as

$$z_{hr}(T) \simeq \frac{1}{3} \left[\sum_{r=1}^{m} \left(e^{-\beta E_r} \right)_{hr} + \sum_{r=m+1}^{\infty} \left(e^{-\beta E_r} \right)_{sfr} \right]. \tag{6.6.6a}$$

Another means for evaluating $z_{hr}(T)$ is to replace the second term of Eq. (6.6.6a) by the difference between the full partition function for the shifted free rotor, given by $z_{sfr}(T) = e^{-\frac{1}{2}\beta V_3}(2\pi I_r k_B T)^{\frac{1}{2}}/\hbar$ and the contribution arising from the first m levels, so that the expression for $z_{hr}(T)$ becomes [44]

$$z_{hr}(T) = \frac{1}{3} \left[\sum_{r=1}^{m} \left(e^{-\beta E_r} \right)_{hr} + e^{-\frac{1}{2}\beta V_3} \left(\frac{\sqrt{2\pi I_r k_B T}}{\hbar} - \sum_{j=-(m-1)/2}^{(m-1)/2} e^{-\hbar^2 j^2/(2I_r k_B T)} \right) \right],$$
$$\tag{6.6.6b}$$

in which the first part of the second term represents $z_{fr}(T)$, given by

$$z_{fr}(T) = \frac{1}{3} \int_{-\infty}^{\infty} e^{-\hbar^2 j^2/(2I_r k_B T)} \, dj. \tag{6.6.6c}$$

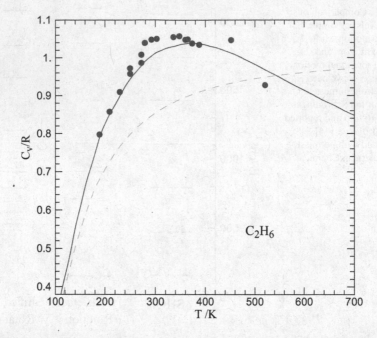

Fig. 6.22 Contribution to the heat capacity at constant volume, $C_V(T)$, due to hindered relative rotational motion of the two methyl groups of ethane (for a rotational barrier height of 289 cm^{-1}) over the temperature range 100 K $\leq T \leq$ 700 K. The 18 experimental points have been obtained using the procedure described in the text. After Ercolani [44], with permission of the American Chemical Society

If we examine Fig. 6.21, we note from a comparison between the degenerate pairs of rotational energy states for C_2H_6 obtained from solution of the hindered rotor Schrödinger equation (with $V_3 \simeq 1032.4$ cm^{-1}) and having energies in excess of the barrier height are slightly higher in energy than the corresponding energy levels for the shifted free oscillator. Moreover, although it is difficult to see from this figure, the difference between these corresponding doubly degenerate levels decreases as the rotational quantum number increases. At some point the two sets of energy levels will cross: calculations [44] for ethane show that the value of the rotational quantum number m for which the hindered rotor and shifted free rotor values cross corresponds to 47.

Pitzer and coworkers [45, 46] have prepared extensive tables based upon $z_{\mathrm{fr}}(T)$ and βV_σ to be utilized for the evaluation of the contributions of hindered rotational motions to the various thermodynamic properties. These tables have been reproduced in a number of textbooks, most notably those by Pitzer [47] and McClelland [48]. The hindered rotational contribution to $C_V(T)$, obtained by employing such tabulations, is shown as the solid curve in Fig. 6.22 for C_2H_6 for temperatures between 100 K and 700 K. In addition, the corresponding contribution to $C_V(T)$ that would have been obtained from a vibrational mode with characteristic vibrational temperature $\Theta_{vib,4} = 415.8\,\text{K}$ (obtained from $\bar{\nu}_4 = 289\,\text{cm}^{-1}$) is shown as a dashed curve, and several experimental values obtained from heat capacity measurements by subtracting the translational, end-over-end rotational, and all vibrational contributions calculated from spectroscopic information are also shown for comparison. The agreement between calculated and experimental values is, on the whole, quite acceptable. More recent computations of the eigenenergies and their employment to compute the hindered rotation contributions to $C_V(T)$, the entropy, $S(T)$, and the enthalpy, $H(T)$, are in excellent agreement [44] with values obtained using the Pitzer tables.

6.7 Problems for This Chapter

1. Show what happens to P, U, S, A, when a constant value b is added to all energy levels E_r occurring in the definition of z_{int} for a molecule.
2. Most molecules in a gas have $\epsilon_{tr} \leq 3k_B T$ and $\epsilon_{tr,x} \leq k_B T$, where $\epsilon_{tr,x}$ is the x-component of the translational energy. For N_2 in a box of volume $8\,\text{cm}^3$ at $0\,°\text{C}$, calculate $\Delta\epsilon_x/(k_B T)$, with $\Delta\epsilon_x$ the spacing between the energies associated with adjacent x translational states when $\epsilon_{tr,x}$ is of order $k_B T$.
3. Spectroscopic intensity measurements for a series of vibrational transitions in a sample of gaseous I_2 give vibrational population ratios $N_1/N_0 = 0.528$ and $N_2/N_0 = 0.279$, in which N_v represents the number of I_2 molecules in a vibrational state characterized by vibrational quantum number v. Show that these results are compatible with the I_2 sample being in thermal equilibrium. Given that the fundamental vibrational frequency for I_2 is $\bar{\nu} = 214.57\,\text{cm}^{-1}$, determine the temperature of the I_2 sample.
4. If a diatomic gas in thermal equilibrium has a $v = 1$ to $v = 0$ population ratio of 0.340, what is the corresponding $v = 3$ to $v = 0$ population ratio?
5. If the fundamental vibrational frequency for N_2 is $\nu_{osc} = 6.9849 \times 10^{13}\,\text{s}^{-1}$, compute the ratio of the $v = 1$ to $v = 0$ populations for (a) $25\,°\text{C}$; (b) $800\,°\text{C}$; (c) $3000\,°\text{C}$.
6. Show, using the SHO approximation for the vibrational motion of a diatomic molecule, that the fractional population of vibrational level v is given by

$$\frac{N_v}{N} = e^{-v\Theta_{vib}/T}(1 - e^{-\Theta_{vib}/T}),$$

with N the total number of molecules. Given that $\Theta_{vib}(N_2) = 3352\,\text{K}$, plot the fractional population of the $v = 1$ vibrational level of N_2 as a function of temperature from $0\,\text{K}$ to $15{,}000\,\text{K}$. Why might you not expect the high-T part of your plot to provide an accurate representation of the real behaviour of N_2?
7. Show that the SHO (vibrational) partition function, $z_{vib}(T)$, given in Eq. (6.2.22) can be rewritten in terms of the hyperbolic cosecant function as

$$z_{vib}(T) = \tfrac{1}{2}\, csch(\Theta_{vib}/2T)\,,$$

and obtain approximations to $z_{vib}(T)$ that are appropriate for temperatures $T \ll \Theta_{vib}$ and $T \gg \Theta_{vib}$.

8. The temperature derivative of $\ln z_{vib}(T)$ determines the average energy for a simple harmonic oscillator. Show that this derivative can be expressed in terms of the hyperbolic cotangent function as

$$\frac{d \ln z_{vib}}{dT} = \frac{\Theta_{vib}}{2T^2}\, coth\left(\frac{\Theta_{vib}}{2T}\right).$$

9. Show that the average energy for a canonical ensemble of N SHOs at temperature T is given by

$$\langle E \rangle_{SHO}(T) = \tfrac{1}{2} N \hbar \omega_{osc}\, coth\left(\tfrac{1}{2}\beta \hbar \omega_{osc}\right)$$

or, equivalently, as

$$\langle E \rangle_{SHO}(T) = \tfrac{1}{2} N \hbar \omega_{osc} + N\, \frac{\hbar \omega_{osc} e^{-\beta \hbar \omega_{osc}}}{1 - e^{-\beta \hbar \omega_{osc}}}\,,$$

with ω_{osc} the (radial) frequency associated with the SHO motion.

10. Obtain expressions in terms of the hyperbolic functions for the vibrational contributions to the Helmholtz energy and the entropy arising from a molar ensemble of one-dimensional simple harmonic oscillators. What is the Gibbs energy for a molar ensemble of such oscillators?

11. Many molecules are found to have quite similar room temperature values of the heat capacity at constant volume. In particular, $\overline{C}_V(Ar)$ and $\overline{C}_V(He)$ both have the value $12.48\,J\,mol^{-1}K^{-1}$ at NTP. Explain why this is so. The value of $\overline{C}_V(N_2)$ at NTP is $20.81\,J\,mol^{-1}K^{-1}$: does this value correspond to what you would predict, given your explanation for the heat capacities of Ar and He? Why or why not?

12. Determine which of the molecules DBr ($\overline{B} = 4.204\,cm^{-1}$), DI ($\overline{B} = 3.22\,cm^{-1}$), CsI ($\overline{B} = 0.0236\,cm^{-1}$), $F^{35}Cl$ ($\overline{B} = 0.514\,cm^{-1}$) will have their rotational contributions to the thermodynamic state functions at temperature $T = 40\,K$ well represented by calculations that employ the high-temperature limit for z_{rot}.

13. Which of H_2, HD, or D_2 will have the largest rotational partition function (use the high-temperature limiting form for z_{rot})? If V and T are fixed, which of these three isotopologues of hydrogen will have the largest value for the translational partition function?

14. Calculate the rotational partition function for $^{16}O_2$ ($\overline{B} = 1.44\,cm^{-1}$) at its boiling point, $90.2\,K$, both by employing the high-temperature approximation

and by discrete summation. Why should only odd values of j be included in your discrete summation?

15. Both ^{12}C and ^{16}O are bosons, each with nuclear spin $I_a = 0$. Obtain an expression for the combined rotational-nuclear spin partition function for a gas of $^{12}C^{16}O_2$ molecules. If the rotational constant for CO_2 is $0.390\,cm^{-1}$, calculate z_{rot} at $T = 298\,K$. Does this have any consequences for the pure rotational Raman spectrum of the $^{12}C^{16}O_2$ molecule? If yes, what are they? If no, why not?

16. In a vibration–rotation spectrum of $H^{35}Cl$ ($I = 2.65 \times 10^{-47}\,kg\,m^2$), the line corresponding to the $j'' = 4$ to $j' = 5$ transition is the most intense. At which temperature has this spectrum been obtained? Which line of this spectrum would have the greatest intensity for temperature $1000\,K$?

17. Determine $\overline{S}^\circ_{298}$, $\overline{C}^\circ_{V,298}$, and $\overline{C}^\circ_{P,298}$ for gaseous HF, given that $\overline{\nu}_{osc}(HF) = 3959$ cm^{-1} and $\overline{B}(HF) = 20.56\,cm^{-1}$.

18. The I_2 molecule has $\nu_{osc} = 6.395 \times 10^{12}\,s^{-1}$, and an equilibrium internuclear separation of 2.67 Å. By making appropriate assumptions, calculate the values of $\overline{U}^\circ_{500} - \overline{U}^\circ_0$, $\overline{H}^\circ_{500} - \overline{U}^\circ_0$, $\overline{S}^\circ_{500}$, and $\overline{G}^\circ_{500} - \overline{U}^\circ_0$ for gaseous I_2.

19. The iodine molecule, I_2, has a moment-of-inertia $I = 750 \times 10^{-47}\,kg\,m^2$ and a characteristic vibrational frequency $\nu_{osc} = 6.40 \times 10^{12}\,s^{-1}$. Calculate the value of the heat capacity $C_V(T)$ and the entropy for one mole of gaseous iodine at $298\,K$ and for a pressure of 1 bar. Calorimetric measurements extended back towards $0\,K$ lead to $\overline{S}^\circ_{298} = 116.7\,J\,mol^{-1}K^{-1}$. Determine the value of $\Delta\overline{S}^\circ_{298}$ for the sublimation of one mole of iodine at $298\,K$. Employ the value $0.305\,mm\,Hg$ for the equilibrium pressure of iodine vapour over solid iodine at $T = 298\,K$ to determine the value of $\Delta\overline{G}^\circ_{298}$ for the sublimation of one mole of iodine at $298\,K$. From your results for $\Delta\overline{S}^\circ_{298}$ and $\Delta\overline{G}^\circ_{298}$, obtain the standard enthalpy of sublimation, $\Delta\overline{H}^\circ_{298}$ for one mole of iodine at $298\,K$ and compare your result with the accepted experimental value of $62.26\,kJ\,mol^{-1}$.

20. For CO_2 the four fundamental vibrational frequencies are given by $\overline{\nu}_{osc} = 1388, 667, 667$, and $2349\,cm^{-1}$; the rotational constant is $\overline{B} = 0.390\,cm^{-1}$. Compute a value for $\overline{S}^\circ_{298}$ for CO_2.

21. A thermodynamics table gives $(\overline{G}^\circ_T - \overline{H}^\circ_0)/T = -257.7\,J\,mol^{-1}K^{-1}$ for $CH_3OH(g)$ at $1000\,K$. Use this information to obtain a value for the canonical partition function $\overline{Z}^\circ$ for $CH_3OH(g)$ at $1000\,K$, assuming ideal gas behaviour.

22. The NO molecule has one unpaired electron, and has an electronic orbital angular momentum that has a projection $\Lambda = 1$ along the NO figure axis, so that its ground electronic term symbol is $^2\Pi$. The total electronic angular momentum J can be $\frac{3}{2}$ or $\frac{1}{2}$. It happens that the ground electronic state is the $^2\Pi_{\frac{1}{2}}$ state, but the first excited electronic $^2\Pi_{\frac{3}{2}}$ state lies only $119.82\,cm^{-1}$ above it. As the $^2\Pi_{\frac{1}{2}}$-$^2\Pi_{\frac{3}{2}}$ separation in energy is far less than the fundamental oscillator frequency $\overline{\nu}_{osc} = 1876.014\,cm^{-1}$, and the characteristic rotational temperatures are very similar for the two electronic states (for which $\overline{B}(^2\Pi_{\frac{1}{2}}) = 1.6634\,cm^{-1}$ and $\overline{B}(^2\Pi_{\frac{3}{2}}) = 1.7111\,cm^{-1}$), the canonical partition function

for the NO molecule can be approximated rather well simply by taking the electronic partition function to have the form

$$z_{\text{el}}(T) = \omega_{\text{e1}} + \omega_{\text{e2}} \exp\{-\epsilon_{\text{e2}}/(k_{\text{B}}T)\},$$

in which $\epsilon_{\text{e2}} = 119.82\,\text{cm}^{-1}$.

Obtain an expression for the contribution, $U_{\text{el}}(T)$, of the excited electronic level to the internal energy $U(T)$ as a function of temperature. From your result for $U_{\text{el}}(T)$ obtain an expression for and discuss the behaviour of the electronic contribution to C_V as a function of temperature. Plot C_V as a function of T for NO, and obtain an expression for the electronic contribution to the entropy, $S_{\text{el}}(T)$. Check that $S_{\text{el}} \to Nk_{\text{B}}\ln 2$ for $T \to 0$ and that $S_{\text{el}} \to Nk_{\text{B}}\ln 4$ for $T \to \infty$.

23. Determine the contribution of the electronic degrees of freedom of the doubly degenerate ground electronic term of the NO molecule to the standard molar entropy of NO. Evaluate $\overline{S}_{\text{el}}(T)$ for temperatures 50 and 298.15 K. Compare your 298.15 K result with $R\ln 4$: why should you expect this to be a meaningful comparison?

24. Starting from Eq. (6.2.59) for the rotational partition function for a heteronuclear diatomic molecule, show that the rotational contribution to the molar heat capacity at constant volume may be written in the form

$$\overline{C}_{V,\text{rot}}(T) = R[\langle\varepsilon_{\text{rot}}^2\rangle(T) - \langle\varepsilon_{\text{rot}}\rangle^2(T)],$$

in which the average $\langle f \rangle$ of $f(j)$ is defined as $\langle f \rangle \equiv \sum_j p_j f(j)$, with p_j given by Eq. (6.2.62b), while ε_{rot} is defined as $\varepsilon_{\text{rot}}(j) \equiv \epsilon_j \Theta_{\text{rot}}/T$, with ϵ_j the rotational energy $\overline{B}j(j+1)$ and $\overline{B} \equiv \hbar^2/(2I)$ is the rotational constant. Compare your computed results with the experimental values 0.2013 at 20 K, 0.9698 at 40 K, 1.0403 at 77.5 K, 1.0201 at 90.2 K, and 1.0000 at 293 K on a plot of $\overline{C}_{V,\text{rot}}(T)/R$ vs. temperature.

25. Compute the rotational partition function for methane (CH_4) at temperatures $T = 50\,\text{K}, 100\,\text{K}, 300\,\text{K}$, and $500\,\text{K}$ by direct summation over rotational states using Eq. (E.2.3) and compare the values obtained with those determined from the classical formula of Eq. (6.3.4). As CH_4 is spherical top molecule, the three principal moments-of-inertia are all equal. The rotational constant for CH_4 is $\overline{B} = 5.2412\,\text{cm}^{-1}$, and the symmetry number for CH_4 is 12. Compare your summations (converged to two decimal places) with the classical (or high temperature) values for $z_{\text{rot}}(T)$.

26. Nuclear spin statistical effects are important for the computation of the intensities of spectroscopic transitions, which depend upon the individual fractional populations of the spectroscopic states. For bulk thermodynamic properties that depend less directly upon the fractional populations and more directly upon the partition function, we should expect the effects of nuclear spin statistics to be relatively small (with the exception of the hydrogen isotopologues).

Obtain a discrete summation expression for $z_{rot}(T)$ that includes the centrifugal-distortion energy, with the rotational energy of a CH_4 molecule given in cm^{-1} units as

$$\epsilon_j = j(j+1)\overline{B} - [j(j+1)]^2\overline{D}.$$

Compute the values of $z_{rot}(T)$ for temperatures 50 K, 100 K, 300 K, and 500 K using $\overline{B}_0 = 5.2412\,cm^{-1}$ and $\overline{D}_0 = 1.23 \times 10^{-4}cm^{-1}$. Compare the results of this calculation with the rigid-rotor values obtained in Problem 25 via $z_{rot-nuc}(T) = 16z_{rot}^{DS}(T)$ with the values $z_{rot-nuc}^{DS}(T)$ 41.993, 116.421, 598.639, and 1287.523 computed for CH_4 at temperatures 50 K, 100 K, 300 K, 500 K, respectively, by Robiette and Dang-Nhu [49] and the closed-form values $z_{rot-nuc}^{CF}(T)$ given via Eq. (6.3.23).

27. Calculate the standard molar entropy of ClO_2 at 298.15 K (NATP). The ClO_2 molecule is a nonlinear triatomic molecule (point symmetry group C_{2v}), and has one unpaired electron (ground electronic term symbol $^2B_1$), so that it is paramagnetic. Its rotational constants have been determined from spectroscopy to be $\overline{B}_A = 1.737\,cm^{-1}$, $\overline{B}_B = 0.332\,cm^{-1}$, and $\overline{B}_C = 0.278\,cm^{-1}$, while the fundamental vibrational oscillator frequencies are $\overline{\nu}_1 = 945.5\,cm^{-1}$, $\overline{\nu}_2 = 447.4\,cm^{-1}$, and $\overline{\nu}_3 = 1110.5\,cm^{-1}$. Remember that standard conditions (NTP) are $P = 1$ bar, $T = 273.15$ K, and $\overline{V}^\circ = 22.711\,dm^3\,mol^{-1}$.

28. Linear triatomic molecules have four independent vibrational degrees of freedom, each characterized by a natural vibrational frequency ν_i ($i = 1, 2, 3, 4$) in Hz [equivalently wave number $\overline{\nu}_i$ in the non-SI unit cm^{-1}] and, in the SHO approximation, a corresponding energy $\epsilon_{vib,i} \equiv (\nu_i + \frac{1}{2})h\nu_i$ [or $\overline{\epsilon}_i = (\nu_i + \frac{1}{2})hc\overline{\nu}_i$], with h the Planck constant and ν_i the vibrational quantum number for vibrational degree of freedom i. The corresponding SHO vibrational partition function for a triatomic linear molecule is hence given by

$$z_{vib}(T) = \prod_{i=1}^{4} \frac{e^{-\Theta_{vib,i}/(2T)}}{1 - e^{\Theta_{vib,i}/T}},$$

in which the characteristic vibrational temperature $\Theta_{vib,i}$ is defined as $\Theta_{vib,i} \equiv \epsilon_{vib,i}/k_B$. Express $z_{vib}(T)$ in terms of hyperbolic functions, and obtain an expression for the temperature derivative of z_{vib}. [Hint: use the technique of logarithmic differentiation.]

29. When the temperature is either very high or very low (by which we mean that k_BT is either much larger than or much smaller than the fundamental oscillator energies or, equivalently, that the temperature T is either much larger or smaller than the characteristic temperatures $\Theta_{vib,}$), then we can introduce approximations for the functions appearing in the expression for $z_{vib}(T)$ given in Problem 28, and thereby obtain appropriate approximate expressions for the vibrational partition function itself. Obtain such expressions for the individual $z_{vib,i}(T)$ valid for these two cases.

30. As it is actually the natural logarithm of the partition function, $\ln z_{\text{vib}}(T)$, for the vibrational degrees of freedom that enters into expressions giving the vibrational contributions to the various thermodynamic state functions, obtain explicit forms for the vibrational contributions, $U_{\text{vib}}(T)$, $A_{\text{vib}}(T)$, and $C_{V,\text{vib}}(T)$ to $U(T, V, N)$, $A(T, V, N)$, and $C_V(T)$ for a gas of N linear triatomic molecules. Show, in particular, that $A_{\text{vib}}(T)$ can be written in the form

$$A_{\text{vib}}(T) = N k_{\text{B}} T \sum_{i=1}^{4} \left[\frac{\hbar \omega_i}{2 k_{\text{B}} T} + \ln(1 - \text{e}^{-\hbar \omega_i / k_{\text{B}} T}) \right].$$

31. Calculate values for $\overline{S}$, $\overline{U}$, $\overline{C}_V$, and $\overline{H}$ for HCN vapour at 1 bar pressure and a temperature of 700 K, given that the spectroscopic values of $\overline{B}$, $\overline{v}_{i,\text{osc}}$ for HCN are $1.478 \, \text{cm}^{-1}$, $713.5 \, \text{cm}^{-1}$, $2096.7 \, \text{cm}^{-1}$, and $3311.5 \, \text{cm}^{-1}$, respectively, while $\mathfrak{D}_\text{e} = 5.65 \, \text{eV}$ (for dissociation into H + CN). Do not forget to identify both the symmetry type (e.g., symmetric stretch, bend, etc.) and number (i.e., degeneracy) of the vibrational motions involved.

32. The nitrous oxide molecule, N_2O, has fundamental vibrational frequencies (in wavenumbers): $\overline{v}_1 = 1284.9 \, \text{cm}^{-1}$, $\overline{v}_2 = 588.8 \, \text{cm}^{-1}$ (doubly degenerate), and $\overline{v}_3 = 2223.8 \, \text{cm}^{-1}$. Calculate the vibrational contributions to A, U, and $C_V / N k_{\text{B}}$ for one mole of N_2O at temperatures of 300 and 1200 K.

33. Combustion models often require the thermodynamic properties of oxygen at high temperatures (1000–5000 K) as input. The lowest *excited* electronic state of the O_2 molecule is a $^1\Delta_g$ state, lying $7918.2 \, \text{cm}^{-1}$ above the ground electronic, $X^3\Sigma_g^-$, state. The $^1\Delta_g$ state has an electronic degeneracy of 2 (associated with what is termed 'lambda-doubling' in molecular spectroscopy). Treat the $X^3\Sigma_g^-$ state as triply degenerate (although this is not strictly true, it is a very good approximation to do so for the present purposes).
By assuming (as usual) that the electronic motion is decoupled from the other internal motions of the O_2 molecule, $z_{\text{el}}(T)$ may be written as

$$z_{\text{el}}(T) = \omega_{\text{eg}} + \omega_{\text{e2}} \text{e}^{-\beta \epsilon_{\text{e2}}},$$

in which ϵ_{e2} and ω_{e2} are the energy and degeneracy, respectively, of the $^1\Delta_g$ state of O_2, while ω_{eg} is the degeneracy of the $X^3\Sigma_g^-$ state. Obtain an expression for the contribution $U_{\text{el}}(T)$ arising from the $^1\Delta_g$ state to U for O_2, and from it obtain an expression for $C_{V\text{el}}(T)$ for O_2.
Obtain expressions for the limiting low and high-temperature behaviours of $C_{V\text{el}}(T)$ for O_2. Obtain an equation governing the temperature, T_{max}, at which $C_{V\text{el}}(T)$ attains its maximum value. Solve this equation for the value of T_{max} at which the contribution of the $^1\Delta_g$ state of O_2 is maximal. [Note that the equation that you obtain will not be an algebraic equation: it belongs to a class of equations referred to as transcendental equations. You may find that the simplest way to solve this equation is by iteration.]

34. Measurements of $C_P(T)$ for O_2 have been reported by Lewis and von Elbe [50]. A selection of their values is: $C_P(200) = 6.951$, $C_P(500) = 7.434$, $C_P(800) = 8.072$, $C_P(1250) = 8.621$, $C_P(1500) = 8.840$, $C_P(2000) = 9.208$, $C_P(2500) = 9.500$ and $C_P(3000) = 9.725$; temperatures in K, heat capacities in $\text{cal}\,\text{mol}^{-1}\text{K}^{-1}$ [based upon $R = 1.98714\,\text{cal}\,\text{mol}^{-1}\text{K}^{-1}$].

From your knowledge of $C_P(T)$ for an ideal gas, utilize the expression obtained in Problem 33 to calculate values of C_P at the experimental temperatures of Lewis and von Elbe, and compare your results with their experimental values. Comment on the quality of agreement that you obtain, and discuss the significance of the role of the $^1\Delta_g$ excited state of O_2 at the temperatures concerned, bearing in mind the results that you obtained in the latter part of Problem 33.

A slightly more recent (than 1933), and relatively complete calculation of the thermodynamic functions for molecular oxygen in the ideal gas state was carried out by Woolley [51]. Compare your values calculated using the rigid-rotor-SHO approximation with the results obtained by Woolley (which go beyond the SHO approximation) and displayed in his Tables 5 and 6. Comment on the appropriateness of the rigid-rotor-SHO approximation that you have employed for your calculations.

35. Just as the possession by the proton (^{1}H) of a nuclear spin of magnitude $\frac{1}{2}$ has significant ramifications for the spectroscopic and thermodynamic properties of H_2, so it has similar consequences for the polyatomic centrosymmetric linear molecule acetylene, $^{12}C_2H_2$. [Note that the ^{12}C nucleus has nuclear spin $I = 0$.]

Given that the electronic ground term of C_2H_2 is $^1\Sigma_g^+$, obtain an expression that is applicable at low temperature for the combined rotational-nuclear spin canonical partition function for the *ortho* and *para* spin modifications of $^{12}C_2H_2$. Discuss the effect of the nuclear spin statistics on the Raman spectrum of $^{12}C_2H_2$. How would the Raman spectrum of $^{12}C_2H_2$ differ from those of its isotopologues $^{12}C_2D_2$ and $^{12}C_2HD$? Determine a value for the residual entropy for each acetylene isotopologue.

36. The near infrared spectrum of $^{12}C_2H_2$ has five fundamental absorption bands, located at 611.7, 729.2, 1973.5, 3294.9, and 3372.5 cm^{-1}. These bands have been identified as having symmetries π_g, π_u, σ_g^+, σ_u^+ and σ_g^+, respectively. Note that vibrations of type $\pi_{u/g}$ are doubly degenerate (cf. individual π MO's), while vibrations of type $\sigma_{u/g}^{\pm}$ are nondegenerate. The rotational constant $\overline{B}_0$ of $^{12}C_2H_2$ has the value 1.1766 cm^{-1}. Calculate $\overline{U}_{298.15}^{\circ} - \overline{U}_0^{\circ}$, $\overline{H}_{298.15}^{\circ} - \overline{H}_0^{\circ}$, $\overline{C}_{V,298.15}$, and $\overline{S}_{298.15}^{\circ}$ for C_2H_2, and compare your values with the accepted literature values [given in the JANAF tables of the American National Institute of Standards and Technology (NIST)].

37. Consider the class of general spherical top molecules, such as silane, SiH_4, with tetrahedral (T_d) symmetry, sulphur hexafluoride, SF_6, with octahedral (O_h) symmetry or buckminsterfullerene, C_{60}, with icosahedral (I_h) symmetry. Obtain an expression for the fractional population p_j for a spherical top

molecule in a rotational level characterized by the rotational quantum number j, and from it derive an expression for the value j_{max} at which the fractional population is maximal. What is the meaning of this expression? Obtain values of j_{max} at temperature $500\,K$ for the CH_4 ($\Theta_{rot} = 7.54\,K$), CF_4 ($\Theta_{rot} = 0.276\,K$), and SF_6 ($\Theta_{rot} = 0.131\,K$) molecules.

38. From spectral data acquired from the atmosphere of Titan, the largest moon of Saturn, an astronomer observes a series of resolved rotational fine-structure lines in an infrared absorption band located at approximately $3010\,cm^{-1}$ that she has identified as the ν_3-fundamental band of CH_4. She notes that, within the accuracy of her astronomical spectral data, the rotational fine-structure transitions show equal maximal intensity for transitions originating from $j = 3$ and $j = 4$. Use this observation to bracket the temperature in the atmosphere of Titan. Are your values consistent with what is known about Titan and its atmosphere?

39. The dimethyl acetylene molecule possesses, in addition to the three end-over-end rotational degrees of freedom, an additional free relative internal rotation of the two end methyl groups. Given that the relevant moment-of-inertia for the relative internal rotation of the two methyl groups is $2.566 \times 10^{-47}\,kg\,m^2$, determine its contribution to the entropy of dimethyl acetylene at $300\,K$, if the symmetry number associated with this internal rotational motion is $\sigma = 3$. As far as the overall (end-over-end) rotational motion of the molecule is concerned, dimethyl acetylene is a symmetric top molecule, with one of its three moments-of-inertia having the value $10.657 \times 10^{-47}\,kg\,m^2$, and the other two each having the value $249.66 \times 10^{-47}\,kg\,m^2$. Construct a sketch showing the axes about which these three rotations occur. What is the entropy contribution due to the overall rotational motion, if the associated symmetry number is $\sigma = 6$?

40. The formaldehyde molecule has a $^1A_1$ ground electronic state. Its infrared spectrum gives the vibrational fundamental frequencies $\bar{\nu}_1(a_1) = 2766\,cm^{-1}$, $\bar{\nu}_2(a_1) = 1746\,cm^{-1}$, $\bar{\nu}_3(a_1) = 1510\,cm^{-1}$, $\bar{\nu}_4(b_1) = 2843\,cm^{-1}$, $\bar{\nu}_5(b_1) = 1251\,cm^{-1}$, and $\bar{\nu}_6(b_2) = 1167\,cm^{-1}$, while the principal moments-of-inertia can be extracted from the far infrared spectrum as $I_{A0} = 2.976 \times 10^{-47}$, $I_{B0} = 21.61 \times 10^{-47}$ and $I_{C0} = 24.68 \times 10^{-47}$, all in units of $kg\,m^2$. The notation a_1, b_1, b_2 accompanying the vibrational fundamental frequencies identify the point-group symmetry species of the corresponding vibrational motion (they are all nondegenerate). Calculate $\overline{U}^\circ_{298.15} - \overline{U}^\circ_0$, $\overline{H}^\circ_{298.15} - \overline{H}^\circ_0$, $\overline{C}_{V,298.15}$, and $\overline{S}^\circ_{298.15}$ for formaldehyde, and compare your values with the accepted literature values (found in the NIST JANAF tables).

41. Give a quantitative explanation of the statement 'At a temperature of $300\,K$, the heat capacity $\overline{C}_V$ of IBr has nearly its classical value, while that for F_2 does not'. The rotational Raman spectrum for F_2 corresponds to a rotational constant $\overline{B} = 0.8833\,cm^{-1}$ and $\bar{\nu}_{osc} = 894.2\,cm^{-1}$, while the infrared spectrum of IBr gives values $\overline{B} = 0.05673\,cm^{-1}$ and $\bar{\nu}_{osc} = 267.01\,cm^{-1}$. By how much do the heat capacities of these two molecular gases differ from the values that would be expected classically?

42. Carry out the steps to obtain Eqs. (6.1.4) from Eq. (6.1.10a).

43. If the rotational and vibrational motions are assumed to be independent for ground electronic state Cl_2 and HCl molecules, so that $\epsilon_{int} = \epsilon_{vib} + \epsilon_{rot}$, how many rotational energy levels of the ground vibrational state can be found in the energy gap between the ground vibrational energy ϵ_0 and the energy of the first vibrational state, i.e., in the interval $(\Delta\epsilon_{vib})_{0\rightarrow 1}$? Assume that the vibrational energy gap for Cl_2 is $554.4\,cm^{-1}$ and that for HCl is $2885.2\,cm^{-1}$. Assume also that both Cl_2 and HCl may be treated as rigid rotors, with $B_e(Cl_2) = 0.240\,cm^{-1}$ and $B_e(HCl) = 10.593\,cm^{-1}$.

44. In Example 6.3, it was mentioned that the molecular oxygen isotopologue $^{17}O_2(X\,^3\Sigma_g^-)$, unlike either $^{16}O_2(X\,^3\Sigma_g^-)$ or $^{18}O_2(X\,^3\Sigma_g^-)$, which have only odd rotational states, has both even and odd rotational states. Explain why this is so. What effect will this have on the pure rotational Raman spectrum of $^{17}O_2(X\,^3\Sigma_g^-)$? How would you expect the pure rotational Raman spectrum of $^{17}O_2(X\,^3\Sigma_g^-)$ to differ from the pure rotational Raman spectra of the homonuclear ^{16}O and ^{18}O diatomic isotopologues or from that of $^{16}O^{18}O(X\,^3\Sigma_g^-)$?

References

1. G. Herzberg, *Molecular Spectra and Molecular Structure. I. Diatomic Molecules*, 2nd edn. (Van Nostrand, New York, 1950), p. 92
2. J.E. Mayer, M. Goeppert-Mayer, *Statistical Mechanics* (Wilcy, New York, 1940)
3. K.P. Huber, G. Herzberg, *Molecular Spectra and Molecular Structure. IV. Constants of Diatomic Molecules* (Van Nostrand, New York, 1979). See also NIST Chemistry WebBook, https://webbook.nist.gov/
4. L.S. Kassel, Chem. Rev. **18**, 277 (1936)
5. D.A. McQuarrie, *Mathematical Methods for Scientists and Engineers* (University Science Books, Sausalito, 2003), pp. 107–112
6. Mulholland was the first to obtain this result, although via a quite different mathematical strategem: see H.P. Mulholland, Proc. Camb. Phil. Soc. **24**, 280 (1928)
7. R.S. McDowell, J. Chem. Phys. **88**, 356 (1988)
8. F.M. Fernández, R.H. Tipping, Spectrochim. Acta **48A**, 1283 (1992)
9. R.S. McDowell, J. Chem. Phys. **39**, 526 (1963)
10. W. Pauli, Phys. Rev. **58**, 716 (1940). The Pauli Exclusion Principle that 'no two electrons can have all four quantum numbers n, ℓ, m_ℓ, m_s the same' was enunciated by Pauli in 1925 [W. Pauli, Z. f. Physik **31**, 765 (1925)], while the more general (symmetry) version was not introduced until 1940
11. A. Eucken, Sitzber. Preuss. Akad. Wiss., **1912**, 41
12. A. Eucken, R. Hiller, Z. f. Phys. Chem. B, **4**, 142 (1929)
13. K. Scheel, W. Heuse, Ann d. Phys. **40**, 473 (1913)
14. I.H. Brinkworth, Proc. R. Soc. A **107**, 510 (1925)
15. R.E. Cornish, E.D. Eastman, J. Am. Chem. Soc. **50**, 627 (1928)
16. K. Klusius, K. Hiller, Z. f. Phys. Chem. B **4**, 158 (1929)
17. D.M. Dennison, Proc. R. Soc. A **115**, 483 (1927)
18. K.F. Bonhöffer, P. Harteck, Z. f. Phys. Chem. B **4**, 113 (1929)
19. J.E. Kilpatrick, Y. Fukuda, S.Y. Larsen, J. Chem. Phys. **43**, 430 (1965)
20. A.R. Gordon, C. Barnes, J. Chem. Phys. **1**, 297 (1933)

21. M.W. Chase, Jr., *NIST-JANAF Thermochemical Tables*, 4th edn. J. Phys. Chem. Ref. Data, Monograph **9** (1998), or http://webbook.nist.gov.chemistry
22. G. Herzberg, *Molecular Spectra and Molecular Structure. III. Electronic Spectra and Electronic Structure of Polyatomic Molecules* (Van Nostrand, New York, 1966)
23. I. Ozier, S.S. Lee, N.F. Ramsey, J. Chem. Phys. **65**, 3985 (1976)
24. K. Fox, J. Quant. Spectrosc. Radiat. Transfer **10**, 1335 (1970)
25. R.S. McDowell, J. Quant. Spectrosc. Radiat. Transfer **38** 337 (1987)
26. A.G. Robiette, M. Dang-Nhu, J. Quant. Spectrosc. Radiat. Transf. **22**, 499 (1979)
27. R.S. McDowell, J. Chem. Phys. **93**, 2801 (1990)
28. J.K.G. Watson, in *Vibrational Spectra and Structure*, vol. 6, ed. by, J.R. Durig (Elsevier, Amsterdam, 1977)
29. L.S. Kassel, J. Chem. Phys. **1**, 576 (1933)
30. J.K.G. Watson, Mol. Phys. **65**, 1377 (1988)
31. R.J. Le Roy, J. Mol. Spectrosc. **192**, 237 (1998)
32. J. Bendtsen, J. Raman Spectrosc. **32**, 989 (2001)
33. J. Bendtsen, F. Rasmussen, J. Raman Spectrosc. **31**, 433 (2000)
34. H.G.M. Edwards, D.A. Long, H.R. Mansour, J. Chem. Soc., Faraday Trans. II **74**, 1203 (1978)
35. H.G.M. Edwards, E.A.M. Good, D.A. Long, J. Chem. Soc. Faraday Trans. II **72**, 927 (1976)
36. H.G.M. Edwards, E.A.M. Good, D.A. Long, J. Chem. Soc. Faraday Trans. II **72**, 865 (1976)
37. J.W. Fleming, J. Cole, Infrared Phys. **14**, 277–292 (1974)
38. W. Demtröder, *Laser Spectroscopy. Basic Concepts and Instrumentation*, 3rd edn. (Springer, Berlin, 2003)
39. H.G.M. Edwards, High-resolution raman spectroscopy of gases, in *Chapter 6 of Essays on Structural Chemistry*, ed. by A.J. Downs, D.A. Long, L.A.K. Staveley (Macmillan, London, 1971)
40. W.F. Giauque, G.W. Stout J. Am. Chem. Soc. **58**, 1144 (1936)
41. G.W. Giauque, M.F. Ashley, Phys. Rev. **43**, 81 (1933)
42. L. Pauling, J. Am. Chem. Soc. **57**, 2680 (1957)
43. K.S. Pitzer, *Quantum Chemistry* (Prentice-Hall, Englewood Cliffs, 1960)
44. G. Ercolani, (a) J. Chem. Ed. **77**, 1495 (2000); (b) J. Chem. Ed. **82**, 1703 (2005)
45. K.S. Pitzer, W.D. Gwinn, J. Chem. Phys. **10**, 428 (1942)
46. J.C.M. Li, K.S. Pitzer, J. Chem. Phys. **60**, 466 (1956)
47. K.S. Pitzer, *Thermodynamics*, 3rd edn. (McGraw-Hill, New York, 1995)
48. B.J. McClelland, *Statistical Thermodynamics* (Chapman-Hall, London, 1973)
49. A.G. Robiette, M. Dang-Nhu, J. Quant. Spectrosc. Rad. Trans. **22**, 499 (1974)
50. B. Lewis and G. von Elbe, J. Am. Chem. Soc. **55**, 511 (1933)
51. H.W. Woolley, J. Res. Nat. Bur. Stand. **40**, 163 (1948)

Chapter 7
Classical Statistical Mechanics

This chapter focusses mainly upon the classical mechanical evaluation of the partition function for a gas of structureless atoms, and begins with a discussion of energy equipartition. The role of interatomic interactions is examined using the grand partition function, as it enables a more convenient separation of the roles played by kinetic and potential energy terms. Expressions are developed for non-ideal contributions to the thermodynamic pressure, entropy, and the internal, Helmholtz, and Gibbs energies of a gas in terms of virial coefficients and their first temperature derivatives. A discussion of distribution functions leads to the classical Liouville and Boltzmann equations, while a consideration of binary collision dynamics enables development of the form for the monatomic Boltzmann collision term. An introduction to nonequilibrium phenomena is provided by the derivation of the traditional transport equations for an atomic gas.

7.1 Introduction

In this chapter we shall consider the classical statistical mechanics for a monatomic gas. We have seen in Chap. 5 that the canonical partition function for a monatomic gas made up of N atoms is given by

$$Z(N, T, V) = \frac{z^N(T, V)}{N!},$$

in which $z(V, T)$ is the canonical partition for a single atom, given as

$$z(T, V) = \left(\frac{2\pi m k_B T}{h^2}\right)^{\frac{3}{2}} V. \tag{7.1.1}$$

© Springer Nature Switzerland AG 2021
F. R. W. McCourt, *Statistical Thermodynamics for Pure and Applied Sciences*,
https://doi.org/10.1007/978-3-030-52006-9_7

As we have seen, $z(T, V)$ is defined quantum mechanically via $z \equiv \sum_r e^{-\beta \epsilon_r}$, with the summation over r a summation over all quantum states r associated with an atom, and hence we might logically expect that the classical expression for z should have a similar form, i.e., $z(T, V) \propto \int \cdots \int e^{-\beta \mathcal{H}(p,q)} \, dp dq$, in which p, q collectively denote all the momenta and coordinates upon which the Hamiltonian (energy) depends. We could write $dp \equiv dp_1 \cdots dp_s$ and $dq \equiv dq_1 \cdots dq_s$, with s giving the number of degrees of freedom of the molecule. For example, a closed-shell atom in its ground electronic state will have $s = 3$ (corresponding to $q_1 = x$, $q_2 = y$, $q_3 = z$; $p_1 = p_x$, $p_2 = p_y$, $p_3 = p_z$). This specific example uses Cartesian coordinates and conjugate momenta: more generally, the momenta and coordinates do not need to be Cartesian. See Appendix G for additional background information on classical mechanics.

The classical Hamiltonian for a single atom of mass m is just its kinetic energy, namely

$$\mathcal{H}_{\text{trans}}(\mathbf{r}, \mathbf{p}) = \frac{1}{2m}\mathbf{p}^2 \, .$$

If we now follow our conjecture in the paragraph above that

$$z(T, V) \propto \iint e^{-\beta \mathcal{H}(\mathbf{r}, \mathbf{p})} \, d\mathbf{p} d\mathbf{r} \, ,$$

we see that

$$z(T, V) \equiv z_{\text{trans}}(T, V) \propto \iint e^{-\beta \mathbf{p}^2/(2m)} \, d\mathbf{p} d\mathbf{r}$$

$$= C_{\text{trans}} V \int_0^\infty \int_0^\pi \int_0^{2\pi} e^{-\beta p^2/(2m)} p^2 \sin \vartheta \, d\varphi d\vartheta dp$$

$$= C_{\text{trans}} V 4\pi \int_0^\infty e^{-\beta p^2/(2m)} p^2 \, dp \, ,$$

with C_{trans} a constant. Upon evaluation of the definite integral, we obtain

$$z_{\text{trans}}(T, V) = C_{\text{trans}} V \left(\frac{2\pi m}{\beta} \right)^{\frac{3}{2}} = C_{\text{trans}} (2\pi m k_B T)^{\frac{3}{2}} V \, . \qquad (7.1.2)$$

Comparison between the classical result (7.1.2) and Eq. (7.1.1) indicates that C_{trans} must equal h^{-3} in order for the two results to be consistent.

In the classical phase-space description of the motion of a (structureless) particle, each point $(\mathbf{r}, \mathbf{p})$ in the six-dimensional phase space represents a possible state of motion for the particle. This leads to the statement that the number of states available to the particle in a specific region of its phase space must be proportional to the (six-dimensional) volume of that region. Thus, for a structureless classical particle having position $\mathbf{r} - \frac{1}{2}d\mathbf{r} \leq \mathbf{r} \leq \mathbf{r} + \frac{1}{2}d\mathbf{r}$ and possessing momentum $\mathbf{p} - \frac{1}{2}d\mathbf{p} \leq$

$\mathbf{p} \leq \mathbf{p} + \frac{1}{2}\mathbf{dp}$, the number of accessible phase-space points will be proportional to the volume of that region of phase space in which the particle is to be found, i.e., to drdp. The argument presented above from the comparison between Eqs. (7.1.1) and (7.1.2) identifies the proportionality constant as h^{-3}.

Moreover, if we carry out classical-type evaluations of the partition functions for a rigid rotor and a simple harmonic oscillator (SHO), we obtain similar results. For a rigid rotor, we may employ generalized coordinates ϑ, φ, and corresponding generalized momenta p_ϑ, p_φ. The classical Hamiltonian for rigid-rotor motion is then given by

$$\mathcal{H}_{\mathrm{rot}} = \frac{1}{2I}\left(p_\vartheta^2 + \frac{p_\varphi^2}{\sin^2\vartheta}\right),$$

from which $z_{\mathrm{rot}}(T)$ is obtained as

$$z_{\mathrm{rot}}(T) = C_{\mathrm{rot}}(8\pi^2 I k_{\mathrm{B}} T),$$

with C_{rot} a constant. Comparison with Eq. (6.2.61b) indicates that $C_{\mathrm{rot}} = h^{-2}$. Similarly, for a SHO, the classical Hamiltonian has both kinetic and potential components, and is given by

$$\mathcal{H}_{\mathrm{vib}} = \frac{p^2}{2\mu} + \frac{1}{2}kx^2.$$

The corresponding classical vibrational partition function is thus

$$z_{\mathrm{vib}}(T) = C_{\mathrm{vib}} \int_{-\infty}^{\infty} \mathrm{d}p \int_{-\infty}^{\infty} \mathrm{d}x\, \mathrm{e}^{-\beta\mathcal{H}} = C_{\mathrm{vib}} \frac{k_{\mathrm{B}} T}{\nu_{\mathrm{osc}}},$$

so that comparison with $z_{\mathrm{vib}}(T) \simeq T/\Theta_{\mathrm{vib}}$ for $T \gg 1$ shows that $C_{\mathrm{vib}} = h^{-1}$.

From these three single-atom/single-molecule comparisons, we may deduce that there will be a factor h^{-1} occurring in the single-molecule partition function for each conjugate coordinate-momentum pair of variables that enter the corresponding expression for the Hamiltonian function. Thus, based upon the preceding observations, let us assume that

$$z = \sum_r \mathrm{e}^{-\beta\epsilon_r} \quad \Rightarrow \quad \frac{1}{h^s} \int \cdots \int \mathrm{e}^{-\beta\mathcal{H}} \prod_{i=1}^{s} \mathrm{d}p_i \mathrm{d}q_i ,$$

a result that we should obtain for an individual molecule (in the RR-SHO approximation). For N independent indistinguishable molecules, we can thus write

$$Z_N(V, T) = \frac{z^N(V, T)}{N!} = \frac{1}{N!} \prod_{j=1}^{N} \left\{ \frac{1}{h^s} \int \cdots \int \mathrm{e}^{-\beta\mathcal{H}_j} \prod_{i=1}^{s} \mathrm{d}p_{ji} \mathrm{d}q_{ji} \right\},$$

in which the Hamiltonian $\mathcal{H}_j$ is defined via $\mathcal{H}_j \equiv \mathcal{H}(p_{j1}, \cdots, p_{js}; q_{j1}, \cdots, q_{js})$.

By relabelling the conjugate momenta and coordinates for the molecules so that the variables p_{nk} and q_{nk} for molecule n (for $k = 1, \cdots, s$) are mapped onto the variables p_i and q_i as p_{ns+k}, q_{ns+k}, we may write $Z_N(V, T)$ as

$$Z_N(T, V) = \frac{1}{N! h^{sN}} \int \cdots \int e^{-\beta \sum_j \mathcal{H}_j} \prod_{i=1}^{sN} dp_i dq_i$$

$$= \frac{1}{N! h^{sN}} \int \cdots \int e^{-\beta \mathcal{H}} \prod_{i=1}^{sN} dp_i dq_i \,,$$

in which $\mathcal{H} \equiv \sum_j \mathcal{H}_j$ is the Hamiltonian for the N-molecule system.

7.1.1 Energy Equipartition in Classical Mechanics

For a structureless classical ideal gas of N point particles, each of mass m, the Hamiltonian $\mathcal{H}$ is typically given as

$$\mathcal{H}\left(\mathbf{r}^N, \mathbf{p}^N\right) = \frac{1}{2m} \sum_{i=1}^{N} \mathbf{p}_i^2 \,, \tag{7.1.3}$$

in which $\mathbf{r}^N = (\mathbf{r}_1, \cdots \mathbf{r}_N)$ and $\mathbf{p}^N = (\mathbf{p}_1, \cdots \mathbf{p}_N)$ are vectors in the $6N$-dimensional classical phase space. The absence of interparticle potential energy terms in expression (7.1.3) is, of course, consistent with the concept of an ideal gas: the point particles are assumed to undergo only elastic collisions with one another via delta-function repulsive interactions. We may note, however, that $\mathcal{H}$ will depend upon $\mathbf{r}^N$ should an *external* potential energy function $V_{\text{ext}}(\mathbf{r}^N)$ be imposed upon the gas.

For the present discussion, we shall focus only upon a structureless classical ideal gas to which no such external field has been applied. When such a gas in a container of volume V is placed in thermal contact with its surroundings and allowed to equilibrate via inelastic collisions with its energy-permeable (but mass-impermeable) walls, we may represent it in terms of a canonical ensemble of N particles that is characterized by volume V and temperature T.

The probability $p(\mathbf{r}^N, \mathbf{p}^N)$ that an N-particle system in thermal equilibrium at a temperature T be in a state in which its N particles have coordinates $\{\mathbf{r}_i\}$ and momenta $\{\mathbf{p}_i\}$, $i = 1, \cdots N$, lying within respective volume elements $\{d\mathbf{r}_i\}$ and $\{d\mathbf{p}_i\}$ is

$$p\left(\mathbf{r}^N, \mathbf{p}^N\right) \prod_{i=1}^{N} d\mathbf{r}_i d\mathbf{p}_i = C e^{-\beta \mathcal{H}(\mathbf{r}^N, \mathbf{p}^N)} \prod_{i=1}^{N} d\mathbf{r}_i d\mathbf{p}_i \tag{7.1.4a}$$

in which the normalization constant C is given by

$$C \equiv \left[\int \cdots \int e^{-\beta \mathcal{H}(\mathbf{r}^N, \mathbf{p}^N)} \prod_{i=1}^{N} d\mathbf{r}_i d\mathbf{p}_i \right]^{-1}, \qquad (7.1.4b)$$

and $\prod_{i=1}^{N} d\mathbf{r}_i d\mathbf{p}_i = d\mathbf{r}_1 \cdots d\mathbf{r}_N d\mathbf{p}_1 \cdots d\mathbf{p}_N$. For a structureless ideal classical gas, the Hamiltonian is simply the sum of quadratic terms in the momentum, typified by the translational energy for the ith particle, namely $a\mathbf{p}_i^2 = a(p_x^2 + p_y^2 + p_z^2)$, as $\mathbf{p}_i^2$ collectively represents three independent degrees of translational freedom. We have introduced $a \equiv 1/(2m)$ for convenience in the following calculation and to emphasize that the proportionality constant for the quadratic dependence ultimately plays no role in the final result.

In the evaluation of the contribution of the ith particle to the translational internal energy $U_{\text{tr}}(T)$ for a structureless ideal gas, integration over all position and momentum coordinates for particles $k \neq i$ contribute factors 1, as does the integration over the position coordinates $\mathbf{r}_i$. The net result is that the remaining probability factor $p_i(\mathbf{p}_i)\, d\mathbf{p}_i$ for the final integration over $\mathbf{p}_i$ can be expressed as

$$p(\mathbf{p}_i)\, d\mathbf{p}_i = \frac{e^{-\beta a(p_x^2 + p_y^2 + p_z^2)} dp_x dp_y dp_z}{\iiint e^{-\beta a(p_x^2 + p_y^2 + p_z^2)} dp_x dp_y dp_z}$$

$$= \frac{e^{-\beta a p_x^2}}{\int e^{-\beta a p_x^2} dp_x} \frac{e^{-\beta a p_y^2}}{\int e^{-\beta a p_y^2} dp_y} \frac{e^{-\beta a p_z^2}}{\int e^{-\beta a p_z^2} dp_z}. \qquad (7.1.5)$$

Note that we have suppressed the subscript i on the right-hand side in order to avoid an unduly complicated notation. It will also be clear that in calculating the contributions to the internal energy from each of the three Cartesian components of the translational energy, the argument employed above may also be applied to the probability (7.1.5), in which case we need simply consider the average over p_x as typical. Our representative calculation is thus reduced to evaluation of the average over the x-component of $\mathbf{p}_i$ using the probability distribution

$$p(p_x) dp_x = \frac{e^{-\beta a p_x^2} dp_x}{\int e^{-\beta a p_x^2} dp_x}, \qquad (7.1.6)$$

in which the integration in the denominator is over all values that may be taken by p_x. This integration may readily be extended from $+\infty$ to $-\infty$, as the integrand is sharply peaked about $p_x = 0$. Thus, if we employ the probability distribution Eq. (7.1.6) to compute the average of the energy ap_x^2, we obtain $\langle ap_x^2 \rangle$ as

$$\langle ap_x^2 \rangle = \frac{\int ap_x^2 e^{-\beta a p_x^2} dp_x}{\int e^{-\beta a p_x^2} dp_x}.$$

Note that we may also utilize parameter differentiation to express $\langle ap_x^2 \rangle$ in the form

$$\langle ap_x^2 \rangle = -\frac{\partial}{\partial \beta} \ln \left(\int e^{-\beta ap_x^2} \, dp_x \right). \tag{7.1.7}$$

A change of variable in Eq. (7.1.7) from p_x to $\xi \equiv \sqrt{\beta} p_x$ then gives $\langle ap_x^2 \rangle$ as

$$\langle ap_x^2 \rangle = -\frac{\partial}{\partial \beta} \ln \left[\frac{1}{\sqrt{\beta}} \int e^{-a\xi^2} d\xi \right]$$

$$= \frac{1}{2\beta},$$

or

$$\langle ap_x^2 \rangle = \tfrac{1}{2} k_B T. \tag{7.1.8}$$

As an analogous result holds for each quadratic contribution to the classical Hamiltonian $\mathcal{H}$, the $\mathbf{p}_i^2/(2m)$ term associated with particle i thus contributes $\tfrac{3}{2}k_B T$ to the per particle translational internal energy $u_{tr}(T)$. It follows, therefore, that the translational internal energy for N ideal gas particles is $U_{tr}(T) = \tfrac{3}{2}Nk_B T$. It is important to note that the contribution (7.1.8) to the internal energy associated with the term ap_x^2 does not depend upon the coefficient a. Thus, even had a depended upon the position or momentum coordinates of particles other than the ith particle (or, for that matter upon the position coordinates of particle i itself), the result (7.1.8) would have been the same. Note that expression (7.1.8) applies specifically to additive contributions to the energy of a particle that depend quadratically upon a generalized coordinate or momentum variable whose domain is unbounded.

The classical equipartition argument was shown by Tolman [1] to hold also for a class of non-quadratic terms in the Hamiltonian. His argument was more or less as follows. If, in the normalization condition

$$\int \cdots \int p\left(\mathbf{r}^N, \mathbf{p}^N\right) \prod_{i=1}^N d\mathbf{r}_i dp_i = \int \cdots \int C e^{-\beta \mathcal{H}(\mathbf{r}^N, \mathbf{p}^N)} \prod_{i=1}^N d\mathbf{r}_i dp_i = 1$$

for the probability $p(\mathbf{r}^N, \mathbf{p}^N)$ given in Eqs. (7.1.4a), an integration by parts over an *arbitrary* generalized coordinate or momentum variable (designated by p_1 for convenience of notation) is carried out, we obtain the expression

$$\iint C \left\{ \left[p_1 e^{-\beta \mathcal{H}} \right]_{p_1=\ell}^{p_1=u} + \int e^{-\beta \mathcal{H}} \beta p_1 \frac{\partial \mathcal{H}}{\partial p_1} \, dp_1 \right\} d\boldsymbol{\pi} d\boldsymbol{\rho} = 1, \tag{7.1.9}$$

with $\int \cdots d\boldsymbol{\pi}$ and $\int \cdots d\boldsymbol{\rho}$ designating integrations over all generalized momenta $\boldsymbol{\pi}$ and coordinates $\boldsymbol{\rho}$ other than the specific variable labelled p_1. For integration over

p_1 in an interval $[\ell, u]$ such that the first term in expression (7.1.9) vanishes, we see that

$$\iint Ce^{-\beta\mathcal{H}} \beta p_1 \frac{\partial \mathcal{H}}{\partial p_1} \prod_{i=1}^{N} d\mathbf{r}_i dp_i \equiv \beta \left\langle p_1 \frac{\partial \mathcal{H}}{\partial p_1} \right\rangle = 1 ,$$

in which $\langle \cdots \rangle$ is defined as

$$\langle \cdots \rangle = \frac{\displaystyle\iint \cdots e^{-\beta\mathcal{H}} \prod_{i=1}^{N} d\mathbf{r}_i dp_i}{\displaystyle\iint e^{-\beta\mathcal{H}} \prod_{i=1}^{N} d\mathbf{r}_i dp_i} . \qquad (7.1.10)$$

We shall re-express this result as

$$\left\langle p_1 \frac{\partial \mathcal{H}}{\partial p_1} \right\rangle = k_B T , \qquad (7.1.11a)$$

which Tolman [1] termed the general equipartition principle. Note that if p_1 appears quadratically in $\mathcal{H}$, then Eq. (7.1.11a) reduces to the usual equipartition statement, namely $\langle p_1^2 \rangle = \frac{1}{2} k_B T$. This generalization of the equipartition result turns out to be quite useful when an external potential, such as that associated with a gravitational field, is applied to a gas. The three-dimensional (3D) version of this generalized equipartition principle is

$$\left\langle \mathbf{p} \cdot \frac{\partial \mathcal{H}}{\partial \mathbf{p}} \right\rangle = 3 k_B T . \qquad (7.1.11b)$$

The generalization (7.1.11a) of the classical equipartition principle is predicated upon a vanishing of the first term in Eq. (7.1.9), most commonly due to unbound-edness of the domains of integration. The result (7.1.11a) makes clear the role played by boundary conditions (via vessel walls) even for the case of a quadratic dependence of the Hamiltonian upon position coordinates. For example, as the Hamiltonian for an ideal gas of 3D simple harmonic oscillators of mass m can be written as

$$\mathcal{H} = \sum_i \mathcal{H}_i , \qquad \text{with} \qquad \mathcal{H}_i = \tfrac{1}{2} k \mathbf{r}_i^2 + \tfrac{1}{2} \mathbf{p}_i^2 / m ,$$

the classical equipartition principle gives

$$\langle \mathcal{H}_i \rangle = \langle \tfrac{1}{2} k \mathbf{r}_i^2 \rangle + \langle \tfrac{1}{2} \mathbf{p}_i^2 / m \rangle = 3 k_B T ,$$

for a container of infinite volume. However, for a container of finite volume, such that the first term in Eq. (7.1.9) cannot be said to vanish, then Eq. (7.1.11a) will fail. Such a failure is averted via the introduction of a wall potential $W(\mathbf{r})$ that goes to infinity at the confining vessel boundaries, and has a range that is small relative to the vessel size. Moreover, this example illustrates the importance of taking into consideration the presence and role, albeit implicit, of system (container) walls.

The role played by wall potential(s) in the classical equipartition principle has been examined by Mello and Rodríguez [2] for a single particle confined to a 1-dimensional potential well. As such an example illustrates the importance and role of confinement, and as generalization to simple three-dimensional geometries is relatively straightforward, we shall follow the same approach here.

The Hamiltonian for a single particle located within the interval $a \leq x \leq b$ and possessing kinetic energy $p^2/(2m)$, potential energy $V_{\text{tot}}(x)$ is given by

$$\mathcal{H}_1 = \frac{p^2}{2m} + V_{\text{tot}}(x) . \tag{7.1.12}$$

We shall consider the potential energy $V_{\text{tot}}(x)$ to have a smoothly-varying component $V_{\text{ext}}(x)$ associated with a source external to $[a, b]$ plus a component $W(x)$ that vanishes for values $a + \delta \leq x \leq b - \delta$ and rises smoothly and rapidly to infinity for $a + \delta < x$ and for $x < b - \delta$, thereby confining the particle to a one-dimensional region of length $L \equiv b - a$. We shall refer to the component $W(x)$ as the 'wall potential' and treat it as the sum of two components, $W_a(x)$ and $W_b(x)$, corresponding to the interval end-points a and b.

We may employ the general equipartition principle (7.1.11b) for x and $V_{\text{tot}}(x)$ to obtain

$$\left\langle x \frac{dV_{\text{tot}}}{dx} \right\rangle \Bigg|_a^b = k_B T . \tag{7.1.13}$$

Following Mello and Rodríguez [2], we shall split Eq. (7.1.13) into its three component terms as

$$\left\langle x \frac{dV_{\text{ext}}}{dx} \right\rangle \Bigg|_a^b + \left\langle x \frac{dW_a}{dx} \right\rangle \Bigg|_a^b + \left\langle x \frac{dW_b}{dx} \right\rangle \Bigg|_a^b = k_B T .$$

If we examine the second term on the left-hand side of this equation, we have

$$\left\langle x \frac{dW_a}{dx} \right\rangle \Bigg|_a^b \equiv \frac{\displaystyle\int_a^b x \frac{dW_a}{dx} e^{-\beta[V_{\text{ext}}(x)+W_a(x)+W_b(x)]} dx}{\displaystyle\int_a^b e^{-\beta V_{\text{tot}}(x)} dx} . \tag{7.1.14}$$

The numerator of Eq. (7.1.14) can be simplified if we note that the derivative of $W_a(x)$ in the integrand will be appreciable only for $x - a \leq \delta$ due to the nature of the potential term $W_a(x)$: this means that we may approximate $xe^{-\beta V_{ext}(x)}$ by $ae^{-\beta V_{ext}(a)}$ and extend the upper limit of integration from b to ∞, thereby giving

$$\int_a^b x \frac{dW_a}{dx} e^{-\beta[V_{ext}(x)+W_a(x)+W_b(x)]} dx \simeq ae^{-\beta V_{ext}(a)} \int_a^\infty \frac{dW_a}{dx} e^{-\beta W_a(x)} dx$$

for the numerator. Parameter differentiation then gives

$$\int_a^b x \frac{dW_a}{dx} e^{-\beta[V_{ext}(x)+W_a(x)+W_b(x)]} dx = ae^{-\beta V_{ext}(a)} \left(-\frac{1}{\beta}\right) \int_a^\infty \frac{\partial e^{-\beta W_a(x)}}{\partial x} dx .$$

Finally, the integral on the right-hand side of this expression may be evaluated directly, to give

$$\int_a^b x \frac{dW_a}{dx} e^{-\beta[V_{ext}(x)+W_a(x)+W_b(x)]} dx = -k_B T a\, e^{-\beta V_{ext}(a)} .$$

For the integral in the denominator of Eq. (1.2.12), we note that the exponential $e^{-\beta V_{tot}(x)}$ only differs from $e^{-\beta V_{ext}(x)}$ in regions that extend only by approximately δ from the walls and, so long as δ is much smaller than $b - a$, we obtain

$$\int_a^b e^{-\beta V_{tot}(x)} dx \simeq \int_a^b e^{-\beta V_{ext}(x)} dx .$$

A similar treatment of the term involving $W_b(x)$ on the left-hand side gives the same denominator, and the numerator as

$$\int_a^b x \frac{dW_b}{dx} e^{-\beta[V_{ext}(x)+W_a(x)+W_b(x)]} dx \simeq k_B T b\, e^{-\beta V_{ext}(b)} .$$

Upon collecting terms, we thus obtain the result

$$\left\langle x \frac{dV_{ext}}{dx} \right\rangle \bigg|_a^b = k_B T \left[1 - \frac{be^{-\beta V_{ext}(b)} - ae^{-\beta V_{ext}(a)}}{\displaystyle\int_a^b e^{-\beta V_{ext}(x)} dx} \right] . \tag{7.1.15}$$

This result shows explicitly that the introduction of a 'wall-potential' that has a range that is short in comparison with typical system dimensions allows us to obtain the generalized equipartition principle (7.1.13) that does not depend upon the detailed form for the wall-potential and, when an external potential acts on the system, to incorporate a correction due to the presence of the boundaries.

Example 7.1 External gravitational field.

A good example of an external field imposed upon a system is afforded by the consideration of the internal energy per particle, $u(T)$, of an ideal gas of structureless particles of mass m under the influence of a gravitational field characterized by an acceleration g acting in the negative z-direction. The single-particle Hamiltonian is then

$$\mathcal{H} = \frac{\mathbf{p}^2}{2m} + mgz.$$

This Hamiltonian has three quadratic momentum terms describing the translational motion of the particle along the three Cartesian directions plus a non-quadratic term, mgz, describing the gravitational contribution to the energy of a particle.

Let us consider a cylindrical column (height H) of this gas, so that the three quadratic momentum terms corresponding to Eq. (7.1.11b) thereby give accordingly a contribution

$$\langle \mathbf{p}^2 \rangle = \tfrac{1}{2} \left\langle \mathbf{p} \cdot \frac{\partial \mathcal{H}}{\partial \mathbf{p}} \right\rangle = \tfrac{3}{2} k_{\mathrm{B}} T$$

to the internal energy $u(T)$. Similarly, the non-quadratic term mgz then corresponds to Eq. (7.1.15), with $a = 0$, $b = H$, and $V_{\mathrm{ext}}(z) = mgz$, so that Eq. (7.1.15) gives

$$\left\langle z \left(\frac{\partial V_{\mathrm{ext}}}{\partial z} \right) \right\rangle \bigg|_0^H = k_{\mathrm{B}} T \left[1 - \beta mg \frac{H e^{-\beta mgH}}{\displaystyle\int_0^H e^{-\beta mgz} \, dz} \right]$$

$$= k_{\mathrm{B}} T \left[1 - \beta mgH \frac{e^{-\beta mgH}}{1 - e^{-\beta mgH}} \right].$$

If we now define a dimensionless parameter η via $\eta \equiv \beta mgH$, then we may write the generalized equipartition result of Tolman [1] as

$$\left\langle z \frac{dV_{\mathrm{ext}}}{dz} \right\rangle \bigg|_0^H = k_{\mathrm{B}} T \left(1 - \frac{\eta}{e^\eta - 1} \right).$$

The sum of these two components thus gives the internal energy per particle, $u(T)$, of a structureless ideal gas as

$$u(T) = \langle \tfrac{1}{2} \mathbf{p}^2 / m \rangle + \left\langle z \frac{dV_{\mathrm{ext}}}{dz} \right\rangle \bigg|_0^H$$

$$= \tfrac{3}{2} k_{\mathrm{B}} T + k_{\mathrm{B}} T \left(1 - \frac{\eta}{e^\eta - 1} \right).$$

If we gather the two terms together, we obtain

$$u(T) = k_B T \left(\frac{5}{2} - \frac{\eta}{e^\eta - 1} \right),$$

which is the same result as that obtained in Example 4.1 of Chap. 4. □

7.1.2 Dealing with Intermolecular Interactions

We may now *conjecture* that the classical limit of Z for systems of interacting molecules has the form

$$Z_N(V, T) = \frac{1}{N! h^{sN}} \int \cdots \int e^{-\beta \mathcal{H}(p,q)} \, dp^N dq^N \,,$$

in which $\mathcal{H}(q, p)$ is now the classical N-molecule Hamiltonian for *interacting* molecules and $dp^N dq^N \equiv \prod_{i=1}^{sN} dp_i dq_i$ provides a convenient short-hand notation for the volume element.

For a monatomic gas, for example, $\mathcal{H}(p, q)$ is given by

$$\mathcal{H}(p, q) = \frac{1}{2m} \sum_{i=1}^{N} \mathbf{p}_i^2 + V(\mathbf{r}_1, \cdots, \mathbf{r}_N). \tag{7.1.16}$$

If we perform the relevant integrations over the momenta of the N molecules we obtain $Z_{class}(N, V, T)$ as

$$Z_{class}(N, V, T) = \frac{1}{N!} \left(\frac{2\pi m k_B T}{h^2} \right)^{\frac{3N}{2}} Z_{NC}, \tag{7.1.17}$$

with Z_{NC} given by

$$Z_{NC} = \int_V e^{-V(\mathbf{r}_1, \cdots, \mathbf{r}_N)} \, d\mathbf{r}_1 \cdots d\mathbf{r}_N, \tag{7.1.18}$$

and referred to as the *classical configuration integral*. In the absence of intermolecular forces, this integral has the value V^N. Almost all research in equilibrium statistical mechanics of imperfect gases and liquids involves various means for evaluating this integral. Equations (7.1.17) and (7.1.18) are thus the basic equations for the study of classical monatomic liquids and imperfect gases.

We have already seen that rotational degrees of freedom can normally be treated classically. Moreover, as most characteristic vibrational temperatures for simple molecules are quite high, typically $\Theta_{\text{vib}} > 1000\,\text{K}$, and electronic excitation energies even higher than vibrational excitation energies, it is not a poor approximation to ignore these degrees of freedom for simple fluids. To include the classical rotational contributions it is only necessary to insert an additional factor z_{rot}^{N} into Eq. (7.1.17): this factor is obtained by carrying out the integration over the conjugate momenta p_{θ} and p_{ϕ} appearing in the rotational Hamiltonian. The classical configuration integral is, however, quite a bit more complicated due to the angular dependencies introduced into the potential energy function $V(\mathbf{r}_1, \cdots, \mathbf{r}_N)$ for molecules, as even the pair interactions become angle-dependent (anisotropic).

7.2 The Virial Equation of State

We shall begin by considering the form of the classical canonical partition function for N molecules, given by

$$
\begin{aligned}
Z_N(V, t) &= \frac{1}{N! h^{3N}} \iint e^{-\beta \mathcal{H}(p,q)} \, d\mathbf{p}^N d\mathbf{q}^N \\
&= \frac{1}{N! h^{3N}} \int e^{-\beta \left(\sum_i^N p_i^2 / 2m \right)} \, d\mathbf{p}^N \int e^{\beta V(\mathbf{r}_1, \cdots, \mathbf{r}_N)} \, d\mathbf{r}^N \\
&= \frac{1}{N! h^{3N}} \prod_{j=1}^{N} \int e^{-\beta p_j^2 / (2m)} \, d\mathbf{p}_i \int e^{-\beta V(\mathbf{r}_1, \cdots, \mathbf{r}_N)} \, d\mathbf{r}^N ,
\end{aligned}
$$

in which the volume elements $d\mathbf{p}^N$ and $d\mathbf{r}^N$ are N-fold integrations over the linear momenta and positions of the N molecules. The integrations over the linear momenta can be performed straightforwardly, with each integration giving rise to a factor $(2\pi m k_B T)^{\frac{3}{2}}$, the final result being

$$
Z_N = \frac{1}{N!} \left(\frac{2\pi m k_B T}{h^2} \right)^{\frac{3N}{2}} Z_{NC} , \tag{7.2.1}
$$

with Z_{NC} the configuration integral

$$
Z_{NC} \equiv \int \cdots \int e^{-\beta V(\mathbf{r}_1, \cdots, \mathbf{r}_N)} \, d\mathbf{r}_1 \cdots d\mathbf{r}_N . \tag{7.2.2}
$$

As it will turn out to be more convenient to obtain the correction to ideal gas behaviour by employing the grand ensemble that we discussed in Chaps. 3 and 4, we shall start with the grand partition function

$$
\Xi(V, T, \mu) = \sum_{N=0}^{\infty} Z(N, V, T) \lambda^N . \tag{7.2.3}
$$

For $N = 0$, there is only one state (the empty state) with zero energy, so that $Z(N = 0, V, T) = 1$: this result allows us to write the expression for $\Xi(V, T, \mu)$ in the convenient form

$$\Xi(V, T, \mu) = 1 + \sum_{N=1}^{\infty} Z_N(V, T)\lambda^N . \tag{7.2.4}$$

Now, we have seen in Chap. 4 that the characteristic function for the grand ensemble is PV [see Eq. (4.2.27)] and that it is related to Ξ via

$$PV = k_B T \ln \Xi . \tag{7.2.5}$$

Further, we have seen that the (average) number of molecules $\overline{N}$ in the system is given by

$$\overline{N} = k_b T \left(\frac{\partial \ln \Xi}{\partial \mu} \right)_{V,T} = \lambda \left(\frac{\partial \ln \Xi}{\partial \lambda} \right)_{V,T} , \tag{7.2.6}$$

[see Eqs. (4.2.30a) and (4.2.34)]. We note from Eq. (7.2.3) that for $\lambda \ll 1$, $\overline{N}$ behaves approximately as

$$\overline{N} = \lambda \left(\frac{\partial \ln \Xi}{\partial \lambda} \right)_{V,T} \simeq \frac{\lambda Z_1(V, T) + \mathcal{O}(\lambda^2)}{1 + \mathcal{O}(\lambda)} \approx \lambda Z_1(V, T) , \tag{7.2.7}$$

from which we see that for $\lambda \ll 1$, the (number) density $\rho \equiv N/V$ approaches $\lambda Z_1/V$: for this reason, we shall define a new activity, ζ, related to the absolute activity λ via

$$\zeta \equiv \frac{\lambda Z_1(V, T)}{V} . \tag{7.2.8}$$

This activity has the property that it approaches the number density ρ when λ is very small. Hence, Eq. (7.2.3) for $\Xi(V, T, \mu)$ may be replaced by a power series in ζ as

$$\Xi(V, T, \mu) = 1 + \sum_{N=1}^{\infty} \left(\frac{Z_N V^N}{Z_1^N} \right) \zeta^N . \tag{7.2.9}$$

Let us further define Z_{NC} as

$$Z_{NC} \equiv N! \left(\frac{V}{Z_1} \right)^N Z_N , \tag{7.2.10}$$

the reason for which will become clear a little later in our development. The definitions for ζ and Z_{NC} now allow us to replace expression (7.2.9) for $\Xi(V, T, \mu)$ by

$$\Xi(V, T, \mu) = 1 + \sum_{N=1}^{\infty} \frac{Z_{NC}(V, T)}{N!} \zeta^N, \qquad (7.2.11)$$

which is a power series representation of the grand partition function in terms of ζ. For $\lambda \ll 1$, this will become a power series expansion of $\Xi(V, T, \mu)$ in terms of the number density ρ.

Let us now *assume* that the pressure can also be expanded in terms of a power series in ζ according to

$$P = k_{\mathrm{B}} T \sum_{j=1}^{\infty} b_j \zeta^j. \qquad (7.2.12)$$

We now wish to determine the unknown coefficients b_j in terms of the configuration integrals Z_{NC}. This is accomplished by substituting expansion (7.2.12) into $\Xi = \exp\{PV/(k_{\mathrm{B}}T)\}$, expanding the exponential, then collecting like powers of ζ and equating the coefficients to those of Eq. (7.2.9), followed by solving for the b_j coefficients in terms of the Z_{NC}. In this way, the first three coefficients b_j in Eq. (7.2.12) may be determined as

$$b_1 = \frac{1}{V} Z_{1C} = 1, \qquad b_2 = \frac{1}{2V} \left(Z_{2C} - Z_{1C}^2 \right),$$

$$b_3 = \frac{1}{3!V} \left(Z_{3C} - 3Z_{2C}Z_{1C} + 2Z_{1C}^3 \right). \qquad (7.2.13)$$

In order to obtain an explicit expression for b_2, we require only the two-molecule and single-molecule configuration integrals Z_{2C} and Z_{1C} [we recall also that the intermolecular potential energy vanishes by definition for the single-molecule case, so that $Z_{1C} \equiv V$ from Eq. (7.2.1)], while b_3 also requires the three-molecule configuration integral Z_{3C}. This procedure effectively reduces our original N-molecule problem to a series of few-molecule problems. This process illustrates the power of working with the grand ensemble and is why we have approached this development using the grand partition function rather than working within the more traditional canonical ensemble.

We are not yet where we wish to be, as our intention is to obtain an expansion of the pressure P in terms of the number density ρ rather than in terms of the activity ζ that we have defined earlier. In order to convert the expansion in the activity ζ to one in terms of ρ, we can begin with our previously-derived expression for the density in terms of Ξ, namely

$$\rho \equiv \frac{N}{V} = \frac{\lambda}{V} \left(\frac{\partial \ln \Xi}{\partial \lambda} \right)_{V,T} = \frac{\zeta}{V} \left(\frac{\partial \ln \Xi}{\partial \zeta} \right)_{V,T}, \qquad (7.2.14)$$

or

$$\rho = \frac{\zeta}{k_{\mathrm{B}}T}\left(\frac{\partial P}{\partial \zeta}\right)_{V,T}. \qquad (7.2.15)$$

This result provides the connection between Eq. (7.2.9), our series expansion for the pressure, and the number density, so that ρ is given by the series expansion

$$\rho = \sum_{j=1}^{\infty} jb_j\zeta^j. \qquad (7.2.16)$$

We may now employ a method known as *regression of series* (see also Appendix B.1) to obtain P as a power series in ρ: we begin by assuming that ζ itself can be written as a power series in ρ as

$$\zeta = a_1\rho + a_2\rho^2 + a_3\rho^3 + \cdots, \qquad (7.2.17)$$

and then substituting this power series into Eq. (7.2.16), followed by a matching of the coefficients of the powers of ρ on the left- and right-hand sides of the resultant equation. This process gives

$$a_1 = 1, \quad a_2 = -2b_2, \quad a_3 = -3b_3 + 8b_2^2. \qquad (7.2.18)$$

Now we find that ζ has the power series representation

$$\zeta = \rho - 2b_2\rho^2 + (8b_2^2 - 3b_3)\rho^3 + \cdots, \qquad (7.2.19)$$

which we may now substitute into Eq. (7.2.12) for P to obtain

$$\begin{aligned}
\frac{P}{k_{\mathrm{B}}T} &= \rho - b_2\rho^2 + \left(4b_2^2 - 2b_3\right)\rho^3 + \cdots \\
&= \rho - \frac{1}{2V}\left(Z_{2C} - Z_{1C}^2\right)\rho^2 \\
&\quad - \frac{1}{3!V}\left[V\left(Z_{3C} - 3Z_{2C}Z_{1C} + 2Z_{1C}^3\right) - 3\left(Z_{2C} - Z_{1C}^2\right)\right]\rho^3 + \cdots.
\end{aligned}$$
$$(7.2.20)$$

If we now compare this expression with the well-known virial expansion of thermodynamics, viz.,

$$\frac{PV}{Nk_{\mathrm{B}}T} = 1 + B_2(T)\rho + B_3(T)\rho^2 + \cdots, \qquad (7.2.21)$$

we see that the second and third virial coefficients $B_2(T)$ and $B_3(T)$ are given in terms of configuration integrals as

$$B_2(T) = -\frac{1}{2V}\left(Z_{2C} - Z_{1C}^2\right) \tag{7.2.22}$$

and

$$B_3(T) = -\frac{1}{3!V}\left[V\left(Z_{3C} - 3Z_{2C}Z_{1C} + 2Z_{1C}^3\right) - 3\left(Z_{2C} - Z_{1C}^2\right)^2\right], \tag{7.2.23}$$

respectively. A word of caution: should the binary intermolecular interaction energy die off with increasing intermolecular separation r_{12} more slowly than r_{12}^{-3}, a virial expansion does not even exist because, as we shall see shortly, the second virial coefficient $B_2(T)$ diverges for such interactions. This means, for example, that in the case of a plasma in which the charged species interact via the Coulomb law (i.e., as r_{12}^{-1}) there can be no virial expansion representation for the pressure.

7.2.1 Second Virial Coefficient for Monatomic Gases

We return to Eq. (7.2.1) for the partition function Z_N. For $N = 1$, we retrieve our previous result for a monatomic gas, namely

$$Z \equiv Z_1 = \left(\frac{2\pi m k_B T}{h^2}\right)^{\frac{3}{2}} V = \frac{V}{\Lambda^3}, \tag{7.2.24}$$

with Λ the thermal de Broglie wavelength. Note that we have utilized the result $Z_{1C} = V$ for the single-molecule case. Moreover, for $N > 1$ we may also write $Z_N(V, T)$ as

$$Z_N(V, T) = \frac{1}{N!}\frac{Z_{NC}(V, T)}{\Lambda^{3N}}. \tag{7.2.25}$$

If we also replace Λ^{-3} in this result by Z_1/V from Eq. (7.2.24), we obtain

$$Z_N(V, T) = \frac{1}{N!}\left(\frac{Z_1}{V}\right)^N Z_{NC}, \tag{7.2.26}$$

which is nothing other than a rearranged version of Eq. (7.2.10).

We wish now to relate the second virial coefficient appearing in Eq. (7.2.8) to the intermolecular interaction. We shall consider only a monatomic gas in order to avoid complications due to angle-dependencies in the intermolecular interaction.

In order to obtain an expression relating the second virial coefficient $B_2(T)$ to the intermolecular interaction, represented by $V(\mathbf{r}_1, \cdots, \mathbf{r}_N)$, we shall require the configuration integrals Z_{1C}, Z_{2C}, which may be obtained as

$$Z_{1C} = \int d\mathbf{r}_1 = V, \tag{7.2.27}$$

$$Z_{2C} = \iint e^{-\beta V(\mathbf{r}_1, \mathbf{r}_2)} \, d\mathbf{r}_1 d\mathbf{r}_2 . \tag{7.2.28}$$

Note that were we desirous of obtaining an expression for the third virial coefficient, $B_3(T)$, in terms of the intermolecular interactions, we would also require the configuration integral Z_{3C}, which is given by

$$Z_{3C} = \iiint e^{-\beta V(\mathbf{r}_1, \mathbf{r}_2, \mathbf{r}_3)} \, d\mathbf{r}_1 d\mathbf{r}_2 d\mathbf{r}_3 . \tag{7.2.29}$$

The intermolecular interaction $V(\mathbf{r}_1, \mathbf{r}_2)$ between a pair of atoms depends only upon the distance $r \equiv |\mathbf{r}_2 - \mathbf{r}_1|$ separating the atoms. If we substitute expressions (7.2.27) and (7.2.28) for Z_{1C} and Z_{2C} into expression (7.2.22) for $B_2(T)$, we obtain

$$B_2(T) = -\frac{1}{2V} \left(Z_{2C} - Z_{1C}^2 \right)$$

$$= -\frac{1}{2V} \iint \left[e^{-\beta V(r)} - 1 \right] d\mathbf{r}_1 d\mathbf{r}_2 .$$

Now, because the integrand, through $V(r)$, depends only upon r, we may change variables from $\mathbf{r}_2$ to $\mathbf{r} = \mathbf{r}_2 - \mathbf{r}_1$, so that $d\mathbf{r}_2 = d\mathbf{r}$, to obtain

$$B_2(T) = -\frac{1}{2V} \int d\mathbf{r}_1 \int \left[e^{-\beta V(r)} - 1 \right] d\mathbf{r}, \tag{7.2.30}$$

following which, integration over $\mathbf{r}_1$ gives a factor V and integration over the angle variables of $\mathbf{r}$ gives an additional factor 4π: the end result for $B_2(T)$ is then

$$B_2(T) = -2\pi \int_0^\infty \left[e^{-\beta V(r)} - 1 \right] r^2 \, dr . \tag{7.2.31}$$

Note that the integral over $\mathbf{r}_1$ gave the volume V of the container, while the integration over r from 0 to ∞ indicates that we have essentially extended the integration beyond the (macroscopic) container. We may do this because the integrand goes quite rapidly to zero as r increases, so that extending the integration from $r_{\max}$, corresponding to the container dimensions, to infinity makes a negligible contribution to the value of B_2. Similar expressions can be obtained for the second virial coefficient for molecular gases [3].

Had we carried out a quantum mechanical derivation of $B_2(T)$ we would have found that the classical expression (7.2.31) for $B_2(T)$ gives by far the dominant contribution to the second virial coefficient at most temperatures. For light atoms (molecules), especially at low temperatures, there will be quantum corrections to the classical expression. A more complete expression for $B_2(T)$ for atoms is [3]

$$B_2(T) = -2\pi \int_0^\infty \left[e^{-\beta V(r_{12})} - 1 \right] r_{12}^2 \, dr_{12}$$

$$+ \frac{h^2}{24\pi m (k_B T)^3} \int_0^\infty e^{-\beta V(r_{12})} \left(\frac{dV}{dr_{12}} \right)^2 r_{12}^2 \, dr_{12} + \mathcal{O}\left(h^3 \right). \qquad (7.2.32)$$

The first term in this expression is the classical contribution and the second term is referred to as the first semiclassical (translational) quantum correction. The second term contributes only a small fraction of the total value for $B_2(T)$. Second and higher order correction terms (representing a series expansion in powers of h^2) also exist, but generally contribute significantly less than the two lowest first-order terms. For ^{4}He, for example, the first translational quantum correction term contributes 4.1% to $B_2(T)$ at 256 K and 17.0% at 83.5 K, while for H_2, the corresponding percentages are 23.0% at 183 K and 3.0% at 592 K. For a heavier atom, such as Ne, we find 6% at 35 K, 9.7% at 95 K, and 0.6% at 392 K. Full quantum mechanical calculations of the second virial coefficient should be carried out for the helium isotopes, the hydrogen isotopologues and, to a lesser extent, the neon isotopes, as the semi-classical expressions are asymptotic expressions, and can fail for sufficiently low temperatures.[1] Figure 7.1 displays results [4] obtained from fully quantum mechanical calculations of the second virial coefficient for ^{4}He. The quantum-mechanical ideal gas contribution for helium is also shown in the inset figure: as may be seen, it has essentially vanished for temperatures in excess of 60 K. The temperature dependence shown in this figure is typical of the behaviour of second virial coefficients.

The extension of these concepts to mixtures of gases is fairly straightforward. For a binary mixture of two chemical species designated by A and B, for example, the second virial coefficient for the mixture can be written as

$$B_2^{\text{mix}}(T) = x_A^2 B_{2,AA} + 2x_A x_B B_{2,AB} + x_B^2 B_{2,BB}, \qquad (7.2.33)$$

with x_A, x_B the mole fractions of A and B in the binary mixture. The second virial coefficients $B_{2,AA}$ and $B_{2,BB}$ are those for the pure gaseous A and B components, respectively, each of which can be associated, through Eq. (7.2.31), with the corresponding intermolecular interaction. The second virial coefficient $B_{2,AB}$, is similar to that for the pure species A and B and, as it depends solely upon the intermolecular

---

[1]For H_2 and He at temperatures below 75 K, 45 K, respectively, an expansion in powers of h^2 no longer converges, and it is actually necessary to carry out full quantum mechanical calculations of $B_2(T)$.

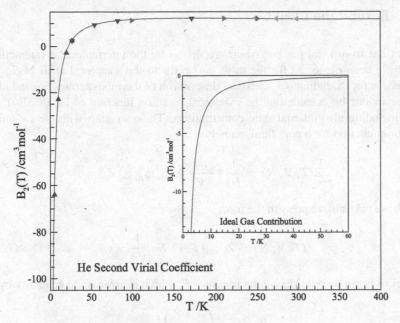

Fig. 7.1 The second virial coefficient for the He–He interaction. A number of representative experimental values of $B_2(T)$ have also been included. Based upon data from Table 2 of [4], with permission of Taylor & Francis

interaction between the unlike chemical species A and B: it is referred to as the *interaction* second virial coefficient. It should be noted, however, that experimental second interaction virial coefficient data are difficult to obtain because they must typically be extracted from second virial coefficient measurements for both pure gases in addition to second virial coefficient measurements for a series of mixture compositions using Eq. (7.2.33).

For molecules, there is another correction term, referred to as the first semi-classical rotational quantum correction which, while generally smaller than the corresponding translational correction, is at the same level of approximation. The relative contributions of the leading correction terms to the interaction second virial coefficient $B_{2,AB}(T)$ arising from the CO_2–He interaction can be seen in Fig. 2 of [3]. The interaction second virial coefficient for binary mixtures in which at least one component is a molecular gas, will consist of a classical contribution plus translational and rotational quantum corrections, to give $B_{2,AB}(T)$ as

$$B_{2,AB}(T) = B_{2,AB}^{class}(T) + B_{2,AB}^{tr}(T) + B_{2,AB}^{rot}(T).$$

The first two contributions listed in this equation correspond to the contributions shown in Eq. (7.2.32) for binary mixtures of atomic gases.

7.3   Beyond the Ideal Gas

Deviations from ideal gas behaviour are driven by the intermolecular interactions that have been neglected for the most part prior to this chapter. Let us begin our discussion for extending our statistical description of thermodynamics beyond ideal gas behaviour by considering the canonical partition function of Eq. (7.2.26), but now including also internal state contributions. Thus, we start with the canonical partition function for a pure fluid, namely

$$Z(T; V, N) = \frac{1}{N!} \left(\frac{z_{\text{trans}}}{V} \right)^N z_{\text{rot}}^N z_{\text{vib}}^N z_{\text{el}}^N Z_{NC},$$

which we may also express in the form

$$Z(T, V, N) = Z_{\text{ideal}}(T, V, N) \frac{1}{V^N} Z_{NC}. \qquad (7.3.1)$$

Let us begin by considering the pressure P, which is given quite generally by

$$P = k_B T \left(\frac{\partial \ln Z}{\partial V} \right)_{T,N}.$$

However, as may be seen from Eqs. (3.2.27) and (3.2.21) for an ideal gas, the factor $Z_{\text{ideal}}(T, V, N)/V^N$ does not depend upon V, so that the pressure is determined by the configuration factor Z_{NC} via

$$P = k_B T \left(\frac{\partial \ln Z_{NC}}{\partial V} \right)_{T,N}. \qquad (7.3.2)$$

This result does not in itself help very much, however, as Z_{NC} involves the intermolecular potential, which in general is a complicated function of all intermolecular separations, thereby making any direct evaluation of Z_{NC} extremely difficult, if not nigh to impossible.

Recall from Eq. (7.1.18) that Z_{NC} is an integral of $e^{-\beta V(\mathbf{r}_1, \cdots, \mathbf{r}_N)}$ over the coordinates $\mathbf{r}_1, \cdots, \mathbf{r}_N$: only for the case in which the general potential energy function can be written as the sum of pairwise spherically symmtric interactions as

$$V(\mathbf{r}_1, \mathbf{r}_2, \cdots, \mathbf{r}_N) = \sum_{i>j} V(|\mathbf{r}_i - \mathbf{r}_j|) \equiv \sum_{i>j} V(r_{ij}) \qquad (7.3.3)$$

can we effect considerable simplification. Pairwise additivity of $V(\mathbf{r}_1, \cdots, \mathbf{r}_N)$ enables the integrand of Z_{NC} to be obtained as

$$e^{-\beta \sum_{i>j} V(r_{ij})} = \prod_{i>j} e^{-\beta V(r_{ij})} = \prod_{i>j} \left[1 + \left(e^{-\beta V(r_{ij})} - 1 \right) \right] ,$$

where we have isolated the role of the two-body interaction terms [note also that the term in parentheses vanishes if $V(r_{ij}) = 0$]. These factors are referred to as Mayer f-functions, and are commonly designated by f_{ij}: they represent deviations from ideal gas behaviour associated with the interactions between pairs of molecules.

We may rewrite the integrand of Z_{NC} in terms of Meyer f-functions as

$$\prod_{i>j} (1 + f_{ij}) = (1 + f_{21})(1 + f_{31}) \cdots (1 + f_{32}) \cdots$$

$$\simeq 1 + f_{21} + f_{31} + \cdots + f_{32} + \cdots ,$$

and retain terms no higher than those linear in the Meyer functions. These linear terms thus represent the pair interaction contributions to Z_{NC}. We then obtain

$$Z_{NC} \simeq \int_V d\mathbf{r}_1 \cdots d\mathbf{r}_N \left(1 + \sum_{i>j}^N f_{ij} \right) = V^N + \int_V d\mathbf{r}_1 \cdots d\mathbf{r}_N \sum_{i>j}^N f_{ij}$$

$$= V^N + V^{N-2} \sum_{i>j}^N \int_V d\mathbf{r}_i d\mathbf{r}_j f_{ij}$$

and, upon recognizing that each integral in this summation has the same value, we see that Z_{NC} is approximately given by

$$Z_{NC} \simeq V^N + V^{N-2} \tfrac{1}{2} N(N-1) \int_V d\mathbf{r}_1 d\mathbf{r}_2 f_{12}(|\mathbf{r}_1 - \mathbf{r}_2|) . \qquad (7.3.4)$$

By making a change of variables from $\{\mathbf{r}_1, \mathbf{r}_2\}$ to $\{\mathbf{r}, \mathbf{r}_1\}$, with $\mathbf{r} \equiv \mathbf{r}_1 - \mathbf{r}_2$, we see that the intermolecular contribution can be obtained as

$$\int_V d\mathbf{r}_1 d\mathbf{r}_2 f_{12}(|\mathbf{r}_1 - \mathbf{r}_2|) = \int_V d\mathbf{r}_1 \int_V d\mathbf{r} f(r) = V \int_V d\mathbf{r} f(r) ,$$

so that Z_{NC} becomes

$$Z_{NC} \simeq V^N - V^{N-1} N^2 \tfrac{1}{2} \int_V d\mathbf{r} \left[1 - e^{-\beta V(r)} \right] , \qquad (7.3.5)$$

in which we have used $N \gg 1$ and $V(r)$ is the interaction potential between a pair of atoms/molecules. Thus, Z_{NC} can be expressed in lowest order in terms of the second virial coefficient $B_2(T)$ given by

$$B_2(T) = \int_V d\mathbf{r} \left[1 - e^{-\beta V(r)} \right].$$

This gives us the leading term in the virial expansion via

$$\ln Z_{NC} \simeq N \ln V + \ln \left[1 - \frac{N^2}{V} B_2(T) \right] \approx N \ln V - \frac{N^2}{V} B_2(T),$$

where in the final step we have employed the Maclaurin expansion for the natural logarithm, namely $\ln(1 - x) \simeq -x + \cdots$. The pressure P is hence given as

$$P = k_B T \left(\frac{\partial \ln Z_{NC}}{\partial V} \right)_{T,N} \simeq \frac{N k_B T}{V} + \frac{N^2 k_B T}{V^2} B_2(T),$$

or, equivalently, as

$$\frac{PV}{N k_B T} \simeq 1 + B_2(T)\rho,$$

with $\rho \equiv N/V$ being the number density. This expression provides the first two terms of the density virial equation of state.

More generally, $\ln Z_{NC}$ can be obtained in terms of the virial expansion as

$$\ln Z_{NC} = N \ln V - N \left[B_2(T)\rho + \tfrac{1}{2} B_3(T)\rho^2 + \tfrac{1}{3} B_4(T)\rho^3 + \cdots \right], \qquad (7.3.6)$$

corresponding to the virial equation of state for the pressure P, i.e.,

$$P = k_b T \left(\frac{\partial \ln Z_{NC}}{\partial V} \right)_{T,N} = \frac{N k_B T}{V} \left[1 + B_2(T)\rho + B_3(T)\rho^2 + \cdots \right]. \tag{7.3.7}$$

This equation is the key equation for determining density corrections to the various thermodynamic state functions. The Helmholtz energy is given, for example, by

$$A(T, V, N) = A_{\text{ideal}}(T, V, N) + N k_B T [B_2(T)\rho + \tfrac{1}{2} B_3(T)\rho + \tfrac{1}{3} B_4(T)\rho^2 + \cdots], \tag{7.3.8}$$

while the internal energy, U, the entropy, S, and the Gibbs energy, G, are found to be given by

$$U(T, V, N) = U_{\text{ideal}}(T, V, N)$$
$$- N k_B T \left[T \frac{dB_2}{dT} \rho + \frac{1}{2} T \frac{dB_3}{dT} \rho^2 + \frac{1}{3} T \frac{dB_4}{dT} \rho^3 + \cdots \right], \tag{7.3.9}$$

$$S(T, V, N) = S_{ideal}(T, V, N) - Nk_B \left[\left(B_2 + T\frac{dB_2}{dT} \right) \rho + \frac{1}{2} \left(B_3 + T\frac{dB_3}{dT} \right) \rho^2 \right.$$

$$\left. + \frac{1}{3} \left(B_4 + \frac{dB_4}{dT} \right) \rho^3 + \cdots \right], \tag{7.3.10}$$

and

$$G(T, V, N) = G_{ideal}(T, V, N) + Nk_B T \left[\frac{2}{1}B_2\rho + \frac{3}{2}B_3\rho^2 + \frac{4}{3}B_4\rho^3 + \cdots \right]. \tag{7.3.11}$$

As the virial expansion is only valid for relatively small departures from ideal gas behaviour, some care must be exercised before determining these corrections for higher densities. It is thus necessary to ensure that the pressure virial equation converges reasonably rapidly prior to employing these expressions.

7.4   An Approximate Description for Dense Fluids

We shall begin with the canonical partition function of Eq. (7.1.17) extended to include the rotational contribution for molecules of mass M, viz.,

$$Z_N(T, V) = \frac{(2\pi M k_B T)^{3N/2}}{N! h^{3N}} z_{rot}^N(T) Z_{NC}. \tag{7.4.1}$$

For a gas of linear molecules, the single-molecule $z_{rot}(T)$ is given by

$$z_{rot}(T) = \frac{8\pi^2 I k_B T}{\sigma h^2} \equiv \frac{T}{\sigma \Theta_{rot}},$$

with σ the symmetry number and Θ_{rot} the characteristic rotational temperature. For nonlinear molecules (see Chap. 6, Sect. 6.3) $z_{rot}(T)$ is given by

$$z_{rot}(T) = \frac{\sqrt{\pi}}{\sigma} \left(\frac{T^3}{\Theta_{rot,A}\Theta_{rot,B}\Theta_{rot,C}} \right)^{\frac{1}{2}},$$

with $\Theta_{rot,A}$, $\Theta_{rot,B}$, $\Theta_{rot,C}$ the rotational temperatures associated with the three (generally different) moments of inertia associated with rotations about the three principal axes of the nonlinear molecule. We note that these two separate cases can be subsumed into a single expression for $z_{rot}(T)$ by defining a mean rotational temperature for nonlinear molecules as

$$\Theta_{rot,m} \equiv (\Theta_{rot,A}\Theta_{rot,B}\Theta_{rot,C})^{\frac{1}{3}},$$

and writing $z_{\text{rot}}(T)$ as

$$z_{\text{rot}}(T) = \frac{\pi^{(w-5)N/2}}{\sigma} \left(\frac{T}{\Theta_{\text{rot},m}}\right)^{(w-3)N/2},$$

with w being the total number of translational plus rotational degrees of freedom in the molecule (w has the value 5 for linear molecules and 6 for nonlinear molecules), with the understanding that for a linear molecule, $\Theta_{\text{rot},m} = \Theta_{\text{rot}}$.

We need to find a means of approximating Z_{NC} by developing a relation for the potential energy associated with the interacting molecules. For dense fluids, such as liquids, a useful approach is to employ the concept of the radial distribution function designated by $g(r)$ and defined so that $4\pi r^2 \rho g(r)\mathrm{d}r$ represents the number density of molecules with centres between r and $r + \mathrm{d}r$ measured relative to a specific molecule in the fluid. As a function of r, $g(r)$ behaves qualitatively as in Fig. 7.2.

This function can be determined either computationally by Monte Carlo simulations using model pair interaction potentials or experimentally via X-ray or neutron scattering: the product of the system mean density and $g(r)\mathrm{d}r$ is often interpreted as giving the local (time-averaged) density in a fluid. For a fluid consisting of nonpolar electrically neutral molecules, it is a reasonable approximation to treat the intermolecular interactions as spherically symmetric, i.e., depending only upon the distance r between molecules. In this context, $4\pi r^2 V(r)\rho g(r)\mathrm{d}r$ represents the potential energy between the central molecule and all molecules whose centres lie between distances r and $r + \mathrm{d}r$ from it. The Lennard-Jones potential form is often utilized for this purpose.

The total potential energy of the fluid is obtained by integrating over all values of r and then multiplying that result by N, since any of the N molecules may serve as the central one. This result must still be divided by two in order that each interacting pair of molecules be counted only once for this calculation. We obtain in this manner the total potential energy

Fig. 7.2 Radial distribution function for Lennard-Jones Ar as a function of the distance from a given atom (located at the origin) [Figure courtesy of Dr. Kevin Bishop]

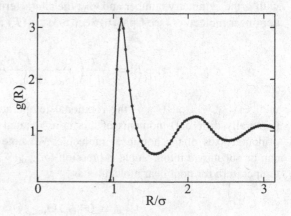

$$\mathcal{V}_N = \tfrac{1}{2} N \int_0^\infty 4\pi r^2 V(r) \rho g(r) \, dr \, . \tag{7.4.2}$$

We shall now employ a relatively crude but nonetheless reasonable approximation for $g(r)$, namely

$$g(r) = \begin{cases} 0, & \text{for } r < D \\ 1, & \text{for } r > D \, , \end{cases} \tag{7.4.3}$$

with D a characteristic molecular diameter (such as twice the Van der Waals radius, for example). With this approximation, we may replace Eq. (7.4.2) for $\mathcal{V}_N$ by the simpler result

$$\mathcal{V}_N = \frac{2\pi N^2}{V} \int_D^\infty r^2 V(r) \, dr \, . \tag{7.4.4}$$

If we define a characteristic parameter a via

$$a \equiv -2\pi \int_D^\infty r^2 V(r) \, dr \, , \tag{7.4.5}$$

we may write $\mathcal{V}_N$ as

$$\mathcal{V}_N = -\frac{aN^2}{V} \, . \tag{7.4.6}$$

Returning now to our evaluation of the configuration integral of Eq. (7.2.2), we may replace $\beta V(\mathbf{r}_1, \cdots, \mathbf{r}_N)$ by $-\beta a N^2 / V$, which enables us to take the exponential outside the integrals to obtain

$$Z_{NC} = e^{aN^2/(Vk_B T)} \int \cdots \int 1 \, d\mathbf{r}_1 \cdots d\mathbf{r}_N \, . \tag{7.4.7}$$

We need to be careful in approximating the trivial-looking integral in this expression. For a dilute gas, it would immediately give a factor V^N. For dense gases (or liquids), however, the molecules themselves occupy a significant fraction of the volume of the physical system. As this occupied space is inaccessible to any given molecule, it should be excluded from the multiple integration. For this reason, a better approximation to our multiple integral will be $(V - Nb)^N$, with b a constant representing the mean volume occupied by an individual molecule in the fluid sample, of order $\pi D^3/6$. Thus, we obtain for Z_{NC} the expression

$$Z_{NC} = e^{aN^2/(Vk_B T)} (V - Nb)^N \, . \tag{7.4.8}$$

The full canonical partition function for this model of a dense fluid is then

$$Z_N(V,T) = \frac{(V-Nb)^N}{N!}\left(\frac{2\pi M k_B T}{h^2}\right)^{3N/2}\frac{\pi^{(w-5)N/2}}{\sigma^N}\left(\frac{T}{\Theta_{\rm rot,m}}\right)^{(w-3)N/2}e^{aN^2/(Vk_B T)}.$$

(7.4.9)

The pressure for this model fluid (equivalently, the thermodynamic equation of state for the fluid) can now be determined, using the usual formula relating the pressure P to the canonical partition function, as

$$P = \frac{Nk_B T}{V-Nb} - \frac{aN^2}{V^2},$$

(7.4.10)

which is the famous Van der Waals equation of state first obtained by Johannes van der Waals in his Ph.D. thesis in 1875.

7.5   The Liouville and Boltzmann Equations

7.5.1   The Liouville Equation

We shall consider a classical system made up of N interacting molecules, each of which has s degrees of freedom: that is, s coordinates will be required to describe each molecule completely, so that sN coordinates, $q_1, \cdots, q_{sN}$, will be needed to describe fully the spatial attributes of the N-molecule system. There will be another sN conjugate momentum coordinates: thus, $2sN$ variables are required for complete specification of the classical mechanical state of an N-body system. When these $2sN$ variables are taken, together with their corresponding equations of motion, the complete past and future courses of the system are determined.

To describe the behaviour of our N-molecule system classically, we may now introduce a Euclidian space of $2sN$ dimensions, with sN pairs of rectangular (q_i, p_i) axes: this space is referred to as the *phase space* for the system, a terminology introduced by Gibbs. The concept of the phase space has been introduced because the state of a classical N-body system at any time t is completely specified by the location of a single point, called a *phase point*, in it. The dynamics of the system as it evolves in time is described by the motion (or *trajectory*) of the phase point through the phase space, and is given by Hamilton's equations of motion,

$$\dot{q}_j = \frac{\partial \mathcal{H}}{\partial p_j}, \quad \dot{p}_j = -\frac{\partial \mathcal{H}}{\partial q_j}, \quad j = 1, \cdots, sN.$$

(7.5.1)

Appendix G contains a brief introduction/review of classical mechanics and Hamilton's equations. Integration of these equations in principle gives $\mathbf{q}(t)$ and $\mathbf{p}(t)$, with $\mathbf{q}$ and $\mathbf{p}$ being sN-dimensional vectors, with the $2sN$ constants of integration determined by the location of the phase point at some initial time $t = t_0$. Of course, this cannot actually be accomplished in practice.

We shall now introduce a classical microcanonical phase space ensemble to aid us in developing the distribution function concept. Consider a large number $\mathcal{N}$ of isolated systems, each having the same values of the macroscopic variables N, V, and $E \equiv U$. As each N-particle system in this ensemble will have a representative phase point in the same phase space, the entire ensemble will make up a cloud of points in phase space, each point tracing out its own (independent) trajectory with time. Note that as each system is isolated in a microcanonical ensemble, this ensures its independence.

The role of the 'equal a priori probability' postulate is to require there to be a representative phase point in phase space for each and every set of coordinates and conjugate momenta that is consistent with the fixed values of N, V, and U. For a microcanonical ensemble, this requirement reduces to a uniform density over a constant energy surface in phase space. The cloud of phase points is thus very dense, which allows us to define a number density $f^{(N)}(\mathbf{p}, \mathbf{q}, t)$ which has the property that the number of systems in the ensemble that have phase points in a volume element $d\mathbf{p}d\mathbf{q}$ about the point $(\mathbf{p}, \mathbf{q})$ at time t is given by $f^{(N)}(\mathbf{p}, \mathbf{q}, t)\,d\mathbf{p}d\mathbf{q}$. We need to impose the *constraint* that the number of members of our ensemble (i.e., the number of N-particle systems) be given by

$$\mathcal{N} = \iint f^{(N)}(\mathbf{p}, \mathbf{q}, t)\,d\mathbf{p}d\mathbf{q}. \tag{7.5.2}$$

As we have seen in Chap. 3, we may define the ensemble average of a quantity $M(\mathbf{p}, \mathbf{q})$ as

$$\overline{M}(t) \equiv \frac{1}{\mathcal{N}} \iint M(\mathbf{p}, \mathbf{q}) f^{(N)}(\mathbf{p}, \mathbf{q}, t)\,d\mathbf{p}d\mathbf{q}. \tag{7.5.3}$$

We may also postulate, as was done originally by Gibbs, that the ensemble average is the same as the corresponding thermodynamic function. To proceed further, we note that as the equations of motion determine the trajectories of the phase points in phase space, they must also determine the density $f^{(N)}(\mathbf{p}, \mathbf{q}, t)$ of the phase points at any time if the dependence of $f^{(N)}(\mathbf{p}, \mathbf{q}, t)$ on $\mathbf{p}, \mathbf{q}$ is known at some initial time t_0.

Let us consider a volume element $\delta\mathbf{p}\delta\mathbf{q} \equiv \delta p_1 \cdots \delta p_{sN}\delta q_1 \cdots \delta q_{sN}$ about $(\mathbf{p}, \mathbf{q})$: the number of phase points $\delta\mathcal{N}$ inside this phase-space volume element will be

$$\delta\mathcal{N} = f^{(N)}(\mathbf{p}, \mathbf{q}, t)\,\delta\mathbf{p}\delta\mathbf{q}. \tag{7.5.4}$$

Now, let us determine the difference between the numbers of phase points entering and leaving this volume element in phase space via a three-step process. In step 1 we shall consider the net change obtained for one variable, in step 2 we shall extend the argument utilized in step 1 to the full set of sN conjugate variables and determine the net change for the full phase-space volume element, and in step 3 we shall equate the result of step 2 with the change in the distribution function $f^{(N)}$ for that phase volume.

1. Let us consider the difference between the number of phase points

$$f^{(N)}(\mathbf{p}, \mathbf{q}, t)\dot{q}_1 \delta q_2 \cdots \delta q_{sN}\delta\mathbf{p} = f^{(N)}(\mathbf{p}, \mathbf{q}, t)\frac{\partial q_1}{\partial t}\delta q_2 \cdots \delta q_{sN}\delta\mathbf{p}$$

at position $(\mathbf{p}, \mathbf{q})$ and those at $(\mathbf{p}, \mathbf{q}')$, with $\mathbf{q}' = (q_1 + \delta q_1, q_2, \cdots, q_{sN})$, which we may express as $f^{(N)}(\mathbf{p}, \mathbf{q}', t)\dot{q}_1' \delta q_2 \cdots \delta q_{sN}\delta\mathbf{p}$. If we expand f and $\dot{q}_1$ to terms linear in δq_1, i.e., upon writing $f^{(N)} + \dfrac{\partial f^{(N)}}{\partial q_1}\delta q_1$ and $\dot{q}_1 + \dfrac{\partial \dot{q}_1}{\partial q_1}\delta q_1$, the net difference $\Delta(q_1)$ between the numbers of phase points at the two values of q_1, which therefore represents a flow in the q_1-direction in phase space, is given by

$$\Delta(q_1) = \left\{ f^{(N)}(\mathbf{p}, \mathbf{q}, t)\dot{q}_1 \right.$$

$$\left. - \left[f^{(N)}(\mathbf{p}, \mathbf{q}, t) + \frac{\partial f^{(N)}}{\partial q_1}\delta q_1 \right]\left[\dot{q}_1 + \frac{\partial \dot{q}_1}{\partial q_1}\delta q_1 \right] \right\} \delta q_2 \cdots \delta q_{sN}\delta\mathbf{p}$$

$$\simeq - \left[\frac{\partial f^{(N)}}{\partial q_1}\dot{q}_1\delta q_1 + f^{(N)}(\mathbf{p}, \mathbf{q}, t)\frac{\partial \dot{q}_1}{\partial q_1}\delta q_1 \right] \delta q_2 \cdots \delta q_{sN}\delta\mathbf{p}$$

$$= - \left(\frac{\partial f^{(N)}}{\partial q_1}\dot{q}_1 + f^{(N)}\frac{\partial \dot{q}_1}{\partial q_1} \right)\delta\mathbf{p}\delta\mathbf{q} \,,$$

upon neglect of the much smaller second-order term. A completely analogous calculation for changes in the conjugate momentum variable p_1 gives

$$\Delta(p_1) = - \left(\frac{\partial f^{(N)}}{\partial p_1}\dot{p}_1 + f^{(N)}\frac{\partial \dot{p}_1}{\partial p_1} \right)\delta\mathbf{p}\delta\mathbf{q}.$$

2. It now becomes clear that the same process that we have just used for the pair of conjugate variables p_1, q_1 would give equivalent expressions for all other conjugate pairs of variables in our representation of the phase space and, consequently, the net change for the full phase-space volume element $\delta\mathbf{p}\delta\mathbf{q}$, which must also represent the change with time of the number of phase points $\delta\mathcal{N}$ in the volume element $\delta\mathbf{q}\delta\mathbf{p}$, will be given by

$$\frac{d(\delta\mathcal{N})}{dt} = - \sum_{j=1}^{sN} \left(\frac{\partial f^{(N)}}{\partial q_j}\dot{q}_j + f^{(N)}\frac{\partial \dot{q}_j}{\partial q_j} + \frac{\partial f^{(N)}}{\partial p_j}\dot{p}_j + f^{(N)}\frac{\partial \dot{p}_j}{\partial p_j} \right)\delta\mathbf{q}\delta\mathbf{p}$$

$$= -\sum_{j=1}^{sN} \left[f^{(N)} \left(\frac{\partial \dot{q}_j}{\partial q_j} + \frac{\partial \dot{p}_j}{\partial p_j} \right) + \frac{\partial f^{(N)}}{\partial q_j} \dot{q}_j + \frac{\partial f^{(N)}}{\partial p_j} \dot{p}_j \right] \delta\mathbf{q}\delta\mathbf{p} \, .$$

We shall now utilize Hamilton's equations (7.5.1) to help simplify this result: the first term on the right-hand side of our net result vanishes, and we can similarly simplify the remaining two terms to give

$$\frac{d(\delta\mathcal{N})}{dt} = -\sum_{j=1}^{sN} \left(\frac{\partial f^{(N)}}{\partial q_j} \frac{\partial \mathcal{H}}{\partial p_j} - \frac{\partial f^{(N)}}{\partial p_j} \frac{\partial \mathcal{H}}{\partial q_j} \right) \tag{7.5.5}$$

for the final result.

3. We are now ready to take the final step in this procedure: the change in the number of phase points passing through the phase volume element $\delta\mathbf{q}\delta\mathbf{p}$ divided by the volume element must represent the change in the phase-point density $f^{(N)}$ with time for that phase volume, i.e.,

$$\frac{\partial f^{(N)}}{\partial t} = -\sum_{j=1}^{sN} \left(\frac{\partial f^{(N)}}{\partial q_j} \frac{\partial \mathcal{H}}{\partial p_j} - \frac{\partial f^{(N)}}{\partial p_j} \frac{\partial \mathcal{H}}{\partial q_j} \right) \, .$$

It is conventional to rearrange this final result slightly and to write it as

$$\frac{\partial f^{(N)}}{\partial t} + \sum_{j=1}^{sN} \left(\frac{\partial f^{(N)}}{\partial q_j} \frac{\partial \mathcal{H}}{\partial p_j} - \frac{\partial f^{(N)}}{\partial p_j} \frac{\partial \mathcal{H}}{\partial q_j} \right) = 0 \, . \tag{7.5.6}$$

Equation (7.5.6), referred to as the Liouville equation, is one of the most fundamental equations of classical statistical mechanics because it provides the starting point for most theories of nonequilibrium statistical mechanics.

In terms of a Cartesian representation of the positions and linear momenta of the molecules in the gas, we may write $\mathbf{q} = \mathbf{r} \equiv (\mathbf{r}_1, \cdots, \mathbf{r}_N)$ for the positions, and $\mathbf{p} \equiv (\mathbf{p}_1, \cdots, \mathbf{p}_N) = m(\dot{\mathbf{r}}_1, \cdots, \dot{\mathbf{r}}_N)$ for the linear momenta of the N molecules. Moreover, as the Hamiltonian for an atomic system is

$$\mathcal{H} = \sum_{i=1}^{N} \frac{\mathbf{p}_i^2}{2m_i} + V(\mathbf{r}_1, \cdots, \mathbf{r}_N) \, , \tag{7.5.7}$$

Hamilton's equations of motion reduce to

$$\frac{\partial \mathcal{H}}{\partial \mathbf{p}_i} = \frac{1}{m_i} \mathbf{p}_i \, , \quad \text{and} \quad \frac{\partial \mathcal{H}}{\partial \mathbf{r}_i} = \frac{\partial V}{\partial \mathbf{r}_i} \equiv -\mathbf{F}_i \, . \tag{7.5.8}$$

Upon substituting Eqs. (7.5.8) into Eq. (7.5.6), our result takes the form

$$\frac{\partial f^{(N)}}{\partial t} + \sum_{i=1}^{N} \frac{1}{m_i} \mathbf{p}_i \cdot \nabla_{\mathbf{r}_i} f^{(N)} + \sum_{i=1}^{N} \mathbf{F}_i \cdot \nabla_{\mathbf{p}_i} f^{(N)} = 0 \,. \tag{7.5.9}$$

In terms of the total (also called *substantial*) time derivative, which is defined as

$$\frac{\mathrm{d}}{\mathrm{d}t} \equiv \frac{\partial}{\partial t} + \sum_{i=1}^{N} \left(\frac{\mathbf{p}_i}{m_i} \cdot \nabla_{\mathbf{r}_i} + \mathbf{F}_i \cdot \nabla_{\mathbf{p}_i} \right) , \tag{7.5.10}$$

the Liouville equation takes an especially simple-looking form, namely

$$\frac{\mathrm{d} f^{(N)}}{\mathrm{d}t} = 0 \,. \tag{7.5.11}$$

What does this result tell us? Recall that $f^{(N)}$ represents the density of phase points in the neighbourhood of any selected moving phase point, so that the Liouville equation stated in this simple form tells us that this 'density' is constant along the trajectory of that phase point. Consequently, the cloud of phase points moves in exactly the same way as an incompressible fluid does: for this reason, Eq. (7.5.11) was said by Gibbs to represent the 'principle of conservation of phase-point density'.

Let us introduce a notation that will allow us to shorten some of the resultant equations in this section. Let us write the N-molecule distribution function $f^{(N)}$ in the form $f^{(N)}(\mathbf{r}_1, \cdots , \mathbf{r}_N, \mathbf{p}_1, \cdots , \mathbf{p}_N, t) = f^{(N)}(\mathbf{x}^N, t)$ with $\mathbf{x}^N \equiv (\mathbf{x}_1, \cdots , \mathbf{x}_N)$ a $6N$-dimensional vector whose components $\mathbf{x}_i \equiv (\mathbf{r}_i, \mathbf{p}_i)$ are each 6-dimensional vectors in the phase space of molecule i: $\mathbf{x}^N$ thus represents a point in the overall $6N$-dimensional phase space Γ of the N molecules. We shall also restrict our considerations to the case of a pure gas, i.e., with all N molecules indistinguishable.

The N-molecule distribution function $f^{(N)}(\mathbf{x}^N, t)$ must be symmetric in the $\mathbf{x}_i$ vectors, and $\int_V f^{(N)}(\mathbf{x}^N, t) \, \mathrm{d}\mathbf{x}^N$ must represent the number of members of the ensembles in the domain (volume V of the macroscopic vessel) while $f^{(N)}(\mathbf{x}^N, t)$ must satisfy the Liouville equation, Eq. (7.5.9) that is,

$$\frac{\partial f^{(N)}}{\partial t} + \sum_{k=1}^{N} \frac{1}{m_k} \mathbf{p}_k \cdot \nabla_{\mathbf{r}_k} f^{(N)} + \sum_{k=1}^{N} \mathbf{F}_k^{(N)} \cdot \nabla_{\mathbf{p}_k} f^{(N)} = 0 \,.$$

We shall now assume that we are dealing with a gaseous system that is sufficiently dilute that essentially only two-body collisions occur between its constituent molecules. Note that the force $\mathbf{F}_k^{(N)}$ may consist of two parts, one of which is an *external* force imposed upon all the molecules in the container, the other being the force exerted upon each molecule by all other molecules in the gas. We shall also treat the potential energy function $V(\mathbf{r}_1, \cdots , \mathbf{r}_N)$ as the sum of two-body potentials, while simultaneously restricting our consideration to central forces, i.e., potential energy functions that depend solely upon the distance between the centres of interacting pairs of molecules. Such potential energy functions are said to be *isotropic*, i.e., independent of angles.

These arguments allow us to write $\mathbf{F}_k^{(N)} \cdot \nabla_{\mathbf{p}_k}$ as

$$\mathbf{F}_k^{(N)} \cdot \nabla_{\mathbf{p}_k} = \mathbf{X}_k \cdot \nabla_{\mathbf{p}_k} + \left[-\nabla_{\mathbf{r}_k} V(\mathbf{r}_1, \cdots, \mathbf{r}_N) \cdot \nabla_{\mathbf{p}_k} \right]$$

or as

$$\mathbf{F}_k^{(N)} \cdot \nabla_{\mathbf{p}_k} = \mathbf{X}_k \cdot \nabla_{\mathbf{p}_k} + \mathbf{F}_k^{\text{int}} \cdot \nabla_{\mathbf{p}_k} \, ,$$

with the intermolecular force $\mathbf{F}_k^{\text{int}}$ acting on molecule k defined as

$$\mathbf{F}_k^{\text{int}} \equiv -\nabla_{\mathbf{r}_k} V(\mathbf{r}_1, \cdots, \mathbf{r}_N) \, ,$$

and representing the force exerted upon molecule k by all the other molecules in the container.

We shall take a slight detour to consider in more detail the nature of the intermolecular potential energy function and the intermolecular forces acting upon the molecules during a binary molecular collision. We begin by examining the potential energy function for the interaction between two molecules, i and j. If this interaction energy depends only upon the distance between the two molecules, then we may write it as

$$V(\mathbf{r}_i, \mathbf{r}_j) = V(|\mathbf{r}_i - \mathbf{r}_j|) \, , \qquad (7.5.12)$$

which allows us to calculate the corresponding intermolecular force exerted upon molecule i by molecule j as

$$\mathbf{F}_{ij}^{\text{int}} = -\nabla_{\mathbf{r}_i} V(|\mathbf{r}_i - \mathbf{r}_j|)$$

$$= -\frac{\mathbf{r}_i - \mathbf{r}_j}{|\mathbf{r}_i - \mathbf{r}_j|} \frac{\partial V}{\partial r_{ij}}$$

$$\equiv \frac{\mathbf{r}_i - \mathbf{r}_j}{|\mathbf{r}_i - \mathbf{r}_j|} F_{ij}^{\text{int}} \, ,$$

in which $r_{ij} \equiv |\mathbf{r}_i - \mathbf{r}_j|$ is the distance between molecules i and j and $F_{ij} \equiv -\dfrac{\partial V}{\partial r_{ij}}$ is the *magnitude* of the force exerted by molecule j upon molecule i. Notice that this force is directed from molecule j towards molecule i [i.e., it is the force exerted upon molecule i by molecule j]. A force that is equal in magnitude but opposite in direction is exerted by molecule i upon molecule j.

By employing the concept of pairwise additivity of the intermolecular potential energy function, we may now introduce the net force on molecule k exerted by all of the other molecules with which it interacts via

$$\mathbf{F}_k^{\text{int}} = -\nabla_{\mathbf{r}_k} \sum_{j=1}^{N} V(|\mathbf{r}_j - \mathbf{r}_k|)$$

$$= \sum_{j=1}^{N} \left[-\nabla_{\mathbf{r}_k} V(|\mathbf{r}_j - \mathbf{r}_k|) \right]$$

$$\equiv \sum_{j=1}^{N} \mathbf{F}_{jk}^{\text{int}} \,.$$

This result allows us to rewrite the Liouville equation as

$$\frac{\partial f^{(N)}}{\partial t} + \sum_{k=1}^{N} \frac{\mathbf{p}_k}{m_k} \cdot \nabla_{\mathbf{r}_k} f^{(N)} + \sum_{k=1}^{N} \mathbf{X}_k \cdot \nabla_{\mathbf{p}_k} f^{(N)} + \sum_{k=1}^{N} \mathbf{F}_k^{\text{int}} \cdot \nabla_{\mathbf{p}_k} f^{(N)} = 0 \,,$$

or by placing the term arising from the intermolecular interactions on the right-hand side, as

$$\frac{\partial f^{(N)}}{\partial t} + \sum_{k=1}^{N} \frac{\mathbf{p}_k}{m_k} \cdot \nabla_{\mathbf{r}_k} f^{(N)} + \sum_{k=1}^{N} \mathbf{X}_k \cdot \nabla_{\mathbf{p}_k} f^{(N)} = -\sum_{j<k} \sum_{k=1}^{N} \mathbf{F}_{jk}^{\text{int}} \cdot \nabla_{\mathbf{p}_k} f^{(N)} \,.$$

$$(7.5.13)$$

Note that the intermolecular interactions (or collision) term can be written explicitly for a pair-wise interaction potential energy function as

$$-\sum_{j<k} \sum_{k=1}^{N} \mathbf{F}_{jk}^{\text{int}} \cdot \nabla_{\mathbf{p}_k} f^{(N)} = \sum_{j<k} \sum_{k=1}^{N} \frac{\partial V}{\partial r_{jk}} \left[\frac{\mathbf{r}_j - \mathbf{r}_k}{r_{jk}} \cdot \nabla_{\mathbf{p}_k} f^{(N)} \right] \,,$$

but cannot readily be simplified further.

If we wish to calculate macroscopic properties of our dilute gas, we only need to know the one-particle distribution function $f^{(1)}$ and (at most) the two-particle distribution function $f^{(2)}$. Let us obtain an equation for $f^{(1)}$ by integrating $f^{(N)}$ over all other molecules. The normalization for $f^{(N)}$ that is most convenient for this purpose is

$$\int f^{(N)}(\mathbf{x}^N, t) \, \mathrm{d}\mathbf{x}^N = N! \,,$$

$$(7.5.14)$$

so that $f^{(1)}(\mathbf{x}, t)$ will be obtained from $f^{(N)}$ as

$$f^{(1)}(\mathbf{x}^1, t) = \frac{1}{(N-1)!} \int f^{(N)}(\mathbf{x}^N, t) \, d\mathbf{x}^{N-1}, \tag{7.5.15}$$

or more generally, $f^{(s)}(\mathbf{x}^s, t)$ is obtained from $f^{(N)}$ as

$$f^{(s)}(\mathbf{x}^s, t) = \frac{1}{(N-s)!} \int f^{(N)}(\mathbf{x}^N, t) \, d\mathbf{x}^{N-s}.$$

With this choice of normalization, we have

$$\int f^{(1)}(\mathbf{x}^1, t) \, d\mathbf{x}^1 = N, \tag{7.5.16}$$

and

$$\int f^{(2)}(\mathbf{x}^2, t) \, d\mathbf{x}^2 = N(N-1). \tag{7.5.17}$$

Thus, from $f^{(1)}(\mathbf{x}^1, t)$ we obtain the number of molecules, N, in the container, while from $f^{(2)}(\mathbf{x}^2, t)$ we obtain the number of pairs, $N(N-1)$, of molecules. We also see that the number density $n(\mathbf{r}, t)$ is given by

$$\int f^{(1)}(\mathbf{x}^1, t) \, d\mathbf{p}_1 = n(\mathbf{r}_1, t). \tag{7.5.18}$$

The corresponding quantity $n^{(2)}(\mathbf{r}_1, \mathbf{r}_2, t)$ obtained from $f^{(2)}(\mathbf{x}^2, t)$ by integrating over the momenta of the two particles is defined via

$$\iint f^{(2)}(\mathbf{x}^2, t) \, d\mathbf{p}_1 d\mathbf{p}_2 = n^{(2)}(\mathbf{r}_1, \mathbf{r}_2, t),$$

and is proportional to n^2.

To obtain the full equation of change for $f^{(1)}$, we shall integrate the terms in the Liouville equation over all but one molecule. We shall, however, find it convenient to deal with this process one term at a time. We find that

$$\frac{1}{(N-1)!} \int \frac{\partial f^{(N)}}{\partial t} \, d\mathbf{x}^{N-1} = \frac{\partial f^{(1)}}{\partial t},$$

$$\frac{1}{(N-1)!} \int \sum_{k=1}^{N} \frac{\mathbf{p}_k}{m_k} \cdot \nabla_{\mathbf{r}_k} f^{(N)} \, d\mathbf{x}^{N-1} \equiv \frac{\mathbf{p}}{m} \cdot \nabla f^{(1)},$$

$$\frac{1}{(N-1)!} \int \sum_{k=1}^{N} \mathbf{X}_k \cdot \nabla_{\mathbf{r}_k} f^{(N)} \, d\mathbf{x}^{N-1} \equiv \mathbf{X} \cdot \nabla_{\mathbf{p}} f^{(1)},$$

for the three terms on the left-hand side of the Liouville equation (7.5.13). Notice that we have also suppressed the subscript '1' in the final results on the right-hand sides of our equations. The intermolecular interaction term on the right-hand side of the Liouville equation is slightly more complicated than the other terms, however, and must be handled more carefully. We begin by integrating over the phase space of $N-2$ of the N molecules and defining in the process a new quantity $\theta(\mathbf{r}_1, \mathbf{r}_2)$ via

$$\frac{1}{(N-2)!} \int \sum_{j<k} \sum_{k=1}^{N} \frac{\partial V}{\partial r_{jk}} \left[\frac{\mathbf{r}_j - \mathbf{r}_k}{r_{jk}} \cdot \nabla_{\mathbf{p}_k} f^{(N)} \right] d\mathbf{x}^{N-2} \equiv \theta(\mathbf{r}_1, \mathbf{r}_2, t) f^{(2)}(\mathbf{x}_1, \mathbf{x}_2, t),$$

so that the integration of this term over $N-1$ particles can be expressed as

$$\frac{1}{(N-1)!} \int \sum_{j<k} \sum_{k=1}^{N} \frac{\partial V}{\partial r_{jk}} \left[\frac{\mathbf{r}_j - \mathbf{r}_k}{r_{jk}} \cdot \nabla_{\mathbf{p}_k} f^{(N)} \right] d\mathbf{x}^{N-1}$$

$$\equiv \int \theta(\mathbf{r}_1, \mathbf{r}_2, t) f^{(2)}(\mathbf{x}_1, \mathbf{x}_2, t) \, d\mathbf{x}_2.$$

This result allows us finally to be able to write down the equation that determines $f^{(1)}$ as

$$\frac{\partial f^{(1)}}{\partial t} + \frac{\mathbf{p}}{m} \cdot \nabla f^{(1)} + \mathbf{X} \cdot \nabla_{\mathbf{p}} f^{(1)} = \int \theta(\mathbf{r}_1, \mathbf{r}_2, t) f^{(2)}(\mathbf{x}_1, \mathbf{r}_2, t) \, d\mathbf{r}_2$$

$$\equiv \left(\frac{\delta f^{(1)}}{\delta t} \right)_{\text{collision}}. \tag{7.5.19}$$

7.5.2   *Interlude on Binary Collision Kinematics*

Let us consider a collision between two molecules possessing masses m_i, velocities $\mathbf{v}_i$, and (internal) rotational angular momenta $\mathbf{j}_i$, with $i = 1, 2$. To distinguish the postcollisional values for the velocities and rotational angular momenta for these attributes from their precollisional values, we shall designate the postcollisional values by primes: thus $\mathbf{v}_i'$ and $\mathbf{j}_i'$ represent the postcollisional velocities and rotational angular momenta of the colliding molecules.[2] As you learned in your first physics

---

[2]In traditional texts and monographs on the kinetic theory of fluids, you will often find precollisional quantities designated by primes and postcollisional quantities unprimed: we are following the normal conventions for classical collision dynamics with respect to the use of primed and unprimed quantities.

course, the laws of conservation of mass, linear momentum, angular momentum, and energy all apply to binary collision dynamics. For a (chemically) nonreactive collision, the conservation of mass has the consequence that both m_1 and m_2 are unchanged after the collision, while for a reactive collision, it means that the total mass, $M = m_1 + m_2$ of the pair of colliding molecules remains unchanged, even though m_1' and m_2' need not equal m_1 and m_2, respectively.

The conservation of linear momentum for a nonreactive binary collision can be expressed in the form

$$m_1\mathbf{v}_1 + m_2\mathbf{v}_2 = m_1\mathbf{v}_1' + m_2\mathbf{v}_2' , \tag{7.5.20}$$

in which precollisional values appear on the left-hand side and postcollisional values appear on the right-hand side of a conservation equation. The conservation of energy can similarly be written as

$$\tfrac{1}{2}m_1v_1^2 + \epsilon_{1,\text{int}} + \tfrac{1}{2}m_2v_2^2 + \epsilon_{2,\text{int}} = \tfrac{1}{2}m_1{v_1'}^2 + \epsilon_{1,\text{int}}' + \tfrac{1}{2}m_2{v_2'}^2 + \epsilon_{2,\text{int}}' , \tag{7.5.21}$$

in which ϵ_{int} represents the molecular internal (normally vibrotational) energy.

To deal with binary collision dynamics, it is customary to define centre-of-mass (CM) and relative velocities by

$$\mathbf{V}_{\text{CM}}' = \frac{m_1}{M}\mathbf{v}_1' + \frac{m_2}{M}\mathbf{v}_2' , \qquad \mathbf{V}_{\text{CM}} = \frac{m_1}{M}\mathbf{v}_1 + \frac{m_2}{M}\mathbf{v}_2 , \tag{7.5.22a}$$

and

$$\mathbf{v}_r' = \mathbf{v}_1' - \mathbf{v}_2' \equiv v_r'\mathbf{e}' , \qquad \mathbf{v}_r = \mathbf{v}_1 - \mathbf{v}_2 \equiv v_r\mathbf{e} , \tag{7.5.22b}$$

in which $\mathbf{e}'$ and $\mathbf{e}$ are unit vectors in the directions of increasing v_r' and v_r, respectively. In terms of these centre-of-mass and reduced velocity variables, the conservation of linear momentum reduces to

$$\mathbf{P} = \mathbf{P}' ; \quad \mathbf{P} = M\mathbf{V}_{\text{CM}}, \ \ \mathbf{P}' = M\mathbf{V}_{\text{CM}}',$$

which reduces to a statement that the CM velocity changes neither in magnitude nor in direction during a binary collision, i.e.,

$$\mathbf{V}_{\text{CM}}' = \mathbf{V}_{\text{CM}} . \tag{7.5.23}$$

In a similar manner, the conservation of energy during a binary collision becomes

$$\tfrac{1}{2}m_r{v_r'}^2 = \tfrac{1}{2}m_rv_r^2 + \Delta\epsilon_{\text{int}} , \tag{7.5.24}$$

in which $m_r \equiv m_1m_2/M$ is the reduced mass of the two molecules, and $\Delta\epsilon_{\text{int}}$ is the difference between the precollisional and postcollisional internal energies of the collision partners, given by

$$\Delta\epsilon_{int} \equiv \epsilon_{1,int} + \epsilon_{2,int} - \epsilon'_{1,int} - \epsilon'_{2,int} \,.$$

For spherically symmetric intermolecular interactions, such as those occurring between atom pairs, the momentum conservation equation is unaffected, but the energy conservation equation reduces to the simple statement that only the direction of the relative velocity vector may change over the course of a binary collision, with the consequence that $v'_r = v_r$.

Inversion of the kinematic equations (7.5.22), combined with Eq. (7.5.23) allows $\mathbf{v}_1$, $\mathbf{v}_2$, $\mathbf{v}'_1$, and $\mathbf{v}'_2$ to be expressed in terms of $\mathbf{V}_{CM}$, $\mathbf{v}_r$, and $\mathbf{v}'_r$ as

$$\mathbf{v}_1 = \mathbf{V}_{CM} + \frac{m_1}{M}\mathbf{v}_r \,, \quad \mathbf{v}_2 = \mathbf{V}_{CM} - \frac{m_2}{M}\mathbf{v}_r \,, \tag{7.5.25a}$$

and

$$\mathbf{v}'_1 = \mathbf{V}_{CM} + \frac{m_1}{M}\mathbf{v}'_r \,, \quad \mathbf{v}'_2 = \mathbf{V}_{CM} - \frac{m_2}{M}\mathbf{v}'_r \,. \tag{7.5.25b}$$

These equations are traditionally referred to as the *kinematic equations for a binary collision*.

From Eqs. (7.5.22a) and (7.5.22b), we may show that Jacobian determinants for the transformation between precollisional and postcollisional CM and relative coordinates are both unity, that is,

$$\frac{\partial(\mathbf{v}_r, \mathbf{V}_{CM})}{\partial(\mathbf{v}_1, \mathbf{v}_2)} = 1 \,, \quad \frac{\partial(\mathbf{v}'_r, \mathbf{V}_{CM})}{\partial(\mathbf{v}'_1, \mathbf{v}'_2)} = 1 \,.$$

These Jacobian values allow us to deduce that

$$d\mathbf{v}_1 d\mathbf{v}_2 = d\mathbf{v}_r d\mathbf{V}_{CM} = v_r^2 dv_r d\mathbf{e} d\mathbf{V}_{CM} \tag{7.5.26a}$$

and

$$d\mathbf{v}'_1 d\mathbf{v}'_2 = d\mathbf{v}'_r d\mathbf{V}_{CM} = v_r^2 dv_r d\mathbf{e}' d\mathbf{V}_{CM} \tag{7.5.26b}$$

for binary collisions between atoms (or even for molecules with a spherically symmetric potential energy function). When we multiply the first of these two identities by $d\mathbf{v}'_r$ and the second by $d\mathbf{v}_r$, their right-hand sides become equal and we obtain the relation

$$d\mathbf{v}'_r d\mathbf{v}_1 d\mathbf{v}_2 = d\mathbf{v}_r d\mathbf{v}'_1 d\mathbf{v}'_2 \,, \tag{7.5.26c}$$

in which it is understood that v_r is to be determined from the conservation of energy as expressed by Eq. (7.5.24). For elastic collisions, this relation reduces to

$$d\mathbf{e}' d\mathbf{v}_1 d\mathbf{v}_2 = d\mathbf{e} d\mathbf{v}'_1 d\mathbf{v}'_2 \,,$$

which is sometimes referred to as the *Liouville theorem for elastic collisions*.

We shall now restrict ourselves to the binary collision of two atoms. The Hamiltonian for a system of two atoms in interaction through a central potential energy function V is given by

$$\mathcal{H}(\mathbf{r}_1, \mathbf{r}_2, \mathbf{p}_1, \mathbf{p}_2) = \frac{\mathbf{p}_1^2}{2m_1} + \frac{\mathbf{p}_2^2}{2m_2} + V(|\mathbf{r}_2 - \mathbf{r}_1|), \qquad (7.5.27)$$

with $\mathbf{r}_1$, $\mathbf{r}_2$ the positions of the atoms relative to some space-fixed origin O, $r \equiv |\mathbf{r}_2 - \mathbf{r}_1|$ is the distance between them ($\mathbf{r} = \mathbf{r}_2 - \mathbf{r}_1$ is the position of atom 1 relative to atom 2, and is called the relative position vector), m_1, m_2 are the atomic masses, and $\mathbf{p}_1$, $\mathbf{p}_2$ are the linear momenta of the atoms.

Panel (a) of Fig. 7.3 depicts the overall geometry associated with a binary elastic collision between two particles, and shows the particle masses, both pre- and post-collisional velocities and relative positions, together with the path taken by their centre-of-mass (CM) during the interaction. The geometry associated with the two

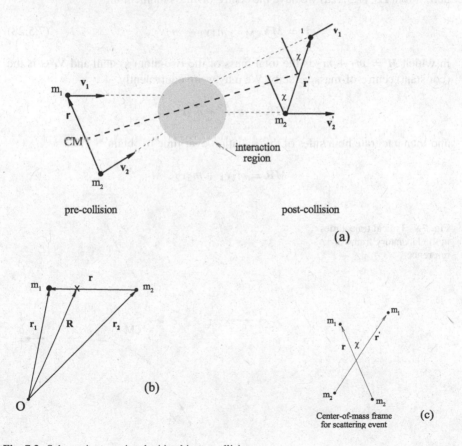

pre-collision

post-collision

(a)

(b)

Center-of-mass frame for scattering event (c)

Fig. 7.3 Schematics associated with a binary collision

particles in interaction is illustrated in Panel (b), with $\mathbf{R}$ the position of the centre-of-mass, $\mathbf{r}_1$ and $\mathbf{r}_2$ the positions of the colliding atoms, all relative to a space-fixed origin O. Panel (c) shows the pre- and post-collisional relative position vectors and the scattering angle χ in the centre-of-mass frame. Because the interaction energy depends only upon the relative distance r between the two atoms, we may anticipate that $\mathcal{H}$ can be simplified if we also express the kinetic energy in terms of the relative and centre-of-mass momenta, by which we understand the linear momenta conjugate to position coordinates $\mathbf{R}$ and $\mathbf{r}$. Because V does not depend upon $\mathbf{R}$ and the kinetic energy terms in $\mathcal{H}$ cannot introduce $\mathbf{R}$ into the Hamiltonian obtained upon transformation into CM and relative coordinates, $\mathcal{H}$ will be cyclic in $\mathbf{R}$, with the consequence that its conjugate momentum $\mathbf{P}$ will be a conserved quantity, i.e., the centre-of-mass momentum $\mathbf{P} = \mathbf{p}_1 + \mathbf{p}_2$ is conserved.

As it has been more traditional to employ a velocity diagram of the type shown in Fig. 7.4 (for a purely repulsive collision between two atoms) rather than linear momenta to illustrate binary collision dynamics, we shall take the same approach here. From Eq. (7.5.22a) we have the centre-of-mass momentum

$$\mathbf{P} = M\mathbf{V}_{\mathrm{CM}} = m_1\mathbf{v}_1 + m_2\mathbf{v}_2 \,, \tag{7.5.28}$$

in which $M = m_1 + m_2$ is the total mass of the two-atom system and $\mathbf{V}_{\mathrm{CM}}$ is the (constant) centre-of-mass velocity. We may write equivalently,

$$M\dot{\mathbf{R}} = m_1\dot{\mathbf{r}}_1 + m_2\dot{\mathbf{r}}_2 \,,$$

and then integrate both sides of this equation over time to obtain

$$M\mathbf{R} = m_1\mathbf{r}_1 + m_2\mathbf{r}_2 \,.$$

Fig. 7.4 Typical trajectories in the laboratory frame of reference

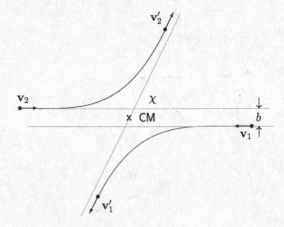

The location $\mathbf{R}$ of the centre-of-mass in the space-fixed coordinates is thus given as

$$\mathbf{R} = \frac{m_1}{M}\mathbf{r}_1 + \frac{m_2}{M}\mathbf{r}_2 . \qquad (7.5.29)$$

The time derivative of the relative position vector $\mathbf{r}$ gives us the relative velocity $\mathbf{v}_{\text{rel}}$ of the atom pair as

$$\mathbf{v}_{\text{rel}} = \dot{\mathbf{r}}_1 - \dot{\mathbf{r}}_2 = \mathbf{v}_2 - \mathbf{v}_1 ,$$

which is related to the momenta $\mathbf{p}_1$ and $\mathbf{p}_2$ as

$$\mathbf{v}_{\text{rel}} = \frac{m_1\mathbf{p}_2 - m_2\mathbf{p}_1}{m_1 m_2} ,$$

or

$$m_1 m_2 \mathbf{v}_{\text{rel}} = m_1\mathbf{p}_2 - m_2\mathbf{p}_1 .$$

Division on both sides of this equation by the total mass M gives

$$\mathbf{p}_{\text{rel}} \equiv \frac{m_1 m_2}{M}\mathbf{v}_{\text{rel}} = \frac{m_1}{M}\mathbf{p}_2 - \frac{m_2}{M}\mathbf{p}_1 \qquad (7.5.30)$$

for the relative linear momentum associated with the interacting atoms: this is the momentum conjugate to the relative position vector $\mathbf{r}$. Equations (7.5.28) and (7.5.30) can be inverted to give $\mathbf{p}_1$ and $\mathbf{p}_2$ in terms of $\mathbf{P}$ and $\mathbf{p}_{\text{rel}}$ as

$$\mathbf{p}_1 = \frac{m_1}{M}\mathbf{P} - \mathbf{p}_{\text{rel}} , \qquad \mathbf{p}_2 = \frac{m_2}{M}\mathbf{P} + \mathbf{p}_{\text{rel}} . \qquad (7.5.31)$$

Using results (7.5.30) and (7.5.31) in Eq. (7.5.27) gives

$$\mathcal{H} = \frac{\mathbf{P}^2}{2M} + \left[\frac{\mathbf{p}_{\text{rel}}^2}{2m_{\text{r}}} + v(r) \right]$$

$$= \mathcal{H}_{\text{CM}} + \mathcal{H}_{\text{rel}} . \qquad (7.5.32)$$

Now, because $\mathbf{P}$ is a constant, so also is $\mathcal{H}_{\text{CM}}$. The two-body problem for the interaction of two atoms has thus been reduced through this argument into an *effective* single-body problem in terms of the motion of a fictitious particle of mass m_{r} in the field of a central potential energy $V(r)$.

The two degrees of freedom associated with the planar motion of a particle of mass m_{r} in the field of a central potential energy function are the distance r and an angle ϕ: we may write the Hamiltonian for such motion [see Section 1 of Appendix G] as

$$\mathcal{H}_{\text{rel}} = \frac{p_r^2}{2m_{\text{r}}} + \frac{p_\phi^2}{2m_{\text{r}}r^2} + V(r). \tag{7.5.33}$$

Notice that the angle ϕ does not itself appear in our expression for $\mathcal{H}$, which means that its conjugate momentum p_ϕ, traditionally called the orbital angular momentum and designated by the symbol L, is a constant of the motion: we shall see later that we may write it in the form $L = m_{\text{r}}gb$ in terms of the relative speed g and a quantity b, called the *impact parameter*,[3] also illustrated in Fig. 7.4. We may thus write the relative energy $E \equiv \mathcal{H}_{\text{rel}}$ as

$$E = \frac{p_r^2}{2m_{\text{r}}} + \frac{L^2}{2m_{\text{r}}r^2} + V(r). \tag{7.5.34}$$

From the first of Hamilton's equations, Eq. (7.5.1), we have

$$\dot{r} = \frac{\partial \mathcal{H}}{\partial p_r} = \frac{p_r}{m_{\text{r}}} ; \quad \dot{\phi} = \frac{\partial \mathcal{H}}{\partial p_\phi} = \frac{p_\phi}{m_{\text{r}}r^2} = \frac{L}{m_{\text{r}}r^2}, \tag{7.5.35}$$

so that we obtain the differentials dr and $d\phi$ for the radial and angular coordinates of the relative motion as

$$dr = \sqrt{\frac{2}{m_{\text{r}}}(E - V) - \frac{L^2}{m_{\text{r}}^2 r^2}}\, dt, \tag{7.5.36a}$$

and

$$d\phi = \frac{L}{m_{\text{r}}r^2}\, dt \quad \Rightarrow \quad dt = \frac{m_{\text{r}}r^2}{L}\, d\phi. \tag{7.5.36b}$$

Upon replacing dt in Eq. (7.5.36b) in terms of dr from Eq. (7.5.36a), we obtain $d\phi$ in terms of r and dr as

$$d\phi = \frac{L}{r^2\sqrt{2m_{\text{r}}(E - V) - L^2/r^2}}\, dr,$$

from which ϕ can be obtained as

$$\phi = L \int \frac{1}{r^2\sqrt{2m_{\text{r}}(E - V) - L^2/r^2}}\, dr + C, \tag{7.5.37}$$

---

[3]Note that because both the relative kinetic energy $E_{\text{rel}} = \frac{1}{2}m_{\text{r}}v_{\text{rel}}^2$ and the orbital angular momentum $L = m_{\text{r}}v_{\text{rel}}b$ are conserved quantities during an elastic collision, b is constant for each trajectory associated with our fictitious particle of mass m_{r}.

Fig. 7.5 A typical binary
collision trajectory in the
centre-of-mass frame

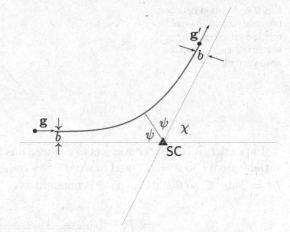

with C the constant of integration.

We think of a binary collision as being typified by three distinct regimes, namely before collision, during collision, after collision. These concepts are illustrated in Fig. 7.5.

These regimes are determined by the nature of the interaction potential energy. If we assign a typical potential energy as having a distance (which we shall designate by r_0 and call the range of the potential) beyond which it has no further influence on the outcome of the collision, then 'before collision' means that the collision partners are approaching one another (though not yet interacting) and $r > r_0$, so that $V(r) = 0$, 'after collision' means that the collision partners are receding from one another and $r > r_0$, so $V(r) = 0$ once again: 'during collision' then refers to the region for which $r < r_0$ and $V(r) \neq 0$.

The scattering event is represented in Fig. 7.5 as the motion of a fictitious particle of mass m_r relative to a scattering centre (SC) located at the origin, and having pre- and post-collisional (relative) velocities $\mathbf{g}$ and $\mathbf{g}'$, respectively. The scattering angle χ, the impact parameter b, and the distance of closest approach (represented by a line from the scattering centre to the trajectory) are all illustrated in this figure. We can also see from Fig. 7.5 that scattering is symmetric about a vector $\mathbf{r}_{\min}$ between the scattering centre and the point of closest approach between the two atoms (whose direction is referred to as defining an 'apse line' for the collision). Alternatively, the relation between $\mathbf{g}$, $\mathbf{g}'$, and χ may be illustrated in the form of a vector diagram, as in Fig. 7.6.

We may also see from Fig. 7.6 that χ is related to the angle ψ between $\mathbf{g}$ and the apse line by $\chi = \pi - 2\psi$. The angle ψ is, in turn, related to the values of the polar coordinate ϕ corresponding to the relative separations r at infinity and at the closest approach of the two atoms, and may be obtained from Eq. (7.5.37) as

$$\psi = \phi(r_\infty) - \phi(r_{\min}) = \int_{r_{\min}}^{\infty} \frac{L/r^2}{\sqrt{2m_r(E - V) - L^2/r^2}} \, dr . \qquad (7.5.38)$$

Fig. 7.6 Relation between
pre- and post-collisional
relative velocities, the
scattering angle χ and the
apse vector **k**

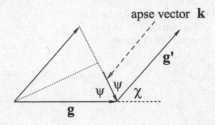

In practice, it is more convenient to evaluate this integral by making a change of variable from r to $u = r^{-1}$ and to write $L^2 = (m_r bg)^2$ in terms of $E = \frac{1}{2} m_r g^2$ as $L^2 = 2m_r b^2 E$, so that the angle ψ is obtained as

$$\psi = \int_0^{u_{\max}} \frac{b}{\sqrt{1 - b^2 u^2 - V(u)/E}} \, du \,, \qquad (7.5.39)$$

with $u_{\max}$ determined from the condition

$$1 - b^2 u_{\max}^2 - \frac{V(u_{\max})}{E} = 0 \,.$$

7.5.3 Scattering Cross Section Concept

Let us consider a beam of atoms with energy E and intensity I (expressed as the number of atoms per square meter per second) incident upon a scattering centre located at the origin. The number of atoms scattered (deflected) per second into an element de' around the solid angle e' will be proportional to the incident intensity I and the element of solid angle (see also Fig. 7.7): if we denote the proportionality factor by σ, and call it the differential scattering cross section, we obtain $I\sigma$ de'. An examination of Fig. 7.7, in which the annular region associated with the impact parameter b in the central part of the figure has been magnified in order to illustrate the relevant volume element for the cylindrical geometry associated with the incoming beam of atoms, shows that this expression must also represent the number of atoms that have passed through an annulus, of width db, of the beam. This then gives the number of deflected atoms as I dϕbdb. Note that in Fig. 7.7, the coordinates for the scattered atoms are spherical polar coordinates and the geometry has been arranged so that the scattering angle χ is the azimuthal angle (i.e., the angle between the final scattered atom and the polar-axis).

The differential cross section σ must therefore also satisfy the relation

$$I\sigma \, \mathrm{d}e' = I \, \mathrm{d}\phi b \mathrm{d}b \,, \qquad (7.5.40)$$

so that

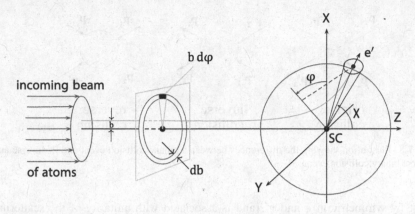

Fig. 7.7 Geometric arrangement associated with scattering of an incoming beam of atoms by a scattering centre (SC). After Figure 3.5 of [5]

$$\sigma \, de' = b \, db \, d\phi . \qquad (7.5.41)$$

From relation (7.5.41) and the element of solid angle $de' = \sin \chi \, d\chi \, d\phi$, we can express the differential scattering cross section as a function of the scattering energy E and the scattering angle χ as

$$\sigma(E, \chi) = \frac{b(E, \chi)}{\sin \chi} \frac{db}{d\chi} . \qquad (7.5.42)$$

The differential scattering cross section is so called because it refers to the number of atoms scattered into a specific direction. If we were to integrate the differential cross section over all directions, we would arrive at the total number of atoms scattered out of the beam of atoms. This quantity is referred to as the 'total cross section' or, more properly, as the 'integral cross section', and is given directly as

$$\sigma_{\text{int}} = \int_{4\pi} \sigma \, de' .$$

We may also write $\sigma_{\text{int}} = \pi r_0^2$ if r_0 is the range of the potential energy function (meaning that the atoms with impact parameters greater than r_0 are not scattered out of the beam): the integral cross section thus represents the obstructional area that the scattering centre presents to the incident beam.

Inverse and Reverse Collisions

For an inverse collision (see Fig. 7.8) to give the same contribution to the scattering cross section as did the original collision, the differential scattering cross section

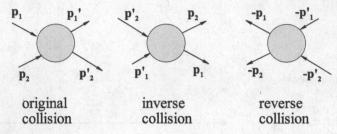

original   inverse   reverse
collision   collision   collision

Fig. 7.8 Illustration showing the differences between a binary collision event and its inverse and reverse binary collision events

must by symmetric in g and g' [and is associated with unitarity of the scattering matrix in QM, for example], while reverse collisions will always contribute the same amount to the scattering process as the original collision because the equations of motion are time-reversal invariant.

7.5.4 The Binary Collision Term and the Boltzmann Equation

We shall write $\left(\dfrac{\delta f^{(1)}}{\delta t}\right)_{\text{coll}}$ as the difference between two terms, $\Gamma^{(+)}$ and $\Gamma^{(-)}$, with $\Gamma^{(+)}$ representing the results of collisions during time interval δt in which an atom enters the phase-space volume element due to collisions with other atoms (it is also known as the 'gain term'), while $\Gamma^{(-)}$ represents the results of collisions during time interval δt in which atoms are ejected from the phase-space volume element by collisions (also known as the 'loss term'). We thus have

$$\left(\frac{\delta f^{(1)}}{\delta t}\right)_{\text{coll}} = \Gamma^{(+)} - \Gamma^{(-)},$$

and now we need to find explicit expressions for $\Gamma^{(+)}$ and $\Gamma^{(-)}$.

The phase-space volume element that we need to consider for this discussion is given by $d\mathbf{r}_1 d\mathbf{p}_1$: to calculate $\Gamma^{(-)}$, we must take into account all binary collisions in which an atom is ejected from the interval $d\mathbf{p}_1$ about $\mathbf{p}_1$. Let us consider two atoms, one located in the phase-space volume element $d\mathbf{r}_1 d\mathbf{p}_1$ about $(\mathbf{r}_1, \mathbf{p}_1)$ with the other in the phase-space volume element $d\mathbf{r}_2 d\mathbf{p}_2$ about $(\mathbf{r}_2, \mathbf{p}_2)$, and determine how many atoms with momentum $\mathbf{p}_2$ in $d\mathbf{r}_2$ undergo collisions with atoms with momentum $\mathbf{p}_1$ in $d\mathbf{r}_1$ in time δt. By definition of the two-particle distribution function $f^{(2)}(\mathbf{x}_1, \mathbf{x}_2, t)$, this must be given by

Fig. 7.9 Determination of
the volume element $d\mathbf{r}_2$ for
the Boltzmann collision
integral. After figure 3.10 of
[5]

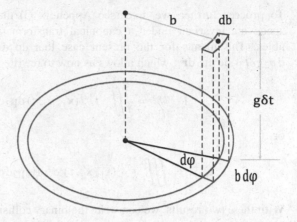

$$\Gamma^{(-)}\delta t = \iiint f^{(2)}(\mathbf{x}_1, \mathbf{x}_2, t)\, d\mathbf{p}_1 d\mathbf{r}_1 d\mathbf{p}_2 d\mathbf{r}_2 \,.$$

All atoms with momentum $\mathbf{p}_2$ that lie within a cylindrical shell of height $g\delta t$ and base area $b\,db\,d\phi$, depicted in Fig. 7.9, will undergo collision with an atom having momentum $\mathbf{p}_1$ in time δt, which means that $d\mathbf{r}_2$ will be given as

$$d\mathbf{r}_2 = g\delta t\, b\,db\,d\phi = g\delta t\sigma\, d\mathbf{e}' \,,$$

with the consequence that the loss term becomes

$$\Gamma^{(-)} = \left[\iiint f^{(2)}(\mathbf{x}_1, \mathbf{x}_2, t)\, d\mathbf{p}_2\, g\sigma\, d\mathbf{e}' \right] d\mathbf{p}_1 d\mathbf{r}_1 \,. \qquad (7.5.43)$$

For the calculation of $\Gamma^{(+)}$, we note that a binary collision that sends a particle labelled 1 into $d\mathbf{p}_1$ about $\mathbf{p}_1$ in time δt is the inverse of the original collision, a process which we can write as

$$(\mathbf{p}_1, \mathbf{p}_2) \quad \longrightarrow \quad (\mathbf{p}_1', \mathbf{p}_2') \,,$$

or

$$(\mathbf{p}_1'', \mathbf{p}_2') \quad \longrightarrow \quad (\mathbf{p}_1', \mathbf{p}_2) \,,$$

so that the gain term can be written down as

$$\Gamma^{(+)}\delta t = \iiint f^{(2)}(\mathbf{x}_1', \mathbf{x}_2', t)\, d\mathbf{p}_2' d\mathbf{r}_2' d\mathbf{p}_1' d\mathbf{r}_1' \delta t \,.$$

To proceed further, we note (see Appendix G) that volume elements in phase space are invariant under a canonical transformation of the phase-space variables. This means for the present case that $d\mathbf{p}_2' d\mathbf{r}_2' d\mathbf{p}_1' d\mathbf{r}_1' = d\mathbf{p}_2 d\mathbf{r}_2 d\mathbf{p}_1 d\mathbf{r}_1 = dp_2 g \delta t \, \sigma \, de' d\mathbf{p}_1 d\mathbf{r}_1$, which allows us now to rewrite the gain term as

$$\Gamma^{(+)} \delta t = \iiint f^{(2)}(\mathbf{x}_1', \mathbf{x}_2', t) \, d\mathbf{p}_2 d\mathbf{r}_2 d\mathbf{p}_1 d\mathbf{r}_1 \delta t$$

or

$$\Gamma^{(+)} = \iiint f^{(2)}(\mathbf{x}_1', \mathbf{x}_2', t) \, \sigma \, de' d\mathbf{p}_2 d\mathbf{p}_1 d\mathbf{r}_1 \,. \tag{7.5.44}$$

With these two results, we can write the binary collision term as

$$\left(\frac{\delta f}{\delta t} \right)_{\text{coll}} = \Gamma^{(+)} - \Gamma^{(-)}$$

$$= \iiint \left[f^{(2)}(\mathbf{x}_1', \mathbf{x}_2', t) - f^{(2)}(\mathbf{x}_1, \mathbf{x}_2, t) \right] g b \, db \, d\phi \, d\mathbf{r}_1 d\mathbf{p}_1 d\mathbf{p}_2 \,.$$

$$\tag{7.5.45}$$

Completion of our derivation of the Boltzmann (transport) equation requires three additional assumptions: the first of these assumptions is called, for want of a better terminology, 'molecular chaos', the second assumption is that of 'uniformity in space', and the third assumption required is that external forces on the system are not sufficiently strong to influence binary collisions between molecules during the rather short time that such collisions typically take. Let us summarize them briefly.

1. **Molecular chaos assumption**: only binary collisions occur, and molecular linear momenta before and after a collision are uncorrelated (i.e., independent); it is based in part upon the observation that for many binary collisions the relative kinetic energy ('collision energy') is sufficiently large that the intermolecular forces in operation are all of a short-ranged nature, so that for distances larger than the range, there are no forces active on the molecules. This also causes the duration of a collision to be very short [typically of order of a few picoseconds (10^{-12}s)]. The end result of this assumption is that we can approximate the two-molecule distribution function as the product of two single-molecule (singlet) distribution functions, viz.,

$$f^{(2)}(\mathbf{x}_1, \mathbf{x}_2, t) \simeq f^{(1)}(\mathbf{x}_1, t) f^{(1)}(\mathbf{x}_2, t) \,. \tag{7.5.46}$$

2. **Uniformity approximation**: this is the assumption that the singlet distribution function is a slowly-changing function of position, so that during a binary collision we can write

$$f^{(1)}(\mathbf{r}_1, \mathbf{p}_1, t) f^{(1)}(\mathbf{r}_2, \mathbf{p}_2, t) \simeq f^{(1)}(\mathbf{r}_1, t; \mathbf{p}_1) f^{(1)}(\mathbf{r}_1, t; \mathbf{p}_2), \qquad (7.5.47)$$

which is sometimes referred to as *localization* of the collision.

3. **Neglect of external forces on the collision process**: this simply means that an applied external field must not be so strong that it contorts the collision trajectories in the time that it takes for a binary collision to occur.

Upon employing these three assumptions in Eq. (7.5.45), we obtain the final form for the Boltzmann equation in which the collision integral is frequently written as

$$\left(\frac{\delta f^{(1)}}{\delta t} \right)_{\text{collision}} = \iint \left[f_1' f_2' - f_1 f_2 \right] g\sigma \, \mathrm{de}' \mathrm{dv}_2, \qquad (7.5.48)$$

with g being the relative speed of the colliding molecules and σ the collision cross section. When the collision is between two atoms, the integration over the element of solid angle de' gives rise to a factor 2π (arising from integration over the azimuthal angle ϕ shown in Fig. 7.9).

The single-particle distribution functions in Eq. (7.5.48) are now considered as functions of velocity rather than of linear momentum and are designated by single subscripts only; moreover, the superscript '(1)' is suppressed, so that f_1 and f_1' are given as

$$f_1 \equiv f^{(1)}(\mathbf{v}_1) = n_1 \left(\frac{m_1}{2\pi k_{\mathrm{D}} T} \right)^{\frac{3}{2}} e^{-m_1 v_1^2 / (2k_{\mathrm{B}} T)},$$

$$f_1' \equiv f^{(1)}(\mathbf{v}_1') = n_1 \left(\frac{m_1}{2\pi k_{\mathrm{B}} T} \right)^{\frac{3}{2}} e^{-m_1 v_1'^2 / (2k_{\mathrm{B}} T)},$$

while f_2 and f_2' are similarly identified as $f_2 \equiv f^{(1)}(\mathbf{v}_2)$ and $f_2' \equiv f^{(1)}(\mathbf{v}_2')$.

Upon combining Eq. (7.5.47) with Eq. (7.5.19), we arrive at the Boltzmann equation for a dilute pure monatomic gas, namely

$$\frac{\partial f_1}{\partial t} + \mathbf{v}_1 \cdot \nabla f_1 + \mathbf{X} \cdot \nabla_v f_1 = \iint \left[f_1' f_2' - f_1 f_2 \right] g\sigma \, \mathrm{de}' \mathrm{dv}_2. \qquad (7.5.49)$$

7.5.5 Nonequilibrium Phenomena and the Boltzmann Equation

We shall focus upon nonequilibrium phenomena in the dilute gas regime, which in practice refers to gases at pressures between about 100 Pa and 1 MPa. If we have N atoms, each with the same mass m_1, in a container of volume V, then the average number of atoms at time t in volume element $\mathrm{dv}_1 \mathrm{dr}_1$ of the classical single-particle phase space is given by $f(\mathbf{v}_1, \mathbf{r}_1, t)\, d\mathbf{r}_1 d\mathbf{v}_1$: this quantity can be interpreted as the probability for finding an atom within the phase-space volume $d\mathbf{r}_1 d\mathbf{v}_1$. The *local* number density $n(\mathbf{r}_1, t)$ at time t and position $\mathbf{r}_1$ is obtained by integrating $f(\mathbf{v}_1, t)$ over $\mathbf{v}_1$, i.e.,

$$n(\mathbf{r}_1, t) = \int f(\mathbf{v}_1, \mathbf{r}_1, t)\, d\mathbf{v}_1 \,. \tag{7.5.50}$$

The total number of atoms in the volume V at time t is then given by

$$N(t) = \int_V n(\mathbf{r}_1, t)\, d\mathbf{r}_1 = \int_V f(\mathbf{v}_1, \mathbf{r}_1, t)\, d\mathbf{v}_1 d\mathbf{r}_1 \,.$$

The local nonequilibrium average $\langle\psi\rangle_{\mathrm{ne}}$ of an arbitrary function $\psi(\mathbf{v}_1)$ is expressed in terms of $f(\mathbf{v}_1, \mathbf{r}_1, t)$ according to

$$n(\mathbf{r}_1, t)\langle\psi\rangle_{\mathrm{ne}}(\mathbf{r}_1, t) = \int f(\mathbf{v}_1, \mathbf{r}_1, t)\psi(\mathbf{v}_1)\, d\mathbf{v}_1 \,. \tag{7.5.51}$$

While these relations need to be generalized slightly in order to apply them to a molecular gas, much of what we obtain for a monatomic gas carries over in a straightforward manner to molecular gases.

Multiplication of both sides of the Boltzmann equation, Eq. (7.5.48), by a function of the velocity, $\psi(\mathbf{v}_1)$, followed by integration over $\mathbf{v}_1$ gives

$$\frac{\partial}{\partial t}\,(n\langle\psi\rangle_{\mathrm{ne}}) + \nabla\cdot(n\langle\mathbf{v}_1\psi(\mathbf{v}_1)\rangle)_{\mathrm{ne}}) = \left(\frac{\delta n\langle\psi\rangle_{\mathrm{ne}}}{\delta t}\right)_{\mathrm{coll}}, \tag{7.5.52}$$

with the right-hand side given as

$$\left(\frac{\delta n\langle\psi\rangle_{\mathrm{ne}}}{\delta t}\right)_{\mathrm{coll}} = \iiint (f_1' f_2' - f_1 f_2)\sigma g\psi(\mathbf{v}_1)\, d\mathbf{e}' d\mathbf{v}_2 d\mathbf{v}_1 \,,$$

and referred to as the *Boltzmann binary collision operator*. We may utilize the symmetry of this binary collision operator to particle labelling, referred to as $1 \leftrightarrow 2$ symmetry, and to pre- and post-collisional velocity interchange, and referred to as primed-unprimed symmetry. This latter symmetry argument is associated with inverse collisions (see Fig. 7.8) to obtain a more directly useful expression for the behaviour of $\langle\psi\rangle_{\mathrm{ne}}$, namely

$$\frac{\partial}{\partial t}\,(n\langle\psi\rangle_{\mathrm{ne}}) + \nabla\cdot(n\langle\mathbf{v}_1\psi(\mathbf{v}_1)\rangle_{\mathrm{ne}}) = \tfrac{1}{4}\iiint (f_1' f_2' - f_1 f_2)\sigma g\Delta\psi(\mathbf{v}_1)\, d\mathbf{e}' d\mathbf{v}_2 d\mathbf{v}_1 \,, \tag{7.5.53}$$

in which $\Delta\psi$ is defined via

$$\Delta\psi \equiv \psi(\mathbf{v}_1) + \psi(\mathbf{v}_2) - \psi(\mathbf{v}_1') - \psi(\mathbf{v}_2') \,. \tag{7.5.54}$$

For atoms there are five fundamental linearly independent dynamical quantities for which $\Delta \psi$ vanishes, so that there are no collisional contributions to Eq. (7.5.54): these are mass, represented by $\psi = 1$, kinetic energy, for which $\psi = \frac{1}{2}mv^2$, and linear momentum, with $\psi = m_1 \mathbf{v}$. These quantities are referred to as the *collisional invariants*. There are no other linearly independent collisional invariants. When $\psi(\mathbf{v})$ is a collisional invariant, the Boltzmann equation gives rise to a conservation law of the general form

$$\frac{\partial}{\partial t}\left(n\langle \psi \rangle_{\mathrm{ne}}\right) + \nabla \cdot \left(n\langle \mathbf{v}\psi \rangle_{\mathrm{ne}}\right) = 0. \qquad (7.5.55\mathrm{a})$$

For $\psi = 1$, Eq. (7.5.53) gives the well-known hydrodynamic equation of continuity for the number density n, viz.,

$$\frac{\partial n}{\partial t} + \nabla \cdot (n\mathbf{v}_0) = 0, \qquad (7.5.55\mathrm{b})$$

with the nonequilibrium average $\mathbf{v}_0 \equiv \langle \mathbf{v} \rangle_{\mathrm{ne}}$ representing the *macroscopic flow velocity*, or *stream velocity*, of the flowing gas. We may also write the equation of continuity in terms of the mass density $\rho = nm$ as

$$\frac{\partial \rho}{\partial t} + \nabla \cdot (\rho \mathbf{v}_0) = \frac{d\rho}{dt} + \rho \nabla \cdot \mathbf{v}_0 = 0,$$

in which the operator $\dfrac{d}{dt} \equiv \dfrac{\partial}{\partial t} + \mathbf{v}_0 \cdot \nabla$ denotes the rate of change for an observer moving with the fluid, and is called the *substantial derivative* in the field of fluid dynamics. The condition $\nabla \cdot \mathbf{v}_0$ is synonymous with the incompressibility condition $\dfrac{d\rho}{dt} = 0$.

The linear momentum conservation law is

$$\frac{\partial}{\partial t}(\rho \mathbf{v}_0) + \nabla \cdot (\rho \langle \mathbf{vv} \rangle_{\mathrm{ne}}) = 0, \qquad (7.5.56\mathrm{a})$$

and is called the equation of motion of the fluid in hydrodynamics. This equation is more commonly written in the form

$$\frac{\partial}{\partial t}(\rho \mathbf{v}_0) = -\nabla \cdot (\rho \mathbf{v}_0 \mathbf{v}_0 + \mathsf{P}), \qquad (7.5.56\mathrm{b})$$

with the dyadic $\rho \mathbf{v}_0 \mathbf{v}_0$ representing the convective flux of momentum through the gas; the pressure tensor P is defined in terms of the nonequilibrium average by

$$\mathsf{P} = \rho \langle \mathbf{VV} \rangle_{\mathrm{ne}}; \qquad (7.5.57)$$

the motion of an individual atom relative to the stream velocity $\mathbf{v}_0$ is represented by

$$\mathbf{V} \equiv \mathbf{V}(\mathbf{r}, t) = \mathbf{v} - \mathbf{v}_0(\mathbf{r}, t), \qquad (7.5.58)$$

and is called the *peculiar velocity* in the kinetic theory of gases.

The equation of motion of the fluid is often expressed in the form of the Navier–Stokes equation,

$$\rho \frac{d\mathbf{v}_0}{dt} = -\nabla \cdot \mathsf{P}, \tag{7.5.59}$$

in which the pressure tensor P is split into two parts as

$$\mathsf{P} = \mathbf{\Pi} + (P + \Pi)\boldsymbol{\delta}, \tag{7.5.60}$$

with $\mathbf{\Pi}$ a symmetric (second rank) tensor, P the equilibrium gas pressure, and Π a nonequilibrium contribution to the scalar pressure (which typically vanishes for a monatomic gas), while $\boldsymbol{\delta}$ is the unit (isotropic second rank) tensor. The energy conservation law is obtained as

$$\frac{\partial}{\partial t}\left(\rho(\tfrac{1}{2}\rho\langle v^2\rangle_{\text{ne}}\right) + \nabla \cdot \left(\tfrac{1}{2}\rho\langle v^2\mathbf{v}\rangle_{\text{ne}}\right) = 0. \tag{7.5.61}$$

It is, however, more commonly presented in the form

$$\frac{\partial}{\partial t}\left[\rho(\tfrac{1}{2}v_0^2 + u)\right] + \nabla \cdot \mathbf{J}^{\mathbf{E}} = 0, \tag{7.5.62}$$

in which u, defined as $u \equiv \tfrac{1}{2}\langle V^2\rangle_{\text{ne}}$, is the specific internal energy (in the thermodynamic sense) and $\mathbf{J}^E$, defined as $\mathbf{J}^E \equiv \tfrac{1}{2}\rho\langle v^2\mathbf{v}\rangle_{\text{ne}}$, represents the total energy flux, namely

$$\mathbf{J}^E = \tfrac{1}{2}\rho v_0^2\mathbf{v}_0 + \rho u\mathbf{v}_0 + \mathsf{P} \cdot \mathbf{v}_0 + \mathbf{q}. \tag{7.5.63}$$

The quantity $\mathbf{q}$ appearing in this equation is defined as

$$\mathbf{q} \equiv \tfrac{1}{2}\rho\langle V^2\mathbf{V}\rangle_{\text{ne}}, \tag{7.5.64}$$

and is known as the *heat flux vector*.

A slightly different, but fully equivalent, version of the energy conservation equation is given in terms of the substantial derivative as[4]

$$\rho \frac{du}{dt} = -\mathsf{P} : \nabla\mathbf{v}_0 - \nabla \cdot \mathbf{q},$$

and is known in hydrodynamics as the *energy balance equation*.

---

[4]The 'double-dot' contraction between two second rank tensors A and B is defined by $\mathsf{A} : \mathsf{B} \equiv \sum_{ij} A_{ij}B_{ij}$, with i, j summed over the three Cartesian directions.

7.6 Problems for this Chapter

1. It is often convenient to express a two-dimensional result in terms of a different set of coordinates, for example, in terms of polar coordinates (r, θ), instead of Cartesian coordinates (x, y). Such coordinate transformations are special cases of a class of transformations referred to as 'contact' transformations, and are often associated with a generating function, such as $F_2(\mathbf{q}, P_i, t) = f_i(\mathbf{q}, t) P_i$ that generates the components Q_i of $\mathbf{Q}$ from the components q_i of $\mathbf{q}$ via the relation $Q_i = (\partial F_2 / \partial P_i) = f_i(\mathbf{q}, t)$. Use the generating function

$$F_2(\mathbf{q}, \mathbf{P}, t) = (q_1^2 + q_2^2)^{\frac{1}{2}} P_1 + P_2 \arctan\left(\frac{q_2}{q_1}\right)$$

   to obtain $Q_1 = r$, $Q_2 = \theta$. Such a transformation of variables is said to be canonical should both the original and the transformed variables satisfy Hamilton's equations, respectively, for a pair of Hamiltonians $\mathcal{H}(\mathbf{q}, \mathbf{p})$ and $\mathcal{K}(\mathbf{Q}, \mathbf{P})$. Show that this condition is satisfied in the present case.

2. Use the generating function

$$F_2(\mathbf{q}, \mathbf{P}) = (q_1^2 + q_2^2 + q_3^2)^{\frac{1}{2}} P_1 + \arctan\left(\frac{q_2}{q_1}\right) P_2 + \arctan\left(\frac{\sqrt{q_1^2 + q_2^2}}{q_3}\right) P_3,$$

   with $Q_i = \partial F_2 / \partial P_i$ to generate the point transformation from Cartesian coordinates (x, y, z) to spherical polar coordinates (r, θ, ϕ).

3. A time-independent transformation may also be shown to be canonical if the condition

$$\sum_i [p_i dq_i - p_i' dq_i'] = h_1(\mathbf{q}, \mathbf{p}) dh_1 + h_2(\mathbf{q}, \mathbf{p}) dh_2 \equiv dh$$

   is satisfied (i.e., the difference is a total differential dh). Use this criterion to show that the transformation $q' = \ln(\frac{1}{p} \sin p)$, $p' = q \cot p$ is a canonical transformation.

4. The radius vector $\mathbf{R}$ to the centre-of-mass for a collection of N mass points $\{m_i | i = 1, \cdots, N\}$ is given by

$$\mathbf{R} \equiv \left(\sum_{i=1}^N m_i \mathbf{r}_i\right)\left(\sum_{i=1}^N m_i\right)^{-1} = \sum_{i=1}^N \frac{m_i}{M} \mathbf{r}_i.$$

   The N particles will have positions $\mathbf{r}_i'$ relative to the centre-of-mass position $\mathbf{R}$ given by $\mathbf{r}_i' \equiv \mathbf{r}_i - \mathbf{R}$. Show that the (translational) kinetic energy, T, of the system is given in terms of these relative coordinates and the centre-of-mass coordinates by

$$T = \tfrac{1}{2} M \dot{R}^2 + \sum_{i=1}^{N} m_i (\dot{\mathbf{r}}_i')^2 \, .$$

5. It has been suggested that an inductively-coupled double-circuit network can be described by a Lagrangian

$$\mathcal{L} = \tfrac{1}{2} \sum_{i=1}^{2} \left[L_i \dot{I}^2 - \frac{I_i^2}{2C_i} + \dot{V}_i(t) I_i \right] + \sum_{j=1}^{2} \sum_{\{k \neq j\}=1}^{2} M_{jk} \dot{I}_j \dot{I}_k \, ,$$

in which the I_i, V_i, L_i, C_i, M_{jk} are, respectively, currents, voltages, inductances, capacitances, and mutual inductances. Obtain the Lagrange equations (see Appendix G.2) and, from them, the Hamiltonian for this double-circuit network.

6. The partition function for an atomic Van der Waals gas is given as

$$Z(T, V, N) = \frac{1}{N!} \left(\frac{2\pi m k_{\mathrm{B}} T}{h^2} \right)^{\frac{3N}{2}} (V - Nb)^N \mathrm{e}^{aN^2/(Vk_{\mathrm{B}}T)} \, .$$

Obtain an expression for the internal energy U, the heat capacity C_V, and the pressure P for this Van der Waals gas. How do the expressions for U and C_V compare with those for an ideal gas?

7. By considering b/V to be small, obtain the density virial expansion for a Van der Waals gas. Express your answer in the form

$$\frac{P}{N k_{\mathrm{B}} T V} = 1 + B_2(T)\rho + B_3(T)\rho^2 + \cdots ,$$

with $\rho \equiv N/V$ the number density of the Van der Waals gas. Obtain explicit expressions for the second and third density virial coefficients $B_2(T)$ and $B_3(T)$. Would your result be any different were you to consider a molecular Van der Waals gas. Explain why or why not.

8. Convert the density virial expansion for a Van der Waals gas into the pressure virial expansion

$$\frac{P}{N k_{\mathrm{B}} T V} = 1 + B_2'(T) P + B_3'(T) P^2 + \cdots ,$$

and give the interrelationships amongst the pressure and density virial coefficients $B_2'(T)$, $B_3'(T)$ and $B_2(T)$, $B_3(T)$. [Hint: consider employing a quadratic approximation for the Van der Waals pressure P to obtain an appropriate expression for V for substitution into the density virial equation.]

9. Consider the following data for O_2 at a temperature $T = 273.15$ K.

P/bar	0.2500	0.5000	0.7500	1.0000
ρ/g dm^{-3}	0.352323	0.704817	1.057481	1.410316

Use these data to determine $B_2(T)$ for O_2 at $T =273.15$ K, given that the molar mass of O_2 is $\overline{M}(O_2) = 31.99880$ g mol^{-1}bar K^{-1}, and $R = 8.31451$ J K^{-1} $= 0.083145$ dm^{3}mol^{-1}bar K^{-1}. [Hint: you will need to obtain $Z(P, T) - 1$ as a function of $\overline{\rho} = \overline{V}^{-1}$. $B_2(T)$ can be obtained in principle from the slope of a plot of $Z(P, T)$ vs. $\overline{\rho}$: however, the slope is not readily determined with sufficient accuracy from a plot based upon so few data. For this reason, make an estimate for $B_2(T)$ by computing the average of the six point-slope values that can be obtained from the rather sparse data set that has been provided].

10. A Van der Waals gas has a compressibility factor $Z = 1.00084$ at temperature $T = 298.15$ K and pressure $P = 1$ bar. The Boyle temperature for the gas (the temperature for which the second density virial coefficient vanishes) is $T_B = 125$ K. Use $Z = 1 + B_2^{VdW}(T)/\overline{V}$, and approximate $\overline{V}$ using the ideal gas law (this is reasonable as $P = 1$ bar is a pressure at which many simple gases behave essentially as ideal gases). Estimate values for the Van der Waals parameters $\overline{a}$ and $\overline{b}$.

References

1. R.C. Tolman, Phys. Rev. **11**, 261 (1918); see also his classic text *The Principles of Statistical Mechanics* (Oxford Univ. Press, Oxford, 1938)
2. P.A. Mello, R.F. Rodríguez, Am. J. Phys. **78**, 820 (2010)
3. F.R.W. McCourt, Virial coefficients, in *Handbook of Molecular Physics and Quantum Chemistry*, ed. by S. Wilson. Molecules in the Physico-Chemical Environment: Spectroscopy, Dynamics and Bulk Properties, vol. 3, chap. 27 (Wiley, Chichester, 2003)
4. R.A. Aziz, F.R.W. McCourt, C.C.K. Wong, Mol. Phys. **61**, 1487 (1987)
5. R.L. Liboff, *Kinetic Theory. Classical, Quantum, and Relativistic Descriptions*, 3rd edn. (Springer, New York, 1996)

Chapter 8
Electric and Magnetic Phenomena

This chapter treats the responses of bulk matter to constant external electric and magnetic fields. The concepts of electric and magnetic dipole moments are reviewed, modifications to thermodynamics needed to accommodate field effects, and the statistical mechanical descriptions of polarizable media in the presence of applied fields are developed. The canonical ensemble for dipolar molecules is employed to obtain an expression for the dielectric constant of a polar gas. The main focus for magnetizable media is on paramagnetism: the statistical mechanics of nuclear spin systems is employed to obtain a statistical thermodynamic expression for the Curie susceptibility of ground level atoms or ions, which is then applied to the lanthanide tripositive ions that play important roles in modern electronic devices. The chapter finishes with a brief consideration of ferromagnetism and the one-dimensional Ising model.

8.1 Responses of Matter to Electric and Magnetic Fields

The response of a material system to an electric and/or magnetic field derives from the responses of the individual constituent atoms or molecules that make up the system under consideration. In principle, electric or magnetic fields will be deemed to be external (or applied) fields if they originate outside the system and will be deemed internal if they are the result of interactions amongst constituents of the system itself. Although, in principle, an applied may either be constant (in time) or oscillatory, we shall focus in the main upon the response of matter to constant external fields.

Two types of electric dipole moment, referred to as either (a) a permanent, or (b) an induced, dipole moment, may be encountered at the atomic/molecular level.

© Springer Nature Switzerland AG 2021
F. R. W. McCourt, *Statistical Thermodynamics for Pure and Applied Sciences*,
https://doi.org/10.1007/978-3-030-52006-9_8

(a) Permanent dipole moments are associated with asymmetric distributions of separated charges $\{\delta q_i\}$ located at corresponding positions $\{r_i\}$ that give rise to a permanent electric dipole moment, μ_d, defined via

$$\mu_d = \sum_i \delta q_i \mathbf{r}_i. \tag{8.1.1a}$$

Ionically-bonded molecules typically possess large permanent electric dipole moments while covalently-bonded molecules typically possess much smaller or, should they have highly symmetric charge distributions, no permanent electric dipole moments. Common examples of covalently-bonded molecules that do possess permanent electric dipole moments range from small nonlinear poly-atomic molecules like water and acetonitrile (CH_3CN), which have relatively large permanent dipole moments, to heteronuclear diatomic molecules like the hydrogen halides, which have fairly small permanent electric dipole moments.

(b) An electric dipole moment that is proportional to the field strength, E, of an applied electric field can be induced in any atom or molecule. This induced electric dipole moment is given by

$$\mu_i = \boldsymbol{\alpha} \cdot \mathbf{E}, \tag{8.1.1b}$$

in which $\boldsymbol{\alpha}$, a second rank tensor known as the polarizability tensor, is a basic property of an electron distribution that is a measure of the ease with which charge separations can be induced by an electric field. Additional information on the microscopic quantum mechanical origins of the polarizability tensor can be found, for example, in Chapter 5 of [1]. Even though many atoms and ions (which often have spherical symmetry) and molecules having a high level of symmetry (corresponding to symmetry point groups T_h, O_h, and I_h) do not possess permanent electric dipole moments, it is possible to create induced electric dipole moments by subjecting these species to an applied constant electric field. For very strong applied electric fields, such as those generated by intense laser radiation, additional (nonlinear) contributions may also need to be taken into account. As electric field manipulations play important roles in many modern devices involving nanomaterials, it is important to outline the role that statistical mechanics/thermodynamics plays in electric-field-dependent phenomena.

Electrons in atoms tend to execute orbital motion about positively-charged nuclei and, in so doing, generate orbital angular momenta $\mathbf{l}\hbar$, with magnitude $\sqrt{l(l+1)}\,\hbar$, $l = 0, 1, 2, \cdots$. In addition, individual electrons also possess an intrinsic angular momentum $\mathbf{s}\hbar$, of magnitude $\frac{\sqrt{3}}{2}\hbar$, referred to as electron spin angular momentum. The resultant total electronic angular momentum, $\hbar\mathbf{J}$, for an individual multi-electron atom or molecule is given by a vector sum of all individual electronic angular momenta. An atom (or molecule) that possesses a net total electronic angular momentum $\hbar\mathbf{J}$ has an electronic magnetic dipole moment, $\mu_d^{(m)}$, given by

$$\mu_d^{(m)} = - g_J \hbar \mu_B \mathbf{J}, \qquad (8.1.2)$$

in which g_J is called the Landé g-factor (whose value depends upon the set of electronic angular momentum quantum numbers characterizing the atom or molecule), while $\mu_B \equiv 9.2740101 \times 10^{-24} \ \mathrm{J \, T^{-1}}$ is the Bohr magneton (the basic unit for electronic magnetic dipole moments).

8.2 Thermodynamics of Polarizable Media

A polarizable medium consists of molecules that possess permanent electric dipole moments or molecules in which an electric dipole moment can be induced by an applied electric field. An appropriate thermodynamic system thus consists of an amount of a polarizable substance bounded by a surface (for example, lying between a pair of flat condenser plates) and subject to a time-independent electric field, $\mathbf{E}(\mathbf{r})$, produced by a conductor carrying charge q.

It has been pointed out by Honig [2] that considerable care should be exercised when dealing with the thermodynamics of dielectric media in the presence of external electric fields. We shall consider a macroscopic volume V of a dielectric material that is surrounded by free space, maintained at a specified electrostatic potential $\phi(\mathbf{r})$, and has a charge density $\rho(\mathbf{r})$ consistent with the Maxwell equation,

$$\nabla \cdot \mathbf{D}(\mathbf{r}) = \rho(\mathbf{r}),$$

determining the electric displacement $\mathbf{D}(\mathbf{r})$ at position $\mathbf{r}$ within the material. To bring an additional incremental charge dq, given by $d\rho(\mathbf{r}) = \nabla \cdot (d\mathbf{D})$, to position $\mathbf{r}$ will thus require incremental work $\delta W'_{\mathrm{elec}}$, given by

$$\delta W'_{\mathrm{elec}} = - \int \phi(\mathbf{r}) \nabla \cdot (d\mathbf{D}) \, dV,$$

to be done on the macroscopic dielectric system. The triple integral in this expression must be evaluated over all regions of space in which $\mathbf{D}$ does not vanish. To make this expression more tractable, we may utilize the vector operator version of the product rule, namely

$$\nabla \cdot (f\mathbf{g}) = \mathbf{g} \cdot \nabla f + f \nabla \cdot \mathbf{g},$$

to obtain

$$\delta W'_{\mathrm{elec}} = - \int \nabla \cdot (\phi d\mathbf{D}) \, dV + \int d\mathbf{D} \cdot \nabla \phi \, dV.$$

By employing the Gauss theorem of vector integral calculus to convert the first integral into a surface integral with a bounding surface sufficiently distant from the region of space occupied by the dielectric, the first component of this expression vanishes. Replacement of $-\nabla\phi$ in the second integral by the electric field $\mathbf{E}$ then gives the incremental work as

$$\delta W'_{\text{elec}} = -\int (\mathbf{E} \cdot d\mathbf{D}) \, dV. \tag{8.2.1}$$

In measuring the properties of dielectric media, it is a common practice to employ a constant electric field that is maintained independently of the dielectric medium, and is in principle *external* to it. We shall denote this constant electric field by $\mathbf{E}_0$. We also know from electromagnetic theory that the energy density associated with an electric field $\mathbf{E}_0$ (often referred to as the self- or vacuum energy) is $\frac{1}{2}\epsilon_0 E_0^2$. This external field is held fixed as the dielectric system is inserted into it.

It is traditional to treat dielectric media in terms of the macroscopic electric dipole moment density $\mathcal{P}$, termed the (electric) polarization, which enters through the electric displacement vector $\mathbf{D}$ introduced by Maxwell via the defining relation

$$\mathbf{D} \equiv \epsilon_0\mathbf{E}_0 + \mathcal{P}, \tag{8.2.2}$$

in which $\epsilon_0 \simeq 8.854188 \times 10^{-12}\,\text{Farad}\,\text{m}^{-1}$ is the permittivity of free space. At the molecular level, $\mathcal{P}$ may be associated with both permanent and field-induced electric dipole moments: however, we must exclude ferro-electric and hysteresis phenomena, as they cannot be described solely in terms of reversible thermodynamic changes.

In order to affect a full separation of the externally applied electric field $\mathbf{E}_0$ and the dielectric thermodynamic system, we may employ an identity,

$$\mathbf{E} \cdot d\mathbf{D} = \mathbf{E}_0 \cdot d\mathbf{D}_0 + (\mathbf{E} \cdot d\mathbf{D}_0 - \mathbf{D} \cdot d\mathbf{E}_0) + (\mathbf{D}_0 \cdot d\mathbf{E}_0 - \mathbf{E}_0 \cdot d\mathbf{D}_0)$$
$$+ \mathbf{E} \cdot (d\mathbf{D} - d\mathbf{D}_0) + (\mathbf{D} - \mathbf{D}_0) \cdot d\mathbf{E}_0, \tag{8.2.3}$$

due to Heine [3], in which $\mathbf{E}_0$ and $\mathbf{D}_0 = \epsilon_0\mathbf{E}_0$ are the electric field and displacement vectors associated with a fixed charge distribution established in free space prior to placing the dielectric medium in the field. Of the five integral terms to be evaluated following substitution of this identity into Eq. (8.2.1), the first term represents the work associated with the establishment of the external $\mathbf{E}_0$ field in free space, and the final three terms can be shown (see Problem 8.3) to vanish. The second term gives

$$\int (\mathbf{E} \cdot d\mathbf{D}_0 - \mathbf{D} \cdot d\mathbf{E}_0) dV = -\int_V \mathcal{P} \, dV \cdot d\mathbf{E}_0,$$

with the final integration now being only over the volume of the thermodynamic system because $\mathcal{P} \equiv 0$ outside the dielectric medium. The work done on the dielectric upon inserting it into an externally-maintained constant electric field $\mathbf{E}_0$ is thus given by the sum of the first two terms as

$$\delta W'_{\text{elec}} = -\int \epsilon_0 \mathbf{E}_0 \cdot d\mathbf{E}_0 \, dV + \left(\int_V \mathcal{P} \, dV \right) \cdot d\mathbf{E}_0$$

$$= -\int d(\tfrac{1}{2}\epsilon_0 E_0^2) \, dV + \left(\int_V \mathcal{P} \, dV \right) \cdot d\mathbf{E}_0.$$

As the first term in this expression is disconnected from the dielectric system, we see that the incremental work δW_{elec} required to create the polarization $\mathcal{P}$ of the dielectric system is thus given simply by

$$\delta W_{\text{elec}} = \left(\int_V \mathcal{P} \, dV \right) \cdot d\mathbf{E}_0. \tag{8.2.4}$$

That $\mathcal{P}$ appears in the expression δW_{elec} for incremental work only in the form of an integral over the volume occupied by the thermodynamic system is suggestive that a simplification can be achieved by working with the incremental work density δw_{elec} related to δW_{elec} by

$$\delta W_{\text{elec}} = \int_V \delta w_{\text{elec}} \, dV. \tag{8.2.5a}$$

Thus δw_{elec}, given by

$$\delta w_{\text{elec}} = \mathcal{P} \cdot d\mathbf{E}_0, \tag{8.2.5b}$$

represents the incremental work density pertinent to the thermodynamic description of a dielectric medium.

For the remainder of this section we shall consider only a thermodynamic system that is both spatially and electrically isotropic, so that the polarization $\mathcal{P}$ induced by the (externally maintained) applied electric field, and thus also $\mathbf{D}$, is parallel to $\mathbf{E}_0$. This allows us to replace the vectors $\mathcal{P}$, $\mathbf{D}$, and $\mathbf{E}_0$ everywhere by the scalars $\mathcal{P}$, D, and E_0, respectively. Moreover, we shall restrict our discussion to uniform dielectric systems, for which

$$\int_V \mathcal{P} \, dV \equiv \mathfrak{D} \tag{8.2.6a}$$

more generally, with $\mathfrak{D}$ the dipole moment of the macroscopic thermodynamic system, or

$$\int_V \mathcal{P} \, dV = \mathcal{P}V, \tag{8.2.6b}$$

more specifically for fixed-volume systems.

We begin with a combined 1st and 2nd Laws expression for a system of isotropic uniform dielectric matter, namely

$$dU_{\text{elec}} = \delta Q + \delta W_{\text{elec}}$$

$$= T\mathrm{d}S - P\mathrm{d}V + \mathfrak{D}\mathrm{d}E_0. \qquad (8.2.7\text{a})$$

For a fixed-volume thermodynamic system, we may work equivalently with an expression given in terms of the internal energy and entropy densities, $u_{\text{elec}}(s, E_0)$ and s, together with the dielectric polarization (i.e., macroscopic dipole moment density), as

$$\mathrm{d}u_{\text{elec}} = T\mathrm{d}s + \mathcal{P}\,\mathrm{d}E_0. \qquad (8.2.7\text{b})$$

We note, however, that the version of the combined 1st and 2nd Laws expression given above does not derive from the traditional form in which each term is represented by the product of an intensive thermodynamic variable and the differential of its corresponding extensive thermodynamic variable. Conversion of Eq. (8.2.7a) into the traditional form may be accomplished by applying the Legendre transform

$$U'(S, V, \mathcal{P}) = U_{\text{elec}}(S, V, E_0) - E_0\mathfrak{D}. \qquad (8.2.8)$$

This Legendre transform leads to a revised 1st and 2nd Laws expression

$$\mathrm{d}U' = T\mathrm{d}S - P\mathrm{d}V - E_0\,\mathrm{d}\mathfrak{D}, \qquad (8.2.9\text{a})$$

from which it is clear that U' is a function of the three extensive thermodynamic variables $S, V, \mathfrak{D}$. We may also consider the corresponding internal energy density u' and constant volume processes, for which the combined 1st and 2nd Laws expression becomes

$$\mathrm{d}u' = T\mathrm{d}s - E_0\mathrm{d}\mathcal{P}, \qquad (8.2.9\text{b})$$

with u' a function of s and $\mathcal{P}$.

The corresponding thermodynamic potentials derived from $U'(S, V, \mathfrak{D})$ are the enthalpy $H'(S, P, \mathfrak{D})$, the Helmholtz energy $A'(T, V, E_0)$, and the Gibbs energy $G'(T, P, E_0)$, given, respectively, by[1]

$$H'(S, P, \mathfrak{D}) \equiv U'(S, V, \mathfrak{D}) + PV, \qquad (8.2.10\text{a})$$

$$A'(T, V, E_0) \equiv U'(S, V, \mathfrak{D}) - TS + E_0\mathfrak{D} \qquad (8.2.10\text{b})$$

---

[1]It is not uncommon to find the enthalpy, rather than the Helmholtz energy defined in terms of the double Legendre transformation of the internal energy. The choice made here for the definition of the Helmholtz energy maintains a desired parallel between the thermodynamic descriptions of dielectric and magnetizable media.

and

$$G'(T, P, E_0) \equiv U'(S, V, \mathfrak{D}) + PV - TS + E_0\mathfrak{D}. \qquad (8.2.10c)$$

These thermodynamic potentials play much the same roles in the thermodynamics of dielectric systems as do their counterparts in the thermodynamics of non-dielectric media.

Thermodynamic relationships associated with a dielectric medium have many analogies with those obtained in terms of pressure and volume, such as expression (2.4.10) relating the partial derivative of the internal energy with respect to system volume to the system pressure and temperature, and expression (2.5.1c) for the difference $C_P - C_V$ between the constant pressure and constant-volume heat capacities for non-dielectric media.

Example 8.1 Dielectric medium equivalent of Eq. (2.4.10).

We shall begin with the extended version of the combined 1st and 2nd Laws expression for a uniform constant-volume thermodynamic system. If we treat the entropy as a function of T, and $\mathfrak{D}$, its total differential dS is given by

$$dS = \left(\frac{\partial S}{\partial T}\right)_{V,\mathfrak{D}} dT + \left(\frac{\partial S}{\partial \mathfrak{D}}\right)_{T,V} d\mathfrak{D}.$$

Upon substituting this equation for dS into Eq. (2.3.9a) and collecting terms, we obtain dU' as

$$dU' = T\left(\frac{\partial S}{\partial T}\right)_{V,\mathfrak{D}} dT + \left[T\left(\frac{\partial S}{\partial \mathfrak{D}}\right)_{T,V} - E_0\right] d\mathfrak{D}$$

$$\equiv \left(\frac{\partial U'}{\partial T}\right)_{V,\mathfrak{D}} dT + \left(\frac{\partial U'}{\partial \mathfrak{D}}\right)_{T,V} d\mathfrak{D}.$$

It is perhaps worth noting that we are now considering the internal energy as a function of T and $\mathfrak{D}$, rather than of S and $\mathfrak{D}$. Comparison of the first expression with the formal mathematical definition of the total differential dU' gives, in particular, the expression

$$\left(\frac{\partial U'}{\partial \mathfrak{D}}\right)_{T,V} = T\left(\frac{\partial S}{\partial \mathfrak{D}}\right)_{T,V} - E_0.$$

To replace the entropy partial derivative in this expression, we shall employ a Maxwell-type relation derivable from the Helmholtz energy $A_{\text{elec}}(T, \mathfrak{D}) \equiv U_{\text{elec}}(S, E_0) - TS - E_0\mathfrak{D}$ obtained from $U_{\text{elec}}(S, E_0)$, namely

$$\left(\frac{\partial S}{\partial \mathfrak{D}}\right)_{T,V} = \left(\frac{\partial E_0}{\partial T}\right)_{V,\mathfrak{D}},$$

followed by replacement of the nonphysical partial derivative of E_0 with respect to T using the $(T, \mathfrak{D}, E_0)$ permutation rule to obtain

$$\left(\frac{\partial U'}{\partial \mathfrak{D}}\right)_{T,V} = -\left[T\left(\frac{\partial \mathfrak{D}}{\partial T}\right)_{V,E_0}\left(\frac{\partial E_0}{\partial \mathfrak{D}}\right)_{T,V} + E_0\right]$$

as the electric field thermodynamic analogue of Eq. (2.4.10).   □

We have seen from the expression for the work done on a dielectric system upon inserting it into an electric field of magnitude E_0 that the macroscopic electric dipole moment $\mathfrak{D}$ of the system plays a role analogous to that of the system volume, while E_0 plays a role analogous to that played by the pressure exerted on the system. We have also seen in Chap. 2 that changes in the internal energy and the enthalpy of a thermodynamic system associated with temperature changes arising from heat (energy) transferred to or from its surroundings are manifested, respectively, through the system heat capacities at constant volume and at constant pressure. Of course, the effects of temperature changes are still partly manifested through C_V and C_P for dielectric materials, as expression (2.5.3) for the heat capacity difference also applies to them. Indeed, it is likely that these heat capacity contributions still play important roles for gas phase dielectrics. For most liquid and essentially all solid dielectric materials, however, field-dipole work makes the dominant contribution to changes in the internal energy (largely because volume changes are very small and pressures are not commensurately high). Thus, the analogous heat capacities C_E and $C_\mathfrak{D}$ may be expected to play significant roles in internal energy and enthalpy changes associated with energy exchange between the dielectric system and its surroundings.

Example 8.2 Heat capacity difference $C_E - C_\mathfrak{D}$ for a dielectric system.
We shall examine the difference between the heat capacities determined for temperature changes made while holding the applied electric field, E_0, constant and those made while maintaining the macroscopic electric dipole moment, $\mathfrak{D}$, at a fixed value. For convenience, we shall consider only constant-volume thermodynamic systems and we shall suppress the subscript '0' on E_0 when referring to the heat capacity obtained under conditions of constant electric field strength $E = E_0$.

If we express the internal energy U as a function of T, and $\mathfrak{D}$, the heat capacity $C_\mathfrak{D}$ is simply given as

$$C_\mathfrak{D} \equiv \left(\frac{\partial U}{\partial T}\right)_\mathfrak{D},$$

while C_E can be obtained from the enthalpy $H'(T, E_0)$ defined in terms of the Legendre transform

$$H'(T, E_0) = U(T, \mathfrak{D}) - \mathfrak{D}E_0$$

of the internal energy $U'(T, \mathfrak{D})$ as

$$C_E \equiv \left(\frac{\partial H'}{\partial T} \right)_{E_0} = \left(\frac{\partial U}{\partial T} \right)_{E_0} - E_0 \left(\frac{\partial \mathfrak{D}}{\partial T} \right)_{E_0}$$

or as

$$C_E = \left(\frac{\partial U}{\partial T} \right)_{\mathfrak{D}} + \left[\left(\frac{\partial U}{\partial \mathfrak{D}} \right)_T - E_0 \right] \left(\frac{\partial \mathfrak{D}}{\partial T} \right)_{E_0}.$$

The difference $C_E - C_{\mathfrak{D}}$ is then given by

$$C_E - C_{\mathcal{P}} = \left[\left(\frac{\partial U}{\partial \mathfrak{D}} \right)_T - E_0 \right] \left(\frac{\partial \mathfrak{D}}{\partial T} \right)_{E_0}$$

which, upon utilizing the expression obtained for the partial derivative of the internal energy with respect to the electric dipole moment in Example 8.1, gives

$$C_E - C_{\mathfrak{D}} = -T \left(\frac{\partial \mathfrak{D}}{\partial T} \right)_{E_0}^2 \left(\frac{\partial E_0}{\partial \mathfrak{D}} \right)_T.$$

This result is more commonly expressed as

$$C_E - C_{\mathfrak{D}} = -T \frac{\left(\frac{\partial \mathfrak{D}}{\partial T} \right)_{E_0}^2}{\left(\frac{\partial \mathfrak{D}}{\partial E_0} \right)_T},$$

as $\mathfrak{D}$ is, in principle, a function of E_0 and T. This expression is analogous to $C_P - C_V$ for non-dielectric systems.   □

8.3 Dielectric Constant of a Polar Gas

8.3.1 Setting the Stage

For simplicity, we shall consider a gas of N heteronuclear diatomic molecules, such as HF, HCl, or HBr in a container of volume V, and subject to an applied electric field **E**. In the absence of an applied electric field, molecules of this type behave in exactly the same way as do homonuclear diatomic molecules like H_2, N_2, or O_2. However, because heteronuclear diatomic molecules do not have equivalent nuclei, there is no longer an inversion centre midway between the nuclei, and consequently it is possible for heteronuclear diatomic molecules to develop noncentrosymmetric charge distributions. For the HCl molecule, for example, we might indicate this by writing a formula of the type $H^{\delta+}Cl^{\delta-}$ to indicate the greater localization of charge around the Cl nucleus: such a charge separation leads naturally to a permanent electric dipole moment μ_d.

The energy of a polar molecule is changed in an applied electric field by an amount that depends upon both the strength of the applied field and the magnitude of the permanent dipole moment of the molecule. The effect of the external electric field on the molecule is known as the Stark effect (and is the electric analogon of the Zeeman effect associated with an external magnetic field). You may have already encountered the Stark effect in a quantum mechanics course: however, if not, all that we really need for our purposes is to know that the interaction energy is, in general, given by

$$E_{\text{field}} = -\boldsymbol{\mu}_d^{(e)} \cdot \mathbf{E} = -\mu_d^{(e)} E \cos\theta, \tag{8.3.1}$$

with the precise form to be taken by $\boldsymbol{\mu}_d^{(e)}$ to be specified later.

8.3.2   Canonical Ensemble Description of a Dielectric Thermodynamic System

Since the energy of an individual molecule depends upon how it is aligned with respect to the field direction, we may ask how such oriented molecules would be distributed when they are at thermal equilibrium in the presence of the applied electric field. To do this, we shall make use of the canonical distribution, which, as we know, gives Boltzmann probabilities, p_r, having the form $p_r \propto e^{-\beta E_r}$ for r in principle discrete. We shall, however, treat this problem classically, consistent with our given for the interaction energy E_{field} expressed in Eq. (8.3.1): what we mean by this is that we shall use the form $\mu_d \cos\theta$, which is a classical expression, rather than employ the z-component (operator), $\widehat{\mu}_{dz}^{(e)}$, of the electric dipole moment vector operator, as would be required were we to wish to obtain a full quantum-mechanical description.

For a classical description, we shall replace p_r (with r discrete) by $f(\theta)$, and call f a distribution function; we shall also replace the summation over r by an integral over the angles needed to specify the direction in space of $\boldsymbol{\mu}_d^{(e)}$ as illustrated in Fig. 8.1 for a charge distribution.

Fig. 8.1 Interaction of a dipolar charge distribution with an applied electric field

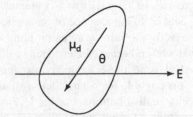

Thus, for our polar gas sample, the partition function for an individual molecule will be

$$z(V, T) = z_{\text{mol}}(V, T) z_{\text{field}}(T),$$ (8.3.2)

and p_r will be given by

$$p_r = \frac{e^{-\beta E_r}}{z(V, T)} = \frac{e^{-\beta E_r^{\text{mol}}} e^{-\beta E_{\text{field}}}}{z_{\text{mol}}(V, t) z_{\text{field}}(T)}$$ (8.3.3)

or

$$p_r = p_r^{\text{mol}} \, p_{\text{field}}.$$ (8.3.4)

Factorization of the probability p_r in this way corresponds to the canonical partition function for N such molecules taking the form

$$Z_N = \frac{z^N}{N!} = \frac{z_{\text{mol}}^N(V, T) z_{\text{field}}^N(T)}{N!} = Z_N^{\text{mol}}(V, T) Z_N^{\text{field}}(T),$$ (8.3.5)

giving rise to the usual additivity result:

$$\ln Z_N = \ln Z_N^{\text{mol}}(V, T) + \ln Z_N^{\text{field}}(T).$$ (8.3.6)

The probability p_r is normally given by

$$p_r = C e^{-\beta E_r},$$

with the constant of proportionality C being determined by the requirement that probabilities are normalized to unity, so that $C = 1/z$, with z the usual canonical partition function,

$$z = \sum_r e^{-\beta E_r}.$$

In the present case we have an analogous statement, with $f(\theta, \phi)$ given as

$$f(\theta, \phi) = C e^{-\beta E_{\text{field}}(\theta, \phi)},$$ (8.3.7)

with C being the reciprocal of the partition function associated with the energies of the precessing oriented dipoles in the electric field. The sum $\sum_r p_r$ over probabilities is replaced here by the normalization requirement

$$\int_0^{2\pi} \int_0^\pi f(\theta, \phi) \sin\theta \, d\theta \, d\phi = 1$$

or

$$C \int_0^{2\pi} \int_0^\pi e^{\beta \mu_d E \cos\theta} \sin\theta \, d\theta \, d\phi = 1,$$

which, if we note that the integrand is independent of ϕ, define x as $x \equiv \cos\theta$, and introduce the parameter $a \equiv \beta\mu_d E$, simplifies to $2\pi C \int_{-1}^{1} e^{ax}\, dx = 1$, from which we may determine C as

$$C = \frac{a}{4\pi \sinh a}. \tag{8.3.8}$$

The electric dipole moment of a closed-shell atom in the presence of an externally applied electric field is given by

$$\boldsymbol{\mu}_d^{(e)} \equiv \boldsymbol{\mu}_i = \alpha \mathbf{E}, \tag{8.3.9}$$

in which α is called the polarizability of the atom. Even for a relatively simple closed-shell molecule, such as HCl, the situation is slightly more complicated, as local charge separations within the molecule not only give rise to both a permanent electric dipole moment, $\boldsymbol{\mu}_d$, but also an anisotropy in the polarizability, making it a second rank tensorial quantity that cannot be represented by a scalar quantity (see [1]).

A molecule in the presence of an electric field $\mathbf{E}$ will have a total electric dipole moment, $\boldsymbol{\mu}_d^{(e)}$, given by the sum of its permanent and electric-field-induced dipole moments, namely

$$\boldsymbol{\mu}_d^{(e)} \simeq \boldsymbol{\mu}_d + \overline{\alpha}\,\mathbf{E}, \tag{8.3.10}$$

in which $\overline{\alpha}$ is referred to as the isotropic polarizability of the molecule. The situation is actually a little more complicated than this (see [1]), as Eq. (8.3.10) ignores the anisotropic components of the polarizability tensor $\boldsymbol{\alpha}$.

The electric polarization $\mathcal{P}$ of a dielectric is defined as the average electric dipole moment per unit volume of the substance, so that if we have N molecules, each with electric dipole moment $\boldsymbol{\mu}_d$, in a volume V, the polarization $\mathcal{P}$ in the presence of an external electric field $\mathbf{E}$ will be

$$\mathcal{P} = \frac{N}{V}\,\langle \boldsymbol{\mu}_d^{(e)} \rangle = \frac{N}{V}\,\langle \boldsymbol{\mu}_d + \overline{\alpha}\,\mathbf{E} \rangle$$

or

$$\mathcal{P} = \frac{N}{V}\,[\overline{\alpha}\,\mathbf{E} + \langle \boldsymbol{\mu}_d \rangle]. \tag{8.3.11}$$

If the electric field lies along the z-axis, then we obtain

$$\mathcal{P}_z = \frac{N}{V}\,[\overline{\alpha}\,E + \langle \mu_d \cos\theta \rangle] \tag{8.3.12}$$

for the z-component of the (electric) polarization, and

$$\mathcal{P}_x = \frac{N}{V} \langle \mu_{\mathrm{d}} \sin\theta \cos\phi \rangle = 0, \quad \mathcal{P}_y = \frac{N}{V} \langle \mu_{\mathrm{d}} \sin\theta \sin\phi \rangle = 0,$$

for the x- and y-components.

We know from Eq. (8.2.2) that the electric polarization $\mathcal{P}$ of the material and the applied electric field $\mathbf{E}$ are related via the electric displacement $\mathbf{D}$ by $\mathbf{D} = \epsilon_0 \mathbf{E} + \mathcal{P}$, with ϵ_0 the electric permittivity of free space. We shall consider here only the case in which the dielectric medium is isotropic. Moreover, we shall limit our discussion to what is called the linear response regime, in which the polarization $\mathcal{P}$ resulting from imposition of the external electric field is proportional to $\mathbf{E}$, i.e.,

$$\mathcal{P} = \chi^{(e)} \mathbf{E}. \tag{8.3.13}$$

The quantity χ^e is a measure of the response of the dielectric medium to the external electric field, and is referred to as the electric susceptibility. This linear relation allows us to write $\mathbf{D}$ in terms of $\mathbf{E}$ as

$$\mathbf{D} = (\epsilon_0 + \chi^{(e)}) \mathbf{E}. \tag{8.3.14}$$

The dielectric constant ε of a dielectric material is then defined through the relation

$$\mathbf{D} \equiv \varepsilon \epsilon_0 \mathbf{E}, \tag{8.3.15}$$

so that from a comparison with our previous result, we see that the dielectric constant can be determined in terms of the electric susceptibility and the permittivity of free space as

$$\varepsilon \equiv 1 + \frac{\chi^{(e)}}{\epsilon_0}. \tag{8.3.16}$$

Having established the connection between the dielectric constant of a dielectric material and its electric susceptibility, let us now return to the calculation of the polarization. We have seen that the only nonzero component of the polarization for an applied electric field in the z-direction is $\mathcal{P}_z$, which is given by

$$\mathcal{P}_z = \frac{N}{V} \overline{\alpha} E + \frac{N}{V} \langle \mu_{\mathrm{d}} \cos\theta \rangle,$$

with

$$\langle \mu_{\mathrm{d}} \cos\theta \rangle = \int_0^{2\pi} \int_0^{\pi} \mu_{\mathrm{d}} \cos\theta \, f(\theta, \phi) \sin\theta \, \mathrm{d}\theta \mathrm{d}\phi.$$

We may evaluate this result in the same manner employed to evaluate the normalization constant C. More specifically, we may write $\mathcal{P}_z$ as

$$\mathcal{P}_z = 2\pi C \mu_{\mathrm{d}} \int_{-1}^{1} e^{ax} x \, dx,$$

in terms of the parameter a, $(x \equiv \cos\theta)$. This integral may be evaluated straightforwardly using integration by parts, to obtain

$$\int_{-1}^{1} e^{ax} x \, dx = \frac{2}{a} \cosh a - \frac{2}{a^2} \sinh a,$$

so that $\langle \mu_{\mathrm{d}} \cos\theta \rangle$ can be rewritten as

$$\langle \mu_{\mathrm{d}} \cos\theta \rangle = \frac{a\mu_{\mathrm{d}}}{4\pi \sinh a} \left[\frac{4\pi}{a} \cosh a - \frac{4\pi}{a^2} \sinh a \right], \qquad (8.3.17)$$

or, more usefully, as

$$\langle \mu_{\mathrm{d}} \cos\theta \rangle = \mu_{\mathrm{d}} \left[\frac{\cosh a}{\sinh a} - \frac{1}{a} \right]. \qquad (8.3.18)$$

The quantity within the square brackets arises in a number of contexts, and was first obtained by Langevin. It is now referred to as the Langevin function, and is defined as

$$\mathcal{L}(x) \equiv \coth x - \frac{1}{x}. \qquad (8.3.19)$$

We may utilize this function to write a more compact version of Eq. (8.3.18), namely

$$\langle \mu_{\mathrm{d}} \cos\theta \rangle = \mu_{\mathrm{d}} \mathcal{L}(a), \quad \text{with } a \equiv \frac{\mu_{\mathrm{d}} E}{k_{\mathrm{B}} T}. \qquad (8.3.20)$$

For a small, i.e., $a \ll 1$, we can expand the Langevin function about zero to give

$$\mathcal{L}(a) \simeq \frac{1 + \frac{1}{2}a^2 + \frac{1}{24}a^4 + \cdots}{a + \frac{1}{6}a^3 + \frac{1}{12}a^5 + \cdots} - \frac{1}{a}$$

$$= \frac{1}{a}\left(1 + \frac{1}{2}a^2 + \frac{1}{24}a^4 + \cdots\right)\left(1 - \frac{1}{6}a^2 - \frac{1}{12}a^4 + \frac{1}{36}a^4 + \cdots\right) - \frac{1}{a}$$

$$= \frac{1}{a}\left(1 + \frac{a^2}{3} - \frac{a^4}{45} + \cdots\right) - \frac{1}{a}$$

or

$$\mathcal{L}(a) \simeq \frac{a}{3} - \frac{a^3}{45}, \quad a \ll 1. \tag{8.3.21}$$

Even though we have obtained an expression that has a term cubic in a, we shall retain only the linear term for the present calculation of the dielectric constant of HCl.

If we employ the linear approximation to the Langevin function in our expression for the polarization, we find that $\mathcal{P}_z$ is given by

$$\mathcal{P}_z = \frac{N}{V} \overline{\alpha} E + \frac{N}{V} \frac{\mu_d^2 E}{3k_B T} \equiv \chi^{(e)} E. \tag{8.3.22}$$

With the electric susceptibility given by

$$\chi^{(e)} = \frac{N}{V} \left(\overline{\alpha} + \frac{\mu_d^2}{3k_B T} \right), \tag{8.3.23}$$

we obtain for the dielectric constant ε the result

$$\varepsilon = 1 + \frac{N}{V} \left(\frac{\overline{\alpha}}{\epsilon_0} + \frac{\mu_d^2}{3\epsilon_0 k_B T} \right). \tag{8.3.24}$$

Example 8.3 Dielectric constant ε of dry HCl gas.

We wish to calculate a value for the dielectric constant ε of dry HCl gas at a temperature of $T = 273$ K. We know that the permanent electric dipole moment of HCl has the value $\mu_d = 1.11$ D, and if we take one mole of HCl, we have $V = \overline{V}^\circ$. and $N = N_0$. We must first convert from the chemical unit debye for the dipole moment into the SI unit using $1\,\mathrm{D} = 3.336 \times 10^{-30}$ C m, so that $\mu_d(\mathrm{HCl}) = 3.70 \times 10^{-30}$ C m. The dielectric permittivity of free space has the value $\epsilon_0 = 8.854188 \times 10^{-12}\,\mathrm{C}^2\mathrm{J}^{-1}\mathrm{m}^{-1}$, and the Loschmidt number $L_0 = N_0/\overline{V}^\circ$ has the value $L_0 = 2.686780 \times 10^{25}\,\mathrm{m}^{-3}$. Finally, the isotropic polarizability of HCl is given in the combination $\overline{\alpha}/\epsilon_0$, which for HCl has the value $\overline{\alpha}(\mathrm{HCl})/\epsilon_0 = 2.61 \times 10^{-30}\,\mathrm{m}^3$.

Using these data we see that the permanent component of the dipole moment of HCl contributes

$$\frac{N_0}{\overline{V}^\circ} \frac{\mu_d^2}{3\epsilon_0 k_B T} \simeq 3.677 \times 10^{-3} \approx 0.0037,$$

to ε, while the electric-field-induced component of the dipole contributes

$$\frac{N_0}{\overline{V}^\circ} \frac{\overline{\alpha}}{\epsilon_0} = (2.687 \times 10^{25})(2.61 \times 10^{-30}) \approx 0.0001,$$

so that the total value of the dielectric constant of HCl obtained in this way is

$$\varepsilon(\text{HCl}) \simeq 1.0038,$$

at temperature $T = 273$ K. If we compare our calculated value with the value $\varepsilon(\text{HCl}) = 1.0039$ given in the standard CRC tables (82nd edition, 2001–2002), we see that our calculation indeed gives a quite reasonable result. □

8.4　Magnetism, I. Paramagnetism

We are all familiar with magnetism in the form of stirrer bar-magnets employed in almost any elementary chemistry laboratory, in the form of the ubiquitous refrigerator magnets in our homes, or from patterns made by iron filings utilized to show the interconnectedness of electricity and magnetism in a typical first-year physics laboratory. These examples all involve materials that have permanent magnetic moments. The nature of magnetic materials is, however, more complex, as there are several distinct types of magnetic material, as characterized by their behaviours in the presence of an external magnetic field, such as that produced by a pole magnet. For example, if a sample of a particular material brought into the magnetic field tends to be repelled by the field, it is said to be diamagnetic, while if it tends to be attracted into the field, it is said to be paramagnetic if it is only weakly attracted, and (normally) to be ferromagnetic if it is strongly attracted (we are speaking here of orders of magnitude differences in the strengths of the attractions). The examples employed to remind us of our familiarity with magnetism all involved ferromagnetic materials, which are readily identified by their abilities to retain their magnetic properties in the absence of an externally applied magnetic field. Neither diamagnetic nor paramagnetic materials can be readily characterized in the absence of an external magnetic field.

Unfortunately, however, the old adage that life is never truly simple also applies to the classification of the magnetic properties of materials. Magnetic properties of paramagnetic and ferromagnetic materials depend in general upon the temperature of the system under consideration. Ferromagnetism is associated with the ability of the constituent atoms/molecules to achieve and maintain alignment of their (magnetic) dipole moments even in the absence of an external magnetic field. The fundamental distinction between paramagnetic and ferromagnetic substances is that in paramagnetism the magnetic moments of the constituent atoms/molecules do not interact with one another, while in ferromagnetism, there is a strong nearest-neighbour interaction. Nonetheless, in both cases, the increase in the thermal motions of the individual constituents as the temperature is increased either progressively destroys the induced alignment of the magnetic dipoles in a paramagnetic system or disrupts the neighbour-interactions in a ferromagnetic system. At sufficiently high temperatures, a paramagnetic system becomes diamagnetic, while a ferromagnetic system undergoes a phase transition at a critical temperature and becomes paramagnetic.

Although we shall investigate paramagnetism is some detail in this and the next section, we shall only provide an introduction to ferromagnetism in Sect. 8.6 via a discussion of the one-dimensional Ising-chain model plus a short summary of some properties of the two-dimensional Ising model solution, as its analytical solution is not mathematically tractable at a typical undergraduate level.

8.4.1 Thermodynamics of Magnetic Dipolar Media

The present treatment of the thermodynamics of a magnetic dipolar medium assumes that the magnetizable material is uniformly distributed over the volume V of the thermodynamic system and magnetized by an externally-maintained constant magnetic field $\mathbf{H}_0$. The corresponding free-space magnetic induction is given by $\mathbf{B}_0 = \mu_0 \mathbf{H}_0$, with μ_0 the permeability of free space defined (in SI units), as precisely $\mu_0 \equiv 4\pi \times 10^{-7}\,\mathrm{H\,m^{-1}}$. The magnetizing field $\mathbf{H}_0$ infuses the thermodynamic system with a field energy, U_{field}, given [4] by

$$U_{\text{field}} = \tfrac{1}{2} V \mathbf{B} \cdot \mathbf{H}_0, \tag{8.4.1}$$

with V the volume of the thermodynamic system and $\mathbf{B}$ the magnetic induction,

$$\mathbf{B} = \mu_0(\mathbf{H}_0 + \mathbf{M}) \equiv \mu \mathbf{H}_0, \tag{8.4.2}$$

that depends upon both the external field $\mathbf{H}_0$ and the internal magnetic field associated with the bulk magnetic moment $\mathbf{M}$. However, because the differential work done by current sources that are part of the surroundings in magnetizing a volume V of a magnetizable substance is given (see Problem 8.7) by $V\mathbf{H}_0 \cdot d\mathbf{B}_0$, it is convenient to treat $\mathbf{B}_0$ as the more fundamental quantity for dealing with the thermodynamics of a magnetizable medium. A careful and thorough analysis of the nature and role played by an external magnetic field has also been given by Callen [5]. The appropriate form giving the incremental energy that is transferred reversibly into a volume dV of a magnetizable substance upon quasistatically immersing it into a magnetic induction field $\mathbf{B}_0$ is thus dU_{field}, which can be shown (see Problem 8.9) to be given by

$$dU_{\text{field}} = \mathbf{B}_0 \cdot d\mathbf{M}. \tag{8.4.3}$$

8.4.2 Magnetic Susceptibility and Adiabatic Demagnetization

In general, the macroscopic magnetic moment of a thermodynamic system is given in terms of the magnetizing field $\mathbf{B}_0$ as

$$\mathbf{M} = \chi^{(m)}(T, \rho, \mathbf{B}_0) \cdot \mathbf{B}_0, \tag{8.4.4a}$$

in which the magnetic susceptibility $\chi^{(m)}$ is a second rank tensor. We shall, however, limit ourselves to the linear response regime, for which $\chi^{(m)}(T, \rho, \mathbf{B}_0)$ becomes a scalar multiple, $\chi^{(m)}(T, \rho)$, of the isotropic second rank tensor $\boldsymbol{\delta}$ (whose Cartesian components are Kronecker deltas δ_{ij}, with $i, j = x, y, z$), so that the macroscopic magnetic moment $\mathbf{M}$ is parallel to $\mathbf{B}_0$, i.e.,

$$\mathbf{M} = \chi^{(m)}(T, \rho)\mathbf{B}_0. \tag{8.4.4b}$$

Note that the magnetic susceptibility may, in principle, still be a function of both the temperature, T, and mass density, ρ, of the thermodynamic system. Within the linear response regime the permittivity μ is given in terms of μ_0 and $\chi^{(m)}$ by

$$\mu = \mu_0(1 + \chi^{(m)}). \tag{8.4.5}$$

The combined 1st and 2nd Laws expression, Eq. (2.3.12a) is extended to include the energy change (8.4.3) induced in the system by the constant external magnetic field, to become

$$dU = dU + dU_{\text{field}} = TdS - PdV + \mathbf{B}_0 \cdot d\mathbf{M} \tag{8.4.6}$$

for a magnetizable thermodynamic system. In the following, we shall focus our discussion on the thermodynamics of a *magnetically isotropic medium*, so that $\mathbf{M}$ lies along the direction defined by $\mathbf{B}_0$, thereby allowing us to replace $\mathbf{M}$ everywhere by M and $\mathbf{B}_0$ everywhere by B_0.

As the magnetic moment of the macroscopic sample depends upon both temperature, T, and (for an isotropic medium) the magnitude, B_0, of the external magnetic flux into which it has been placed, the total differential dM of the macroscopic magnetic moment may be expressed as

$$dM = \left(\frac{\partial M}{\partial T}\right)_{B_0} dT + \left(\frac{\partial M}{\partial B_0}\right)_T dB_0. \tag{8.4.7}$$

The response of a macroscopic collection of magnetic dipoles to a change in the magnitude of the applied magnetic flux, B_0, under isothermal conditions is known as the isothermal magnetic susceptibility, and is denoted by $\chi^{(m)}(T)$. From expression (8.4.7), we see that $\chi^{(m)}(T)$ is thus given by

$$\chi^{(m)}(T) = \left(\frac{\partial M}{\partial B_0}\right)_T. \tag{8.4.8}$$

As pointed out by Pippard [6], the thermodynamic results traditionally obtained for the internal energy U as a function of pressure P and volume V, as well as the expressions for the constant pressure and constant-volume heat capacities and their difference, as given in Eq. (2.5.1c) can be transformed into equivalent expressions for B_0 and M via the transformation $P \leftrightarrow B_0$, $V \leftrightarrow -M$. In particular, we note that the relevant permutation rule for a magnetic system,

$$\left(\frac{\partial B_0}{\partial T}\right)_M \left(\frac{\partial M}{\partial B_0}\right)_T \left(\frac{\partial T}{\partial M}\right)_{B_0} = -1, \tag{8.4.9}$$

may be employed to eliminate the (often formal) nonphysical partial derivative $(\partial B_0 / \partial T)_M$ from a thermodynamic expression. Moreover, an expression analogous to Eq. (2.4.10) is

$$\left(\frac{\partial U}{\partial M}\right)_T = -T \left(\frac{\partial B_0}{\partial T}\right)_M + B_0, \tag{8.4.10}$$

while heat capacities C_M and C_B are given, analogously to Eqs. (2.3.15c) and (2.4.1c) in terms of the thermodynamic internal energy U and enthalpy H as

$$C_M \equiv \left(\frac{\partial Q_{rev}}{\partial T}\right)_M = \left(\frac{\partial U}{\partial T}\right)_M, \tag{8.4.11}$$

and

$$C_B \equiv \left(\frac{\partial Q_{rev}}{\partial T}\right)_{B_0} = \left(\frac{\partial H}{\partial T}\right)_{B_0}. \tag{8.4.12}$$

Similarly, an expression for $C_B - C_M$ can immediately be obtained from Eq. (2.5.1c) for fluids as

$$C_B - C_M = T \left(\frac{\partial B_0}{\partial M}\right)_T \left(\frac{\partial M}{\partial T}\right)_{B_0}^2. \tag{8.4.13a}$$

A physically more appealing form for this expression is

$$C_B - C_M = T \frac{\left(\frac{\partial M}{\partial T}\right)_{B_0}^2}{\left(\frac{\partial M}{\partial B_0}\right)_T}, \tag{8.4.13b}$$

as M is a function of B_0 and T.

A magnetic susceptibility χ_{mag} is often defined phenomenologically via the relation

$$M = \chi_{mag}(B_0, T, \rho) B_0, \tag{8.4.14a}$$

relating the bulk magnetic moment M to the applied magnetic induction B_0. We note that the isothermal magnetic susceptibility $\chi^{(m)}(T)$ defined in Eq. (8.4.8), and also commonly referred to as the isothermal differential susceptibility, is related to the phenomenological magnetic susceptibility of Eq. (8.4.14a) via the expression

$$\chi^{(m)}(B_0, T, \rho) \equiv \left(\frac{\partial M}{\partial B_0}\right)_T = \left[B_0\left(\frac{\partial \chi_{mag}}{\partial B_0}\right)_T + \chi_{mag}(B_0, T, \rho)\right]. \qquad (8.4.14b)$$

This expression becomes

$$\chi^{(m)}(T) = \chi_{mag}(T) \qquad (8.4.14c)$$

should χ_{mag} depend only upon temperature.

For a magnetic field-independent magnetic susceptibility, which is the only type considered herein,[2] the partial derivative in the numerator of Eq. (8.4.13b) becomes

$$\left(\frac{\partial M}{\partial T}\right)_{B_0} = B_0\left(\frac{d\chi_{mag}}{dT}\right),$$

and the difference $C_B - C_M$ is well approximated by

$$C_B - C_M \simeq \frac{TB_0^2}{\chi_{mag}(T)}\left(\frac{d\chi_{mag}}{dT}\right)^2. \qquad (8.4.15)$$

If we recall that substances for which $\chi^{(m)}(T) < 0$ are called diamagnetic, while those for which $0 < \chi^{(m)}(T) \lesssim 1$ are called paramagnetic, then this expression tells us, perhaps not surprisingly, that $C_B = C_M$ for $B_0 = 0$, that $C_B \geq C_M$ for paramagnetic materials, and that $C_B \leq C_M$ for diamagnetic materials.

From the combined 1st and 2nd Laws expression for a magnetic system, the heat δQ transferred during a reversible magnetization is obtained as

$$dQ_{rev} = \left(\frac{\partial U}{\partial T}\right)_M dT + \left(\frac{\partial U}{\partial M}\right)_T dM - B_0 dM$$

$$= C_M dT + \left[\left(\frac{\partial U}{\partial M}\right)_T - B_0\right]dM.$$

Now, upon employing Eq. (8.4.10) and treating M as a function of B_0 and T, δQ_{rev} may also be expressed as

---

[2]As noted by Pippard (footnote on pp. 65, 66 of [6]), this excludes ferromagnetic materials (because they exhibit hysteresis), high magnetic field strengths and, in some cases, magnetic materials at very low temperatures. Moreover, as noted earlier, the present discussion is also limited to isotropic materials in the linear response regime, for which $\chi_{mag}(T, \rho)$ is a scalar.

$$dQ_{rev} = \left[C_M + T \left(\frac{\partial B_0}{\partial M} \right)_T \left(\frac{\partial M}{\partial T} \right)_{B_0}^2 \right] dT + T \left(\frac{\partial M}{\partial T} \right)_{B_0} dB_0.$$

This result may be further simplified by utilizing Eq. (8.4.13a) to obtain

$$dQ_{rev} = C_B dT + T \left(\frac{\partial M}{\partial T} \right)_{B_0} dB_0.$$

Finally, we may utilize the phenomenological equation of state (8.4.14a) for the magnetization to give dQ_{rev} as

$$dQ_{rev} = C_B dT + B_0 T \frac{d\chi_{mag}}{dT} dB_0. \tag{8.4.16}$$

Hence, for a reversible adiabatic process (for which $\delta Q_{rev} = 0$), we obtain a differential temperature change dT of

$$dT = - \frac{B_0 T}{C_{B_0}} \frac{d\chi}{dT} dB_0. \tag{8.4.17}$$

As B_0, T, and C_B are all positive, while $dB_0 < 0$ for demagnetization, the sign of dT is the sign of the temperature derivative of $\chi(T)$. For a paramagnetic solid that obeys the Curie law of magnetization, viz.,

$$\chi_{mag}(T) = \frac{N\mu_m^2}{3k_B T}, \tag{8.4.18}$$

we thus see that $d\chi_{mag}/dT < 0$, so that the temperature is decreased upon adiabatic demagnetization of a paramagnetic sample. This is, in fact, the procedure that was employed for many years to obtain temperatures significantly lower than the temperature of liquid helium, typically in the temperature range 0.1 K to 0.001 K.

Interlude, Energy Transfer Between System and Surroundings

The nature of the energy transferred between a mechanical thermodynamic system and its surroundings is typically termed 'work' when the energy transfer results in volume or pressure changes in the system and is typically termed 'heat' when the energy transfer leads to changes in the temperature or entropy of the system. In both cases, the transferred energy is redistributed through the thermodynamic system in a manner that is consistent with the macroscopic change to the system. For a magnetizable thermodynamic system in an applied magnetic field, however, the situation is somewhat different, in that the magnetizable components of the thermodynamic system interact with the field to create new energy states that enable the system to store the energy supplied by the field. However, unlike the oft-considered example of an ideal gas, in which there is a mechanism, namely collisions, for redistributing the energy (heat) transferred to the ideal gas from the surroundings or the work done on the system by the surroundings into other dynamical modes (such as the translational motion or internal states) of the constituent particles of the thermodynamic system, no mechanism exists for transferring the stored (magnetic) energy into other dynamical modes of the thermodynamic system. That the sole outcome arising from placing a magnetizable thermodynamic system into a magnetic field is a reduction in entropy due to the creation of preferential alignment of spins (or other angular momenta) along the field direction (thereby increasing the order of the system) is a potential source of confusion about the proper analogy to be drawn between the intensive-extensive B_0-M pair of thermodynamic variables and the P-V or T-S pairs employed to generate the enthalpy or Helmholtz energy.

The most common analogy made between traditional thermodynamic systems and magnetizable thermodynamic systems is that between pressure–volume work, and the field-moment 'work', $B_0 dM$, in which $\mathbf{B}_0$ plays the role of pressure, P, and $-\mathbf{M}$ plays the role of volume, V [4–7]. Consistent with this manner of looking upon the magnetizable system in terms of the internal energy U_{mag} as a function of $S, V, \mathbf{M}$ is the introduction of a magnetic enthalpy, $H_{\mathrm{mag}}(S, P, \mathbf{B}_0) \equiv U(S, V, \mathbf{M}) + PV - \mathbf{B}_0 \cdot \mathbf{M}$, a magnetic Helmholtz energy $A_{\mathrm{mag}}(T, V, \mathbf{M}) \equiv U(S, V, \mathbf{M}) - TS$, and a magnetic Gibbs energy $G_{\mathrm{mag}}(T, P, \mathbf{B}_0) \equiv U(S, V, \mathbf{M}) - TS + PV - \mathbf{B}_0 \cdot \mathbf{M}$. Of course, the internal energy U remains exclusively a function of the three extensive thermodynamic variables $S, V, \mathbf{M}$, while the Gibbs energy remains exclusively a function of the three intensive thermodynamic variables T, P, and $\mathbf{B}_0$.

The analogy between the $\mathbf{B}$–$\mathbf{M}$ pair of complementary variables and the corresponding T-S pair would leave the enthalpy unchanged as $H'_{\mathrm{mag}}(S, P, \mathbf{M}) \equiv U(S, V, \mathbf{M}) + PV$, would introduce a new Helmholtz energy $A'_{\mathrm{mag}}(T, V, \mathbf{B}_0) \equiv U(S, V, \mathbf{M}) - TS - \mathbf{B}_0 \cdot \mathbf{M}$, and would give the same Gibbs energy $G_{\mathrm{mag}}(T, P, \mathbf{B}_0)$ as that obtained using the P-V analogy. However, as neither of these two interpretations leads to full compatibility between the thermodynamic and the statistical mechanical description of a magnetic system given in the following subsection, we shall follow a description provided in §5.8 of [2].   □

8.4.3   Thermodynamic Potentials for Magnetizable Systems

Because it is experimentally very difficult to control the magnetization $\mathcal{M}$ (equivalently, the magnetic moment $\mathbf{M}$) of a macroscopic system, while it is relatively easy to control an external magnetic field $\mathbf{H}_0$ (equivalently, a magnetic induction field $\mathbf{B}_0 \equiv \mu_0 \mathbf{H}_0$), it is useful to work with the Legendre-transformed internal energy U_{mag}, defined as

$$U_{\mathrm{mag}}(S, V, \mathbf{B}_0) \equiv U(S, V, \mathbf{M}) - \mathbf{B}_0 \cdot \mathbf{M}, \tag{8.4.19a}$$

in which $U(S, V, \mathbf{M})$ corresponds to Eq. (4.1.6), while $\mathbf{M}$, given by

$$\mathbf{M} \equiv \int_V \mathcal{M} \, d^3 r, \qquad (8.4.19b)$$

represents the macroscopic magnetic moment of a magnetizable system with variable volume V and magnetization $\mathcal{M}$ (for integrand 1, the integration simply gives the system volume V). From expression (8.4.19a), we obtain the differential of U_{mag} as

$$dU_{\text{mag}} = dU - d(\mathbf{B}_0 \cdot \mathbf{M})$$
$$= T \, dS - P \, dV + \mathbf{B}_0 \cdot d\mathbf{M} - d(\mathbf{B}_0 \cdot \mathbf{M}).$$

As the volume V is variable, the integral (8.4.19b) has variable limits (see Appendix B.1): this results in d $(\mathbf{B}_0 \cdot \mathbf{H})$ taking the form

$$d(\mathbf{B}_0 \cdot \mathbf{M}) = \int_V \frac{1}{V} d(V \mathcal{M} \cdot \mathbf{B}_0) \, d^3 r$$
$$= \int_V \frac{\mathcal{M} \cdot \mathbf{B}_0}{V} \, dV + \int_V [\mathbf{B}_0 \cdot d\mathcal{M} + \mathcal{M} \cdot d\mathbf{B}_0] \, d^3 r.$$

With this result, dU_{mag} becomes

$$dU_{\text{mag}} = T \, dS - \left[P + \int_V \frac{\mathcal{M} \cdot \mathbf{B}_0}{V} \, d^3 r \right] dV - \left(\int_V \mathcal{M} \, d^3 r \right) \cdot d\mathbf{B}_0$$

$$= T \, dS - P_{\text{mag}} \, dV - \mathbf{M} \cdot d\mathbf{B}_0, \qquad (8.4.20a)$$

in which the modified pressure P_{mag} is defined as

$$P_{\text{mag}} \equiv P + \int_V \frac{\mathcal{M} \cdot \mathbf{B}_0}{V} \, d^3 r. \qquad (8.4.20b)$$

Enthalpy, Helmholtz energy, and Gibbs energy functions consistent with this procedure are given as $H = U + P_{\text{mag}} V$, $A = U - TS$, and $G = U + P_{\text{mag}} V - TS$, from which we obtain the set of differentials

$$dH = T \, dS + V \, dP_{\text{mag}} - \mathbf{M} \cdot d\mathbf{B}_0, \qquad (8.4.21a)$$

$$dA = -S \, dT - P_{\text{mag}} \, dV - \mathbf{M} \cdot d\mathbf{B}_0, \qquad (8.4.21b)$$

$$dG = -S \, dT + V \, dP_{\text{mag}} - \mathbf{M} \cdot d\mathbf{B}_0. \qquad (8.4.21c)$$

Equation (8.4.20a) together with Eqs. 8.4.21a, 8.4.21b, and 8.4.21c represents a set of total differentials for the thermodynamic functions of state appropriate to experiments performed on magnetizable systems, be they gaseous, liquid, or solid. For a thermodynamic system that has a uniform mass density and is magnetically isotropic, **M** becomes simply

$$\mathbf{M} = \int_V \mathcal{M} \, d^3 r = V\mathcal{M},$$ (8.4.22)

while $\mathcal{M}$ (and hence **M**) and $\mathbf{B}_0$ are collinear.

8.5 Paramagnetism in Spin Systems

8.5.1 Setting the Stage

Because magnetic moments are typically associated with intrinsic angular momenta (traditionally referred to as 'spin') possessed by fundamental quantum mechanical particles, and as there is no reason to assume that all readers of the following sections will be familiar with the fundamental properties of such spins, we shall set the stage for an examination of the statistical mechanics associated with typical spin systems with a primer on the quantum mechanics of nuclear spins, thereby ensuring that everyone possesses the same level of basic knowledge needed to achieve an understanding the statistical thermodynamics of simple magnetic systems.

Many nuclei possess intrinsic angular momenta, commonly designated by **I** and referred to as 'nuclear spin'. These angular momenta are in many ways analogous to the intrinsic 'spin-$\frac{1}{2}$' angular momentum (with magnitude $\frac{\sqrt{3}}{2}\hbar$) possessed by electrons. The fundamental nuclear particles (i.e., protons and neutrons) that make up atomic nuclei, like electrons, also possess intrinsic 'spin-$\frac{1}{2}$' angular momenta. Atomic nuclei, which are made up from protons and neutrons, possess what are referred to as 'composite angular momenta' or 'nuclear spins' that derive from coupling of the intrinsic angular momenta of their constituent protons and neutrons. The only exception is the H atom nucleus with its single proton. The D atomic nucleus, for example, consists of one proton and one neutron, and has a resultant 'spin-1' angular momentum (with magnitude $\sqrt{2}\hbar$, that corresponds to the vector operator $\widehat{\mathbf{I}}$ with $I = 1$.

An individual nucleus that possesses a (composite or net) nuclear spin described by a corresponding nuclear spin (vector) operator $\widehat{\mathbf{I}}$ has an associated nuclear magnetic moment (vector) operator $\widehat{\boldsymbol{\mu}}_I$, given by $\widehat{\boldsymbol{\mu}}_I = g_I \mu_N \widehat{\mathbf{I}}$. The constant of proportionality $g_I \mu_N$ is a product of a factor g_I that has a specific value for a given nucleus and a factor μ_N that represents the magnitude of a typical nuclear magnetic moment: g_I is simply referred to as the nuclear g-factor, while μ_N, defined by $\mu_N \equiv \hbar e/(2m_p)$, with e the proton charge and m_p the proton mass, is known as the nuclear magneton, and has the (approximate) value $5.0508 \times 10^{-27}\,\mathrm{JT}^{-1}$ (Joules per Tesla).

In the absence of a magnetic field (induction) **B** (units: Tesla), the energy levels associated with a given nuclear magnetic moment are all degenerate, but in the presence of an external magnetic field, they are split via the Zeeman interaction in the Hamiltonian, namely $\mathcal{H} = -\widehat{\boldsymbol{\mu}}_I \cdot \mathbf{B} = -\widehat{\mu}_{Iz}B$, with $\widehat{\mu}_{Iz} \equiv g_I\mu_\mathrm{N}\widehat{I}_z$. Only the z-component $\widehat{I}_z$ of the nuclear angular momentum vector operator appears in the expression for the Zeeman Hamiltonian because we normally choose to associate the z-direction with the direction of the (constant) applied (or external) magnetic field **B**.

Because the spin (or magnetic, or Zeeman) Hamiltonian $\mathcal{H}_\mathrm{spin}$ is proportional to the $\widehat{I}_z$ operator, it is convenient to choose simultaneous eigenstates ψ_spin of the two (commuting) operators $\widehat{I}^2$ and $\widehat{I}_z$ as the eigenstates of the spin Hamiltonian. Recall that (if not, consider what is written here as the defining relations), as simultaneous eigenstates of these two operators, these states will have the properties

$$\widehat{I}^2\psi_\mathrm{spin} = I(I+1)\psi_\mathrm{spin}\,, \quad \widehat{I}_z\psi_\mathrm{spin} = M_I\psi_\mathrm{spin}, \tag{8.5.1}$$

with $-I \leq M_I \leq +I$.

Thus, if we utilize these simultaneous eigenfunctions ψ_spin as eigenfunctions of the magnetic Hamiltonian, then we may write

$$\mathcal{H}_\mathrm{spin}\psi_\mathrm{spin} = E_\mathrm{spin}\psi_\mathrm{spin}. \tag{8.5.2a}$$

Moreover, if we utilize the form

$$\mathcal{H}_\mathrm{spin} = -\widehat{\mu}_z B = -g_I\mu_\mathrm{N}B\widehat{I}_z \tag{8.5.2b}$$

for the spin Hamiltonian, this then allows us to obtain the spin-energies, for example, for a spin-$\frac{1}{2}$ nucleus (in the presence of a constant magnetic field of strength B) as

$$E_\mathrm{spin} = \mp\tfrac{1}{2}g_I\mu_\mathrm{N}B, \tag{8.5.2c}$$

depending upon whether m_I has the value $+\frac{1}{2}$ or $-\frac{1}{2}$; (note that the spin-'up' value $M_I = +\frac{1}{2}$ has the lower energy). It is also useful to note that $\widehat{\mu}_z\psi_\mathrm{spin} = \mu_z\psi_\mathrm{spin}$ gives the corresponding (eigen)values for the magnetic moment, i.e.,

$$\mu_{z\pm} = \pm\tfrac{1}{2}g_I\mu_\mathrm{N}. \tag{8.5.3}$$

8.5.2 Statistical Mechanics of Spin Systems

Although the focus of the statistical mechanical treatment of paramagnetism expounded in this section will be upon nuclear spin systems, the same concepts apply, with appropriate adaptation, to electronic spin systems.

Much of the behaviours characterizing spin systems can be illustrated by an examination of the simplest exemplar of a spin system, namely a spin-$\frac{1}{2}$ species, such as a ^{3}He or ^{129}Xe atom,[3] which has nuclear spin angular momentum $\mathbf{I}$ of magnitude $\frac{\sqrt{3}}{2}\hbar$, corresponding to the nuclear angular momentum quantum number $I = \frac{1}{2}$. We may use the results obtained in Sect. 8.5.1 to treat such a system.

The canonical partition function, $z_{\text{spin}}(T)$, for an individual atom with nuclear spin-$\frac{1}{2}$ in the presence of a constant magnetic induction field of magnitude B_0 is given by

$$z_{\text{spin}}(T, B_0) \equiv \sum_r e^{-\beta E_r} = e^{\frac{1}{2}\beta g_I \mu_N B_0} + e^{-\frac{1}{2}\beta g_I \mu_N B_0},$$

or more compactly in terms of the hyperbolic cosine, as

$$z_{\text{spin}}(T, B_0) = 2 \cosh\left(\tfrac{1}{2}\beta g_I \mu_N B_0\right). \tag{8.5.4}$$

The associated probability factors for the two spin states are then given by

$$p_+ = \frac{e^{\frac{1}{2}\beta g_I \mu_N B_0}}{z_{\text{spin}}(\beta)} \;; \quad p_- = \frac{e^{-\frac{1}{2}\beta g_I \mu_N B_0}}{z_{\text{spin}}(\beta)}, \tag{8.5.5}$$

with the $+$ sign associated with $M_I = \frac{1}{2}$ and the $-$ sign associated with $M_I = -\frac{1}{2}$. These probabilities allow us to determine the ensemble average magnetic moment $\langle \mu_z \rangle$ associated with an individual spin particle as

$$\begin{aligned}
\langle \mu_z \rangle &= p_+ \mu_{z+} + p_- \mu_{z-} \\
&= \frac{\frac{1}{2}g_I \mu_N e^{\frac{1}{2}\beta g_I \mu_N B_0}}{2\cosh\left(\frac{1}{2}\beta g_I \mu_N B_0\right)} + \frac{\left(-\frac{1}{2}g_I \mu_N\right) e^{-\frac{1}{2}\beta g_I \mu_N B_0}}{2\cosh\left(\frac{1}{2}\beta g_I \mu_N B_0\right)} \\
&= \tfrac{1}{2}g_I \mu_N \frac{e^{\frac{1}{2}\beta g_I \mu_N B_0} - e^{-\frac{1}{2}\beta g_I \mu_N B_0}}{2\cosh\left(\frac{1}{2}\beta g_I \mu_N B_0\right)},
\end{aligned}$$

which can be more compactly written as

$$\langle \mu_z \rangle(T) = \tfrac{1}{2}g_I \mu_N \tanh\left(\tfrac{1}{2}\beta g_I \mu_N B_0\right). \tag{8.5.6}$$

---

[3]Of course, we could equally well consider a number of other nuclear spin-$\frac{1}{2}$ (multi-electron) atoms, such as ^{89}Y (yttrium), ^{107}Ag, ^{109}Ag, ^{169}Tm (thulium), or ^{207}Pb, for example. However, the noble gas atoms ^{3}He and ^{129}Xe provide slightly simpler exemplars, as they have zero net electronic angular momentum due to their fully completed electronic shell structures.

For high temperatures, by which we mean temperatures such that $\beta g_I \mu_N B_0 \ll 1$, we may expand the exponentials and retain only the first two terms for each of them, to give

$$\tanh\left(\tfrac{1}{2}\beta g_I \mu_N B_0\right) \simeq \frac{\left(1 + \tfrac{1}{2}\beta g_I \mu_N B_0 + \cdots\right) - \left(1 - \tfrac{1}{2}\beta g_I \mu_N B_0 + \cdots\right)}{\left(1 + \tfrac{1}{2}\beta g_I \mu_N B_0 + \cdots\right) + \left(1 - \tfrac{1}{2}\beta g_I \mu_N B_0 + \cdots\right)}$$

$$\simeq \tfrac{1}{2}\beta g_I \mu_N B_0.$$

As a consequence, the average magnetic moment behaves at high temperatures as

$$\langle \mu_z \rangle_{\text{high } T} \simeq \frac{\left(\tfrac{1}{2}g_I \mu_N\right)^2 B_0}{k_B T}, \qquad (8.5.7a)$$

i.e., it varies inversely with T. This is as we might anticipate intuitively in that, the higher the temperature the more randomly oriented the magnetic moment will become, and thus the smaller its average will be.

For temperatures sufficiently low that $\beta \mu_N B_0 \gg 1$, we have $e^{\frac{1}{2}\beta g_I \mu_N B_0} \gg e^{-\frac{1}{2}\beta g_I \mu_N B_0}$, with a corresponding average magnetic moment given by

$$\langle \mu_z \rangle_{\text{low } T} \simeq \tfrac{1}{2}g_I \mu_N \frac{e^{\frac{1}{2}\beta g_I \mu_N B_0}}{e^{\frac{1}{2}\beta g_I \mu_N B_0}} = \tfrac{1}{2}g_I \mu_N, \qquad (8.5.7b)$$

and is constant, i.e., independent of T. Thus, if T is low enough, the value of $\langle \mu_z \rangle$ is simply $\mu_z = \tfrac{1}{2}g_I \mu_N$, since the spin particle remains in the ground state E_+, for which the energy is minimal.

Before proceeding further, it will be useful to pause and consider more carefully the connection between the statistical mechanical description of our spin system in terms of a canonical ensemble of a macroscopically large number N of spin-particles and the thermodynamics of a magnetizable system as described in the previous section. We saw in Chap. 2 that two pairs of conjugate variables, namely $\{T, S\}$ and $\{P, V\}$, enter into the description of a traditional thermodynamic system, and that any pair of these variables, one chosen from each set, may be employed to describe the system. When we established the canonical ensemble for a fixed number N of particles in Chap. 4, we saw that the relevant variables characterizing it were the volume V and temperature T and that the corresponding thermodynamic potential for the system energy treated as a function of T and V is the Helmholtz energy, A. It is thus appropriate to take some care to ensure that we make the proper assignment of the characteristic function for a canonical ensemble consisting of N spins so that no contradiction arises between the thermodynamic and statistical mechanical descriptions of a macroscopic spin system. To accomplish this goal, we begin by noting the introduction of the magnetic internal energy $U_{\text{mag}}(S, V, \mathbf{B}_0)$

in Sect. 8.4.3 via Eq. (8.4.19a) to aid in the description of a magnetizable system brought quasistatically into an externally-maintained constant magnetic induction field. For present purposes, we shall prefer to refer to this internal energy as U_{spin}.

If we now consider N *isolated* (i.e., distinguishable by virtue of being at specified locations in a solid lattice) $I = \frac{1}{2}$ atoms, the macroscopic magnetization $\overline{\mathbf{M}}$ will be given by

$$\overline{\mathbf{M}} = N\langle\boldsymbol{\mu}\rangle, \tag{8.5.8a}$$

in which $\langle\boldsymbol{\mu}\rangle$ represents the macroscopic average of the magnetic dipole moments of the N spin-$\frac{1}{2}$ atoms. As both μ_x and μ_y precess about the z-axis (with a typical NMR frequency), the equilibrium values of $\overline{M}_x$ and $\overline{M}_y$ average to zero on a typical NMR time scale, and only the z-component of $\overline{\mathbf{M}}$, given via Eq. (8.5.6) as

$$\overline{M}_z = N\langle\mu_z\rangle = \tfrac{1}{2}Ng_I\mu_N \tanh(\tfrac{1}{2}\beta g_I\mu_N B_0), \tag{8.5.8b}$$

survives. We shall hereinafter denote $\overline{M}_z$ as M.

The characteristic function [8] for a canonical ensemble[4] of N distinguishable nuclear spin-$\frac{1}{2}$ atoms is the Helmholtz energy, A_{spin}, given by

$$A_{\text{spin}}(T, B_0) = -k_B T \ln Z_{\text{spin}}(N, T, B_0), \tag{8.5.9a}$$

in terms of the canonical partition function $Z_{\text{spin}}(T, B_0)$ for N atoms, or as

$$A_{\text{spin}}(T, B_0) = -N k_B T \ln z_{\text{spin}}(T, B_0), \tag{8.5.9b}$$

in terms of the single-nucleus partition function $z_{\text{spin}}(T, B_0)$ given by Eq. (8.5.4). In this manner, the Helmholtz energy for a spin-$\frac{1}{2}$ particle in a magnetic induction field B_0 is obtained as an explicit function of T and B_0, namely

$$A_{\text{spin}}(T, B_0) = -N k_B T \ln[2\cosh(\tfrac{1}{2}\beta\mu_N g_I B_0)]. \tag{8.5.10}$$

It is perhaps worth noting that the extensivity of the Helmholtz energy is here represented specifically by the factor N.

The thermodynamic expression for the Helmholtz energy is, according to the development in Sect. 8.4.3,

$$\begin{aligned} A_{\text{spin}}(T, V, B_0) &\equiv U_{\text{spin}}(S, V, B_0) - TS \\ &= U(S, V, M) - TS - B_0 M. \end{aligned} \tag{8.5.11a}$$

---

[4]The importance of making an appropriate identification of the characteristic function for the canonical ensemble and an unambiguous connection with macroscopic thermodynamics has been emphasized by Castellano [8], who points out, in particular, that this connection is not a minor issue, as the identification must also hold for systems of interacting spins.

The total differential for A_{spin} will thus be

$$dA_{\text{spin}} = -S dT - P_{\text{mag}} dV - M dB_0, \qquad (8.5.11b)$$

from which we see that the entropy, modified pressure P_{mag} defined by Eq. (8.4.20b), and magnetization M may be determined directly from the expression for A_{spin} as

$$S_{\text{spin}} = -\left(\frac{\partial A_{\text{spin}}}{\partial T}\right)_{V, B_0}, \quad P_{\text{mag}} = -\left(\frac{\partial A_{\text{spin}}}{\partial V}\right)_{T, B_0}, \quad M_{\text{spin}} = -\left(\frac{\partial A_{\text{spin}}}{\partial B_0}\right)_{T, V}. \tag{8.5.12}$$

For a solid macroscopic sample of N (distinguishable) spin particles, $Z_{\text{spin}}(N, T) = z_{\text{spin}}^N(T)$, and the thermodynamic internal energy is given by

$$U_{\text{spin}} = k_B T^2 \left(\frac{\partial \ln Z_{\text{spin}}}{\partial T}\right)_N = N k_B T^2 \frac{1}{z_{\text{spin}}} \frac{dz_{\text{spin}}}{dT}. \tag{8.5.13a}$$

Evaluation of the right-hand side of this expression gives

$$U_{\text{spin}} = -N \left(\tfrac{1}{2} g_I \mu_N B_0\right) \tanh\left(\tfrac{1}{2}\beta g_I \mu_N B_0\right), \tag{8.5.13b}$$

which can be seen, using Eq. (8.5.8b), to equal $-B_0 M$. Upon taking note of the defining relation $U_{\text{spin}} \equiv U - B_0 M$, we obtain the intriguing (and perhaps somewhat surprising) result that $U = 0$ for a nuclear spin system.

Of course, we may also utilize the three relations (8.5.12) connecting the entropy, modified pressure, and macroscopic magnetic moment to the Helmholtz energy to obtain $S_{\text{spin}}(M)$, P_{mag}, and $M(T, B_0)$ as

$$S_{\text{spin}}(N, M) = N k_B \left\{ \ln\left[2\cosh\left(\tfrac{1}{2}\beta \mu_N g_I B_0\right)\right] - \frac{\mu_N g_I B_0}{2 k_B T} \tanh(\tfrac{1}{2}\beta \mu_N g_I B_0) \right\}, \tag{8.5.14}$$

$$M(T, B_0) = \tfrac{1}{2} N \mu_N g_I \tanh(\tfrac{1}{2}\beta \mu_N g_I B_0), \tag{8.5.15}$$

and $P_{\text{mag}} = \mathcal{M} B_0$.

It is useful to compare expression (8.5.14) for the entropy of the spin system with its thermodynamic definition in terms of the spin contributions to the internal and Helmholtz energies as

$$S_{\text{spin}}(N, T) = \frac{U_{\text{spin}}(N, T) - A_{\text{spin}}(N, T)}{T} = \frac{U_{\text{spin}}(N, T)}{T} + N k_B \ln z_{\text{spin}}(T).$$

A corresponding expression for a heat capacity, C_B, can be obtained as

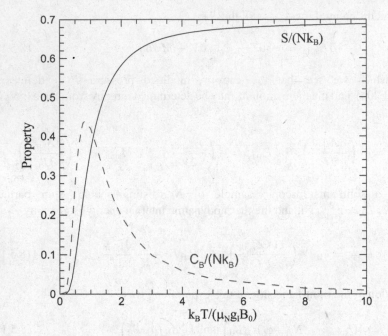

Fig. 8.2 The temperature dependence of the per-particle contributions to the entropy and heat capacity associated with a collection of N noninteracting and distinguishable spin-$\frac{1}{2}$ nuclei in a magnetic induction field of strength B_0

$$C_B = \left(\frac{\partial U_{\text{spin}}}{\partial T}\right)_{N,B_0} = -N\,\tfrac{1}{2}g_I\mu_N B_0 \left(\frac{\partial \tanh(\frac{1}{2}\beta g_I \mu_N B_0)}{\partial T}\right)_{B_0},$$

which we may write in a more compact form as

$$C_B = N k_B \left(\frac{g_I \mu_N B_0}{2 k_B T}\right)^2 \text{sech}^2\left(\tfrac{1}{2}\beta g_I \mu_N B_0\right). \qquad (8.5.16)$$

The temperature dependences of $C_B(N,T)/(Nk_B)$ and $S_{\text{spin}}(N,T)/(Nk_B)$ for N spin-$\frac{1}{2}$ particles are shown in Fig. 8.2.

The behaviour of $C_B(T)$ illustrated in Fig. 8.2 is characteristic for a spin system, and is referred to as a Schottky heat capacity (or specific heat) anomaly. Notice that by $k_B T/(\mu B) \simeq 10$, the value of C_B has dropped below approximately 5 % of its maximum value, and similarly, S_{spin} has very nearly reached its high-temperature limiting value. Moreover, we see that the temperature regime over which a spin system contributes to the heat capacity is determined by the ratio of μ_N to k_B, i.e., by a (small) multiple of $3.6581 \times 10^{-4}\,\text{K}\,\text{T}^{-1}$: for a fixed field strength $B_0 = 10\,\text{T}$ (approximately the magnetic field strength for a 500 MHz NMR spectrometer), the maximum in C_B would occur at a temperature $T_{\text{max}} = 0.0078$ K for a sample of ^{3}He.

It may be shown (see Problem 8.5) that the macroscopic magnetic moment of a magnetizable thermodynamic system may be obtained from the Helmholtz energy $A_{\text{mag}}(T, V, \mathbf{B}_0)$ given by the Legendre transform

$$A_{\text{mag}}(T, V, \mathbf{B}_0) = U_{\text{mag}}(S, V, \mathbf{M}) - TS - \mathbf{B}_0 \cdot \mathbf{M} \tag{8.5.17a}$$

as

$$\mathbf{M} = -\left(\frac{\partial A_{\text{mag}}}{\partial \mathbf{B}_0}\right)_{T,V}. \tag{8.5.17b}$$

Thus, we may obtain the magnetic moment $\overline{\mathbf{M}}(\mathbf{B}, T)$ associated with the system of N spins under consideration here from the fundamental canonical ensemble relation,

$$A_{\text{spin}}(N, T) = -k_{\text{B}}T \ln Z_{\text{spin}}(N, T),$$

in general as

$$\overline{\mathbf{M}}(\mathbf{B}_0, T) = Nk_{\text{B}}T\left(\frac{\partial z_{\text{spin}}}{\partial \mathbf{B}_0}\right)_{T,V}. \tag{8.5.18}$$

More specifically, for an isotropic medium subjected to a uniaxial magnetic field, so that $\mathbf{B}_0$, $\mathbf{M}$ are collinear and $\mathbf{B}_0 \cdot \mathbf{M} \equiv B_0 M$, as

$$\overline{M}(B_0, T) = Nk_{\text{B}}T\,\frac{1}{z_{\text{spin}}}\left(\frac{\partial z_{\text{spin}}}{\partial B_0}\right)_{T,V}. \tag{8.5.19}$$

We may utilize relation (8.4.8) to obtain a formal expression for $\chi_T^{(\text{m})}(N, B, T)$ as

$$\chi_T^{(\text{m})}(N, B, T) = Nk_{\text{B}}T\left[-\frac{1}{z_{\text{spin}}^2}\left(\frac{\partial z_{\text{spin}}}{\partial B_0}\right)^2 + \frac{1}{z_{\text{spin}}}\left(\frac{\partial^2 z_{\text{spin}}}{\partial B_0^2}\right)\right].$$

The first term on the right-hand side of this expression is clearly equal to $-\beta^2 \overline{m}^2 \equiv -\beta^2 \overline{M}^2/N^2$, with $\overline{m}$ the ensemble average magnetic moment per particle, while an examination of the second term gives $\beta^2 \overline{m^2} \equiv \beta^2 \overline{M^2}/N^2$, so that $\chi_T^{(\text{m})}$ is given by

$$\chi_T^{(\text{m})} = \frac{1}{Nk_{\text{B}}T}\left[\overline{M^2} - \overline{M}^2\right]. \tag{8.5.20}$$

This result shows that the magnetic susceptibility is determined by fluctuations in the magnetization. Both expressions (8.4.8) and (8.5.20) are general. We may test the consistency of these general expressions explicitly for our spin-$\frac{1}{2}$ system: from expression (8.5.8b) for $\overline{M}_z$, we see that Eq. (8.4.8) gives

$$\chi_T^{(m)} = \left(\frac{\partial \overline{M}_z}{\partial B_0}\right)_T = N(\tfrac{1}{2}g_I\mu_N)^2\beta\,\mathrm{sech}^2(\tfrac{1}{2}\beta g_I\mu_N B_0),$$

while expression (8.5.20) gives

$$\chi_T^{(m)} = \frac{\beta}{N}(\tfrac{1}{2}g_I\mu_N)^2[1 - \tanh^2(\tfrac{1}{2}\beta g_I\mu_N B_0)]$$

which, upon application of the identity $1 - \tanh^2 x = \mathrm{sech}^2 x$, gives the same final result.

Example 8.4 Thermodynamic properties of a spin-1 species.

The thermodynamic properties of a spin-1 magnetic system in the solid state are exemplified by pD_2 (similarly, oH_2). We know that the nuclear spin angular momentum of a pD_2 molecule interacts with a constant (homogeneous) magnetic field $\mathbf{B}_0$ lying along the z-axis through the Zeeman Hamiltonian (8.5.2b), so that the relevant spin operator appearing there is the z-component $-g_I\mu_N B_0\widehat{I}_z$ of $\widehat{\boldsymbol{\mu}}^{(m)}$ $= -g_I\mu_N\widehat{\mathbf{I}}$. The corresponding energy eigenvalue equation is

$$\begin{aligned}\mathcal{H}|IM_I\rangle &= -g_I\mu_N B_0\widehat{I}_z|IM_I\rangle \\ &= -g_I\mu_N B_0 M_I|IM_I\rangle \\ &\equiv E_{M_I}|IM_I\rangle.\end{aligned}$$

For a spin-one ($I = 1$) species, such as para-deuterium molecules, there will thus be three energy states, one each corresponding to $M_I = -1, 0, 1$: we may therefore write $E_{M_I} = -\mu B_0 M_I$, with $\mu = g_I\mu_N$. Associated with the values $M_I = -1, 0, 1$, we have three corresponding magnetic moments $\mu_- = \mu$, $\mu_0 = 0$, and $\mu_+ = -\mu$, together with their three magnetic energies, $E_- = \mu B_0$, $E_0 = 0$, and $E_+ = -\mu B_0$. These values may be utilized to construct the canonical partition function for a single pD_2 molecule as

$$z_{\mathrm{spin}}(\beta) \equiv \sum_{\mathrm{states}} e^{-\beta\epsilon_{\mathrm{states}}} = \sum_{M_I=-1}^{+1} e^{-\beta E_{M_I}} = e^{-\beta E_+} + 1 + e^{-\beta E_-}$$

which, for convenience, we may rewrite in terms of the hyperbolic cosine function as

$$z_{\mathrm{spin}}(\beta) = 2\cosh(\beta\mu B_0) + 1.$$

The relevant probability factors required for the calculation of canonical ensemble averages for an individual pD_2 molecule are thus given by

$$p_{+1} = \frac{e^{\beta\mu B_0}}{z_{\text{spin}}(\beta)}, \qquad p_0 = \frac{1}{z_{\text{spin}}(\beta)}, \qquad p_{-1} = \frac{e^{-\beta\mu B_0}}{z_{\text{spin}}(\beta)}.$$

As the magnetization of a bulk sample of N spin-one species is given by N times the ensemble average of the magnetic moment of a single spin-1 species, we shall start with this calculation: the definition of the ensemble average magnetic moment for an individual spin-1 species is

$$\langle\mu\rangle \equiv p_+\mu_+ + p_0\mu_0 + p_-\mu_- = p_+\mu_+ + p_-\mu_-,$$

as $\mu_0 \equiv 0$. Upon employing the forms that we have just obtained for the probabilities, we obtain

$$\langle\mu\rangle = \frac{\mu e^{\beta\mu B_0}}{2\cosh(\beta\mu B_0) + 1} + \frac{-\mu e^{-\beta\mu B_0}}{2\cosh(\beta\mu B_0) + 1},$$

which we may simplify to

$$\langle\mu\rangle = \mu\,\frac{2\sinh(\beta\mu B_0)}{2\cosh(\beta\mu B_0) + 1}.$$

The bulk magnetization $\overline{M}$ for a collection of N pD_2 molecules is thus given as

$$\overline{M} = N\mu\,\frac{2\sinh(\beta\mu B_0)}{2\cosh(\beta\mu B_0) + 1}.$$

For high temperatures, i.e., for $\beta\mu B_0 \ll 1$, we see that $\cosh(\beta\mu B_0) \simeq 1$, $\sinh(\beta\mu B_0) \simeq \beta\mu B_0$, so that

$$\overline{M} = N\langle\mu\rangle \simeq \frac{2}{3}\beta\mu^2 B_0 = \frac{2\mu^2 B_0}{3k_B T},$$

which tells us that the magnetization dies off inversely with temperature. For low temperatures, for which $\beta\mu B_0 \gg 1$, we see that $\cosh(\beta\mu B) \simeq \sinh(\beta\mu B_0) \simeq \frac{1}{2}e^{\beta\mu B_0} \gg 1$, so that the magnetization behaves like

$$\overline{M} = N\langle\mu\rangle \simeq N\mu\,\frac{2\left(\frac{1}{2}e^{\beta\mu B_0}\right)}{2\left(\frac{1}{2}e^{\beta\mu B_0}\right) + 1} \simeq N\mu,$$

a constant.

The canonical partition function for the N pD_2 molecules in our solid sample is given in terms of the partition function z_{spin} that we obtained above as

$$Z_N^{\text{spin}}(T) = [z_{\text{spin}}(T)]^N,$$

so that the internal energy U_{mag} associated with our spin system is given by

$$
\begin{aligned}
U_{\mathrm{spin}}(T) &= Nk_{\mathrm{B}}T^2 \left(\frac{\partial \ln z_{\mathrm{spin}}}{\partial T} \right)_V \\
&= -Nk_{\mathrm{B}}T^2 \frac{2\mu B_0 \sinh(\beta\mu B_0)}{k_{\mathrm{B}}T^2[2\cosh(\beta\mu B_0)+1]} \\
&= -N\mu B_0 \frac{\tanh(\beta\mu B_0)}{1+\frac{1}{2}\mathrm{sech}(\beta\mu B_0)}.
\end{aligned}
$$

Notice that, if we compare this result for $U_{\mathrm{spin}}(T)$ with the expression for the magnetization $\overline{M}$, we find that $U_{\mathrm{spin}}(T) = -\overline{M}B_0$, which is as we should expect. We may directly write down an expression for the entropy of our spin system using our statistical thermodynamic expression

$$
S_{\mathrm{spin}} = \frac{U_{\mathrm{spin}}}{T} + k_{\mathrm{B}} \ln z_{\mathrm{spin}}^N,
$$

to obtain

$$
S_{\mathrm{spin}} = Nk_{\mathrm{B}} \left\{ \ln[2\cosh(\beta\mu B_0)+1] - \frac{\mu B_0}{k_{\mathrm{B}}T} \frac{\tanh(\beta\mu B_0)}{1+\frac{1}{2}\mathrm{sech}(\beta\mu B_0)} \right\}.
$$

Although the expression for the heat capacity for a spin-1 system is rather complicated, it is instructive to see that it does not behave as a function of temperature any differently than does the simpler case of a spin-$\frac{1}{2}$ system. The heat capacity at constant volume can be obtained as the partial derivative with respect to temperature of $U_{\mathrm{spin}}(T)$, viz.,

$$
C_B = Nk_{\mathrm{B}} \left(\frac{\mu B_0}{k_{\mathrm{B}}T} \right)^2 \frac{\mathrm{sech}(\beta\mu B_0)\left[\mathrm{sech}(\beta\mu B_0)+\frac{1}{2}\right]}{\left[1+\frac{1}{2}\mathrm{sech}(\beta\mu B_0)\right]^2}.
$$

Plots of the spin entropy S_{spin} and C_B are shown in Fig. 8.3.

We see from a comparison between Figs. 8.2 and 8.3 that the behaviours of the spin entropy, $S_{\mathrm{spin}}(T)$, and heat capacity, $C_{\mathrm{spin}}(T)$, are qualitatively the same for the spin-$\frac{1}{2}$ and spin-1 systems, differing only in the magnitude of the high-temperature entropy and the heat capacity maximum. □

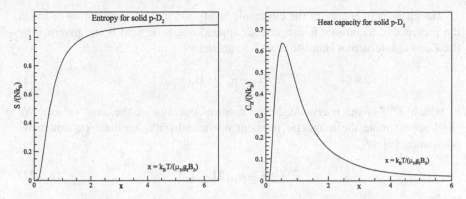

Fig. 8.3 The temperature per-particle dependence of the spin magnetization contribution to the entropy and heat capacity associated with a molecular crystal formed from the spin-1 species pD_2. Precisely the same result would be obtained for an oH_2 crystal

8.5.3 Multi-Electron Atoms/Ions: Setting the Stage

Compounds of the transition metals involve various ions with open-shell configurations. As is discussed in greater detail in Appendix D, the inclusion of inter-electron repulsions in the electronic Hamiltonian leads to the requirement that the electronic orbital angular momenta of the valence electrons (i.e., those electrons external to filled electronic shells) be coupled vectorially to give well-defined total electronic angular momenta, $\mathbf{L}$, for the multi-electron atom or ion. This requirement, when coupled with the Pauli Symmetry Principle for indistinguishable electrons, then requires that the individual electron spins be coupled vectorially to give well-defined total spin angular momenta, $\mathbf{S}$, for the multi-electron atom or ion. The total electronic spin and orbital angular momenta then interact via what is called the spin–orbit interaction term, $\hbar^2 A_{so}\mathbf{S} \cdot \mathbf{L}$, in the electronic Hamiltonian for the multi-electron atom. The presence of the spin–orbit coupling term in the electronic Hamiltonian requires that $\mathbf{S}$ and $\mathbf{L}$ be coupled vectorially to give well-defined total electronic angular momenta, $\mathbf{J}$, for the multi-electron atom or ion. As is also discussed in Appendix D, a pair of values (S, L) corresponds to an electronic term, and each set of values (S, L, J) corresponds to an electronic level within an electronic term: these levels are designated by a term symbol consisting of an upper case letter indicating the value of L, a pre-superscript giving the value of $2S+1$, and called the multiplicity, and a post-subscript giving the value of J for the level. Each resultant total electronic angular momentum thus corresponds to a distinct $(2J+1)$-fold degenerate (energy) level within an electronic term. We may characterize the energy eigenstates of the electronic Hamiltonian $\mathcal{H}_0$ at this level by $|\alpha, L, S, J, M_J\rangle$ in Dirac notation, with (α, S, L) representing quantum numbers characterizing the term energy, and the allowed values of J characterizing the energy levels within a given term.

The changes in energy of the electronic levels of a multi-electron atom or ion in the presence of a constant homogeneous applied magnetic field **B** are governed by the Zeeman interaction Hamiltonian, $\mathcal{H}_Z$, given by

$$\mathcal{H}_Z = -\widehat{\boldsymbol{\mu}}^{(m)} \cdot \mathbf{B}, \tag{8.5.21}$$

in which $\widehat{\boldsymbol{\mu}}^{(m)}$ is the magnetic dipole moment operator of the atom or ion. We shall approximate the magnetic moment operator by its dominant paramagnetic component [9] as[5]

$$\widehat{\boldsymbol{\mu}}^{(m)} = -\mu_B(2\widehat{\mathbf{J}} - \widehat{\mathbf{L}}), \tag{8.5.22}$$

in which **J** is the total angular momentum, **L** is the total orbital angular momentum for the multi-electron atom or ion, while μ_B, defined by

$$\mu_B \equiv \frac{\hbar|e|}{2m_e}, \tag{8.5.23}$$

with $\hbar$ the Planck constant divided by 2π, $|e|$ the magnitude of the electronic charge, and m_e the mass of the electron, is the Bohr magneton. The Bohr magneton is the fundamental unit of electronic magnetic moments, and has the approximate value[6] $9.27401 \times 10^{-24}\,\mathrm{J\,T^{-1}}$. A much smaller diamagnetic component of the electronic magnetic moment is too small to play a role in our present considerations.

As the Zeeman interaction energy is small in comparison with the field-free electronic energy, we may treat the Zeeman Hamiltonian as a perturbation to the field-free Hamiltonian, $\mathcal{H}_0$, with eigenenergies $E_0(\alpha, L, S, J)$. According to perturbation theory, the energy due to the Zeeman interaction can therefore be expressed in a power series in the magnetic field strength B as

$$E(\alpha, L, S, J, M_J) = E_0(\alpha, L, S, J) + W_1(L, S, J, M_J)B$$

$$+ W_2(L, S, J, M_J)B^2 + \cdots, \tag{8.5.24}$$

in which $W_1(L, S, J, M_J)$ and $W_2(L, S, J, M_J)$ are known as the first- and second-order Zeeman energies of the atom or ion in an energy level characterized by α, L, S, J. The field-free energy $E_0(\alpha, L, S, J)$ can be split into two parts as

$$E_0(\alpha, L, S, J) = E_0(\alpha, L, S) + E_{so}(L, S, J), \tag{8.5.25}$$

---

[5]For a derivation of this form for the magnetic dipole moment of a multi-electron atom see, for example, pp. 128–129 of [9].

[6]That the Bohr magneton is much larger than the nuclear magneton that we encountered earlier is simply due to the much smaller mass of the electron. For a more precise value of the Bohr magneton, see the NIST website https://physics.nist.gov/PhysRefData.

with $E_0(\alpha, L, S)$ representing the term energy, characterized by specific values of L, S, and with $E_{so}(L, S, J)$ giving the spin–orbit coupling energy

$$E_{so}(S, L, J) = \tfrac{1}{2}\hbar^2 A_{so}[J(J+1) - S(S+1) - L(L+1)], \qquad (8.5.26)$$

for the energy of an individual level, characterized by J, within the term.

We shall work with the standard time-independent perturbation theory expression given, for example, in [10, 11] for the energy associated with a perturbed Hamiltonian $\mathcal{H} = \mathcal{H}^{(0)} + \mathcal{H}^{(1)}$. If we work in terms of the set of eigenstates $\{|n\rangle\}$ of the unperturbed Hamiltonian $\mathcal{H}^{(0)}$, then the perturbed energy E_n is given by

$$E_n \simeq E_n^{(0)} + H_{nn}^{(1)} + \sum_{n' \neq n} \frac{|H_{nn'}^{(1)}|^2}{E_n^{(0)} - E_{n'}^{(0)}},$$

to second order in the perturbation. The energies $E_n^{(0)}$ are the eigenvalues of the unperturbed Hamiltonian (assumed to be known exactly), and the matrix elements $H_{nn'}^{(1)}$ of $\mathcal{H}^{(1)}$ are between eigenstates of $\mathcal{H}^{(0)}$.

For the case of a multi-electron atom or ion, the unperturbed eigenstates of $\mathcal{H}^{(0)}$ are the $|\alpha, L, S, J, M_J\rangle$ states, so that the first-order perturbation energy $W_1 B$ arising from the Zeeman interaction is given by

$$W_1 B = \mu_B B \langle \alpha, L, S, J, M_J | 2\widehat{J}_z - \widehat{L}_z | \alpha, L, S, J, M_J \rangle$$

$$= \mu_B B (2M_J - \langle \alpha, L, S, J, M_J | \widehat{L}_z | \alpha, L, S, J, M_J \rangle). \qquad (8.5.27)$$

The diagonal elements of the $\widehat{L}_z$ operator are given via the Wigner–Eckart theorem [12] as

$$\langle \alpha, L, S, J, M_J | \widehat{L}_z | \alpha, L, S, J, M_J \rangle = \frac{J(J+1) + L(L+1) - S(S+1)}{2J(J+1)} M_J, \qquad (8.5.28)$$

so that the first-order perturbation energy $W_1 B$ becomes

$$W_1 B = \mu_B B M_J \left[2 - \frac{J(J+1) + L(L+1) - S(S+1)}{2J(J+1)} \right],$$

more commonly written in the form

$$W_1(L, S, J, M_J)B = \mu_B B M_J g_J(L, S) M_J, \qquad (8.5.29)$$

with $g_J(L, S)$ given by

$$g_J(L, S) = 1 + \frac{J(J+1) + S(S+1) - L(L+1)}{2J(J+1)}, \qquad (8.5.30)$$

and known as the Landé g-factor for a multi-electron atom or ion in a specific (L,S,J) energy level. Note that we could also write the first-order Zeeman perturbation energy as $W_1(L, S, J, M_J)B = \mu_{JM_J}(L, S)B$, with $\mu_{JM_J}(L, S) \equiv \mu_B g_J(L, S)M_J$ representing the magnetic moment of the multi-electron atom or ion in the energy state $E(\alpha, L, S, J, M_J)$.

To determine the second-order Zeeman contribution to the energy, we note that the component of $\widehat{\mu}_z^{(m)}$ that is proportional to $\widehat{J}_z$ is diagonal in both the J and M_J quantum numbers (because it commutes with both $\widehat{J}_z$ and $\widehat{J^2}$), while the component proportional to $\widehat{L}_z$, which commutes with $\widehat{J}_z$ but not with $\widehat{J^2}$, although diagonal in M_J, is not diagonal in J. The total orbital angular momentum operator $\widehat{\mathbf{L}}$ happens, however, to belong to a class of vector operators $\widehat{\mathbf{V}}$ whose Cartesian components satisfy the nine commutation relations represented collectively by the equation

$$[\widehat{J}_\ell, \widehat{V}_m]_- = i\varepsilon_{\ell mn}\widehat{V}_n,$$

in which the subscripts ℓ, m, n take on values x, y, z, and the symbol $\varepsilon_{\ell mn}$ is +1 for ℓmn an even permutation of xyz, −1 for ℓmn an odd permutation of xyz, and 0 otherwise. The Wigner–Eckart theorem applies precisely to this class of operators, so that it may be utilized to compute the non-vanishing off-diagonal elements of $\widehat{L}_z$. According to the Wigner–Eckart theorem, such a vector operator has off-diagonal elements between the J and $J' = J \pm 1$ eigenstates of $\widehat{J^2}$. For present purposes, the relevant eigenstates are the coupled eigenstates $|\alpha, L, S, J, M_J\rangle$, and the appropriate matrix elements for the operator $\widehat{L}_z$ are

$$\langle\alpha, L, S, J - 1, M_J|\widehat{L}_z|\alpha, L, S, J, M_J\rangle = \langle\alpha, L, S, J, M_J|\widehat{L}_z|\alpha, L, S, J - 1, M_J\rangle$$

$$= -\sqrt{f(L, S, J)(J^2 - M_J^2)},$$

$$(8.5.31)$$

and

$$\langle\alpha, L, S, J + 1, M_J|\widehat{L}_z|\alpha, L, S, J, M_J\rangle = \langle\alpha, L, S, J, M_J|\widehat{L}_z|\alpha, L, S, J + 1, M_J\rangle$$

$$= -\sqrt{f(L, S, J + 1)[(J + 1)^2 - M_J^2]},$$

$$(8.5.32)$$

with $f(L, S, J)$ defined as

$$f(L, S, J) \equiv \frac{[J^2 - (L - S)^2][(L + S + 1)^2 - J^2]}{4J^2(4J^2 - 1)}.$$

$$(8.5.33)$$

We may obtain the second-order perturbation energy $W_2 B^2$ from these results, with W_2 given as

$$W_2(L, S, J, M_J)$$

$$= \frac{\mu_B^2}{\hbar^2 A_{so}} \left[\frac{f(L, S, J)(J^2 - M_J^2)}{2J} - \frac{f(L, S, J+1)[(J+1)^2 - M_J^2]}{2(J+1)} \right].$$

$$(8.5.34)$$

8.5.4 Magnetic Susceptibility for Ground Level Atoms or Ions

An atom or ion found in a specific energy level characterized by total electronic angular momentum J will also possess a corresponding magnetic moment $\mu_J^{(m)}$. In an applied magnetic field $\mathbf{B}$, the magnetic moment $\mu_J^{(m)}$ can assume any of the values $g_J(L, S)\mu_B M_J$ $(-J \leq M_J \leq +J)$ as its component along the field direction. As we have seen, the energy of an atom or ion in a $\mathbf{B}$-field is changed by the Zeeman interaction $\mathcal{H}_Z = -\mu^{(m)} \cdot \mathbf{B}$: in particular, it lifts the $(2J + 1)$-fold degeneracy of the M_J-states of the J-level. If the Zeeman interaction Hamiltonian may be treated as a perturbation on the field-independent electronic Hamiltonian (often a good approximation), then the term energy $E_J(S, L)$ becomes

$$E(L, S, J, M_J) = E_J(L, S) + g_J(L, S)\mu_B B M_J, \qquad (8.5.35)$$

with the second term representing the first-order Zeeman energy.

In the presence of an applied magnetic field of strength B, the canonical partition function for a multi-electron atom or ion that has no thermally accessible excited electronic levels at the temperature T of the sample containing it will be given by

$$z(T, B) = \sum_{M_J=-J}^{+J} e^{-\beta[E_J(L,S) + g_J(L,S)\mu_B M_J B]}$$

$$= e^{-\beta E_J(L,S)} \sum_{M_J=-J}^{+J} e^{-\beta g_J(L,S)\mu_B M_J B}$$

$$= e^{-\beta E_J(L,S)} z_{mag}(T, B). \qquad (8.5.36)$$

The magnetic partition function, $z_{mag}(T, B)$, appearing in the expression for $z(T, V)$ is defined as

$$z_{mag}(T, B) \equiv \sum_{M_J=-J}^{+J} e^{-\beta g_J(L,S)\mu_B M_J B}, \qquad (8.5.37)$$

and β is, as usual, $(k_B T)^{-1}$. Corresponding fractional populations $p_{JM_J}(T, B)$ for the (no longer degenerate) magnetic states are then given accordingly by

$$p_{JM_J}(T, B) = \frac{e^{-\beta g_J(L,S)\mu_B BM_J}}{z_{\text{mag}}(T, B)}. \tag{8.5.38}$$

We note, in passing, that the fractional populations of Eq. (8.5.38) do not involve the original term energy $E_J(L, S)$: this is another manifestation of the arbitrariness of the zero of energy in the determination of fractional populations. Note also that much of the following derivation does not require the Zeeman energy to be small relative to $k_B T$.

To obtain an expression for the magnetization for an individual atom or ion, let us begin by calculating $\langle \mu_J \rangle(T, B)$ as

$$\langle \mu_J \rangle(T, B) \equiv \sum_{M_J=-J}^{+J} \mu_{JM_J}(L, S) \, p_{JM_J}(T, B). \tag{8.5.39}$$

Upon employing Eq. (8.5.38) for the fractional populations for the Zeeman states, we obtain

$$\langle \mu_J \rangle(T, B) = \frac{1}{z_{\text{mag}}(T, B)} \sum_{m_J=-J}^{+J} g_J(L, S) M_J \mu_B e^{-\beta g_J(L,S)\mu_B M_J B}. \tag{8.5.40}$$

By noting that the prefactor in the summation can be generated by taking the partial derivative with respect to the field strength B of the exponential, we obtain $\langle \mu_J \rangle(T, B)$ as

$$\langle \mu_J \rangle(T, B) = \frac{1}{z_{\text{mag}}(T, B)} \sum_{M_J=-J}^{+J} \left[-k_B T \frac{\partial}{\partial B} \left(e^{-\beta g_J \mu_B M_J B} \right) \right]$$

$$= -\frac{k_B T}{z_{\text{mag}}(T, B)} \frac{\partial}{\partial B} \left(\sum_{M_J=-J}^{+J} e^{-\beta g_J \mu_B M_J B} \right). \tag{8.5.41}$$

Once we recognize that the quantity within parentheses is simply $z_{\text{mag}}(B, T)$, we see that we have the rather pleasing result

$$\langle \mu_J \rangle(T, B) = -k_B T \left(\frac{\partial \ln z_{\text{mag}}(T, B)}{\partial B} \right)_T \tag{8.5.42}$$

for the average magnetic moment for an isolated multi-electron atom or ion.

We shall find it convenient to rewrite expression (8.5.37) as

$$z_{\text{mag}}(T, B) = e^{\beta g_J(L,S)\mu_B BJ} \sum_{\ell=0}^{2J} \left(e^{-\beta g_J(L,S)\mu_B B} \right)^{\ell}, \tag{8.5.43}$$

which better illustrates that the summation in Eq. (8.5.37) is a finite geometric progression. By starting from this expression for $z_{mag}(T, B)$, we may, with a little manipulation, obtain a closed-form expression for it, namely

$$z_{mag}(T, B) = e^{\beta g_J \mu_B BJ} \frac{(e^{-\beta g_J \mu_B B})^{2J+1} - 1}{e^{-\beta g_J \mu_B B} - 1}. \tag{8.5.44}$$

We may further rearrange this expression into the more symmetric form

$$z_{mag}(T, B) = \frac{(e^{\beta g_J \mu_B B})^{J+\frac{1}{2}} - (e^{\beta g_J \mu_B B})^{-J-\frac{1}{2}}}{(e^{\beta g_J \mu_B B})^{\frac{1}{2}} - (e^{\beta g_J \mu_B B})^{-\frac{1}{2}}}, \tag{8.5.45}$$

which specifically suggests that we may employ the definition of the hyperbolic sine function to rewrite it as

$$z_{mag}(T, B) = \frac{\sinh[\frac{1}{2}\beta(2J + 1)g_J \mu_B B]}{\sinh(\frac{1}{2}\beta g_J \mu_B B)}.$$

Magnetization in a bulk sample is simply the net magnetic moment associated with the sample, i.e., it is given by the average magnetic moment of a single paramagnetic atom or ion multiplied by the number of such ions in the bulk sample. Thus, we may write the electronic magnetization M for a bulk sample as

$$M(T, B) = N\langle \mu_J \rangle(T, B) = -Nk_B T \left(\frac{\partial \ln z_{mag}(T, B)}{\partial B} \right)_T. \tag{8.5.46}$$

If we are working with a solid sample, the atoms or ions will be localized (at lattice sites in the solid), and hence distinguishable, so that the appropriate form for the canonical partition function for a solid sample containing N atoms or ions is

$$Z_N(T, B) = [z_{mag}(T, B)]^N = \frac{\sinh^N[\frac{1}{2}\beta(2J + 1)g_J \mu_B B]}{\sinh^N(\frac{1}{2}\beta g_J \mu_B B)}. \tag{8.5.47}$$

We can now make use of this result to rewrite our expression for the electronic bulk magnetization $M(N, T, B)$ as[7]

$$M(N, T, B) = -k_B T \left(\frac{\partial \ln Z_N(T, B)}{\partial B} \right)_{T,N}. \tag{8.5.48}$$

---

[7]Note that the negative sign here arises specifically because we have explicitly taken the sign of the electronic magnetic dipole moment into account in Eqs. (8.5.36) and (8.5.43). For nuclear spin magnetization the overall sign in Eq. (8.5.48) would be positive.

This is a fundamental relation for magnetic systems. Essentially the entire thermodynamics of magnetic systems can be derived from this rather important result, taken together with Eq. (8.5.37) for the canonical partition function.

Let us now employ the specific form of Eq. (8.5.47) for $Z_N(T, B)$ to obtain a working expression for the magnetization. To accomplish this task, we begin with the expression for the magnetization of our bulk sample, written in the form

$$M = k_B T \frac{\partial}{\partial B} \ln \left\{ \frac{\sinh^N \left[\dfrac{(2J + 1)g_J \mu_B B}{2k_B T} \right]}{\sinh^N \left[\dfrac{g_J \mu_B B}{2k_B T} \right]} \right\}$$

$$= N k_B T \frac{\partial}{\partial B} \left\{ \ln \sinh \left[\frac{(2J + 1)g_J \mu_B B}{2k_B T} \right] - \ln \sinh \left[\frac{g_J \mu_B B}{2k_B T} \right] \right\}, \quad (8.5.49)$$

or

$$M = N k_B T \left\{ \frac{(2J + 1)g_J \mu_B}{2k_B T} \coth \left[\frac{(2J + 1)g_J \mu_B B}{2k_B T} \right] - \frac{g_J \mu_B}{2k_B T} \coth \left[\frac{g_J \mu_B B}{2k_B T} \right] \right\}.$$
$$(8.5.50)$$

It is traditional to write this rather complicated-looking result as

$$M(T, B) = N g_J \mu_B J \mathcal{B}_J \left(\frac{g_J \mu_B J B}{k_B T} \right), \quad (8.5.51)$$

in which $\mathcal{B}_J(x)$, known as a Brillouin function, is defined as

$$\mathcal{B}_J(x) \equiv \frac{2J + 1}{2J} \coth \left(\frac{2J + 1}{2J} x \right) - \frac{1}{2J} \coth \left(\frac{x}{2J} \right). \quad (8.5.52)$$

For our application, x is given by $x = g_J(L, S)\mu_B J B/(k_B T)$. The behaviours of the Brillouin functions are illustrated in Fig. 8.4.

As in the dielectric example of the previous section of this chapter, it is of interest to examine what form our expression takes when x is small, i.e., when $g_J(L, S)\mu_B J B \ll k_B T$. In this case, the argument of the Brillouin function is small, and we can use a Maclaurin-like expansion for $\coth y$, namely

$$\coth y \simeq \frac{1}{y} \left(1 + \frac{1}{3} y^2 + \cdots \right),$$

so that $\mathcal{B}_J(x)$ can be approximated as

$$\mathcal{B}_J(x) \simeq \frac{2J + 1}{2J} \left(\frac{2J}{(2J + 1)x} + \frac{1}{3} \frac{(2J + 1)x}{2J} \right) - \frac{1}{2J} \left(\frac{2J}{x} + \frac{1}{3} \frac{x}{2J} \right)$$

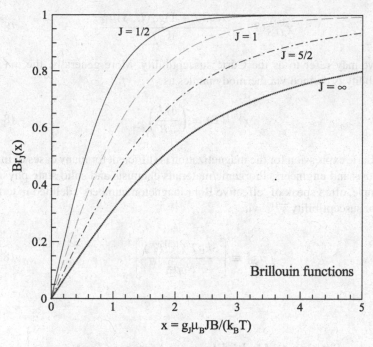

Fig. 8.4 The temperature dependence of the Brillouin functions

or as

$$\mathcal{B}_J(x) \simeq \frac{J+1}{3J}\, x = \frac{J+1}{3}\,\frac{g_J(L,S)\mu_B B}{k_B T}. \tag{8.5.53}$$

Thus, for temperatures T such that $k_B T \gg g_J(L,S)\mu_B J B$, we can obtain an expression for the magnetization M from Eqs. (8.5.51) and (8.5.53), namely

$$M(T,B) \simeq N\,\frac{J(J+1)g_J^2(L,S)\mu_B^2 B}{3k_B T}. \tag{8.5.54}$$

This expression is commonly found in the magnetism of paramagnetic materials, where it is known as the Curie magnetization. It also appears in the nuclear magnetic resonance literature.

The magnetic susceptibility $\chi^{(m)}(T)$ is defined via

$$M(T,B) = \chi^{(m)}(T)B \tag{8.5.55}$$

in much the same way as the electric susceptibility was defined. A comparison between Eqs. (8.5.54) and (8.5.55) gives the magnetic susceptibility for a collection of N multi-electron atoms or ions as

$$\chi_{JLS}^{(m)}(T) = \frac{NJ(J+1)g_J^2(L,S)\mu_B^2}{3k_B T},$$ (8.5.56)

which we may refer to as the Curie susceptibility. More generally, the magnetic susceptibility is defined via thermodynamics as

$$\chi^{(m)}(T) \equiv \left(\frac{\partial M}{\partial B}\right)_T.$$ (8.5.57)

The Curie expression for the magnetization is sufficient for many cases of interest to scientists and engineers. Inorganic/materials chemists and solid state physicists, for example, often speak of 'effective Bohr magneton numbers', defined in terms of the molar susceptibility $\overline{\chi}^{(m)}$ via

$$\mu_{\text{eff}} \equiv \left[\frac{3k_B T \overline{\chi}^{(m)}(T)}{N_0 \mu_B^2}\right]^{\frac{1}{2}}.$$ (8.5.58)

8.5.5 Application to Multi-Electron Atoms or Ions

Expression (8.5.56) for the magnetic susceptibility of a multi-electron atom or ion applies when the environment of the atom or ion is spherically symmetric and only a single level of its ground term is thermally accessible. Strictly speaking, a fully spherically symmetric environment occurs for these atoms or ions only in the gas phase, corresponding to temperatures for which it is likely that all levels of the ground term plus even some of the excited terms would be thermally populated. Transition metal ions in solids are almost always strongly affected by their surroundings, which always have symmetries lower than spherical. The nearest-neighbours at the molecular level are often chemical moieties referred to as ligands, and the resultant electronic environment is correspondingly referred to as a ligand field. A ligand field shifts and, more importantly, splits the term and level energies of the multi-electron atom or ion, thereby making a more detailed examination necessary before meaningful comparisons can be made with experiment. To consider the effects of ligand fields on the transition metal ions would divert us from the theme of statistical mechanics. A number of excellent yet readable treatments of this subject, such as those of Schläfer and Gliemann [13] or Figgis and Hitchman [14], are available.

There is a class of compounds for which the details of the environment are much less important because the electronic shell being filled as one moves across the periodic table is the $4f$-shell, whose electrons penetrate deep into the interior of the atom or ion (because of the nature of the $4f$ orbitals). This deep penetration

Table 8.1 Properties of the free lanthanide tripositive ions

Ln^{3+} ion	La^{3+}	Ce^{3+}	Pr^{3+}	Nd^{3+}	Pm^{3+}	Sm^{3+}	Eu^{3+}	Gd^{3+}
Ground level	$^1S_0$	$^2F_{\frac{5}{2}}$	$^3H_4$	$^4I_{\frac{9}{2}}$	$^5I_4$	$^6H_{\frac{5}{2}}$	$^7F_0$	$^8S_{\frac{7}{2}}$
$\hbar^2 A_{so}/cm^{-1}$	0	640	375	300	–	236	227	0
μ_{eff}(expt)	0	2.46	3.61	3.63	–	1.66	3.61	7.97
g_J	1	$\frac{6}{7}$	$\frac{4}{5}$	$\frac{8}{11}$	$\frac{3}{5}$	$\frac{2}{7}$	$\frac{3}{2}$	2
μ_{eff}^J(Curie)	0	2.535	3.578	3.618	2.683	0.845	0	7.937
Ln^{3+} ion	Tb^{3+}	Dy^{3+}	Ho^{3+}	Er^{3+}	Tm^{3+}	Yb^{3+}	Lu^{3+}	
Ground level	$^7F_6$	$^6H_{\frac{15}{2}}$	$^5I_6$	$^4I_{\frac{15}{2}}$	$^3H_6$	$^2F_{\frac{7}{2}}$	$^1S_0$	
$\hbar^2 A_{so}/cm^{-1}$	−270	−364	−520	−823	−1375	−2950	0	
μ_{eff}(expt)	9.81	10.6	10.5	9.46	7.51	4.47	0	
g_J	$\frac{3}{2}$	$\frac{4}{3}$	$\frac{5}{4}$	$\frac{6}{5}$	$\frac{7}{6}$	$\frac{8}{7}$	1	
μ_{eff}^J(Curie)	9.721	10.646	10.61	9.581	7.561	4.436	0	

into the atomic interior effectively shields the $4f$ electrons from the environment created by surrounding ligands in the solid state. This class of compounds is referred to as the rare earth compounds.[8] Because the $4f$ electrons are well shielded from surrounding ligands in solid lanthanide salts, the ligand fields in these solids can be treated as only minor perturbations to the electronic Hamiltonians for electrons in the valence shells of lanthanide ions. This shielding of the $4f$ electrons allows a meaningful comparison to be made between results obtained for free (i.e., spherically symmetric) lanthanide atoms or ions and experiment.

Let us now consider specifically the Lanthanide series of tripositive ions. The Curie magnetization for this series of ions can be calculated directly once the atomic term symbols are known. The relevant information has been entered into Table 8.1, together with the experimental values for this series of ions.

The Curie magnetization for a collection of paramagnetic ions (such as Gd^{3+}), all found in unique levels (for Gd^{3+}, the $^8S_{\frac{7}{2}}$ level), is given by

$$M_J(T, B) \simeq N \frac{J(J+1)g_J^2\mu_B^2 B}{3k_B T} = \chi_J^{(m)}(T)B, \tag{8.5.59}$$

in which case μ_{eff}^J is obtained from Eq. (8.5.58) as

$$\mu_{eff}(L, S, J) = g_J(L, S)\sqrt{J(J+1)}. \tag{8.5.60}$$

---

[8] See, for example, the recent article 'Rare earths in a nutshell' [J. Zhou and G. A. Fiete, *Physics Today* **73**, 66 (2020)] on the importance of the rare earths for many technological products.

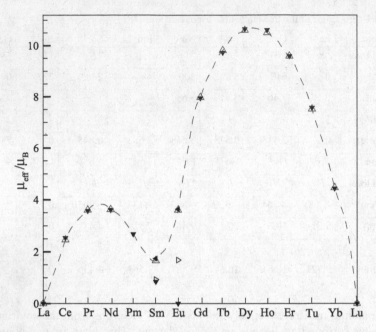

Fig. 8.5 Effective magnetic moments of the lanthanide tripositive ions: ▼, Curie values; Δ, experiment; ▷, including excited levels; ◄, including Zeeman second approximation

Values obtained with this expression are referred to as the 'Curie values' for the magnetization. This expression provides a good representation for most atomic ions, and is commonly employed by inorganic/materials chemists.

Experimental values for the effective magnetic moments of the tripositive lanthanide (Ln) ions are compared with the corresponding Curie values obtained from Eq. (8.5.58) in Fig. 8.5. Apart from the Sm^{3+} and Eu^{3+} ions, the level of agreement between the calculated Curie values and experiment is generally quite acceptable. To see why Sm^{3+}, and especially Eu^{3+}, give experimental magnetic moments that are so dramatically at odds with the Curie values, we need to examine the structures of the atomic terms for the lanthanide series of ions more closely, as the Curie magnetization formula of Eq. (8.5.58) implicitly assumes that all atoms or ions are confined to their ground electronic energy levels.

Whether or not a multi-electron atom or ion will be found in one or more of its excited electronic energy levels at a given temperature depends upon the magnitude of the 'spin–orbit splitting' of the different energy levels belonging to that particular term relative to the thermal energy $k_B T$. The spin–orbit coupling energy for a level specified by J within a given term is given by

$$E_{so}(S, L, J) = \tfrac{1}{2}\hbar^2 A_{so}[J(J + 1) - S(S + 1) - L(L + 1)]. \tag{8.5.61}$$

How far above the ground level energy an excited level lies is determined by the difference between the spin–orbit energies $E_{so}(S, L, J)$ of the excited and ground levels of the ion. Because the values of L and S for a given term do not change, this difference may be simplified to

$$\Delta E_{so}(J, J_{min}) = \tfrac{1}{2}\hbar^2 A_{so}[J(J+1) - J_{min}(J_{min}+1)]. \quad (8.5.62)$$

If we employ Eq. (8.5.62) and the data in Table 8.2 for the Ln^{3+} ions Ce^{3+}, Pr^{3+}, Nd^{3+}, Sm^{3+}, and Eu^{3+} we find, for example, that the single excited $^2F_{\frac{7}{2}}$ level of the ^2F term for Ce^{3+} lies 2240 cm^{-1} above the ground $^2F_{\frac{5}{2}}$ level, and the two excited levels, $^3H_5$ and $^3H_6$, of Pr^{3+} lie 1875 cm^{-1} and 4125 cm^{-1}, respectively, above the ground $^3H_4$ level. For convenience, the excitation energies for this set of five Ln^{3+} ions have been collected into Table 8.2 and illustrated in Fig. 8.6.

Figure 8.6 illustrates the spin–orbit splitting of the energy levels in the ground terms for five tripositive lanthanide ions from the first half of the sequence of chemically very similar rare earth elements. Gd^{3+}, with a half-filled $4f$ shell and an electron configuration $[Xe](4f)^7$, has a total electronic orbital angular momentum L of zero, so that its resultant ^8S term has only a single level with $J = \frac{7}{2}$. Its nearest excited energy level belongs to a ^6P term, and lies approximately 32,000 cm^{-1} above the ground ^8S term. As can be seen from Table 8.1, the magnetization for Ln^{3+} ions beyond Gd^{3+}, i.e., in the second half of the rare earth sequence, are well described by the Curie magnetization expression. The reason for this is that a negative spin–orbit coupling constant gives rise to an inversion of the energy ordering for the levels within a term. More specifically, this means that for a given term the level with the largest total electronic angular momentum lies lowest in energy, and consequently, the largest spin–orbit level splittings occur between

Table 8.2 Level energies and Landé g-factors for tripositive lanthanide ions

Ion	Excited level	g_J	$\Delta E_{so}/\text{cm}^{-1}$	Ion	Excited level	g_J	$\Delta E_{so}/\text{cm}^{-1}$
Ce^{3+}	$^2F_{\frac{7}{2}}$	$\frac{8}{7}$	2240	Sm^{3+}	$^6H_{\frac{13}{2}}$	$\frac{254}{195}$	4720
Pr^{3+}	$^3H_5$	$\frac{31}{30}$	1875		$^6H_{\frac{15}{2}}$	$\frac{344}{255}$	6490
	$^3H_6$	$\frac{4}{3}$	4125	Eu^{3+}	$^7F_1$	$\frac{3}{2}$	227
Nd^{3+}	$^4I_{\frac{11}{2}}$	$\frac{138}{143}$	1650		$^7F_2$	$\frac{3}{2}$	681
	$^4I_{\frac{13}{2}}$	$\frac{216}{195}$	3600		$^7F_3$	$\frac{3}{2}$	1862
	$^4I_{\frac{15}{2}}$	$\frac{306}{255}$	5850		$^7F_4$	$\frac{3}{2}$	2270
Sm^{3+}	$^6H_{\frac{7}{2}}$	$\frac{52}{63}$	826		$^7F_5$	$\frac{3}{2}$	3405
	$^6H_{\frac{9}{2}}$	$\frac{106}{99}$	1888		$^7F_6$	$\frac{3}{2}$	4767
	$^6H_{\frac{11}{2}}$	$\frac{172}{143}$	3186				

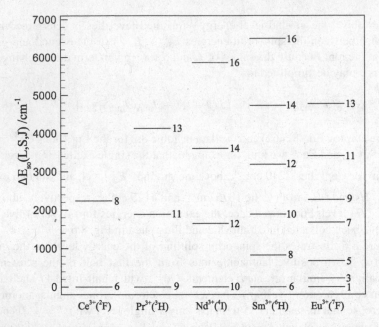

Fig. 8.6 Energies and degeneracies of the levels of the electronic ground terms of the lanthanide tripositive ions, Ln^{3+}, having electronic configurations $[Xe](4f)^i$, with $i = 1, 2, 3, 5, 6$ and [Xe] representing the electronic configuration of a xenon atom. The degeneracies $\omega_J = 2J + 1$ are shown to the right of the corresponding energy levels

the lowest, rather than the highest, levels in the ground term, making thermal accessibility of excited levels within the term negligible.

In order to have an appreciable fraction of atoms in an excited electronic level of its ground term, the assembly of atoms must form an equilibrium ensemble at a temperature T for which $k_B T$ is of the order of the excitation energy for that level. Thus, to find an appreciable fraction of a sample of Ce^{3+} ions in the excited $^2F_{\frac{7}{2}}$ level, the assembly of Ce^{3+} ions would need to be at a temperature of order 2500 K. It is therefore very unlikely that the $^2F_{\frac{7}{2}}$ excited level of Ce^{3+} will contribute to the magnetic moment of a sample of Ce^{3+} ions at temperatures below about 2000 K. Similarly, neither Pr^{3+} nor Nd^{3+} ions at room temperature should have any excited level contributions to their magnetic moments. However, Sm^{3+} ions, and especially Eu^{3+} ions, have relatively low-lying excited electronic levels, with the lowest, $^6H_{\frac{7}{2}}$, excited level for the Sm^{3+} ion lying $826\,cm^{-1}$ above its $^6H_{\frac{5}{2}}$ ground level, and the lowest, $^7F_1$, excited level for the Eu^{3+} ion lying a mere $227\,cm^{-1}$ above its $^7F_0$ ground level. It is thus quite likely that some of the excited levels for these two ions will have significant thermal populations even at room temperature, and it will thus be necessary to include contributions from those ions found in the excited levels of these two species in order to explain the deviations between experiment and the Curie values for the effective magnetic moments for the Sm^{3+} and Eu^{3+} ions.

8.5.6 Magnetic Susceptibilities of Sm³⁺ and Eu³⁺ Ions

To include the additional low-lying atomic levels for the Sm^{3+} and Eu^{3+} ions, we need to generalize our expression for the magnetization to include all levels of the ground term, rather than just its lowest-lying energy level. We shall restrict our discussion to cases for which energies associated with the applied magnetic field are small relative to $k_B T$ as well as to the spin–orbit coupling energy E_{so} for the ground term.

We shall also include the second-order Zeeman energy terms as well as take spin–orbit coupling energies into account in the following. Starting from Eq. (8.5.24) for the electronic energy of a multi-electron atom or ion, we see that its magnetic moment $\mu^{(m)}$ can be obtained from the energy expression as

$$\mu_{LS}^{(m)}(J, M_J) = -\frac{\partial E}{\partial B} = -W_1(L, S, J, M_J) - 2W_2(L, S, J, M_J)B + \cdots .$$
(8.5.63)

The first-order Zeeman energy, W_1, is linear in M_J, and the second-order Zeeman energy, W_2, has a quadratic dependence on M_J [see Eqs. (8.5.29) and (8.5.34)]. The magnetization associated with N multi-electron atoms or ions, all found in a given electronic level characterized by J, will then be

$$M_{LSJ}(T, B) = \frac{1}{\sum\limits_{M_J=-J}^{+J} e^{-\beta E_{LSJ}(M_J)}} N \sum_{M_J=-J}^{+J} \mu_{LS}(J, M_J) e^{-\beta E_{LSJ}(M_J)}.$$
(8.5.64)

The corresponding magnetic susceptibility $\chi_{LSJ}^{(m)}(T) \equiv M_{LSJ}(T, B)/B$ is then obtained from

$$\chi_{LSJ}^{(m)}(T) = -\frac{N}{Bz_J(T)} \sum_{M_J=-J}^{+J} (W_1 + 2W_2 B) e^{-\beta(E_0 + W_1 B + W_2 B^2 + \cdots)},$$
(8.5.65)

with $z_J(T)$ the canonical partition function for the Zeeman states, viz.,

$$z_J(T) \equiv \sum_{M_J=-J}^{+J} e^{-\beta(E_0 + W_1 B + W_2 B^2 + \cdots)}.$$
(8.5.66)

For W_1 and W_2 small relative to $k_B T$, we may write $\chi_{LSJ}^{(m)}(T)$ as

$$\chi_{LSJ}^{(m)}(T) \simeq -\frac{N}{Bz_J(T)} \sum_{M_J=-J}^{+J} e^{-\beta E_0}(W_1 + 2W_2 B)(1 - \beta W_1 B - \beta W_2 B^2 + \cdots)$$

$$= \frac{N}{z_J(T)} e^{-\beta E_0} \sum_{M_J=-J}^{+J} [\beta W_1^2(M_J) - 2W_2(M_J)], \tag{8.5.67}$$

where we have made use of the fact that terms that are overall odd in M_J vanish upon summation over M_J, and we have neglected contributions from quartic or higher-order terms in M_J. In the same vein, $z_J(T)$ is given by $z_J(T) \simeq (2J+1)e^{-\beta E_0}$, so that $\chi_{LSJ}^{(m)}(T)$ becomes

$$\chi_{LSJ}^{(m)}(T) \simeq \frac{N}{2J+1} \sum_{M_J=-J}^{+J} [\beta W_1^2(M_J) - 2W_2(M_J)], \tag{8.5.68}$$

with $W_1^2(M_J)$ and $W_2(M_J)$ given, respectively, by

$$W_1^2(L, S, J, M_J) = g_J(L, S)^2 \mu_B^2 M_J^2, \tag{8.5.69}$$

and

$$W_2(L, S, J, M_J)$$
$$= \frac{\mu_B^2}{\hbar^2 A_{so}} \left\{ \frac{f(L, S, J)(J^2 - M_J^2)}{J} - \frac{f(L, S, J+1)[(J+1)^2 - M_J^2]}{J+1} \right\}. \tag{8.5.70}$$

The summation over M_J^2 required in these two expressions is given by

$$\sum_{M_J+-J}^{+J} M_J^2 = \tfrac{1}{3} J(J+1)(2J+1),$$

so that the second-order Zeeman contribution to the susceptibility becomes

$$\sum_{M_J=-J}^{+J} W_2(M_J) = \frac{\mu_B^2}{3\hbar^2 A_{so}} [f(L, S, J)(2J-1)$$
$$- f(L, S, J+1)(2J+3)](2J+1), \tag{8.5.71}$$

while the squared first-order Zeeman contribution gives

$$\sum_{M_J+-J}^{+J} W_1^2(M_J) = g_J^2(L, S)\mu_B^2 \tfrac{1}{3} J(J+1)(2J+1). \tag{8.5.72}$$

The latter contribution to the magnetic susceptibility gives, as it should, our previous result (8.5.56) for a single level atom or ion. We may now combine the above results for the first- and second-order Zeeman contributions to obtain

$$\chi_{LSJ}^{(m)}(T) = N \left[\frac{g_J^2(L,S)\mu_B^2 J(J+1)}{3k_B T} + \alpha(J) \right], \tag{8.5.73}$$

with the second-order Zeeman contribution $\alpha(J)$ expressed as

$$\alpha(J) \equiv \frac{\mu_B^2}{6\hbar^2 A_{so}(2J+1)} [f_2(L,S,J+1) - f_2(L,S,J)], \tag{8.5.74}$$

in which $f_2(L,S,J)$ is defined via

$$f_2(L,S,J) \equiv \frac{1}{J^2} [J^2 - (L-S)^2][(L+S+1)^2 - J^2]. \tag{8.5.75}$$

For a term containing several levels, each with a characteristic total electronic angular momentum J and a corresponding degeneracy $2J+1$, the fractional populations of the levels within that term will be given by

$$p_J(T) = \frac{(2J+1)e^{-\beta E_{so}(L,S,J)}}{\sum_J (2J+1)e^{-\beta E_{so}(L,S,J)}}, \tag{8.5.76}$$

with the summation over J running from $J_{min} = |L-S|$ to $J_{max} = L+S$. The resultant susceptibility for the electronic term will then be

$$\begin{aligned}
\chi_{LS}^{(m)}(T) &= N \sum_{J=J_{min}}^{J_{max}} p_J \chi_{LSJ}^{(m)}(T) \\
&= \frac{N}{z_{so}(T)} \sum_{J=J_{min}}^{J_{max}} (2J+1)e^{-\beta E_{so}(L,S,J)} \chi_{LSJ}^{(m)}(T).
\end{aligned} \tag{8.5.77}$$

For convenience, we have defined the ground term canonical partition function $z_{so}(T)$ as

$$z_{so}(T) \equiv \sum_{J=J_{min}}^{J_{max}} (2J+1)e^{-\beta E_{so}(L,S,J)}. \tag{8.5.78}$$

The final expression for $\chi^{(m)}$ is thus given by

$$\chi^{(m)} = \frac{N_0}{z_{so}} \sum_{J=J_{min}}^{J_{max}} \left[\frac{g_J^2 J(J+1)\mu_B^2}{3k_B T} + \alpha(J) \right] (2J+1)e^{-\beta E_{so}(S,L,J)}. \tag{8.5.79}$$

An expression for μ_{eff}^2 when more than one J level is thermally accessible can now be defined in terms of $\chi_{LS}^{(m)}(T)$ via

$$
\mu_{\text{eff}}^2 \equiv \frac{3k_{\text{B}}T\overline{\chi}_{LS}^{(m)}(T)}{N_0\mu_{\text{B}}^2}
$$

$$
= \frac{1}{z_{\text{so}}} \sum_{J=J_{\min}}^{J_{\max}} \left[g_J^2(L,S)J(J+1) + \frac{3k_{\text{B}}T\alpha(J)}{\mu_{\text{B}}^2} \right] (2J+1)e^{-\beta E_{\text{so}}(L,S,J)}.
$$

$$(8.5.80)$$

This expression for μ_{eff}^2 was first obtained by Van Vleck [15], and was evaluated in detail for the Sm^{3+} and Eu^{3+} ions by Frank [16] in her M.Sc. thesis.

Except for the Sm^{3+} and Eu^{3+} ions, for which inclusion of the excited levels of the ground electronic term leads to significant increases over their simple Curie values for μ_{eff}, no meaningful change from the Curie value is obtained for any other triply-charged lanthanide ion. This lack of change in the calculated values of μ_{eff} at a temperature of 293 K for the lanthanide tripositive ions other than Sm^{3+} and Eu^{3+} is due to the large energy separations between the levels in the ground terms, as can be seen from Table 8.2. Not only do larger separations between levels for these ions lead to very small contributions from excited states in the first term of Eq. (8.5.79), they also lead to much smaller (even negligible) second-order Zeeman contributions because of the denominator that appears in the expression for the second-order Zeeman energies. For the Sm^{3+} ion, thermal accessibility of the first excited ^6H level contributes an additional 0.135 to μ_{eff}, while the second-order Zeeman energy term contributes an additional 0.765 to μ_{eff}, to give a final calculated value $\mu_{\text{eff}}(Sa^{3+}) = 1.729$ for $T = 293$ K. For the Eu^{3+} ion, the contributions from the first two excited levels of its ground term raise $\mu_{\text{eff}}(Eu^{3+})$ from 0 to 1.679, and the second-order Zeeman contributions bring the calculated value for $\mu_{\text{eff}}(Eu^{3+})$ to 3.6351 for $T = 293$ K. Both the $\mu_{\text{eff}}(Sa^{3+})$ and $\mu_{\text{eff}}(Eu^{3+})$ values are thus in accord with their experimental values once the role of both the excited ground levels and second-order external field effects have been taken into account, as can be seen from Fig. 8.5 and from the summary contained in Table 8.3.

Table 8.3 Summary of the first- and second-order Zeeman contributions to the effective magnetic moments of the Sm^{3+} and Eu^{3+} ions

$\mu_{\text{eff}}(Ln^{3+})$	Contribution	1st-order only	Including 2nd-order	Increase
$\mu_{\text{eff}}(Sm^{3+})$	Ground term only	0.8451	1.6624	0.8173
	Including excited levels	0.9270	1.7374	0.8104
$\mu_{\text{eff}}(Eu^{3+})$	Ground term only	0	4.6553	–
	Including excited levels	1.6791	3.6351	1.9560

Note that the value $\mu_{eff} = 4.6553$ given in Table 8.3 for the ground level 2nd-order Zeeman contribution to the effective magnetic moment for Eu^{3+} is actually a considerable overestimate of μ_{eff}, as it assumes a fractional population 1 for the ground term, rather than only about 0.46, which is the relevant fractional population for the Eu^{3+} ground level at temperature $T = 293$ K. A more realistic estimate, $\mu_{eff}(Eu^{3+}) = 3.1524$, for the 2nd-order Zeeman contribution is obtained upon employing the full $z_{so}(T)$ value of 2.1808 for Eu^{3+} at $T = 293$ K.

8.6 Magnetism, II. Ferromagnetism

8.6.1 Setting the Stage

It has long been known in thermodynamics that changes in phase of a substance, such as the gas-liquid or liquid-solid phase transitions that occur at specific temperatures (commonly referred to as the boiling point and the freezing point) for the substance and at which two or more phases co-exist typically give rise to analytic discontinuities in some of the thermodynamic functions for the system, especially in the neighbourhood of what is called the critical point. These discontinuities may become quite distinct in the vicinity of a critical point. No such non-analytic behaviour is seen for the ideal gas, whose constituent particles do not interact, and whose success is in principle limited to those ranges of temperature and pressure in which the actual interparticle interactions play only minor roles. Interparticle interactions play essential roles, however, in phase transitions, such as gas condensation, melting of solids, or phase co-existence phenomena. A characteristic feature of the ideal gas model is that energy levels of a macroscopic collection of noninteracting particles (atoms, ions, molecules) can be expressed directly in terms of the energy levels of the individual particles themselves. For macroscopic systems of interacting particles, however, it happens that under certain circumstances, large numbers of the constituent particles are able to interact with one another in a cooperative manner that assumes a macroscopic significance at a particular temperature (referred to as a critical temperature, T_c, for the system) that corresponds to the onset of a change of phase for the macroscopic substance. A proper treatment of such cooperative phenomena can prove to be formidable, so that it is often necessary to invent simple, mathematically tractable, models in order to obtain an understanding of them. We shall be interested here specifically in attempting to reach some understanding of the phenomenon of ferromagnetism and the ferromagnetic to paramagnetic phase transition. For the interested reader, more complete and detailed expositions of these interesting phenomena can be found in [17–21].

When we examined monatomic solids in Chap. 5 and the paramagnetism of solid inorganic salts in previous sections in the present chapter, we treated the atoms/ions as sets of noninteracting particles located at fixed lattice sites defining the solid crystal. We have also examined the role of interatomic/intermolecular interactions briefly in Chap. 7 to reach some understanding of corrections to the thermodynamic state functions when binary interactions between the atoms or molecules can no longer be neglected. In Sect. 7.4, we also examined briefly the concept of the critical point, particularly the concept of a critical temperature and phase co-existence within a Van der Waals model fluid. In this section, we shall examine the role played by nearest-neighbour interactions amongst the particles occupying lattice sites of a crystal.

The best-known examples of ferromagnetic materials involve atoms or ions of the ferrous metals iron, cobalt, and nickel. The relevant atoms, Fe, Co, Ni, are all transition metals with electron configurations $[Ar](4s)^2(3d)^n$ electronic configurations, with $n = 6, 7, 8$, and ground electronic terms ^5D, ^4F, and ^3F, respectively. They thus possess both spin and orbital electronic angular momenta, coupled to give total electronic angular momenta $J = L + S, L + S - 1, \cdots, |L - S|$, in which L designates the total electronic orbital angular momentum quantum number and S the total electronic spin angular momentum quantum number. Nonetheless, it has been experimentally well established that ferromagnetism appears to be associated only with the electronic spins.

We have seen that a collection of N noninteracting spins gives rise to a macroscopic magnetization $\overline{\mathbf{M}}$ in the presence of a constant external magnetic field. From Eq. (8.5.8b) for $\overline{M}_z$, for example, it is clear that for $B = 0$, the magnetization will vanish. It is then reasonable to ask what happens when we consider a collection of N interacting spins. This question turns out to be difficult to answer in general, as the mathematics becomes virtually intractable for the most relevant case, namely that of a three-dimensional solid. Although the two-dimensional case with nearest- and next-nearest-neighbour remains intractable, that with nearest-neighbour plus magnetic field interactions becomes numerically accessible, while the two-dimensional case with nearest-neighbour interactions in zero magnetic field is analytically solvable [22], though essentially beyond the level of the present treatment.[9]

A solution of the simplest model, namely that of a one-dimensional chain of magnetic dipole moments, each of which may be either aligned in the direction established by an applied magnetic field of strength B or aligned opposite to that direction, and interacts with its nearest-neighbours, thereby establishing a linking of the chain elements, can be attained relatively straightforwardly. This rather important, though simple, model is known as the one-dimensional Ising model. A history of the Ising model has been given by Brush [23], and a more detailed treatment of the Ising model, its generalizations, and its many applications has been given by Pathria [24]. In its simplest version, it may be represented physically through 'expressing the Hamiltonian $\mathcal{H}_N$ of the collection of N 'spins' as

---

[9] An exact analytical solution to the two-dimensional nearest-neighbour interaction in the absence of an external magnetic field was first given in 1944 by Onsager [22].

$$\mathcal{H}_N = -J \sum_{\text{n.n.}} \hat{s}_{iz}\hat{s}_{jz} - \mu B \sum_{i=1}^{N} \hat{s}_{iz}, \tag{8.6.1a}$$

in which $\hat{s}_{iz}, \hat{s}_{jz}$ have eigenvalues ± 1, J is referred to as the exchange energy, and the notation n.n. stands for 'nearest neighbours'. If we note that our model requires that we take each interaction between a nearest-neighbour pair of spins into account only once, then the summation over nearest-neighbours that appears in expression (8.6.1a) can be replaced by an equivalent summation that explicitly takes this requirement into account to give

$$\mathcal{H}_N = -\sum_{k=1}^{N-1} [J\hat{s}_k\hat{s}_{k+1} + \mu B\hat{s}_k] \tag{8.6.1b}$$

wherein, for convenience of notation, $\hat{s}$ now stands for $\hat{s}_z$. If we work in a basis consisting of the spin eigenstates of $\hat{s}_z$, then we may replace expression (8.6.1b) for the Hamiltonian by the corresponding energy expression

$$E_N = -\sum_{k=1}^{N-1} [Js_ks_{k+1} + \mu Bs_k], \tag{8.6.2}$$

in which s_k ($\{k = 1, 2, \cdots, N\}$) now explicitly represents an eigenvalue (i.e., ± 1) of the operator $\hat{s}_k$ (i.e., $\equiv \hat{s}_{kz}$).

The true quantum mechanical nature of a spin system is masked in this version of the Ising model (which actually predates quantum mechanics), as we consider the spins to be either aligned parallel or anti-parallel to a fixed direction (normally considered to be that associated with a constant homogeneous external magnetic field): thus, for a set of quantum mechanical spin-particles, we consider only the z-components of the spins, so that there are no commutation problems, and a full quantum mechanical treatment becomes unnecessary. For a set of N spin-$\frac{1}{2}$ particles, expression (8.6.1b) is essentially complete: for a spin particle for which $s > \frac{1}{2}$, only the largest M_s value will be considered in order to avoid complications that arise from inclusion of the other eigenvalues of $\hat{s}_z$. In any case, we shall consider that the exchange energy J subsumes all factors relating electron spins to electronic spin magnetic moments. An electron spin magnetic moment μ has also been included in the field term so that μB represents the magnitude of the Zeeman interaction energy between a spin particle and the external magnetic field. We shall consider only interactions between nearest-neighbours and then only for the $J > 0$ case. Hence, according to Eq. (8.6.2), the exchange energy will be $-J$ if the spins are parallel and $+J$ if they are anti-parallel. The way in which Eq. (8.6.1b) has been written allows an interpretation of the net number of 'up' (i.e., $s = +1$) spins as the magnetization arising from this model.

Because the total number of spins has been fixed, the natural ensemble for describing a system of N such spins is the canonical ensemble. We note, however, that even though the spins are distinguishable by virtue of their locations at distinct lattice sites, because they interact with one another we are no longer able to express the canonical partition function $Z_N(T)$ for our assembly of N spins as the canonical partition function $z(T)$ for a single spin particle raised to the power N. We are thus reduced in principle to explicitly performing the sum over all 2^N microstates accessible to such a one-dimensional collection of N spins. However, as the 2^N energies involved in the partition function sum will be too numerous to take into account directly for $N \gg 1$, we must either be able to obtain $Z_N(T)$ from its defining expression via some mathematical trick, or we shall be reduced to deducing the mathematical form for $Z_N(T)$ by an essentially inductive procedure.

We note also that while boundary conditions are irrelevant for the thermodynamic limit in which $N \to \infty$, $V \to \infty$, and the number density N/V is held constant, they do matter for finite N. There are two obvious choices for boundary conditions for a one-dimensional system. These are: either (a) allow the ends to be free, in which case two of the spins (one located at each end of the chain) will have only a single nearest-neighbour, while the remaining $N-2$ spins each have two nearest-neighbours, or (b) place the N spins equidistant from one another on a circle, with the Nth spin next to the first spin, so that all spins are equivalent, thereby eliminating the boundary. We shall examine the one-dimensional Ising model firstly in the absence, thereafter in the presence, of an external constant and homogeneous magnetic field. The first type of boundary condition (referred to as free boundary conditions) will prove useful in the former case, while the elimination of the boundary via the second condition (referred to as the toroidal boundary condition) will aid us in working through the latter case.

8.6.2 One-Dimensional Ising Chain in the Absence of an External Magnetic Field

We shall find it to be helpful to choose free boundary conditions when there is no external magnetic field, so that the energy of our Ising spin-chain is given simply by

$$E_N = -J \sum_{k=1}^{N-1} s_k s_{k+1}. \tag{8.6.3}$$

In order to obtain a closed-form expression for the partition function $Z_N(T)$ associated with this Hamiltonian, we note firstly that the 2^N energy values, E_{r_N}, associated with the eigenstates of $\mathcal{H}_N$ arise from the product $s_k s_{k+1}$ when s_k and s_{k+1} are independently assigned their allowed values ± 1. The 2^N values for r_N may thus be determined via the multiple two-term summations of $s_1, s_2 \cdots s_N$, so that the summation over r_N from 1 to 2^N that appears in the defining relation

$$Z_N(T) \equiv \sum_{r_N=1}^{2^N} e^{-\beta E_{r_N}} \tag{8.6.4}$$

is expressed equivalently by this multiple summation over the allowed values taken by the s_k. This replacement will allow us to employ the defining relation directly to evaluate the canonical partition function by making use of the structure of Eq. (8.6.3) for the energy to write $Z_N(T)$ as

$$Z_N(T) = \sum_{r_N=1}^{2^N} \prod_{k=1}^{N-1} \left(e^{\beta J s_k s_{k+1}} \right)$$

$$= \sum_{s_1=\pm 1} \sum_{s_2=\pm 1} e^{\beta J s_1 s_2} \sum_{s_3=\pm 1} e^{\beta J s_2 s_3} \cdots \sum_{s_N=\pm 1} e^{\beta J s_{N-1} s_N}, \tag{8.6.5a}$$

in which the summation over the 2^N spin microstates is now represented by the product of summations of the set of spins $\{s_k\}$ over their allowed values ± 1. We may unravel this nested summation by starting with the summation over s_N, and observing that it can occur only once, as s_N lies at one of the free ends of the chain: this summation gives $e^{\beta J} + e^{-\beta J} = 2 \cosh \beta J$ independently of the value taken by s_{N-1}. Precisely the same situation then occurs sequentially for the next N-2 spins, with each summation providing an additional factor $2 \cosh \beta J$. The final summation over $s_1 = \pm 1$ reduces to $\sum_{s_1} 1 = 2$ (because a single spin has no interaction). In this way, $Z_N(T)$ assumes the closed form

$$Z_N(T) = 2(2 \cosh \beta J)^{N-1}, \tag{8.6.5b}$$

from which the Helmholtz energy $A_{\text{spin}}(N, T)$ for an Ising chain (in the absence of an applied magnetic field) is obtained as

$$A_{\text{spin}}(N, T) = -k_B T \ln Z_N(T)$$

$$= -k_B T [\ln 2 + (N - 1) \ln(2 \cosh \beta J)]$$

$$\simeq -N k_B T \ln(2 \cosh \beta J), \tag{8.6.6}$$

for $N \gg 1$.

We may also utilize the usual canonical ensemble expressions relating the internal energy, the entropy, and the heat capacity for our one-dimensional Ising chain to the canonical partition function $Z_N(T)$, to obtain the expressions

$$U_{\text{spin}}(N, T) = -N J \tanh \beta J, \tag{8.6.7}$$

$$S_{\text{spin}}(N, T) = N k_B \left[\ln(e^{2\beta J} + 1) - \frac{2\beta J}{1 + e^{-\beta J}} \right], \tag{8.6.8}$$

and

$$C_{\text{spin}}(N, T) = Nk_{\text{B}}[\beta J \,\text{sech}\, \beta J]^2. \quad\quad (8.6.9)$$

8.6.3 One-Dimensional Ising Chain in the Presence of an External Magnetic Field

Free boundary conditions do not permit a straightforward solution of the one-dimensional Ising chain in the presence of an external magnetic field. In this case, it is convenient to employ the toroidal boundary condition, for which the nearest-neighbours for the spin $\hat{s}_N$ are $\hat{s}_{N-1}$ and $\hat{s}_1$. For the toroidal topology, we are thus able to re-express our Hamiltonian (8.6.1b) in a symmetric manner as

$$\mathcal{H}_N = -J \sum_{k=1}^{N-1} \hat{s}_k \hat{s}_{k+1} - \tfrac{1}{2}\mu B \sum_{k=1}^{N-1} (\hat{s}_k + \hat{s}_{k+1}). \quad\quad (8.6.10)$$

The corresponding energy expression is

$$E_N = -J \sum_{k=1}^{N-1} s_k s_{k+1} - \tfrac{1}{2}\mu B \sum_{k=1}^{N-1} (s_k + s_{k+1}), \quad\quad (8.6.11)$$

in which the s_k take values ± 1 in the multiple summation version of the sum-over-states that then defines $Z_N(B, T)$.

In order to obtain a closed analytic form for the summation that appears in the expression for $Z_N(B, T)$ when the energy of the spin chain is given by Eq. (8.6.11), we shall employ a generalization [25] of the procedure that we employed in the previous subsection for $Z_N(T)$ in the absence of an external magnetic field **B**.[10] As the presence of the external field term in the energy expression (8.6.11) greatly complicates the form for the final result obtained from carrying out the summations over the spin values s_k, we shall draw upon our knowledge of matrix algebra to simplify the mathematics.

Interlude: Transfer Matrices

To introduce this procedure, let us consider briefly the essentially trivial two-spin case, in which s_1 and s_2 are one another's nearest-neighbours (i.e., $s_{N+1} \equiv s_3$ is

---

[10]This procedure was introduced by Kramers and Wannier [25] in order to study the two-dimensional Ising model. Although they referred to their procedure as the eigenvalue method, it is now commonly called the transfer matrix method.

simply s_1). For this simplest possible case, the canonical partition function $Z_2(B, T)$ may be obtained upon inspection as

$$Z_2(B, T) = \left[e^{\beta(J+B\mu)}\right]^2 + 2\left(e^{-\beta J}\right)^2 + \left[e^{\beta(J-B\mu)}\right]^2,$$

which we may readily recognize as the trace (i.e., the sum of the diagonal elements) of T^2, with T being the 2×2 matrix

$$T = \begin{pmatrix} e^{\beta(J+B\mu)} & e^{-\beta J} \\ e^{-\beta J} & e^{\beta(J-B\mu)} \end{pmatrix}, \tag{8.6.12a}$$

commonly referred to as a *transfer matrix* (hence the notation T). From matrix algebra, we know that we may employ a similarity transformation matrix S (and its inverse S^{-1}) made up from the normalized eigenvectors of T to transform T into its diagonal form, viz.,

$$T_{diag} = S \cdot T \cdot S^{-1} = \begin{pmatrix} \lambda_+ & 0 \\ 0 & \lambda_- \end{pmatrix}, \tag{8.6.12b}$$

in which λ_+ and λ_- are the eigenvalues of T. We obtain the eigenvalues of T from $T = \lambda 1$ by setting the determinant $|T - \lambda 1|$ to zero and solving for λ. Thus, using expression (8.6.12a) for T, its eigenvalue equation is

$$\lambda^2 - 2\lambda e^{\beta J} \cosh(\beta\mu B) + 2\sinh(2\beta J) = 0. \tag{8.6.12c}$$

The two roots of this eigenvalue equation, namely

$$\lambda_\pm = e^{\beta J} \cosh(\beta\mu B) \pm \sqrt{e^{-2\beta J} + e^{2\beta J}\sinh^2(\beta\mu B)}, \tag{8.6.12d}$$

may then be substituted into expression (8.6.12b). □

Let us return now to the more general problem of evaluating the canonical partition function $Z_N(B, T)$ for a chain of N Ising spins, expressed in the form

$$Z_N(B, T) = \sum_{s_1=\pm 1} \sum_{s_2=\pm 1} \cdots \sum_{s_N=\pm 1} \exp\left\{\beta J \sum_{k=1}^N s_k s_{k+1} + \tfrac{1}{2}\beta\mu B \sum_{k=1}^N (s_k + s_{k+1})\right\}, \tag{8.6.13a}$$

with $s_{N+1} \equiv s_1$. It will be clear from what we have seen in the interlude that $Z_N(B, T)$ can be expressed in the form of a matrix product as

$$Z_N(B, T) = \sum_{s_1=\pm 1} \sum_{s_2=\pm 1} \cdots \sum_{s_N=\pm 1} T_{s_1 s_2} T_{s_2 s_3} \cdots T_{s_{N-1} s_N} T_{s_N s_1}.$$

The summation over $s_2, \cdots, s_N$ constitutes, according to the rules of matrix multiplication, the $s_1 s_1$ diagonal matrix element of the matrix product T^N, so that $Z_N(B,T)$ is thus given as

$$Z_N(B,T) = \sum_{s_1=\pm 1} (\mathsf{T}^N)_{s_1 s_1} \equiv \mathrm{Tr}\, \mathsf{T}^N. \tag{8.6.13b}$$

If we replace each transfer matrix vector by its similarity-transformed diagonal version, we can see that $\check{Z}_N(B,T)$ is given quite simply in terms of the eigenvalues $\lambda_\pm$ of T as

$$Z_N(B,T) = \lambda_+^N + \lambda_-^N = \lambda_+^N \left[1 + \left(\frac{\lambda_-}{\lambda_+}\right)^N\right]. \tag{8.6.13c}$$

From Eq. (8.6.12d) giving the eigenvalues λ_+ and λ_-, it will be clear that $\lambda_+ > \lambda_-$, and hence $(\lambda_+/\lambda_-)^N \ll N$ for $N \gg 1$, in which case, it is λ_+ that determines the physical properties in the thermodynamic limit of a macroscopically large Ising chain of spins.

As all thermodynamic properties of our spin system are obtained in terms of $\ln Z_N(B,T)$ and/or it derivatives, we shall specifically be concerned with the behaviour of $N \ln \lambda_+$ in the thermodynamic limit. The Helmholtz energy $A(B,T)$ is thus given as

$$A(B,T) = -NJ - Nk_\mathrm{B}T \ln\left\{\cosh(\beta\mu B) + \sqrt{e^{-4\beta J} + \sinh^2(\beta\mu B)}\right\}, \tag{8.6.14a}$$

while the internal energy $U(B,T)$ is given as

$$U(B,T) = -N\left\{J + \frac{\mu B \sinh(\beta\mu B)}{\sqrt{e^{-4\beta J} + \sinh^2(\beta\mu B)}}\right.$$

$$\left. - \frac{2Je^{-4\beta J}}{[\cosh(\beta\mu B) + \sqrt{e^{-4\beta J} + \sinh^2(\beta\mu B)}]\sqrt{e^{-4\beta J} + \sinh^2(\beta\mu B)}}\right\}. \tag{8.6.14b}$$

The magnetization $M(B,T)$ can be obtained from the Helmholtz energy according to Eq. (8.5.12), giving

$$M(B,T) = \frac{N\mu \sinh(\beta\mu B)}{\sqrt{e^{-4\beta J} + \sinh^2(\beta\mu B)}} \equiv Nm(B,T), \tag{8.6.14c}$$

with $m(B,T)$ the magnetization per spin.

The heat capacity may be obtained as the partial derivative of $U(B, T)$ with respect to temperature, while the susceptibility $\chi(B, T)$ may be obtained as the partial derivative of $M(B, T)$ with respect to field strength via Eq. (8.4.4b), from which we obtain

$$\chi^{(m)}(T) = \left(\frac{\partial M}{\partial B}\right)_{T,N} = N\left(\frac{\partial m}{\partial B}\right)_{\beta,N},$$

so that $\chi^{(m)}(T)$ is given as

$$\chi^{(m)}(T) = N\beta\mu^2 \frac{\cosh(\beta\mu B)e^{-4\beta J}}{[e^{-4\beta J} + \sinh^2(\beta\mu B)]^{\frac{3}{2}}}. \tag{8.6.14d}$$

From Eq. (8.6.14c) for the magnetization of our N-spin chain, we see that $\lim_{B \to 0} M(B, T) = 0$, so that a one-dimensional Ising model (or Ising chain) does not allow for spontaneous magnetization, and there is hence no phase transition from a ferromagnetic phase to a paramagnetic phase at any finite temperature, in spite of the existence of nearest-neighbour interactions between the spins. If, however, we examine the limiting case for $T \to 0$, we find that

$$\lim_{T \to 0} M(B, T) = N\mu \tag{8.6.15}$$

for arbitrary nonzero values B of the external magnetic field. The one-dimensional Ising model does, in principle, have a phase transition at a critical temperature $T_c = 0$. Moreover, if we examine the behaviour of the heat capacity $C_{B=0}(T)$, given by

$$C_0(T) = Nk_B(\beta J)^2 \operatorname{sech}^2(\beta J), \tag{8.6.16}$$

we see that although it passes through a maximum, it remains a smooth function of temperature and does not show a discontinuity, which would be a typical characteristic feature of a thermodynamic phase change. A more extensive discussion of this aspect of the Ising chain can be found in Chapter 11 of Pathria's text [24].

8.6.4 Correlation Between Spins: 1-D Ising Model

As the difference between ferromagnetism and paramagnetism hangs upon the existence or nonexistence, respectively, of long-range order in the absence of an externally-imposed magnetic field, we shall be interested in the tendency of the spins in the Ising chain to be aligned with one another. To address the tendency of a pair of spins in the Ising chain to be aligned or not, we shall introduce a function $g_{kk'}(B, T)$ defined as

$$g_{kk'}(B, T) \equiv \langle s_k s_{k'} \rangle(B, T) - \langle s_k \rangle(B, T)\langle s_{k'} \rangle(B, T). \tag{8.6.17a}$$

For toroidal boundary conditions (for which all N spins are equivalent), the average $\mu \langle s_k \rangle (B, T)$ gives the magnetization per spin, m, which does not depend upon the site label k [see Eq. (8.6.14c)], so that $g_{kk'}(B, T)$ depends only upon the number of spin sites $|k' - k|$ separating spin s_k from spin $s_{k'}$ for fixed values of magnetic field strength B and temperature T. We may thus replace Eq. (8.6.17a) by

$$g_{|k'-k|}(B, T) = \langle s_k s_{k'} \rangle (B, T) - \langle s_k \rangle^2 (B, T). \qquad (8.6.17b)$$

Should $k' = k$, Eq. (8.6.17b) gives

$$g_0(B, T) = \langle s_k^2 \rangle (B, T) - \langle s_k \rangle^2 (B, T), \qquad (8.6.17c)$$

which is proportional to the susceptibility, $\chi^{(m)}(B, T)$, for a system of N equivalent magnetic dipole moments [see Eq. (8.6.14d)]. For $k' \neq k$, however, $g_{|k'-k|}(B, T)$ provides information on how the alignment of spin s_k is affected by spin $s_{k'}$ as it is sited farther and farther away from s_k, thereby providing a measure of how well the motions of spins s_k and $s_{k'}$ are correlated. This is the source of the terminology 'correlation function' for $g_{|k'-k|}(B, T)$. In particular, if $g_{|k'-k|}(B, T) = 0$, we say that the spins s_k and $s_{k'}$ are uncorrelated.

From Eq. (8.6.11), we see that the interaction energy for an Ising chain represents a competition between the interaction energy J between adjacent spins and the energy of interaction between the spins and the applied magnetic field of strength B. For any reasonable field strength B, the Zeeman interaction dominates over the spin–spin interaction, and gives rise to a macroscopic paramagnetic state of the spin chain, much as it does for a set of noninteracting spins.

Hence, because ferromagnetism will only be present if the magnetization of the set of spins in the absence of an applied magnetic field is non-vanishing, we shall focus our interest upon the correlation function for zero applied magnetic field, i.e., upon

$$g_{|k'-k|}(0, T) = \langle s_k s_{k'} \rangle (0, T) - \langle s_k \rangle^2 (0, T). \qquad (8.6.18a)$$

As $\langle s_k \rangle (B, T)$ is proportional to the magnetization per spin, $m(B, T)$, we see from Eq. (8.6.14c) that $m(0, T)$, and hence $\langle s_k \rangle (0, T)$, vanishes for $B = 0$, so that $g_{|k'-k|}(0, T)$ is given simply by

$$g_{|k'-k|}(0, T) = \langle s_k s_{k'} \rangle (0, T).$$

It will prove to be slightly more transparent to express this correlation function in terms of the number of sites r separating the two spins, or as

$$g_r(0, T) = \langle s_k s_{k+r} \rangle (0, T). \qquad (8.6.18b)$$

Our calculation will be greatly simplified by the assumption of free boundary conditions for the $B = 0$ case.

In the absence of an applied magnetic field, we may express the allowed energies for an Ising chain of N spins in the slightly generalized form

$$E_N = -\sum_{k=1}^{N-1} J_k s_k s_{k+1},$$

in which the spin–spin interaction energy between nearest-neighbour spins s_k and s_{k+1} are allowed to be site-dependent. We may, therefore, express the correlation function $g_{|k'-k|}(0, T)$ as

$$g_{|k'-k|}(0, T) = \langle s_{k'} s_k \rangle(0, T)$$

$$= \frac{1}{Z_N(T)} \sum_{s_1=\pm 1} \cdots \sum_{s_N=\pm 1} s_{k'} s_k \exp\left\{-\sum_{i=1}^{N-1} \beta J_i s_i s_{i+1}\right\},$$

with $Z_N(T; J_1, \cdots, J_{N-1})$ given by

$$Z_N(T; J_1, \cdots, J_{N-1}) = \sum_{s_1=\pm 1} \cdots \sum_{s_N=\pm 1} \prod_{i=1}^{N-1} e^{\beta J_i s_i s_{i+1}}. \tag{8.6.19a}$$

Because s_N occurs only once in this expression for Z_N for free boundary conditions, we may perform the summation over s_N explicitly as

$$\sum_{s_N=\pm 1} e^{\beta J_{N-1} s_{N-1} s_N} = 2\cosh(\beta J_{N-1} s_{N-1})$$

$$= 2\cosh(\beta J_{N-1}),$$

which is independent of the values ± 1 assumed by s_{N-1}. Utilization of this result in Eq. (8.6.19a) leads to a recurrence relation for the canonical partition function for the Ising chain, namely

$$Z_N(T; J_1, \cdots, J_{N-1}) = 2\cosh(\beta J_{N-1}) Z_{N-1}(T; J_1, \cdots, J_{N-2}). \tag{8.6.19b}$$

Iteration of this recurrence relation then gives for $Z_N(T)$ the result

$$Z_N(T; J_1, \cdots, J_{N-1}) = \left(\sum_{s_1=\pm 1} 1\right) \prod_{i=1}^{N-1} [2\cosh(\beta J_i)]$$

$$= 2^N \prod_{i=1}^{N-1} \cosh(\beta J_i) \tag{8.6.19c}$$

for the canonical partition function for an Ising chain of N spins.

We are now poised to obtain an expression for the nearest-neighbour (n.n.) spin–spin correlation function $g_k(\text{n.n.}) \equiv \langle s_k s_{k+1} \rangle$. Because $Z_N(T)$ is parameterized by the spin–spin interaction parameters J_i $(i = 1, \cdots, N - 1)$, we see that $g_k(\text{n.n.})$ may be obtained in terms of the parameter-derivative of $Z_N(T)$ with respect to J_k as

$$g_k(\text{n.n.}) = \frac{1}{Z_N(T)} \left(\frac{1}{\beta} \frac{\partial Z_N}{\partial J_k} \right). \tag{8.6.20}$$

If we now utilize Eq. (8.6.19c) for $Z_N(T)$, we obtain directly the result

$$g_k(\text{n.n.}) = \tanh(\beta J_k). \tag{8.6.21}$$

To extend this result to the spin–spin correlation function $g_k(r)$ for a pair of spins s_k, $s_{k'}$ separated by r lattice sites, and defined in Eq. (8.6.18b), we firstly note that it can be expressed as the ensemble average of a sequence of r nearest-neighbour factors, namely as

$$g_k(r) = \langle s_k s_{k+r} \rangle$$
$$= \langle (s_k s_{k+1})(s_{k+1} s_{k+2}) \cdots (s_{k+r-1} s_{k+r}) \rangle, \tag{8.6.22a}$$

as $s_i^2 = 1$ $(i = 1, \cdots, N - 1)$. Each pair of spin values appearing in this expression can be generated by parameter-differentiation of $Z_N(T)$, so that $g_k(r)$ is given, analogously to Eq. (8.6.20) for $g_k(\text{n.n.})$, by

$$g_k(r) = \frac{1}{Z_N} \left(\frac{1}{\beta} \frac{\partial}{\partial J_k} \right) \left(\frac{1}{\beta} \frac{\partial}{\partial J_{k+1}} \right) \cdots \left(\frac{1}{\beta} \frac{\partial}{\partial J_{k+r-1}} \right) Z_N. \tag{8.6.22b}$$

Upon utilizing Eq. (8.6.19c) for $Z_N(T)$, we find that $g_k(r)$ is obtained as

$$g_k(r) = \prod_{i=k}^{k+r-1} \tanh(\beta J_i). \tag{8.6.22c}$$

When all nearest-neighbour spin–spin interactions are equal, i.e., $J_i \equiv J$ $(i = 1, \cdots, N - 1)$, the spin–spin correlation function becomes

$$g(r) = \tanh^r(\beta J), \tag{8.6.22d}$$

which is site-independent.

We note that expression (8.6.20d) can be rewritten in terms of an exponential as

$$g(r) = e^{-r/\xi}, \tag{8.6.23}$$

in which ξ, referred to as the correlation length, is defined formally as the reciprocal of $\ln[\coth(\beta J)]$. When the correlation function is written in this format, it becomes clear that for the temperature approaching zero (i.e., for $\beta J \gg 1$), the correlation length ξ can be approximated by

$$\xi \approx \tfrac{1}{2}e^{2\beta J},$$

which diverges in the limit $T \to 0$. This divergence of the correlation length supports the comments made at the end of Sect. 8.6.3, namely that although there is a critical temperature for which the one-dimensional Ising chain undergoes spontaneous magnetization in zero external field, that critical temperature is $T_c = 0$.

8.6.5 Some Comments on the Two-Dimensional Ising Model

Exact quantitative results were first obtained for the two-dimensional Ising model by Kramers and Wannier [25] in 1941. Shortly thereafter, Onsager [22] employed their transfer matrix method to obtain an analytic expression for the Helmholtz energy in the absence of an external magnetic field, and from it established that there was an anomaly in the heat capacity at the critical temperature obtained earlier by Kramers and Wannier. Because the Onsager solution lies beyond the scope of the present treatment, we shall be content here to provide only an outline of the solution to this problem along the lines of that provided earlier by Pathria [24].

The partition function for the square-lattice two-dimensional Ising model is obtained as [24]

$$Z_N(T) = 2^{N/2}\cosh(2\beta J)\exp\left\{\frac{N}{\pi}\int_0^{\pi/2}\ln[1 + \sqrt{1 - \kappa^2\sin^2\phi}\,]\,d\phi\right\},$$
$$(8.6.24a)$$

in which κ is defined as

$$\kappa \equiv 2\tanh(2\beta J)\,\mathrm{sech}\,(2\beta J). \qquad (8.6.24b)$$

Differentiation of Eq. (8.6.24a) with respect to $-\beta$ gives the zero-field internal energy $U_0(T)$ as

$$U_0(T) = -2NJ\tanh(2\beta J) + \frac{N}{\pi}\left(\kappa\frac{d\kappa}{d\beta}\right)\int_0^{\pi/2}\frac{\sin^2\phi}{[1 + \sqrt{1 + \kappa^2\sin^2\phi}\,]\sqrt{1 - \kappa^2\sin^2\phi}}.$$

Upon rationalizing the integrand on the right-hand side of this expression, $U_0(T)$ can be written as

$$U_0(T) = -2NJ\tanh(2\beta J) + \frac{1}{\pi}\left(\frac{1}{\kappa}\frac{d\kappa}{d\beta}\right)\left[-\frac{\pi}{2} + K_1(\kappa)\right],$$

in which $K_1(\kappa)$ is the complete elliptic integral of the first kind, namely

$$K_1(\kappa) = \int_0^{\pi/2} \frac{d\phi}{\sqrt{1 - \kappa^2 \sin^2 \phi}},$$

with κ known as the modulus of the elliptic integral. Logarithmic differentiation of κ with respect to β gives

$$\frac{1}{\kappa}\left(\frac{d\kappa}{d\beta}\right) = 2J[\coth(2\kappa) - 2\tanh(2\kappa)],$$

so that $U_0(T)$ simplifies to

$$U_0(T) = -NJ\coth(2\beta J)\left[1 + \frac{2\kappa'}{\pi}K_1(\kappa)\right], \tag{8.6.25}$$

in which κ', defined as $\kappa' \equiv \sqrt{1 - \kappa^2}$, is referred to as the complementary modulus.

Differentiation of expression (8.6.25) for $U_0(T)$ with respect to T, followed by utilization of the expressions

$$\frac{d\kappa}{d\beta} = -\frac{\kappa'}{\kappa}\left(\frac{d\kappa'}{d\beta}\right) = -\frac{\kappa'}{\kappa}8J\tanh(2\beta J)[1 - \tanh^2(2\beta J)]$$

and

$$\frac{dK_1}{d\kappa} = \frac{1}{\kappa'^2\kappa}[E_1(\kappa) - \kappa'^2 K_1(\kappa)],$$

in which $E_1(\kappa)$ is the complete elliptic integral of the second kind,

$$E_1(\kappa) = \int_0^{\pi/2} \sqrt{1 - \kappa^2 \sin^2 \phi}\, d\phi,$$

results in the expression

$$C_0(T) = Nk_B\frac{2}{\pi}(\beta J)^2\coth(2\beta J)\{2[K_1(\kappa) - E_1(\kappa)] - (1 - \kappa')[\tfrac{1}{2}\pi + \kappa' K_1(\kappa)]\}, \tag{8.6.26}$$

for the heat capacity $C_0(T)$.

Because the complete elliptic integral $K_1(\kappa)$ has a singularity at $\kappa = 1$ (equivalently, for $\kappa' = 0$) that behaves as

$$K_1(\kappa) \approx \ln\left(\frac{4}{|\kappa'|}\right),$$

while $E_1(\kappa)$ is approximately unity, the heat capacity $C_0(T)$ has a logarithmic singularity at a temperature T_c determined from $\kappa_c = 1$ ($\kappa'_c = 0$). This condition gives $T_c \approx 2.431 J/k_B$ for the square lattice, which has a coordination number (i.e., number of nearest-neighbours) q of 4. Two-dimensional lattices with different coordination numbers have different critical temperatures, the higher the coordination number, the higher will T_c be: for example, for $q = 3$, $T_c \approx 1.517 J/k_B$, while for $q = 6$, $T_c \approx 3.640 J/k_B$. In the vicinity of $T = T_c$ the heat capacity $C_0(T)$ for a square Ising lattice behaves as

$$C_0(T) \simeq -0.4945 \ln \left| 1 - \frac{T}{T_c} \right| - 0.3063, \qquad (8.6.27)$$

which clearly shows a logarithmic divergence at $T = T_c$, and hence signifies a phase transition from paramagnetic to ferromagnetic behaviour as the temperature of the lattice is lowered below T_c.

8.7 Problems for This Chapter

1. Employ the combined 1st and 2nd Laws expression (8.2.7a) for dU_{elec} for a uniform-density dielectric thermodynamic system, with $\mathfrak{D} \equiv \int_V \mathcal{P} \, dV$, to obtain the well-known first Maxwell relation and two Maxwell-like relations.

2. Employ the combined 1st and 2nd Laws expression (8.2.9a) for a dielectric medium subjected to an electric field E_0, coupled with the enthalpy $H'(S, P, \mathfrak{D}) = U'(S, V, \mathfrak{D}) + PV$ and Helmholtz energy $A'(T, V, E_0) = U'(S, V, \mathfrak{D}) - TS + E_0\mathfrak{D}$ to obtain a set of Maxwell and Maxwell-like relations relevant to the dielectric medium.

3. Show, using the basic properties of $\mathbf{E}$-fields and $\mathbf{D}$-fields, together with the Gauss divergence theorem, that the integrals over all space of the three terms $(\mathbf{D}_0 \cdot d\mathbf{E}_0 - \mathbf{E}_0 \cdot d\mathbf{D}_0)$, $\mathbf{E} \cdot (d\mathbf{D} - d\mathbf{D}_0)$, and $(\mathbf{D} - \mathbf{D}_0) \cdot d\mathbf{E}_0$ appearing in the Heine identity, Eq. (8.2.3), all vanish.

4. Calculate the dielectric constant ε for 1 mole of dry HX (X = F, Br, I) under 1 bar pressure and at a temperature of 25 °C, given that the relevant permanent electric dipole moments, $\mu_d(HX)$, have values 1.8262 ± 0.0001 D, 0.8272 ± 0003 D, and 0.448 ± 0.006 D, while the average (isotropic) dipole polarizabilities, $\bar{\alpha}(HX)/\epsilon_0$, have values $0.83 \times 10^{-30} \, \text{m}^3$, $3.61 \times 10^{-30} \text{m}^3$, $5.44 \times 10^{-30} \text{m}^3$, respectively, for X = F, Br, I.

5. Show that the macroscopic magnetic moment, $\mathbf{M}$, of a magnetizable thermodynamic system of volume V is given by the partial derivative with respect to the magnetic induction, $\mathbf{B}$, of the Helmholtz energy, A_{mag}, obtained from the internal energy, $U_{\text{mag}}(S, V, \mathbf{M})$, by an appropriate Legendre transform.

6. For a uniform and magnetically isotropic magnetizable dipolar medium subjected to an applied magnetic field (induction) $\mathbf{B}_0$, determine the set of

Maxwell and Maxwell-like relations arising from the combined 1st and 2nd Laws expression (8.4.6) in conjunction with the Gibbs energy $G(T, P, B_0) \equiv U(S, V, M) + PV - TS - MB_0$.

7. Show that the entropy, S_{spin}, of a system of N distinguishable spin-$\frac{1}{2}$ particles in an applied magnetic induction field of strength B_0 and given by Eq. (8.5.14) may also be written as

$$S_{\text{spin}}(N, T, B_0) = \tfrac{1}{2} N k_{\text{B}} \left\{ \ln \left[\frac{4(N\mu_N g_I)^2}{(N\mu_N g_I)^2 - 4M^2(T, B_0)} \right] \right.$$
$$\left. - \frac{4M(T, B_0)}{N\mu_N g_I} \tanh^{-1} \left(\frac{2M(T, B_0)}{N\mu_N g_I} \right) \right\}.$$

What consequence does this result imply for C_M?

8. Use Eq. (8.4.13b) to establish that the heat capacity at constant magnetic moment, C_M, vanishes for a system of N distinguishable spin-$\frac{1}{2}$ particles.

9. Starting from expression (8.4.1) for the magnetic field contribution to the internal energy, show that its contribution to the combined 1st and 2nd Laws expression for dU is given by $dU_{\text{field}} = \mathbf{B}_0 \cdot d\mathbf{M}$ if the magnetizable thermodynamic system is of fixed volume V and has fixed permeability μ. What condition(s) is(are) thus placed upon the magnetic susceptibility of the material?

10. Starting from the combined 1st and 2nd laws expression $dU = T\,dS + B\,dM$, and by considering both the internal energy U and the entropy S to be functions of temperature T and magnetic moment M, show that the heat transferred between a macroscopic system and its surroundings is given by

$$\delta Q_{\text{rev}} = C_B(T)\,dT + T\left(\frac{\partial M}{\partial T}\right)_B dB.$$

11. If the equation of state for a magnetizable sample is given by $M = \chi(T)B$ with $\chi(T)$ known as the magnetic susceptibility, show that for an *adiabatic demagnetization* process (in which the magnetic induction is decreased by an amount dB), the temperature change, dT, is given by

$$dT = -\frac{BT}{C_B}\frac{d\chi}{dT}\,dB.$$

Many paramagnetic crystals have a susceptibility given by the Curie law,

$$\chi(T) = \frac{N\mu_{\text{m}}^2}{3k_{\text{B}}T},$$

in which N is the number of paramagnetic molecules or ions in the crystal, and μ_m represents the magnetic moment of an individual molecule or ion. Determine whether the temperature increases or decreases for such paramagnetic crystals when they are demagnetized adiabatically.

12. The thermodynamic Helmholtz energy $A_{\text{spin}}(T, V, B_0)$ appropriate to a canonical ensemble of N distinguishable spin-$\frac{1}{2}$ nuclei is related to the internal energy $U(S, V, M)$ via Eq. (8.5.11a). Evaluate $U(S, V, M)$ explicitly for this system, and show that $U \equiv 0$. Comment upon why this result is actually appropriate for this spin system.

13. Obtain expressions for the variance, $\langle (\Delta M)^2 \rangle$, and the standard deviation, σ_M, associated with magnetic moment fluctuations to which a solid macroscopic sample containing N independent spin-$\frac{1}{2}$ nuclei in a magnetic induction field of strength B_0 is subject.

14. The standard deviation σ_M for (also referred to as the fluctuation in) the macroscopic magnetic moment M of a thermodynamic magnetic system is defined as the square root of its variance $\overline{M^2} - \overline{M}^2$. Obtain an expression for $\sigma_M / \overline{M}$ in terms of the magnetic susceptibility $\chi^{)m)}$. Use your result to justify the statement that the thermodynamic magnetic moment M is the same as $\overline{M}$.

15. Expressions (8.4.13a) were obtained by appealing to the observation that $-M$ corresponds to V and B_0 corresponds to P in terms of their roles in thermodynamics. By starting from Eq. (8.4.6) for P and V constant, show that the same final expression for the heat capacity difference $C_B - C_M$ is attained directly. Consider a homogeneous magnetically isotropic thermodynamic system, so that $\mathbf{M}$ and $\mathbf{B}_0$ may be replaced by scalars M and B_0.

16. When a sample of a magnetizable substance is magnetized by bringing it into a magnetic field of strength B while maintaining thermal contact with its surroundings at temperature T, the magnetization is hence inherently a function of both temperature and applied field strength. It thus makes sense to think in terms of energy exchange, δQ, between a (thermodynamic) system and its surroundings under the condition of constant applied magnetic field strength, B, rather than at constant magnetization. Obtain an expression for the difference between $C_B(T)$ and $C_M(T)$, and discuss its analogy with the difference between $C_P(T)$ and $C_V(T)$ obtained in Chap. 2.

17. The combined 1st and 2nd laws expression for a magnetizable substance may be written as

$$dU = T\,dS + B\,dM - P\,dV,$$

with B the magnetic induction and M the magnetic moment of the substance, so that $B\,dM$ represents the work involved in changing the sample magnetic moment by dM. Obtain an expression for the heat, δQ_{rev}, exchanged between the sample and its surroundings, and show that for $U \equiv U(T, M)$, δQ_{rev} can be written as

$$\delta Q_{\text{rev}} = C_M(T)\,dT + \left[\left(\frac{\partial U}{\partial M} \right)_T - B \right] dM,$$

with $C_M(T)$ the heat capacity of the solid sample at constant magnetization.

18. Show that the correlation function $g(r)$ can be expressed in the form given in Eq. (8.6.18b) and show that the correlation length ξ behaves as $e^{2\beta J}/2$ for temperatures approaching zero.

References

1. P.W. Atkins, R.S. Friedman, *Molecular Quantum Mechanics*, 3rd edn. (Oxford University Press, Oxford, 1997), Chapter 12. See also A.J. Stone, *Theory of Intermolecular Forces*, 2nd edn. (Clarendon Press, Oxford, 2013)
2. J.M. Honig, *Thermodynamics*, 2nd edn. (Academic Press, San Diego, 1999), p. 493. More detailed discussions of the thermodynamics of electric and magnetic systems can be found in §§1.6(f,g), §1.7, and §§5.6–5.8 of this reference
3. V. Heine, Proc. Camb. Phil. Soc. **52**(3), 546 (1956)
4. Y. Zimmels, Phys. Rev. E **52**, 1452 (1995)
5. H.B. Callen, *Thermodynamics and an Introduction to Thermostatics*, 2nd edn. (Wiley, New York, 1985). See especially Appendix B
6. A.B. Pippard, *The Elements of Classical Thermodynamics* (Cambridge University Press, Cambridge, 1957)
7. F. Mandl, *Statistical Physics*, 2nd edn. (Wiley, Chichester, 1988)
8. G. Castellano, J. Magn. Magn. Mater. **260**, 146 (2003)
9. J.S. Griffith, *The Theory of Transition-Metal Ions* (Cambridge University Press, Cambridge, 1961)
10. D.W. Griffiths, *Introduction to Quantum Mechanics* (Prentice Hall, Upper Saddle River, 1994)
11. D.A. McQuarrie, *Quantum Chemistry*, 2nd edn. (University Science Books, Sausalito, 2008)
12. The Wigner-Eckart theorem lies beyond the scope of our discussion, but is developed in most standard advanced quantum mechanics texts and in all specialized texts on angular momentum: see, for example, R.N. Zare, *Angular Momentum. Understanding Spatial Aspects in Chemistry and Physics* (Wiley, New York, 1988)
13. H.L. Schläfer, G. Gliemann, *Basic Principles of Ligand Field Theory* (Wiley-Interscience, London, 1969)
14. B.N. Figgis, M.A. Hitchman, *Ligand Field Theory and Its Applications* (Wiley-VCH, New York, 2000)
15. J.H. Van Vleck, Phys. Rev. **31**, 587 (1928)
16. A. Frank, Phys. Rev. **39**, 119 (1932)
17. A. Münster, *Statistical Thermodynamics*, vol. II (Springer-Verlag, New York, 1974), Chapter XIV. This work is often considered to be the 'bible' of statistical thermodynamics
18. R.J. Baxter, *Exactly Solved Models in Statistical Mechanics* (Dover Publications, New York, 2007), Chapter 7. Square-Lattice Ising Model
19. C.J. Thompson, *Mathematical Statistical Mechanics* (Princeton University Press, Princeton, 1972)
20. B.M. McCoy, T.T. Wu, *The Two-Dimensional Ising Model* (Harvard University Press, Cambridge, 1973)
21. G.A. Baker, Jr., *Quantitative Theory of Critical Phenomena* (Academic Press, New York, 1990)
22. L. Onsager, Phys. Rev. **65**, 117 (1944)
23. S.G. Brush, Revs. Mod. Phys. **39**, 883 (1967)
24. R.K. Pathria, *Statistical Mechanics*, 2nd edn., (Butterworths-Heinemann, Oxford, 1996). See especially Chapters 11 and 12
25. H.A. Kramers, G.H. Wannier, Phys. Rev. **60**, 252, 263 (1941)

Chapter 9
Chemical Equilibrium

This chapter is concerned with chemical equilibrium. The canonical ensemble is employed to obtain a statistical thermodynamic expression for the Gibbs energy of a pure polyatomic gas. This expression is then employed to obtain the change in the Gibbs energy for a reacting gas mixture and an expression for the equilibrium constant for a chemical reaction at constant pressure and temperature, thereby giving a statistical mechanical working relation for it. Four extended examples are examined in order to illustrate the use of the statistical expression. The relation between the constant-pressure and constant-concentration equilibrium constants is established. The chapter finishes with an application to the chemical kinetic rate constant for a chemical reaction that proceeds via a model two-step mechanism that employs an 'activated complex' or 'transition state' to facilitate completion of the reaction.

9.1 Setting the Stage

We shall begin by looking carefully at the form of the Gibbs energy for a pure gas. In order to do this, let us recall that we have seen in Eqs. (3.2.22) and (6.2.7) that the canonical partition function Z_N may be well approximated by

$$Z_N = \frac{1}{N!}(z)^N = \frac{1}{N!} z_{\text{tr}}^N z_{\text{rot}}^N z_{\text{vib}}^N z_{\text{el}}^N .$$ (9.1.1)

© Springer Nature Switzerland AG 2021

F. R. W. McCourt, *Statistical Thermodynamics for Pure and Applied Sciences*,
https://doi.org/10.1007/978-3-030-52006-9_9

The Helmholtz energy for a pure gas is therefore given by

$$A = -k_B T \ln Z_N$$

$$= -k_B T \ln \left(\frac{z^N}{N!} \right)$$

$$= -k_B T \ln \left(z^N \right) + k_B T \ln N!, \tag{9.1.2}$$

which, if we utilize the Stirling approximation for $N!$, can be approximated by

$$A \simeq -N k_B T - N k_B T \ln(z/N). \tag{9.1.3}$$

On a molar basis, we would hence obtain

$$\overline{A} + RT = -RT \ln(z/N_0), \tag{9.1.4}$$

in which $\overline{A}$ is the Helmholtz energy per mole of gas, and R is the universal gas constant, which is related to the Avogadro number N_0 and the Boltzmann constant k_B by $R = N_0 k_B$.

We can now write the Gibbs energy per mole as

$$\overline{G} \equiv \overline{A} + P\overline{V}; \qquad P\overline{V} = N_0 k_B T = RT. \tag{9.1.5}$$

The molar Gibbs energy is thus given in terms of the molecular partition function by

$$\overline{G} = -RT \ln(\overline{z}/N_0). \tag{9.1.6}$$

The natural variables for the Gibbs energy are temperature T and pressure P. The pressure dependence of $\overline{G}$ arises from $z_{tr}(V, T)$ through $\overline{V}$, so that

$$\overline{G} = -RT \ln \left\{ \left[\frac{2\pi M k_B T}{h^2} \right]^{\frac{3}{2}} \frac{\overline{V}}{N_0} z_{rot} z_{vib} \omega_{el} \right\}, \tag{9.1.7}$$

which, upon substituting for $\overline{V}$, becomes

$$\overline{G} = -RT \ln \left\{ \left[\frac{2\pi M k_B T}{h^2} \right]^{\frac{3}{2}} \frac{RT}{N_0 P^\circ} z_{rot} z_{vib} \omega_{el} \right\} + RT \ln \left(\frac{P}{P^\circ} \right), \tag{9.1.8}$$

with the standard pressure, $P^\circ \equiv 1$ bar. In all cases with which we shall be concerned, molecules will be in their ground electronic states; we have therefore replaced $z_{el}(T)$ by the corresponding ground electronic state degeneracy, ω_{el}.

As chemical equilibria are discussed in thermodynamics in terms of molar Gibbs energy changes under standard conditions, in particular, $P = P° = 1$ bar, we shall be concerned in this chapter largely with $\overline{G}°$, which is given by

$$\overline{G}° = -RT \ln \left\{ \left[\frac{2\pi M k_B T}{h^2} \right]^{\frac{3}{2}} \frac{RT}{N_0} z_{rot} z_{vib} \omega_{el} \right\} \equiv -RT \ln \left(\frac{\overline{z}°}{N_0} \right). \quad (9.1.9)$$

This quantity is referred to in thermodynamics as the *standard molar Gibbs energy*. As G is an extensive quantity, we can obtain $G°$ for ν moles of the substance from $\overline{G}°$ via

$$G° = \nu\overline{G}° = -\nu RT \ln(\overline{z}°/N_0). \quad (9.1.10)$$

9.2 Derivation of the Expression for K_P

We know from thermodynamics that the change in the Gibbs energy (evaluated at a standard pressure of 1 bar) is related to the equilibrium constant K_P for a chemically reacting mixture by

$$\Delta G° = -RT \ln K_P, \quad (9.2.1)$$

in which $\Delta G°$ is the difference between the standard Gibbs energies of the products and reactants, viz., $\Delta G° \equiv \sum G°_{prod} - \sum G°_{react}$.

If, for example, we are dealing with the reaction

$$\nu_A A + \nu_B B \quad \rightleftharpoons \quad \nu_C C + \nu_D D, \quad (9.2.2)$$

then $\Delta G°$ can be obtained from

$$\Delta G° = \nu_C \overline{G}°_C + \nu_D \overline{G}°_D - \nu_A \overline{G}°_A - \nu_B \overline{G}°_B. \quad (9.2.3)$$

If we now combine expression (9.1.9) for $\overline{G}°$ with the thermodynamic expression (9.2.1) relating $\Delta G°$ to the equilibrium constant K_P, we obtain the result

$$\Delta G° = -RT \ln \left\{ \frac{(\overline{z}°_C/N_0)^{\nu_C} (\overline{z}°_D/N_0)^{\nu_D}}{(\overline{z}°_B/N_0)^{\nu_B} (\overline{z}°_A/N_0)^{\nu_A}} \right\} = -RT \ln K_P. \quad (9.2.4)$$

When we solve Eq. (9.2.4) for K_P, we have the statistical mechanical expression

$$K_P = \frac{(\overline{z}°_C/N_0)^{\nu_C} (\overline{z}°_D/N_0)^{\nu_D}}{(\overline{z}°_B/N_0)^{\nu_B} (\overline{z}°_A/N_0)^{\nu_A}}. \quad (9.2.5)$$

Thus far, we have not had to specify the origin for the internal energies of the molecules involved in the chemical reaction. For a given temperature, all factors in expression (9.2.5) are constants, since all $\overline{z}^\circ$ factors are calculated for the *standard pressure* of 1 bar and not for the equilibrium pressure.

We must now consider the form for the molecular partition function $\overline{z}^\circ$, given as

$$\overline{z}^\circ = \left[\frac{2\pi M k_B T}{h^2} \right]^{\frac{3}{2}} \frac{RT}{P^\circ} z_{\mathrm{rot}} z_{\mathrm{vib}} z_{\mathrm{el}} \,. \tag{9.2.6}$$

We have seen earlier that there are three basic possibilities that can be considered as energy origins for diatomic molecules: the SSA zero, the bottom of the potential well, and the ground vibrational energy. We need to be concerned about this, as the choice of energy origin affects the way in which we represent the vibrational and electronic partition functions in our expression for K_P:

For the SSA energy origin, we have seen in Chap. 6 that $z_{\mathrm{el}}(T)$ and $z_{\mathrm{vib}}(T)$ are given by $z_{\mathrm{el}}(T) = \omega_{\mathrm{el}} e^{\beta \mathfrak{D}_e}$ (rather than simply ω_{el}) and $z_{\mathrm{vib}}(T) = z_{\mathrm{vib}}^{\mathrm{SHO}}(T)$, so that the energy origin for the vibrational partition function is the bottom of the potential well. We shall see, however, that it is more convenient to employ the ground vibrational energy as the zero of energy for the calculation of the vibrational partition function, with the consequence being that $\overline{z}^\circ$ has the form

$$\overline{z}^\circ = \widetilde{z}^\circ e^{\beta D_0} \,, \tag{9.2.7a}$$

in which D_0 is the dissociation energy, and $\widetilde{z}^\circ$ is given by

$$\widetilde{z}^\circ = \overline{z}_{\mathrm{tr}}^\circ z_{\mathrm{rot}} \widetilde{z}_{\mathrm{vib}} z_{\mathrm{el}} \,, \tag{9.2.7b}$$

with $\widetilde{z}_{\mathrm{vib}}$ given by Eq. (5.5.6), and $z_{\mathrm{el}} = \omega_{\mathrm{el}}$. Our expression for K_P must therefore be revised accordingly to

$$K_P = \frac{(\overline{z}_C^\circ/N_0)^{\nu_C} (\overline{z}_D^\circ/N_0)^{\nu_D}}{(\overline{z}_B^\circ/N_0)^{\nu_B} (\overline{z}_A^\circ/N_0)^{\nu_A}} \, e^{\beta \left(\nu_C D_0^C + \nu_D D_0^D - \nu_A D_0^A - \nu_B D_0^B \right)} \,, \tag{9.2.8}$$

with it now being understood that $\overline{z}^\circ$ is evaluated with $\widetilde{z}_{\mathrm{vib}}(T)$ and not with $z_{\mathrm{vib}}^{\mathrm{SHO}}(T)$.

Let us now consider a reaction as it would (hypothetically) occur with all reactants and products in standard states at 0 K, so that

$$\epsilon_{\mathrm{tr}}^i = 0 \,, \qquad \epsilon_{\mathrm{rot}}^i = 0 \,, \qquad \epsilon_{\mathrm{vib}} = -D_0^i \,.$$

Let us also imagine that the reactants are dissociated into component atoms which, at 0 K, would be in precisely the state that we have employed to define zero on the SSA scale. This process can be constructed as follows:

- the dissociation of ν_A molecules of A and ν_B molecules of B requires an energy *input* $\nu_A D_0^A + \nu_B D_0^B$;
- the recombination of the atoms at the zero on the SSA scale into product molecules has an energy *output* $-(\nu_C D_0^C + \nu_D D_0^D)$.

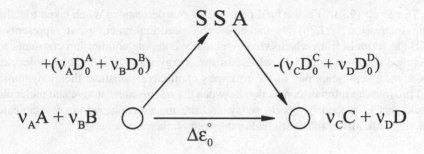

Fig. 9.1 Hess law cycle for reaction of two diatomic molecules at 0 K

If we now consult the Hess law cycle shown in Fig. 9.1, we see that

$$\Delta\epsilon_0^\circ \equiv \left(\nu_A D_0^A + \nu_B D_0^B\right) - \left(\nu_C D_0^C + \nu_D D_0^D\right)$$
$$= -\left(\nu_C D_0^C + \nu_D D_0^D - \nu_A D_0^A - \nu_B D_0^B\right), \qquad (9.2.9)$$

with the latter version written in the conventional products minus reactants format.

Equation (9.2.9) gives the result should the reaction occur on a truly microscopic scale. If we convert this reaction to the macroscopic scale, by allowing ν_A moles of A to react with ν_B moles of B to form ν_C moles of C and ν_D moles of D, then the net change in the macroscopic (or thermodynamic internal) energy U_0° is defined as

$$\Delta U_0^\circ \equiv N_0 \Delta\epsilon_0^\circ. \qquad (9.2.10a)$$

Equivalently, we may express $\Delta\epsilon_0^\circ$ in thermodynamic terms as

$$\Delta\epsilon_0^\circ = \frac{\Delta U_0^\circ}{N_0}, \qquad (9.2.10b)$$

for the relation connecting microscopic and macroscopic energies. This latter result then implies that

$$\frac{\nu_C D_0^C + \nu_D D_0^D - \nu_A D_0^A - \nu_B D_0^B}{k_B T} = -\frac{\Delta\epsilon_0^\circ}{k_B T} = -\frac{\Delta U_0^\circ}{RT}, \qquad (9.2.11)$$

as $U_0^\circ \equiv N_0 \epsilon_0^\circ$ and $R \equiv N_0 k_B$. Hence, we obtain

$$K_P = \frac{(\overline{z}_C^\circ/N_0)^{\nu_C}(\overline{z}_D^\circ/N_0)^{\nu_D}}{(\overline{z}_A^\circ/N_0)^{\nu_A}(\overline{z}_B^\circ/N_0)^{\nu_B}} e^{-\Delta\overline{U}_0^\circ/(RT)} \qquad (9.2.12)$$

as a working expression for $K_P(T)$.

The result (9.2.12) is the final outcome of our derivation. When taken together with expression (9.2.7b) for the molecular partition functions, it represents a working formula from which we may actually compute equilibrium constants for chemical reactions from a knowledge of the energy states of individual molecules. As such, it represents one of the triumphs of modern statistical thermodynamics and provides an intimate connection between the microscopic atomic and molecular world that is probed by spectroscopy and the macroscopic world of chemically reacting bulk matter that is studied in the field of chemical kinetics.

9.2.1 Some Practical Matters

In the particular case that reactant and product molecules are monatomic or diatomic, which is often the case, we can provide compact working expressions for the individual partition functions $z°$. Let us consider specifically the case in which all participants in the reaction are diatomic molecules. We shall examine the form taken by $\overline{z}°/N_0$, which can be written as

$$\frac{\overline{z}°}{N_0} = \frac{\overline{z}°_{\text{trans}}(T)}{N_0} z_{\text{rot}}(T) \, \widetilde{z}_{\text{vib}}(T) \, \omega_{\text{el}} \, .$$

If we recall that the translational component, $z_{\text{trans}}(T, V)$, has the form $V/\Lambda^3 = (2\pi M k_B T/h^2)^{\frac{3}{2}} V$, then for standard pressure, $P°$, we may write

$$\frac{\overline{z}°_{\text{tr}}(T)}{N_0} = \left(\frac{2\pi M k_B T}{h^2}\right)^{\frac{3}{2}} \frac{\overline{V}°}{N_0}$$

$$= \frac{(2\pi \overline{M} k_B T)^{\frac{3}{2}}}{h^3 N_0^{\frac{3}{2}}} \frac{\overline{V}°}{N_0} \, ,$$

which is a function of temperature T only, as $\overline{V}°$ is given according to the ideal gas law by

$$\overline{V}° = \frac{RT}{P°} \, .$$

Note also that $\overline{M}$ is the molecular weight expressed in kg mol^{-1} and that $R/P°$ has the value 83.1446×10^{-6} m^3 mol^{-1} K^{-1} for $P° = 1$ bar. Employment of this expression for $\overline{V}°$ allows $\overline{z}°_{\text{tr}}/N_0$ to be obtained as

$$\frac{\overline{z}°_{\text{tr}}}{N_0} = \frac{(2\pi k_B)^{\frac{3}{2}}}{h^3 N_0^{\frac{5}{2}}} \frac{R}{P°} \overline{M}^{\frac{3}{2}} T^{\frac{5}{2}} \, . \tag{9.2.13a}$$

Upon utilizing the values, $h = 6.62607015 \times 10^{-34}$ J s for the Planck constant, $k_B = 1.380649 \times 10^{-23}$ J K^{-1} for the Boltzmann constant, $N_0 = 6.02214076 \times 10^{23}$ mol^{-1} for the Avogadro number, and $R \simeq 0.0831446$ dm^3 bar K^{-1}mol^{-1} for the universal gas constant, the collection of fundamental constants can be evaluated once and for all, to give 820.5094 in SI units, so that the translational partition function is given by

$$\frac{\overline{z}^\circ_{tr}}{N_0} = 820.5094 \overline{M}^{\frac{3}{2}} T^{\frac{5}{2}} . \tag{9.2.13b}$$

This form for the translational partition function then allows us to express $\overline{z}^\circ/N_0$ for diatomic molecules generically as

$$\frac{\overline{z}^\circ}{N_0} = 820.5094 \, \overline{M}^{\frac{3}{2}} \, T^{\frac{5}{2}} \, \frac{T}{\sigma \Theta_{rot}} \, \frac{1}{1 - e^{-\Theta_{vib}/T}} \, \omega_{el} . \tag{9.2.14}$$

The characteristic temperatures Θ_{rot} and Θ_{vib} are defined in terms of the molecular moment-of-inertia I and characteristic vibrational frequency ν_{osc} by

$$\Theta_{rot} \simeq \frac{40.2753 \times 10^{-47}}{I} \, \text{K} , \quad \text{or} \quad \Theta_{rot} = \frac{\overline{B}_0}{k_B} \, \text{K} , \tag{9.2.15a}$$

and

$$\Theta_{vib} = \frac{\nu_0}{2.08366 \times 10^{10}} \, \text{K} \quad \text{or} \quad \Theta_{vib} = \frac{\overline{\nu}_0}{k_B} \, \text{K} , \tag{9.2.15b}$$

with $\overline{B}_0$, $\overline{\nu}_0$, and $\overline{k}_B$ all in cm^{-1}, and ν_0 in s^{-1}; equivalently, they may be obtained from tabulations, such as Table 6.1, or directly from the experimental spectroscopic constants.

9.2.2 Relation Between $K_P(T)$ and $K_c(T)$

Equilibrium constants are often written in physical chemistry in terms of the molar concentrations of the reacting species, rather than in terms of pressure. For ideal gases and indeed, for gases in general, it proves to be more useful to deal with the equilibrium constant $K_P(T)$ rather than with $K_c(T)$. It will nonetheless be useful to consider the relationship between these two equilibrium constants for ideal gases.

We may write the ideal gas equation in terms of concentration as

$$P = \frac{n}{V} RT \equiv cRT ,$$

with c being the concentration in moles per unit volume (normally cubic decimetres, dm^3). The equilibrium constant $K_P(T)$ for reaction (9.2.2) becomes

$$K_P \equiv \frac{P_C^{\nu_C} P_D^{\nu_D}}{P_A^{\nu_A} P_B^{\nu_B}} = \frac{c_C^{\nu_C} c_D^{\nu_D}}{c_A^{\nu_A} c_B^{\nu_B}} \left(\frac{RT}{P^\circ}\right)^{\nu_C + \nu_D - \nu_A - \nu_B}. \tag{9.2.16}$$

In this expression, the concentrations must be related to some standard concentration c° (which plays the same role for concentrations that P° plays for pressures). Thus, $K_P(T)$ will be related to $K_c(T)$ by

$$K_P(T) = K_c(T) \left(\frac{c^\circ RT}{P^\circ}\right)^{\nu_C + \nu_D - \nu_A - \nu_B}, \tag{9.2.17a}$$

with $K_c(T)$ given as

$$K_c(T) = \frac{(c_C/c^\circ)^{\nu_C}(c_D/c^\circ)^{\nu_D}}{(c_A/c^\circ)^{\nu_A}(c_B/c^\circ)^{\nu_B}}. \tag{9.2.17b}$$

Both K_P and K_c are unitless: the choice for P° or c° determines which units are to be associated with the universal gas constant R. For example, P° is typically chosen to be 1 bar, while the most common choice for c° is $1\,mol\,dm^{-3}$ ($1\,dm^3 \equiv$ 1 litre): for these choices of P° and c°, the factor $c^\circ RT/P^\circ$ becomes RT in unit $dm^3\,bar\,mol^{-1}$, so that R must be given as $R \simeq 0.083145\,dm^3\,bar\,mol^{-1}\,K^{-1}$. It is not uncommon to find the relation (9.2.17a) expressed as

$$K_P(T) = K_c(T)(RT)^{\nu_C + \nu_D - \nu_A - \nu_B} \tag{9.2.17c}$$

with it being implicitly understood that $c^\circ = 1\,dm^3\,mol^{-1}$ and $P^\circ = 1$ bar. For an ideal gas, the volume dependence of a molecular partition function is contained entirely in the translational component, so that $z(T, V)/V$ is independent of volume and depends only upon temperature.

At constant T and V, the total differential for the Helmholtz energy may be written as

$$dA = \mu_A dn_A + \mu_B dn_B + \mu_C dn_C + \mu_D dn_D.$$

From the perspective of the extent of reaction, in which the numbers of moles of reactants and products can be written as $n_A = n_A^{(0)} - \nu_A \xi$, $n_B = n_B^{(0)} - \nu_B \xi$, $n_C = n_C^{(0)} + \nu_C \xi$, and $n_D = n_D^{(0)} + \nu_D \xi$, we find that for chemical equilibrium, for which $dA = 0$, we have

$$dA = [\nu_C \mu_C + \nu_D \mu_D - \nu_A \mu_A - \nu_B \mu_B]d\xi = 0,$$

or

$$\nu_C \mu_C + \nu_D \mu_D - \nu_A \mu_A - \nu_B \mu_B = 0. \tag{9.2.18}$$

The chemical potential for species k is given by

$$\mu_k = -RT \left(\frac{\partial \ln Z}{\partial N_k} \right)_{N_j \neq N_k, T, V}$$

$$= -RT \ln \left(\frac{z_k(T, V)}{N_k} \right)_{T, V}.$$

This expression for μ_k is required by the thermodynamic definition of the chemical potential μ_k as the partial derivative of the Helmholtz energy with respect to the number of moles of chemical species k at constant temperature, constant volume, and constant numbers of moles of chemical species other than species k. Its employment for the chemical potentials $\mu_k(T, V)$ in the equilibrium requirement (9.2.18) then gives

$$\frac{N_C^{\nu_C} N_D^{\nu_D}}{N_A^{\nu_A} N_B^{\nu_B}} = \frac{(z_C)^{\nu_C} (z_D)^{\nu_D}}{(z_A)^{\nu_A} (z_B)^{\nu_B}}.$$

If we convert this expression into the corresponding ratio of number densities[1] $\rho_k \equiv N_k/V$ of the equilibrium mixture components, we obtain K_c as

$$K_c(T) = \frac{\rho_C^{\nu_C} \rho_D^{\nu_D}}{\rho_A^{\nu_A} \rho_B^{\nu_B}} = \frac{(z_C/V)^{\nu_C} (z_D/V)^{\nu_D}}{(z_A/V)^{\nu_A} (z_B/V)^{\nu_B}}. \tag{9.2.19}$$

That this expression for $K_c(T)$ is, in fact, the same as K_c given in Eq. (9.2.17b) can be seen from the relation $c_k = \rho_k/N_0$. Moreover, for an ideal gas, $K_c(T)$ is indeed a function of temperature only.

9.3 Four Extended Examples

Example 9.1 Dissociation of a homonuclear diatomic molecule into its constituent atoms.

The dissociation of a gas phase homonuclear diatomic molecule into its constituent atoms is represented by the chemical formula

$$X_2(g) \rightleftharpoons 2X(g). \tag{9.3.1}$$

---

[1] The symbol ρ has been employed here for number density (rather than for mass density) instead of n because n has already been utilized to designate the number of moles.

The equilibrium constant K_p for this gas-phase reaction can be written simply as

$$K_P = \left(\frac{\bar{z}_X^\circ}{N_0}\right)^2 \frac{\bar{z}_{X_2}^\circ}{N_0}\, e^{-\Delta \bar{U}_0^\circ / RT}\,, \tag{9.3.2}$$

with $\Delta \bar{U}_0^\circ$ given by

$$\Delta \bar{U}_0^\circ = N_0 \left[\sum D_0(\text{prod}) - \sum D_0(\text{react})\right] = N_0 D_0(X_2)\,.$$

Note that this expression indicates that our convention is that $D_0 \equiv 0$ for atomic species.

In this case, the partition functions are given by

$$\frac{\bar{z}_X^\circ}{N_0} = 820.5094\, \overline{M}_X^{\frac{3}{2}} T^{\frac{5}{2}}\, \omega_{\text{el}}^X\,, \tag{9.3.3}$$

$$\frac{\bar{z}_{X_2}^\circ}{N_0} = 820.5094\, \overline{M}_{X_2}^{\frac{3}{2}} T^{\frac{5}{2}}\, \frac{T}{\sigma\,\Theta_{\text{rot}}}\, \frac{e^{\beta D_0}}{1 - e^{-\Theta_{\text{vib}}/T}}\, \omega_{\text{el}}^{X_2}\,, \tag{9.3.4}$$

respectively, for the monatomic and diatomic species. When these expressions for the partition functions are substituted into Eq. (9.2.12) for K_P, we obtain

$$K_P = 820.5094\, \frac{\overline{M}_X^3 T^{\frac{3}{2}} \left(\omega_{\text{el}}^X\right)^2}{\overline{M}_{X_2}^{\frac{3}{2}}\, \omega_{\text{el}}^{X_2}}\, \sigma\,\Theta_{\text{rot}} \left(1 - e^{-\Theta_{\text{vib}}/T}\right) e^{-\beta D_0(X_2)}\,. \tag{9.3.5}$$

This example also affords us the opportunity to study the temperature dependence of the equilibrium constant from first principles and to make yet another connection with thermodynamics.

We can write the internal energy change at temperature T in terms of the internal energy change at 0 K together with the internal energy changes associated with the various degrees of freedom for the constituents of the chemical reaction. This can be illustrated readily from a Hess cycle type of diagram, as seen in Fig. 9.2.

By utilizing this Hess cycle-type diagram for the dissociation of X_2, we may obtain the change ΔU_T° for the dissociation reaction at temperature T in terms of ΔU_0° at 0 K, plus the reactant and product internal energies.

The internal energy U_T for X_2 can be expressed as

$$U_T(X_2) = U_{T,\text{trans}}(X_2) + U_{T,\text{rot}}(X_2) + U_{T,\text{vib}}(X_2) + U_{T,\text{el}}(X_2)$$

$$\overline{U}_T(X_2) = \tfrac{3}{2}RT + RT + \frac{R\Theta_{\text{vib}}}{e^{\Theta_{\text{vib}}/T} - 1} + 0\,,$$

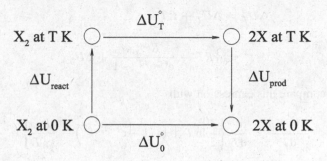

Fig. 9.2 Hess law diagram for dissociation of a homonuclear diatomic molecule

the final zero appears because the ground electronic state for the I_2 molecule is a 'singlet-sigma' state, specifically $X\ ^1\Sigma_g^+$, with $\omega_0 = 1$. Thus, for the reaction $X_2(g)\ \rightleftharpoons\ 2X(g)$, we obtain

$$\Delta U_T^\circ = \Delta U_0^\circ + 2U_T(X) - U_T(X_2)$$

$$= \Delta U_0^\circ + \tfrac{1}{2}RT - \frac{R\Theta_{vib}}{e^{\Theta_{vib}/T} - 1}.$$

Chemists, for example, typically speak in terms of enthalpy changes rather than internal energy changes: from the thermodynamic definition of enthalpy as

$$H = U + PV\,,$$

we may write the corresponding enthalpy change as

$$\Delta \overline{H}_T^\circ = \Delta \overline{U}_T^\circ + \Delta(PV)_T$$

$$= \Delta \overline{U}_T^\circ + RT\Delta\nu\,,$$

in which the subscript 'T' on the product PV indicates that we are considering an *isothermal reaction* involving nominally ideal gases. This is, of course, the most common condition for experiments in chemical kinetics. For nominally ideal gases, the quantity $\Delta(PV)_T$ is given simply by $RT\Delta\nu$, in which $\Delta\nu$ represents the *net change in the number of moles* in the balanced chemical reaction. The enthalpy change ΔH_T° can therefore be written as

$$\Delta H_T^\circ = \Delta U_T^\circ + RT\Delta\nu\,. \tag{9.3.6}$$

For the particular example with which we are working, i.e., $X_2\ \rightleftharpoons\ 2X$, $\Delta\nu = 1$, and

$$\Delta H_T^\circ = \Delta U_T^\circ + RT$$

$$= \tfrac{3}{2}RT - \frac{R\Theta_{\text{vib}}}{e^{\Theta_{\text{vib}}/T} - 1} + \Delta U_0^\circ . \tag{9.3.7}$$

Let us now compare this expression with

$$\frac{\mathrm{d}\ln K_P}{\mathrm{d}T} = \frac{\mathrm{d}}{\mathrm{d}T}\left\{ \frac{3}{2}\ln T + \ln\left(1 - e^{-\Theta_{\text{vib}}/T}\right) - \frac{D_0}{k_{\text{B}}T} \right\}$$

$$= \frac{3}{2T} - \frac{\Theta_{\text{vib}}}{T^2}\frac{1}{e^{\Theta_{\text{vib}}/T} - 1} + \underbrace{\frac{D_0}{k_{\text{B}}T^2}}_{\Delta U_0^\circ/RT^2} , \tag{9.3.8}$$

so that we obtain the relation

$$\frac{\mathrm{d}\ln K_P}{\mathrm{d}T} = \frac{\Delta H_T^\circ}{RT^2} , \tag{9.3.9}$$

which is known from classical thermodynamics as the Van 't Hoff law of chemical kinetics.   □

Example 9.2 Temperature dependence of the dissociation of I_2.

Our second example will be a more explicit calculation of the equilibrium constant for the dissociation of iodine, as a particular reaction for which we can readily carry out a complete detailed calculation in order to compare our calculated results with experiment. We shall carry out our first calculation at a temperature of 1000 °C (or 1273 K). The chemical reaction is

$$I_2(g) \quad \rightleftharpoons \quad 2I(g) .$$

The SI atomic weight of iodine is $\overline{M}_I = 0.127\,\text{kg mol}^{-1}$, while the SI molecular weight of I_2 is $\overline{M}_{I_2} = 0.254\,\text{kg mol}^{-1}$. The iodine atom has a $^2P_{\frac{3}{2}}$ ground state, so that $\omega_{\text{e}1}^I = 4$. The iodine molecule, with its $^1\Sigma_g^+$ ground electronic state, has $\omega_{\text{e}1}^{I_2} = 1$, while the symmetry number σ is 2 because I_2 is a homonuclear diatomic molecule.

The spacing between the lines in the pure rotational Raman spectrum gives us the required information to obtain the moment-of-inertia of the I_2 molecule: the experimental value is $I = 750 \times 10^{-47}\,\text{kg m}^2$, so that $\Theta_{\text{rot}} = 0.0537$ K. Similarly, from the vibrational Raman spectrum, we may obtain the fundamental oscillator frequency $\nu_{\text{osc}} = 6.41 \times 10^{12}\,\text{s}^{-1}$, from which we obtain $\Theta_{\text{vib}} = 308$ K. Finally, from spectroscopy, we also obtain the value of D_0 which, for molecular iodine, is $2.4712 \times 10^{-19}\,\text{J mol}^{-1}$.

We may now employ these data in Eqs. (9.2.13) and (9.2.12) to calculate the value of K_P for the dissociation of molecular iodine at temperature $T = 1000\,°C$ (or 1273 K). The exponential factor in Eq. (9.2.12) is calculated at $T = 1273$ K to be

$$e^{\beta D_0(I_2)} \simeq e^{-2.4712 \times 10^{-19}/(1.38066 \times 10^{-23} \times 1273)} = e^{-14.060},$$

or

$$e^{-\beta D_0(I_2)} \simeq 7.828 \times 10^{-7}.$$

Hence, we find that K_P for this reaction at $T = 1273$ K becomes

$$K_P = 820.5094 \, \frac{(0.127)^3 (1273)^{\frac{3}{2}}}{(0.254)^{\frac{3}{2}}} \left(\frac{4^2}{1}\right) 2(0.0537)(0.215)\left(7.85 \times 10^{-7}\right)$$

$$\simeq 1.725 \times 10^{-1}$$

$$\approx 0.172.$$

This result may be compared with the experimental value 0.167 obtained for $K_P(T = 1273\,\text{K})$.

We should note here that, as we have made no allowance in our calculations for nonideality of the gases, anharmonicity of vibrational motion, partial coupling of vibrational and rotational motions, centrifugal distortion, and so on, we should consider agreement to within 3% to be quite satisfactory.

Let us now consider the temperature variation of K_P. We have seen earlier that

$$\frac{d \ln K_P}{dT} = \frac{3}{2T} - \frac{\Theta_{vib}}{T^2} \frac{e^{-\Theta_{vib}/T}}{1 - e^{-\Theta_{vib}/T}} + \frac{D_0}{k_B T^2} = \frac{\Delta H_T^\circ}{RT^2}. \tag{9.3.10}$$

We may obtain a fairly simple expression for the temperature dependence of K_P provided that the variation of $\Delta \overline{H}_T^\circ$ with T can be neglected. This expression is simply

$$\ln \left\{ \frac{K_P(T_2)}{K_P(T_1)} \right\} = -\frac{\Delta H_T^\circ}{R} \left(\frac{1}{T_2} - \frac{1}{T_1} \right). \tag{9.3.11}$$

We stress that this result holds only when the variation of ΔH_T° with T may be neglected (i.e., when we have taken $\Delta H_T^\circ = \Delta H^\circ$ as constant, or approximately so, over a temperature interval containing T_1 and T_2). Notice that for the case that we are presently considering, we have an explicit expression for ΔH_T° in terms of spectroscopically determined molecular constants, namely

$$\Delta H_T^\circ = N_0 \left\{ \frac{3}{2} k_B T - \frac{k_B \Theta_{vib} e^{-\Theta_{vib}/T}}{1 - e^{-\Theta_{vib}/T}} + D_0 \right\}, \tag{9.3.12}$$

and that over the temperature range 800–1300 K, ΔH_T° varies by no more than 1%, so that taking it to be constant does not introduce a large error into our calculation of K_P.

Let us take $T = 1273$ K as a representative temperature for the temperature range 800–1500 K, and calculate $\Delta \overline{H}_T^\circ$ at this temperature via

$$\Delta \overline{H}_T^\circ = \tfrac{3}{2}RT - \frac{R\Theta_{\mathrm{vib}}}{e^{\Theta_{\mathrm{vib}}/T} - 1} + \Delta \overline{U}_0^\circ. \qquad (9.3.13)$$

For this, we also need $\Delta \overline{U}_0^\circ$, which may be determined from the spectroscopic dissociation energy D_0 for I_2 as $\Delta \overline{U}_0^\circ = 6.02214 \times 10^{23} \times 2.4712 \times 10^{-19}\,\mathrm{J\,mol^{-1}}$ $= 148{,}819\,\mathrm{J\,mol^{-1}}$. Substitution of the relevant data into Eq. 9.3.13 then gives us the result

$$\Delta \overline{H}_T^\circ = \tfrac{3}{2}(8.31447 \times 1273) - \frac{8.31447 \times 208}{e^{308/1273} - 1} + 148{,}819$$

$$= 15{,}876.5 - 9{,}355.5 + 148{,}819\,\mathrm{J\,mol^{-1}},$$

or $\Delta \overline{H}_T^\circ = 155{,}340\,\mathrm{J\,mol^{-1}}$.

Upon taking the value $\Delta H_T^\circ = 2.57948 \times 10^{-19}$ J/particle, we obtain

$$\log_{10} \frac{K_P(T_2)}{0.173} = -\frac{\Delta \overline{H}_T^\circ \log_{10} e}{R}\left(\frac{1}{T_2} - \frac{1}{1273}\right)$$

$$= -\frac{0.43429 \Delta \overline{H}_T^\circ}{R}\left(\frac{1}{T_2} - \frac{1}{1273}\right)$$

$$= -\frac{8113.9}{T_2} + 6.3781$$

for the temperature dependence of K_P. Thus, at $T = 1200\,^\circ\mathrm{C}$ (1473 K), we obtain

$$\log_{10}\left[\frac{K_P(T = 1473\,\mathrm{K})}{0.173}\right] = -5.5084 + 6.3781 = 0.8696_8,$$

from which we see that $K_P(T = 1473\,\mathrm{K}) = 1.28$. Unfortunately, there is no experimental value with which we may compare our calculated result at this temperature. However, based upon our comparison with the experimental result at 1273 K, we should expect a difference of the order of 4–5% due to the various approximations that we have made. At $T = 800\,^\circ\mathrm{C}$ (1073 K), we obtain

$$\log_{10}\left[\frac{K_P(T = 1073\,\mathrm{K})}{0.173}\right] = -7.5619 + 6.3781 = -1.1838,$$

from which we obtain the value 0.0113 for $K_P(T = 1073\,\text{K})$. In this case, we do have an experimental value, $K_P = 0.0114$, with which we may compare our computed result. If we recall that many of the approximations that we have utilized in obtaining our calculated result for K_P are worse the higher we go in temperature, this improved level of agreement between theory and experiment indicates that it is indeed likely that much of the difference between our calculated value at 1073 K and the experimental value is due to the approximations that we have made in obtaining our calculated result. □

Example 9.3 Ionization of an alkali metal atom.

Consider the ionization of an alkali metal atom M to give a free electron plus the M^+ ion,

$$ M\!\left(^2S_{\frac{1}{2}}\right) \;\overset{I_p}{\rightarrow}\; M^+\!\left(^1S_0\right) + e^- , $$

in which I_p is the ionization potential for the reaction. The free electron is a spin-$\frac{1}{2}$ fermion which has only translational energy; M and M^+ will have nuclear spin degeneracies in common, and if we ignore the difference in mass between M and M^+, the translational partition functions for M and M^+ will cancel out in the expression

$$ K_P = \frac{z^\circ_{M^+} z^\circ_{e^-}}{z^\circ_M}\, \frac{1}{N_0}\, e^{-\beta I_p} \tag{9.3.14} $$

for K_P. We then obtain

$$ K_P = \frac{2(2\pi m_e k_B T)^{\frac{3}{2}}\,\overline{V}^\circ}{h^3}\, \frac{z_{\text{el,M}^+}}{z_{\text{el,M}}}\, \frac{1}{N_0}\, e^{-\beta I_p} . \tag{9.3.15} $$

Now, since $\omega_{\text{el}}(M) = 2$, and $\omega_{\text{el}}(M^+) = 1$, while $\overline{V}^\circ = N_0 k_B T/P^\circ$, we have

$$ K_P = \frac{(2\pi m_e k_B T)^{\frac{3}{2}}}{h^3}\, \frac{k_B T}{P^\circ}\, e^{-\beta I_p} $$

$$ = \frac{1}{h^3}\,(2\pi)^{\frac{3}{2}} k_B^{\frac{5}{2}} m_e^{\frac{3}{2}} T^{\frac{5}{2}} e^{-\beta I_p} $$

$$ \simeq \frac{(15.75)(7.08 \times 10^{-58})}{2.909 \times 10^{-100}\, P^\circ}\, m_e^{\frac{3}{2}} T^{\frac{5}{2}} e^{-\beta I_p} , \tag{9.3.16} $$

or

$$K_P \simeq 38.35 \times 10^{42} \, m_e^{\frac{3}{2}} \, T^{\frac{5}{2}} e^{-\beta I_p}$$

$$= 3.334 \times 10^{-2} \, T^{\frac{5}{2}} e^{-\beta I_p} . \tag{9.3.17}$$

Our final form for K_P is then the expression

$$K_P = 0.03334 \, T^{\frac{5}{2}} \, e^{-\beta I_p} . \tag{9.3.18}$$

K_P for the process $M \to M^+ + e^-$ may also be written (as in first year chemistry) in terms of partial pressures as

$$K_P = \frac{(P_{M^+}/P^\circ)(P_{e^-}/P^\circ)}{(P_M/P^\circ)} . \tag{9.3.19}$$

Naturally, the thermal ionization process is, in general, not 100% efficient, and only a fraction of the atoms will be ionized at a given temperature T. The expression for the ionization process can then be written in terms of the fraction α of the M atoms that are ionized as

$$(1 - \alpha)M \quad \to \quad \alpha M^+ + \alpha e^- , \tag{9.3.20}$$

from which we obtain $P_{M^+} = \alpha P$, $P_{e^-} = \alpha P$, and $P_M = (1 - \alpha)P$, with P the total gas pressure. In this way, we find that K_P and α are related by

$$K_P = \frac{\alpha^2}{1 - \alpha} \frac{P}{P^\circ} \simeq \alpha^2 \frac{P}{P^\circ} , \qquad \text{for } \alpha \ll 1 . \tag{9.3.21}$$

From this result, we obtain the expression

$$\alpha = K_P^{\frac{1}{2}} \left(\frac{P^\circ}{P} \right)^{\frac{1}{2}} = 0.1826 \left(\frac{P^\circ}{P} \right)^{\frac{1}{2}} T^{\frac{5}{4}} e^{-\beta I_p/2} \tag{9.3.22}$$

for the degree of ionization α of M atoms. This equation is known in plasma physics as the Saha equation.

If we consider the specific example of Cs, we have $I_p = 6.237 \times 10^{-19}$ J/atom, and βI_p is given by

$$\beta I_p = \frac{6.237 \times 10^{-19} \, \text{J}}{1.3807 \times 10^{-23} \, \text{J K}^{-1} \, \text{T}} \simeq \frac{4.518 \times 10^4 \, \text{K}}{T} . \tag{9.3.23}$$

Thus, for this particular alkali metal, we obtain α as

$$\alpha = 0.1826 \left(\frac{P^\circ}{P}\right)^{\frac{1}{2}} T^{\frac{5}{4}} e^{-2.259 \times 10^4 / T} \tag{9.3.24}$$

for arbitrary temperature T. At 1500 K and a total gas pressure of 0.01313 bar, for example, we obtain $\alpha(\text{Cs}) = 0.0136$. ☐

Example 9.4 A simple isotopic exchange reaction.

Isotopic exchange reactions are particularly simple examples of chemical reactions that still illustrate the utility of the statistical mechanical expression for the equilibrium constant. We shall obtain an expression for the equilibrium constant, $K_P(T)$, for the isotopic exchange reaction between two homonuclear chlorine isotopologues, specifically,

$$^{35}\text{Cl}_2 + {}^{37}\text{Cl}_2 \rightleftharpoons 2\,{}^{35}\text{Cl}^{37}\text{Cl}.$$

The starting point for our determination of a final expression for K_P for this isotopic exchange reaction is Eq. (9.2.12). If we identify A, B, and C $\equiv$ D as $^{35}\text{Cl}_2$, $^{37}\text{Cl}_2$, and $^{35}\text{Cl}^{37}\text{Cl}$, respectively, we obtain

$$K_P = \frac{\left[\overline{z}^\circ\left({}^{35}\text{Cl}^{37}\text{Cl}\right)\right]^2}{\overline{z}^\circ\left({}^{35}\text{Cl}_2\right)\overline{z}^\circ\left({}^{37}\text{Cl}_2\right)} e^{-\Delta\overline{U}_0^\circ/(RT)}.$$

Upon recalling (1), that we may write each of the molecular partition functions $\overline{z}^\circ$ as a product of the individual translational, rotational, vibrational, and electronic partition functions ($\overline{z}^\circ_{\text{trans}}$, z_{rot}, z_{vib}, and z_{el}) and (2), that in obtaining Eq. (9.2.12), we have employed the electronic partition functions and the zero-point vibrational energies of the participating molecules to obtain the exponential factor involving $\Delta\overline{U}_0^\circ$, and we see that $K_P(T)$ takes the form

$$K_P(T) = \left(\frac{\left[\overline{M}\left({}^{35}\text{Cl}^{37}\text{Cl}\right)\right]^2}{\overline{M}\left({}^{35}\text{Cl}_2\right)\overline{M}\left({}^{37}\text{Cl}_2\right)}\right)^{\frac{3}{2}} \frac{\sigma\left({}^{35}\text{Cl}_2\right)\Theta_{\text{rot}}\left({}^{35}\text{Cl}_2\right)\sigma\left({}^{37}\text{Cl}_2\right)\Theta_{\text{rot}}\left({}^{37}\text{Cl}_2\right)}{\left[\sigma\left({}^{35}\text{Cl}^{37}\text{Cl}\right)\Theta_{\text{rot}}\left({}^{35}\text{Cl}^{37}\text{Cl}\right)\right]^2}$$

$$\times \frac{\left[\overline{z}_{\text{vib}}\left({}^{35}\text{Cl}^{37}\text{Cl}\right)\right]^2}{\overline{z}_{\text{vib}}\left({}^{35}\text{Cl}_2\right)\overline{z}_{\text{vib}}\left({}^{37}\text{Cl}_2\right)} \frac{\left[\omega_{\text{el}}\left({}^{35}\text{Cl}^{37}\text{Cl}\right)\right]^2}{\omega_{\text{el}}\left({}^{35}\text{Cl}_2\right)\omega_{\text{el}}\left({}^{37}\text{Cl}_2\right)} e^{-\Delta\overline{U}_0^\circ/(RT)},$$

in which $\overline{M}$ represents molecular weight, and $\overline{z}_{\text{vib}}$ is defined in Eq. (5.5.6).

Although spectroscopic results for Θ_{rot}, Θ_{vib}, and D_0 are readily available for $^{35}\text{Cl}_2$, they are generally more difficult to obtain for the other chlorine isotopologues. It will suffice here for us to work within the spirit of the Born–Oppenheimer approximation, in which the equilibrium separation r_e, the molecular force constant κ, and chemical dissociation energy D_e are the same for all isotopologues, so that we may therefore deduce the values of the characteristic rotational and vibrational temperatures for the heteronuclear and $^{37}Cl_2$ homonuclear isotopologues of Cl_2 from the experimental values for the $^{35}Cl_2$ isotopologue. We could, in principle, also deduce values for the spectroscopic dissociation energy, D_0, for these isotopologues. However, it is relatively difficult to obtain very accurate values for D_0 without carrying out an elaborate fit of the potential energy function to an extensive set of accurate spectroscopic data. As even a small error in the value of $D_0(^{35}Cl_2)$ could lead to a significant error in the values for $K_P(T)$, we shall attempt to avoid doing this here.

To this end, we recall that the characteristic rotational temperature Θ_{rot} for a linear molecule A is defined as

$$\Theta_{rot}(A) \equiv \frac{\hbar^2}{2m_{r,A}r_e^2(A)},$$

with $m_{r,A}$ denoting the reduced mass of A. Because all three isotopologues have the same value of r_e, the product $m_r\Theta_{rot} = \hbar^2/(2r_e)$ is also fixed for the other isotopologues, consistent with the Born–Oppenheimer approximation. Thus, from a knowledge of the reduced masses of the various isotopologues and the value of Θ_{rot} for one of them, the others may be determined from

$$\Theta_{rot}(B) = \frac{m_{r,A}}{m_{r,B}} \Theta_{rot}(A).$$

In a similar manner, given that the characteristic vibrational temperature for isotopologue A is given by $\Theta_{vib}(A) \equiv (2\pi)^{-1}\sqrt{\kappa/m_{r,A}}$, we have a means of determining Θ_{vib} for the other isotopologues from that for isotopologue A, namely,

$$\Theta_{vib}(B) = \sqrt{\frac{m_{r,A}}{m_{r,B}}} \Theta_{vib}(A).$$

Finally, as the spectroscopic dissociation energy D_0 and the chemical dissociation energy $\mathfrak{D}_e$ (i.e., the depth of the potential well) are related to Θ_{vib} by $-D_0 = -\mathfrak{D}_e + \frac{1}{2}k_B\Theta_{vib}$, we also have a relation connecting the spectroscopic dissociation energies of two isotopologues and their respective characteristic vibrational temperatures, namely,

$$D_0(B) = D_0(A) + \tfrac{1}{2}k_B[\Theta_{vib}(A) - \Theta_{vib}(B)].$$

However, rather than employing the values of D_0 evaluated in this fashion, then substituting them into the expression for $\Delta\overline{U}_0^\circ$, we see that we may directly obtain a Born–Oppenheimer expression for $\Delta\overline{U}_0^\circ$ for our isotopologues reaction as

Table 9.1 Characteristic values for Cl_2 isotopologues.

Molecule	Θ_{rot}/K	Θ_{vib}/K	D_0/eV	m_r /amu
$^{35}Cl_2$	0.3511	798.18	2.4794	17.4844
$^{37}Cl_2$	0.3711	776.33	2.4804	18.4826
$^{35}Cl^{37}Cl$	0.3608	787.33	2.4799	17.9696

$$\Delta \overline{U}_0^\circ = 2D_0(C) - D_0(A) - D_0(B)$$
$$= \tfrac{1}{2}k_B[2\Theta_{vib}(C) - \Theta_{vib}(A) - \Theta_{vib}(B)].$$

The three relations giving Θ_{rot}, Θ_{vib}, and D_0 from values measured for the $^{35}Cl_2$ isotopologue provide us with the values shown in Table 9.1.

These tabulated values have been obtained from the experimental values $D_0 = 2.4794\,eV$, $\overline{\nu}_{osc} = 556.38\,cm^{-1}$, and $\overline{B}_e = 0.2440\,cm^{-1}$ for the mass number 35 homonuclear isotopologue of Cl_2, together with the isotopic masses $m(^{35}Cl) = 34.9688\,amu$ and $m(^{37}Cl) = 36.9651\,amu$. Thus, we find for the chlorine isotopic exchange reaction that the translational and rotational factors in $K_P(T)$ have the values 1.0021 and 4(1.0009), with the factor 4 deriving from the product of the symmetry factors σ for the isotopologues. The temperature dependence is thus associated with the vibrational factor and the exponential of $\Delta \overline{U}_0^\circ/(RT)$: both of these factors are only weakly temperature-dependent, with the latter given by $e^{-0.1084/T}$, and the former varying, for example, from 0.9998 for $T = 300\,K$ to 0.9997 for $T = 800\,K$. The net result is that $K_P(T)$ is very nearly constant, with values 4(1.0024) and 4(1.0026) at these two temperatures. In fact, this behaviour is typical of isotopic exchange reactions between two homonuclear isotopologues, apart from the hydrogen isotopologues, for which the mass factors play a significantly larger role. $\quad\square$

9.4 Application to Chemical Kinetics

The 'activated complex theory' approach to chemical kinetics contrasts with the 'collision theory' approach, as it is based upon a more macroscopic (or 'thermodynamic') viewpoint of reaction rate constants. Specifically, it is founded upon an assumption that *all* reactions proceed by an effective two-step mechanism, namely,

$$A + B \;\overset{k_f}{\rightleftharpoons}\; (AB)^{\ddagger} \tag{1}$$

$$(AB)^{\ddagger} \;\overset{k}{\longrightarrow}\; products \tag{2}$$

to give

$$A + B \;\overset{k_2}{\longrightarrow}\; products,$$

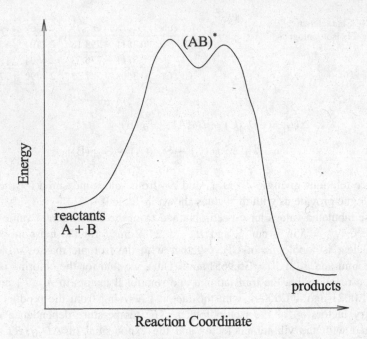

Fig. 9.3 Typical reaction coordinate path for a chemical reaction

in which the first step may be treated as an equilibrium reaction between the 'activated complex' $(AB)^{\ddagger}$ (represented as $(AB)^{*}$ in Fig. 9.3) and the reactants A and B, while the second step consists of a unimolecular decay of a particular activated complex to give the products.[2] The final equation in the sequence above thus represents the net reaction of A and B to give products, with a kinetic rate constant k_2, where in this case, the subscript '2' indicates that the overall reaction order is 2.

The equilibrium assumption in step (1) above is governed by an equilibrium constant $K_{eq}^{\ddagger}$ given by

$$K_{eq}^{\ddagger} = \frac{[(AB)^{\ddagger}]}{[A][B]}. \tag{9.4.1}$$

We also know from chemical kinetics that if the two reactions making up the equilibrium reactions of step (1) are both elementary reactions, then

---

[2]Note that the formation of an activated complex is a reversible process in which many of the intermediate species formed by the reactants dissociate back into reactants, and some of them dissociate into the product molecules of the net reaction. An activated complex represents the set of configurations that can be formed by the reactant species as they proceed through a reaction sequence, while the transition state represents a particular molecular configuration that occurs in the vicinity of the maximum in a reaction profile.

$$K_{eq}^{\ddagger} = \frac{k_f}{k_b} \, ,$$

in which k_f and k_b are the rate constants for the forward and backward reactions. Finally, we know from statistical mechanical considerations [see Eq. (9.2.12)] that we may also write $K_{eq}^{\ddagger}$ in terms of appropriate canonical partition functions as

$$K_{eq}^{\ddagger} = \frac{Z((AB)^{\ddagger}) e^{-\beta E^{\ddagger}}}{Z(A) Z(B)} \, , \tag{9.4.2}$$

in which $E^{\ddagger}$ is the energy required to form the activated complex $(AB)^{\ddagger}$.

The application of basic chemical kinetics arguments to step (2) in the above reaction sequence gives rise to an expression for the net rate, namely,

$$\text{Net rate} = \frac{d[\text{products}]}{dt} = k[(AB)^{\ddagger}] = k K_{eq}^{\ddagger} [A][B] \, ,$$

while from the net reaction, the net rate may be expressed as

$$\frac{d[\text{products}]}{dt} = k_2 [A][B] \, .$$

The equivalence of these two expressions then requires that the *effective bimolecular rate constant* k_2 be given by

$$k_2 = k K_{eq}^{\ddagger} \, . \tag{9.4.3}$$

Now let us consider the breakup of $(AB)^{\ddagger}$ to give the products of the reaction as occurring when *one* of the vibrational modes of what is referred to as the 'activated complex' $(AB)^{\ddagger}$ receives a large surge of energy that causes fragmentation of the complex into products. If we can consider the complex as being made up of two stable fragments weakly bound together, then one of the vibrational modes of the complex will approximate diatom stretching, and we can ascribe the force constant for the vibrational motion with the bound energy for that mode.

It is really only at this point that statistical mechanics enters. Let us designate the canonical partition function for chemical species α by $Z(\alpha)$. We have seen that there are two basic principles governing the distribution of molecules amongst various states, these being

 (i) tendency to maximize entropy: all states of a system having equal energy have equal probability of being achieved;
(ii) tendency to minimize energy: two states having different energies E_1 and E_2 will have a population ratio of

$$\frac{p_1}{p_2} = \frac{\omega_1 e^{-\beta E_1}}{\omega_2 e^{-\beta E_2}} = \frac{\omega_1}{\omega_2} \, e^{-\beta(E_1 - E_2)} \, .$$

Because of these two principles, the canonical partition function $Z(\alpha)$, defined as

$$Z(\alpha) \equiv \sum_i \omega_i \, e^{-\beta E_i(\alpha)} \,,$$

represents a 'weighted counting' of the number of states of the system. We can express $Z_{\mathrm{tot}}(\alpha)$ in the form that we have earlier been considering, i.e., as

$$Z_{\mathrm{tot}}(\alpha) = Z_{\mathrm{tr}}(\alpha) Z_{\mathrm{rot}}(\alpha) Z_{\mathrm{vib}}(\alpha) Z_{\mathrm{el}}(\alpha) \,.$$

9.4.1 The ACT Model

In this model, the 'activated complex' $(AB)^{\ddagger}$ is considered to lie in a small potential hollow in the reaction coordinate, so that progress along the reaction path is equivalent to vibrational motion (stretching) along this degree of freedom: the motion is assigned a 'mode frequency' ν_{rc}, in which the subscript 'rc' stands for 'reaction coordinate'. We shall utilize the notation $\bar{z}^{\circ}$ that we have developed here rather than that traditionally employed in the (older) chemical kinetics literature.

Unimolecular decay of $(AB)^{\ddagger}$ implies that the rate of reaction v is given by

$$v = \nu_{\mathrm{rc}}\big[(AB)^{\ddagger}\big] \,, \tag{9.4.4a}$$

so that upon using the kinetic information contained in step (1) of the proposed kinetic mechanism, we obtain

$$v = \nu_{\mathrm{rc}} K_{\mathrm{eq}}^{\ddagger}[A][B]$$

$$= \frac{k_{\mathrm{B}}T}{h} \, \frac{\bar{z}^{\circ\prime}((AB)^{\ddagger})}{\bar{z}_{\mathrm{tot}}^{\circ}(A)\bar{z}_{\mathrm{tot}}^{\circ}(B)} \, e^{-\beta E^{\ddagger}} \, [A][B] \,, \tag{9.4.4b}$$

with $\bar{z}^{\circ\prime}((AB)^{\ddagger})$ and $\bar{z}_{\mathrm{tot}}^{\circ}((AB)^{\ddagger})$ related by

$$\bar{z}_{\mathrm{tot}}^{\circ}\big((AB)^{\ddagger}\big) = \bar{z}^{\circ\prime}\big((AB)^{\ddagger}\big)\tilde{z}_{\mathrm{vib}}(\nu_{\mathrm{rc}}) = \bar{z}^{\circ\prime}\big((AB)^{\ddagger}\big) \, \frac{1}{1 - e^{-\beta h \nu_{\mathrm{rc}}}} \,. \tag{9.4.4c}$$

It is almost always the case that $\beta h \nu_{\mathrm{c}} \gg 0$, so that $\bar{z}_{\mathrm{tot}}^{\circ}((AB)^{\ddagger})$ can meaningfully be approximated as

$$\bar{z}_{\mathrm{tot}}^{\circ}\big((AB)^{\ddagger}\big) \simeq \frac{k_{\mathrm{B}}T}{h\nu_{\mathrm{rc}}} \, \bar{z}^{\circ\prime}\big((AB)^{\ddagger}\big) \,. \tag{9.4.4d}$$

Notice that no explicit assumptions about the local minimum have been required, as the reaction coordinate frequency ν_{rc} cancels out from the final expressions, so that the net result for $k_2(T)$ is

$$k_2(T) = \left[\frac{k_B T}{h} \frac{\overline{z}^{\circ\prime}\left((AB)^{\ddagger}\right)}{\overline{z}^{\circ}_{tot}(A)\overline{z}^{\circ}_{tot}(B)} \right] e^{-\beta E^{\ddagger}}. \tag{9.4.5}$$

Let us estimate the values for typical molecules at $T = 300$ K, bearing in mind that we are only interested at this time in order-of-magnitude results. For the overall reaction sequence

$$A + B \quad \rightleftharpoons \quad (AB)^{\ddagger} \quad \longrightarrow \quad \text{products},$$

we shall consider that each A molecule has N_A atoms and that each B molecule has N_B atoms.

If each chemical species involved in the above reaction is a nonlinear molecule, it will prove convenient for the purpose of obtaining an order-of-magnitude estimate of the temperature dependence of $k_2(T)$ to express the canonical partition functions for A, B, and $(AB)^{\ddagger}$ as

$$\overline{z}^{\circ}_{tot}(A) = \overline{z}^{\circ}_{tr}\zeta^3_{rot}\widetilde{z}^{3N_A-6}_{vib}, \qquad \overline{z}^{\circ}_{tot}(B) = \overline{z}^{\circ}_{tr}\zeta^3_{rot}\widetilde{z}^{3N_B-6}_{vib},$$

$$\overline{z}^{\circ\prime}\left((AB)^{\ddagger}\right) = \overline{z}^{\circ}_{tr}\zeta^3_{rot}\widetilde{z}^{3N_A+3N_B-7}_{vib}.$$

The only significant differences between the translational partition functions for the three species arise from their different masses appearing in the expression for their de Broglie wavelengths, as the molar volumes are essentially the same. Of course, the characteristic vibrational temperatures associated with the various modes of molecular vibration will differ, but typically by less than an order of magnitude, so that it makes sense to employ some representative average value for each molecule in our calculation. Representation of the rotational partition functions is slightly more complicated, as there will, in principle, be three different moments of inertia for a nonlinear molecule, as opposed to two equal moments of inertia for a linear molecule, so that $z_{rot}(T)$ can be written as the square of two rotational factors $\zeta_{rot}(T)$ for a linear molecule or the product of three (slightly different) rotational factors $\zeta_{rot}(T)$ for a nonlinear molecule. It will suffice, however, for present purposes of obtaining an order-of-magnitude estimate of k_2 to treat the ζ_{rot} factors as roughly independent of the chemical species. Table 9.2 summarizes the relevant information for our estimate.

Upon employing the order-of-magnitude values given in Table 9.2, we may estimate k_2 as

$$k_2 \simeq \frac{k_B T}{h} \frac{\widetilde{z}^5_{vib}}{z^{\circ}_{tr}\zeta^3_{rot}} e^{-\beta E^{\ddagger}} \approx \frac{4 \times 10^{-24}}{7 \times 10^{-34}} \frac{(\sqrt{10})^5 e^{-\beta E^{\ddagger}}}{(10^{30})(\sqrt{1000})^3}$$

Table 9.2 Order-of-magnitude room temperature contributions to the ACT bimolecular rate constant from the various molecular degrees of freedom

Mode of freedom	Statistical thermodynamics		Order of magnitude[a]
Translation	$\Lambda(T)$ [Eq. (3.2.22)]	$z_{\mathrm{tr}}^{\circ}(T, V) = \dfrac{V^{\circ}}{\Lambda^3(T)}$	$\approx 10^{30}$
Rotation	$\begin{cases} \zeta_{\mathrm{rot}}^3 \text{(nonlinear)} \\[1em] \zeta_{\mathrm{rot}}^2 \text{(linear)} \end{cases}$	$\zeta_{\mathrm{rot}}(T) = \left(\dfrac{8\pi^2 I k_{\mathrm{B}} T}{h^2}\right)^{\frac{1}{2}}$	$\approx 10\text{--}100/\mathrm{df}$
Vibration	$\begin{cases} \tilde{z}_{\mathrm{vib}}^{3N-6} \text{(nonlinear)} \\[1em] \tilde{z}_{\mathrm{vib}}^{3N-5} \text{(linear)} \end{cases}$	$\tilde{z}_{\mathrm{vib}}(T) = \dfrac{1}{1 - e^{-\beta h \nu_{\mathrm{osc}}}}$	$\approx 1\text{--}10/\mathrm{df}$

[a] 'df' stands for 'degree of (rotational or vibrational) freedom'.

$$\approx 10^{-20}\, e^{-\beta E^{\ddagger}}.$$

Note that we have employed geometric mean values for the relevant vibrational and rotational factors in the expression for $k_2(T)$. Moreover, we must be very careful with the units! If $k_2(T)$ is in molecular units, then expression (9.4.5) can be used, provided that the concentrations [A], [B], and [(AB)‡] are expressed in molecules/unit volume. If, however, [A], [B], and [(AB)‡] are in the usual laboratory unit of mol/L, then we should ultimately incorporate factors $N_0 V$ into our expression for $k_2(T)$ to give

$$k_2(T) = \frac{k_{\mathrm{B}} T}{h} \frac{\bar{z}^{\circ\prime}(\mathrm{AB}^{\ddagger})/N_0 V}{(\bar{z}_{\mathrm{A}}^{\circ}/N_0 V)(\bar{z}_{\mathrm{B}}^{\circ}/N_0 V)}\, e^{-\Delta U_0^{\circ}/(RT)}. \tag{9.4.6}$$

Example 9.5 Reaction between I and Br.

Let us consider the simple atom–atom reaction

$$\mathrm{I}^*\left({}^2 P_{\frac{1}{2}}\right) + \mathrm{Br}\left({}^2 P_{\frac{3}{2}}\right) \quad \longrightarrow \quad \mathrm{I}\left({}^2 P_{\frac{3}{2}}\right) + \mathrm{Br}^*\left({}^2 P_{\frac{1}{2}}\right)$$

between I and Br. From the atomic term symbols, we may deduce that $\omega_{\mathrm{el}}(\mathrm{I}^*) = 2$, $\omega_{\mathrm{el}}(\mathrm{Br}) = 4$, $\omega_{\mathrm{el}}(\mathrm{I}) = 4$, and $\omega_{\mathrm{el}}(\mathrm{Br}^*) = 2$.

This equation describes the exchange of electronic energy between excited iodine atoms, I*, and ground-state bromine atoms, Br. The reaction proceeds via a 'transition state' that is a ${}^3 \Pi_1$ excited electronic state of the IBr molecule, with an equilibrium bond length of 0.45 nm. The ground electronic state degeneracy for a ${}^3 \Pi$ term is $\omega_{e1} = 2$. The barrier to reaction has a height $\Delta E = 9.7 \times 10^{-20}$ J. ACT (or 'transition-state theory' [TST]) gives

$$k_{\mathrm{TST}} = \frac{k_{\mathrm{B}} T}{h} \frac{z_{\mathrm{AB}}^{\ddagger \circ}}{\bar{z}_{\mathrm{A}}^{\circ} \bar{z}_{\mathrm{B}}^{\circ}}\, e^{-\beta E_{\mathrm{A}}}$$

$$= \frac{k_B T}{h} \frac{z_{rot}^{IBr} z_{tr}^{IBr} z_{el}^{IBr}}{z_{el}^{I*} z_{tr}^{I*} z_{el}^{Br} z_{tr}^{Br}}$$

for this reaction. More specifically, we obtain

$$k_{TST} = \frac{k_B T}{h} \left(\frac{T}{\Theta_{rot}^{IBr}} \right) \frac{(2\pi M k_B T/h^2)^{\frac{3}{2}} 2V^\circ e^{-\beta E_A}}{(2\pi m_I k_B T/h^2)^{\frac{3}{2}} 2V^\circ (2\pi m_{Br} k_B T/h^2)^{\frac{3}{2}} 4V^\circ}$$

for k_{TST}, or even more specifically,

$$k_{TST} = \frac{k_B T}{h} \left(\frac{T}{\Theta_{rot}^{IBr}} \right) \frac{1}{4} \left(\frac{M}{m_I m_{Br}} \right)^{\frac{3}{2}} \frac{h^3}{(2\pi k_B T)^{\frac{3}{2}} V^\circ} e^{-\beta E_A}$$

$$= \frac{k_B T}{h} \frac{1}{4} \left(\frac{T}{\Theta_{rot}^{IBr}} \right) \frac{1}{(2\pi \mu_{IBr} k_B T/h^2)^{\frac{3}{2}} V^\circ} e^{-\beta E_A},$$

in Hz $\equiv$ s^{-1}. Of course, M in this expression represents the mass of the intermediate IBr, so that $M = m_I + m_{Br}$. As the IBr molecule has only a single vibrational degree of freedom, the first factor in our final expression for k_{TST} for this reaction represents $\tilde{z}_{vib}$, and the penultimate factor represents z_{tr}° for the relative motion of the I and Br species. $\square$

9.4.2 From TST Expression to Experiment

For the overall chemical reaction

$$A + B \longrightarrow \text{products},$$

with reaction rate v given by $v = k_2[A][B]$, we have the dimensional result $\left[\frac{mol}{volume} \right] \frac{1}{s} = k_2 \left(\frac{mol}{volume} \right)^2$, which means that k_2 has units volume mol^{-1} s^{-1}.

Because k_{TST} is in s^{-1}, we see that we must first multiply k_{TST} by $\overline{V}^\circ N_0$ in order to obtain a rate constant k_2 that corresponds to a bulk gas experimental value. Thus, for the reaction of iodine atoms with bromine atoms, $k_2(T)$ takes the form

$$k_2(T) = k_{TST}(T) \overline{V}^\circ N_0 = \frac{k_B T}{h} \frac{1}{4} \frac{T}{\Theta_{rot}^{IBr}} \frac{N_0}{[2\pi \mu_{IBr} k_B T/h^2]^{\frac{3}{2}}} e^{-E_A/(k_B T)},$$

with the characteristic rotational temperature $\Theta_{rot}^{IBr} = 0.0244$ K and the reduced mass for IBr having the value $\mu_{IBr} = 49.03$ amu. Thus, we find that the temperature dependence of k_2 is given by

$$k_2(T) = \frac{(6.62608 \times 10^{-34})^2}{4(2\pi)^{\frac{3}{2}}(1.3807 \times 10^{-23})^{\frac{1}{2}}} \frac{\sqrt{T}(6.02214 \times 10^{23})}{(49.03 \times 1.6605 \times 10^{-27})^{\frac{3}{2}}} \frac{e^{-7025.4/T}}{0.0244},$$

or

$$k_2(T) = 2.16_3 \times 10^7 \ \mathrm{m^3 s^{-1} mol^{-1}} \sqrt{T} e^{-7025.4/T}$$

$$= 2.16_3 \times 10^{10} \ \mathrm{L\, s^{-1} mol^{-1}} \ \sqrt{T} \ e^{7025.4/T}.$$

Thus, for $T = 650\,\mathrm{K}$, $k_2 = 1.115 \times 10^7\,\mathrm{L\,mol^{-1}s^{-1}}$, and for $T = 1000\,\mathrm{K}$, $k_2 = 6.08_6 \times 10^8\,\mathrm{L\,mol^{-1}s^{-1}}$.

9.5   Problems for This Chapter

1. Sodium vapour exists in two forms, monatomic sodium, Na, and diatomic sodium, $\mathrm{Na_2}$. These species are related via the dissociation reaction

$$\mathrm{Na_2}(g) \quad \rightleftharpoons \quad 2\mathrm{Na}(g).$$

   Given that $\Theta_{\mathrm{vib}}(^{23}\mathrm{Na_2}) = 228.56$ K, $\Theta_{\mathrm{rot}}(^{23}\mathrm{Na_2}) = 0.2220\,\mathrm{K}$, $\overline{M} = 0.02299\,\mathrm{kg\,mol^{-1}}$, and that the electronic term symbols for Na and $\mathrm{Na_2}$ are $X^2S_{\frac{1}{2}}$ and $X^1\Sigma_g^+$, respectively, obtain an expression for the equilibrium constant for the dissociation of sodium and evaluate $K_{\mathrm{eq}} = K_P$ for sodium at temperatures of 298 and 1163 K. Note that the boiling temperature for liquid Na metal is $T_\mathrm{b} = 1163\,\mathrm{K}$.

2. Tritium (the mass three isotopes of hydrogen) is important in the nuclear power industry, as it is one of the by-products generated by some nuclear reactors. Molecular tritium, $\mathrm{T_2}$, undergoes an isotopic exchange reaction with molecular hydrogen, $\mathrm{H_2}$, to give the mixed isotopologue HT according to the equilibrium reaction

$$\mathrm{H_2}(^1\Sigma_g^+) + \mathrm{T_2}(^1\Sigma_g^+) \rightleftharpoons 2\mathrm{HT}(^1\Sigma^+).$$

   Experimental data pertinent to the hydrogen isotopologues $\mathrm{H_2}$, $\mathrm{T_2}$, and HT are given in the following table:

Isotopologue	$\overline{M}/\mathrm{kg\,mol^{-1}}$	$\Theta_{\mathrm{rot}}/\mathrm{K}$	$\Theta_{\mathrm{vib}}/\mathrm{K}$
H_2	2.018	85.30	6215
T_2	6.032	28.51	3593
HT	4.024	56.90	5076

From the relation between the depth $(-D_e)$ of the potential well for a diatomic molecule, the zero-point vibrational energy $(\frac{1}{2}h\nu_{\mathrm{osc}})$, and the experimentally accessible dissociation energy $D_0(H_2) = 431.79\,\mathrm{kJ\,mol^{-1}}$ determine D_0 values for the T_2 and HT hydrogen isotopologues. Obtain an expression for evaluating the equilibrium constant, $K_P(T)$, for the tritium–hydrogen exchange reaction, and employ it to evaluate $K_P(T)$ at temperatures $T = 300\,\mathrm{K}, 500\,\mathrm{K}$, and $1000\,\mathrm{K}$.

3. The equilibrium constant K_{eq} characterizing the isotopic exchange of a deuterium atom between a deuterated acid and water may be evaluated using spectroscopic data for the molecules involved in the reaction. Calculate values of the equilibrium constant at temperatures of 300 and 1000 K for the gas-phase isotopic exchange reaction

$$DCl(g) + H_2O(g) \rightleftharpoons HDO(g) + HCl(g)\,,$$

using the following approximate values for relevant spectroscopic data (in $\mathrm{cm^{-1}}$):

- H_2O: $\overline{\nu}_1(a_1) = 3657, \overline{\nu}_2(a_1) = 1595, \overline{\nu}_3(b_2) = 3756, \overline{A}_0 = 27.9, \overline{B}_0 = 14.5,$ $\overline{C}_0 = 9.28.$
- HDO: $\overline{\nu}_1(a') = 2727, \overline{\nu}_2(a') = 1402, \overline{\nu}_3(a') = 3707, \overline{A}_0 = 23.4, \overline{B}_0 = 9.10,$ $\overline{C}_0 = 6.41.$
- HCl: $\overline{\nu}(a) = 2886, \overline{B}_0 = 10.4$; DCl: $\overline{\nu}(a) = 2091, \overline{B}_0 = 5.39.$

All vibrational modes are nondegenerate.

4. Obtain an expression for the equilibrium constant $K_{\mathrm{eq}}(T)$ for the isotopic exchange reaction

$$CD_4(g) + HCl(g) \rightleftharpoons CHD_3(g) + DCl(g)\,.$$

Make a plot of $K_{\mathrm{eq}}(T)$ vs. T to illustrate the temperature dependence of $K_{\mathrm{eq}}(T)$ in the temperature range $700\,\mathrm{K} \le T \le 1000\,\mathrm{K}$ using the relevant approximate values for the vibrational frequencies and rotational constants (in $\mathrm{cm^{-1}}$):

- CD_4: $\overline{\nu}_1(a_1) = 2109, \overline{\nu}_2(e) = 1092, \overline{\nu}_3(f_2) = 2259, \overline{\nu}_4(f_2) = 996, \overline{B}_0 = 2.63.$
- CHD_3: $\overline{\nu}_1(a_1) = 2293, \overline{\nu}_2(a_1) = 2142, \overline{\nu}_3(a_1) = 1003, \overline{\nu}_4(e) = 2263,$ $\overline{\nu}_5(e) = 1291, \overline{\nu}_6(e) = 1036, \overline{A}_0 = 2.63, \overline{B}_0 = 3.28.$
- HCl: $\overline{\nu}(a) = 2886, \overline{B}_0 = 10.4$; DCl: $\overline{\nu}(a) = 2091, \overline{B}_0 = 5.39.$

Vibrations having symmetry designations a, e, and f have degeneracies 1, 2, and 3, respectively.

5. Find the values for the equilibrium constant for the dissociation of molecular bromine at temperatures of 300 K, 500 K, and 1000 K. The dissociation reaction is

$$Br_2(g) \rightleftharpoons 2Br(g) .$$

The term symbol for ground level molecular bromine is $X^1\Sigma_g^+$, its spectroscopic dissociation energy is $D_0 = 1.9707\,eV$, and $\bar{\nu}_{osc} = 323.166\,cm^{-1}$, $\bar{B}_0 = 0.08148\,cm^{-1}$; use reduced mass $\mu = 39.4592$ for Br_2. The ground term for atomic bromine is $X^2P_{\frac{3}{2}}$.

6. Calculate values for the equilibrium constant for the formation of $^{84}Kr_2$ Van der Waals molecules in a sample of krypton gas at temperatures of 200 K, 300 K, and 500 K. The ^{84}Kr atom has a $^1S_0$ ground electronic term, with no low-lying excited electronic terms. The relevant spectroscopic constants for the $^{84}Kr_2$ molecule are $D_0 = 126.9\,cm^{-1}$, $\bar{\nu}_{osc} = 23.022\,cm^{-1}$, and $\bar{B}_0 = 0.02439\,cm^{-1}$; the Kr_2 molecule also has no low-lying electronic states. Compare your values for $K_{eq}(T)$ calculated for temperatures of 300 K and 500 K for the Kr–Kr_2 system with corresponding values of $K_{eq}(T)$ obtained in Problem 5 for the Br–Br_2 system.

7. Compare the equilibrium constants for the reactions

$$H_2(g) + I_2(g) \rightleftharpoons 2HI(g) , \qquad D_2(g) + I_2(g) \rightleftharpoons 2DI(g) ,$$

using the data provided in the accompanying table.

Molecule	Term symbol	M $10^3 kg\,mol^{-1}$	D_0/eV	$\bar{\nu}_{osc}/cm^{-1}$	$\bar{B}_0/cm^{-1}$
H_2	$^1\Sigma_g^+$	2.1057	4.4781	4158.87	59.052
D_2	$^1\Sigma_g^+$	4.0282	4.5563	2993.39	29.904
$^{127}I_2$	$^1\Sigma_g^+$	253.8090	1.5424	213.285	0.0368
$H^{127}I$	$^1\Sigma^+$	127.9123	3.054	2229.66	6.341
$D^{127}I$	$^1\Sigma^+$	128.9186	3.095	1519.76	3.223

8. The transition state for the reaction $D + H_2 \rightleftharpoons DH + H$ is the linear molecule D–H–H. By locating your origin at the D atom, with the D–H and H–H separations both given by $d = 0.930 \times 10^{-10}$ m, obtain an expression for the moment-of-inertia of the DH_2 molecule in terms of the masses and bond distances. Calculate the moment-of-inertia I_b for rotation about an axis perpendicular to the bond axis and passing through the centre-of-mass of the DH_2 molecule. Use this result, together with the moment-of-inertia for H_2 and symmetry numbers $\sigma(H_2) = 2$, $\sigma^{\ddagger} = \sigma(DH_2) = 1$ to calculate the ratio $z_{rot}^{\ddagger}/z_{rot}(H_2)$ for temperatures of 450 K and 600 K.

9. Calculate the equilibrium constant K_P for the gaseous reaction

$$2HI(g) \quad \rightleftharpoons \quad H_2(g) + I_2(g)$$

using the following spectroscopic data [K. P. Huber and G. Herzberg [1]]:

Molecule	$H_2\left({}^1\Sigma_g^+\right)$	$I_2\left({}^1\Sigma_g^+\right)$	$HI\left({}^1\Sigma^+\right)$
$\overline{B}/\mathrm{cm}^{-1}$	59.052	0.0368	6.3415
$\overline{\nu}_{\mathrm{osc}}/\mathrm{cm}^{-1}$	4158.87	213.28	2229.66
D_0/eV	4.47813	1.54238	3.0541

Plot $K_P(T)$ as a function of temperature for the temperature range 300 K $\leq$ $T \leq$ 3000 K. In particular, compare your calculated values with the experimental values reported by Taylor and Crist [2].

T/K	$10^2\,K_P$	T/K	$10^2\,K_P$
666.8	1.639 ± 0.010	730.8	2.018 ± 0.006
	1.645 ± 0.009		2.007 ± 0.008
698.6	1.812 ± 0.007	763.8	2.192 ± 0.007
	1.811 ± 0.002		2.172 ± 0.004

References

1. K.P. Huber, G. Herzberg, *Molecular Spectra and Molecular Structure. IV. Constants of Diatomic Molecules* (Van Nostrand–Reinhold, New York, 1979). Also found online at https://webbook. nist.gov/
2. A.H. Taylor, Jr., R.H. Crist, J. Amer. Chem. Soc. **63**, 1377 (1941)

Chapter 10
Quantum Statistics

This chapter examines the statistics imposed by quantum mechanical symmetry restrictions upon wavefunctions for systems of two types of inherently indistinguishable quantum mechanical particles, called bosons and fermions. The distinctions between fermions and bosons, symmetry requirements that apply to them, and the natures of their statistical descriptions are elaborated using the grand ensemble. Each type of quantum statistics reduces under appropriate conditions to classical statistics. The ^{3}He isotope, conduction electrons in metals, and free electrons in semiconductors are treated as ideal fermion gases, while the ^{4}He isotope and photons are treated as ideal boson gases. Discussions of strongly degenerate fermion and boson gases are presented. Both ideal fermion and ideal boson gases have nonzero second virial coefficients at low temperatures. The chemical potentials for fermion and boson gases show distinctly different temperature dependences at low temperatures, but merge with the classical mechanical chemical potential at high temperatures.

10.1   Setting the Stage

We have thus far considered systems in which we have not had to worry at a fundamental level about quantum mechanical symmetry restrictions on the wavefunctions for a molecular system, and have consequently restricted our considerations to cases in which the number of available states by far exceeds the number of particles. In this way situations in which more than one particle is to be found in a given state are extremely rare, i.e., on average only one particle occupies a particular quantum mechanical state of the N-particle system. At a more fundamental level, however, we should take into consideration the fact that if we are dealing with fermions (particles with an intrinsic half-odd-integer spin), the N-body wavefunction must be antisymmetric to the interchange of two identical fermions, while for bosons

© Springer Nature Switzerland AG 2021
F. R. W. McCourt, *Statistical Thermodynamics for Pure and Applied Sciences*,
https://doi.org/10.1007/978-3-030-52006-9_10

(particles possessing an intrinsic integer or zero spin), the N-body wavefunction must be symmetric to the interchange of two identical bosons. The statistics obeyed by these two basic types of particle are quite different, and are known, respectively, as Fermi–Dirac (FD) and Bose–Einstein (BE) statistics.

Fermi–Dirac and Bose–Einstein statistics are, in fact, the only exact statistics, and are hence the only appropriate statistics to use, especially for the treatment of indistinguishable species. Of course, classical (or Maxwell–Boltzmann) statistics applies for systems of distinguishable species, and we have argued heuristically in Chap. 3 (Sect. 3.2.3) that, as first proposed by Gibbs, the canonical partition function $Z_N(T, V)$ for N indistinguishable particles can be obtained from that for N distinguishable particles via division by $N!$. Indeed, we shall see in Sect. 10.3 that both the exact Fermi–Dirac and Bose–Einstein quantum-statistical partition functions reduce under appropriate conditions to the Gibbs version of the Maxwell–Boltzmann canonical partition function for a system of N indistinguishable particles.

Fermions
- A fundamental QM symmetry requirement for fermions is that the total wavefunction Ψ_T for an N-particle system be antisymmetric under the interchange of any two of the identical fermions, viz.,

$$\Psi_T(\cdots i \cdots j \cdots) = -\Psi_T(\cdots j \cdots i \cdots) \, ;$$

  this does not lead to a new state of the gas, so that the particles must be genuinely indistinguishable in enumerating the distinct states of the gas.
- Suppose now that two particles i and j, both in the *same* single-particle state s, are interchanged, then necessarily

$$\Psi_T(\cdots i \cdots j \cdots) = \Psi_T(\cdots j \cdots i \cdots) \, .$$

  Because both this statement and the previous one can be true simultaneously only if $\Psi_T \equiv 0$, this implies that there are *no* states of a fermi gas for which two or more particles are in the *same* single-particle state. This is the origin of the Pauli Exclusion Principle.

Bosons
- A fundamental QM symmetry requirement for bosons is that the total wavefunction Ψ_T for an N-particle system be symmetric under the interchange of any two of the identical bosons, viz.,

$$\Psi_T(\cdots i \cdots j \cdots) = \Psi_T(\cdots j \cdots i \cdots) \, ,$$

  and we see that in this case, there is *no* restriction on how many bosons may be in any given one-particle state.

Example 10.1 Illustration of differences between FD, BE, MB statistics.

Consider now 2 particles, and 3 states, and ask the question 'in how many ways can the 2 particles be arranged in the 3 states using each of Maxwell–Boltzmann, Fermi–Dirac and Bose–Einstein statistics?'.

Case 1: Maxwell–Boltzmann statistics (distinguishable particles).

State		
1	2	3
1, 2	–	–
–	1, 2	–
–	–	1,2
1	2	–
2	1	–
1	–	2
2	–	1
–	1	2
–	2	1
Total number of states = 9		

Case 2: Fermi–Dirac statistics (indistinguishable particles).

State		
1	2	3
1	1	–
1	–	1
–	1	1
Total number of states = 3		

Case 3: Bose–Einstein statistics (indistinguishable particles).

State		
1	2	3
1, 1	–	–
–	1, 1	–
–	–	1, 1
1	1	–
1	–	1
–	1	1
Total number of states = 6		

Let us now examine the ratio r of the probability that the two particles are found in the same state to the probability that the two particles are found in different states for each of the three types of statistics that we have been discussing. That is, we shall calculate the quantity

$$r \equiv \frac{\text{probability that the 2 particles are found in same state}}{\text{probability that the 2 particles are found in different states}}$$

We obtain the answers:

$$r_{\text{MB}} = \frac{3/9}{6/9} = \frac{1}{2},$$

$$r_{\text{FD}} = \frac{0/3}{3/3} = 0,$$

$$r_{\text{BE}} = \frac{3/6}{3/6} = 1,$$

from which we may deduce that the BE case has a greater tendency for particles to bunch up, while the FD case has a greater tendency for particles to be in different states. □

Literal enumeration of the number of microstates accessible to systems consisting of large numbers of Bose–Einstein or Fermi–Dirac particles along the lines that we have just considered for two such particles in Example 10.1 would quickly become a formidable task. We may, however, utilize combinatoric formulae to determine the numbers of possible BE or FD microstates accessible to such multi-particle systems. The number of microstates obtained for n_b bosons distributed over n_s states is given by

$$N_{\text{BE}} = \frac{(n_b + n_s - 1)!}{n_b! \, (n_s - n_b)!}, \tag{10.1.1a}$$

while the number of microstates obtained for n_f fermions distributed over n_s states is given by

$$N_{\text{FD}} = \frac{n_s!}{n_f! \, (n_s - n_f)!}. \tag{10.1.1b}$$

For comparison, we note that the expression for the number of microstates available to a system of distinguishable particles corresponding to classical Maxwell–Boltzmann statistics, with n_s states available to n_{mb} particles, is given by

$$N_{\text{MB}} = (n_s)^{n_{\text{mb}}}. \tag{10.1.1c}$$

Let us now consider directly the effect on enumeration of the many-body states available to a macroscopic system of particles. The energy for a given state is

$$E_{ijkl...} = \epsilon_i + \epsilon_j + \epsilon_k + \epsilon_l + \dots , \tag{10.1.2}$$

while the canonical partition function is given by

$$Z(N, T, V) = \sum_{ijkl...} e^{-\beta(\epsilon_i + \epsilon_j + \dots)} . \tag{10.1.3}$$

When the molecules are indistinguishable, we cannot *in principle* sum over $i, j, k, l, \dots$ separately, as we could have done had they been distinguishable.

We have seen that there are two cases to consider:

(a) **fermions**: severely restricted occupation of available single-particle energy states, as the antisymmetry of the wavefunction requires that no two identical fermions occupy the same single-particle state. This means that *any terms in which two or more indices are the same must be excluded from the summation* in an expression for the canonical partition function $Z(N, T, V)$.

(b) **bosons**: unrestricted occupation of available single-particle energy states is allowed. However, there are just as many problems for bosons as for fermions. For example, if we consider the case in which all indices except one are the same, we have $\epsilon_i + \epsilon_j + \epsilon_j + \dots + \epsilon_j, i \neq j$, and this is the same as $\epsilon_j + \epsilon_i + \epsilon_j + \dots + \epsilon_j, i \neq j$, and so on, so that this case will be included N times in the summation over all states if the sums are done independently.

In both cases, if all particles are in different states, an unrestricted summation will have included $N!$ terms where only one should have been allowed.

These facts make a direct summation to obtain $Z(N, T, V)$ extremely difficult to perform, both for fermions and for bosons. The real problem occurs here when we cannot simply ignore those terms in which two or more indices are the same in comparison with those terms for which all indices are different. We have already seen that when such terms can be ignored, it is sufficient to sum over all indices unrestrictedly, and then divide by $N!$ to correct for the overcounting made thereby. We have said earlier that this will be valid when the number of N-body energy states available for the system of N particles greatly exceeds the number of particles.

We can be more explicit for translational states, for which we have obtained an expression, Eq. (3.2.18a), giving the number of translational states, $\Phi(V, \epsilon)$, having energies $\epsilon \geq 0$. If we express $\Phi(V, \epsilon)$ as

$$\Phi(V, \epsilon) = \frac{V}{\Lambda^3(T)} \frac{2}{6\sqrt{\pi}} \left(\frac{\epsilon}{k_B T} \right)^{\frac{3}{2}},$$

in terms of the thermal de Broglie wavelength $\Lambda(T)$, defined as

$$\Lambda(T) \equiv \frac{h}{(2\pi m k_B T)^{\frac{1}{2}}} \tag{10.1.4}$$

Table 10.1 When are
quantum statistics needed?

Substance	T/K	$N\Lambda^3/V$
Liquid ^{4}He	4	2.2
^{4}He gas	4	0.15
^{4}He gas	20	2.8×10^{-3}
^{4}He gas	100	4.8×10^{-5}
Liquid Ne	27	1.5×10^{-2}
Ne gas	27	1.1×10^{-4}
Ne gas	100	4.3×10^{-6}
Liquid Ar	86	7.0×10^{-4}
Ar gas	86	2.2×10^{-6}
Liquid Kr	127	7.5×10^{-5}
Kr gas	127	2.8×10^{-7}
e^- in Na	300	2025
Liquid H_2	20	0.44
H_2 gas	20	7.0×10^{-3}

then the condition $\Phi(V,\epsilon) \gg N$ becomes equivalent to

$$\frac{V}{\Lambda^3(T)} \gg N \quad \Rightarrow \quad \frac{N\Lambda^3(T)}{V} = \frac{\Lambda^3}{V/N} \ll 1, \tag{10.1.5a}$$

in which V/N, which represents the system volume per particle, is the reciprocal of the number density $n \equiv N/V$, while $\Lambda^3(T)$ provides a measure of the quantum mechanical volume of a system particle. Notice that condition (10.1.5a) is favoured by large particle mass, high temperature, and low number density. Values of $N\Lambda^3(T)/V$ are given in Table 10.1 for a representative selection of substances. Thus, we may interpret condition (10.1.5a) as requiring that, on average, if the volume V/N available to a particle in the system under consideration is much greater than $\Lambda^3(T)$, then a Gibbs classical indistinguishability description will be valid. It then also follows that if

$$\frac{\Lambda^3(T)}{V/N} \gtrsim 1, \tag{10.1.5b}$$

then an appropriate quantum-statistical treatment will be needed. Such a quantum-statistical treatment is thus favoured by small particle mass, low temperature, and high number density.

From the values for $N\Lambda^3/V$ given in Table 10.1 for a number of common substances we see that we should require a fully quantum-statistical treatment especially for: electrons in metals (FD), liquid ^{3}He (FD), ^{3}He gas below 6 K (FD), liquid ^{4}He (BE), ^{4}He gas below 6 K (BE), and liquid H_2 (BE). At normal earth temperatures we see that, except for an electron gas (e.g. electrons in metals), our condition for MB statistics is typically satisfied for all substances in Table 10.1. Indeed, electrons in metals remain highly quantum mechanical even at very high temperatures. We have also seen that at sufficiently high temperatures we can treat the energy of our system as essentially classical, so that MB statistics and the classical limit are essentially equivalent.

10.2 Two Types of Quantum Statistics

There are two cases to consider in the evaluation of the partition function for sets of indistinguishable particles. The resultant distribution for fermions is known as the 'Fermi–Dirac distribution', while that for bosons is known as the 'Bose–Einstein distribution'. We should note, moreover, that these distributions are the only *exact* distributions.

N-particle wavefunctions are complicated to construct, as the requirement of antisymmetrization (fermions) or symmetrization (bosons) may place strong restrictions on the accessibility of states at the specified energy to the particles. However, because we are dealing with inherently indistinguishable particles, the only meaningful statistical data are the numbers of particles that occupy the allowed quantum states. Thus, rather than attempting to construct sets of N-particle wavefunctions, it will prove more useful to describe the system as a whole in terms of 'occupation numbers', n_k, of the available/accessible energy states. In quantum chemistry, this is referred to as the Fock basis. Each single-particle state corresponds to a distinct energy ϵ_k, and the energy states are ordered by increasing energy. We may introduce the notation $|n_1, n_2, \ldots, n_k, \ldots\rangle$ to designate the N-particle state for which n_1 particles occupy the state of energy ϵ_1, n_2 that of energy ϵ_2, and so forth. Of course, there must be a constraint on the set of n_k-values, since the total number of particles must be N, i.e., the n_k are required to satisfy the condition

$$\sum_k n_k = N \,, \tag{10.2.1}$$

while at the same time the total energy E_N of the N-particle state must be given by

$$\sum_k n_k \epsilon_k = E_N \,. \tag{10.2.2}$$

If we wished to work in terms of the canonical ensemble, we would have to write $Z(N, V, T)$ as

$$Z(N, V, T) = \sum_j e^{-\beta E_j} = \sum_{\{n_k\}}^{*} \exp\left\{-\beta \sum_i n_i \epsilon_i\right\} \,, \tag{10.2.3}$$

in which the asterisk on the outer summation is to remind us that only those n_k such that $\sum_k n_k = N$ are to be included in it. This restriction is extremely difficult to incorporate mathematically. We may avoid this problem, however, if we work in the grand ensemble, in which N can vary. For this reason we shall use the grand partition function,

$$\Xi(T, V, \mu) = \sum_{N=0}^{\infty} e^{\beta\mu N} Z(N, T, V), \qquad (10.2.4)$$

from this point onwards.

Let us examine some of the equivalent forms in which Ξ can be written. The ones of interest to us at the moment are

$$\Xi(T, V, \lambda) = \sum_{N=0}^{\infty} \lambda^N \sum_{\{n_k\}}^{*} \exp\left\{-\beta \sum_i \epsilon_{n_i}\right\}$$

$$= \sum_{N=0}^{\infty} \sum_{\{n_k\}}^{*} \lambda^{\Sigma_i n_i} \exp\left\{-\beta \sum_i \epsilon_i n_i\right\}$$

$$= \sum_{N=0}^{\infty} \sum_{\{n_k\}}^{*} \prod_k \left(\lambda e^{-\beta\epsilon_k}\right)^{n_k}, \qquad (10.2.5)$$

in which we have written as earlier Eq. (4.2.21) $\lambda \equiv e^{\beta\mu}$.

Now comes the crucial step in our reasoning. Because we are summing over all values of N in the grand ensemble, each n_k will range over all its possible values, and we can rewrite for the grand partition function as

$$\Xi(T, V, \lambda) = \sum_{n_1=0}^{n_1^{\max}} \sum_{n_2=0}^{n_2^{\max}} \cdots \prod_k \left(\lambda e^{-\beta\epsilon_k}\right)^{n_k}$$

$$= \sum_{n_1=0}^{n_1^{\max}} \left(\lambda e^{-\beta\epsilon_1}\right)^{n_1} \sum_{n_2=0}^{n_2^{\max}} \left(\lambda e^{-\beta\epsilon_2}\right)^{n_2} \cdots$$

$$= \prod_{k=1}^{\infty} \sum_{n_k=0}^{n_k^{\max}} \left(\lambda e^{-\beta\epsilon_k}\right)^{n_k}. \qquad (10.2.6)$$

(a) For fermions we know that the occupation numbers are restricted to $n_k = 0, 1$, so that $n_k^{\max} = 1$, with the consequence that the partition function expression is given as

$$\Xi_{\mathrm{FD}}(T, V, \lambda) = \prod_k \left(1 + \lambda e^{-\beta\epsilon_k}\right) \qquad (10.2.7a)$$

or, in terms of the chemical potential μ, as

$$\Xi_{\text{FD}}(T, V, \mu) = \prod_k \left(1 + e^{\beta(\mu - \epsilon_k)}\right). \qquad (10.2.7b)$$

From Eq. (10.2.7a) we see that $\ln \Xi$ is thus given by

$$\ln \Xi_{\text{FD}}(T, V, \lambda) = \sum_k \ln(1 + \lambda e^{-\beta \epsilon_k}). \qquad (10.2.8)$$

(b) For bosons the occupation number can take on any integer value, i.e., $n_k = 0, 1, 2, \ldots$, and hence $n_k^{\max} = \infty$, with the consequence that the partition function is given by

$$\Xi_{\text{BE}}(T, V, \lambda) = \prod_k \sum_{n_k=0}^{\infty} \left(\lambda e^{-\beta \epsilon_k}\right)^{n_k}$$

$$= \prod_k (1 - \lambda e^{-\beta \epsilon_k})^{-1}. \qquad (10.2.9a)$$

Note that the summation over the boson occupation number, unlike the summation over the fermion occupation number in Eqs. (10.2.7), has introduced a constraint, as the summation will not converge to $(1 - \lambda e^{-\beta \epsilon_k})^{-1}$ unless $\lambda e^{-\beta \epsilon_k} < 1$. If we rewrite expression (10.2.9a) in terms of the chemical potential μ as

$$\Xi_{\text{BE}}(T, V, \mu) = \prod_k \left(1 - e^{\beta(\mu - \epsilon_k)}\right)^{-1}, \qquad (10.2.9b)$$

we see that this result requires that the chemical potential μ must be (algebraically) less than all energies ϵ_k (including the ground state energy) in order for expression (10.2.9b) to be valid. Should $\epsilon_{\text{ground}} = 0$, the chemical potential cannot be positive. From Eq. (10.2.9a), we obtain the BE equivalent of Eq. (10.2.8), namely

$$\ln \Xi_{\text{BE}}(T, V, \lambda) = -\sum_k \ln(1 - \lambda e^{-\beta \epsilon_k}). \qquad (10.2.10)$$

We may collect these two cases together into a single formula for Ξ, namely

$$\Xi_{\substack{\text{FD} \\ \text{BE}}}(T, V, \lambda) = \prod_k \left(1 \pm \lambda e^{-\beta \epsilon_k}\right)^{\pm 1}. \qquad (10.2.11)$$

What can we say generally regarding quantum systems? Let us begin with our expression for the grand partition function $\Xi_{FD}^{BE}(T, V, \lambda)$ of Eq. (10.2.11). If we recall [see Eq. (4.2.31)] that we may define a grand potential $\Phi(T, V, \mu)$, equivalently $\Phi(T, V, \lambda)$, via

$$\Phi(T, V, \lambda) \equiv -k_B T \ln \Xi(T, V, \lambda),$$

then $\Phi(T, V, \lambda)$ will be given as

$$\Phi(T, V, \lambda) = \mp k_B T \sum_k \ln \left(1 \pm \lambda e^{-\beta \epsilon_k}\right), \tag{10.2.12}$$

with it now being understood that the upper signs apply to FD and the lower signs to BE systems.

We shall approximate the sum over energy states associated with translational motion by an integration over continuum energy using the prescription (3.2.19), namely

$$\sum_k \cdots \quad \rightarrow \quad \int \Omega(\epsilon) d\epsilon \cdots, \tag{10.2.13a}$$

in which $\Omega(\epsilon)$ is the density-of-states,

$$\Omega(\epsilon) = (2I_a + 1) 2\pi \left(\frac{2m}{h^2}\right)^{\frac{3}{2}} V \sqrt{\epsilon}, \tag{10.2.13b}$$

with I_a the magnitude of any *intrinsic* angular momentum possessed by the quantum particles under consideration. This allows us to write $\Phi(T, V, \mu)$ as

$$\Phi(T, V, \mu) = \mp k_B T \int_0^\infty \Omega(\epsilon) \ln \left(1 \pm e^{\beta(\mu - \epsilon)}\right) d\epsilon, \tag{10.2.14a}$$

or, more explicitly, upon substituting Eq. (10.2.13b) for $\Omega(\epsilon)$, and with $x \equiv \beta \epsilon$, $\alpha \equiv \beta \mu$, as

$$\Phi(T, V, \mu) = \mp \frac{(2I_a + 1)V}{\Lambda^3(T)} \frac{2k_B T}{\sqrt{\pi}} \int_0^\infty \sqrt{x} \ln(1 \pm e^{-(x - \alpha)}) \, dx. \tag{10.2.14b}$$

Integration by parts gives

$$\Phi(T, V, \mu) = -\frac{(2I_a + 1)V}{\Lambda^3(T)} \frac{4k_B T}{3\sqrt{\pi}} \int_0^\infty \frac{x^{\frac{3}{2}}}{e^{x - \alpha} \pm 1} \, dx. \tag{10.2.15}$$

The integrals in expression (10.2.15), defined by

$$g_\nu^\pm(\lambda) \equiv \frac{1}{\Gamma(\nu)} \int_0^\infty \frac{x^{\nu-1}}{\frac{1}{\lambda}e^x \pm 1}\, dx \,, \tag{10.2.16}$$

with the upper sign applying to Fermi–Dirac gases, the lower sign to Bose–Einstein gases, are special cases of generalized zeta functions, and

$$\Gamma(\nu) \equiv \int_0^\infty e^{-t} t^{\nu-1} dt \tag{10.2.17}$$

is the gamma function. The grand potential may thus be expressed in terms of these functions as

$$\Phi_{\mathrm{BE}}(T, V, \lambda) = -\frac{(2I_a + 1)Vk_{\mathrm{B}}T}{\Lambda^3(T)}\, g_{\frac{5}{2}}^-(\lambda) \,, \tag{10.2.18}$$

for a Bose–Einstein gas, and as

$$\Phi_{\mathrm{FD}}(T, V, \lambda) = -\frac{(2I_a + 1)Vk_{\mathrm{B}}T}{\Lambda^3(T)}\, g_{\frac{5}{2}}^+(\lambda) \,, \tag{10.2.19}$$

for a Fermi–Dirac gas.

Individual members of the grand ensemble are themselves canonical ensembles, each with a fixed number N of molecules (particles) such that $N = \sum_k n_k$ is the sum over the occupation numbers associated with the set of energy states $\{\epsilon_k\}$. The most probable number $\overline{N}$ for the grand ensemble corresponds to the thermodynamic (macroscopic) equilibrium value N_{thermo}, and must therefore be given by

$$N_{\text{thermo}} \equiv \overline{N} = \sum_k \overline{n}_k \,. \tag{10.2.20}$$

We shall employ this result to identify the mean (or equilibrium) populations $\overline{n}_k$ of individual energy states ϵ_k. We may obtain useful expressions for the mean occupation numbers by noting that Eq. (4.2.32) established that

$$\overline{N} = \lambda \left(\frac{\partial \ln \Xi}{\partial \lambda}\right)_{T,V} = \sum_k \frac{\lambda e^{-\beta \epsilon_k}}{1 \pm \lambda e^{-\beta \epsilon_k}} \tag{10.2.21}$$

for Fermi–Dirac and Bose–Einstein systems. A comparison between Eqs. (10.2.20) and (10.2.21) shows that the $\overline{n}_k$ can therefore be expressed as

$$\overline{n}_k = \frac{\lambda e^{-\beta \epsilon_k}}{1 \pm \lambda e^{-\beta \epsilon_k}} \,. \tag{10.2.22}$$

The mean occupation numbers $\bar{n}_k(\epsilon_k)$ thus provide us with a means of calculating averages of quantities that depend upon the energies ϵ_k of the quantum particles. The simplest such calculation is that of the internal energy, which is given by

$$U(T, V, \lambda) \equiv \sum_k \bar{n}_k(\epsilon_k)\epsilon_k = \sum_k \frac{\lambda \epsilon_k e^{-\beta \epsilon_k}}{1 \pm \lambda e^{-\beta \epsilon_k}}. \tag{10.2.23}$$

As for our calculation of the grand potential $\Phi(V, T, \lambda)$, we shall replace the summation over k by an integration over ϵ:

$$U(T, V, \mu) = \frac{(2I_a + 1)V}{\Lambda^3(T)} \frac{2k_B T}{\sqrt{\pi}} \int_0^\infty \frac{x^{\frac{3}{2}}}{e^{x-\alpha} \pm 1} \, dx, \tag{10.2.24}$$

so that a comparison between Eqs. (10.2.15) and (10.2.24), recalling also that $\Phi = -PV$, shows us that for both BE and FD gases we have

$$PV = \tfrac{2}{3} U, \tag{10.2.25}$$

which is the same relation that we obtained for a classical gas.

Example 10.2 Equation of state for structureless particles in a gravitational field.

We have seen in Example 3.4 of Chap. 3 that the density-of-states for structureless particles subject to a gravitational field characterized by an acceleration g is given by

$$\Omega(x, \eta) = \frac{4\pi V \beta}{3\eta} \left(\frac{2m}{h^2 \beta}\right)^{\frac{3}{2}} \left[x^{\frac{3}{2}} - (x - \eta)^{\frac{3}{2}} H(\eta/x)\right],$$

in which $x \equiv \beta\epsilon$, $\eta \equiv \beta m g H$, and $H(y)$ is the Heaviside function. The effect of the gravitational field is parameterized by η. Equation (10.2.14a) for the grand potential $\Phi(T, V, \mu) = -PV$ then gives the equation of state as

$$PV = \pm k_B T \int_0^\infty \Omega(x, \eta) \ln(1 \pm e^{\alpha-x}) \, dx$$

$(\alpha \equiv \beta\mu)$, so that PV is given by

$$PV = \pm \frac{4V\sqrt{\pi}}{3\eta\Lambda^3} k_B T \int_0^\infty \left[x^{\frac{3}{2}} - (x - \eta)^{\frac{3}{2}} H(\eta/x)\right] \ln(1 \pm e^{\alpha-x}) \, dx.$$

The integral appearing in the first term, namely

$$I(\alpha) = \int_0^\infty x^{\frac{3}{2}} \ln(1 \pm e^{\alpha-x}) \, dx$$

may be simplified by means of an integration-by-parts in terms of $x^{\frac{3}{2}}\,dx$ and $\ln(1 \pm e^{\alpha-x})$ to give

$$\int_0^\infty x^{\frac{3}{2}} \ln(1 \pm e^{\alpha-x})\,dx = \pm \int_0^\infty \frac{x^{\frac{5}{2}}}{e^{x-\alpha} \pm 1}\,dx$$

$$= \pm \tfrac{2}{5}\Gamma\left(\tfrac{7}{2}\right) g_{\frac{5}{2}}^{\pm}(\alpha),$$

with $\Gamma(\tfrac{7}{2}) = 15\sqrt{\pi}/8$. The same integration-by-parts technique may be employed to evaluate the integral in the second term following a change of variables from x to $y = x - \eta$. Combining the two results then gives [1]

$$PV = k_B T \, \frac{V}{\Lambda^3 \eta}\left[g_{\frac{5}{2}}^{\pm}(\alpha) - g_{\frac{5}{2}}^{\pm}(\alpha - \eta)\right],$$

for the equation of state of a quantum ideal gas in a gravitational field. $\qquad\qquad\square$

From expression (10.2.15) for $\Phi(V, T, \mu)$ we note that if we rewrite the integral over ϵ in terms of the dimensionless variable $x \equiv \beta\epsilon$, we may also express $\Phi(V, T, \alpha)$ as

$$\Phi(T, V, \alpha) = -\frac{(2I_a+1)V}{\Lambda^3(T)} \, \frac{4k_B T}{3\sqrt{\pi}} \int_0^\infty \frac{x^3}{e^{x-\alpha} \pm 1}\,dx$$

$$\equiv -VT^{\frac{5}{2}}\varphi(\alpha), \qquad\qquad\qquad (10.2.26)$$

in which φ is a function of the dimensionless variable α only. This expression for the grand potential allows us to express the system pressure P as

$$P = T^{\frac{5}{2}}\varphi(\alpha) \qquad\qquad\qquad (10.2.27)$$

in terms of T and α.

Because the entropy S is obtained as a partial derivative of Φ with respect to the temperature T, with μ, V held fixed [see Eq. (4.2.30b)], we may obtain the entropy as

$$S(T, V, \alpha) = VT^{\frac{3}{2}}\left[\tfrac{5}{2}\varphi(\alpha) - \alpha\varphi'(\alpha)\right], \qquad\qquad (10.2.28)$$

in which $\varphi'(\alpha)$ is the derivative of $\varphi(\alpha)$ with respect to α. Similarly, as $\overline{N}$ is given in terms of the grand potential $\Phi(T, V, \alpha)$ via Eq. (4.2.30a) as

$$\overline{N} = -\left(\frac{\partial\Phi}{\partial\mu}\right)_{T,V},$$

we obtain

$$\overline{N}(T, V, \alpha) = V T^{\frac{3}{2}} \frac{1}{k_B} \varphi'(\alpha).$$ (10.2.29)

We are now set to consider an adiabatic process, for which both the number of particles $\overline{N}$ and the entropy S are held constant. In particular, this means that the ratio

$$\frac{S}{\overline{N} k_B} = \frac{5}{2} \frac{\varphi(\alpha)}{\varphi'(\alpha)} - \alpha$$

must be constant which, in turn, requires that α, and hence $\varphi(\alpha)$ and $\varphi'(\alpha)$, be constant. The constancy of $\varphi(\mu/T)$, taken together with Eq. (10.2.27) for the pressure, then requires that $P T^{-\frac{5}{2}}$ be constant. Similarly, the constancy of $\overline{N}$, together with the constancy of $\varphi'(\mu/T)$ and Eq. (10.2.29) for $\overline{N}$, requires that $V T^{\frac{3}{2}}$ be constant. Hence, if we now raise $V T^{\frac{3}{2}}$ to the power 5/3, and multiply it by $P T^{-\frac{5}{2}}$ (in order to eliminate temperature) that product must also be constant: we thus have the requirement

$$P V^{\frac{5}{3}} = \text{constant}$$ (10.2.30)

for an adiabatic process in a quantum mechanical gas. This requirement is the same as that for an adiabatic process in a classical gas. Note, however, that in this case, it does not represent $P V^{\gamma}$, with $\gamma = C_P/C_V$, as $C_P/C_V \neq 5/3$ even for ideal BE and FD quantum gases.

10.3   Connection to Maxwell–Boltzmann Statistics

We have seen that when $\overline{N} \Lambda^3 / V$ is small we may apply the classical Maxwell–Boltzmann statistics and that when $\overline{N} \Lambda^3 / V$ is of order 1 or larger we must apply one of the quantum statistics. In particular, we note that $\overline{N} \Lambda^3(T)/V$ will be small for

- fixed temperature when $\overline{N}/V$ is small: if the container is also of fixed volume, then $\overline{N}/V$ small implies that $\overline{N}$ is small, which in turn implies via Eq. (10.2.20) that the individual $\overline{n}_k$ values must be very small, i.e.,

$$\overline{n}_k = \frac{1}{e^{\beta(\epsilon_k - \mu)} \pm 1} \ll 1.$$

This condition holds when $e^{\beta(\epsilon_k - \mu)} \gg 1$ or, equivalently, when $\epsilon_k \gg \mu$, in which case $\overline{n}_k$ can be approximated as

$$\overline{n}_k \simeq e^{-\beta(\epsilon_k - \mu)} = \lambda e^{-\beta \epsilon_k}.$$

- fixed $\overline{N}/V$ when $\Lambda(T)$ is small or, equivalently, when T is high: from Eq. (10.2.22), we thus see that $\lambda e^{-\beta\epsilon_k} \ll 1$, as for $\overline{N}/V$ is small.

Thus, $\overline{n}_k \to 0$ may be achieved by taking $\lambda \to 0$. This result is consistent with our knowledge that for classical (or Maxwell–Boltzmann) statistics to apply, there must be many more quantum N-body states than there are particles to occupy them, so that on average most states are unoccupied.

For Bose–Einstein particles (bosons), the mean occupation number $\overline{n}_k$ is given by the expression

$$\overline{n}_k^{BE} = \frac{\lambda e^{-\beta\epsilon_k}}{1 - \lambda e^{-\beta\epsilon_k}} = \frac{1}{\frac{1}{\lambda}e^{\beta\epsilon_k} - 1},$$

while for Fermi–Dirac particles (fermions) $\overline{n}_k$ is given by the expression

$$\overline{n}_k^{FD} = \frac{\lambda e^{-\beta\epsilon_k}}{1 + \lambda e^{-\beta\epsilon_k}} = \frac{1}{\frac{1}{\lambda}e^{\beta\epsilon_k} + 1},$$

and in the limit in which $\lambda e^{-\beta\epsilon_k} \ll 1$, both $\overline{n}_k^{BE}$ and $\overline{n}_k^{FD}$ reduce to a common result, namely

$$\overline{n}_k^{BE} = \overline{n}_k^{FD} \simeq \lambda e^{-\beta\epsilon_k} \ll 1. \tag{10.3.1a}$$

This expression gives the Maxwell–Boltzmann result, namely

$$\overline{n}_k^{MB} = \lambda e^{-\beta\epsilon_k}. \tag{10.3.1b}$$

In the Maxwell–Boltzmann limit (10.3.1b) for the mean occupation number $\overline{n}_k^{MB}$, expression (10.2.20) for the thermodynamic $\overline{N}$ takes the form

$$\overline{N} \simeq \sum_k \lambda e^{-\beta\epsilon_k} = \lambda \sum_k e^{-\beta\epsilon_k}. \tag{10.3.2}$$

If we focus upon a single well-defined value of $\overline{N}$, say N, we would have

$$N = \lambda \sum_k e^{-\beta\epsilon_k} = \lambda z, \tag{10.3.3}$$

in which z is the usual canonical partition function for a single particle. From expression (10.3.3) we see that the absolute activity λ is thus related to z and N by

$$\lambda = \frac{N}{z}. \tag{10.3.4}$$

Thus, for $\lambda \ll 1$, the Maxwell–Boltzmann mean occupation number $\bar{n}_k^{\text{MB}}$ can be written as

$$\bar{n}_k^{\text{MB}} = \frac{N}{z} e^{-\beta \epsilon_k} , \tag{10.3.5a}$$

from which we see that the classical probability p_k^{MB} is related to the mean occupation number via

$$p_k^{\text{MB}} \equiv \frac{e^{-\beta \epsilon_k}}{z} = \frac{\bar{n}_k^{\text{MB}}}{N} . \tag{10.3.5b}$$

In this sense, for both FD and BE statistics, the mean occupation number $\bar{n}_k$ reduces to the classical probability p_k^{MB}.

We note, in passing, that as the average number of particles, $\overline{N}$, and the average energy $\overline{E}$, are given, respectively, by

$$\overline{N} = \sum_k \bar{n}_k \simeq \sum_k \lambda e^{-\beta \epsilon_k} \tag{10.3.6}$$

and

$$\overline{E} \simeq \sum_k \lambda \epsilon_k e^{-\beta \epsilon_k} , \tag{10.3.7}$$

the average energy per particle, $\bar{\epsilon} \equiv \overline{E}/\overline{N}$, becomes

$$\bar{\epsilon} = \frac{\displaystyle\sum_k \lambda \epsilon_k e^{-\beta \epsilon_k}}{\displaystyle\sum_k \lambda e^{-\beta \epsilon_k}} = \frac{1}{z} \sum_k \epsilon_k e^{-\beta \epsilon_k} , \tag{10.3.8}$$

and the absolute activity λ (equivalently, the chemical potential μ) no longer plays a role in the statistical expressions.

In this same limit, the equation of state becomes

$$PV = \pm k_{\text{B}} T \sum_k (\pm) \lambda e^{-\beta \epsilon_k} . \tag{10.3.9}$$

Upon utilizing Eq. (10.2.11) for Ξ and employing the approximation $\ln(1+x) \simeq x$ for $x \ll 1$, we may also express PV as

$$PV = \lambda k_{\text{B}} T \sum_k e^{-\beta \epsilon_k}$$

$$= k_{\text{B}} T \lambda z . \tag{10.3.10}$$

Now, upon recalling [see Eq. (4.2.31)] that $PV/(k_B T) = \ln \Xi$, we see that Ξ may also be obtained as

$$\Xi(T, V, \lambda) = e^{\lambda z(T,V)} . \qquad (10.3.11a)$$

If we compare Eq. (4.2.4), namely

$$\Xi(T, V, \lambda) = \sum_{N=0}^{\infty} \lambda^N Z_N(T, V),$$

relating the canonical partition function $Z_N(T, V)$ to the grand partition function $\Xi(T, V, \lambda)$ with the expanded form of Eq. (10.3.11a), namely

$$\Xi(T, V, \lambda) = \sum_{N=0}^{\infty} \frac{[\lambda z(T, V)]^N}{N!} , \qquad (10.3.11b)$$

we may conclude that $Z_N(T, V)$ is given in terms of the single particle canonical partition function $z(T, V)$ by

$$Z_N(T, V) = \frac{z^N(T, V)}{N!} . \qquad (10.3.12)$$

This procedure has provided us with a more rigorous derivation of our earlier result, Eq. (3.2.27), for indistinguishable particles/molecules using Maxwell–Boltzmann statistics and based upon the Gibbs rationalization for division of the power z^N by $N!$.

Let us now briefly reconsider the mean occupation number, $\bar{n}_k$, which is given by

$$\bar{n}_k = \frac{\lambda e^{-\beta \epsilon_k}}{1 \pm \lambda e^{-\beta \epsilon_k}} = \frac{1}{\frac{1}{\lambda} e^{\beta \epsilon_k} \pm 1} . \qquad (10.3.13)$$

If we recall that $\lambda = e^{\mu/k_B T}$, we see that $\lambda e^{-\beta \epsilon} = 1$ when $\mu = \epsilon$. This result has strong implications for each of the statistics: from Eq. (10.2.22), $\bar{n}_k$ goes to infinity for bosons, takes the value $\frac{1}{2}$ for fermions, and takes the value 1 for the limiting case of Maxwell–Boltzmann statistics as ϵ_k approaches μ. This can also be seen from Fig. 10.1, which compares mean occupation numbers for the three cases. We note also that for large values of ϵ (i.e., $\epsilon \gg \mu$) the quantum states of the system are rarely occupied, if at all, and the probability of multiple occupancy becomes essentially nil.

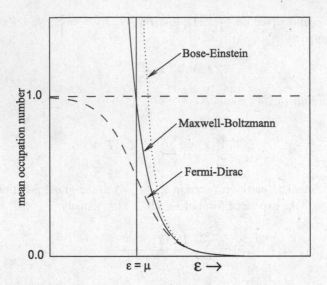

Fig. 10.1 Comparison between BE, FD, and MB statistics

We have also seen that $\overline{n}_k$ is the quantum mechanical counterpart of the probability p_k that we used in our classical (or Maxwell–Boltzmann) statistical description. The internal energy $U \equiv \overline{E}$ is given in terms of $\overline{n}_k$ by

$$U = N\overline{\epsilon} = \sum_k \overline{n}_k \epsilon_k = \sum_k \frac{\lambda \epsilon_k e^{-\beta \epsilon_k}}{1 \pm \lambda e^{-\beta \epsilon_k}}, \qquad (10.3.14)$$

while the equations of state are given by

$$(PV)_{\text{FD}} = k_{\text{B}} T \ln(\Xi)_{\text{FD}}(V, T, \lambda) = k_{\text{B}} T \sum_k \ln\left(1 + \lambda e^{-\beta \epsilon_k}\right) \qquad (10.3.15)$$

for a Fermi–Dirac system, and by

$$(PV)_{\text{BE}} = k_{\text{B}} T \ln(\Xi)_{\text{BE}}(V, T, \lambda) = -k_{\text{B}} T \sum_k \ln\left(1 - \lambda e^{-\beta \epsilon_k}\right) \qquad (10.3.16)$$

for a Bose–Einstein system. We shall discuss these results further for certain special cases yet to be defined.

Equations (10.3.15) and (10.3.16) are exact expressions for the equations of state for Fermi–Dirac and Bose–Einstein particles. They are not, however, the most convenient representations for the equations of state due to their dependence on the chemical potential μ (equivalently, the absolute activity λ).

10.4 Fermi–Dirac Statistics

10.4.1 The Ideal Fermi Gas

As has been pointed out in the introductory section of this chapter, the classical Gibbs treatment of indistinguishable particles is valid for $N\Lambda^3(T)/V \ll 1$. The ^{3}He isotope of helium is a fermion (due to its nuclear spin $I_a = \frac{1}{2}$) and is the lightest element for which the atom is the naturally-occurring form on our planet.

If we consider ^{3}He at temperature $T = 300\,\mathrm{K}$, for example, it has a thermal de Broglie wavelength, $\Lambda(T)$, of approximately 0.45 nm. For a sample of ^{3}He at a pressure $P^\circ = 1\,\mathrm{bar}$ and molar volume $\overline{V}^\circ = 24.943\,\mathrm{L}$, we obtain $\overline{V}^\circ/N_0 \simeq (6.44\,\mathrm{nm})^3$, so that $N_0\Lambda^3(T = 300\,\mathrm{K})/\overline{V}^\circ(300\,\mathrm{K}) \simeq 3.42 \times 10^{-4}$, which is indeed quite small. This means that even for ^{3}He, we do not really require the use of Fermi–Dirac statistics to describe properties of an ideal gas with this mass at a temperature of 300 K. However, if we carry out the same estimate for a sample of ^{3}He at a temperature of 5 K, for example, we obtain $N_0\Lambda^3(5\,\mathrm{K})/\overline{V}^\circ(5\,\mathrm{K}) \simeq 61.4$: even for a temperature $T = 50\,\mathrm{K}$, this ratio has a value of approximately 0.108, so that it becomes clear that a quantum-statistical description of ^{3}He will be necessary for ^{3}He at any temperature significantly below room temperature.

In the following we shall focus upon a gas of spin-$\frac{1}{2}$ fermions, so that

$$\Omega_{\mathrm{FD}}(\epsilon) = 4\pi \left(\frac{2m}{h^2}\right)^{\frac{3}{2}} V\epsilon^{\frac{1}{2}}, \tag{10.4.1}$$

since the number of spin-$\frac{1}{2}$ fermion states is twice the number of classical translational states [note that for arbitrary spin I_a, we would require a factor $2I_a + 1$ rather than simply 2, see Eq. (10.2.13b)].

From Eq. (10.2.12), together with the relation $\Phi = -PV$, we obtain the equation of state for the Fermi–Dirac gas as

$$\left(\frac{PV}{k_\mathrm{B}T}\right)_{\mathrm{FD}} = \sum_k \ln\left(1 + \lambda e^{-\beta\epsilon_k}\right).$$

We begin by replacing this discrete variable expression by the continuum approximation

$$\left(\frac{PV}{k_\mathrm{B}T}\right)_{\mathrm{FD}} \simeq \int_0^\infty \Omega_{\mathrm{FD}}(\epsilon) \ln\left(1 + \lambda e^{-\beta\epsilon}\right) \mathrm{d}\epsilon$$

$$= 4\pi \left(\frac{2m}{h^2}\right)^{\frac{3}{2}} V \int_0^\infty \epsilon^{\frac{1}{2}} \ln\left(1 + \lambda e^{-\beta\epsilon}\right) \mathrm{d}\epsilon. \tag{10.4.2}$$

Further, from Eq. (10.2.21) we obtain the number of Fermi–Dirac particles

$$\overline{N}_{\mathrm{FD}} = \sum_k \frac{\lambda e^{-\beta\epsilon}}{1 + \lambda e^{-\beta\epsilon}} ,$$

which we shall similarly approximate using a continuum expression, namely

$$\overline{N}_{\mathrm{FD}} \simeq \int_0^\infty \Omega_{\mathrm{FD}}(\epsilon) \frac{\lambda e^{-\beta\epsilon}}{1 + \lambda e^{-\beta\epsilon}} \, d\epsilon$$

$$= 4\pi \left(\frac{2m}{h^2}\right)^{\frac{3}{2}} V \int_0^\infty \frac{\lambda \epsilon^{\frac{1}{2}} e^{-\beta\epsilon}}{1 + \lambda e^{-\beta\epsilon}} \, d\epsilon . \qquad (10.4.3)$$

Because these integrals cannot be evaluated in closed form, we shall be content for small λ to use the following expansions in order to generate a series solution:

$$\ln\left(1 + \lambda e^{-\beta\epsilon}\right) = \sum_{j=1}^\infty \frac{(-1)^{j-1}}{j} \lambda^j e^{-\beta j\epsilon} , \qquad (10.4.4)$$

and

$$\frac{1}{1 + \lambda e^{-\beta\epsilon}} = \sum_{j=0}^\infty (-1)^j \lambda^j e^{-\beta j\epsilon} . \qquad (10.4.5)$$

Substitution of these expressions into Eqs. (10.4.2) and (10.4.3), followed by term-by-term integration, gives

$$\frac{P}{k_{\mathrm{B}}T} = \frac{2}{\Lambda^3} \sum_{j=1}^\infty \frac{(-1)^{j-1}\lambda^j}{j^{\frac{5}{2}}} = \frac{2}{\Lambda^3}\left(\lambda - \frac{\lambda^2}{2^{\frac{5}{2}}} + \frac{\lambda^3}{3^{\frac{5}{2}}} - \cdots\right) , \qquad (10.4.6)$$

and

$$\frac{\overline{N}}{V} = \rho = \frac{2}{\Lambda^3} \sum_{j=1}^\infty \frac{(-1)^{j-1}\lambda^j}{j^{\frac{3}{2}}} = \frac{2}{\Lambda^3}\left(\lambda - \frac{\lambda^2}{2^{\frac{3}{2}}} + \frac{\lambda^3}{3^{\frac{3}{2}}} - \cdots\right) . \qquad (10.4.7)$$

We may assume that λ in Eqs. (10.4.6) and (10.4.7) is itself a power series in the number density ρ. From Eq. (10.4.7) we see that the power series expansion of λ must be such that

$$\lambda = \frac{\Lambda^3 \rho}{2} + a_1 \rho^2 + a_2 \rho^3 + \cdots , \qquad (10.4.8)$$

so that if we now take powers of λ and substitute them back into Eq. (10.4.7) and equate powers of ρ on the LH and RH sides of the resultant equation, we can readily obtain expressions for a_1 and a_2, namely

$$a_1 = \frac{1}{4}\frac{\Lambda^6}{2^{\frac{3}{2}}}, \qquad a_2 = \frac{1}{8}\left(\frac{\Lambda^9}{4} - \frac{\Lambda^9}{3^{\frac{3}{2}}}\right). \tag{10.4.9}$$

As a consequence of these results, the expansion for λ in terms of ρ becomes

$$\lambda = \frac{\Lambda^3 \rho}{2} + \frac{1}{4}\frac{\Lambda^6}{2^{\frac{3}{2}}}\rho^2 + \frac{\Lambda^9}{8}\left(\frac{1}{4} - \frac{1}{3^{\frac{3}{2}}}\right)\rho^3 + \dots. \tag{10.4.10}$$

If we substitute this expansion for λ into Eq. (10.4.6), we obtain for the equation of state for our ideal FD gas the result

$$\frac{P}{k_B T} = \rho + \frac{\Lambda^3 \rho^2}{2^{\frac{7}{2}}} + \Lambda^6 \rho^3 \left(\frac{1}{8} - \frac{2}{3^{\frac{5}{2}}}\right) + \dots. \tag{10.4.11}$$

If we now compare the result (10.4.11) with the traditional virial expansion for the pressure of an ideal spin-$\frac{1}{2}$ fermion gas,

$$\frac{P}{k_B T} = \rho + B_2(T)\rho^2 + B_3(T)\rho^3 + \dots, \tag{10.4.12}$$

we see that

$$B_2(T) = \frac{\Lambda^3}{2^{\frac{7}{2}}}. \tag{10.4.13}$$

and

$$B_3(T) = \frac{\Lambda^6}{4}\left(\frac{1}{8} - \frac{2}{3^{\frac{5}{2}}}\right). \tag{10.4.14}$$

It is important to note that these expressions for virial coefficients are for quantum mechanical particles that correspond to ideal gas classical particles, i.e., particles in the absence of normal interparticle interactions. These virial contributions are thus truly quantum mechanical in origin, and can be associated with effective interactions between the particles that are imposed upon them in the Fermi–Dirac case considered above by the *requirement* of a fully antisymmetrized N-body wavefunction. This effective interaction between ideal fermions is *repulsive*, as shown by the positive leading correction term, B_2, to the perfect quantum gas (corresponding to the classical ideal gas) limit. The effective (repulsive) interaction between identical fermions is sometimes referred to as a 'pairing repulsion' or as the 'Pauli repulsion'.

Table 10.2 Second Virial
Coefficient for ^{3}He at Low
Temperatures

T/K	Contributions		Total
	B_{ideal}	$B_{interaction}$	$B_2(T)$
0.5	152.95	−508.08	−355.13
1.0	54.08	−286.53	−232.45
2.0	19.12	−147.85	−128.73
3.0	10.41	−95.31	−84.90
4.0	6.41	−68.01	−61.25
5.0	4.84	−51.36	−46.52
6.0	3.68	−40.16	−36.48
7.0	2.92	−32.12	−29.20
8.0	2.39	−26.05	−23.66
9.0	2.00	−21.31	−19.31
10.0	1.71	−17.51	−15.80

Table 10.2 shows the quantum mechanical contribution $B_{2,\,ideal}(T)$ given by
Eq. (10.4.13) (labelled as B_{ideal}) plus, for comparison, the nonideal (i.e., normal
interparticle interaction) contribution $B_{interaction}$, to $B_2(T)$ for ^{3}He gas over the
temperature range $0.5\,K \le T \le 10.0\,K$ [2]. All values of B_2 are given in the
traditional units of $cm^3\,mol^{-1}$. Note, in particular, that the quantum contribution
has almost vanished by temperature 10 K.

10.4.2   The Chemical Potential: Fermi Gas

We have seen in Sect. 5.2 of Chap. 5, given that the characteristic temperature T_0
at which the chemical potential μ changes sign is of the order of 0.01 K for a
typical classical ideal gas, that μ is, as illustrated in Fig. 5.1, inherently negative.
The temperature dependence of the chemical potential of an ideal Fermi–Dirac
gas is similarly illustrated in Fig. 10.2, where μ/ϵ_F has been plotted vs. T/T_F.
The characteristic temperature T_F at which μ changes sign is called the Fermi
temperature and is defined as

$$T_F \equiv \frac{\epsilon_F}{k_B}.$$
(10.4.15)

However, as T_F for electrons in a metal is typically of the order of thousands of
kelvin, is this behaviour reasonable, given what we have seen for the ideal classical
gas? To see why the behaviour of μ exhibited in Fig. 10.2 is indeed appropriate, let
us consider briefly what is happening for an ideal Fermi gas [3, 4].

At $T = 0$ we know, essentially from the definition of the Fermi energy ϵ_F, both
that all single-particle states lying below ϵ_F are filled, in accordance with the Pauli
Principle, and that all single-particle states having energies in excess of ϵ_F are empty.
As there can be only one microstate corresponding to this situation, the entropy S

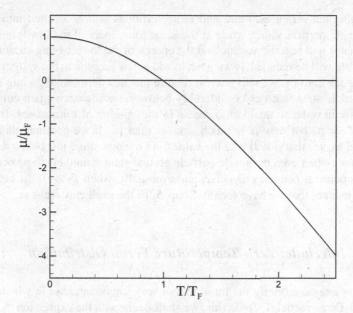

Fig. 10.2 The temperature dependence of the chemical potential for a Fermi–Dirac gas. Figure provided by Dr. Lee Huntington

will necessarily be zero, in full agreement with the usual statement of the 3rd Law of thermodynamics. Let us consider now what happens when another fermion is added to the system at $T = 0$. This additional fermion must be placed into the single-particle state that lies immediately above the Fermi level, thereby increasing the total internal energy of the system by $\Delta U = \epsilon_F$. According to the thermodynamic definition of the chemical potential, which we recall is $\mu \equiv \left(\dfrac{\partial U}{\partial N} \right)_{S,V}$, the change ΔU in the internal energy associated with this action must equal the chemical potential μ. As $\epsilon_F > 0$, the chemical potential that we have obtained is thus positive and, as there is still only one microstate available to the system, the entropy remains unchanged at zero.

Let us now consider what happens when the temperature T is nonzero. We note firstly that the (total) internal energy of the system will increase when some of the fermions begin to occupy excited single-particle states, and secondly, that the entropy of the system will also increase as more microstates become available. If we consider the effect of adding one more fermion at a very low temperature (i.e., $T \ll T_F$), we must focus upon adding the fermion in such a way that the system entropy remains unchanged: the only way to accomplish this is to place the new fermion into a vacant single-particle state that lies slightly below the Fermi energy, while simultaneously cooling the gas slightly in order to avoid an increase in the number of accessible microstates. The change ΔU in the internal energy will again give μ, but in this case, μ, while positive, will be slightly smaller than ϵ_F.

At higher temperatures, more and more fermions will be excited into higher-energy single-particle states, while at the same time, more of the low-lying single-particle states will become vacant, and the energy of the lowest-lying vacant single-particle state will be reduced. If we wish to add a new fermion to the system without increasing the entropy, we will have to require this new fermion to go into a vacant single-particle state that lies considerably below ϵ_F, while once again cooling the gas slightly in order to avoid an increase in the number of microstates. In such a case, $\Delta U = \mu$, but now μ is much smaller than ϵ_F. If we continue this line of reasoning, we see that μ will take the value 0 for a temperature just below the Fermi temperature (when even the single-particle ground state is unlikely to be occupied). After this point, μ becomes negative, and eventually, when $T \gg T_F$, μ behaves in the same manner that we have seen in Chap. 5 for the ideal classical gas.

10.4.3   Interlude: Zero-Temperature Fermi Distribution

Let us now examine briefly the limiting, but very important, case in which $T = 0$ for Fermi–Dirac particles. To do this, we shall begin with the expression

$$\overline{n}_k^{\mathrm{FD}} = \frac{1}{e^{\beta(\epsilon_k - \mu)} + 1} \, ,$$

and take the limit that T goes to zero to obtain

$$\lim_{T \to 0} \overline{n}_k^{\mathrm{FD}} = \lim_{\beta \to \infty} \overline{n}_k^{\mathrm{FD}} = \lim_{\beta \to \infty} \frac{1}{e^{\beta(\epsilon_k - \mu)} + 1} \, .$$

We must consider two separate cases for this limit, depending upon whether ϵ_k is greater than or less than μ: now, if $\epsilon_k > \mu$, the result is

$$\lim_{\beta \to \infty} \frac{1}{e^{\beta(\epsilon_k - \mu)} + 1} = 0 \, ,$$

while if $\epsilon_k < \mu$, the result is

$$\lim_{\beta \to \infty} \frac{1}{e^{\beta(\epsilon_k - \mu)} + 1} = 1 \, .$$

Finally, when $\epsilon_k = \mu$, we see that $\overline{n}_k = 0.5$. Hence, for absolute zero temperature, the graph of $\overline{n}_k$ for Fermi–Dirac particles is a discontinuous function with the form shown in Fig. 10.3.

All states with energy ϵ_k less than $\epsilon_F \equiv \mu(0)$ are occupied, and all states with energy $\epsilon_k > \mu(0)$ are unoccupied: ϵ_F is called the Fermi energy, and its value on an energy scale is referred to as the Fermi level.

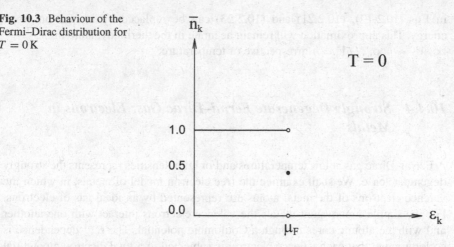

Fig. 10.3 Behaviour of the Fermi–Dirac distribution for $T = 0\,\mathrm{K}$

For Bose–Einstein particles we know that $\bar{n}_k$ cannot be negative, and that $\bar{n}_k^{BE}$ is given by

$$\bar{n}_k^{BE} = \frac{\lambda e^{-\beta\epsilon_k}}{1 - \lambda e^{-\beta\epsilon_k}} = \frac{1}{\frac{1}{\lambda}e^{\beta\epsilon_k} - 1}.$$

The ground state energy for a Bose–Einstein particle is ϵ_0, which we can take to define the zero of energy, i.e., we may set $\epsilon_0 \equiv 0$, so that the mean occupation number for bosons in the ground state becomes

$$\bar{n}_0^{BE} = \frac{\lambda}{1 - \lambda}; \quad 0 \le \lambda < 1.$$

We have characterized the ideal gas by the absence of interparticle (or inter-molecular) forces, and we have indicated that the existence of such interparticle forces classically must be taken into consideration whenever the second virial coefficient (in the virial equation of state) becomes important. The particles in a classical ideal gas can thus be considered to be independent of one another. For ideal quantum gases, however, the particles are not really independent of one another, because of the occupation number rules, which do not permit the doubling up of fermions in the same spin state, but do permit it for bosons. Because of these rules, and the concomitant subtle correlations between particles, the properties of an ideal quantum gas cannot be stated solely in terms of a single-particle partition function. Let us therefore examine the equation of state to be obtained for a weakly degenerate quantum gas (this is the case for which $\lambda \ll 1$).

For the ensuing discussions we shall restrict ourselves to cases of gases of quantum particles with no internal structure apart from spin, so that the summations in Eqs. (10.2.12), (10.2.21) and (10.2.23) can be replaced by integrations over energy. This approximation will remain accurate in the thermodynamic limit $\overline{N} \to \infty$, $V \to \infty$, $\overline{N}/V = n$, irrespective of temperature.

10.4.4 Strongly Degenerate Fermi–Dirac Gas: Electrons in Metals

A Fermi–Dirac gas at low temperatures and/or high densities represents the strongly degenerate case. We shall examine the free electron model of metals, in which the valence electrons of the metal atoms are represented by an ideal gas of electrons. For the simple monovalent metals the valence electrons interact with one another and with the atomic cores through a Coulombic potential. The R^{-1} dependence is so long-range that each valence electron is subjected to a total electronic potential that is almost constant from location to location in the metallic crystal. The resulting, and essentially free, electrons are referred to as conduction electrons. In fact, as we shall see, many of the physically observable properties of metals can be associated more readily with quantum-statistical effects than with the details of the electron-electron and electron-ionic core interactions.

The electron occupation number $\overline{n}_k$ for the state with energy ϵ_k is given by Eq. (10.3.13), with the + sign, namely

$$\overline{n}_k = \frac{\lambda e^{-\beta \epsilon_k}}{1 + \lambda e^{-\beta \epsilon_k}} = \frac{1}{1 + e^{\beta(\epsilon_k - \mu)}} . \tag{10.4.16}$$

Now, if we assume, as in the previous section, a continuous distribution of energy states, we can write this in the form

$$f(\epsilon) = \frac{1}{1 + e^{\beta(\epsilon - \mu)}} , \tag{10.4.17}$$

in which $f(\epsilon)$ represents the probability that a given state is occupied (see Fig. 10.4). For $T = 0$ the distribution is unity for $\epsilon < \mu$ and zero for $\epsilon > \mu$. All states having energy less than $\epsilon_F \equiv \mu(0)$ are occupied, while those having energy greater than ϵ_F are unoccupied: in this sense, ϵ_F serves as a cut-off energy. As the number of states lying between ϵ and $\epsilon + d\epsilon$ is given by Eq. (10.4.1), we are able to determine ϵ_F directly from the fact that all states lying below the energy ϵ_F are occupied, while all states lying above ϵ_F remain unoccupied.

We shall obtain the equation of state for the conduction electrons by obtaining firstly an expression for their internal energy, $U_0 \equiv U_{FD}(T = 0)$, then utilizing the thermodynamic relation

$$P = - \left(\frac{\partial U}{\partial V} \right)_{S,N}$$

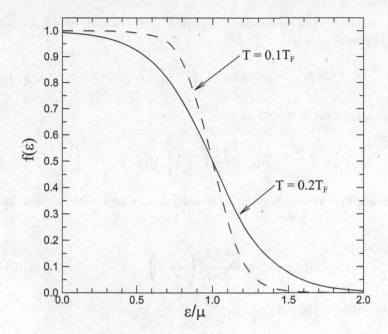

Fig. 10.4 Fermi–Dirac distribution for finite temperature

relating the thermodynamic pressure and the internal energy, to obtain an expression for $P_{ce} \equiv P(T = 0)$. We begin with the defining relation,

$$U = \sum_k \overline{n}_k \epsilon_k ,$$

for the thermodynamic internal energy, and convert the summation over the discrete energy states into an integration over continuum energy states via Eqs. (10.2.13) with $I_a = \frac{1}{2}$. The internal energy is then given in general as

$$U \simeq 4\pi \left(\frac{2m_e}{h^2}\right)^{\frac{3}{2}} V \int_0^\infty f(\epsilon) \epsilon^{\frac{3}{2}} \, d\epsilon . \tag{10.4.18}$$

Because the Fermi–Dirac distribution for temperature $T = 0$ is a step function, expression (10.4.18) simplifies considerably to yield for U_0 the result

$$U_0 \simeq 4\pi \left(\frac{2m_e}{h^2}\right)^{\frac{3}{2}} V \int_0^{\epsilon_F} \epsilon^{\frac{3}{2}} \, d\epsilon$$

$$= \frac{8\pi}{5} \left(\frac{2m_e}{h^2}\right)^{\frac{3}{2}} V \epsilon_F^{\frac{5}{2}} . \tag{10.4.19}$$

The total number of conduction electrons, $\overline{N}_{ce}$, in a metallic crystal is similarly given, in general as

$$\overline{N}_{FD} \equiv N_{ce} = \int_0^\infty \Omega_{FD}(\epsilon)\, d\epsilon \qquad (10.4.20)$$

or, more specifically, at $T = 0$, as

$$\overline{N}_{ce} = \frac{8\pi}{5} \left(\frac{2m_e}{h^2}\right)^{\frac{3}{2}} V\epsilon_F^{\frac{3}{2}}. \qquad (10.4.21)$$

This result allows us to express the Fermi energy, ϵ_F, in terms of the number density, $\overline{N}_{ce}/V$, as

$$\epsilon_F = \frac{h^2}{8m_e} \left(\frac{3\overline{N}_{ce}}{\pi V}\right)^{\frac{2}{3}} \qquad (10.4.22a)$$

or, upon evaluating the constants, as

$$\epsilon_F = 5.8423 \times 10^{-38} n_{ce}^{\frac{2}{3}} \text{ J}, \qquad (10.4.22b)$$

in terms of the number density $n_{ce} \equiv \overline{N}_{ce}/V$. We may also define a characteristic temperature, T_F, for the conduction electron gas via the Fermi energy as

$$T_F = \frac{\epsilon_F}{k_B} \qquad (10.4.23a)$$

or, expressed directly in terms of n_{ce} as

$$T_F = 4.2315 \times 10^{-15} n_{ce}^{\frac{2}{3}} \text{ K}. \qquad (10.4.23b)$$

Let us note in passing that the internal energy, U_0, may also be expressed quite simply in terms of $\overline{N}_{ce}$ and ϵ_F as

$$U_0 = \tfrac{3}{5}\overline{N}_{ce}\epsilon_F. \qquad (10.4.24)$$

Example 10.3 Conduction electrons in sodium metal.

As the sodium atom has an unpaired $3s$ valence electron, each atom in a sample of metallic sodium contributes one electron tö the sodium metal conduction band. To determine the number density of these sodium conduction electrons in a sample of the metal, given by

$$n_{ce}(Na) = \frac{\overline{N}_{ce}(Na)}{V(Na)},$$

we require values for the number of conduction electrons, $\overline{N}_{ce}$, and the volume V of the sample. For convenience, we shall determine the value for $n_{ce}(Na)$ for a molar sample of metallic sodium at the standard temperature of 25 °C (298.15 K). For this, we need values for the molecular weight (molar mass) and the mass density, $\rho(Na)$. These quantities are available from any standard reference source, such as the NIST website http://webbook.nist.gov.chemistry or the *CRC Handbook of Chemistry and Physics* [5], for example: $\overline{M}(Na) = 22.99\,\text{g mol}^{-1}$ $(0.02299\,\text{kg mol}^{-1}$ in SI units) and $\rho(Na) = 0.97\,\text{g cm}^{-3}$ at 298.15 K $(970\,\text{kg m}^{-3}$ in SI units). The molar volume of sodium metal at $T = 298.15$ K is thus obtained from $\overline{V} = \overline{M}/\rho$ as

$$\overline{V}(Na) = \frac{0.02299\,\text{kg mol}^{-1}}{970\,\text{kg m}^{-3}}$$

$$= 2.37_0 \times 10^{-5}\,\text{m}^3\,\text{mol}^{-1}.$$

Note that the overbars on $\overline{V}$ and $\overline{M}$ denote molar amounts, unlike the overbar on $\overline{N}$ that we have employed to distinguish the thermodynamic quantity N_{thermo} from the variable value of N for members of the grand ensemble.

The number density of conduction electrons in metallic sodium at $T = 298.15$ K is thus given by

$$n_{ce}(Na) = \frac{N_0}{\overline{V}(Na)} = \frac{6.022 \times 10^{23}\,\text{mol}^{-1}}{2.370 \times 10^{-5}\,\text{m}^3\,\text{mol}^{-1}}$$

$$= 25.40_9 \times 10^{27}\,\text{m}^{-3}.$$

We shall also require a value for $n_{ce}(Na)$ at 0 K, which we may estimate from $n_{ce}(Na)$ at 298.15 K by making use of the coefficient of thermal expansion, $\alpha(Na) = 7.1 \times 10^{-5}\,\text{K}^{-1}$ (from the CRC Handbook of Chemistry and Physics) to estimate the relative change in the molar volume as $\Delta\overline{V}(Na)/\overline{V}(Na) \approx \alpha(Na)\Delta T$, with $\Delta T = -298.15$ K, to give

$$\overline{V}_{T=0}(Na) \simeq (1 - 0.0212)\overline{V}_{T=298}(Na)$$

$$= 2.31_5 \times 10^{-5}\,\text{m}^3\,\text{mol}^{-1}.$$

This gives a conduction electron number density of

$$n_{ce}(T = 0) = 25.96 \times 10^{27}\,\text{m}^{-3},$$

from which the Fermi energy $\epsilon_F(Na)$ is obtained via Eq. (10.4.21) as

$$\epsilon_F(Na) = \left(\frac{5n_{ce}(T = 0)}{8\pi}\right)^{\frac{2}{3}} \frac{h^2}{8m_e}$$

$$= 5.12_2 \times 10^{-19}\,\text{J}$$

or, in units of eV, as

$$\epsilon_F \simeq 3.19_7 \, \text{eV} \, ,$$

compared with 3.24 eV given in the CRC Handbook of Chemistry and Physics. The corresponding Fermi temperature, T_F, is then

$$T_F \simeq 37,000 \, \text{K} \, . \quad \square$$

We see from this example that the Fermi temperature is typically of the order of several tens of thousands of degrees kelvin, which means that for a typical solid metal sample, the thermal energy $k_B T$ will be only a small fraction of the Fermi energy. For this reason, the distribution $f(\epsilon)$ behaves essentially like a step function, so that approximating it as

$$f(\epsilon) \approx f_0(\epsilon) = \begin{cases} 1, & \epsilon < \epsilon_F \\ 0, & \epsilon > \epsilon_F \end{cases} , \qquad (10.4.25)$$

is actually an excellent approximation even at room temperature. The Pauli Exclusion Principle thus implies that a FD gas has a large mean energy even at absolute zero.

The result (10.4.4) for U_0 implies that the contribution of the conduction electrons to the heat capacity, $\overline{C}_V$, is zero, which differs sharply from the value $\frac{3}{2} k_B$ per electron that would be predicted by classical equipartition arguments [see, for example, Sect. 5.5.2]. Physically, this is explained by the fact that ϵ_F is so high that only a very small fraction of all electrons will lie within $k_B T$ of the top of the distribution, that is, near vacant quantum states to which they may be excited thermally in order to contribute to $\overline{C}_V$. For this reason, only a very small number of the conduction electrons can contribute to $\overline{C}_V$, and hence the experimental heat capacity associated with the conduction electrons will be very nearly zero.

A zero contribution from the conduction electrons to the heat capacity of a metal will strictly be true only when we have precisely the 0 K distribution. How, then, will $\overline{C}_V$ behave for $f(\epsilon)$ differing slightly from $f_0(\epsilon)$? To have some idea of what to expect, we shall pursue the physical argument a little further. Those electrons whose energies lie no more than $k_B T$ below ϵ_F can be thermally excited to at most $k_B T$ above ϵ_F, since electrons deeper below the Fermi surface cannot be thermally excited (as there will be no *empty* levels into which they can be excited). The number of electrons, $\overline{N}'_{ce}$, so excited will thus be

$$\overline{N}'_{ce} \simeq \left(\frac{k_B T}{\epsilon_F} \right) \overline{N}_{ce} \, , \qquad (10.4.26)$$

and their average excitation energy will be $k_B T$, so that their contribution, ΔU, to the internal energy will be

$$\Delta U \simeq \overline{N}'_{ce} k_B T = \frac{\overline{N}_{ce} k_B{}^2 T^2}{\epsilon_F} . \tag{10.4.27}$$

The electronic contribution to the heat capacity will thus be

$$\Delta C_V \simeq 2\overline{N}_{ce} \frac{k_B{}^2 T}{\epsilon_F} = 2\overline{N}_{ce} k_B \frac{T}{T_F} , \tag{10.4.28}$$

which is linear in temperature, and very small at room temperature, given the value for the Fermi temperature, T_F, for a typical metal.

A. The Sommerfeld Expansion

The physical argument presented above indicates, as was recognized by Sommerfeld, the need for a low-temperature/high-density expansion of the thermodynamic properties to obtain an expression for $\Delta C_V(T)$. To this end, let us define an appropriate expansion parameter η as

$$\eta \equiv \frac{k_B T}{\epsilon_F} = (\beta \epsilon_F)^{-1} . \tag{10.4.29}$$

Let us recall that for a continuum distribution of energy states, the summation over $\overline{n}_k$ for a discrete distribution of energy states is replaced by an integration over ϵ of the summand weighted by the density of states. Thus, we make the replacements

$$\sum_{\overline{n}_k} \cdots \leftrightarrow \int_0^\infty d\epsilon \, \Omega_{FD}(\epsilon) \cdots , \tag{10.4.30}$$

and

$$\overline{n}_k(\epsilon) \leftrightarrow f(\epsilon) , \tag{10.4.31}$$

to convert expression (10.2.21) for $\overline{N}_{ce}$ into

$$\overline{N}_{ce} = \int_0^\infty f(\epsilon) \Omega_{ce}(\epsilon) \, d\epsilon , \tag{10.4.32}$$

with $\Omega_{ce}(\epsilon)$ given by Eq. (10.2.13b) with $I_a = \frac{1}{2}$, as

$$\Omega_{ce}(\epsilon) = 4\pi \left(\frac{2m_e}{h^2} \right)^{\frac{3}{2}} V \epsilon^{\frac{1}{2}} \tag{10.4.33a}$$

$$\equiv 2\Omega_0(V) \epsilon^{\frac{1}{2}} , \tag{10.4.33b}$$

with $2I_a + 1 = 2$, $\Omega_0(V) \equiv 2\pi(2m_e/h^2)^{\frac{3}{2}}V$, so that $\Omega_0(V)\epsilon^{\frac{1}{2}}$ represents the density-of-states for a classical (i.e., spinless) particle.

In the same way, we may write the internal energy U as

$$U = \int_0^\infty f(\epsilon)u(\epsilon)\,\mathrm{d}\epsilon\,, \tag{10.4.34}$$

with $u(\epsilon)$ given by $\epsilon\Omega(\epsilon)$ as

$$u(\epsilon) \equiv 4\pi \left(\frac{2m_e}{h^2}\right)^{\frac{3}{2}} V\epsilon^{\frac{3}{2}}\,. \tag{10.4.35}$$

The integral (10.4.34) for U could be evaluated numerically, but more can be understood of the behaviour of U if we attempt to evaluate it analytically. To accomplish this task, we shall integrate by parts, using $f(\epsilon)\mathrm{d}\epsilon$ as '$\mathrm{d}v$', and $u(\epsilon)$ as 'u' in the integration-by-parts formula $\int u\mathrm{d}v = uv - \int v\mathrm{d}u$. Thus, upon carrying out an integration by parts, we obtain the internal energy as

$$U = \int_0^\infty f(\epsilon)u(\epsilon)\,\mathrm{d}\epsilon = -\int_0^\infty f'(\epsilon)U(\epsilon)\,\mathrm{d}\epsilon\,, \tag{10.4.36}$$

with $U(\epsilon)$ given by

$$U(\epsilon) \equiv \int_0^\epsilon u(x)\,\mathrm{d}x = \frac{8\pi}{5}\left(\frac{2m_e}{h^2}\right)^{\frac{3}{2}} V\epsilon^{\frac{5}{2}}\,. \tag{10.4.37}$$

Because $f'(\epsilon)$ exhibits a very sharp spike at $\epsilon = \mu$, most of the contribution to such integrals comes from the region of ϵ around μ, and we may usefully make a Taylor expansion around this point, and then carry out the integration term-by-term. Thus, we expand $U(\epsilon)$ as

$$U(\epsilon) = U(\mu) + U'(\mu)(\epsilon - \mu) + \frac{1}{2!}U''(\mu)(\epsilon - \mu)^2 + \ldots, \tag{10.4.38}$$

which, when substituted into Eq. (10.4.36) gives

$$U = -\int f'(\epsilon)U(\epsilon)\,\mathrm{d}\epsilon$$

$$= -\int f'(\epsilon)\left[U(\mu) + U'(\mu)(\epsilon - \mu) + \frac{1}{2}U''(\mu)(\epsilon - \mu)^2 + \ldots\right]\mathrm{d}\epsilon$$

$$= -\left[U(\mu)\int f'(\epsilon)\,\mathrm{d}\epsilon + U'(\mu)\int f'(\epsilon)(\epsilon - \mu)\,\mathrm{d}\epsilon\right.$$

$$+ \frac{1}{2} U''(\mu) \int f'(\epsilon)(\epsilon - \mu)^2 \, d\epsilon + \dots \Bigg]$$

$$= U(\mu) L_0 + U'(\mu) L_1 + U''(\mu) L_2 + \dots, \tag{10.4.39}$$

with the L_i given, in general, by

$$L_i \equiv - \int_0^\infty (\epsilon - \mu)^i f'(\epsilon) \, d\epsilon. \tag{10.4.40}$$

For the time being we need only L_0, L_1 and L_2: their values can be shown to be 1, 0, and $\pi^2 (k_B T)^2 / 3$, respectively, so that U can be written as

$$U = U(\mu) + \frac{\pi^2}{6} (k_B T)^2 U''(\mu) + \dots. \tag{10.4.41}$$

By applying the same procedure to expression (10.4.27), i.e., carrying out an integration by parts as in Eq. (10.4.36), but with

$$\overline{N}_{ce}(\epsilon) = \frac{8\pi}{3} \left(\frac{2m_e}{h^2} \right)^{\frac{3}{2}} V \epsilon^{\frac{3}{2}} \tag{10.4.42}$$

replacing $U(\epsilon)$ in Eq. (10.4.39), we arrive at a similar expression for $\overline{N}$, namely

$$\overline{N}_{ce} = \overline{N}_{ce}(\mu) + \frac{\pi^2}{6} (k_B T)^2 \overline{N}''_{ce}(\mu) + \dots. \tag{10.4.43}$$

From this last result, together with expression (10.4.42) and its derivatives with respect to ϵ, we obtain for $\overline{N}_{ce}$ the expression

$$\overline{N}_{ce} = \frac{8\pi}{3} \left(\frac{2m_e}{h^2} \right)^{\frac{3}{2}} V \mu^{\frac{3}{2}} \left[1 + \frac{\pi^2}{8} (\beta\mu)^{-2} + \dots \right]. \tag{10.4.44}$$

If we now rearrange this result by multiplying both sides of the equation by $(3/8\pi)(h^2/2m_e)^{\frac{3}{2}}/V$, we obtain

$$\frac{\overline{N}}{V} \left(\frac{3}{8\pi} \right) \left(\frac{h^2}{2m_e} \right)^{\frac{3}{2}} = \mu^{\frac{3}{2}} \left[1 + \frac{\pi^2}{8} (\beta\mu)^{-2} + \dots \right], \tag{10.4.45}$$

which becomes, upon employing expression (10.4.22a) for the Fermi energy ϵ_F,

$$\epsilon_F^{\frac{3}{2}} = \mu^{\frac{3}{2}} \left[1 + \frac{\pi^2}{8} (\beta\mu)^{-2} + \dots \right]. \tag{10.4.46}$$

We may employ this result to give ϵ_F in terms of the chemical potential μ as

$$\epsilon_F = \mu \left[1 + \frac{\pi^2}{8}(\beta\mu)^{-2} + \ldots \right]^{\frac{2}{3}} \simeq \mu \left[1 + \frac{\pi^2}{12}(\beta\mu)^{-2} + \ldots \right] \quad (10.4.47)$$

or, perhaps more usefully, to show the deviation of the chemical potential from the Fermi energy via

$$\mu \simeq \epsilon_F \left[1 - \frac{\pi^2}{12}(\beta\mu)^{-2} + \ldots \right]. \quad (10.4.48a)$$

Finally, as this expression shows that μ has order-of-magnitude ϵ_F, it will still be a good approximation to replace μ in the second term on the RHS of Eq. (10.4.48a) by ϵ_F to give the chemical potential as

$$\mu \approx \epsilon_F \left[1 - \frac{\pi^2}{12}(\beta\epsilon_F)^{-2} + \ldots \right]. \quad (10.4.48b)$$

Expressed in terms of the smallness parameter η defined in Eq. (10.4.29), we have

$$\mu = \epsilon_F \left[1 - \frac{\pi^2}{12}\eta^2 + \mathcal{O}\left(\frac{T^4}{T_F^4}\right) \right], \quad (10.4.49)$$

from which it is clear that μ changes rather slowly with temperature. Consequently, we may meaningfully approximate μ by ϵ_F throughout the entire solid-state temperature range of a metal.

Let us return now to consideration of U. Following the same steps that we have used to arrive at Eq. (10.4.44) for $\overline{N}_{ce}$, we obtain the sequence of equations listed below:

$$U = \frac{8\pi}{5}\left(\frac{2m_e}{h^2}\right)^{\frac{3}{2}} V\mu^{\frac{5}{2}} \left[1 + \frac{5\pi^2}{8}(\beta\mu)^{-2} + \ldots \right]$$

$$= U_0 \left(\frac{\mu}{\epsilon_F}\right)^{\frac{5}{2}} \left[1 + \frac{5\pi^2}{8}(\beta\mu)^{-2} + \ldots \right]$$

$$\simeq U_0 \left[1 + \frac{5\pi^2}{12}\eta^2 + \ldots \right], \quad (10.4.50)$$

with U_0 given by

$$U_0 \equiv \frac{8\pi}{5}\left(\frac{2m_e}{h^2}\right)^{\frac{3}{2}} V\epsilon_F^{\frac{5}{2}} = \tfrac{3}{5}\overline{N}_{ce}\epsilon_F. \quad (10.4.51)$$

We may also express the grand potential $\Phi = -2U/3$ as

$$\Phi = -\frac{16}{15}\left(\frac{2m_e}{h^2}\right)^{\frac{3}{2}} V \epsilon_F^{\frac{5}{2}}\left[1 + \frac{5\pi^2}{12}\eta^2 + \dots\right]$$

$$= -\frac{2}{5}\overline{N}_{ce}\epsilon_F\left[1 + \frac{5\pi^2}{12}\eta^2 + \dots\right], \tag{10.4.52}$$

with the final expression utilizing the defining relation for the Fermi energy ϵ_F. We can also obtain a corresponding expression for the entropy from $TS = U + PV - \mu N$ as

$$TS = \frac{5}{2} PV - \mu \overline{N}_{ce}$$

$$= \overline{N}_{ce}\epsilon_F\left[1 + \frac{5\pi^2}{12}\eta^2 - \left(1 - \frac{\pi^2}{12}\eta^2\right) + \mathcal{O}(\eta^4)\right]$$

so that the entropy is given by

$$S = \overline{N}_{ce}k_B\,\frac{\pi^2}{2}\,\frac{T}{T_F} + \mathcal{O}\left(\frac{T^3}{T_F^3}\right). \tag{10.4.53}$$

Finally, we can obtain the electronic contribution to the heat capacity C_V arising from the electron gas in the metal by differentiating expression (10.4.50) for the internal energy contribution with respect to T. The electronic contribution to C_V is thus given as

$$C_V = U_0 \frac{5\pi^2}{12}(2\eta)\frac{d\eta}{dT}$$

$$= \frac{1}{2}\pi^2\overline{N}_{ce}k_B\,\frac{T}{T_F}, \tag{10.4.54}$$

which is linear in T, as expected upon physical grounds. As $C_V(T)$ depends linearly upon T/T_F, and T_F is typically very large, the electronic contribution to C_V in a metallic sample may be observed only should the contributions from all other sources be extremely small. Within the temperature range of a solid metal this can occur only at temperatures so low that even the lattice vibrational contributions (given by the Debye formula, Eq. (5.6.22b), for example) have decreased sufficiently that they no longer dominate over the electronic contribution. A plot of $\overline{C}_V/T$ vs. T^2 should then give an intercept from which the Fermi temperature for the metal may be determined, while the slope should allow a determination of the Debye temperature. A typical plot is shown in Fig. 10.5: the dashed line gives the lattice contribution according to the Debye formula, while the solid line gives the lattice plus electronic contributions.

Fig. 10.5 Heat capacity of
metallic silver at very low
temperatures. Experimental
values: Corak et al. [6]

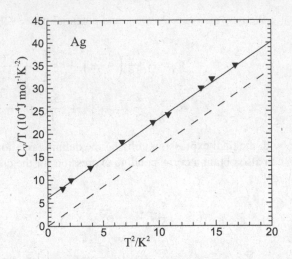

B. Pauli Paramagnetism

The density-of-states for the translational energies of electrons confined to a potential box (traditionally employed to model conduction electrons in a metal) is given by Eqs. (10.4.33). For $T = 0$, all translational energy eigenstates with $\epsilon \leq \epsilon_F$ will be occupied and all translational energy eigenstates for which $\epsilon > \epsilon_F$ will be unoccupied. We know that, according to the Pauli Exclusion Principle, no two (indistinguishable) fermions may occupy the same energy eigenstate. However, if we note that, as each electron possesses an *intrinsic* angular momentum **s** (termed 'spin') of magnitude $s = \frac{1}{2}$, it will either be in a state with $m_s = +\frac{1}{2}$ or $m_s = -\frac{1}{2}$, so that an electron in a state characterized by a positive spin-projection quantum number m_s will thus be distinguishable from an electron in a state characterized by a negative spin-projection quantum number. Each translational energy eigenstate may therefore accommodate two electrons, provided that they are 'spin-paired'. We may visualize this situation pictorially by constructing a diagram in which the density-of-states for electrons with $m_s = +\frac{1}{2}$ (as a function of translational energy ϵ along the vertical-axis) faces to the right along the horizontal-axis while the density-of-states for electrons with $m_s = -\frac{1}{2}$ (again, as a function of translational energy ϵ along the vertical-axis) faces to the left along the horizontal-axis, as illustrated in Panel (a) of Fig. 10.6.

In the presence of a magnetic field **B** that interacts, via the Zeeman Hamiltonian $\mathcal{H}_{mag} \equiv -\boldsymbol{\mu}^{(m)} \cdot \mathbf{B}$, with the electronic spin magnetic dipole moment $\boldsymbol{\mu}_e^{(m)} = -g_e \beta_0 \mathbf{s}$, each conduction electron acquires an additional energy $E_{mag} = m_s g_e \beta_0 B$, with $g_e \simeq 2.0023$ the electronic g-factor for a free electron and $\beta_0 \equiv |e|\hbar/(2m_e) \simeq 9.274 \times 10^{-24}\,\mathrm{J\,T^{-1}}$ the Bohr magneton (which serves as the fundamental unit for electronic magnetic moments). Thus all electrons with $m_s = \frac{1}{2}$ are raised in energy by $\mu_0 B \equiv \frac{1}{2} g_e \beta_0 B$, while all electrons with $m_s = -\frac{1}{2}$ are lowered in energy by $\mu_0 B$ (we recall that positive magnetic moments align with the direction

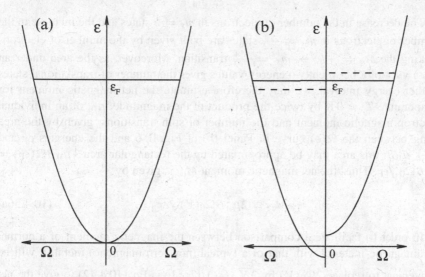

Fig. 10.6 Density-of-states for conduction electrons in the (**a**) absence and (**b**) presence of a magnetic field at temperature $T = 0\,\mathrm{K}$

of an applied magnetic field in order to attain a lower energy state), as illustrated in Panel (b) of Fig. 10.6.

The Fermi–Dirac distribution, $f(\epsilon)$, remains effectively unchanged by the introduction of the magnetic field. However, as the density-of-states is given in terms of translational energy relative to zero, a downward shift of all energies for $m_s = -\frac{1}{2}$ electrons by $\mu_0 B$ means that the density-of-states becomes

$$\Omega(\epsilon, V) = \Omega_0(V)(\epsilon + \mu_0 B)^{\frac{1}{2}}, \qquad (10.4.55\text{a})$$

rather than $\Omega(\epsilon, V) = \Omega_0(V)\epsilon^{\frac{1}{2}}$. Similarly, we see that an upward shift of all $m_s = +\frac{1}{2}$ electron energies by $\mu_0 B$ results in a density-of-states

$$\Omega(\epsilon, V) = \Omega_0(V)(\epsilon - \mu_0 B)^{\frac{1}{2}}. \qquad (10.4.55\text{b})$$

The Fermi energy ϵ_F is indicated in each panel of Fig. 10.6 by a solid horizontal line, and the energies corresponding to the maximal energies attained by $m_s = \pm\frac{1}{2}$ electrons are indicated by dashed lines in Panel (b). As an electron in an $m_s = +\frac{1}{2}$ state may only attain a lower energy by making a $\Delta m_s = -1$ transition into an equivalent, but unoccupied, $m_s = -1$ electron translational state, it will be clear from Panel (b) of Fig. 10.6 that only those $m_s = +\frac{1}{2}$ electrons lying above the Fermi energy ϵ_F will have nearby unoccupied accessible $m_s = -\frac{1}{2}$ translational states lying below the Fermi energy.

The decrease in the number of electrons in $m_s = \frac{1}{2}$ states and the increase in the number of electrons in $m_s = -\frac{1}{2}$ states are both given by the number of electrons making the $m_s = +\frac{1}{2} \rightarrow m_s = -\frac{1}{2}$ transition. Moreover, as the area under an $\Omega(\epsilon)$ vs. ϵ plot between two energy values gives the number of translational states in that energy interval, we may therefore estimate the net magnetic moment for temperature $T = 0\,\mathrm{K}$ by twice the product of the magnitude, μ_0, of an individual electron magnetic moment and the number of spin transitions, given by the area lying between the $\Omega(\epsilon)$ curve in Panel (b) of Fig. 10.6 and the energies ϵ_F and $\epsilon_F + \mu_0 B$: this area may be approximated by the rectangular area $\frac{1}{2}\mu_0 B\Omega(\epsilon_F) = \mu_0 B\Omega_0\sqrt{\epsilon_F}$. Thus, the net magnetic moment M_{ce} is given by

$$M_{\mathrm{ce}} \simeq 2\mu_0^2 B\Omega_0(V)\sqrt{\epsilon_F} . \tag{10.4.56a}$$

In order to facilitate a comparison between the magnetic moment of a normal paramagnetic material with that of a typical (non-ferromagnetic) metal, it will be convenient to replace $2\Omega_0(V)$ by $3\overline{N}_{\mathrm{ce}}(\epsilon_F)/[2\epsilon_F^{\frac{3}{2}}]$ via Eq. (10.4.42) to give the net magnetic moment due to the conduction electrons as

$$M_{\mathrm{ce}} \simeq \mu_0^2 B \frac{3\overline{N}_{\mathrm{ce}}(\epsilon_F)}{2\epsilon_F} . \tag{10.4.56b}$$

Upon comparing this expression for M_{ce} with the Curie law expression,

$$M \equiv \chi^{(m)}(T, \rho)B = \frac{N\mu_m^2 B}{3k_B T} ,$$

of Eqs. (8.4.14a) and (8.4.8) for the magnetic moment of a normal paramagnetic material, we see that these two expressions are indeed rather similar, with the principal difference between them being replacement of the thermal energy $k_B T$ by the Fermi energy ϵ_F. It will thus already be clear from this $0\,\mathrm{K}$ result for M_{ce} that the magnetic moment induced in a metal by a magnetic field is, by virtue of the difference between a typical thermal energy and ϵ_F, many orders of magnitude smaller than the magnetic moment induced in a bulk substance whose constituents contain unpaired electrons. Physically, the difference can be associated with the fact that the spins of all unpaired electrons in a normal paramagnetic substance (by virtue of their association with individual atoms at distinct spatial locations) are able to align along the direction of the applied magnetic field, while only the spins of those conduction electrons associated with energy levels that lie no more than $\mu_0 B$ below the Fermi level of a metal may ultimately be both unpaired and aligned along the direction of the applied magnetic field.

For nonzero temperatures, the number of continuum translational energy states for $m_s = -\frac{1}{2}$ electrons is given by

$$N(m_s = -\tfrac{1}{2}) = \Omega_0(V) \int_{-\mu_0 B}^{\infty} (\epsilon + \mu_0 B)^{\frac{1}{2}} f(\epsilon)\, d\epsilon , \tag{10.4.57a}$$

the number of continuum translational states for $m_s = +\frac{1}{2}$ electrons is given by

$$N\left(m_s = +\tfrac{1}{2}\right) = \Omega_0(V) \int_{-\mu_0 B}^{\infty} (\epsilon - \mu_0 B)^{\frac{1}{2}} f(\epsilon)\, d\epsilon, \qquad (10.4.57b)$$

and the net magnetic moment is given by

$$M_{\text{ce}} = \mu_0 \left[N(m_s = -\tfrac{1}{2}) - N\left(m_s = +\tfrac{1}{2}\right) \right]$$

$$= \mu_0 \Omega_0(V) \left[\int_{-\mu_0 B}^{\infty} (\epsilon + \mu_0 B)^{\frac{1}{2}} f(\epsilon)\, d\epsilon - \int_{\mu_0 B}^{\infty} (\epsilon - \mu_0 B)^{\frac{1}{2}} f(\epsilon)\, d\epsilon \right].$$
$$(10.4.58)$$

Evaluation of the integrals in expression (10.4.58) involves a number of steps, the first of which is to carry out an integration by parts to convert them into integrals having the form

$$I(\mu, \mu_0 B) = - \int_{\mp \mu_0 B}^{\infty} (\epsilon \pm \mu_0 B)^{\frac{3}{2}} f'(\epsilon)\, d\epsilon,$$

in which $f'(\epsilon) = \beta e^{\beta(\epsilon - \mu)}/[1 + e^{\beta(\epsilon - \mu)}]^2$. Then, following an expansion of $(\epsilon \pm \mu_0 B)^{\frac{3}{2}}$ in a Taylor series around $\epsilon = \mu$, a transformation of the integration variable from ϵ to $x \equiv \beta(\epsilon - \mu)$, and an extension of the lower limit of integration to $-\infty$, the leading contributions to the two integrals give

$$M_{\text{ce}} = \tfrac{2}{3} \mu_0 \Omega_0(V) \left[(\mu + \mu_0 B)^{\frac{3}{2}} - (\mu - \mu_0 B)^{\frac{3}{2}} \right.$$

$$\left. + \frac{(k_B T)^2}{8(\mu + \mu_0 B)^{\frac{1}{2}}} - \frac{(k_B T)^2}{8(\mu - \mu_0 B)^{\frac{1}{2}}} \right]$$

for the magnetic moment associated with conduction electrons. Binomial expansions of $(\mu \pm \mu_0 B)^{\frac{3}{2}}$ and $(\mu \pm \mu_0 B)^{-\frac{1}{2}}$ then give M_{ce} as

$$M_{\text{ce}} \simeq 2\Omega_0(V)\mu_0^2 B \left[\mu^{\frac{1}{2}} - \frac{\pi^2 (k_B T)^2}{24\mu^{\frac{3}{2}}} \right] \qquad (10.4.59)$$

in terms of the chemical potential μ for the electrons. We may employ Eq. (10.4.49), which gives the temperature dependence of the chemical potential μ as

$$\mu \simeq \epsilon_F \left[1 - \frac{\pi^2}{12} \left(\frac{k_B T}{\epsilon_F} \right)^2 + \ldots \right],$$

to obtain an expression for M_{ce} in terms of the 0 K result (10.4.56b) plus a correction term for the temperature dependence. In particular, we can see from Eq. (10.4.49) that we may safely approximate μ by ϵ_F in the second term on the right-hand side of Eq. (10.4.59), but we must be more careful with the $\mu^{\frac{1}{2}}$ term, by approximating it as

$$\mu^{\frac{1}{2}} \simeq \epsilon_F^{\frac{1}{2}} \left[1 - \frac{\pi^2}{12} \left(\frac{k_B T}{\epsilon_F} \right)^2 + \dots \right]^{\frac{1}{2}} \approx \epsilon_F^{\frac{1}{2}} \left[1 - \frac{\pi^2}{24} \left(\frac{k_B T}{\epsilon_F} \right)^2 \right],$$

in order not to neglect an additional contribution of order T^2 to M_{ce}. Thus, we obtain the leading temperature dependence of the magnetic moment of the conduction electrons as

$$M_{ce}(T) \simeq 2\mu_0^2 B \Omega_0(V) \epsilon_F^{\frac{1}{2}} \left[1 - \frac{\pi^2}{24} \left(\frac{k_B T}{\epsilon_F} \right)^2 - \frac{\pi^2}{24} \left(\frac{k_B T}{\epsilon_F} \right)^2 \right]$$

$$= \frac{3 \overline{N}_{ce} \mu_0^2 B}{2 \epsilon_F} \left[1 - \frac{\pi^2}{12} \left(\frac{k_B T}{\epsilon_F} \right)^2 \right], \tag{10.4.60}$$

in which the leading term is the $T = 0\,\text{K}$ magnetic moment. Note also that the leading temperature-dependent term is indeed very small.

10.5   Electrons in Semiconductors

We have seen that the density of states for a free electron gas can be written as

$$\Omega(\epsilon) = \frac{8\pi V}{h^3} \left(2m^3 \epsilon \right)^{\frac{1}{2}}. \tag{10.5.1}$$

When atoms form a crystal (solid), some of the electrons no longer belong to particular atoms, but are shared in some sense by all atoms in the solid, giving those electrons a dual nature of atomic and free electrons; there is a broadening of the discrete atomic orbitals to form bands in the solid. These two situations are illustrated in Fig. 10.7. See also Appendix F.

The probability that a one-particle state i is not occupied can be represented by $\overline{n}_i'$, with

$$\overline{n}_i' \equiv 1 - \overline{n}_i = \frac{1}{e^{-\beta(\epsilon_i - \mu)} + 1}; \tag{10.5.2}$$

if we take the condition in which all the electronic states are occupied as the standard, then we say that the state i is occupied by a *hole* when it is not occupied by

Fig. 10.7 Electron densities of state for free electrons, [panel (**a**)], and for an intrinsic semiconductor, [panel (**b**)]

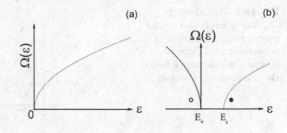

an electron. We shall see that holes may be treated as particles provided that they are fermions. The treatment in this subsection has been largely inspired by Chapter 4 of [7].

An electron (charge $-e$) occupying a one-electron state is assigned a linear momentum $\mathbf{p}$ and an energy $\varepsilon(\mathbf{p})$, while a hole (with an effective charge e) occupying a one-electron state will have a linear momentum $\mathbf{p}' = -\mathbf{p}$, an energy $\varepsilon'(\mathbf{p}') = K - \varepsilon(\mathbf{p})$, with K a constant, and a Fermi potential $\mu' = C - \mu$, with C another constant. We may therefore represent $\overline{n}'_{p'}$ by

$$\overline{n}'_{\mathbf{p}'} = \frac{1}{e^{\beta[\varepsilon'(\mathbf{p}') - \mu]} + 1} \, .$$

An *intrinsic semiconductor* has a filled valence band and an empty conduction band at $T = 0$ K. For $T \neq 0$, electrons are excited from the valence band into the conduction band, leaving a corresponding number of holes in the valence band. Both the electrons and the holes may play roles in electrical conduction. The temperature at which electrical conduction appears depends upon the magnitude of the energy gap, E_g, between the valence and conduction bands.

An *impurity semiconductor* is a solid that has a relatively large energy gap, but may be able to exhibit good conductivity at relatively low temperatures due to the presence of a particular impurity in its makeup. Impurity semiconductors are of two types:

- n-type semiconductors, in which an impurity level is filled by electrons at $T = 0$ K, but for $T \neq 0$ is such that the impurity (donor) level electrons are excited into the conduction band, where they give rise to electrical conduction: germanium (Ge) containing phosphorus (P) and Ge containing arsenic (As) are examples of n-type semiconductors.
- p-type semiconductors, in which there are impurity levels that are not occupied by electrons for $T = 0$ K, but at $T \neq 0$ are able to accommodate excited electrons from the valence band (acceptor levels), thereby producing holes in the valence band that give rise to electrical conduction: Ge containing boron (B) and Ge containing aluminium (Al) are examples of p-type semiconductors.

The occupation of energy levels in typical impurity semiconductors is illustrated in Fig. 10.8.

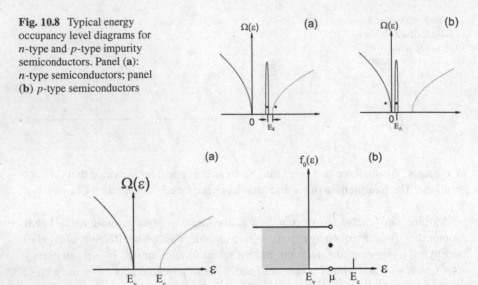

Fig. 10.8 Typical energy occupancy level diagrams for n-type and p-type impurity semiconductors. Panel (**a**): n-type semiconductors; panel (**b**) p-type semiconductors

Fig. 10.9 Density-of-states [panel (**a**)] and $T = 0\,\mathrm{K}$ Fermi distribution [panel (**b**)] for intrinsic semiconductors

We shall first examine intrinsic semiconductors in greater detail, after which we shall take a brief foray into the world of impurity semiconductors.

10.5.1   Intrinsic Semiconductors

At temperature T the energy of a promoted conduction electron can often be approximated by

$$\epsilon(\mathbf{p}) = E_{\mathrm{g}} + \frac{1}{2m_{\mathrm{e}}^*}\,\mathbf{p}^2 \,, \tag{10.5.3}$$

while that of the 'hole' left behind in the valence band may be approximated by

$$\epsilon'(\mathbf{p}') = -\frac{1}{2m_{\mathrm{h}}^*}\,\mathbf{p}'^2 \,, \tag{10.5.4}$$

with m_{e}^* and m_{h}^* being effective masses of the electron and hole [see Appendix F.3]. We note in passing that these are the masses required for the electrons and holes to behave like free particles; they often have values that are only fractions of the electron rest mass. These relationships are illustrated in Fig. 10.9.

We see that for an intrinsic semiconductor, $\Omega(\epsilon)$ has a large gap, of magnitude $E_g \equiv E_c - E_v$, which has the consequence that the definition of the number of electrons in a crystal (solid), given by

$$N = \int_0^\infty f(\epsilon) n(\epsilon)\, d\epsilon\,, \qquad (10.5.5)$$

splits naturally into two parts, one each for the valence and conduction bands. Thus, we may write

$$N = \int_0^{E_v} f(\epsilon) n_v(\epsilon)\, d\epsilon + \int_{E_c}^\infty f(\epsilon) n_c(\epsilon)\, d\epsilon\,, \qquad (10.5.6)$$

in which $n_v(\epsilon)$ and $n_c(\epsilon)$ represent the densities of states in the valence and conduction bands, respectively.

Let us now make use of the fact that for $T = 0$, we have

$$
\begin{aligned}
N &= \int_0^\infty f_0(\epsilon) n(\epsilon)\, d\epsilon \\
&= \int_0^\mu n(\epsilon)\, d\epsilon \qquad (10.5.7)\\
&= \int_0^{E_v} n_v(\epsilon)\, d\epsilon
\end{aligned}
$$

so that our expression (10.5.6) for N can now be rewritten as

$$\int_0^{E_v} [1 - f(\epsilon)] n_v(\epsilon)\, d\epsilon = \int_{E_c}^\infty f(\epsilon) n_c(\epsilon)\, d\epsilon\,, \qquad (10.5.8)$$

which suggests that the number of electrons excited into the conduction band is equal to the number of holes left in the valence band (a result that would seem to be more or less intuitively obvious), i.e.,

$$N_h = N_{ce}\,. \qquad (10.5.9)$$

It will be convenient to determine conduction electron energies relative to E_c, and to determine hole energies relative to the top of the valence band, E_v; we shall accordingly employ variables $\epsilon_{ce} \equiv \epsilon - E_c$ and $\epsilon_h \equiv -(\epsilon - E_v)$ for the energies of the conduction electrons and holes, respectively [see Fig. 10.10a].

Notice that we may write

$$1 - f(\epsilon - \mu) = 1 - \frac{1}{e^{\beta(\epsilon - \mu)} + 1} = \frac{1}{e^{-\beta(\epsilon - \mu)} + 1} = f(\epsilon_h + \mu_h)\,, \qquad (10.5.10)$$

Fig. 10.10 Coordinate systems for describing valence and conduction band energies in an intrinsic semiconductor. Panel (**a**) illustrates the energies of the valence and conduction bands relative to an arbitrary energy origin, while panel (**b**) illustrates the variables utilized for electrons and holes in the text

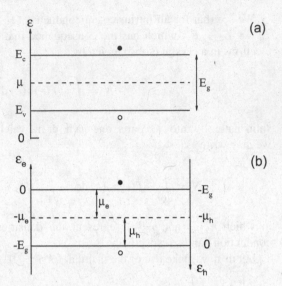

with $\epsilon_h + \mu_h = -(\epsilon - \mu)$: the probability for finding a hole in an energy level lying $|\epsilon - \mu|$ *below* the Fermi level[1] μ is the same as the probability for finding an electron at energy $|\epsilon - \mu|$ *above* the Fermi level. Thus holes, like the electrons whose removal created them, obey Fermi–Dirac statistics, but as if their energies were measured downwards [see Fig. 10.10b]. Equation (10.5.8) can now be rewritten as

$$\int_0^\infty f(\epsilon_h + \mu_h) n_v(\epsilon_h)\, d\epsilon_h = \int_0^\infty f(\epsilon_{ce} + \mu_{ce}) n_c(\epsilon_{ce})\, d\epsilon_{ce}, \qquad (10.5.11)$$

with the constraint that μ_{ce} and μ_h should in fact correspond to the same overall Fermi level μ. Note that because μ_{ce} and μ_h are measured from energy origins that are separated by E_g, we must have

$$\mu_{ce} + \mu_h = E_g. \qquad (10.5.12)$$

Let us turn now to the evaluation of N_h and N_{ce} near the top of the valence band and near the bottom of the conduction band, respectively, where we can utilize the approximate expressions given in Eqs. (10.5.3) and (10.5.4) for ϵ_{ce} and ϵ_h. Thus, we have

$$\frac{N_{ce}}{V} \simeq \frac{2}{h^3} \int \frac{d\mathbf{p}}{e^{\beta[\epsilon(\mathbf{p}) - \mu]} + 1} = \frac{2}{h^3} \int \frac{d\mathbf{p}}{e^{\beta(E_g - \mu + \frac{1}{2}\mathbf{p}^2/m_{ce}^*)} + 1}$$

---

[1] The chemical potential for electrons in solids is often referred to as the Fermi level, and approximated by the Fermi energy ϵ_F ($\equiv \mu_F$).

or

$$\frac{N_{\mathrm{ce}}}{V} \approx \frac{2}{h^3} \int e^{-\beta\left(E_{\mathrm{g}}-\mu+\frac{1}{2}\mathbf{p}^2/m_{\mathrm{ce}}^*\right)} \, d\mathbf{p}$$

$$= \frac{2}{h^3} 4\pi e^{-\beta(E_{\mathrm{g}}-\mu)} \int_0^\infty p^2 e^{-\beta\frac{1}{2}p^2/m_{\mathrm{ce}}^*} \, dp$$

$$= \frac{2}{h^3} 4\pi e^{-\beta(E_{\mathrm{g}}-\mu)} (2m_{\mathrm{ce}}^* k_{\mathrm{B}} T)^{\frac{3}{2}} \int_0^\infty x^2 e^{-x^2} \, dx .$$

Upon recognizing that the integral in the final line of this sequence has the value $\sqrt{\pi}/4$, we obtain for N_{ce} the expression

$$N_{\mathrm{ce}} = 2 \left(\frac{2m_{\mathrm{ce}}^* \pi k_{\mathrm{B}} T}{h^2} \right)^{\frac{3}{2}} V \, e^{-\beta(E_{\mathrm{g}}-\mu)} . \tag{10.5.13}$$

Similarly, for N_{h} we obtain the expression

$$\frac{N_{\mathrm{h}}}{V} \simeq \frac{2}{h^3} \int \frac{d\mathbf{p}'}{e^{\beta[\mu-\varepsilon'(\mathbf{p}')]}+1} = \frac{2}{h^3} \int \frac{d\mathbf{p}'}{e^{\beta\left(\mu+\frac{1}{2}p'^2/m_{\mathrm{h}}^*\right)}+1}$$

$$\simeq \frac{2}{h^3} \int e^{-\beta\left(\mu+\frac{1}{2}p'^2/m_{\mathrm{h}}^*\right)} \, d\mathbf{p}'$$

or

$$N_{\mathrm{h}} \simeq 2 \left(\frac{2\pi m_{\mathrm{h}}^* k_{\mathrm{B}} T}{h^2} \right)^{\frac{3}{2}} V \, e^{-\beta\mu} . \tag{10.5.14}$$

Equality of N_{ce} and N_{h} allows us to solve for $e^{\beta\mu}$ as

$$(m_{\mathrm{ce}}^*)^{\frac{3}{2}} e^{-\beta(E_{\mathrm{g}}-\mu)} = (m_{\mathrm{h}}^*)^{\frac{3}{2}} e^{-\beta\mu} ,$$

or

$$e^{2\beta\mu} = \left(\frac{m_{\mathrm{h}}^*}{m_{\mathrm{ce}}^*} \right)^{\frac{3}{2}} e^{\beta E_{\mathrm{g}}} ,$$

from which we find that

$$e^{\beta\mu} = \left(\frac{m_{\mathrm{h}}^*}{m_{\mathrm{ce}}^*} \right)^{\frac{3}{4}} e^{\frac{1}{2}\beta E_{\mathrm{g}}} . \tag{10.5.15}$$

This gives us an expression for the Fermi level in terms of the semiconductor energy gap E_g and the effective masses of the conduction electrons and holes, viz.,

$$\mu = \tfrac{1}{2}E_g + \tfrac{3}{4}k_B T \ln\left(\frac{m_h^*}{m_{ce}^*}\right). \tag{10.5.16}$$

Thus, only when the effective masses of the conduction electrons and holes are equal does the Fermi level lie precisely at the midpoint of the energy gap.

We may employ relation (10.5.15) in our expression above for N_{ce} to obtain

$$N_{ce} = 2\left(\frac{2\pi m_{ce}^* k_B T}{h^2}\right)^{\frac{3}{2}} V\, e^{-\beta E_g}\left(\frac{m_h^*}{m_{ce}^*}\right)^{\frac{3}{4}} e^{\frac{1}{2}\beta E_g},$$

or

$$N_{ce} = 2\left(\frac{2\pi\sqrt{m_{ce}^* m_h^*}\,k_B T}{h^2}\right)^{\frac{3}{2}} V\, e^{-\frac{1}{2}\beta E_g}, \tag{10.5.17}$$

and similarly,

$$N_h = 2\left(\frac{2\pi\sqrt{m_{ce}^* m_h^*}\,k_B T}{h^2}\right)^{\frac{3}{2}} V\, e^{-\frac{1}{2}\beta E_g}, \tag{10.5.18}$$

i.e., $N_h = N_{ce}$. This equation is often used to determine the free-carrier concentration in intrinsic semiconductors.

Example 10.4 Elemental tellurium (Te).

Elemental tellurium has an energy gap $E_g = 0.33$ eV (2661.61 cm^{-1}) and, at room temperature, may be considered to be an intrinsic semiconductor. The effective mass for conduction electrons in Te is $m_{ce}^* = 0.68 m_e$, while the effective mass for the holes in the valence band is $m_h^* = 0.19 m_e$. If we employ $T = 300$ K as room temperature, we see that the Fermi level μ_F is given by

$$\mu_F = \tfrac{1}{2}(0.33 \text{ eV}) + \tfrac{3}{4}k_B T \ln\left(\frac{0.19 m_e}{0.68 m_e}\right)$$

$$= \tfrac{1}{2}(0.33)(8065.479) \text{ cm}^{-1} + \tfrac{3}{4}(0.695 \text{ cm}^{-1}\text{K}^{-1})(300 \text{ K}) \ln\left(\frac{0.19}{0.68}\right)$$

$$= 1330.81 - 199.39 \text{ cm}^{-1},$$

or

$$\mu_F = 1131.42 \text{ cm}^{-1},$$

which may be compared with the thermal energy $k_B T = 208.50\,\mathrm{cm}^{-1}$ for $T = 300\,\mathrm{K}$. Thus, we see that for Te, μ_F is much greater than $k_B T$ at room temperature. This side calculation justifies the simplification that we utilized in order to obtain formal expressions for N_{ce} and N_h.

Now we may calculate the number density for the conduction electrons as

$$\frac{N_{ce}}{V} = 2\left[\frac{2\pi\,(0.3594)\,(9.1095\times10^{-31})\,(1.38066\times10^{-23})\,(300)}{(6.62618\times10^{-34})^2}\right]^{\frac{3}{2}} e^{-1330.81/208.50}$$

$$\simeq 9.14\times10^{23}\,\mathrm{m}^{-3}\ .$$

This value may be compared with the Loschmidt number, which is approximately $L_0 \simeq 2.6516\times10^{25}\,\mathrm{m}^{-3}$, so that

$$\frac{\overline{N}_{ce}}{L_0} \simeq 3.45\times10^{-2}\ .$$

We have employed the Loschmidt number as an estimate of the number density of electrons in the valence band of the semiconductor.   □

10.5.2 n-Type Impurity Semiconductors

We shall now consider donor levels lying at an energy E_D below the bottom of the conduction band: $\overline{N}_D$ is the number of donors per unit volume, $\overline{N}_{de}$ is the number of donor electrons per unit volume, and $\overline{N}_{ce}$ is the number of conduction electrons per unit volume. We shall consider only the case in which a donor level may not simultaneously be occupied by more than one electron and the conduction electrons are nondegenerate.

We begin with an approximate expression for the Helmholtz energy associated with the electrons in the donor levels given [7] by

$$A \simeq -N_{de}E_D - k_B T \ln\left\{\frac{N_D!}{N_{de}!(N_D - N_{de})!}\,2^{N_{de}}\right\}, \qquad (10.5.19)$$

in which the second term is a configurational term associated with the total number of configurations for N_{de} electrons (having spin 'up' or spin 'down') distributed over the donor levels associated with the N_D impurity sites. The number density of conduction electrons $\overline{N}_{ce}$ will be given, as for an intrinsic semiconductor, by

$$\overline{N}_{ce} = \frac{2}{h^3}\int\frac{d\mathbf{p}}{e^{\beta(\epsilon-\mu)}+1}, \qquad \epsilon = \frac{p^2}{2m_{ce}^*}. \qquad (10.5.20)$$

We may employ the Stirling approximation for the factorials in our expression for the Helmholtz energy in order to obtain a simpler expression, namely

$$A \simeq -N_{\text{de}} E_{\text{D}} - k_{\text{B}} T \ln \left\{ \frac{N_{\text{D}}^{N_{\text{D}}}}{N_{\text{de}}^{N_{\text{de}}} (N_{\text{D}} - N_{\text{de}})^{(N_{\text{D}} - N_{\text{de}})}} 2^{N_{\text{de}}} \right\} , \qquad (10.5.21)$$

and we may utilize this expression to obtain the chemical potential (Fermi level) μ as

$$\mu = \left(\frac{\partial A}{\partial N_{\text{de}}} \right)_{N_{\text{ce}}, V, T} = -E_{\text{D}} - k_{\text{B}} T \frac{\partial}{\partial N_{\text{de}}} \{ N_{\text{de}} \ln 2 + N_{\text{D}} \ln N_{\text{D}} - N_{\text{de}} \ln N_{\text{de}}$$

$$-(N_{\text{D}} - N_{\text{de}}) \ln(N_{\text{D}} - N_{\text{de}}) \}$$

$$= -E_{\text{D}} - k_{\text{B}} T \{ \ln 2 - \ln N_{\text{de}} - 1 + 1 + \ln(N_{\text{D}} - N_{\text{de}}) \} ,$$

or

$$\mu = -E_{\text{D}} - k_{\text{B}} T \left[\ln 2 - \ln \left(\frac{N_{\text{de}}}{N_{\text{D}} - N_{\text{de}}} \right) \right] . \qquad (10.5.22)$$

We may also rewrite this result in the form

$$\frac{\mu + E_{\text{D}}}{k_{\text{B}} T} + \ln 2 = \ln \left(\frac{N_{\text{de}}}{N_{\text{D}} - N_{\text{de}}} \right) , \qquad (10.5.23)$$

or, upon exponentiating both sides of the equation, in the form

$$\frac{N_{\text{de}}}{N_{\text{D}} - N_{\text{de}}} = 2 e^{\beta(\mu + E_{\text{D}})} . \qquad (10.5.24)$$

Upon rearranging this result, we obtain an expression for N_{de}, viz.,

$$N_{\text{de}} = \frac{2 N_{\text{D}} e^{\beta(\mu + E_{\text{D}})}}{1 + 2 e^{\beta(\mu + E_{\text{D}})}} = \frac{N_{\text{D}}}{1 + \frac{1}{2} e^{-\beta(E_{\text{D}} + \mu)}} , \qquad (10.5.25)$$

while we obtain an expression for $\overline{N}_{\text{ce}}$ via relation (10.5.20):

$$\overline{N}_{\text{ce}} = \frac{2}{h^3} \int \frac{d\mathbf{p}}{e^{\beta(\epsilon - \mu)} + 1} , \qquad \epsilon = \frac{p^2}{2 m_{\text{ce}}^*}$$

$$= 2 \left(\frac{2 \pi m_{\text{ce}}^* k_{\text{B}} T}{h^2} \right)^{\frac{3}{2}} e^{\beta \mu} . \qquad (10.5.26)$$

Here, however, we shall prefer to write this result as

$$\overline{N}_{ce} = \overline{N}_c e^{\beta\mu} \; ; \qquad \overline{N}_c \equiv 2 \left(\frac{2\pi m_{ce}^* k_B T}{h^2} \right)^{\frac{3}{2}} , \qquad (10.5.27)$$

so that

$$e^{\beta\mu} = \frac{\overline{N}_{ce}}{\overline{N}_c} = \frac{N_{ce}}{N_c} . \qquad (10.5.28)$$

We may now combine this last result with our result Eq. (10.5.24) to obtain

$$\frac{N_{de}}{N_D - N_{de}} = 2 e^{\beta E_D} e^{\beta\mu} = 2 \frac{N_{ce}}{N_c} e^{\beta E_D} , \qquad (10.5.29)$$

from which we obtain the relation

$$\frac{N_{ce}(N_D - N_{de})}{N_{de}} = \tfrac{1}{2} N_c e^{-\beta E_D} . \qquad (10.5.30)$$

This expression may be interpreted in terms of a dissociation equilibrium for the donors D via

$$D \;\; \rightleftharpoons \;\; D^+ + e^- . \qquad (10.5.31)$$

If, for example, we recall the relationship between number density $n_A \equiv \overline{N}_A$ and concentration [A] of a chemical species A as

$$n_A = N_0 [A] , \qquad (10.5.32)$$

with N_0 being the Avogadro number, $N_0 = 6.02214 \times 10^{23} \text{ mol}^{-1}$, we have

$$\frac{n_D}{N_0} = [D] , \quad \frac{1}{N_0}(n_D - n_{de}) = [D^+] , \quad \frac{n_{ce}}{N_0} = [e^-] . \qquad (10.5.33)$$

The equilibrium constant K_{eq} for process (10.5.31) is defined as

$$K_{eq}(T) \equiv \frac{[D^+][e^-]}{[D]} , \qquad (10.5.34)$$

with the temperature dependence of the equilibrium constant $K_{eq}(T)$ given explicitly as

$$K_{eq}(T) = \tfrac{1}{2} n_c e^{-\beta E_D} . \qquad (10.5.35)$$

As for an intrinsic semiconductor, the number of conduction electrons N_{ce} will be equal to the number of vacancies in the donor levels, which is given by $N_D - N_{de}$ (the difference between the number of donors and the number of donor electrons remaining in the donor levels): this is referred to as the condition of electrical neutrality, expressed in our case by $n_{ce} = n_D - n_{de}$. With the condition of electrical neutrality, Eq. (10.5.30) becomes

$$\frac{n_{ce}}{n_D - n_{de}} = \tfrac{1}{2} n_c e^{-\beta E_D} , \tag{10.5.36}$$

or, equivalently,

$$n_{ce}^2 + \tfrac{1}{2} n_c e^{-\beta E_D} n_{ce} - \tfrac{1}{2} n_D n_c e^{-\beta E_D} = 0 . \tag{10.5.37}$$

We may solve this latter quadratic equation in n_{ce} for n_{ce} as

$$n_{ce} = -\frac{n_c e^{-\beta E_D}}{4} \pm \tfrac{1}{2}\sqrt{\left(\tfrac{1}{2}n_c e^{-\beta E_D}\right)^2 + 2n_D n_c e^{-\beta E_D}}$$

$$= \frac{n_c e^{-\beta E_D}}{4}\left\{ -1 \pm \sqrt{1 + \frac{8n_D}{n_c}e^{\beta E_D}} \right\} .$$

It will be clear that the lower sign of the $\pm$ before the square root expression in the equation above is extraneous, as n_{ce} cannot be negative. Thus the number density of conduction electrons becomes

$$n_{ce} = \frac{1}{4}\, n_c e^{-\beta E_D}\left\{ -1 + \sqrt{1 + \frac{8n_D}{n_c}e^{\beta E_D}} \right\} , \tag{10.5.38}$$

and, upon using Eq. (10.5.28), the Fermi level (chemical potential) becomes

$$\mu = k_B T \ln\left(\frac{n_{ce}}{n_c}\right) = -E_D + k_B T \ln\left[\frac{1}{4}\left\{ -1 + \sqrt{1 + \frac{8n_D}{n_c}\, e^{\beta E_D}} \right\} \right] . \tag{10.5.39}$$

Let us now take a brief look at the high- and low-temperature behaviour of the density of conduction electrons in such a doped semiconductor:

- *Low Temperatures*: We have $e^{\beta E_D} \gg \dfrac{n_c}{8n_D}$ $\left(\text{or} \ \dfrac{8n_D}{n_c}e^{\beta E_D} \gg 1\right)$, and

$$n_{ce} \simeq \frac{n_c}{4}\, e^{-\beta E_D}\sqrt{\frac{8n_D}{n_c}\, e^{\beta E_D}} = \left(\frac{n_c n_D}{2}\right)^{\tfrac{1}{2}} e^{-\beta E_D} ,$$

or

$$n_{ce} = n_D \left(\frac{n_c}{2n_D}\right)^{\frac{1}{2}} e^{-\beta E_D}.$$ (10.5.40)

From this last result we see that the number density of conduction electrons will be much smaller than the number of donor levels, i.e.,

$$n_{ce} \ll n_D,$$ (10.5.41)

and that the Fermi level will be given by

$$\mu_F = -\tfrac{1}{2}E_D + \tfrac{1}{2} k_B T \ln\left(\frac{n_D}{2n_c}\right)^{\frac{1}{2}} \approx -\tfrac{1}{2} E_D.$$ (10.5.42)

This result shows us that the Fermi level lies approximately at the middle of the gap between the bottom of the conduction band and the donor levels, thus corresponding to a slight ionization of the donors.

- *High Temperatures*: If the temperature is sufficiently high, then $e^{\beta E_D} \approx 1$, $\frac{8n_D}{n_c} e^{\beta E_D} \ll 1$, and we can employ the Maclaurin expansion $(1 + x)^{\frac{1}{2}} \approx 1 + \tfrac{1}{2}x - \tfrac{1}{8}x^2 + \ldots$, so that

$$n_{ce} \simeq \frac{n_c}{4} e^{-\beta E_D} \left\{ 1 + 1 + \frac{1}{2}\frac{8n_D}{n_c}e^{\beta E_D} - \frac{1}{8}\left(\frac{8n_D}{n_c}e^{\beta E_D}\right)^2 + \ldots \right\}$$

$$= n_D - \frac{2n_D^2}{n_c}e^{\beta E_D} + \ldots$$

or

$$n_{ce} \simeq n_D \left(1 - \frac{2n_D}{n_c}e^{\beta E_D} + \ldots\right),$$ (10.5.43)

while the Fermi level becomes

$$\mu_F = -E_D + k_B T \ln\left\{ \frac{1}{4}\left[\frac{4n_D}{n_c}e^{\beta E_D} - \frac{1}{8}\left(\frac{8n_D}{n_c}e^{\beta E_D}\right)^2 + \ldots\right] \right\}$$

$$= -E_D + k_B T \ln\left(\frac{n_D}{n_c}\right) + E_D + k_B T \ln\left(1 - \frac{2n_D}{n_c}e^{\beta E_D} + \ldots\right),$$

or

$$\mu_F \simeq k_B T \left[\ln\left(\frac{n_D}{n_c}\right) - \frac{2n_D}{n_c} e^{\beta E_D} + \ldots\right].$$ (10.5.44)

The electrical conductivity σ of a semiconductor can be shown to have the form

$$\sigma = \frac{e^2 n_{ce}}{m^*}\, \tau\,, \tag{10.5.45}$$

in which τ is a relaxation time characterizing the nonequilibrium situation that occurs when a weak electric field is applied to the semiconductor. The above expressions for n_{ce} suggest that in the low and high temperature limits, the electrical conductivity behaves as

$$\sigma \simeq \frac{e^2 \tau}{m^*} \left(\frac{2\pi m^* k_B T}{h^2}\right)^{\frac{3}{4}} n_D^{\frac{1}{2}}\, e^{-\beta E_D/2}\,, \tag{10.5.46}$$

for low temperatures, and as

$$\sigma \simeq \frac{e^2 \tau}{m^*}\, n_D \tag{10.5.47}$$

for high temperatures.

10.6 Bose–Einstein Statistics

10.6.1 The Ideal Bose Gas

Just as the ^{3}He isotope of helium serves as the smallest-mass element whose naturally-occurring form is as atoms that are (composite) fermions, its ^{4}He isotope serves as the smallest-mass element whose naturally-occurring form is as atoms that are (composite) bosons (essentially due to its $I_a = 0$ nuclear spin). As the essential distinction between ^{3}He and ^{4}He is the small difference between the masses of the two isotopes, a treatment of a gas of ^{4}He atoms via the Gibbs prescription for (classically) indistinguishable particles at temperatures of the order of $T = 300$ K is justified by precisely the same arguments that we considered for ^{3}He. By the same token, ^{4}He must also be treated via Bose–Einstein quantum statistics for temperatures significantly below room temperature.

From Eqs. (10.2.12) and (10.2.21) the sums that are to be changed to integrals are now

$$\left(\frac{PV}{k_B T}\right)_{BE} = -\sum_k \ln\left(1 - \lambda e^{-\beta \epsilon_k}\right) \tag{10.6.1}$$

and

$$\overline{N}_{\mathrm{BE}} = \sum_k \frac{\lambda e^{-\beta \epsilon_k}}{1 - \lambda e^{-\beta \epsilon_k}}, \tag{10.6.2}$$

for Bose–Einstein statistics.

Both summands diverge when $\lambda e^{-\beta \epsilon_0} \rightarrow 1$ for the energy of the lowest translational state. From the particle-in-a-box model we have $\epsilon_0 = 3h^2/(8mV^{\frac{2}{3}})$, but we may for convenience measure all our energy levels relative to this one, and thereby effectively set $\epsilon_0 = 0$, as no physical results can depend upon where the zero of energy states is set. We must hence separate out in the sums (10.6.1) and (10.6.2) the $k = 0$ term, and then replace the rest of the series by an integral. This procedure differs from how we proceeded for fermions. Moreover, for bosons we shall treat only the spinless case, so that

$$\Omega_{\mathrm{BE}}(\epsilon) = 2\pi \left(\frac{2m}{h^2} \right)^{\frac{3}{2}} V \epsilon^{\frac{1}{2}}, \tag{10.6.3}$$

and we obtain

$$\frac{P}{k_{\mathrm{B}}T} = -\frac{1}{V} \ln(1 - \lambda) - 2\pi \left(\frac{2m}{h^2} \right)^{\frac{3}{2}} \int_{\epsilon>0}^{\infty} \epsilon^{\frac{1}{2}} \ln \left(1 - \lambda e^{-\beta \epsilon} \right) d\epsilon, \tag{10.6.4}$$

and

$$\frac{\overline{N}}{V} = \rho = \frac{\lambda}{V(1 - \lambda)} + 2\pi \left(\frac{2m}{h^2} \right)^{\frac{3}{2}} \int_{\epsilon>0}^{\infty} \frac{\lambda \epsilon^{\frac{1}{2}} e^{-\beta \epsilon}}{1 - \lambda e^{-\beta \epsilon}} d\epsilon. \tag{10.6.5}$$

The terms in $1/V$ are needed only for very degenerate situations in which $\lambda \rightarrow 1$, and may be ignored in the present small-λ case: for V of macroscopic dimension these terms may be discarded unless $\lambda \rightarrow 1$, and we shall do so here.

If we proceed as we did for the fermion case in Sect. 10.4.1 using expansions analogous to Eqs. (10.4.4) and (10.4.5) to obtain Eqs. (10.4.6) and (10.4.7), we obtain the corresponding Bose–Einstein equations

$$\left(\frac{P}{k_{\mathrm{B}}T} \right)_{\mathrm{BE}} = \frac{1}{\Lambda^3} \sum_{j=1}^{\infty} \frac{\lambda^j}{j^{\frac{5}{2}}}, \tag{10.6.6}$$

$$\left(\frac{\overline{N}}{V} \right)_{\mathrm{BE}} = \rho_{\mathrm{BE}} = \frac{1}{\Lambda^3} \sum_{j=1}^{\infty} \frac{\lambda^j}{j^{\frac{3}{2}}}, \tag{10.6.7}$$

and

$$\lambda = \Lambda^3 \rho - \frac{\Lambda^6}{2^{\frac{3}{2}}} \rho^2 + \left(\frac{1}{4} - \frac{1}{3^{\frac{3}{2}}}\right) \Lambda^9 \rho^3 + \dots . \tag{10.6.8}$$

Substitution of Eq. (10.6.8) into Eq. (10.6.6), followed by equating the powers of ρ_{BE} on the two sides of the resulting equation, gives the pressure for a dilute (ideal) Bose–Einstein gas as [see also Table 10.3.]

$$\left(\frac{P}{k_B T}\right)_{BE} = \rho - \frac{\Lambda^3}{2^{\frac{5}{2}}} \rho^2 + \Lambda^6 \left(\frac{1}{8} - \frac{2}{3^{\frac{5}{2}}}\right) \rho^3 + \dots , \tag{10.6.9}$$

with

$$B_2(T) = -\frac{\Lambda^3}{2^{\frac{5}{2}}} , \tag{10.6.10}$$

$$B_3(T) = \left(\frac{1}{8} - \frac{2}{3^{\frac{5}{2}}}\right) \Lambda^6 . \tag{10.6.11}$$

The effective interaction in this case is *attractive*, and is just the opposite to what would be obtained for a gas of spinless fermions (could such a thing exist!).

Table 10.3 Contributions to the second virial coefficient of ^{4}He at low temperatures[a]

T/K	Contributions to B			Total
	B_{ideal}	$B_{interaction}$	B_{bound}	$B_2(T)$
0.5	−200.08	−1172.00	−5.20	−1377.30
1.0	−70.74	−402.61	−0.92	−474.26
2.0	−25.01	−167.36	−0.16	−192.53
3.0	−13.61	−105.12	−0.12	−118.79
4.0	−8.84	−75.09	−0.06	−83.96
5.0	−6.33	−57.07	−0.03	−63.42
6.0	−4.81	−44.98	−0.02	−49.80
7.0	−3.82	−36.28		−40.11
8.0	−3.13	−29.72		−32.85
9.0	−2.62	−24.59		−27.21
10.0	−2.24	−20.46		−22.70

[a]Contributions to the second virial coefficient of ^{4}He. All values of B_2 are given in cm^3 mol^{-1}. Data from Ref. [2], with permission of Taylor and Francis

10.6.2 The Chemical Potential: Bose Gas

The behaviour of the chemical potential for an ideal Bose gas is illustrated in Fig. 10.11: the behaviour at low temperatures is clearly different from either that of an ideal Fermi gas (see Fig. 10.2) or an ideal classical gas. For high temperatures, μ_{BE} behaves similarly to μ for an ideal classical gas, as might be expected. The difference between μ_{BE} and μ_{FD} at low temperatures is directly due to the difference in statistics associated with the two types of quantum particle.

For temperatures T near 0, bosons begin to accumulate in the lowest energy eigenstates available, i.e., the ground state $\epsilon = 0$, as bosons, unlike Fermions, do not obey an exclusion principle. At $T = 0$, all bosons would be in the ground state with energy zero: there is thus only one available microstate for the system, so that the entropy $S = k_B \ln \Omega = k_B \ln 1 = 0$. What happens, then, if we add another boson to the system, while keeping the system entropy and volume constant? To keep the entropy constant, the boson must also go into the ground state, so that $\Delta U = 0$, and hence $\mu = \Delta U = 0$.

Now let us examine what will happen at a very low temperature $T > 0$: the internal energy of the gas will have increased, with a few bosons in excited states, most of them remaining in the ground state, and the entropy will also have increased due to the concomitant increase in the number of available system microstates.

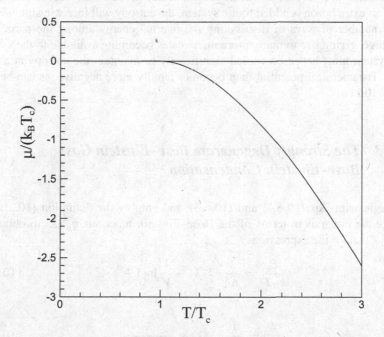

Fig. 10.11 Temperature dependence of the chemical potential for a Bose–Einstein gas. Figure provided by Dr. Lee Huntington

What happens now, if we add another boson to the system? Even if we add the boson to the ground state population, the number of microstates available to the system will increase slightly (because the number of ways of distributing the internal energy of the system amongst the particles normally increases with the addition of another particle) and consequently, the internal energy of the gas should be allowed to decrease slightly (by cooling the system) in order that the number of available microstates not increase, thereby holding the system entropy fixed. The small change ΔU in U will thus be negative, so that μ will be very small in magnitude, and negative.

By extending this argument, we can see that the chemical potential of a Bose gas remains close to zero, and negative, until a temperature T_c, which characterizes the temperature for which the occupancy of the ground state becomes vanishingly small, is reached. As we have already seen for spinless bosons, this critical temperature is given by

$$T_c = \frac{1}{\left[g_{\frac{3}{2}}^-(1) \right]^{\frac{2}{3}}} \frac{h^2}{2\pi m k_B} \left(\frac{N}{g_I V} \right)^{\frac{2}{3}} \tag{10.6.12}$$

for a gas of bosons with spin I ($g_I \equiv 2I + 1$ is the relevant spin degeneracy). For temperatures $T > T_c$, there will no longer be any ground state bosons present, so that if an extra boson is added to the system, the entropy will increase significantly, as the number of ways of distributing the internal energy among the particles is now large, giving rise to many more microstates becoming available to the system. The system must hence be cooled significantly to maintain the entropy at a fixed value. The chemical potential then becomes rapidly more negative, as can be seen in Fig. 10.11.

10.6.3   The Strongly Degenerate Bose–Einstein Gas: Bose–Einstein Condensation

We begin with Eqs. (10.6.4) and (10.6.5), and employ the definition (10.2.16) to replace the integrals in terms of the Bose–Einstein functions $g_n^-(\lambda)$ to obtain for $P/(k_B T)$ and ρ the expressions

$$\frac{P}{k_B T} = \frac{1}{\Lambda^3} g_{\frac{5}{2}}^-(\lambda) - \frac{1}{V} \ln(1 - \lambda), \tag{10.6.13}$$

and

$$\rho = \frac{1}{\Lambda^3} g_{\frac{3}{2}}^-(\lambda) + \frac{\lambda}{V(1 - \lambda)}. \tag{10.6.14}$$

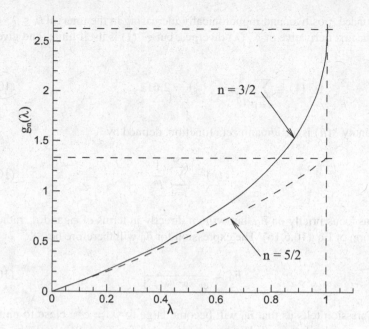

Fig. 10.12 Behaviour of the Bose–Einstein functions $g_{\frac{3}{2}}^-(\lambda)$ and $g_{\frac{5}{2}}^-(\lambda)$ for $0 \leq \lambda \leq 1$

As we have separated out the terms associated with $\lambda \rightarrow 1$, we may represent the Bose–Einstein functions $g_l^-(\lambda)$ in terms of the power series obtained from the integral expression (10.2.16) upon expanding the integrand as a power series in λe^{-x}, obtaining ultimately

$$g_l^-(\lambda) \equiv \sum_{k=1}^{\infty} \frac{\lambda^k}{k^l}. \tag{10.6.15}$$

For additional details on these interesting functions, see Pathria [8]. We note for future use that all of these functions behave as λ for sufficiently small values of λ. The behaviours of $g_{\frac{3}{2}}^-(\lambda)$ and $g_{\frac{5}{2}}^-(\lambda)$ are illustrated in Fig. 10.12.

Bose–Einstein degeneracy is associated with λ taking values near unity: this means that, in principle, we can no longer simply ignore the V^{-1} terms in Eqs. (10.6.13) and (10.6.14). From Eq. (10.2.22) the average number of particles in the ground state is

$$\bar{n}_0 = \frac{\lambda}{1 - \lambda}, \tag{10.6.16}$$

from which we see that λ is restricted to the interval $0 \leq \lambda < 1$ (cf. FD statistics, for which λ need only be positive-definite, i.e., $0 \leq \lambda < \infty$). The function $g_{\frac{3}{2}}^-(\lambda)$ is thus bounded, positive, and monotonically increasing in the interval $0 \leq \lambda < 1$; at $\lambda = 1$ the first derivative of $g_{\frac{3}{2}}^-(\lambda)$ diverges, but $g_{\frac{3}{2}}^-(1)$ is itself finite, and given by

$$g_{\frac{3}{2}}^-(1) = \sum_{l=1}^{\infty} \frac{1}{l^{\frac{3}{2}}} = \zeta\left(\tfrac{3}{2}\right) = 2.612\ldots \tag{10.6.17}$$

The quantity $\zeta(n)$ is a Riemann zeta function, defined by

$$\zeta(n) \equiv \sum_{l=1}^{\infty} \frac{1}{l^n}. \tag{10.6.18}$$

Let us focus briefly on $\overline{n}_0$, but written directly in terms of ϵ_0 and μ, rather than in the form of Eq. (10.6.16). The expression for $\overline{n}_0$ will therefore be

$$\overline{n}_0 = \frac{1}{e^{\beta(\epsilon_0 - \mu)} - 1}. \tag{10.6.19}$$

This expression tells us that $\overline{n}_0$ will become large for $\lambda = e^{\beta\mu}$ close to unity: this can happen only if $\mu < 0$. We have argued that for $T = 0$, μ will, by its very definition, be small in magnitude whenever there is a significant number of bosons in the ground state; hence, for $\overline{n}_0$ to be large, the temperature must also be 'small enough' that the exponential in Eq. (10.6.19) be close to unity. If we expand the exponential, and retain only the first two terms in its expansion, we see that $\overline{n}_0$ will be very large when

$$\overline{n}_0 \simeq \frac{k_B T}{\epsilon_0 - \mu}.$$

We also can see from this expression that the chemical potential μ must be equal to ϵ_0 for $T = 0$, and just slightly less than ϵ_0 for $T \neq 0$, as $\overline{n}_0$ cannot be negative. Recall that we have also seen that ϵ_0 may be chosen arbitrarily, and that the usual choice is to set $\epsilon_0 = 0$, which means that μ must be small and negative, all of which is consistent with our picture of the chemical potential for a BE quantum gas. An obvious question that arises from this discussion is 'How low must the temperature be in order that the vast majority of the bosons are found in the ground state?'

To obtain the equation of state, as we did for the FD electron gas, we need somehow to invert Eq. (10.6.14) so as to obtain λ as a function of ρ, then substitute that result into Eq. (10.6.13) for $P/k_B T$. However, as Eq. (10.6.14) cannot be solved analytically, it must be solved either graphically or numerically.

It is clear from Eqs. (10.6.13) and (10.6.14) that $\lambda = 1$ defines some sort of critical point for a BE gas. Let us therefore utilize the non-divergent term in Eq. (10.6.14) to define a critical temperature corresponding, for a given particle density ρ, to $\lambda = 1$: i.e., we define T_c via

$$\rho = \frac{g_{\frac{3}{2}}^-(1)}{\Lambda^3(T_c)} . \tag{10.6.20}$$

The critical temperature T_c thus depends upon the density ρ of the BE gas. Note also that for $\lambda < 1$, when the second term of Eq. (10.6.14) can be neglected, $g_{\frac{3}{2}}^-(\lambda)$ $= N\Lambda^3/V$ implicitly defines λ. Moreover, this expression correctly reduces to $\lambda = N\Lambda^3/V$, as it must for a classical gas (for which $\lambda \ll 1$).

It will be more convenient for the purposes of the following discussion to focus our attention on the number N of bosons, rather than upon their density $\rho = N/V$. Thus, we shall rewrite Eq. (10.6.14) for ρ as

$$N = \frac{N}{\Lambda^3(T)} \frac{1}{\rho} g_{\frac{3}{2}}^-(\lambda) + \frac{\lambda}{1-\lambda} . \tag{10.6.21}$$

We shall also replace ρ in this equation by Eq. (10.6.20), to obtain

$$N = N \frac{\Lambda^3(T_c)g_{\frac{3}{2}}^-(\lambda)}{\Lambda^3(T)g_{\frac{3}{2}}^-(1)} + \frac{\lambda}{1-\lambda} ,$$

which we may also write as

$$N = N \left(\frac{T}{T_c}\right)^{\frac{3}{2}} \frac{g_{\frac{3}{2}}^-(\lambda)}{g_{\frac{3}{2}}^-(1)} + \frac{\lambda}{1-\lambda} \tag{10.6.22a}$$

$$\equiv N_{\text{ex}} + n_0 . \tag{10.6.22b}$$

Recall that N represents the total number of bosons in the system, so that N_{ex}, which is defined via the first term of Eq. (10.6.22b) as

$$N_{\text{ex}} = N \left(\frac{T}{T_c}\right)^{\frac{3}{2}} \frac{g_{\frac{3}{2}}^-(\lambda)}{g_{\frac{3}{2}}^-(1)} \equiv \sum_{k \neq 0} \bar{n}_k , \tag{10.6.23a}$$

represents the total number of bosons in excited states, while $\bar{n}_0$, given by

$$\bar{n}_0 \equiv \frac{\lambda}{1-\lambda} , \tag{10.6.23b}$$

is the number of bosons in the ground state. It is for this reason that the second term in Eq. (10.6.14) cannot be ignored when λ approaches unity.

We may similarly enquire as to the role played by the $\ln(\lambda - 1)$ term in the pressure equation (10.6.13). Firstly, we note that because we have identified $\lambda/(1 - \lambda)$ with $\bar{n}_0$ in Eq. (10.6.23b), we may invert it to express λ as $\lambda = \bar{n}_0/(\bar{n}_0 + 1)$. Thus,

$V^{-1}\ln(1-\lambda)$ becomes $V^{-1}\ln(\overline{n}_0+1)$, so that in the thermodynamic limit it is at most of order $N^{-1}\ln N$, and may therefore be neglected for all values of λ.

Two temperature regimes must be considered separately, namely those temperature regimes that lie either above or below the critical temperature, T_c.

(a) For temperatures $T > T_c$, we have $\lambda < 1$, so that the term $\lambda/(1-\lambda)$ is finite, and may be neglected relative to N. The value of λ may be determined either graphically or numerically by setting the coefficient of N in expression (10.6.23a) to 1, i.e., by solving the equation

$$g_{\frac{3}{2}}^{-}(\lambda) = g_{\frac{3}{2}}^{-}(1)\left(\frac{T}{T_c}\right)^{\frac{3}{2}} \tag{10.6.24}$$

for λ.

(b) As we know that $\lambda \to 1$ for temperatures $T < T_c$, we may replace λ in the first term of expression (10.6.22a) by 1, thereby obtaining N as

$$N \simeq N\left(\frac{T}{T_c}\right)^{\frac{3}{2}} + \frac{\lambda}{1-\lambda} = N\left(\frac{T}{T_c}\right)^{\frac{3}{2}} + \overline{n}_0,$$

from which we obtain $\overline{n}_0$ as

$$\overline{n}_0 = N\left[1 - \left(\frac{T}{T_c}\right)^{\frac{3}{2}}\right]. \tag{10.6.25}$$

For $T > T_c$ the fraction of bosons in the ground state is essentially zero. This is the usual situation, in which the bosons are distributed smoothly over the many quantum states that are available to each particle. However, we note that as the temperature is lowered below T_c, the ground state suddenly begins to be appreciably populated, with its population increasing, until at $T = 0$ all particles are in the ground state. This is analogous to a phase transition, or 'condensation' of the particles into their ground states. It is, however, not a condensation in physical space, but is rather one in what can be referred to as 'momentum space'. This process is referred to as the *Bose–Einstein condensation*, and is one of the means of describing the occurrence of superfluidity in liquid ^{4}He when the critical temperature of 2.18 K is reached.

Let us now define the Bose condensate mole fraction X_0 via the thermodynamic limit, '$N \to \infty$, $V \to \infty$, N/V = constant'. In this limit, we can express X_0 as

$$X_0 = \begin{cases} 0, & T > T_c \\ 1 - (T/T_c)^{\frac{3}{2}}, & T < T_c \end{cases}. \tag{10.6.26}$$

The quantity $\sqrt{X_0}$ is traditionally referred to as the 'order parameter', and goes to zero as $(T_c - T)^{\frac{1}{2}}$ in the neighbourhood of T_c. The behaviours of X_0 and $\sqrt{X_0}$ are illustrated in Fig. 10.13.

We have argued that the second term in Eq. (10.6.13) may be neglected relative to the first term for all values of λ. We have also seen that there is a critical temperature, T_c, for a Bose–Einstein gas below which λ can be set to unity, so that we may write the pressure in a form reflecting this difference in behaviour above and below T_c as

$$P = \begin{cases} \dfrac{k_B T}{\Lambda^3}\, g_{\frac{5}{2}}^{-}(\lambda)\,, & T > T_c \\[2ex] \dfrac{k_B T}{\Lambda^3}\, g_{\frac{5}{2}}^{-}(1)\,, & T < T_c \end{cases} \qquad (10.6.27)$$

We see from this expression that $P \propto T^{\frac{5}{2}}$ for $T < T_c$ and that for $T > T_c$ we may write the pressure as

$$P = \frac{N k_B T}{V}\, \frac{g_{\frac{5}{2}}^{-}(\lambda)}{g_{\frac{3}{2}}^{-}(\lambda)} \qquad (10.6.28)$$

by employing Eq. (10.6.20) for ρ when $T > T_c$. Recall that for sufficiently small values of λ, we have $g_{\frac{5}{2}}^{-}(\lambda) \simeq g_{\frac{3}{2}}^{-}(\lambda) \approx \lambda$, so that we thereby retrieve the ideal classical gas equation of state.

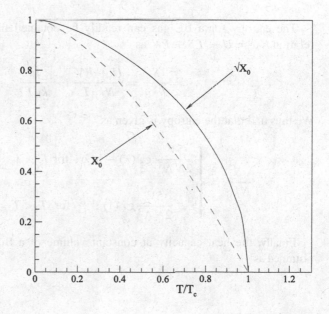

Fig. 10.13 The order parameter and the Bose–Einstein condensate fraction X_0 as functions of the ratio T/T_c

We have seen that, quite generally, the internal energy U can be obtained from the grand partition function Ξ as

$$U = -\left(\frac{\partial}{\partial \beta} \ln \Xi\right)_{\lambda, V} = k_B T^2 \left\{\frac{\partial}{\partial T}\left(\frac{PV}{k_B T}\right)\right\}_{\lambda, V}$$

and, as

$$\left(\frac{PV}{k_B T}\right)_{BE} = \frac{1}{\Lambda^3(T)} g_{\frac{5}{2}}(\lambda)$$

for $T > T_c$, we obtain U as

$$U = k_B T^2 g_{\frac{5}{2}}(\lambda) \frac{d}{dT}\left(\frac{1}{\Lambda^3}\right)$$

$$= \frac{3}{2} k_B T \frac{V}{\Lambda^3} g_{\frac{5}{2}}(\lambda)$$

for $T > T_c$. To obtain U for $T < T_c$, we need only replace λ by 1, so that the two expressions for the internal energy can be merged to give

$$U = \begin{cases} \dfrac{3k_B T V}{2\Lambda^3} g_{\frac{5}{2}}(\lambda) \,, \ T > T_c \\[3mm] \dfrac{k_B T}{\Lambda^3} g_{\frac{5}{2}}(1) \,, \quad T < T_c \end{cases} \qquad (10.6.29)$$

The entropy for a BE gas can readily be obtained from the thermodynamic relation $N\mu = U - TS + PV$ as

$$\frac{S}{Nk_B} = \frac{U + PV}{Nk_B T} - \frac{\mu}{k_B T} .$$

We thus find that the entropy is given as

$$\frac{S}{Nk_B} = \begin{cases} \dfrac{5V}{2N\Lambda^3} g_{\frac{5}{2}}(\lambda) - \ln \lambda \,, \ \text{for } T > T_c \\[3mm] \dfrac{5V}{2N\Lambda^3} g_{\frac{5}{2}}(1) \,, \qquad \text{for } T < T_c \end{cases} \qquad (10.6.30)$$

Finally, the heat capacity at constant volume of a Bose–Einstein gas can be obtained as

$$C_V(T) = \begin{cases} \dfrac{15 g_{\frac{5}{2}}^-(\lambda)}{4 g_{\frac{3}{2}}^-(\lambda)} - \dfrac{9 g_{\frac{3}{2}}^-(\lambda)}{4 g_{\frac{1}{2}}^-(\lambda)}\,, & \text{for } T > T_c \\[4mm] \dfrac{15 V}{4 N \Lambda^3}\, g_{\frac{5}{2}}^-(1)\,, & \text{for } T < T_c \end{cases} \qquad (10.6.31)$$

We note that at $T = T_c$, where C_V exhibits a cusp, $C_V/(N k_B)$ is given by

$$\frac{C_V(T_c)}{N k_B} = \frac{15}{4} \frac{g_{\frac{5}{2}}^-(1)}{4 g_{\frac{3}{2}}^-(1)} \simeq 1.925\,,$$

which lies above the classical value of 1.5.

Only the excited states contribute to the entropy and internal energy of a BE gas. The condensate entropy is zero, as is its internal energy. At $T = T_c$, the heat capacity of a BE gas has a discontinuous first derivative at T_c, and corresponds to what is termed a 1st-order phase transition.

10.6.4 The Photon Gas: Blackbody Radiation

Blackbody radiation (which we may regard as a photon gas) satisfies the Planck distribution, which takes the form

$$\bar{n}_{\text{photon}} = \frac{1}{e^{\beta h \nu} - 1}\,. \qquad (10.6.32)$$

This is the same Planck distribution considered in a typical quantum mechanics course. The photon is a boson that possesses an intrinsic spin $\hbar$ but, unlike other spin-1 bosons, has a spin-degeneracy of 2, rather than 3; the nonexistence of the $m_I = 0$ projection of the photon spin is associated with the photon having rest-mass zero. The $m_I = \pm 1$ states correspond to right- and left-circular polarization states of the photon. The energy ϵ_k and linear momentum p_k associated with an individual photon are related by $\epsilon_k = p_k c$, with c the speed of light. Because ϵ_k is simultaneously related to the frequency ν_k of the light wave with which it is associated by $\epsilon_k = h \nu_k$, with h the Planck constant, we also have the de Broglie relation $p_k = h \nu_k / c$.

If we compare the Planck distribution with the general BE distribution, as represented by the boson mean occupation number

$$\bar{n}_{\text{BE}} = \frac{1}{e^{\beta(\epsilon - \mu)} - 1}\,, \qquad (10.6.33)$$

we see that the two formulae are compatible only if the chemical potential μ for a photon gas is identically zero for all temperatures. This may also appear to be another astounding result. Why is it so? A number of explanations can be given, but they all reduce to the fundamental reason being that photons, unlike most other particles, can be created (emitted) and destroyed (absorbed) by matter, so that their number is not conserved. Recall that we introduced the chemical potential μ (through $\gamma = -\beta\mu$) in the grand ensemble as a constraint on the total number of particles [characterizing material contact between our system of interest and the 'reservoir']. If the total number of particles is not conserved, then there is no need for μ to be assigned a specific value, leaving it arbitrary, so that we may set it to zero.

The internal energy U for a photon gas is given by

$$U = \sum_i \bar{n}_i \epsilon_i = \sum_i \frac{h\nu_i}{e^{\beta h\nu_i} - 1}, \tag{10.6.34}$$

which, upon replacing the discrete summation by an integration over continuous frequency, becomes

$$U = \int_0^\infty \Omega(\nu) \frac{h\nu}{e^{\beta h\nu} - 1} \, d\nu, \tag{10.6.35}$$

with the density of photon states $\Omega(\nu)$ given by

$$\Omega(\nu) = \frac{8\pi V \nu^2}{c^3}, \tag{10.6.36}$$

in which ν is the photon frequency and c is the speed of light. Thus, the internal energy can be written as

$$U = V \int_0^\infty \frac{8\pi h\nu^3}{c^3(e^{\beta h\nu} - 1)} \, d\nu \tag{10.6.37}$$

or, as written in terms of the radiation Planck density $\rho(\nu, T)$,

$$U = V \int_0^\infty \rho(\nu, T) \, d\nu. \tag{10.6.38}$$

We shall make use of the geometric progression in reverse to write the integrand of Eq. (10.6.37) as

$$\frac{\nu^3}{e^{\beta h\nu} - 1} = \frac{\nu^3 e^{-\beta h\nu}}{1 - e^{\beta h\nu}} = \nu^3 e^{-\beta h\nu} \sum_{n=0}^\infty (e^{-\beta h\nu})^n$$

$$= \nu^3 \sum_{n=1}^\infty (e^{-\beta h\nu})^n.$$

Thus, following the introduction of the dimensionless variable $x \equiv \beta h\nu$, the internal energy for the blackbody radiation is given by

$$U = \frac{8\pi V (k_B T)^4}{c^3 h^3} \sum_{n=1}^{\infty} \int_0^{\infty} x^3 e^{-nx} \, dx = \frac{48\pi V}{c^3 h^3} (k_B T)^4 \zeta(4) , \tag{10.6.39}$$

in which $\zeta(4)$ is the zeta function

$$\zeta(4) \equiv \sum_{n=1}^{\infty} \frac{1}{n^4} = \frac{\pi^4}{90} . \tag{10.6.40}$$

In this way, we obtain the internal energy U of a photon gas as

$$U = \frac{8\pi^5 k_B^4 V}{15 c^3 h^3} T^4 = \frac{4\sigma V}{c} T^4 , \tag{10.6.41}$$

with $\sigma = 2\pi^5 k_B^4 / (15 c^2 h^3) \simeq 5.67037 \times 10^{-8} \, \mathrm{J\,m^{-2}\,s^{-1}\,K^{-4}}$ the Stefan-Boltzmann constant of thermodynamics. An expression for the heat capacity at constant volume associated with thermal radiation may be obtained directly from expression (10.6.41) as

$$C_V = \left(\frac{\partial U}{\partial T} \right)_V = \frac{16\sigma V}{c} T^3 . \tag{10.6.42}$$

Let us now examine the radiation pressure P, which can be obtained through the equation of state

$$\frac{PV}{k_B T} = -\sum_i \ln(1 - e^{-\beta h\nu_i})$$

$$= -\int_0^{\infty} \Omega(\nu) \ln(1 - e^{-\beta h\nu}) \, d\nu$$

$$= -\frac{8\pi V}{c^3} \int_0^{\infty} \nu^2 \ln(1 - e^{-\beta h\nu}) \, d\nu . \tag{10.6.43}$$

This expression can be simplified by carrying out an integration by parts, to give

$$P = \frac{8\pi h}{3c^3} \int_0^{\infty} \frac{\nu^3}{e^{\beta h\nu} - 1} \, d\nu = \frac{U}{3V} . \tag{10.6.44}$$

Radiation thus exerts a pressure given by

$$P = \frac{4\sigma}{3c} T^4 . \tag{10.6.45}$$

Given that $\sigma/c \simeq 1.8915 \times 10^{-16}\,\mathrm{Pa\,K^{-4}}$, radiation pressure will thus be negligible under terrestrial conditions. However, in stellar interiors, where temperatures can reach several million degrees, radiation pressure can, and does, play a significant role.

We may obtain an explicit expression for the dependence of the number of thermal photons from our expression for $\overline{N}$ by converting the discrete sum over photon energy states into an integral over a continuous frequency distribution having the density of photon states $\Omega(\nu)$ given by Eq. (10.6.36). This gives $\overline{N}$ as

$$
\begin{aligned}
\overline{N} &= \frac{8\pi V}{(hc\beta)^3} \int_0^\infty \frac{x^2}{e^x - 1}\, \mathrm{d}x \\
&= \frac{16\pi k_B^3}{h^3 c^3}\, \zeta(3)\, V T^3 ,
\end{aligned}
\tag{10.6.46a}
$$

or, upon evaluating the constant, as

$$
\overline{N} = 2.0286 \times 10^7\, V T^3 .
\tag{10.6.46b}
$$

In particular, let us note that the number of photons in a photon gas at thermal equilibrium at temperature T depends linearly upon the volume V of its black-body container.

Example 10.5 Comparison of the pressures of a thermal photon gas and an ideal classical gas at a temperature of 1000 K.

Let us consider a tungsten-walled container of volume $V = 10\,\mathrm{m}^3$ at temperature $T = 1000\,\mathrm{K}$. Thus, for thermal photons, the radiation pressure is given by Eq. (10.6.45) as

$$
P = 2.5220 \times 10^{-16}\, T^4 ,
$$

giving $P_{\text{radiation}}(T = 1000\,\mathrm{K}) \simeq 2.5 \times 10^{-4}\,\mathrm{Pa}\,[1\,\mathrm{J\,m^{-3}} \equiv 1\,\mathrm{Pa}]$ or $2.5 \times 10^{-9}\,\mathrm{bar}$, and the number of photons in the enclosure, given by Eq. (10.6.46b) will be of order 2×10^{17}. As the radiation pressure, P_{rad}, is independent of the system volume (i.e., the volume of the container), we shall compare it with the pressure exerted by an ideal gas that represents what might be the laboratory situation generated by starting with an ideal gas in a container of volume V at room temperature (say $T = 300\,\mathrm{K}$), then heated to $T = 1000\,\mathrm{K}$, so that both N and V are fixed. The resultant ideal gas pressure will thus be

$$
P_{1000\,\mathrm{K}} = P_{300\,\mathrm{K}} \left(\frac{1000}{300} \right) k_B \simeq 3.33\,\mathrm{bar}
$$

for $P_{300\,\mathrm{K}} = 1\,\mathrm{bar}$, corresponding (for $V = 10\,\mathrm{m}^3$) to $\overline{N} \simeq 2.4 \times 10^{21}$ particles. Thus, the radiation contributes a partial pressure $P_{\text{rad}} \approx 3 \times 10^{-4}\,\mathrm{Pa}$, versus the ideal gas contribution of $P \approx 3.3 \times 10^5$ Pa to the total pressure in the container at a temperature of 1000 K. $\square$

As the chemical potential μ is identically zero for a photon gas, the Gibbs energy vanishes (i.e., $G = N\mu = 0$), and hence the entropy S associated with radiation may be obtained from the relation

$$G = U - TS + PV = 0$$

as

$$S = \frac{U + PV}{T} = \frac{4}{3} \frac{U}{T} = \tfrac{1}{3} C_V. \qquad (10.6.47a)$$

Another means of checking the consistency of zero chemical potential is to utilize the Helmholtz energy A, which is given by

$$A(T, V) \equiv A = U - TS = \left(4 - \tfrac{16}{3}\right) \frac{\sigma V}{c} T^4 = -\frac{4\sigma V}{3c} T^4, \qquad (10.6.47b)$$

to determine the chemical potential using the thermodynamic relation

$$\mu = \left(\frac{\partial A}{\partial N}\right)_{T,V} = 0.$$

Indeed, expression (10.6.47b) for the Helmholtz energy is consistent with the chemical potential being identically zero, as we see that A is independent of N for thermal photons.

That the chemical potential vanishes identically for a photon gas in thermal equilibrium at temperature T requires us to look more carefully into some of the general results that were obtained for the grand ensemble in Chap. 4. For example, in discussing fluctuations in the number of particles, we saw in Eq. (4.2.43) that the relative standard deviation in the number of particles took the form

$$\left(\frac{\sigma_N}{\overline{N}}\right)^2 = \frac{k_B T}{V} \kappa_T,$$

in which κ_T, defined as

$$\kappa_T \equiv -\frac{1}{V} \left(\frac{\partial V}{\partial P}\right)_T,$$

is the isothermal compressibility of the substance. However, as we have seen from Eq. (10.6.44) that the pressure for a thermal photon gas depends only upon the temperature, T, and is independent of V, the partial derivative $(\frac{\partial P}{\partial V})_T$ is zero. Normally, $(\frac{\partial P}{\partial V})_T$ and $(\frac{\partial V}{\partial P})_T$ are related by Eq. (B.1.12) as reciprocals of one another. We might thus be tempted to conclude that this relation then means that $\left(\frac{\partial V}{\partial P}\right)_T$ is infinite: this would, however, be an incorrect deduction. The proper interpretation of the vanishing of $\left(\frac{\partial P}{\partial V}\right)_T$ is that the partial derivative $\left(\frac{\partial V}{\partial P}\right)_T$ does not exist, so that Eq. (4.2.43) does not apply to a thermal photon gas, a point that has been established quite convincingly by Leff [9]. This conclusion is also supported by examining expression (4.2.35), which gives $\overline{N}$ in general in terms of the grand partition function $\Xi(T, V, \mu)$ as

$$\overline{N} = k_B T \left(\frac{\partial \ln \Xi}{\partial \mu}\right)_{T,V}.$$

As $\Xi(T, V, \mu)$ is effectively a function only of T and V for the thermal photon gas, this expression relating $\Xi(T, V, \mu)$ to $\overline{N}$ also does not apply to thermal photons, and we must therefore consider the behaviour of the number of thermal photons more carefully.

To examine fluctuations in the number of photons in a thermal photon gas, we return to the expression for the grand partition function $\Xi(T, V)$, which we may write explicitly for the photon gas in terms of the occupation numbers, $\{n_s\}$, and photon energies, $\{\epsilon_s\}$, of the photon energy states $\{s\}$ as

$$\Xi(T, V) = \sum_{\{n_s\}} e^{-\beta \sum_s n_s \epsilon_s}, \qquad (10.6.48)$$

from which the mean occupation numbers $\overline{n}_s$, given by

$$\overline{n}_s = -k_B T \left(\frac{\partial \ln \Xi}{\partial \epsilon_s}\right)_{T,V},$$

are thus obtained as

$$\overline{n}_s = \frac{\displaystyle\sum_{n_s=0}^{\infty} n_s e^{-\beta n_s \epsilon_s}}{\displaystyle\sum_{n_s=0}^{\infty} e^{-\beta n_s \epsilon_s}}. \qquad (10.6.49)$$

Straightforward differentiation of expression (10.6.48) for $\Xi(T, V)$ shows that the variance in the mean occupation number of photons in photon energy state s is given by

$$\overline{n_s^2} - \overline{n}_s^2 = \frac{1}{\beta^2} \left(\frac{\partial^2 \ln \Xi}{\partial \epsilon_s^2}\right)_{T,V}.$$

Examination of the second partial derivative in this expression allows us to rewrite it as

$$-k_{\mathrm{B}}T\,\frac{\partial}{\partial\epsilon_s}\left[-k_{\mathrm{B}}T\left(\frac{\partial\ln\Xi}{\partial\epsilon_s}\right)_{T,V}\right]_{T,V}=-k_{\mathrm{B}}T\left(\frac{\partial\overline{n}_s}{\partial\epsilon_s}\right)_{T,V},$$

so that the variance in the mean occupation number of photons in energy state ϵ_s is given as

$$\overline{n_s^2}-\overline{n}_s^2=-k_{\mathrm{B}}T\left(\frac{\partial\overline{n}_s}{\partial\epsilon_s}\right)_{T,V}. \tag{10.6.50}$$

From Eq. (10.6.33), we see that $\overline{n}_s$ for bosons is given (for $\lambda=1$) by

$$\overline{n}_s=\frac{1}{e^{\beta\epsilon_s}-1}.$$

Thus, the right-hand side of Eq. (10.6.50) becomes

$$-k_{\mathrm{B}}T\,\frac{\partial\overline{n}_s}{\partial\epsilon_s}=\frac{e^{\beta\epsilon_s}}{(e^{\beta\epsilon_s}-1)^2}$$

$$=\left[\frac{1}{e^{\beta\epsilon_s}-1}+\frac{1}{\left(e^{\beta\epsilon_s}-1\right)^2}\right]$$

$$=\overline{n}_s(1+\overline{n}_s).$$

With this result, the variance in $\overline{n}_s$ then takes the form

$$\overline{n_s^2}-\overline{n}_s^2=\overline{n}_s(1+\overline{n}_s). \tag{10.6.51}$$

Ultimately, however, we are interested in fluctuations in the macroscopic number $\overline{N}$ of photons in a thermal photon gas. To examine these fluctuations, we begin by noting that we have previously defined $\overline{N}$ in terms of the mean occupation numbers $\{\overline{n}_s\}$ via Eqs. (10.2.1) and (10.2.20) as $\overline{N}=\sum_s\overline{n}_s$. This enables us to express the variance in $\overline{N}$ as

$$\overline{N^2}-\overline{N}^2=\overline{\left(\sum_s n_s\right)^2}-\left(\sum_s\overline{n}_s\right)^2$$

or, equivalently, as the double summation

$$\overline{N^2}-\overline{N}^2=\sum_{s,s'}\left[\overline{n_s n_{s'}}-\overline{n}_s\,\overline{n}_{s'}\right].$$

Because we may treat n_s and $n_{s'}$ as statistically independent when $s \neq s'$, we obtain $\overline{n_s n_{s'}} = \overline{n}_s \overline{n}_{s'}$ for $s \neq s'$, and only the terms for which $s = s'$ remain in the double sum, thereby giving

$$\overline{N^2} - \overline{N}^2 = \sum_s \overline{n}_s (1 + \overline{n}_s) \qquad (10.6.52a)$$

for the variance in $\overline{N}$ (i.e., in the thermodynamic N). We may also express the variance in $\overline{N}$ in terms of the photon state energies ϵ_s as

$$\overline{N^2} - \overline{N}^2 = \sum_s \frac{e^{\beta \epsilon_s}}{(e^{\beta \epsilon_s} - 1)^2} , \qquad (10.6.52b)$$

from which it is also clear that the variance in $\overline{N}$ is a function of temperature alone.

Evaluation of the sum over photon states on the right-hand side of Eq. (10.6.52b) gives the variance explicitly as

$$\overline{N^2} - \overline{N}^2 = \frac{8\pi^3 k_B^3}{3h^3 c^3} VT^3$$

which, by employing Eq. (10.6.52b), may finally be written in terms of $\overline{N}$ as

$$\overline{N^2} - \overline{N}^2 = \frac{\pi^2}{6\zeta(3)} \overline{N} . \qquad (10.6.53)$$

This expression may, in turn, be utilized to establish that the relative rms deviation $\sigma_{\overline{N}}/\overline{N}$, with $\sigma_{\overline{N}}$ defined via $\sigma_{\overline{N}}^2 \equiv [\overline{N^2} - \overline{N}^2]^{\frac{1}{2}}$, thus behaves as

$$\frac{\sigma_{\overline{N}}}{\overline{N}} = \frac{1}{\sqrt{\overline{N}}} \frac{\pi}{\sqrt{6\zeta(3)}}$$

$$\simeq \frac{1.170}{\sqrt{\overline{N}}} . \qquad (10.6.54)$$

10.7   Density Matrix and Ensemble Averages

In the derivation of the Liouville Equation given in Sect. 7.5.1, we introduced the classical density of phase-space points, designated by $f^{(N)}(\mathbf{p}, \mathbf{q})$, associated with a system of N particles. The equivalent quantity in a fully quantum mechanical description is referred to as the *density matrix* operator, and is traditionally designated by ρ. The form taken by ρ is most readily illustrated using Dirac notation, in which the members of a set of quantum states $\{\phi | i = 1, \ldots, \}$ are designated $\{|i\rangle; i = 1, \ldots, \}$, with $|i\rangle$ known as a 'ket'. The adjoint, $|i\rangle^\dagger$, of a ket is referred to as a 'bra', and is designated $\langle i|$. In terms of these quantities, the inner product between the two quantum states represented by $|i\rangle$ and $|j\rangle$ is given by the 'bracket' $\langle i|j\rangle$. If a quantum mechanical system can be described in terms of a single quantum state, it is said to be in a *pure state*; otherwise, it is said to be in a *mixed state*, that is, a state that is represented by a linear combination of the $|\varphi_\alpha\rangle$, namely

$$|\psi\rangle = \sum_\alpha p_\alpha |\varphi_\alpha\rangle,$$

in which the p_α represent (relative) probabilities satisfying

$$\sum_\alpha p_\alpha = 1.$$

An ensemble of pure quantum states is thus necessarily represented by a mixed quantum state.

We shall employ Roman letters to identify eigenstates of the Hamiltonian and Greek letters to label the members of an ensemble containing N subsystems, each of which is in a pure quantum state $|\varphi_\alpha\rangle$ that has been normalized to 1, i.e., $\langle \varphi_\alpha | \varphi_\alpha \rangle = 1; \alpha = 1, \ldots, N$. If we further choose a complete orthonormal basis set $\{|i\rangle; i = 1, \ldots, \}$ (such as the eigenfunctions of the system Hamiltonian), then each system eigenfunction $|\varphi_\alpha\rangle$ may be expressed in terms of this basis set as

$$|\varphi_\alpha\rangle = \sum_i c_i^{(\alpha)} |i\rangle \equiv \sum_i \langle \varphi_\alpha | i \rangle |i\rangle. \qquad (10.7.1)$$

Let us identify the expectation value of an operator $\mathcal{A}$ in the pure state $|\varphi_\alpha\rangle$ associated with an individual member of the ensemble as

$$\langle \mathcal{A} \rangle_\alpha \equiv \langle \varphi_\alpha | \mathcal{A} | \varphi_\alpha \rangle = \sum_{i,j} c_i^{(\alpha)*} c_i^{(\alpha)} \langle j | \mathcal{A} | i \rangle \qquad (10.7.2a)$$

or, in terms of Dirac notation,

$$\langle \mathcal{A} \rangle_\alpha = \sum_{i,j} \langle i | \varphi_\alpha \rangle \langle \varphi_\alpha | j \rangle \langle j | \mathcal{A} | i \rangle. \qquad (10.7.2b)$$

We may now express the ensemble average $\langle \mathcal{A} \rangle$ as

$$\langle \mathcal{A} \rangle \equiv \sum_{\alpha=1}^{N} p_\alpha \langle \mathcal{A} \rangle_\alpha, \qquad (10.7.3)$$

in which p_α $(0 \le p_\alpha \le 1$, and $\sum_{\alpha=1}^{N} p_\alpha = 1)$ is the probability that the system is found in quantum state $|\varphi_\alpha\rangle$.

Substitution of Eq. (10.7.2b) into Eq. (10.7.3) gives the ensemble average $\langle \mathcal{A} \rangle$ of the operator $\mathcal{A}$ as

$$\langle \mathcal{A} \rangle = \sum_{\alpha=1}^{N} \sum_{i,j} \langle i | \varphi_\alpha \rangle \, p_\alpha \, \langle \varphi_\alpha | j \rangle \, \langle j | \mathcal{A} | i \rangle \,.$$

Upon defining a density matrix operator ρ as

$$\rho \equiv \sum_{\alpha} | \varphi_\alpha \rangle \, p_\alpha \, \langle \varphi_\alpha | , \qquad (10.7.4a)$$

with matrix elements ρ_{ij} given by

$$\rho_{ij} \equiv \langle i | \rho | j \rangle = \sum_{\alpha=1}^{N} \langle i | \varphi_\alpha \rangle \, p_\alpha \, \langle \varphi_\alpha | j \rangle \,, \qquad (10.7.4b)$$

then gives the ensemble average of $\mathcal{A}$ as

$$\langle \mathcal{A} \rangle = \sum_{i,j} \langle i | \rho | j \rangle \, \langle j | \mathcal{A} | i \rangle \equiv \sum_{i} \langle i | \rho \mathcal{A} | i \rangle \,, \qquad (10.7.5)$$

upon utilizing the completeness of the (orthonormal) basis set. Thus, the ensemble average $\langle \mathcal{A} \rangle$ of the operator $\mathcal{A}$ is obtained as the sum of the diagonal elements, or trace Tr the product of operator $\rho \mathcal{A}$, or

$$\langle \mathcal{A} \rangle = Tr\{ \rho \mathcal{A} \} \,. \qquad (10.7.6)$$

As the density matrix ρ is by definition an Hermitian operator, it follows that the density matrix corresponding to a system eigenfunction ϕ that is normalized to unity must satisfy the condition

$$Tr \, \rho = \sum_{i} \langle i | \rho | i \rangle = 1 \,.$$

Diagonal elements of ρ may thus be interpreted as probabilities of occurrence of the energy eigenstates of $\mathcal{H}$ in the ensemble, thereby providing a basis for utilizing density matrix calculations of ensemble averages.

Although a time-independent density matrix suffices for the description of equilibrium systems because their observable properties are typically independent of time, a time dependent density matrix would be needed for the description of observable properties in a nonequilibrium system. For additional material on the density matrix and its applications, see Bransden and Joachain [10].

Interlude   Time Dependence of the Density Matrix
The time evolution of the density matrix is given in general as

$$i\hbar \frac{\partial \rho}{\partial t} = [\mathcal{H}, \rho]_-$$

in terms of the commutator $[\mathcal{H}, \rho]_- \equiv \mathcal{H}\rho - \rho\mathcal{H}$ of the system Hamiltonian and the density matrix. For a time-independent Hamiltonian, the time evolution of $\rho(t)$ is given in terms of the initial density matrix $\rho(0)$ as

$$\rho(t) = \mathrm{e}^{-i\mathcal{H}t/\hbar} \rho(0) \mathrm{e}^{i\mathcal{H}t/\hbar},$$

in which case, the matrix elements of $\rho(t)$ may be written explicitly as

$$\langle i|\rho(t)|j \rangle = \mathrm{e}^{i(E_j - E_i)t/\hbar} \langle i|\rho(0)|j \rangle.$$

This last result shows, in particular, that for a time-independent Hamiltonian the diagonal elements of $\rho(t)$ are constant, while the off-diagonal elements of $\rho(t)$ oscillate in time with frequencies $\nu_{ij} = (E_j - E_i)/h$. This expression is the source of the so-called random phase approximation, whereby there is assumed to be no phase correlation between the members of an ensemble.   □

The elements of the density matrix for an equilibrium system, given by

$$\langle i|\rho|j \rangle = \overline{|c_i|^2} \delta_{ij} \tag{10.7.7}$$

correspond to diagonal elements equal to the probabilities for finding the system in a given energy eigenstate of $\mathcal{H}$ and all off-diagonal elements equal to zero. We have seen that for a microcanonical ensemble, each of the $\Omega(E)$ accessible microstates of a system having an energy that lies in an interval δE about energy E has the same probability, namely $1/\Omega(E)$, so that the relevant density matrix elements are given by

$$\langle i|\rho|j \rangle = \frac{1}{\Omega(E)} \delta_{ij}. \tag{10.7.8}$$

The density matrix is thus simply a multiple of the identity matrix. Similarly, we have seen that for the canonical ensemble thermal contact between the members of the ensemble and a heat bath determines a reciprocal temperature parameter $\beta \equiv (k_B T)^{-1}$. The elements of the density matrix for a canonical ensemble are thus given by

$$\langle i|\rho|j \rangle = \frac{1}{Z(T)} \mathrm{e}^{-\beta E_i} \delta_{ij}, \tag{10.7.9a}$$

in terms of the partition function

$$Z(T) \equiv \sum_i e^{-\beta E_i} , \qquad (10.7.9b)$$

with the sum being over the energy *states* of the system. Thus, for the canonical ensemble the density matrix takes the form

$$\rho = \frac{e^{-\beta \mathcal{H}}}{Tr\{e^{-\beta \mathcal{H}}\}} = \frac{1}{Z} e^{-\beta \mathcal{H}} . \qquad (10.7.9c)$$

The ensemble average of the expectation value of an operator $\mathcal{A}$ is thus given in the canonical ensemble by

$$\langle \mathcal{A} \rangle = \frac{1}{Z(T)} Tr\{e^{-\beta \mathcal{H}} \mathcal{A}\} . \qquad (10.7.10)$$

Most applications of the density matrix approach involve the canonical ensemble version. For completeness, however, the density matrix operator for the grand ensemble, has been shown by Greiner et al. [11] to have the form

$$\rho = \frac{1}{\Xi} e^{-\beta(\mathcal{H} - \mu \mathcal{N})} , \qquad (10.7.11)$$

in which Ξ is the grand partition function, μ is the chemical potential, and $\mathcal{N}$ is the particle number operator.

10.8   Problems for This Chapter

1. In how many ways may N identical, yet distinguishable, objects be arranged in M different containers, with no limitation on the number of objects in any given container? Such arrangements apply to a classical, or Maxwell–Boltzmann, system consisting of distinguishable particles.

2. In how many ways may N identical, indistinguishable objects be arranged in M different containers ($N < M$) such that no more than a single object occupies any given container? This process corresponds to the conditions that apply to a system of Fermi–Dirac particles.

3. In how many ways may N identical, indistinguishable objects be arranged in M different containers, with no limitation placed upon the number of objects in a given container? This process corresponds to the conditions that apply to a system of Bose–Einstein particles.

4. Employ the combinatoric formulae (10.1.1a) and (10.1.1b) for the numbers of microstates accessible to multi-particle Bose–Einstein and Fermi–Dirac sys-

tems to determine the number of microstates obtained for 3 bosons distributed over 10 states and for 3 fermions distributed over 10 states. Carry out the same calculations for 3 bosons (fermions), but now distributed over 100 states and comment upon any trend that may appear to be developing.

5. To explore briefly the evaluation of the canonical partition function for systems of quantum particles, we shall consider a simple model that consists of a box that supports ten single-particle ground-energy states which, for convenience, we shall set to zero. Determine the number of two-particle states that can be formed, both when the particles are classical (distinguishable or indistinguishable) and when they are quantum (bosonic or fermionic), and compare your results with the corresponding value for the canonical partition function. Determine the probabilities for finding both particles in the same 1-particle state.

6. If five indistinguishable particles are placed into a container that supports a set of equally-spaced nondegenerate energy states and supplied with sufficient energy to form their ground state configuration, what form will that configuration assume for MB (i.e., classical), FD, and BE particles? If one additional unit of energy is made available for distribution amongst the five particles, how will these configurations change? What forms will the relevant configurations take upon supplying another one, two, additional units of energy to the system? How might we distinguish between systems of indistinguishable and distinguishable particles?

7. The system internal energy U and average number of particles $\overline{N}$ are given in terms of the mean occupation numbers $\overline{n}_k$ defined in Eq. (10.2.22) for Fermi–Dirac (FD) and Bose–Einstein (BE) systems by Eqs. (10.2.23) and (10.2.21), respectively. Expression (10.2.22) for the $\overline{n}_k$ is also commonly rewritten in the convenient form $\overline{n}_k = [e^{\beta(\epsilon_k - \mu)} \pm 1]^{-1}$. Employ Eqs. (10.2.21) and (10.2.23) for $\overline{N}$ and U to aid in interpreting the meanings of T and μ for systems consisting of quantum mechanical particles treated in principle in terms of microcanonical, canonical, or grand ensembles.

8. Consider a system of fermions at temperature $T = 77.4$ K, (i.e., liquid nitrogen temperature). Compute the probability for occupation of single-particle states lying at energies 1 eV, 0.1 eV, 0.02 eV, and 0.01 eV above the value of the chemical potential μ and, similarly, lying at energies -0.01 eV, -0.02 eV, -0.1 eV, and -1.0 eV below μ. Compare the results obtained for $T = 77.4$ K with those obtained for $T = 500$ K. Given that the chemical potential for electrons in metals is typically of the order of 500 eV, what does this behaviour say about the nature of the Fermi–Dirac distribution for electrons in metals?

9. Consider two single-particle fermion states h, k such that state h lies below the chemical potential μ by an amount a and state k lies above the chemical potential by the same amount, i.e., $\epsilon_h = \mu - a$ and $\epsilon_k = \mu + a$. How are the probabilities $\overline{n}_k$ and $\overline{n}_k$ for states h and k related?

10. For a system of bosons at the temperature of liquid nitrogen (i.e., $T = 77.4$ K), determine the average occupancy for a single-particle state having energy ϵ and obtain an expression giving the probability for this bosonic state to contain 0, 1,

2, or 3 bosons. Compute the probabilities for states lying at energies 10^{-6} eV, 10^{-3} eV, 0.01 eV, 0.1 eV, and 1 eV above μ.

11. In Chap. 4 we obtained an expression for the chemical potential for a system that obeys MB statistics. Suppose that you knew the distribution function, $\overline{n}_{MB} = e^{-\beta(\epsilon-\mu)}$, but didn't know the expression for μ. You could still determine μ by requiring that the total number of particles, summed over all states be N. Carry out this calculation to rederive the formula $\mu = -k_B T \ln(z/N)$. This is the actual procedure employed to determine μ in quantum statistics.

12. How large must $\epsilon - \mu$ be before the FD, BE, and MB distributions are within 1% of one another? Is it possible for this condition to be violated for gases at temperatures normally accessible on Earth?

13. Obtain an expression for $\overline{N}$ for an ideal quantum gas subject to a gravitational field with acceleration g. (Hint: see Example 10.2.)

14. Use Eq. (4.2.10) giving the internal energy $U(T, V, \lambda)$ together with Eq. (10.2.8) for the natural logarithm of the grand partition function to show that

$$U(T, V, \lambda) = \sum_k \frac{\epsilon_k \lambda e^{-\beta\epsilon_k}}{1 \pm \lambda e^{-\beta\epsilon_k}}$$

and, by utilizing the continuum approximation to the summation over the energy states for an ideal quantum gas in the presence of a gravitational field, show that $U(T, V, \lambda)$ is given by

$$U(T, V, \lambda) = k_B T \frac{V}{\Lambda^3(T)} \left\{ \frac{5}{2\eta} \left[g_{\frac{5}{2}}^{\pm}(\alpha) - g_{\frac{5}{2}}^{\pm}(\alpha - \eta) \right] - g_{\frac{3}{2}}^{\pm}(\alpha - \eta) \right\}.$$

15. Employ the results obtained in Example 10.2 and Problems 13 and 14 to show that the entropy $S/(Nk_B)$ associated with an ideal quantum gas in a gravitational field is given by

$$\frac{S}{Nk_B} = \frac{7}{2\eta} \left[g_{\frac{5}{2}}^{\pm}(\alpha) - g_{\frac{5}{2}}^{\pm}(\alpha - \eta) \right] - \frac{\alpha}{\eta} \left[g_{\frac{3}{2}}^{\pm}(\alpha) - g_{\frac{3}{2}}^{\pm}(\alpha - \eta) \right] - g_{\frac{3}{2}}^{\pm}(\alpha - \eta).$$

16. Obtain Eq. (10.2.28) for the entropy $S(T, V, \mu)$ from Eq. (4.2.30b).

17. Temperatures at the cores of stars typified by our sun are of the order of 10^7 K, so that all electrons have been stripped from the atomic nuclei to form an electron gas with a number density $\overline{N}/V$ of the order of 10^{32} m^{-3}. Would it be a valid approximation to treat these electrons as a degenerate Fermi gas?

18. The mass density, ρ, of platinum metal is $\rho(Pt) = 21.5$ g cm^3 and that of europium is $\rho(Eu) = 5.24$ g cm^3, while the corresponding molar masses are $\overline{M}(Pt) = 0.19508$ kg mol^{-1} and $\overline{M}(Eu) = 0.15196$ kg mol^{-1}. If each atom in a metallic sample is assumed to contribute a single electron to the conduction band, what are the values for their Fermi temperatures? Would it be justifiable

to treat conduction electrons in these metals as degenerate Fermi gases at room temperature?

19. By considering that the conduction electrons of Pt and Eu form degenerate Fermi gases, compute room temperature (300 K) values for the degeneracy pressures associated with the conduction electrons in these metals, and evaluate their contributions to the heat capacities (at constant volume) of the two metals.

20. Provide the missing steps to obtain Eq. (10.4.56b) for the magnetic moment of the conduction electrons at temperature T in a typical metal from Eq. (10.4.56a).

21. A simple model for the density of states, $\Omega_{cb}(\epsilon)$, near the bottom of the conduction band of a semiconductor employs the functional form obtained for a free Fermi gas, but with the energy, ϵ_{cb}, of the bottom of the conduction band serving as the appropriate energy origin. The density of states, $\Omega_{vb}(\epsilon)$, near the top of the valence band is then modeled by the mirror image of this function to give

$$\Omega_{cb}(\epsilon) = \frac{3\overline{N}}{2\sqrt{\epsilon_F}}(\epsilon - \epsilon_{cb})^{\frac{1}{2}} \quad \text{and} \quad \Omega_{vb}(\epsilon) = \frac{3\overline{N}}{2\sqrt{\epsilon_F}}(\epsilon_{vb} - \epsilon)^{\frac{1}{2}},$$

with $\Omega_{cb}(\epsilon) \equiv 0$ for $\epsilon < \epsilon_{cb}$ and $\Omega_{vb}(\epsilon) \equiv 0$ for $\epsilon > \epsilon_{vb}$ (i.e., for energies within the gap). Why must the chemical potential for this model fall precisely in the centre of the energy gap separating the valence and conduction bands? Obtain an expression for the number of conduction electrons per unit volume as a function of temperature and gap-width.

22. For Si near room temperature, the energy gap is approximately 1.11 eV. Use the expression obtained in Problem 17 above to estimate the order-of-magnitude of the number of conduction electrons per cm^3 for Si at a temperature of 300 K. How does this compare with the number of conduction electrons for a similar amount of Pt? Explain why a semiconductor becomes a much better conductor of electrons as the temperature is increased. Support your argument quantitatively. Estimate the gap-width between valence and conduction bands necessary for a material to be an insulator rather than a semiconductor.

23. The particle-in-a-box model for particles of mass m confined to a square potential box with sides of length L has eigenstates with energies $\epsilon_{n_x,n_y} = h^2(n_x^2 + n_y^2)/(8mL^2)$ in terms of the two quantum numbers n_x and n_y that may assume only positive integer values. What is the maximal number N_{FD} of indistinguishable spin-I_a fermions that can be placed into a square potential box at temperature $T = 0$ if the highest occupied state corresponds to (n_x^{max}, n_y^{max})? How does that number differ, if at all, from the maximal number of spin-I_a fermions that can occupy the energy states of a square potential box? Obtain an expression for the Fermi energy, ϵ_F. Express your results in terms of the effective quantum number n_{max} defined via $n_{max}^2 \equiv (n_x^{max})^2 + (n_y^{max})^2$.

24. Obtain an expression for the internal energy, U, for spin-I_a fermions in a two-dimensional square potential box, and show that the internal energy per fermion is one-half the Fermi energy.

25. Show that the density of states, $\Omega_{FD}(\epsilon)$, for particles in a 2D square potential box is constant, and express it in terms of $\overline{N}$ and ϵ_F.

26. Consider an electron gas confined to a 2D square potential box. Employ the constancy of the density of states for particles in such a 2D potential box (as found in Problem 25) to obtain an expression for $\overline{N}$ for the electron gas, and show that the chemical potential for such an electron gas is given by $\mu = k_B T \ln(e^{\beta \epsilon_F} - 1)$.

27. For temperatures T such that $k_B T \gg \epsilon_F$, compare the expression obtained in Problem 26 for the chemical potential of a two-dimensional electron-gas with the expression for the chemical potential μ Eq. (5.1.8) for the translational states of a classical gas and comment.

28. Obtain an expression for the heat capacity, $C_V(T)$, for spin-I_a fermions confined to a two-dimensional square potential box.

29. Determine the efficiency, $\eta_{\text{photon gas}}$, of a Carnot engine for which the working substance is a thermalized photon quantum gas (i.e., blackbody radiation) and compare your expression for $\eta_{\text{photon gas}}$ with that for an ideal classical gas.

30. Set up physical descriptions to illustrate free expansions of classical and quantum ideal (blackbody radiation) gases and utilize them to compare and contrast the two free expansions.

References

1. P.T. Landsberg, J. Dunning-Davies, D. Pollard, Amer. J. Phys. **62**, 712 (1994)
2. R.A. Aziz, C.C.K. Wong, F.R.W. McCourt, Mol. Phys. **61**, 1487 (1987)
3. G. Cook, R.H. Dickerson, Amer. J. Phys. **63**, 737 (1995)
4. V.C. Aguilera-Nevarro, G.A. Estevez, Amer. J. Phys. **56**, 456 (1988)
5. W.M. Haynes (ed.), *Handbook of Physics and Chemistry*, 97th edn. (CRC Press, Baton Rouge, 2016)
6. W.S. Corak, M.P. Garfunkel, C.B. Satterthwaite, A. Wexler, Phys. Rev. **98**, 1699 (1955)
7. R. Kubo, H. Ichimura, T. Usai, N. Hashizume, *Statistical Mechanics. An Advanced Course with Problems and Solutions* (North-Holland, Amsterdam, 1965)
8. R.K. Pathria, *Statistical Mechanics* (Pergamon, London, 1972)
9. H.S. Leff, Amer. J. Phys. **83**, 362 (2015)
10. B.H. Bransden, C.J. Joachain, *Quantum Mechanics*, , Chap. 14, 2nd edn. (Prentice Hall, London, 2000)
11. W. Greiner, J. Neise, H. Stöker, *Thermodynamics and Statistical Mechanics*, Chap. 10 (Springer, New York, 1995)

Appendix A
Aspects of Combinatorial Analysis

1. *Fundamental Principle of Counting*: if some event can occur in n_1 different ways and if, following this event, a second event can occur in n_2 different ways, and so on, $\cdots$, then the number of ways that the events can occur in that order is $n_1 n_2 \cdots$.

2. *Permutations*: Any arrangement of a set of n objects in a given order is called a permutation. Any arrangement of any $m \leq n$ of the objects in a given order is called an m-permutation, or a permutation of n objects taken m at a time. The number of permutations of n objects taken m at a time is

$$P(n, m) = \frac{n!}{(n - m)!}.$$
(A.1)

3. *Combinations*: A combination of n objects taken m at a time irrespective of the order is given by

$$C(n, m) = \frac{P(n, m)}{m!} = \frac{n!}{m!(n - m)!}.$$
(A.2)

A.1 Some Basic Probability Concepts

Consider 1 marble and 2 boxes, and ask 'How can we distribute the marble between the two boxes?' Obviously, it can be done in two ways, i.e., in one box, or in the other box.

© Springer Nature Switzerland AG 2021
F. R. W. McCourt, *Statistical Thermodynamics for Pure and Applied Sciences*,
https://doi.org/10.1007/978-3-030-52006-9

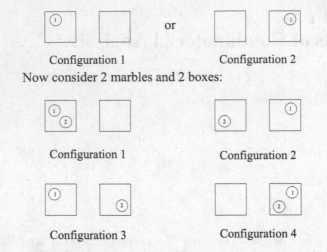

Were we to continue this process, we would find 2^3 configurations for 3 marbles and 2 boxes, 2^4 configurations for 4 marbles and 2 boxes, $\cdots$, 2^N configurations for N marbles and 2 boxes. From such illustrations, we can see that the number of times that *all* marbles are either in the one box *or* in the other box is

- 1 out of 2 when $N = 1$
- 1 out of 4 when $N = 2$
- 1 out of 8 when $N = 3$
- 1 out of 16 when $N = 4$

  $\vdots$

- 1 out of 2^N when N is arbitrary.

Note that we have made the *assumption* that the likelihood of a marble being in a box is unaffected by the presence of other marbles in it. This is true so long as the total volume occupied by the marbles is negligibly small in comparison to the volume of the boxes into which the marbles are being placed.

We attach a special name to the relative frequency of finding all N marbles in one box or the other: we call it the *probability* for the occurrence of such an event, and write

$$p_N = \frac{1}{2^N}. \tag{A.1.1}$$

Since the probability of finding all the marbles in one box is the same as the probability of finding no marbles in the other box, we see that

$$p_0 = \frac{1}{2^N}. \tag{A.1.2}$$

A natural question to ask is therefore: 'What is the probability that n of the N marbles are found in one of the two boxes?'. Our figures tell us that

$$\text{for } N = 2: \quad p_0 = \frac{1}{4}; \quad p_1 = \frac{2}{4}; \quad p_2 = \frac{1}{4}$$

$$\text{for } N = 4: \quad p_0 = \frac{1}{16}; \quad p_1 = \frac{4}{16}; \quad p_2 = \frac{6}{16}; \quad p_3 = \frac{4}{16}; \quad p_4 = \frac{1}{16}.$$

Thus, more generally then, if we let $C(n)$ be the number of ways of distributing the marbles between the boxes so that n of them are found in one (knowing which), then the probability of finding n marbles in that particular box is

$$p_n = \frac{C(n)}{2^N}. \tag{A.1.3}$$

We can also infer from our diagrams that if N is large, as long as n is either close to N or close to zero, then $C(n) \ll 2^N$, so that $p_n \ll 1$. Now, if p_n is small, then it is very unlikely that all of the marbles (or none, as the case may be) are to be found in one particular box. Such situations are *very rarely realized*: we refer to such situations as *non-random*, or *orderly*. We also can see that $C(n)$ is maximal when $n \simeq n' \approx \frac{N}{2}$ and, in this case, p_n is larger, i.e., such configurations are more probable: these are called *random*, or *disorderly* situations.

We have just examined probability concepts based upon two possible outcomes (i.e., RHS or LHS of the box for a marble). An event in general need not have just two outcomes: for example a 'fair' die will have 6 possible outcomes, corresponding to the six faces of the die, and each outcome will have a probability $\frac{1}{6}$ (rather than $\frac{1}{2}$).

The theory of probability is a tool that allows us to make better guesses in a consistent manner.

Let us work with what we shall refer to as a system $\mathcal{A}$, which we can take for the purposes of illustration to be a coin. We shall then perform an experiment, or observation, on it (for the present example, toss the coin), and record the outcome (for the present example, H or T). We must assume that the situation is time-independent, i.e., the conditions at the end of the experiment are the same as those at the beginning, and we then repeat the experiment many times, e.g., $\mathcal{N} = 500$, and record all the outcomes. Having carried out the set of observations, we then ask the question 'If I toss the coin just once, can you predict whether the outcome will be H or T?'. In order to attempt to answer this question we might, for example, examine the outcomes for the first 100 experiments, and perhaps find that $\mathcal{N}_H = 42$, $\mathcal{N}_T = 58 \, (\mathcal{N} = 100)$, so that $p_H = 0.42$, $p_T = 0.58$. This answer we would, of course, not find particularly satisfactory, since we would intuitively know (from our lifelong experience) that the real answer should be closer to $p_H = p_T = 0.50$. What do we do? Well, we analyse yet more of our data. If we examined 200 experiments, we would perhaps find $\mathcal{N}_H = 106$, $\mathcal{N}_T = 94 \, (\mathcal{N} = 200)$, which implies that $p_H = \frac{166}{200} = 0.53$ and $p_T = \frac{94}{200} = 0.47$. We see immediately that the probabilities have changed with the number of experiments performed. This tells us that we have not performed enough experiments for our results to be converged. Let us then use all the data that we have acquired. We would perhaps find that $\mathcal{N}_H = 245, \mathcal{N}_T = 255, (\mathcal{N} = 500)$, so that $p_H = 0.49$ and $p_T = 0.51$. We are still seeing a change in the values of the probabilities with the number of experiments! Let us now examine the sequence of results that we have obtained.

$\mathcal{N}$	p_H	p_T
100	0.42	0.58
200	0.53	0.47
500	0.49	0.51
.	.	.
.	.	.
∞	0.50	0.50

The answer to the original question that we posed is therefore: 'All that I can tell you is that the probability of obtaining H is $\frac{1}{2}$, i.e., a 50% chance.'

In a sense we have made a guess, but it was a *reasoned* guess, based upon the assumptions that the coin is an 'honest' coin (i.e., that the centre-of-mass of the coin coincides with the geometrical centre), and that there are only two possible outcomes for our coin toss (H or T). We have seen that in the long run the numbers $\mathcal{N}_H$ and $\mathcal{N}_T$ will tend to be equal, i.e., $\mathcal{N}_H = \mathcal{N}_T = \frac{1}{2}\mathcal{N}$

$$p_H = \frac{\mathcal{N}_H}{\mathcal{N}} \xrightarrow[\text{in the long run}]{} \frac{\frac{1}{2}\mathcal{N}}{\mathcal{N}} = \frac{1}{2} \qquad (A.1.4)$$

We can speak of probabilities only when the number of experiments or trials is extremely large—equivalently, we can only make proper statistical analyses on large quantities of data. In the case just discussed in detail we have the *normalization condition* $p_H + p_T = 1$, since we know for certainty that the outcome of any trial will be either H or T. More generally, for experiments leading to M mutually exclusive outcomes, we have seen that the normalization condition is

$$\sum_{r=1}^{M} p_r = 1. \qquad (A.1.5)$$

1. Let us now ask about the probability of occurrence of *either* outcome r or outcome s.

   • $\mathcal{N}_r$ systems in the ensemble exhibit outcome r

- $\mathcal{N}_s$ systems in the ensemble exhibit outcome s
- $\mathcal{N}_r + \mathcal{N}_s$ systems in the ensemble exhibit either r or s

Therefore $p(r \ \ or \ \ s) = (\mathcal{N}_r + \mathcal{N}_s)/\mathcal{N} = p_r + p_s$. This result can be generalized. Note that *either* $\cdots$ *or* excludes the possibility of r and s together.

2. What is the probability of the *joint* occurrence of both outcomes r *and* s? If the probability for s is unaffected by the occurrence of r (we say that outcomes r and s are *statistically independent* or *uncorrelated*), then if $\mathcal{N}_r \ \rightarrow \ r$ and $\mathcal{N}_{rs}$ of these systems also exhibit s, then

$$p_s = \frac{\mathcal{N}_{rs}}{\mathcal{N}_r} \ \Rightarrow \ \mathcal{N}_{rs} = \mathcal{N}_r p_s \qquad (A.1.6)$$

and hence

$$p_{rs} \equiv \frac{\mathcal{N}_{rs}}{\mathcal{N}} = \frac{\mathcal{N}_r}{\mathcal{N}} p_s = p_r p_s . \qquad (A.1.7)$$

Thus, if outcomes r and s are statistically independent, then $p_{rs} = p_r p_s$.

Example Let us consider an ideal gas consisting of N atoms enclosed in a container of volume V_T. Let us now divide the box into two parts, of volumes V and V', such that $V + V' = V_T$. We treat the atoms as statistically independent, and construct an ensemble of such containers of gas. Let p be the probability that a given atom is found in volume V, and $q = 1 - p$ be the probability that a given atom is found in the remaining volume V'. At equilibrium the atoms tend to be *uniformly distributed* throughout the volume V_0, so that

$$p = \frac{V}{V_T} ; \qquad q = \frac{V'}{V_T} .$$

We now ask the question: 'What is the probability $p(n)$ that n of the N atoms are found in the volume V, while the remaining $N - n$ atoms are found in the volume V'?'.

$$\underbrace{ppp\cdots p}_{n \text{ factors}} \underbrace{qqq\cdots q}_{N-n \text{ factors}} = p^n q^{N-n} .$$

Each set of n atoms out of a total N atoms can be chosen in

$$\binom{N}{n} = C(N, n) = \frac{N!}{n!(N - n)!} \qquad (A.1.8)$$

ways. Hence we see that

$$p(N, n) = \binom{N}{n} p^n q^{N-n} . \qquad (A.1.9)$$

$\square$

This result is often written in the form

$$p(n) = \frac{N!}{n!(N-n)!} \, p^n q^{N-n} \,, \qquad\qquad \text{(A.1.10)}$$

and is known as the *binomial distribution*. This name is based upon the well-known binomial theorem

$$(x+y)^n = \sum_{m=0}^{n} \binom{n}{m} x^m y^{n-m} \,. \qquad\qquad \text{(A.1.11)}$$

The Gaussian distribution

$$p(N,n) = p(N,\overline{n}) \, e^{-(n-\overline{n})^2/(2Npq)} \,, \qquad \int_0^N dn \, p(N,n) = 1 \,, \qquad \text{(A.1.12)}$$

is the continuous variable equivalent of the binomial distribution. The value of $p(N,\overline{n})$ is obtained from the normalization condition, which gives

$$p(N,\overline{n}) \int_0^N dn \, e^{-(n-\overline{n})^2/(Npq)} = 1 \,.$$

Evaluation of this integral proceeds by changing variables from n to $y = n - \overline{n}$, so that the integral over n from 0 to N is replaced by an integral over y from $-\overline{n}$ to $N - \overline{n}$. For N sufficiently large (say, of order 10^6 or more), it can be replaced by

$$\int_{-\overline{n}}^{N-\overline{n}} dy \, e^{-y^2/(2Npq)} \simeq \int_{-\infty}^{\infty} dy \, e^{-y^2/(2Npq)}$$

$$= 2 \int_0^{\infty} dy \, e^{-y^2/(2Npq)} \,,$$

with the latter step arising from the even symmetry under inversion (i.e., $y \to -y$) on the interval $[-a, a]$ over which the integration is being performed. The definite integral appearing in the final expression has the form $\int_0^{\infty} e^{-ax^2} \, dx$, which can be obtained from a table of definite integrals.[1] Similar arguments may be employed for the evaluation of the mean value, $\langle n \rangle$, of n and the variance, $\langle (n-\overline{n})^2 \rangle$, for n.

---

[1] It may also be evaluated directly by noting that its square can be written as the product of the same integral over the interval $[0,\infty)$ for x and for y, i.e., as the double integral over the positive quadrant of the xy-plane of the integrand $e^{a(x^2+y^2)}$: this double integral can then be converted into polar coordinates and readily evaluated as $\pi/(4a)$, whose square root gives $\int_0^{\infty} e^{-ax^2} \, dx = \frac{1}{2}\sqrt{\pi/a}$.

Appendix B
Review of Traditional Calculus

B.1 Derivatives

B.1.1 Functions of a Single Variable

Statistical thermodynamics and statistical mechanics typically involve the use of derivatives, both of functions of a single variable and of multivariable functions. In the latter case, we shall typically not examine functions of more than two independent variables, as all of the important properties that will be needed can be illustrated in terms of functions of three variables. We begin simply by listing some of the most important (or, for our purposes, useful) results taught in typical first year calculus courses.

(i) One of the most important rules is the *product rule*, which gives the derivative of the product of two functions of a single variable in terms of products of these functions and their derivatives. It takes the form

$$\frac{d}{dx}\{f(x)g(x)\} = \frac{df}{dx}g(x) + f(x)\frac{dg}{dx}$$

$$\equiv f'_x\,g(x) + f(x)\,g'_x. \tag{B.1.1}$$

The latter version employs the shorthand notation f'_x for the derivative of $f(x)$ with respect to x.

(ii) Another very important rule of differentiation is the *chain rule* for obtaining the derivative of a composite function $F(x) = f(g(x))$ in terms of simpler derivatives. This rule may be expressed as

© Springer Nature Switzerland AG 2021
F. R. W. McCourt, *Statistical Thermodynamics for Pure and Applied Sciences*,
https://doi.org/10.1007/978-3-030-52006-9

$$\frac{d}{dx} F(x) = \frac{d}{dx} f(g(x))$$

$$= \frac{df}{dg} \cdot \frac{dg}{dx} \equiv f_g' \cdot g_x' \,. \tag{B.1.2}$$

(iii) Another rule that occurs with some frequency is the *quotient rule*, which may be recalled in the rather complicated, but explicit, form

$$\frac{d}{dx} \left(\frac{f(x)}{g(x)} \right) = \frac{g(x) f_x' - f(x) g_x'}{g^2(x)} \,. \tag{B.1.3}$$

It may, however, be obtained fairly straightforwardly as a special case of the product rule taken together with the quotient rule in which $F(x)$ is given by $F(x) = 1/g(x)$, so that the quotient f/g is considered as the product fF of the functions f and F.

(iv) Derivatives of *non-algebraic functions*: a number of simple non-algebraic functions that are encountered in typical statistical mechanical problems are the exponential, natural logarithmic, trigonometric, and hyperbolic functions, for which the most relevant derivatives are summarized below.

$$\frac{d}{dx} e^x = e^x \,, \qquad\qquad \frac{d}{dx} e^{f(x)} = e^{f(x)} \frac{df}{dx} \,,$$

$$\frac{d}{dx} \ln x = \frac{1}{x} \,, \qquad\qquad \frac{d}{dx} \ln f(x) = \frac{1}{f(x)} \frac{df}{dx}$$

$$\frac{d}{dx} \sin x = \cos x \,, \qquad\qquad \frac{d}{dx} \cos x = -\sin x \,,$$

$$\frac{d}{dx} \sinh x = \cosh x \,, \qquad\qquad \frac{d}{dx} \cosh x = \sinh x \,.$$

B.1.2   Multivariate Functions

(i) Definition of *Partial Derivatives*. Consider a (continuous) function of the two independent variables x and y. We shall firstly establish the defining relations for what is called a *partial derivative* by employing an explicit form, $z = f(x, y)$, for the manner in which z depends on the two independent variables x and y even though, more often than not, the dependence of z upon x and y can only be provided implicitly via an expression of the form $f(x, y, z) = 0$. For an implicit relation, we shall see that it is a matter of choice as to which of the three variables involved in the expression is to be considered to be the dependent variable.

The definitions of the two partial derivatives for $z = f(x, y)$ may now be given. For y held fixed we have

$$\frac{\partial z}{\partial x} \equiv \frac{\partial f}{\partial x} = \lim_{\Delta x \to 0} \frac{f(x + \Delta x, y) - f(x, y)}{\Delta x} \,, \tag{B.1.4a}$$

while for x held fixed, we have

$$\frac{\partial z}{\partial y} \equiv \frac{\partial f}{\partial y} = \lim_{\Delta y \to 0} \frac{f(x, y + \Delta y) - f(x, y)}{\Delta y} \,. \tag{B.1.4b}$$

It is worth noting that although we may express the derivative $f'(x)$ of a function f of the single variable x as the quotient of two differentials, df and dx because the defining relation $df(x_0) = f'(x_0)dx$ provides a $1 \leftrightarrow 1$ relationship between the differential df at x_0 and the independent differential dx for a given value x_0 of x, while from the corresponding relationship,

$$df(x_0, y_0) = \left(\frac{\partial f}{\partial x}\right)_y (x_0, y_0)dx + \left(\frac{\partial f}{\partial y}\right)_x (x_0, y_0)dy \,,$$

for a pair of given values (x_0, y_0) of (x, y), it is no longer possible to establish such $1 \leftrightarrow 1$ relationships between the partial derivative functions and the independent differentials df and dx or df and dy. For this reason, partial derivatives must be treated as functions that cannot be obtained as the quotients of dependent and independent differentials, and it makes no sense to write ∂f, ∂x, or ∂y on its own.

Higher-order partial derivatives: Note from Eqs. (B.1.4a) and (B.1.4b), that if $z = f(x, y)$ is a function of the independent variables x and y, then the partial derivatives $\frac{\partial z}{\partial x}$ and $\frac{\partial z}{\partial y}$ are also functions of x and y. Hence, these (first) partial derivatives can be differentiated in turn with respect to x and y. There are thus four second partial derivatives, namely,

$$\frac{\partial^2 z}{\partial x^2}, \quad \frac{\partial^2 z}{\partial y^2}, \quad \frac{\partial^2 z}{\partial x \partial y}, \quad \frac{\partial^2 z}{\partial y \partial x}; \tag{B.1.5a}$$

note that in general the mixed second partial derivatives are equal, i.e.,

$$\frac{\partial^2 z}{\partial x \partial y} = \frac{\partial^2 z}{\partial y \partial x} \,. \tag{B.1.5b}$$

Chain rule for multivariate functions: Extension of the chain rule (B.1.2) to functions of two variables is reasonably straightforward. If z is a function of the two variables x and y, and x, y are in turn functions of the two variables u and v, then the chain rule for the partial derivatives of z with respect to u and v is

$$\frac{\partial z}{\partial u} = \frac{\partial z}{\partial x}\frac{\partial x}{\partial u} + \frac{\partial z}{\partial y}\frac{\partial y}{\partial u}, \tag{B.1.6a}$$

$$\frac{\partial z}{\partial v} = \frac{\partial z}{\partial x}\frac{\partial x}{\partial v} + \frac{\partial z}{\partial y}\frac{\partial y}{\partial v}. \tag{B.1.6b}$$

If x and y both depend upon the same single parameter t, the Eqs. (B.1.6a) and (B.1.6b) reduce to

$$\frac{dz}{dt} = \frac{\partial z}{\partial x}\frac{dx}{dt} + \frac{\partial z}{\partial y}\frac{dy}{dt} \tag{B.1.6c}$$

giving the derivative of z with respect to t.

(ii) *Differentials and the Total Differential.*

The usual statement for a differential, as seen in single-variable calculus, is

$$y = f(x) \quad \Rightarrow \quad dy = f'(x)\Delta x = f'(x)dx, \tag{B.1.7}$$

sometimes also expressed in the form

$$\frac{dy}{dx}\left(\frac{dx}{dt}dt\right), \quad \text{or as} \quad dy = y'(x)x'(t)\,dt, \tag{B.1.8}$$

when f is a composite function, i.e., when $y = f(x)$ and x can also be expressed as a function of t as $x = g(t)$.

If we have a function of two variables, $z = f(x, y)$, then the *total differential*, dz, of (the dependent variable) z is given by the linear combination

$$dz = \left(\frac{\partial z}{\partial x}\right)_y dx + \left(\frac{\partial z}{\partial y}\right)_x dy, \tag{B.1.9}$$

in terms of the partial derivatives with respect to the independent variables x, y, and their respective (independent) differentials dx and dy. We have placed parentheses with subscripts around the partial derivatives, as is commonly done in the physical sciences, especially thermodynamics (where a plethora of choices of independent and dependent variables can be made). This physical science convention often proves to be useful for reminding us which physical variables are being considered to be independent.

(iii) In mathematics, the functional relation $f(x, y, z) = 0$ represents an implicit function of x, y, z. Sometimes we are able to isolate one of the variables in terms of the others: for example, if we could isolate z from $f(x, y, z) = 0$ in the form $z = F(x, y)$, then because z can be expressed as an explicit function of x and y, we may identify z as the dependent variable, in which case, x and y are the independent variables. We shall restrict ourselves in the following to functions of three variables: note, however, that even when we cannot express one of the three variables explicitly in terms of the other two variables, we may, because of the implicit function relation $f(x, y, z) = 0$, consider any one of the variables x, y, z as a function of the other two.

(iv) *Two Important Relations for Partial Derivatives.*

If we consider x as a function of y and z, we may write the total differential dx for x as

$$dx = \left(\frac{\partial x}{\partial y}\right)_z dy + \left(\frac{\partial x}{\partial z}\right)_y dz. \qquad (B.1.10a)$$

If, equivalently, we consider y to be a function of x and z, then we may write a total differential dy of y as

$$dy = \left(\frac{\partial y}{\partial x}\right)_z dx + \left(\frac{\partial y}{\partial z}\right)_x dz. \qquad (B.1.10b)$$

We shall examine what happens when we substitute Eq. (B.1.10b) for dy into Eq. (B.1.10a) for dx: upon making this substitution, we obtain

$$dx = \left(\frac{\partial x}{\partial y}\right)_z \left[\left(\frac{\partial y}{\partial x}\right)_z dx + \left(\frac{\partial y}{\partial z}\right)_x dz\right] + \left(\frac{\partial x}{\partial z}\right)_y dz$$

$$= \left(\frac{\partial x}{\partial y}\right)_z \left(\frac{\partial y}{\partial x}\right)_z dx + \left[\left(\frac{\partial x}{\partial y}\right)_z \left(\frac{\partial y}{\partial z}\right)_x + \left(\frac{\partial x}{\partial z}\right)_y\right] dz. \qquad (B.1.11)$$

Only two of x, y, z can be independent variables because of the implicit function relation $f(x, y, z) = 0$. If we choose x, z to be the independent variables, then Eq. (B.1.11) must hold for *arbitrary* choices of dx and dz.

For $dz = 0$, $dx \neq 0$, we thus see that

$$\left(\frac{\partial x}{\partial y}\right)_z \left(\frac{\partial y}{\partial x}\right)_z = 1,$$

from which we obtain the reciprocal relation

$$\left(\frac{\partial x}{\partial y}\right)_z = \frac{1}{\left(\dfrac{\partial y}{\partial x}\right)_z} \qquad (B.1.12)$$

between $\left(\dfrac{\partial x}{\partial y}\right)_z$ and $\left(\dfrac{\partial y}{\partial x}\right)_z$.

For $dx = 0$, $dz \neq 0$, we find similarly, that

$$\left(\frac{\partial x}{\partial y}\right)_z \left(\frac{\partial y}{\partial z}\right)_x + \left(\frac{\partial x}{\partial z}\right)_y = 0,$$

or

$$\left(\frac{\partial x}{\partial y}\right)_z \left(\frac{\partial y}{\partial z}\right)_y = -\left(\frac{\partial x}{\partial z}\right)_y,$$

which is often written in the equivalent form

$$\left(\frac{\partial x}{\partial y}\right)_z \left(\frac{\partial y}{\partial z}\right)_x \left(\frac{\partial z}{\partial x}\right)_y = -1. \tag{B.1.13}$$

This is an important relation amongst the partial derivatives, and is sometimes referred to in the physical sciences as the 'permutation rule' for partial derivatives.

(v) *Equivalence of Variables.*

If f is a function of x, y, z, and a relation exists amongst x, y, z, then f itself may be regarded as a function of *any two* of x, y, z. This is guaranteed by the relation $f(x, y, z) = 0$. Were we to choose x and y as the variables upon which f depends, we could write the total differential df of f as

$$df = \left(\frac{\partial f}{\partial x}\right)_y dx + \left(\frac{\partial f}{\partial y}\right)_x dy.$$

A simple rearrangement of this equation into the form

$$\left(\frac{\partial f}{\partial x}\right)_y dx = df - \left(\frac{\partial f}{\partial y}\right)_x dy,$$

allows dx to be expressed as

$$dx = \frac{1}{\left(\dfrac{\partial f}{\partial x}\right)_y} df - \frac{\left(\dfrac{\partial f}{\partial y}\right)_x}{\left(\dfrac{\partial f}{\partial x}\right)_y} dy$$

$$= \left(\frac{\partial x}{\partial f}\right)_y df - \left(\frac{\partial x}{\partial f}\right)_y \left(\frac{\partial f}{\partial y}\right)_x dy,$$

where we have made use of property (B.1.12). Now we may employ the permutation rule (B.1.13) in the form

$$\left(\frac{\partial x}{\partial f}\right)_y \left(\frac{\partial f}{\partial y}\right)_x \left(\frac{\partial y}{\partial x}\right)_f = -1,$$

or equivalently, in the form

$$\left(\frac{\partial x}{\partial f}\right)_y \left(\frac{\partial f}{\partial y}\right)_x = -\frac{1}{\left(\frac{\partial y}{\partial x}\right)_f} = -\left(\frac{\partial x}{\partial y}\right)_f,$$

to obtain dx as

$$dx = \left(\frac{\partial x}{\partial f}\right)_y df + \left(\frac{\partial x}{\partial y}\right)_f dy. \qquad (B.1.14)$$

Expression (B.1.14) tells us that we may now regard x as a function of f and y.

Similarly, we may regard y as a function of f and z, so that dy may be written as

$$dy = \left(\frac{\partial y}{\partial f}\right)_z df + \left(\frac{\partial y}{\partial z}\right)_f dz.$$

Substitution of this expression for dy into expression (B.1.14) for dx gives us

$$dx = \left(\frac{\partial x}{\partial f}\right)_y df + \left(\frac{\partial x}{\partial y}\right)_f \left[\left(\frac{\partial y}{\partial f}\right)_z df + \left(\frac{\partial y}{\partial z}\right)_f dz\right]$$

$$= \left[\left(\frac{\partial x}{\partial f}\right)_y + \left(\frac{\partial x}{\partial y}\right)_f \left(\frac{\partial y}{\partial f}\right)_z\right] df + \left(\frac{\partial x}{\partial y}\right)_f \left(\frac{\partial y}{\partial z}\right)_f dz.$$

This result should now be compared with the *definition* for the total differential dx for x expressed as a function of f and z, namely,

$$dx \equiv \left(\frac{\partial x}{\partial f}\right)_z df + \left(\frac{\partial x}{\partial z}\right)_f dz.$$

Comparison of the components of the differential dz gives

$$\left(\frac{\partial x}{\partial z}\right)_f = \left(\frac{\partial x}{\partial y}\right)_f \left(\frac{\partial y}{\partial z}\right)_f,$$

which may otherwise be expressed in the form

$$\left(\frac{\partial x}{\partial y}\right)_f \left(\frac{\partial y}{\partial z}\right)_f \left(\frac{\partial z}{\partial x}\right)_f = +1; \qquad (B.1.15)$$

this is also an important relation amongst partial derivatives. Notice, however, the distinction between this relation and the permutation rule (B.1.13) obtained earlier.

(vi) *Some Properties of Exact Differentials.*

Mathematically, if a differential df can be written in the form

$$df = P(x, y)\, dx + Q(x, y)\, dy$$

for all (x, y) in some region of the xy-plane and for all values of dx and dy, we say that $P(x, y)\, dx + Q(x, y)\, dy$ is an exact differential provided that $P(x, y)$, $Q(x, y)$, $\left(\dfrac{\partial P}{\partial y}\right)_x$, and $\left(\dfrac{\partial Q}{\partial x}\right)_y$ are all continuous on a rectangular region S of the xy-plane and that $\left(\dfrac{\partial P}{\partial y}\right)_x = \left(\dfrac{\partial Q}{\partial x}\right)_y$ for all (x, y) contained in S. This requirement turns out to be both necessary and sufficient. An immediate consequence of this condition is that if for a given function $f(x, y)$, both its two first partial derivatives and its two mixed second partial derivatives are continuous at (x_0, y_0), then

$$\frac{\partial^2 f}{\partial x \partial y} = \frac{\partial^2 f}{\partial y \partial x}$$

at (x_0, y_0). This equality of the mixed second partial derivatives has very important consequences for thermodynamics, as it is the source of the famous Maxwell relations.

(vii) *Homogeneous Function Theorem.*

Let us consider briefly the Euler theorem for homogeneous functions, which can be stated as follows. If $\lambda f(x, y, z) = f(\lambda x, \lambda y, \lambda z)$, then

$$f = x\, \frac{\partial f}{\partial x} + y\, \frac{\partial f}{\partial y} + z\, \frac{\partial f}{\partial z} \, .$$

This particular theorem is relatively easy to prove, and it is perhaps useful to see how the proof works.

We begin by defining a function $F(x, y, z)$ via

$$\lambda f(x, y, z) = F(x, y, z) \equiv f(\lambda x, \lambda y, \lambda z)$$

or $F(x, y, z) = f(x', y', z')$, in which $x' = \lambda x$, $y' = \lambda y$, $z' = \lambda z$. Then, the derivative of $F(x, y, z)$ with respect to λ is given by

$$f(x, y, z) = \frac{\partial}{\partial \lambda}[\lambda f(x, y, z)]$$

$$= \frac{\partial F}{\partial \lambda}$$

or

$$f(x, y, z) = \frac{\partial f}{\partial x'} \frac{dx'}{d\lambda} + \frac{\partial f}{\partial y'} \frac{dy'}{d\lambda} + \frac{\partial f}{\partial z'} \frac{dz'}{d\lambda}$$

$$= x \frac{\partial f}{\partial x} + y \frac{\partial f}{\partial y} + z \frac{\partial f}{\partial z},$$

which is what we set out to establish.

(viii) *Derivatives of integrals with variable limits.*[1]

It sometimes happens in thermodynamics, for example, that a function may be defined in terms of an integration over the volume of the system being examined. Consider a multivariate function $g(X, Y, V)$ defined in terms of an integral over the system volume of the multivariate function $f(X, Y, V)$, namely

$$g(X, Y, V) \equiv \int_V f(X, Y, V) \, d^3r.$$

Of course, should we wish to form the partial derivative of $g(X, Y, V)$ with respect to one of the variables X or Y, we are able to interchange the order of operation of that partial derivative and the integration over the volume of the system. However, should we wish to evaluate the partial derivative of $g(X, Y, V)$ with respect to the volume V, we need to have a means of dealing with the fact that the limits on the integral in general depend upon the form of the region of space that contains the system.

We shall illustrate the procedure for a system that is bounded by a cube that is centred at the origin and has sides of length L. Let us consider the partial derivative of $g(X, Y, V)$ with respect to V (which requires that X and Y be held fixed), for which we shall write

$$\left(\frac{\partial g}{\partial V} \right)_{X,Y} = \frac{\partial}{\partial V} \int_V f(X, Y, V) \, d^3r$$

$$= \frac{\partial}{\partial V} \left[\int_{-\frac{L}{2}}^{+\frac{L}{2}} \int_{-\frac{L}{2}}^{+\frac{L}{2}} \int_{-\frac{L}{2}}^{+\frac{L}{2}} f(X, Y, V) \, d^3r \right]_{X,Y}.$$

For a cubic region of volume L^3, the (partial) derivative with respect to V may also be represented as

$$\frac{\partial}{\partial V} = \frac{dL}{dV} \frac{\partial}{\partial L} = \frac{1}{3L^2} \frac{\partial}{\partial L}.$$

---

[1] J. M. Honig, *Thermodynamics*, 2nd Ed., (Academic Press, San Diego, 1999).

By carrying out a transformation of coordinates from Cartesian coordinates x, y, z to 'fractional coordinates' $\xi = x/L, \eta = y/L, \zeta = z/L$, with Jacobian $J = L^3$, the volume element $dxdydz$ is replaced by the volume element $Jd\xi d\eta d\zeta = L^3 d\xi d\eta d\zeta$, so that the partial derivative becomes

$$\left(\frac{\partial g}{\partial V}\right)_{X,Y} = \frac{1}{3L^2}\frac{\partial}{\partial L}\left[\int_{-\frac{1}{2}}^{+\frac{1}{2}}\int_{-\frac{1}{2}}^{+\frac{1}{2}}\int_{-\frac{1}{2}}^{+\frac{1}{2}} L^3 f(X,Y,L)\,d\xi d\eta d\zeta\right]_{X,Y}.$$

Because the limits on the triple integral are now fixed, the differentiation with respect to L may be taken inside the triple integral to give

$$\left(\frac{\partial g}{\partial V}\right)_{X,Y} = \int_{-\frac{1}{2}}^{+\frac{1}{2}}\int_{-\frac{1}{2}}^{+\frac{1}{2}}\int_{-\frac{1}{2}}^{+\frac{1}{2}} f(X,Y,L)\,d\xi d\eta d\zeta$$

$$+ \int_{-\frac{1}{2}}^{+\frac{1}{2}}\int_{-\frac{1}{2}}^{+\frac{1}{2}}\int_{-\frac{1}{2}}^{+\frac{1}{2}} \frac{L}{3}\left(\frac{\partial f}{\partial L}\right)_{X,Y} d\xi d\eta d\zeta.$$

Upon making the inverse transformation from ξ, η, ζ to x, y, z, we obtain the result

$$\left(\frac{\partial g}{\partial V}\right)_{X,Y} = \frac{1}{L^3}\int_{-\frac{L}{2}}^{+\frac{L}{2}}\int_{-\frac{L}{2}}^{+\frac{L}{2}}\int_{-\frac{L}{2}}^{+\frac{L}{2}} f(X,Y,V)\,dxdydz$$

$$+ \int_{-\frac{L}{2}}^{+\frac{L}{2}}\int_{-\frac{L}{2}}^{+\frac{L}{2}}\int_{-\frac{L}{2}}^{+\frac{L}{2}} \frac{1}{3L^2}\left(\frac{\partial f}{\partial L}\right)_{X,Y} dxdydz$$

or, as $L^3 = V$ and $\frac{1}{3L^2}\frac{\partial}{\partial L} = \frac{\partial}{\partial V}$,

$$\left(\frac{\partial g}{\partial V}\right)_{X,Y} = \frac{1}{V}\int_V f(X,Y,V)\,d^3r + \int_V \left(\frac{\partial f}{\partial V}\right)_{X,Y} d^3r.$$

Forming the differential of such an integral may be handled in the same manner. For example, to obtain the differential dg of $g(X,Y,V)$ involves the following steps. Firstly, a coordinate transformation from Cartesian x, y, z coordinates to fractional ξ, η, ζ coordinates is carried out, thereby obtaining the differential dg as

$$dg \equiv d\left(\int_V f(X,Y,V)\,d^3r\right)$$

$$= d\left(\int_{-\frac{1}{2}}^{+\frac{1}{2}}\int_{-\frac{1}{2}}^{+\frac{1}{2}}\int_{-\frac{1}{2}}^{+\frac{1}{2}} Vf(X,Y,V)\,d\xi d\eta d\zeta\right),$$

with $V = L^3$ for a cubic region. Next, the operations of forming the differential and carrying out the triple integration over the region may be interchanged, to give dg as

$$dg = \int_{-\frac{1}{2}}^{+\frac{1}{2}} \int_{-\frac{1}{2}}^{+\frac{1}{2}} \int_{-\frac{1}{2}}^{+\frac{1}{2}} d(Vf)\, d\xi\, d\eta\, d\zeta \,,$$

or equivalently, as

$$dg = \int_{-\frac{L}{2}}^{+\frac{L}{2}} \int_{-\frac{L}{2}}^{+\frac{L}{2}} \int_{-\frac{L}{2}}^{+\frac{L}{2}} \frac{1}{L^3} d(Vf)\, dx\, dy\, dz \,.$$

Finally, this expression for dg may be rewritten in the form

$$dg = \int_V \frac{1}{V} d(Vf)\, d^3r \,.$$

B.1.3 Inverse Differential Operator: Antiderivative

Amongst the operators defined in finite-difference calculus are:[2] the difference operator, Δ, the shifting operator, $\mathcal{E}$, the differential operator, $\mathcal{D}$, and the integral operator, $\mathcal{J}$. These four operators are defined in terms of their effects on a function $f(x)$, specifically, via

$$\mathcal{E}f(x) = f(x+h)\,, \qquad\qquad (\text{B.1.16a})$$

$$\Delta f(x) = f(x+h) - f(x)\,, \qquad\qquad (\text{B.1.16b})$$

$$\mathcal{D}f(x) = f'(x)\,, \qquad\qquad (\text{B.1.16c})$$

$$\mathcal{J}f(x) = \int_x^{x+h} f(x)\, dx \,. \qquad\qquad (\text{B.1.16d})$$

Positive integer powers of these operators are defined by *iteration*, and the zeroth power of any operator is the identity operator, 1 , that leaves any function unchanged. All four of these operators are linear operators, which means that if $\mathcal{L}$ is a linear operator, then $\mathcal{L}[af(x) + bg(x)] = a\mathcal{L}f(x) + b\mathcal{L}g(x)$.

Two operators $\mathcal{O}_1$ and $\mathcal{O}_2$ are said to be equal if $\mathcal{O}_1 f(x) = \mathcal{O}_2 f(x)$ for all functions $f(x)$ for which the operations are defined. Each of the four linear operators defined above possesses the commutative, distributive, and associative

---

[2]See, for example, F. Hildebrand, *Introduction to Numerical Analysis*, 2nd Ed. (McGraw–Hill, New York, 1974).

properties that are shared by all real numbers, and they satisfy the exponent law $\mathcal{L}^m \mathcal{L}^n = \mathcal{L}^{m+n}$. The commutation of $\mathcal{D}$ and $\mathcal{J}$ serves to illustrate these properties: we have

$$\mathcal{D}\mathcal{J} f(x) \equiv \frac{d}{dx} \int_x^{x+h} f(t)\, dt = f(x+h) - f(x) = \int_x^{x+h} \frac{d f(t)}{dt}\, dt = \mathcal{J}\mathcal{D} f(c),$$

from which we also may note that the operators $\mathcal{J}$, $\mathcal{D}$, and Δ are interrelated through $\mathcal{J}\mathcal{D} = \mathcal{D}\mathcal{J} = \Delta$.

The proper inverse operator, $\mathcal{L}^{-1}$, of an operator $\mathcal{L}$ is related to $\mathcal{L}$ by $\mathcal{L}^{-1}\mathcal{L} = \mathcal{L}\mathcal{L}^{-1} = 1$. Some caution, however, does need to be applied to the inverses of Δ, $\mathcal{J}$, and $\mathcal{D}$, as they annihilate, respectively, functions of period h, derivatives of functions of period h, and constants. As we shall not normally be concerned here with functions of period h, the only inverse operator with which we shall need to exercise caution is $\mathcal{D}^{-1}$. In this case, the cautionary note leads to the well-known 'constant of integration' associated with antiderivatives, and discussed in every first year calculus course. Stated otherwise, $\mathcal{D}^{-1}$ corresponds to the formation of an indefinite integral (or antiderivative) and, while the derivative of the antiderivative gives precisely the original function, the antiderivative of the derivative of that function introduces an arbitrary additive constant, called the 'constant of integration'.

In order to avoid this problem, we may note that $\Delta \mathcal{D}^{-1} f(x)$ is uniquely determined, as the difference operator Δ annihilates the arbitrary constant. From the expression $\mathcal{D}^{-1} f(x) = F(x) + C$, we see that the application of Δ gives

$$\Delta \mathcal{D}^{-1} f(x) = F(x+h) - F(x) = \mathcal{J} f(x),$$

which means that an equivalent operator equation is $\Delta \mathcal{D}^{-1} = \mathcal{J}$. We can also see that the shifting operator, $\mathcal{E}$, the difference operator $\mathcal{D}$, and the integral operator, $\mathcal{J}$, are related via $\mathcal{D}\mathcal{J} = \mathcal{J}\mathcal{D} = \mathcal{E} - 1$.

B.1.4   Taylor and Maclaurin Series

We shall have frequent need to make use of these concepts. However, we shall not need to go beyond their development in a first year calculus course. You will recall that the fundamental expressions for Taylor and Maclaurin series expansion are

$$f(x) = f(a) + f'(a)(x-a) + \frac{1}{2!} f''(a)(x-a)^2 + \cdots \qquad \text{(B.1.17)}$$

for the Taylor series expansion of a function $f(x)$ about the point $x = a$, and

$$f(x) = f(0) + f'(0)x + \frac{1}{2!} f''(0)x^2 + \cdots \qquad \text{(B.1.18)}$$

for the Maclaurin series expansion of a function $f(x)$ about $x = 0$. You will also recall that the Maclaurin series is just a special case of the Taylor series, for $a = 0$. Those expansions most commonly encountered are the Maclaurin expansions of the functions e^x, $\sin x$, $\cos x$, $\sinh x$, $\cosh x$, $\tan^{-1} x$, and $(1 - x)^{-1}$. These expansions are given by

$$e^x = 1 + x + \frac{x^2}{2!} + \cdots = \sum_{i=0}^{\infty} \frac{x^i}{i!}, \tag{B.1.19a}$$

$$\sin x = x - \frac{x^3}{3!} + \frac{x^5}{5!} + \cdots = \sum_{i=0}^{\infty} \frac{(-1)^i x^{2i+1}}{(2i + 1)!}, \tag{B.1.19b}$$

$$\cos x = 1 - \frac{x^2}{2!} + \frac{x^4}{4!} + \cdots = \sum_{i=0}^{\infty} \frac{(-1)^i x^{2i}}{(2i)!}, \tag{B.1.19c}$$

$$\sinh x = x + \frac{x^3}{3!} + \frac{x^5}{5!} + \cdots = \sum_{i=0}^{\infty} \frac{x^{2i+1}}{(2i + 1)!}, \tag{B.1.19d}$$

$$\cosh x = 1 + \frac{x^2}{2!} + \frac{x^4}{4!} + \cdots = \sum_{i=0}^{\infty} \frac{x^{2i}}{(2i)!}, \tag{B.1.19e}$$

$$\tan^{-1} x = x - \frac{x^3}{3} + \frac{x^5}{5} - \cdots = \sum_{i=0}^{\infty} \frac{(-1)^i x^{2i+1}}{2i + 1}, \tag{B.1.19f}$$

$$\frac{1}{1 - x} = 1 + x + x^2 + \cdots = \sum_{i=0}^{\infty} x^i. \tag{B.1.19g}$$

Note that, although the first five of these expansions hold for all values of x, the latter two expansions are valid (i.e., converge) only for $-1 \le x \le 1$ for the inverse tangent function, and $-1 < x < 1$ for the reciprocal of $1 - x$.

Example (Bernoulli Numbers) an application of infinite series representation.

Bernoulli numbers, B_k, are obtained by employing the infinite series representation of the function $x/(e^x - 1)$ to define them via

$$\frac{x}{e^x - 1} \equiv \sum_{k=0}^{\infty} \frac{B_k}{k!} x^k. \tag{B.1.20}$$

It will be clear that the Bernoulli numbers may be obtained formally by carrying out repeated differentiation of the two sides of this defining relation, followed by setting x to zero, to give the B_k as

$$B_k = \left[\frac{\mathrm{d}^k}{\mathrm{d}x^k} \left(\frac{x}{e^x - 1} \right) \right]_{x=0} .$$

It will also be fairly obvious that this is not necessarily a particularly practical means for determining these numbers, especially for larger integers. A more practical means for generating values for the B_k can be obtained from a rearrangement of Eq. (B.1.20) into the form

$$\frac{1}{x} (e^x - 1) \sum_{k=0}^{\infty} \frac{B_k x^k}{k!} = 1 ,$$

followed by expansion of e^x, multiplication of the two resultant series expansions, and utilization of the linear independence of the powers of x to evaluate the B_k coefficients sequentially, starting from $k = 0$.[3] This process readily gives $B_0 = 1$, $B_1 = -\frac{1}{2}$, $B_2 = \frac{1}{6}$, $B_3 = 0$, $B_4 = -\frac{1}{30}$, $B_5 = 0$, $B_6 = \frac{1}{42}$, $B_7 = 0$, $B_8 = -\frac{1}{30}$, and so on. It turns out that $B_{2k+1} = 0$, for $k \geq 1$. Note also that Bernoulli numbers may now readily be evaluated by either of the above methods upon utilizing modern symbolic computation packages.   □

B.1.5   The Euler–Maclaurin Summation Formula

We follow here a heuristic treatment by Gautschi[4] to obtain the Euler–Maclaurin summation formula by starting from the well-known Taylor series expression for a function $f(x)$, namely,

$$f(x) = f(a) + f'(a)(x - a) + \frac{1}{2!} f''(a)(x - a)^2 + \cdots .$$

For x near in value to a, we may write $x = a + h$, so that the Taylor series expression for $f(x)$ may be expressed in terms of the differential operator $\mathcal{D} \equiv \dfrac{d}{da}$ and rearranged into the form

$$f(h + a) - f(a) = \sum_{k=1}^{\infty} \frac{h^k \mathcal{D}^k}{k!} f(a) .$$

Substitution of x for a then produces a version of the Taylor expansion of $f(x)$ in operator format, i.e.,

---

[3] See, e.g., G. Arfken, *Mathematics for Physicists*, 2nd Ed. (Academic Press, New York, 1970).

[4] W. Gautschi, *Numerical Analysis: An Introduction* (Birkhäuser, Boston, 1997).

$$f(x+h) - f(x) = \sum_{k=1}^{\infty} \frac{h^k \mathcal{D}^k}{k!} f(x) \equiv (e^{h\mathcal{D}} - 1)f(x) ,$$

with the final step resulting from the definition of the exponential of an operator. We may solve this equation formally for $f(x)$, to obtain

$$f(x) = (e^{h\mathcal{D}} - 1)^{-1} [f(x+h) - f(x)] ,$$

in which $(e^{h\mathcal{D}} - 1)^{-1}$ is an inverse operator. Upon utilizing the series expansion of the operator function $(e^{\mathcal{D}} - 1))^{-1}$ in the form

$$(e^{\mathcal{D}} - 1)^{-1} = \sum_{k=0}^{\infty} \frac{B_k}{k!} \mathcal{D}^{k-1} ,$$

we obtain $f(x)$ in the form

$$f(x) = \sum_{r=0}^{\infty} \frac{B_r}{r!} (h\mathcal{D})^{r-1} [f(x+h) - f(x)] .$$

If we now separate out the first two terms of the expansion, we find that $f(x)$ is obtained as

$$f(x) = \left[(h\mathcal{D})^{-1} - \tfrac{1}{2} + \sum_{r=2}^{\infty} \frac{B_r}{r!} (h\mathcal{D})^{r-1} \right] [f(x+h) - f(x)]$$

$$= (h\mathcal{D})^{-1} [f(x+h) - f(x)] - \tfrac{1}{2} [f(x+h) - f(x)]$$

$$+ \sum_{r=2}^{\infty} \frac{B_r}{r!} h^{r-1} \left[f^{(r-1)}(x+h) - f^{(r-1)}(x) \right] ,$$

in which $f^{(n)}(x)$ represents the nth derivative of $f(x)$ with respect to x.

By utilizing the antiderivative nature of the inverse differential operator $\mathcal{D}^{-1}$, together with the property $B_{2k+1} = 0, k \geq 1$, of the Bernoulli numbers, we obtain the result

$$f(x) = \frac{1}{h} \int_x^{x+h} f(t) \, dt - \tfrac{1}{2} [f(x+h) - f(x)]$$

$$+ \sum_{k=1}^{\infty} \frac{B_{2k}}{(2k)!} h^{2k-1} \left[f^{(2k-1)}(x+h) - f^{(2k-1)}(x) \right] ,$$

which we may rewrite as

$$\frac{h}{2}[f(x+h)+f(x)] = \int_x^{x+h} f(t)\,dt$$

$$+ \sum_{k=1}^{\infty} \frac{B_{2k}}{(2k)!} h^{2k} \left[f^{(2k-1)}(x+h) - f^{(2k-1)}(x) \right].$$

Upon setting $x = a + ih$, and summing over i from 0 to $n-1$, we obtain

$$\frac{1}{2}h \sum_{i=0}^{n-1} [f(a+(i+1)h) + f(a+ih)] = \int_a^{a+nh} f(t)\,dt$$

$$+ \sum_{k=1}^{\infty} \frac{B_{2k}}{(2k)!} h^{2k} \left[f^{(2k-1)}(a+nh) - f^{(2k-1)}(a) \right]. \qquad (\text{B.1.21})$$

Now, as the left-hand side of this expression can be reduced to $h \sum_{i=0}^{n} f(a+ih)$ $-\frac{1}{2}h[f(a) + f(a+nh)]$, Eq. (B.1.21) becomes

$$h \sum_{i-0}^{n} f(a+ih) = \int_a^{a+nh} f(t)\,dt + \frac{1}{2}h[f(a) + f(a+nh)]$$

$$+ \sum_{k=1}^{\infty} \frac{B_{2k}}{(2k)!} h^{2k} \left[f^{(2k-1)}(a+nh) - f^{(2k-1)}(a) \right].$$

$$(\text{B.1.22})$$

Substitution of $h = 1$ into Eq. (B.1.22) gives the summation of a function of integers in the form

$$\sum_{i=0}^{n} f(a+i) = \int_a^{a+n} f(t)\,dt + \frac{1}{2}[f(a) + f(a+n)]$$

$$+ \sum_{k=1}^{\infty} \frac{B_{2k}}{(2k)!} \left[f^{(2k-1)}(a+n) - f^{(2k-1)}(a) \right].$$

Upon setting a to 0 in the above result, we obtain an expression for the sum over integers of a function that is defined on the real line in terms of the corresponding definite integral over that function plus correction terms that depend upon the derivatives of the function: explicitly, we find that[5]

---

[5]Note that the Bernoulli numbers B_{2k}, $k \geq 2$ alternate in sign: this sign alternation is often taken into account in versions of the Euler–Maclaurin summation formula by introducing a factor $(-1)^k$

$$\sum_{i=0}^{n} f(i) = \int_{0}^{n} f(t)\,dt + \tfrac{1}{2}[f(0) + f(n)] - \sum_{k=1}^{\infty} \frac{B_{2k}}{(2k)!} \left[f^{(2k-1)}(0) - f^{(2k-1)}(n) \right].$$

$$(B.1.23)$$

This result is referred to as the Euler–Maclaurin summation formula.

B.1.6 Fundamental Theorem of the Calculus

The fundamental theorem of calculus, as expressed in first year calculus courses, is

$$\frac{d}{dx} \int_{a}^{x} f(t)\,dt = f(x). \qquad (B.1.24)$$

The fundamental theorem can be generalized simply to deal with composite functions by utilizing the chain rule for derivatives, as follows:

$$\frac{d}{dx} \int_{a}^{g(x)} f(t)\,dt = \frac{dg}{dx} \frac{d}{dg} \int_{a}^{g} f(t)\,dt = f(g(x)) \frac{dg}{dx}. \qquad (B.1.25)$$

We sometimes encounter functions in statistical thermodynamics that are defined according to this theorem. You will see this in particular in the Debye model for the heat capacity of a solid.

There will also be occasion for us to utilize a result, known as the Leibniz parameter differentiation rule, that arises when we consider functions of the form

$$F(T) = \int_{a(T)}^{b(T)} f(t, T)\,dt, \qquad (B.1.26)$$

in which the upper and/or lower limit of the integral is a function of the parameter (or variable) for which the derivative is needed. What is its derivative with respect to T? Let us work it out. Formally, it can be written as

$$\frac{dF}{dT} = \frac{d}{dT} \int_{a(T)}^{b(T)} f(t, T)\,dt. \qquad (B.1.27)$$

What we need to do to proceed with evaluating this derivative is to convert our expression into one that will allow us to utilize the fundamental theorem of the calculus. We may do this by introducing a constant c into the limit of the integral as follows:

---

into the higher-derivatives summation term and employing a redefined set of Bernoulli numbers that are solely positive.

$$\frac{\mathrm{d}F}{\mathrm{d}T} = \frac{\mathrm{d}}{\mathrm{d}T}\left\{\int_c^{b(T)} f(t,T)\,\mathrm{d}t - \int_c^{a(T)} f(t,T)\,\mathrm{d}t\right\}, \tag{B.1.28}$$

then employing the chain rule version of the fundamental theorem to give

$$\frac{\mathrm{d}F}{\mathrm{d}T} = \frac{\partial}{\partial b}\int_c^b f(t,T)\,\mathrm{d}t\,\frac{\mathrm{d}b}{\mathrm{d}T} + \frac{\partial}{\partial T}\int_c^b f(t,T)\,\mathrm{d}t$$

$$- \frac{\partial}{\partial a}\int_c^a f(t,T)\,\mathrm{d}t\,\frac{\mathrm{d}a}{\mathrm{d}T} - \frac{\partial}{\partial T}\int_c^a f(t,T)\,\mathrm{d}t, \tag{B.1.29}$$

which can be simplified to give the final Leibniz rule as

$$\frac{\mathrm{d}}{\mathrm{d}T}\int_{a(T)}^{b(T)} f(t,T)\,\mathrm{d}t = \int_{a(T)}^{b(T)} \frac{\partial f}{\partial T}\,\mathrm{d}t + \frac{\mathrm{d}b}{\mathrm{d}T} f[b(T),T] - \frac{\mathrm{d}a}{\mathrm{d}T} f[(a(T),T]. \tag{B.1.30}$$

B.1.7 Regression of Series Concept

Let us consider that we may express a quantity x as a power series in the variable y, viz.,

$$x = \sum_{i=1}^{\infty} a_i y^i, \tag{B.1.31}$$

in which the coefficients a_i are known. However, if what we would actually like to have is y expressed as a power series in x, how do we go about obtaining the appropriate expression? We shall have occasion to ask this question a couple of times in the course of our study of statistical mechanics.

To answer this query, we shall begin by *assuming* that y can be written as a power series in x, namely,

$$y \equiv \sum_{j=1}^{\infty} b_j x^j, \tag{B.1.32}$$

in which the coefficients b_j are for the moment unknown (and hence to be determined). To see how this procedure works, we shall obtain expressions for b_1, b_2, b_3, and b_4 by substituting Eq. (B.1.32) into Eq. (B.1.31), followed by collecting like powers of x on the right-hand side of the equation. This procedure results in x being expressed as

$$x = a_1 y + a_2 y^2 + a_3 y^3 + a_4 y^4 + \cdots$$
$$= a_1[b_1 x + b_2 x^2 + b_3 x^3 + b_4 x^4 + \cdots] + a_2[b_1 x + b_2 x^2 + b_3 x^3 + \cdots]^2$$
$$+ a_3[b_1 x + b_2 x^2 + \cdots]^3 + a_4[b_1 x + \cdots]^4 + \cdots$$

in which we have retained specifically only the number of terms in the assumed expansion of y as a function of x needed to generate all terms having powers of x up to and including x^4 on the right-hand side of the resulting equation.

If we now collect powers of x on the right-hand side of this equation, we arrive at the identity

$$x \equiv a_1 b_1 x + [a_1 b_2 + a_2 b_1^2]x^2 + [a_1 b_3 + 2a_2 b_1 b_2 + a_3 b_1^3]x^3$$
$$+ [a_1 b_4 + a_2 b_2^2 + 2a_2 b_1 b_3 + 3b_1^2 b_2 a_3 + a_4 b_1^4]x^4 + \mathcal{O}(x^5),$$

from which we obtain the set of relations

$$a_1 b_1 = 1$$
$$a_1 b_2 + a_2 b_1^2 = 0$$
$$a_1 b_3 + 2a_2 b_1 b_2 + a_3 b_1^3 = 0$$
$$a_1 b_4 + a_2 b_2^2 + 2a_2 b_1 b_3 + 3a_3 b_1^2 b_2 + a_4 b_1^4 = 0,$$

that must be satisfied by the coefficients b_j. Note that we have expressed the b_j in terms of the coefficients a_i and b_k, with $k < j$ in order to keep our expressions relatively manageable. These equations can be solved sequentially for the b_j coefficients in terms of the a_i coefficients and the b_k coefficients for k-values already obtained in the solution process. We thus obtain the first four coefficients b_j as

$$b_1 = \frac{1}{a_1},$$

$$b_2 = -a_2 b_1^3 = -\frac{a_2}{a_1^3},$$

$$b_3 = -b_1^2(2a_2 b_2 + a_3 b_1^2) = \frac{1}{a_1^5}(2a_2^2 - a_3 a_1),$$

$$b_4 = -\frac{1}{a_1^5}(a_4 + 5a_2^3 b_1^2 - 5a_3 a_2 b_1) = \frac{1}{a_1^7}[5a_1 a_2 a_3 - 5a_2^3 - a_1^2 a_4].$$

B.2   Legendre Transforms

In the simplest (i.e., one-dimensional) case, the Legendre transform provides the procedure whereby the relation $y = y(x)$, which we may represent as a curve in the Cartesian xy-plane with the tangent lines to it at points (x, y) having slopes given by the derivative $y'(x)$, can be interpreted as the envelope of a set of tangent lines. Elimination of the variable x between $y(x)$ and $y'(x)$ gives y as a function of y'.

We may express the tangent line $t(x)$ passing through the point (x, y) on the curve generated by $y = y(x)$ as

$$t(x) = y'x + t_i , \tag{B.2.1a}$$

with y' its slope and t_i its y-intercept. By writing the slope y' explicitly in the form

$$y' = \frac{y - t_i}{x - 0} , \tag{B.2.1b}$$

we may generate the set of y-intercepts associated with the tangent lines to the curve $y = y(x)$ from the expression

$$t_i(x) = y - y'x . \tag{B.2.2a}$$

We shall refer to this expression for t_i as the Legendre transform of y. This set of tangent lines is equivalent to the curve $y = y(x)$, as we shall now illustrate explicitly with a very simple example.

Let us examine the one-dimensional example, $y(x) = x^2 + 2x + 3$, to give us a clearer understanding of the role of the Legendre transform. The first derivative of $y(x)$ is $y'(x) = 2x + 2$. Two members of the family of curves $y(x) = x^2 + 2x + y_0$, with $y_0 = 3, 4, 5$, are illustrated in Fig. B.1, together with a set of tangent lines to the curve $y(x) = x^2 + 2x + 3$.

Fig. B.1 Illustration of the concept of a Legendre transform. Members of a family of parabolas $y = x^2 + 2x + y_0$ with $y_0 = 3, 5$, together with a subset of the family of tangent lines to the parabola $y = x^2 + 2x + 3$

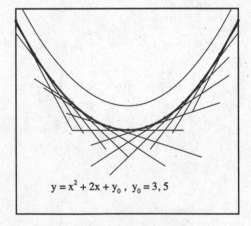

$$y = x^2 + 2x + y_0 , \; y_0 = 3, 5$$

We note in passing that the expression $y(x) = x^2 + 2x + y_0$ represents a family of parabolas, all having a common derivative function $y'(x)$. In other words, $y(x)$ is the general solution to the first-order linear ordinary differential equation $y'(x) = 2x + 2$, and y_0 thus represents the constant of integration for its solution $y(x) = \int y' dx + y_0$. Notice also, that for the specific example with $y_0 = 3$, we may solve $y'(x) = 2x + 2$ for x to obtain $x(y') = \frac{1}{2}y' - 1$ which, upon substitution into $y(x)$, gives y as a function of y' as $y(y') = \frac{1}{4}(y')^2 + 2$. Alternatively, for the present example, we could explicitly invert $y(x) = x^2 + 2x + 3$ by solving the quadratic equation $x^2 + 2x + 3 - y = 0$ for x to obtain $x = -1 \pm \sqrt{y - 2}$, which corresponds to two separate functions of y on the domain $y \in [3, \infty)$, since for x to qualify as a function of y, it must be single-valued on its domain of definition. You might wish to check that this also gives the same result for $y(y')$.

We see from Fig. B.1 that the set of tangent lines is indeed equivalent to the curve mapped out by the set of points (x, y) in the sense that $y = y(x)$ represents the envelope of the family of tangent lines. This illustrates what is referred to as the duality between the point geometry with which we are most familiar and the line geometry introduced by Plücker in the nineteenth century. This duality may also be expressed by noting that in the more traditional point geometry representation, the relevant information is conveyed by the set of points (x, y), with y given by an expression $y = y(x)$, while in the Plücker line geometry the same information is conveyed by a set of points (y', t_i) giving the slopes and y-intercepts of the set of tangent lines whose envelope corresponds to the expression $y = y(x)$.

Nothing that has been said here precludes us from extending the Legendre transform concept to multivariable functions. As discussed by Callen,[6] $z = f(x, y)$ represents a surface that can be viewed as being mapped out by the set of points (x, y, z) or as the envelope of the set of tangent planes to the surface. The most general Legendre transform for z would then be

$$t_i(x) = z - \left(\frac{\partial f}{\partial x}\right)_y x - \left(\frac{\partial f}{\partial y}\right)_x y, \qquad (\text{B.2.2b})$$

involving both independent variables x and y. Note, however, that we may still carry out a partial Legendre transform that involves only x or only y, while leaving the other variable unchanged.

The total differential df for a function of two variables $f \equiv f(x, y)$ is given as

$$df = u\, dx + v\, dy; \qquad u \equiv \left(\frac{\partial f}{\partial x}\right)_y ; v \equiv \left(\frac{\partial f}{\partial y}\right)_x. \qquad (\text{B.2.3})$$

---

[6]H. B. Callen, *Thermodynamics and an Introduction to Thermostatics*, 2nd Edition (Wiley, New York, 1985), §§5.2, 5.3, and 12.4.

Should we wish to change the basis of our description of f from the two variables (x, y) to two different variables (u, y), for example, then we can achieve this goal by *defining* a function $g(u, y)$ such that $g = f - ux$. With this definition of g, we can write its total differential dg as

$$dg = df - u\,dx - x\,du = (u\,dx + v\,dy) - u\,dx - x\,du,$$

or, upon simplification,

$$dg = -x\,du + v\,dy. \tag{B.2.4}$$

This total differential tells us that g is a function of the two independent variables (u, y), and that its partial derivatives with respect to these variables are

$$-x = \left(\frac{\partial g}{\partial u}\right)_y, \quad v = \left(\frac{\partial g}{\partial y}\right)_u.$$

We shall now consider application of the Legendre transform to thermodynamics.

A thermodynamic expression that corresponds to Eq. (B.2.3) is the differential expression for the combined first and second laws, viz.,

$$dU = T\,dS - P\,dV,$$

in which the internal energy U is represented as a function of two independent extensive thermodynamic variables, specifically the entropy S and the volume V, of the (macroscopic) system. Two different Legendre transforms of $U(S, V)$ can be considered.

(a) If $f(x, y)$ is identified with $U(x, y)$, x is identified with V, y with S, and u, v are identified, respectively, with $-P, T$, then $g \equiv H = U + PV$ is a Legendre transform of $U(S, V)$ with a total differential $dH = T\,dS + V\,dP$, so that the enthalpy H is a function of the independent variables S and P, one an extensive, the other an intensive thermodynamic variable.

(b) Similarly to what we have just considered, if we instead identified x with S, y with V, and u with T, v with $-P$, then $g \equiv A = U - TS$ is also a Legendre transform of the internal energy U, with differential $dA = -S\,dT - P\,dV$, so that the Helmholtz energy A is a function of the independent variables T and V, again one intensive and one extensive thermodynamic variable.

Notice that in example (a) we replaced the extensive thermodynamic variable V by the intensive thermodynamic variable P, while in example (b) we replaced the extensive thermodynamic variable S by the intensive thermodynamic variable T. It will now likely be clear that were we to make the identifications $x \equiv V$, $y \equiv T$, $u \equiv -S$, $v \equiv -P$, corresponding to $f \equiv A$, then introduce the Legendre transform $G \equiv A + PV$ to obtain the total differential $dG = -S\,dT + V\,dP$, the resultant Gibbs energy G would be a function of two intensive thermodynamic variables,

T and P, which are referred to in thermodynamics as the 'natural variables' for G. We could, of course, arrive at precisely the same new thermodynamic function via the Legendre transform $G \equiv H - TS$ of the enthalpy H. Moreover, we could also have transformed both extensive independent thermodynamic variables simultaneously into intensive independent thermodynamic variables by carrying out a double Legendre transform, $G \equiv U - TS + PV$, of the internal energy U, corresponding to $g = f - ux - vy$, so that the resultant total differential becomes $dg = -xdu - ydv$.

It is perhaps worthwhile to take a closer look at the Legendre transform to obtain the thermodynamic enthalpy H from the internal energy U. To do this, we note that from the combined first and second law expression $dU = TdS - PdV$, we have the relations

$$T \equiv \left(\frac{\partial U}{\partial S}\right)_V, \qquad P \equiv -\left(\frac{\partial U}{\partial V}\right)_S,$$

so that the defining relation for the enthalpy, namely,

$$H(P, S) = U(V, S) + PV,$$

indeed corresponds to our general definition of the Legendre transform

$$t_1 = y - y' x$$

upon identifying $H(P, S)$ with t_1, $U(V, S)$ with y, $-P$ with y', and V with x. This expression also makes it clear that the enthalpy correlates with the internal energy treated as a function of the pressure P and entropy S, rather than as a function of the volume V and entropy S. Similar considerations show that the Helmholtz energy correlates with the internal energy as a function of the temperature T and volume V, while the Gibbs energy can be viewed as the internal energy treated as a function of temperature T and pressure P.

B.3 Calculus of Variations: But Briefly

We shall examine the problem of finding a path $y = y(x)$ between two values x_1 and x_2 of x which has the property that the integral of some function $f(y, y', x)$ is an extremum. We use the notation $y' \equiv \dfrac{dy}{dx}$ in the following. Further, we shall restrict ourselves to dealing only with a one-dimensional version of the calculus of variations.

We begin by desiring to find the particular path $y = y(x)$ that makes the line integral

$$I \equiv \int_{x_1}^{x_2} f(y, y', x)\, dx \tag{B.3.1}$$

between the points (x_1, y_1) and (x_2, y_2) either a maximum or a minimum.

This idea is illustrated in Fig. B.2, with each path between the points (x_1, y_1) and (x_2, y_2) labelled by a different value of a parameter α. We shall let $\alpha = 0$ coincide with the path (or paths) giving an extremum for the line integral. In such a case, y would be a function both of x and of α, so that

$$y(x, \alpha) = y(x, 0) + \alpha \eta(x) , \qquad (B.3.2)$$

in which $\eta(x)$ is any function of x that vanishes at both $x = x_1$ and $x = x_2$. Note that Eq. (B.3.2) represents a parametric family of curves in the xy-plane. The use of any parametric representation $y = y(x, \alpha)$ makes I a function of the parameter α, so that we may express I as

$$I(\alpha) = \int_{x_1}^{x_2} f[y(x, \alpha), y'(x, \alpha), x] \, dx. \qquad (B.3.3)$$

The condition for an extremum of I with respect to α will thus be

$$\left(\frac{\partial I}{\partial \alpha} \right)_{\alpha=0} = 0 . \qquad (B.3.4)$$

Let us now see how to express this appropriately. We begin by considering the partial derivative of I with respect to α, which we may write in the form

$$\frac{\partial I}{\partial \alpha} = \int_{x_1}^{x_2} \left\{ \frac{\partial f}{\partial y} \frac{\partial y}{\partial \alpha} + \frac{\partial f}{\partial y'} \frac{\partial y'}{\partial \alpha} \right\} \, dx . \qquad (B.3.5)$$

To proceed further, let us utilize an integration by parts to simplify the integral of the second term in the integrand: we find that

$$\int_{x_1}^{x_2} \frac{\partial f}{\partial y'} \frac{\partial y'}{\partial \alpha} \, dx = \int_{x_1}^{x_2} \frac{\partial f}{\partial y'} \frac{\partial^2 y}{\partial \alpha \partial x} \, dx$$

$$= \frac{\partial f}{\partial y'} \frac{\partial y}{\partial \alpha} \Big|_{x_1}^{x_2} - \int_{x_1}^{x_2} \left[\frac{d}{dx} \left(\frac{\partial f}{\partial y'} \right) \right] \frac{\partial y}{\partial x} \, dx ,$$

Fig. B.2 Illustration of the concept of the calculus of variations. Four paths between two points in the xy-plane corresponding to specific values of a parameter α

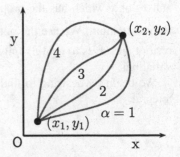

in which each curve is constrained to pass through the points (x_1, y_1) and (x_2, y_2). This constraint on the path to be taken means that $y(x, \alpha)$ is independent of α at those points, and hence $\left. \dfrac{\partial y}{\partial \alpha} \right|_{(x_1,y_1)} = \left. \dfrac{\partial y}{\partial \alpha} \right|_{(x_2,y_2)} = 0$. We are therefore able to write Eq. (B.3.5) as

$$\frac{\partial I}{\partial \alpha} = \int_{x_1}^{x_2} \left(\frac{\partial f}{\partial y} - \frac{d}{dx} \frac{\partial f}{\partial y'} \right) \frac{\partial y}{\partial \alpha} \, dx \,. \tag{B.3.6}$$

If we now multiply Eq. (B.3.6) by the differential $d\alpha$ and evaluate all partial derivatives with respect to α at $\alpha = 0$, we obtain

$$\left(\frac{\partial I}{\partial \alpha} \right)_{\alpha=0} d\alpha = \int_{x_1}^{x_2} \left(\frac{\partial f}{\partial y} - \frac{d}{dx} \frac{\partial f}{\partial y'} \right) \left(\frac{\partial y}{\partial \alpha} \right)_{\alpha=0} dx d\alpha \,. \tag{B.3.7}$$

We shall define the quantity on the left-hand side of this equation as δI, i.e.,

$$\left(\frac{\partial I}{\partial \alpha} \right)_{\alpha=0} d\alpha \equiv \delta I \,, \tag{B.3.8}$$

and call it the *variation* of I; similarly, we define $\left(\dfrac{\partial y}{\partial \alpha} \right)_{\alpha=0} d\alpha \equiv \delta y$, with δy thus representing some *arbitrary* variation of $y(x)$ obtained by variation of the parameter α about $\alpha = 0$. Because δy is arbitrary, we can only have

$$\delta I = \int_{x_1}^{x_2} \left(\frac{\partial f}{\partial y} - \frac{d}{dx} \frac{\partial f}{\partial y'} \right) dx \delta y = 0 \tag{B.3.9}$$

if the integrand is itself zero, i.e.,

$$\frac{\partial f}{\partial y} - \frac{d}{dx} \frac{\partial f}{\partial y'} = 0 \,. \tag{B.3.10}$$

Example the shortest distance between two points in a plane.

We shall begin with the standard first year calculus expression for the differential ds of arc length: $ds = \sqrt{(dx)^2 + (dy)^2}$, from which we may determine the length of an arc between two points in the (x, y)-plane as the integral

$$I \equiv \int_{x=x_1}^{x=x_2} ds = \int_{x_1}^{x_2} \sqrt{1 + \left(\frac{dy}{dx} \right)^2} \, dx \,.$$

Thus, the variational principle requires that the integral I be a *minimum* for the shortest path. We have $f = \sqrt{1 + y'^2}$, from which we see that

$$\frac{\partial f}{\partial y} = 0; \qquad \frac{\partial f}{\partial y'} = \frac{y'}{\sqrt{1 + y'^2}}.$$

Consequently, $\dfrac{df}{dy} = \dfrac{d}{dx}\dfrac{\partial f}{\partial y'}$ reduces to the statement that

$$\frac{d}{dx}\left(\frac{y'}{\sqrt{1 + y'^2}}\right) = 0.$$

This result will hold when $y'/\sqrt{1 + y'^2}$ is a constant, say c; this then allows us to determine y' in terms of c as $y' = c/\sqrt{1 - c^2}$. This result, in turn, tells us that y' is itself a constant, which we shall designate as a. Finally, from $y' = a$, we see that the equation for the shortest path between the two points (x_1, y_1) and (x_2, y_2) in the xy-plane is the straight line

$$y = ax + b,$$

with a given by $a = (y_2 - y_1)/(x_2 - x_1)$ and b by $b = y_1 - ax_1$, a result that we know intuitively from high school Euclidian geometry.   □

B.4   Jacobians

A function $y = f(x_1, x_2, x_3)$ of three variables has an associated total differential

$$dy = f_{,x_1}\, dx_1 + f_{,x_2}\, dx_2 + f_{,x_3}\, dx_3, \qquad (B.4.1)$$

which is a *linear function* of dx_1, dx_2, dx_3, with coefficients given by the partial derivatives $f_{,x_1}, f_{,x_2}, f_{,x_3}$ of f evaluated at the point (x_1, x_2, x_3).

If we now consider, e.g., three functions y_1, y_2, y_3 of the variables x_1, x_2, x_3, which we may represent by the equations

$$y_i = f_i(x_1, x_2, x_3), \qquad i = 1, 2, 3, \qquad (B.4.2)$$

then we can write the corresponding set of three total differentials as

$$\begin{aligned}
dy_1 &= f_{1,x_1}\, dx_1 + f_{1,x_2}\, dx_2 + f_{1,x_3}\, dx_3, \\
dy_2 &= f_{2,x_1}\, dx_1 + f_{2,x_2}\, dx_2 + f_{2,x_3}\, dx_3, \\
dy_3 &= f_{3,x_1}\, dx_1 + f_{3,x_2}\, dx_2 + f_{3,x_3}\, dx_3.
\end{aligned} \qquad (B.4.3)$$

We may also write this equation in matrix format as

$$\begin{pmatrix} dy_1 \\ dy_2 \\ dy_3 \end{pmatrix} = \begin{pmatrix} f_{1,x_1} & f_{1,x_2} & f_{1,x_3} \\ f_{2,x_1} & f_{2,x_2} & f_{2,x_3} \\ f_{3,x_1} & f_{3,x_2} & f_{3,x_3} \end{pmatrix} \begin{pmatrix} dx_1 \\ dx_2 \\ dx_3 \end{pmatrix},$$ (B.4.4)

or in the form

$$dy_i = \sum_{j=1}^{3} f_{i,j} \, dx_j, \qquad i = 1, 2, 3.$$ (B.4.5)

Notice that the set of Eqs. (B.4.2) describes a *mapping* of a domain of R^3 into R^3, while Eqs. (B.4.3) describe a *linear mapping* which approximates the (in general nonlinear) mapping (B.4.2) near a chosen point.

The 3×3 matrix of partial derivatives appearing in relation (B.4.4) is called the *Jacobian matrix* of the set of functions $\{f_1, f_2, f_3\}$ of the set of three variables $\{x_1, x_2, x_3\}$. Its entries are the partial derivatives of the functions f_1, f_2, f_3 evaluated at a chosen point in their (common) domain. The determinant of this matrix is referred to as the *Jacobian determinant*, and is given the symbol J:

$$J = \begin{vmatrix} \dfrac{\partial f_1}{\partial x_1} & \dfrac{\partial f_1}{\partial x_2} & \dfrac{\partial f_1}{\partial x_3} \\[2mm] \dfrac{\partial f_2}{\partial x_1} & \dfrac{\partial f_2}{\partial x_2} & \dfrac{\partial f_2}{\partial x_3} \\[2mm] \dfrac{\partial f_3}{\partial x_1} & \dfrac{\partial f_3}{\partial x_2} & \dfrac{\partial f_3}{\partial x_3} \end{vmatrix} \equiv \dfrac{\partial(f_1, f_2, f_3)}{\partial(x_1, x_2, x_3)}.$$ (B.4.6)

One important application of the Jacobian determinant is in the transformation of variables in multiple integrals. For functions of a single variable, we are familiar with the application of the chain rule (B.1.2) to give the rule for a change of integration variable in a definite integral, viz.

$$\int_{x_1}^{x_2} f(x) dx = \int_{u_1}^{u_2} f[x(u)] \frac{dx}{du} \, du \equiv \int_{u_1}^{u_2} g(u) \, du,$$ (B.4.7)

with $g(u) = f[x(u)]D_u x$. The equivalent expression for double integrals is

$$\int_{R_{xy}} \int f(x, y) \, dx dy = \int_{R_{uv}} \int f[x(u, v), y(u, v)] \left| \frac{\partial(x, y)}{\partial(u, v)} \right| \, du dv,$$ (B.4.8)

involving the absolute value of the Jacobian determinant. Technically, the functions $x(u, v)$, $y(u, v)$ are assumed to be defined and have continuous (partial) derivatives in a region R_{uv} of the uv-plane. The corresponding points (x, y) lie in the region R_{xy}

of the xy-plane, and it is assumed that the inverse functions $u = u(x, y)$, $v = v(x, y)$ are defined and continuous in R_{xy}, so that the correspondence between R_{xy} and R_{uv} is one-to-one. It is also assumed that the Jacobian is either positive throughout R_{uv} or negative throughout R_{uv}.

Appendix C
Stirling Approximations

C.1 Lowest-Level Approximation to $n!$

We begin with the expression for $n!$, which is

$$n! = n(n-1)(n-2) \cdots (2)(1). \qquad (\text{C.1.1})$$

The natural logarithm of $n!$ can be written (see Fig. C.1)

$$\ln n! = \ln n + \ln(n-1) + \cdots + \ln 2 + \ln 1$$

$$= \sum_{m=1}^{n} \ln m. \qquad (\text{C.1.2})$$

The spirit of the Stirling approximation is to replace the summation in the last line of Eq. (C.1.2) by an integral, so that

$$\ln n! \simeq \int_{1}^{n} dx \, \ln x. \qquad (\text{C.1.3})$$

This integral can be integrated by parts, as in your first year calculus course, to give

$$\ln n! = \left[x \ln x - x \right]_{1}^{n},$$

or

$$\ln n! \simeq n \ln n - n + 1, \qquad (\text{C.1.4})$$

© Springer Nature Switzerland AG 2021
F. R. W. McCourt, *Statistical Thermodynamics for Pure and Applied Sciences*,
https://doi.org/10.1007/978-3-030-52006-9

Fig. C.1 Illustration of the Stirling approximation for $\ln n!$

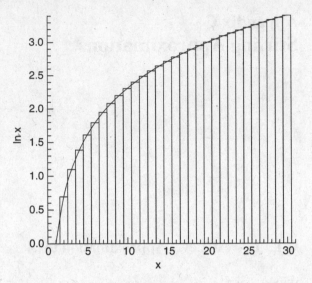

which is Stirling's approximation for $\ln n!$. If we recognize that n is large enough that we can neglect 1 in comparison to n, then we can also rewrite the Stirling approximation as

$$n! \simeq \left(\frac{n}{e}\right)^n . \tag{C.1.5}$$

We shall refer to this as the lowest-order Stirling approximation to $n!$. Moreover, it is the one that we shall employ for the most part in this course.

We can also obtain directly an expression in the same spirit for the derivative of $\ln n!$: if m is sufficiently small relative to n, we can approximate the derivative of $\ln n!$ by the finite-difference expression

$$\frac{d}{dn} \ln n! \simeq \frac{\ln(n+m)! - \ln n!}{m}$$

which becomes, using the properties of the (natural) logarithm

$$\frac{d}{dn} \ln n! = \frac{1}{m} \ln \left[\frac{(n+m)!}{n!} \right]$$

$$= \frac{1}{m} \ln \left[\frac{(n+m)(n+m-1)\cdots(n+1)n!}{n!} \right] .$$

Now, recalling that the spirit of Stirling's approximation is for $n \gg 1$, and recalling further that $m \approx 1$, we obtain

$$\frac{d}{dn} \ln n! = \frac{1}{m} \ln[(n+m)(n+m-1)\cdots(n+1)]$$

$$\simeq \frac{1}{m} \ln\left(n^m\right) = \ln n$$

C.2 Improved Stirling Approximation to n!

Euler's gamma function, denoted $\Gamma(x)$, is a continuous function that gives the values for the discrete function $n!$ for positive integer arguments. This function is defined as

$$\Gamma(x) \equiv \int_0^\infty t^{x-1} e^{-t}\, dt\,. \tag{C.2.1}$$

If we begin with $\Gamma(x+1) = \int_0^\infty t^x e^{-t}\, dt$ and integrate by parts, we obtain $\Gamma(x+1) = x\Gamma(x)$ for arbitrary values of x. If we now let x be n, we see that we can iterate the integration-by-parts process to obtain

$$\Gamma(n+1) = n\Gamma(n) = \cdots = n(n-1)(n-2)\cdots 2 \cdot \Gamma(1)\,,$$

with

$$\Gamma(1) = \int_0^\infty t^{1-1} e^{-t}\, dt = 1\,,$$

so that

$$\Gamma(n+1) = n!\,. \tag{C.2.2}$$

If n is very large, then the major contribution to the integral defining $n!$ comes from those values of t lying near $t = n$, and extending to either side of $t = n$ by $\pm\sqrt{n}$. By substituting $t = n + y\sqrt{n}$ into the defining integral for $n!$ we obtain

$$n! = \int_0^\infty t^n e^{-t}\, dt = \sqrt{n} \int_{-\sqrt{n}}^\infty e^{-(n+y\sqrt{n})} \left(n + y\sqrt{n}\right)^n dy\,,$$

which we can rewrite as

$$n! = \sqrt{n} \left(\frac{n}{e}\right)^n \int_{-\sqrt{n}}^\infty e^{-y\sqrt{n}} \left(1 + \frac{y}{\sqrt{n}}\right)^n dy\,. \tag{C.2.3}$$

Let us call the integrand in Eq. (C.2.3) $I(y)$ and let us examine its behaviour as a function of y around $y = 0$: the derivative of $I(y)$ is

$$I'(y) = \frac{d}{dy}\left[e^{-y\sqrt{n}}\left(1+\frac{y}{\sqrt{n}}\right)^n\right] = -ye^{-y\sqrt{n}}\left(1+\frac{y}{\sqrt{n}}\right)^{n-1},$$

from which we see that $I'(0) = 0$, so that $I(y)$ has an extremum at $y = 0$. We may anticipate that this extremum is a maximum, but we may also examine the second derivative directly, given as it is by

$$I''(y) = -(1-y^2)e^{-y\sqrt{n}}\left(1+\frac{y}{\sqrt{n}}\right)^{n-2},$$

from which $I''(0) = -1$, so that the extremum is indeed a maximum. To demonstrate the rapid falloff of $I(y)$ from its value at $y = 0$, it is sufficient to examine a few typical values of $I(y)$ in the interval $[-\sqrt{n}, \sqrt{n}]$.

Let us now obtain a Gaussian function, which we shall call $I_G(y)$, centred at $y = 0$ to approximate $I(y)$. To accomplish this task, we shall start with $\ln I(y)$, viz.,

$$\ln I(y) = \ln\left[e^{-y\sqrt{n}}\left(1+\frac{y}{\sqrt{n}}\right)^n\right]$$
$$= -y\sqrt{n} + n\ln\left(1+\frac{y}{\sqrt{n}}\right).$$

For n large, $x = y/\sqrt{n}$ will be small, so that we may employ the Maclaurin expansion for $\ln(1+x) = x - \frac{1}{2}x^2 + \frac{1}{3}x^3 - \frac{1}{4}x^4 + \cdots$ to obtain

$$\ln I(y) \simeq -y\sqrt{n} + n\left[\frac{y}{\sqrt{n}} - \frac{y^2}{2n} + \frac{y^3}{3n\sqrt{n}} - \frac{y^4}{4n^2} + \cdots\right]$$
$$= -\frac{y^2}{2} + \frac{y^3}{3\sqrt{n}} - \frac{y^4}{4n} + \cdots.$$

This result gives $I(y)$ as

$$I(y) = e^{-\left(\frac{y^2}{2} - \frac{y^3}{3\sqrt{n}} + \frac{y^4}{2n} + \cdots\right)} = I_G(y)e^{\left(\frac{y^3}{3\sqrt{n}} - \frac{y^4}{2n} + \cdots\right)}, \tag{C.2.4}$$

which is in the form of a Gaussian function $I_G(y) = e^{-y^2/2}$ times an exponential correction factor. If n is sufficiently large, we may ignore the correction factor (i.e., replace it by 1) to obtain a Gaussian approximation to $n!$ given by

$$n! \simeq \sqrt{n} \left(\frac{n}{e}\right)^n \int_{-\sqrt{n}}^{\infty} e^{-y^2/2} \, dy.$$

The same arguments that allow us to make a Gaussian approximation to $I(y)$ allow us to extend the lower limit of our integral from $-\sqrt{n}$ to $-\infty$, so that

$$n! \simeq \sqrt{n} \left(\frac{n}{e}\right)^n \int_{-\infty}^{\infty} e^{-y^2/2} \, dy,$$

or

$$n! \simeq \sqrt{2\pi n} \left(\frac{n}{e}\right)^n, \qquad\qquad (C.2.5)$$

as the definite integral has the value $\sqrt{2\pi}$. Further improvements to the approximation for $n!$, and comments on them have been given by Mermin.[1]

---

[1] N. D. Mermin, *Amer. J. Phys.* **52**, 362 (1984).

Appendix D
Atomic and Molecular Term Symbols

D.1 Term Symbols for Multi-electron Atoms

The set of quantum numbers n, l, m_l, s, m_s associated with the set of quantum mechanical operators $\{\mathcal{H}, \widehat{l^2}, \widehat{l_z}, \widehat{s^2}, \widehat{s_z}\}$ apply to the individual electrons of a multi-electron atom. In particular, the values taken by the orbital angular momentum quantum number l (referred to as the 'azimuthal' quantum number in the old Bohr theory of the hydrogen atom) are used to assign labels to the atomic orbitals in which electrons are placed in the electron configuration model of atomic structure. Each atomic orbital in this model is designated by the value of a principal quantum number n, together with a letter that correlates with one of the allowed values for the magnitude of the orbital angular momentum that is consistent with that principal quantum number value. These letter designations are shown in Table D.1.

One of the main objectives of the quantum mechanical treatment of multi-electron atoms is to determine both the energies of the various states that can be occupied by such atoms and an appropriate way of identifying/labelling them. As energies are the eigenvalues associated with the eigenstates of the multi-electron Hamiltonian, and are obtained by solving the multi-electron Schrödinger equation, and as the energy eigenfunctions will simultaneously be eigenfunctions of any operator that commutes with the (appropriate) Hamiltonian, we may utilize the eigenvalues associated with a set of such operators that also commute mutually with one another to label both the eigenstates of the Hamiltonian and their corresponding energy eigenvalues.

There are different levels at which the electrons of a multi-electron atom can be considered. The scheme described in this appendix is based upon bringing in the various terms making up the full Hamiltonian one at a time in the order in which they contribute to the total energy eigenvalues of $\mathcal{H}$, starting with the largest contribution and proceeding stepwise to the smallest.

The simplest level is that of the so-called independent electron model, which corresponds to the Hamiltonian having the structure $\mathcal{H} = \mathcal{H}_{cf}$ (the central-field

© Springer Nature Switzerland AG 2021
F. R. W. McCourt, *Statistical Thermodynamics for Pure and Applied Sciences*,
https://doi.org/10.1007/978-3-030-52006-9

Table D.1 Atomic electron
orbital labels

Value of l	0	1	2	3	4	$\cdots$
Atomic orbital	s	p	d	f	g	$\cdots$

Hamiltonian), so that each electron moves in the field set up by the nucleus and all the other electrons taken together:

- From the operator point of view, each operator $\widehat{l}_i^2, \widehat{l}_{zi}, \widehat{s}_i^2, \widehat{s}_{zi}$ commutes with $\mathcal{H}_{cf}$ and mutually with one another.
- Corresponding to this set of mutually commuting operators, we speak of electron configurations, such as $(1s)^2(2s)^2(2p)^3$ for the N atom, which has filled $1s$ and $2s$ orbitals, plus a half-filled set of three $2p$ orbitals.

At the next level the most common procedure is to take inter-electron repulsions into account (this process is called Russell–Saunders coupling) as the next most important set of terms in $\mathcal{H}$, so that $\mathcal{H} = \mathcal{H}_{cf} + \mathcal{H}_{ee}$.

From the operator point of view we find that the individual $\widehat{l}_{iz}$ operators no longer commute with $\mathcal{H}$, because they do not commute with $\mathcal{H}_{ee}$. This is because the inter-electron repulsion terms involve the position operators $\widehat{\mathbf{r}}_i$ and $\widehat{\mathbf{r}}_j$ [through the Coulombic potential $V(|\widehat{\mathbf{r}}_i - \widehat{\mathbf{r}}_j|)$], while the orbital angular momentum operators $\widehat{\mathbf{l}}_k (k = i, j)$ for electrons i and j have the form $\widehat{\mathbf{r}}_k \times \widehat{\mathbf{p}}_k$, with the consequence that neither $\widehat{\mathbf{l}}_i$ nor $\widehat{\mathbf{l}}_j$ commutes with $|\widehat{\mathbf{r}}_i - \widehat{\mathbf{r}}_j|$ and hence with $\mathcal{H}_{ee}$, but the commutators $[\mathcal{H}_{ee}, \widehat{\mathbf{l}}_i]_-$ and $[\mathcal{H}_{ee}, \widehat{\mathbf{l}}_j]_-$ are related by $[\mathcal{H}_{ee}, \widehat{\mathbf{l}}_i]_- = -[\mathcal{H}_{ee}, \widehat{\mathbf{l}}_j]_-$, so that their vector sum $\widehat{\mathbf{l}}_i + \widehat{\mathbf{l}}_j$ commutes with $\mathcal{H}_{ee}$. This means that the total orbital angular momentum operator $\widehat{\mathbf{L}} = \sum_i \widehat{\mathbf{l}}_i$ commutes with $\mathcal{H}_{ee}$. We may therefore choose $\widehat{L^2}$, together with any one of its three space-fixed components, such as $\widehat{L}_z = \sum_i \widehat{l}_{iz}$, to provide the corresponding good quantum numbers L, M_L.

The orbital angular momentum eigenstates corresponding to specific values of L, M_L are made up of linear combinations of outer products of the eigenstates of the set of operators $\{\widehat{l}^2_i, \widehat{l}_{iz}\}$. The set of $2L+1$ such linear combination states corresponding to a given value of L (one for each value of M_L) all have the same specific symmetry with regard to the interchange of pairs of electrons: they are either all symmetric or all antisymmetric to such interchanges. For a complete state description of a multi-electron atom, however, we need to adjoin the electron spin dependence to the orbital angular momentum dependence to give a total wavefunction $\Psi_{total} = \psi_{obital}\psi_{spin}$ for the atom. Moreover, as the Pauli Principle requires that the total wavefunction for a multi-electron atom be antisymmetric to the interchange of any pair of electrons in the atom, and as the total orbital angular momentum eigenstates have specific interchange symmetries, it is then also necessary that we combine total orbital angular momentum states $\psi_{orbital}$ with total spin states ψ_{spin} that have electron interchange symmetry opposite to that of $\psi_{orbital}$.

Total electronic spin angular momenta having specific electron interchange symmetries can be constructed from the primitive spin states in the same manner that was employed for the total orbital angular momentum states, giving $2S + 1$ (S, M_S)-states all having the same total spin S: these total spin states correspond

Table D.2 Multi-electron atom term labels

Value of L	0	1	2	3	4	$\cdots$
Term label	S	P	D	F	G	$\cdots$

to the total spin angular momentum operators $\widehat{\mathbf{S}} = \sum_i \widehat{\mathbf{s}}_i$ and $\widehat{S}_z = \sum_i \widehat{s}_{iz}$. This process gives us the set of operators $\{\mathcal{H}, \widehat{L^2}, \widehat{L}_z, \widehat{S^2}, \widehat{S}_z\}$, with corresponding good quantum numbers $\{E, L, M_L, S, M_S\}$, and we speak of atomic terms (hence the name 'Term Symbol').

By analogy with the symbols employed for orbitals at the electron configuration stage, an upper case Roman letter corresponding to the magnitude of L is assigned to designate each term, as shown in Table D.2. The value of S that has been matched up with a specific value of L is conveyed by placing the value for $2S+1$ at the upper left side of the letter designating the value of L to give a term symbol: for example, for $S = 0$ (so that $2S + 1 = 1$) and $L = 2$, we would have a corresponding term symbol ^1D, while for $S = 1$ and $L = 1$, we would obtain the term symbol ^3P. The value $2S + 1$ is called the 'term multiplicity': specifically, for our example we have 'singlet D' and 'triplet P' terms, respectively.

D.2 Extended Example: The He Atom

The simplest multi-electron atom is the He atom, with two electrons. Its ground state has the electronic configuration $(1s)^2$, and its two electrons can be ascribed the sets of values $\left\{1, 0, 0, +\frac{1}{2}\right\}$ and $\left\{1, 0, 0, -\frac{1}{2}\right\}$ for the four quantum numbers $\{n, \ell, m_\ell, m_s\}$ that characterize individual (hydrogenic) electrons. The two He electrons have the same values $n = 1$ for the principal quantum number, $\ell = 0$, $m_\ell = 0$ for the orbital angular momentum quantum number and that of its z-component, and hence according to the Pauli Principle for electrons in the form that 'No two electrons may have all four quantum numbers n, ℓ, m_ℓ, and m_s the same', m_s must have the value $+\frac{1}{2}$ for one of its $1s$ electrons and the value $-\frac{1}{2}$ for the other.

As angular momentum is a vector quantity, the total angular momentum operator $\mathbf{S}$ associated with the two $1s$-electrons must be the vector sum of the individual electron spin angular momentum operators $\mathbf{s}_1$ and $\mathbf{s}_2$: this means that the Cartesian components must also add and, in particular, it means that $S_z = s_{1z} + s_{2z}$, from which the corresponding spin magnetic quantum number relation $M_S = m_{1s} + m_{2s}$ is obtained. Now, because m_{1s} and m_{2s} for the $(1s)^2$ configuration are necessarily $+\frac{1}{2}$ and $-\frac{1}{2}$ (or vice versa), the only possible value for M_S is zero, which in turn implies that only $S = 0$ is allowed for this configuration. The total orbital angular momentum operator can be written similarly as $\mathbf{L} = \mathbf{l}_1 + \mathbf{l}_2$, so that $M_L = m_{\ell_1} + m_{\ell_2}$: for $\ell_1 = \ell_2 = 0$, it is clear that only $L = 0$ is allowed. All of this information is communicated via a quantity referred to as an electronic term symbol and made up from an upper case letter that designates the magnitude assigned to the quantum number L, a superscript at the upper left of the letter designating the value $2S + 1$

associated with the total electronic spin angular momentum quantum number S, and a subscript at the lower right of the letter designating the value of the total electronic angular momentum quantum number J associated with the total electronic angular momentum (vector) operator $\mathbf{J} = \mathbf{L} + \mathbf{S}$ (in the simplest angular momentum coupling scheme, referred to as Russell–Saunders coupling). The unique atomic term symbol associated with the ground electronic configuration $(1s)^2$ is thus ${}^1S_0$.

The simplest excited electronic configurations of the He atom are those in which one electron remains in the $1s$ (hydrogenic) atomic orbital and the second He electron is in an (excited) higher-energy orbital. The lowest such excited configuration is the $(1s)^1(2s)^1$ configuration, for which $n_1 = 1$ and $n_2 = 2$, so that the Pauli Principle is already satisfied for this configuration. The orbital angular momentum quantum numbers are $\ell_1 = \ell_2 = 0$ (because both electrons lie in s-orbitals), so that $L = 0$ only is allowed, as for the ground electronic configuration. Because ℓ_1 and ℓ_2 are both 0, the quantum numbers m_{ℓ_1} and m_{ℓ_2} are both 0 also. However, as the Pauli Principle has already been satisfied, the m_s-values for the two electrons are no longer required to have opposite signs, and there will be four possible pair combinations $\left(m_{s_1}, m_{s_2}\right)$, namely $\left(+\frac{1}{2}, +\frac{1}{2}\right)$, $\left(+\frac{1}{2}, -\frac{1}{2}\right)$, $\left(-\frac{1}{2}, +\frac{1}{2}\right)$, $\left(-\frac{1}{2}, -\frac{1}{2}\right)$. Two of these four pairs of m_{s_1}, m_{s_2}-values clearly correspond to a total spin angular momentum quantum number $S = 1$ (namely, the pairs for which $M_S = m_{s_1} + m_{s_2} = 1, -1$): the other two pairs correspond to $M_S = 0$, and must somehow be associated with the $S = 1, M_S = 0$ and $S = 0, M_S = 0$ states. A more general version of the Pauli Principle for electrons states that the total wavefunction for a system of electrons must be antisymmetric to the interchange of any pair of electrons. This version of the Pauli Principle allows us to determine that the $S = 1, M_S = 0$ wavefunction must then be the sum of the wavefunctions corresponding to the primitive pairs $(m_{s_1}, m_{s_2}) = \left(+\frac{1}{2}, -\frac{1}{2}\right)$ and $\left(-\frac{1}{2}, +\frac{1}{2}\right)$, while the $S = 0, M_S = 0$ wavefunction must be the difference between the wavefunctions for these two primitive pairs, and the four linearly independent spin states for the $(1s)^1(2s)^1$ He configuration then correspond to a symmetric triplet of spin pair-states and an antisymmetric singlet spin pair-state. $L = 0$ and $S = 1$ give $J = L + S = 1$, while $L = 0, S = 0$ gives $J = L + S = 0$, so that we have two electronic terms, ${}^3S_1$ and ${}^1S_0$, associated with the $(1s)^1(2s)^1$ He configuration.

Similar arguments applied to the more highly excited configurations give rise by the same arguments that we have utilized for the $(1s)^1(2s)^1$ configuration to pairs of terms $({}^3S_1, {}^1S_0)$ for other $(1s)^1(ns)^1$ configurations, $({}^3P_{2,1,0}, {}^1P_1)$ for $(1s)^1(np)^1$ configurations, $({}^3D_{3,2,1}, {}^1D_2)$ for $(1s)^1(nd)^1$ configurations, and so on. The set of energies thereby obtained can be displayed in an energy level diagram referred to as a Grotrian diagram: a typical diagram of this type for the He atom can be found, for example, on p. 65 of G. Herzberg, *Atomic Spectra and Atomic Structure* (Dover, New York, 1944). The Herzberg figure also includes the set of electric dipole-allowed spectroscopic transitions, with selection rules $\Delta S = 0$, $\Delta L = \pm 1$: the electronic spin selection rule $\Delta S = 0$ means that triplet and singlet terms cannot interconvert by absorption or emission of photons, and hence the singlet and triplet electronic spin states form spectroscopically distinct species. This distinctness was acknowledged in the early part of the twentieth century by referring to the set of metastable triplet terms as *ortho*-helium (meaning high-spin) and the set of singlet terms as *para*-helium.

At the third (and final stage for this discussion), the spin–orbit interaction is taken into account by adding the term $\mathcal{H}_{so} = A\widehat{\mathbf{S}} \cdot \widehat{\mathbf{L}}$ to the Hamiltonian. It can be shown that $\widehat{\mathbf{S}} \cdot \widehat{\mathbf{L}}$ commutes neither with $\widehat{\mathbf{L}}$, nor with $\widehat{\mathbf{S}}$, but it does commute with their vector sum $\widehat{\mathbf{J}} = \widehat{\mathbf{L}} + \widehat{\mathbf{S}}$. The good quantum numbers are now only J, M_J, corresponding to the operators $\widehat{J^2}$, $\widehat{J_z}$ (however, we retain L, S so that we know which level is the origin for a particular value of J appearing in our labelling procedure). The quantum number J has the set of values determined by $J = L+S, L+S-1, \cdots, |L-S|$. The J values characterize what are referred to as levels, and each pair of (J, M_J)-values designates a specific state within a level. The value of J is typically given as a subscript to the term symbol from which it was deduced. Thus, for the two examples given at the previous stage of our description, namely the ^1D and ^3P terms, we would have levels designated by $^1D_2$ in the first case (because there is only one possible value for J when $L = 2$ and $S = 0$), and $^3P_2$, $^3P_1$, $^3P_0$ arising from the original ^3P term. Hence, we see that the 'singlet D' term indeed has only one (fivefold degenerate) level, while the 'triplet P' term gives rise to three levels, with degeneracies 5, 3, and 1, respectively. For atoms, the term symbol tells us the degeneracy of the energy level to which the symbol corresponds: this degeneracy is always given by the value of $2J + 1$.

For atomic configurations with nonequivalent valence electrons, such as the excited electronic $(1s)^2(2s)^1(2p)^1$ configuration for Be, which has a total of 12 possible states (consisting of all possible combinations of the two available states for the $2s$ orbital and the six available states for the $2p$ orbital), all possible terms are present (as the Pauli Principle has already been satisfied once the electrons have different values for their azimuthal quantum numbers). The example of the Be excited electronic configuration $\text{Be}[(1s)^2(2s)^1(2p)^1]$ serves to illustrate this point. We have $l_1 = 0$ and $l_2 = 1$, which gives $L = l_1 + l_2 = 1$ as the only total orbital angular momentum possible, and this, in turn, leads us to P terms. Each electron has spin $\frac{1}{2}$, so that the total electron spin S can have values 1 and 0, which leads to ^3P and ^1P terms. For the triplet term, we have $L = 1$, $S = 1$, from which we obtain $J = 2, 1, 0$, while for ^1P, we have $L = 1$ and $S = 0$, so that $J = 1$ only. Thus, we arrive finally at a $^1P_1$ level, with 3 states, plus the $^3P_2$, $^3P_1$, $^3P_0$ levels arising from the ^3P term, with 9 states (made up of 5, 3, 1 state(s), respectively, for the three levels) for a total of $3 + 9 = 12$ total electronic states for this configuration. This example also illustrates the principle that the number of states available to an atom in a given configuration remains the same even when the various electronic coupling mechanisms are taken into account. This principle can be utilized as a check on the final results obtained by the Russell–Saunders coupling scheme employed in the present discussion or, indeed, for any other coupling scheme that may be necessitated.

For the example of a Si atom considered in §IV.2, where we found that the $(1s)^2(2s)^2(2p)^6(3s)^2(3p)^2$ ground electronic configuration gave rise to three terms, namely ^1S, ^1D, and ^3P, we see that $\mathbf{J} = \mathbf{L} + \mathbf{S}$ gives rise only to $J = 0$ for the ^1S term, to $J = 2$ for the ^1D term, and to $J = 2, 1, 0$ for the ^3P term. The total number of (J, M_J)-states contained in the levels belonging to these terms is thus 15: 5 from the $^1D_2$ level, 1 from the $^1S_0$ level, and 9 from the $^3P_{2,1,0}$ levels. This is precisely the number $6 \cdot 5/2$ of states for a $(ns)^2(np)^2$ electronic configuration once the Pauli Principle has been taken into account.

D.3   Term Symbols for Linear Molecules

The determination of term symbols for linear molecules proceeds similarly to what we have just seen for atoms. We shall only draw the appropriate analogies here, rather than go through a detailed derivation. The main idea is to use a new set of designators: to this end, we note that we will use the correspondence

Atoms (roman letters) $\longleftrightarrow$ Molecules (Greek letters)

Individual electrons are assigned to molecular, rather than atomic, orbitals; the individual molecular orbitals are assigned orbital angular momentum projection quantum numbers λ as shown in Table D.3.

Note that molecular orbitals (MOs) are complicated by the fact that we have both $s\sigma$ and $p\sigma$ molecular orbitals, as well as $p\pi$ orbitals, and so on. There is also a more complex nodal structure for these orbitals, because of the two nuclear centres, with the consequence that we now have to deal also with both what are referred to as 'bonding' and 'antibonding' orbitals, depending upon whether or not the orbital has a nodal plane lying between the two nuclei and perpendicular to the figure axis. Antibonding MOs are typically designated by an asterisk, e.g., $2p_x\pi^*$.

Unlike atoms, in which the electrons move in a spherically symmetric force field, the electrons in a linear molecule move in the cylindrically symmetric force field associated with the two separated nuclei. As a consequence, the total electronic orbital angular momentum $\mathbf{L}$ is no longer a constant of the motion as for a spherically symmetric atom: for linear molecules, only the M_L component of $\mathbf{L}$ lying along the symmetry axis (the bond axis) of the molecule is a constant of the motion. This difference between a linear molecule and an atom is due to the strong electrostatic field generated by the two positively charged nuclei in the linear molecule: this electrostatic field causes a precessional motion of the orbital angular momentum vector about the figure axis, thereby averaging out the components of

Table D.3 Electron Orbital Labels for Linear Molecules

Value of λ	0	1	2	3	$\cdots$
Orbital type	σ	π	δ	ϕ	$\cdots$

Table D.4 Electron term labels for linear molecules

Value of Λ	0	1	2	3	$\cdots$
Term Label	Σ	Π	Δ	Φ	$\cdots$

L that are perpendicular to it. The electrostatic field also affects the energies of the orbital states via the Stark effect, in which the energy levels are split according to the values of M_L^2, so that each Stark level is doubly degenerate. For this reason, electronic states of diatomic molecules are traditionally classified by the value of $\Lambda = |M_L|$ rather than by the value of L. A corresponding angular momentum vector $\mathbf{\Lambda}$ then represents the component of the resultant total electronic orbital angular momentum along the internuclear axis.

The quantum number Λ can have values $\Lambda = 0, 1, 2, \ldots, L$, with each value corresponding to an electronic term: there will thus be $L + 1$ distinct terms, all with different energies. The values taken by Λ then play the role that the L-values do for atoms: the symbols utilized for identifying the terms for linear molecules are upper case Greek letters, as illustrated in Table D.4.

Spin angular momentum, unlike orbital angular momentum, is unaffected by an electric field. However, the orbital motions of electrons in the linear molecules for Π, Δ, $\cdots$ terms generate an internal magnetic field directed along the molecular figure axis. This field then serves as the quantization axis for the total electron spin angular momentum, so that $\mathbf{S}$ also precesses about the figure axis, with the consequence that only the component $M_S \equiv \Sigma$, with $2S + 1$ allowed values, $-S \leq \Sigma \leq +S$, remains. We note that Σ is undefined for $\Lambda = 0$ (because there is no orbitally created internal magnetic field to define a quantization axis in this case). The total electronic angular momentum for a linear molecule is denoted by $\Omega \equiv \Lambda + \Sigma$ in the same way that $\mathbf{J} = \mathbf{L} + \mathbf{S}$ for atoms, but the addition process is algebraic rather than vectorial.

The quantity $2S + 1$ is still known as the term multiplicity, as it was for multi-electron atoms: in addition, it gives the true multiplicity of the molecular terms in all cases. The value for $2S + 1$ is placed, as for atoms, at the upper left-hand corner of the term symbol, e.g., $^2\Pi$, $^3\Delta$, and so on.

Let us consider an accounting of the number of states available to a diatomic molecule. The nitric oxide molecule, NO, is perhaps the best known example of a molecule with a $^2\Pi$ ground electronic term. It has the ground electronic molecular orbital configuration

$$\text{NO} : [\text{fc}](2p_x\pi^*)^1(2p_y\pi)^0(2p_z\sigma)^0,$$

with [fc] representing the filled set of core molecular orbitals, i.e., $[\text{fc}] \equiv (1s\sigma)^2(1s\sigma^*)^2(2s\sigma)^2(2s\sigma^*)^2(2p_x\pi)^2(2p_y\pi)^2(2p_z\sigma)^2$, so that $\lambda = \Lambda = 1$, and $S = \frac{1}{2}$, so that $\Sigma = \pm\frac{1}{2}$. From $\Lambda = 1$, $\Sigma = \pm\frac{1}{2}$, we obtain $\Omega = |\Lambda + \Sigma| = \frac{3}{2}, \frac{1}{2}$, and our molecular term is $^2\Pi$, split into two levels, $^2\Pi_{\frac{3}{2}}$, $^2\Pi_{\frac{1}{2}}$, each of which is doubly degenerate, corresponding to states having total angular momentum projection quantum numbers $\pm\Omega$. The first excited electronic configuration for the NO molecule is

$$NO : [fc](2p_x\pi^*)^0(2p_y\pi)^0(2p_z\sigma)^1 \, .$$

In this case, $\lambda = \Lambda = 0$, giving a Σ term; moreover, as $S = \frac{1}{2}$, as for the ground electronic configuration, we have a $^2\Sigma$ term.

For the ground electronic configuration NO molecule, we note that we have a single unpaired electron in a π-orbital, corresponding to four possible states associated with an electron that can be either 'spin-up' or 'spin-down' in either a $2p_x\pi$ or a $2p_y\pi$ bonding molecular orbital. If we examine the electronic levels $^2\Pi_{\frac{1}{2}}$ and $^2\Pi_{\frac{3}{2}}$ (each lying at a different energy), we see that two of the four possible states are associated with the $^2\Pi_{\frac{1}{2}}$ level with $\Omega = \frac{1}{2}$ (corresponding to $\Sigma = \pm\frac{1}{2}$) while the other two states are associated with the $^2\Pi_{\frac{3}{2}}$ level (corresponding to $\Sigma = \pm\frac{3}{2}$).

D.4   Term Symbols for Nonlinear Molecules

The term symbols for linear molecules correspond to the notation employed for the irreducible representations of the linear molecule point symmetry groups $C_{\infty v}$ (for noncentrosymmetric linear molecules) and $D_{\infty h}$ (for centrosymmetric linear molecules). Point symmetry groups in general consist of the set of all symmetry operations, such as rotation axes, reflection planes, rotation-reflection axes containing the centre-of-mass of the molecule, as well as inversion through the centre-of-mass for centrosymmetric molecules (plus the identity operation) that leave the molecule after the symmetry operation indistinguishable from what it was before the symmetry operation. In particular, the angular momentum states must belong to an irreducible representation of the molecular point symmetry group.

If we were to re-examine our assignment of electronic terms for linear molecules, we would notice that the one-dimensional irreducible representation of the $C_{\infty v}$ and $D_{\infty h}$ point symmetry groups is labelled Σ, while the irreducible representations labelled by Π, Δ, Φ, $\cdots$, are all two-dimensional. For nonlinear molecules, apart from the spherical group (the group that applies to atoms) and the icosahedral group (the group appropriate for some boranes and for buckminsterfullerene) for which higher-dimensional irreducible representations exist (i.e., of dimension 4 or higher), and to a lesser extent, the tetrahedral and octahedral groups (in which three-dimensional irreducible representations are found), the highest dimension for irreducible representations is typically two. Note that only one-dimensional representations exist for many of the simpler point symmetry groups corresponding to molecules with relatively low symmetries.

The method for determining the appropriate irreducible representations for the electronic states of a nonlinear polyatomic molecule involves first determining the irreducible representations to which the molecular orbitals to be occupied by the electrons belong, and then by using group outer products of these representations, determine the irreducible representation for the orbital angular momentum for the multi-electron molecule. This symmetry label is then identified by the appropriate upper case Roman letter. One-dimensional irreducible representations are commonly labelled by symmetry types A or B (sometimes, though rarely, C or D), while two-dimensional irreducible representations are typically labelled by E, and three-dimensional ones by T. As there may be, and often is, more than one independent irreducible representation of a particular symmetry type, the various occurrences may be distinguished by subscripts $1, 2, \ldots$, as well as by primed and unprimed letters.

The total electronic spin for the ground term in most nonlinear polyatomic molecules is most commonly zero (all spins paired). Occasionally, doublets or even triplets (relatively uncommon) also occur, corresponding to unpaired spins in the molecules. Electronic term symbols like $^2A_1$, $^3B_{1g}$, $^3T_{1u}$, $^2E'$, and so on can be found: everything but the spin multiplicity (in the upper left-hand corner of the term symbol) is related to point-group symmetry information on the irreducible representation.

Two Examples The H_2O and NH_2 molecules both have the same symmetry point group, C_{2v}, with four one-dimensional irreducible representations, $\{A_1, A_2, B_1, B_2\}$. They also have very similar ground state electronic configurations, namely,

$$H_2O: \ (1a_1)^2(a_1)^2(1b_2)^2(3a_1)^2(1b_1)^2 \ ,$$

$$NH_2: \ (1a_1)^2(a_1)^2(1b_2)^2(3a_1)^2(1b_1)^1 \ .$$

Note that, as for atoms and linear molecules, the orbitals are designated by lower case Roman letters; the numbers 1, 2, 3, preceding the letters identify the number of independent orbitals having the same symmetry type (basically the same role as played by the principal quantum number for atoms and linear molecules). A configuration in general represents a reducible representation, and we wish to know which irreducible representations of the point symmetry group it contains. One method which we can use is the method of outer products within a group. As we are working with a group that has only one-dimensional irreducible representations, we do not need to worry too much about the formalities: all that we really need to know for the moment is that the product of two one-dimensional irreducible representations in the same group is again a one-dimensional irreducible representation of that group, that the product of any one-dimensional irreducible representation with itself is always the totally symmetric a_1 or equivalent irreducible representation in that symmetry point group, and that the product of the a_1 representation with any other irreducible representation gives that representation (which is another way of saying that the a_1 representation is the identity for the outer product process). This means, for example, that in every case in which there are two electrons in a given orbital (i.e., a filled orbital), it will behave as the a_1 representation. For the H_2O molecule, this clearly leads to an overall a_1 irreducible representation within the C_{2v} symmetry point group: we label this as an A_1 term according to the spectroscopic convention for naming terms. We can also see that all electrons are paired in this configuration, so that the total electronic spin S will be zero, and hence the term is a singlet term. This gives us the term symbol $^1A_1$ for the H_2O molecule. For the NH_2 radical (molecule), everything is the same except that the highest energy MO has only one electron in it: hence, the electron configuration up to the half-occupied $1b_1$ MO gives a_1 symmetry, so that the outer product of this a_1 symmetry with the singly occupied $1b_1$ MO gives b_1 symmetry, which we assign as the term B_1. There is a single unpaired valence electron for NH_2, so that the total electronic spin is $\frac{1}{2}$ and we have a doublet molecule, with term symbol $^2B_1$.

Appendix E
Spherical/Symmetric Top Molecules

E.1 Linear Molecules: A Brief Review

We have seen (section "Heteronuclear Diatomic Molecules") that the classical expression for the rotational partition function for a linear molecule may be obtained by approximating the sum over the rotational angular quantum number j of the rotational energy states by treating j as a continuous variable, and integrating over it. More generally, we have

$$z_{\mathrm{rot}}(T) = \frac{1}{\sigma} \sum_{j,m} \mathrm{e}^{-E_{\mathrm{rot}}(j,m)/(k_{\mathrm{B}}T)}, \tag{E.1.1}$$

with $E_{\mathrm{rot}}(j,m) = \hbar^2 j(j+1)/2I$, and σ a symmetry number associated with the molecular symmetry group. Upon defining the characteristic rotational temperature Θ_{rot} by $\Theta_{\mathrm{rot}} \equiv \hbar^2/(2Ik_{\mathrm{B}})$, we obtain

$$z_{\mathrm{rot}}(T) = \frac{1}{\sigma} \sum_{j,m} \mathrm{e}^{-j(j+1)\Theta_{\mathrm{rot}}/T}$$

or

$$z_{\mathrm{rot}}(T) = \frac{1}{\sigma} \sum_{j=0}^{\infty} (2j+1)\mathrm{e}^{-j(j+1)\Theta_{\mathrm{rot}}/T}. \tag{E.1.2}$$

The classical approximation to $z_{\mathrm{rot}}(T)$ is thus given by

$$z_{\mathrm{rot}}(T) = \frac{1}{\sigma} \int_0^{\infty} (2j+1)\mathrm{e}^{-j(j+1)\Theta_{\mathrm{rot}}/T}\, \mathrm{d}j,$$

© Springer Nature Switzerland AG 2021
F. R. W. McCourt, *Statistical Thermodynamics for Pure and Applied Sciences*,
https://doi.org/10.1007/978-3-030-52006-9

which can be integrated directly by setting $j(j+1) \equiv x$ and realizing that $(2j+1)\,dj = d[j(j+1)]$, so that

$$z_{\text{rot}}(T) = \frac{1}{\sigma} \int_0^\infty e^{-\Theta_{\text{rot}} x / T}\, dx = \frac{T}{\sigma\, \Theta_{\text{rot}}}. \tag{E.1.3}$$

Recall that for centrosymmetric linear molecules $\sigma = 2$, while for noncentrosymmetric linear molecules $\sigma = 1$.

E.2   Spherical Top Molecules

For a spherical top rotor (which has all three moments-of-inertia equal) we have three quantum numbers that are relevant, namely, j, k, m, with m the usual projection quantum number associated with a space-fixed z-axis, while k is a second projection quantum number, associated with a body-fixed z-axis. Both k and m have values ranging between $-j$ and $+j$. The energy for a spherical top molecule is given by

$$E_{\text{rot}}(j, k, m) = \frac{\hbar^2}{2I} j(j+1), \tag{E.2.1}$$

exactly as for a linear molecule. The quantum partition function, however, is

$$z_{\text{rot}}(T) = \frac{1}{\sigma} \sum_{j=0}^\infty \sum_{k=-j}^{+j} \sum_{m=-j}^{+j} e^{-j(j+1)\Theta_{\text{rot}}/T}, \tag{E.2.2}$$

with Θ_{rot} defined in precisely the same manner as for a linear molecule, while σ is given by the order of the subgroup of all pure rotations that leave the molecule indistinguishable. The sums over m and k can be performed trivially, each sum giving a factor $(2j+1)$:

$$z_{\text{rot}}(T) = \frac{1}{\sigma} \sum_0^\infty (2j+1)^2\, e^{-j(j+1)\Theta_{\text{rot}}/T}, \tag{E.2.3}$$

and the classical equivalent is simply

$$z_{\text{rot}}(T) = \frac{1}{\sigma} \int_0^\infty (2j+1)^2\, e^{-j(j+1)\Theta_{\text{rot}}/T}\, dj. \tag{E.2.4}$$

In this case, however, recognizing that $(2j+1)dj = d[j(j+1)]$ does not help us, because of the second $(2j+1)$ factor. We must therefore be slightly more clever than before in order to evaluate this integral.

In this case we note that $j(j+1)$ can be written as

$$j(j+1) = \left(j+\frac{1}{2}\right)^2 - \frac{1}{4}$$

$$\approx \left(j+\frac{1}{2}\right)^2 \tag{E.2.5}$$

for j sufficiently large, so that we can approximate $z_{\text{rot}}(T)$ as

$$z_{\text{rot}}(T) \simeq \frac{4}{\sigma} \int_0^\infty \left(j+\frac{1}{2}\right)^2 e^{-(j+\frac{1}{2})^2 \Theta_{\text{rot}}/T} \, d(j+\tfrac{1}{2}). \tag{E.2.6}$$

In order to evaluate this integral it will be convenient to make a change of variable from $j+\frac{1}{2}$ to y, defined via

$$y \equiv \left(j+\frac{1}{2}\right)\left(\frac{\Theta_{\text{rot}}}{T}\right)^{\frac{1}{2}}, \tag{E.2.7}$$

which gives

$$z_{\text{rot}}(T) = \frac{4}{\sigma}\left(\frac{T}{\Theta_{\text{rot}}}\right)^{\frac{3}{2}} \int_0^\infty y^2 e^{-y^2} \, dy. \tag{E.2.8}$$

The value of the integral can either be looked up in a mathematical handbook, or evaluated through integration by parts, as

$$\int_0^\infty y^2 e^{-y^2} \, dy = -\frac{1}{2} y e^{-y^2}\Big|_0^\infty + \int_0^\infty e^{-y^2} \, dy = \frac{\sqrt{\pi}}{4}, \tag{E.2.9}$$

and hence the rotational partition function for a spherical top molecule is given by

$$z_{\text{rot}}(T) \simeq \frac{\sqrt{\pi}}{\sigma}\left(\frac{T}{\Theta_{\text{rot}}}\right)^{\frac{3}{2}}. \tag{E.2.10}$$

E.3 Symmetric Top Molecules

Evaluation of the rotational partition function for symmetric top molecules is more difficult than for spherical top molecules, because of the additional contribution to the rotational energy of the molecule depending upon the quantum number k. A typical symmetric top molecule is one belonging to the point symmetry group C_{3v}, as illustrated in Fig. E.1.

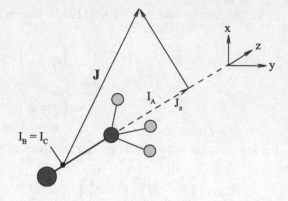

Fig. E.1 Symmetric top molecule and body-fixed axis system

The energy of the rigid rotational motion of a symmetric top molecule is given by

$$E_{\text{rot}} = \frac{\hbar^2}{2I_B}(J_x^2 + J_y^2) + \frac{\hbar^2}{2I_A}J_z^2$$

$$= \frac{\hbar^2}{2I_B}(J_x^2 + J_y^2 + J_z^2) - \frac{\hbar^2}{2I_B}J_z^2 + \frac{\hbar^2}{2I_A}J_z^2 . \tag{E.3.1}$$

The rotational Hamiltonian corresponding to this energy is then given by

$$\mathcal{H}_{\text{rot}} = \frac{\hbar^2}{2I_B}\widehat{J^2} + \left(\frac{\hbar^2}{2I_A} - \frac{\hbar^2}{2I_B}\right)\widehat{J_z^2} . \tag{E.3.2}$$

The eigenstates of this rigid body rotational Hamiltonian can be designated by $|jkm\rangle$, and correspond to eigenenergies E_{jkm}:

$$\mathcal{H}_{\text{rot}}|jkm\rangle = E_{jkm}|jkm\rangle . \tag{E.3.3}$$

The quantum states $|jkm\rangle$ are also simultaneous eigenstates of the set of (mutually commuting) operators $\widehat{J^2}, \widehat{J_z}, \widehat{J_Z}$, so that the energy eigenvalues for this molecule are

$$E_{jkm} = \frac{\hbar^2}{2I_B}j(j+1) + \left(\frac{\hbar^2}{2I_A} - \frac{\hbar^2}{2I_B}\right)k^2 . \tag{E.3.4}$$

The quantity that appears in the exponent of the rotational partition function expression is $E_{jkm}/(k_B T)$. For convenience this will be written as

$$\frac{E_{jkm}}{k_B T} = \alpha j(j+1) + \beta k^2 , \tag{E.3.5}$$

in which

$$\alpha \equiv \frac{\hbar^2}{2I_B k_B T}\,; \quad \beta \equiv \frac{\hbar^2}{2k_B T}\left(\frac{1}{I_B} - \frac{1}{I_A}\right), \quad \Rightarrow \quad \frac{\hbar^2}{2I_A k_B T} = \alpha + \beta\,. \quad \text{(E.3.6)}$$

From the foregoing, we see that the rotational partition function is given by

$$z_{\text{rot}}(T) = \frac{1}{\sigma}\sum_{j=0}^{\infty}\sum_{k=-j}^{+j}(2j+1)e^{-\alpha j(j+1)-\beta k^2} \quad \text{(E.3.7)}$$

$$\simeq \frac{2}{\sigma}\sum_{j=0}^{\infty}\sum_{k=-j}^{+j} j e^{-\alpha j^2 - \beta k^2}$$

$$\approx \frac{2}{\sigma}\int_0^{\infty}\int_{-j}^{+j} j e^{-\alpha j^2 - \beta k^2}\,\mathrm{d}j\mathrm{d}k\,. \quad \text{(E.3.8)}$$

Here we make a change of variables of the form: $k' = \sqrt{\beta}\,k$, $j' = \sqrt{\alpha}\,j$, so that

$$0 \le j < \infty \quad \Longrightarrow \quad 0 \le j' < \infty, \quad \text{(E.3.9)}$$

and the limits $-j \le k \le +j$ imply, for example, that if $k = j$, then the corresponding value of k' is $k' = \sqrt{\beta}\,j$, and as $j = j'/\sqrt{\alpha}$, we see that k' is related to j' by $k' = \sqrt{\beta/\alpha}\,j'$. With these transformations, the required integral is

$$z_{\text{rot}} = \frac{2}{\sigma(\sqrt{\alpha})^2\sqrt{\beta}}\int_0^{\infty}\int_{-\sqrt{\beta/\alpha}\,j'}^{\sqrt{\beta/\alpha}\,j'} j' e^{-j'^2 - k'^2}\,\mathrm{d}k'\mathrm{d}j'\,. \quad \text{(E.3.10)}$$

We cannot proceed further with the integral in Eq. (E.3.10) unless we can find a transformation that will bring it into a form that we can integrate directly. If we look carefully at the nature of the function (the complication is in the form of the exponential), the fact that we have the sum of two squared quantities suggests that a transformation to something like a polar representation should help us to simplify this double integral. To this end let us introduce the transformation

$$k' \equiv R\sin\theta\,, \quad j' \equiv R\cos\theta\,, \quad \text{(E.3.11)}$$

for which the Jacobian of the transformation is

$$J = \begin{vmatrix} \dfrac{\partial k'}{\partial R} & \dfrac{\partial k'}{\partial \theta} \\[2ex] \dfrac{\partial j'}{\partial R} & \dfrac{\partial j'}{\partial \theta} \end{vmatrix}$$

$$= \begin{vmatrix} \sin\theta & R\cos\theta \\ \cos\theta & -R\sin\theta \end{vmatrix} \tag{E.3.12}$$

$$= R.$$

We have also to take care of the limits on the k' integration: since $k' = \sqrt{\beta/\alpha}\, j'$, we see that

$$\frac{k'}{j'} = \left(\frac{\beta}{\alpha}\right)^{\frac{1}{2}} = \tan\theta\,, \tag{E.3.13}$$

so that θ is given by

$$\theta = \tan^{-1}\sqrt{\frac{\beta}{\alpha}}\,. \tag{E.3.14}$$

Upon making this substitution, and recognizing that the argument of the k' integration is an even function of k', we obtain

$$z_{\mathrm{rot}}(T) = \frac{4}{\sigma\alpha\sqrt{\beta}} \int_0^\infty \int_0^{\sqrt{\frac{\beta}{\alpha}}\,j'} j'\mathrm{e}^{-j'^2 - k'^2}\, \mathrm{d}k'\mathrm{d}j'$$

$$= \frac{4}{\sigma\alpha\sqrt{\beta}} \int_0^\infty \int_0^{\tan^{-1}\sqrt{\frac{\beta}{\alpha}}} R\cos\theta\, \mathrm{e}^{-R^2\cos^2\theta - R^2\sin^2\theta}\, R\,\mathrm{d}\theta\mathrm{d}R\,,$$

which simplifies to

$$z_{\mathrm{rot}}(T) = \frac{4}{\sigma\alpha\sqrt{\beta}} \int_0^\infty \int_0^{\tan^{-1}\sqrt{\frac{\beta}{\alpha}}} R^2\mathrm{e}^{-R^2}\cos\theta\, \mathrm{d}\theta\mathrm{d}R \tag{E.3.15}$$

$$= \frac{4}{\sigma\alpha\sqrt{\beta}} \int_0^\infty R^2\mathrm{e}^{-R^2}\, \mathrm{d}R \int_0^{\tan^{-1}\sqrt{\frac{\beta}{\alpha}}} \cos\theta\, \mathrm{d}\theta\,, \tag{E.3.16}$$

so that, finally, we obtain the result

$$z_{\mathrm{rot}}(T) = \frac{\sqrt{\pi}}{\sigma}\, \frac{1}{\alpha\sqrt{\alpha+\beta}}$$

$$= \frac{\sqrt{\pi}}{\sigma} \left(\frac{2I_B k_{\mathrm{B}}T}{\hbar^2}\right) \left(\frac{2I_A k_{\mathrm{B}}T}{\hbar^2}\right)^{\frac{1}{2}}\,. \tag{E.3.17}$$

Appendix F
Review of Solid State Concepts

As there are some aspects of the solid state that need to be understood when dealing with semiconductors and their properties, we shall review briefly a few of the relevant concepts needed for our discussion. It is not our intent here to provide a rigorous treatment of the subject, but simply to attempt to give a more or less qualitative overview of some important aspects of the solid state. We shall do this by utilizing a model approach introduced by Feynman and known as the coupled modes approach. The essence of the Feynman approach is to divide a system into its components and study what happens to the system if the components are weakly coupled to one another. Thus, it is a type of perturbation approach (similar to the concept discussed in an introductory quantum mechanics course).

F.1 The Feynman Model and Energy Bands

We shall introduce the concept and the model by considering a simple homonuclear diatomic species and the formation of molecular energy levels associated with the molecular orbital description of molecules that you will have encountered in an introductory general chemistry course.

The simplest homonuclear diatomic molecule is the hydrogen molecular ion, H_2^+, which consists of two protons and one electron. It is simple enough for it to be solved exactly, and is the molecular analogon of the H atom for atoms. We begin with the time-dependent Schrödinger equation, which is

$$\left(-\frac{\hbar^2}{2m}\nabla^2 + \widehat{V} \right) \Psi(\mathbf{r}, t) = i\hbar\, \frac{\partial \Psi}{\partial t} \,, \tag{F.1.1}$$

© Springer Nature Switzerland AG 2021
F. R. W. McCourt, *Statistical Thermodynamics for Pure and Applied Sciences*,
https://doi.org/10.1007/978-3-030-52006-9

or, in operator notation using the Hamiltonian operator $\mathcal{H}$,

$$\mathcal{H}\Psi(\mathbf{r}, t) = \hbar \, \frac{\partial \Psi}{\partial t} \, . \tag{F.1.2}$$

We shall employ a slightly modified separation of variables approach, in which we write $\Psi(\mathbf{r}, t)$ as

$$\Psi(\mathrm{r}, t) = \sum_j w_j(t)\psi_j(\mathbf{r}) \, , \tag{F.1.3}$$

with $\Psi(\mathbf{r}, t)$ represented as a time-dependent superposition of a set of solutions of the related time-independent Schrödinger equation.

For our purposes, we shall ask what the probability is that an electron, say, is in state j at time t, focussing upon only the temporal variation and essentially ignoring the spatial variation. To do this, we eliminate the spatial dependence in the following way. We substitute Eq. (F.1.3) into the time-dependent Schrödinger equation to obtain

$$\sum_j w_j(t)\psi_j(\mathbf{r}) = i\hbar \sum_j \psi_j(\mathbf{r}) \, \frac{\mathrm{d}w_j}{\mathrm{d}t} \, , \tag{F.1.4}$$

then multiply both sides of this equation by $\psi_k^*(\mathbf{r})$ and integrate over all space:

$$\sum_j w_j(t) \int \psi_k^*(\mathbf{r})\mathcal{H}\psi_j(\mathbf{r}) \, \mathrm{d}\mathbf{r} = i\hbar \sum_j \frac{\mathrm{d}w_j}{\mathrm{d}t} \int \psi_k^*(\mathbf{r})\psi_j(\mathbf{r}) \, \mathrm{d}\mathbf{r} \, . \tag{F.1.5}$$

Recall from your quantum mechanics course that as $\psi_j(\mathbf{r})$ and $\psi_k(\mathbf{r})$ are eigensolutions of the corresponding time-independent Schrödinger equation with different energy eigenvalues, they are necessarily orthogonal, i.e., $\int \psi_k^*(\mathbf{r})\psi_j(\mathbf{r}) \, \mathrm{d}\mathbf{r} = \delta_{jk}$, provided that the functions $\{\psi_i(\mathbf{r})\}$ are properly normalized. Recall also that $\int \psi_k^*(\mathbf{r})\mathcal{H}\psi_j(\mathbf{r}) \, \mathrm{d}\mathbf{r} \equiv H_{kj}$ is a matrix element of the Hamiltonian operator [i.e., once we choose a basis set, operators are replaced by their matrix representations in that basis]. By doing this, we convert the Schrödinger equation into a set of differential equations for the $w_k(t)$ functions, namely,

$$i\hbar \, \frac{\mathrm{d}w_k}{\mathrm{d}t} = \sum_j H_{kj}w_j \, , \qquad k = 1, 2, 3, \dots \, . \tag{F.1.6}$$

In Feyman's approach, as we shall see in a moment, we do not actually need to determine the spatial component functions $\psi_i(\mathbf{r})$ nor do we have to evaluate the matrix elements H_{kj}; these latter quantities will be guessed at, based purely upon physical grounds.

We shall require a minimum of two states in order that we can speak of a coupling between states: let us therefore work with just such a minimum number of states for simplicity. We consider, therefore, the two equations

$$i\hbar \frac{dw_1}{dt} = H_{11}w_1 + H_{12}w_2\,,\qquad\qquad(\text{F.1.7})$$

$$i\hbar \frac{dw_2}{dt} = H_{21}w_1 + H_{22}w_2\,.\qquad\qquad(\text{F.1.8})$$

We note firstly that if $H_{12} = H_{21} = 0$, then states 1 and 2 are not coupled, and the solutions for them would be simply

$$w_k = C_k e^{-iH_{kk}t/\hbar}\,;\qquad k = 1, 2\,,\qquad\qquad(\text{F.1.9})$$

with corresponding probabilities for our system being in state k given by $|w_k(t)|^2 = |C_k|^2$. Hence, once the system is in state k, it will remain there forever (as the C_k are simply constants in this case).

For the H_2^+ molecular ion, these two solutions correspond pictorially to $\odot + $ and $+ \odot$, where $+$ represents a bare proton and $\odot$ represents a hydrogen atom, with its electron in one of the hydrogenic states [we shall also pare this description down to the H atom in its $1s$ state, again for the sake of simplicity]. What is meant by saying that the two states are uncoupled? Simply that if the electron is at proton 1 at the start, it will always be with proton 1, and similarly for the electron if it starts out with proton 2. For the two protons separated by a large distance, this is a realistic possibility: however, when the proton and H atom are brought closer and closer together, we know from quantum mechanics that the electron can tunnel through the potential barrier separating the two protons, so that as the proton and H atom approach one another, there is an increasing probability that the electron will be able to pass from the H atom to the bare proton: these two states are then said to be *coupled*, as they are no longer entirely independent of one another.

Weak coupling means that it remains meaningful to talk about one or the other state, because even though the two states influence one another, they still essentially preserve their separate identities. For H_2^+ we have $H_{11} = H_{22} = E_0$ and $H_{12} = H_{21} = -A$, so that our pair of differential equations becomes

$$i\hbar \frac{dw_1}{dt} = E_0 w_1 - A w_2\,,\qquad\qquad(\text{F.1.10})$$

$$i\hbar \frac{dw_2}{dt} = -A w_1 + E_0 w_2\,.\qquad\qquad(\text{F.1.11})$$

Let us try a solution of the form $w_k = C_k e^{-iEt/\hbar}$, $k = 1, 2$, and solve the resulting pair of algebraic equations for E (and also, in principle, for C_1 and C_2; however, we shall not need them here). We obtain

$$C_1 E = E_0 C_1 - A C_2 \,, \tag{F.1.12}$$

$$C_2 E = -A C_1 + E_0 C_2 \tag{F.1.13}$$

which can have a nontrivial solution for C_1 and C_2 only if the determinant of their coefficients vanishes, i.e., provided that

$$\begin{vmatrix} E_0 - E & -A \\ -A & E_0 - E \end{vmatrix} = 0 \tag{F.1.14}$$

or

$$(E - E_0)^2 - A^2 = 0 \quad \Rightarrow \quad E = E_0 \pm A \,. \tag{F.1.15}$$

From this result we see that the original energy E_0 of the uncoupled states has been replaced in the presence of the coupling by a symmetric splitting of the energy into the pair of energies $E_0 \pm A$. Whenever coupling occurs, the energy splits. For our H_2^+ example, if the H atom electron is a $1s$ electron, the two energy levels generated for H_2^+ will be what are referred to as the $1s\sigma_g$ bonding and $1s\sigma_u$ antibonding molecular orbital energy levels that you may have discussed in a general chemistry course.

Now we ask how A and E_0 behave for our model system. We know that the larger the coupling (i.e., magnitude of A), the larger the energy splitting will be, so that A must somehow be related to the probability that the electron can tunnel through the barrier between the two protons. Now, as we know that tunneling probabilities typically vary exponentially with the distance into the barrier, we may expect that A varies as on the left-hand side of Fig. F.1. Similarly, as E_0 consists of the potential plus kinetic energies of the electron, together with the potential energies of the two protons (for any given fixed distance between them), we may argue that when the two protons are far apart, their potential energies are essentially zero (or very very small), while the electron energy, because it is *bound* to one of the protons, is quite negative (recall that a $1s$-electron in an H atom is bound by 13.6 eV). This means that E_0 is negative for large interproton separations, but rises rapidly when the interproton separation is less than the average distance of the (fluctuating) electron from the protons: E_0 then behaves as on the left-hand side of Fig. F.1. If we form the two linear combinations $E = E_0 \pm A$ and display these values as functions of the interproton separation, we obtain the results displayed on the right-hand side of Fig. F.1.

F.2 Generalization of the Feynman Model to Solids

We shall now utilize the Feynman model to see how band structures occur in solids. We begin by asking what happens when N atoms are brought close together. Generalizing upon our discussion of the H_2^+ molecular ion, it is not unreasonable to expect that there will be an N-fold split in the energy levels: we should then

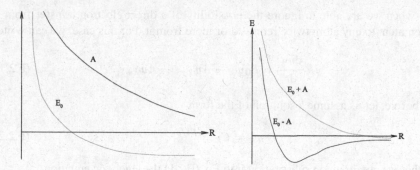

Fig. F.1 Behaviour of the coupling parameter A, the energy E_0, and the combinations $E_0 \pm A$ as functions of the interproton separation for H_2^+

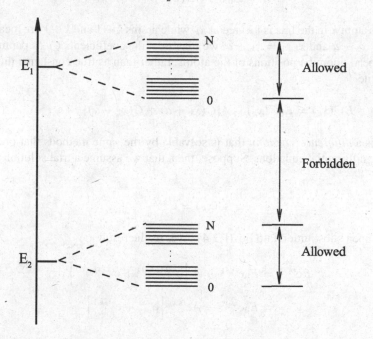

Fig. F.2 Formation of energy bands in an N-atom solid

expect to find clusters of energy levels, and we can see that it is reasonable that these clusters of levels would form allowed energy bands separated by forbidden energy zones (gaps) between them, as illustrated in Fig. F.2.

In an array of atoms, as occurs in a solid or, more simply, in a line of atoms (1D-case), we may anticipate that there is a finite probability that an electron can jump (transfer) from atom k to atom $k + 1$ or to atom $k - 1$, each separated by a distance a from atom k, with equal probability, but it is much less likely for that same electron to jump from atom k to atom $k + 2$ or to atom $k - 2$, and much much less likely for the electron to jump to atoms that are even farther away. Let us look at what we can say when we are able to ignore the possibility of a direct electron transfer from a given atom to any atom twice removed or more from it. For this case, we can write

$$i\hbar \frac{dw_k}{dt} = E_0 w_k - A w_{k-1} - A w_{k+1} .$$ (F.2.1)

As before, let us assume a solution of the form

$$w_k = C_k e^{-iEt/\hbar} ,$$ (F.2.2)

so that we obtain, upon substitution into Eq. (F.2.1) the algebraic equation

$$E C_k = E_0 C_k - A(C_{k-1} + C_{k+1}) .$$ (F.2.3)

Now, as atom k in the line is located at x_k while atoms $k-1$ and $k+1$ are located at $x_{k-1} = x_k - a$ and $x_{k+1} = x_k + a$, we may treat the coefficients C_k as parameters that depend upon the positions of the atoms, rather than as true constants: thus, we may write

$$E C(x_k) = E_0 C(x_k) - A[C(x_k + a) + C(x_k - a)] ,$$ (F.2.4)

which is a *difference equation* that is solvable by the same methods that one uses to solve differential equations. Suppose, then, that we assume a trial solution of the form

$$C(x_k) = e^{ikx_k}$$ (F.2.5)

which, upon substitution into Eq. (F.2.4) gives us the result

$$E e^{ikx_k} = E_0 e^{ikx_k} - A \left\{ e^{ik(x_k+a)} + e^{ik(x_k-a)} \right\}$$
$$= E_0 e^{ikx_k} - A e^{ikx_k} \left[e^{ika} + e^{-ika} \right]$$

or

$$E = E_0 - 2A \cos ka .$$ (F.2.6)

This result is shown in Fig. F.3, and we see that indeed the allowed energies fall into an energy band centred on energy E_0 and $4A$ wide: energies below $E_0 - 2A$ and above $E_0 + 2A$ are forbidden to the electrons. Notice that this argument could equally well apply to holes, and to other types of particles. Moreover, it is relatively readily generalizable to three-dimensional solids: rectangular structure, with distances a, b, c in the directions of the coordinate axes x, y, z, respectively. If we designate translations along the three axes by distance a, b, and c by **a**, **b**, **c**, we may also take advantage of the conciseness of employing vector notation.

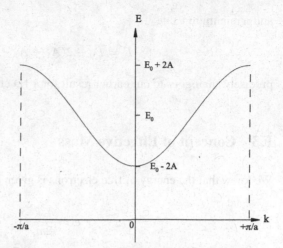

Fig. F.3 Behaviour of the electron energy as a function of wave number k for a one-dimensional crystal

By analogy with w_k for the 1D crystal, we have $w(\mathbf{r})$, and the probability that an electron is attached to the atom at point $(x, y, z) \equiv \mathbf{r}$ is given by $|w(\mathbf{r})|^2$. We then obtain the result

$$i\hbar \, \frac{\partial w(\mathbf{r}, t)}{\partial t} = E_0 w(\mathbf{r}, t) - A_x[w(\mathbf{r} + \mathbf{a}, t) + w(\mathbf{r} - \mathbf{a}, t)]$$

$$- A_y[w(\mathbf{r} + \mathbf{b}, t) + w(\mathbf{r} - \mathbf{b}, t)] - A_z[w(\mathbf{r} + \mathbf{c}, t) + w(\mathbf{r} - \mathbf{c}, t)] \,.$$

Now let us attempt a trial solution of the type

$$w(\mathbf{r}, t) = e^{-iEt/\hbar} e^{i\mathbf{k}\cdot\mathbf{r}} \,.$$

Substitution of this assumed trial solution into our differential equation for $w(\mathbf{r}, t)$ gives

$$i\hbar \left(-\frac{iE}{\hbar}\right) = E_0 w(\mathbf{r}, t) - A_x \left(e^{i\mathbf{k}\cdot\mathbf{a}} + e^{-i\mathbf{k}\cdot\mathbf{a}}\right) w(\mathbf{r}, t)$$

$$- A_y \left(e^{i\mathbf{k}\cdot\mathbf{b}} + e^{-i\mathbf{k}\cdot\mathbf{a}}\right) w(\mathbf{r}, t) - A_x \left(e^{i\mathbf{k}\cdot\mathbf{c}} + e^{-i\mathbf{k}\cdot\mathbf{c}}\right) w(\mathbf{r}, t)$$

or, upon eliminating the common factors from the two sides,

$$E = E_0 - 2A_x \cos \mathbf{k} \cdot \mathbf{a} - 2A_y \cos \mathbf{k} \cdot \mathbf{b} - A_z \cos \mathbf{k} \cdot \mathbf{c} \,.$$

This result tells us that for an electron in a 3D solid, its energy is restricted to lie between the maximum value

$$E = E_0 + 2(A_x + A_y + A_z)$$

and a minimum value

$$E = E_0 - 2(A_x + A_y + A_z) \,,$$

precisely analogous to our earlier result for a 1D crystal.

F.3 Concept of Effective Mass

We know that the energy of free electrons is given by

$$\epsilon = \frac{\hbar^2 k^2}{2m_e} \,, \tag{F.3.1}$$

from which we see that the coefficient of k^2 determines the curvature of the ϵ vs. k curves. As the energy for free electrons is simply quadratic, the curvature is constant and has the value $\hbar^2/m_e$. More generally, we may restate this by saying that the reciprocal mass, $1/m_e$, determines the curvature at any given point in k-space, and require that a statement of this type applies also to electrons in conduction bands of a solid, even though they are no longer free electrons. We have seen that according to the Feynman model, the energy of an electron in a band in a solid is, in the weak-coupling limit, given by

$$\epsilon = \epsilon_0 - 2A \cos(ka) \,, \tag{F.3.2}$$

with A the coupling constant. The relationship between the energy of a wave and its wave vector is called the dispersion relation: the dispersion relation for free electrons is thus simply $\epsilon = \hbar^2 k^2/(2m_e)$, while that for an electron in an energy band in a solid (in the weak-coupling case) is as given above in Eq. (F.3.2). Let us see what this form for the dispersion relation can tell us.

The curvature of the energy vs. wave number plot for electrons in the energy band will be given from Eq. (F.3.2) as

$$\frac{d^2\epsilon}{dk^2} = 2a^2 A \cos(ka) \,, \tag{F.3.3}$$

for any value of k lying between $\pm\frac{\pi}{a}$. Let us focus for the moment upon what happens near the bottom of the energy band, where we may approximate the energy of a conduction electron by

$$\epsilon \simeq \epsilon_0 - 2A \left(1 - \tfrac{1}{2}k^2 a^2 + \cdots \right)$$

$$= (\epsilon_0 - 2A) + Aa^2 k^2 + \cdots \tag{F.3.4}$$

for $ka \ll 1$. The term in brackets just identifies the bottom of the conduction band, so we can replace $\epsilon_0 - 2A$ by E_c, so that we have

$$\epsilon = E_c + Aa^2k^2 + \cdots , \tag{F.3.5}$$

from which we can see that conduction electrons near the bottom of the band would behave like free electrons, but with an *effective mass* m^* defined as

$$\frac{\hbar^2}{2m^*} \equiv Aa^2 . \tag{F.3.6}$$

We might wonder whether this effective mass m^* is different from the value m_e for a free electron. We can get some idea of how different it can be by making an order-of-magnitude estimate using values for A and a that are reasonable for a typical metal or semiconductor. If we take A to be of order 10^{-20} J to 10^{-21} J, and a a typical lattice distance of order 3×10^{-10} m, then we can estimate the effective mass m^* as

$$m^* = \frac{\hbar^2}{Aa^2} \simeq \frac{(1.055 \times 10^{-34})^2 \mathrm{J}^2 \mathrm{s}^2}{2 \times (5 \times 10^{-20}\,\mathrm{J}) \times (3 \times 10^{-10}\,\mathrm{m})^2}$$

$$\approx \frac{1.12 \times 10^{-68}\ \mathrm{kg\,m}^2}{10^{-19} \times 9 \times 10^{-20}\,\mathrm{m}^2} \approx 1.2 \times 10^{-30}\,\mathrm{kg} ,$$

which can be compared with the free electron mass $m_e \simeq 9.1 \times 10^{-31}$ kg. For this particular calculation, we see that the effective mass is about $1.3m_e$: in practice, the effective masses of electrons in energy bands in solids can be anywhere from 100 times smaller to 100 times greater than m_e! More generally, the effective mass of an electron in a band is defined in terms of the curvature of the dispersion relation via

$$m^* = \hbar^2 / \left(\frac{\mathrm{d}^2\epsilon}{\mathrm{d}k^2} \right) . \tag{F.3.7}$$

Thus, by analogy with a free electron which has a well-defined mass, m_e, and which, when subjected to an applied electric field $\mathbf{E}$ is accelerated according to Newtonian mechanics, an electron within a crystalline solid can also be considered to be accelerated by an electric field in a Newtonian manner, but with a mass m^*, rather than the free electron mass m_e.

We know that if we think of an electron as a particle with a well-defined location (remember, that we say that such an electron is localized), then we must in quantum mechanics necessarily think of the electron as being represented by a wave packet, i.e., a superposition of a very large number of waves with rather similar de Broglie wavelengths. The electron travels with the group velocity associated with the wave packet, rather than with any of the phase velocities associated with the individual waves that make up the wave packet. While the group velocity is traditionally defined by the derivative of the angular frequency with respect to wave number via

$$v_g \equiv \frac{d\omega}{dk} \,, \qquad \omega = \frac{\epsilon}{\hbar} \,,$$

it can be written for our purposes as

$$v_g = \frac{1}{\hbar} \frac{d\epsilon}{dk} \,. \tag{F.3.8}$$

If we combine Newton's second law

$$\mathbf{F} = \frac{d\mathbf{p}}{dt} \,,$$

with the relation $\mathbf{p} = \hbar\mathbf{k}$ between linear momentum and wave number, we have

$$\mathbf{F} = \hbar \frac{d\mathbf{k}}{dt} \,, \tag{F.3.9}$$

as a relevant expression of Newton's law. Now, let us consider how the group velocity that we have associated with an electron in a solid behaves: we have, upon taking the time derivative of Eq. (F.3.8)

$$\frac{dv_g}{dt} = \frac{1}{\hbar} \frac{d}{dt} \left(\frac{d\epsilon}{dk} \right) = \frac{1}{\hbar} \left(\frac{d^2\epsilon}{dk^2} \right) \left(\frac{dk}{dt} \right) \,, \tag{F.3.10}$$

so that upon replacing dk/dt by F, we obtain

$$\frac{dv_g}{dt} = \left(\frac{1}{\hbar^2} \frac{d^2\epsilon}{dk^2} \right) F \,,$$

or

$$F = \frac{\hbar^2}{(d^2\epsilon/dk^2)} \frac{dv_g}{dt} \,. \tag{F.3.11}$$

This is a form of Newton's second law provided that we may define an effective mass m^* such that

$$F = m^* \frac{dv_g}{dt} \,. \tag{F.3.12}$$

This is precisely the same criterion that we have used earlier to define an effective mass for electrons in an energy band in a solid.

We may summarize much of what we have said above by returning briefly to an electron in a 1D lattice, for which we have seen that

$$\epsilon(k) = \epsilon_0 - 2A \cos(ka) \,, \tag{F.3.13}$$

so that the electron group velocity v_g is given by

$$v_g(k) = \frac{1}{\hbar} \frac{d\epsilon}{dk} = \frac{2Aa}{\hbar} \sin(ka), \quad\quad\quad (F.3.14)$$

with a corresponding reduced mass m^*, which is also a function of k, given by

$$m^*(k) = \frac{\hbar^2}{2Aa^2} \frac{1}{\cos(ka)}. \quad\quad\quad (F.3.15)$$

Let us briefly take a side trip to see how electrons in an energy band in a solid react to the application of an electric field of magnitude E in order to determine the effective number of electrons available for carrying an electric current. An applied electric field will impose a force eE on each electron, so that the equation of motion (F.3.11) for an electron will become

$$\frac{dv_g}{dt} = \frac{1}{\hbar} \frac{d^2\epsilon}{dk^2} eE. \quad\quad\quad (F.3.16)$$

The effect of accelerating all available electrons, each carrying charge e gives the rate of change of electric current I_0 that flows initially when an electric field is applied.[1] We may therefore write figuratively

$$\frac{dI_0}{dt} = \sum \frac{d}{dt} (ev_g)$$

$$= \frac{1}{\hbar^2} e^2 E \sum \frac{d^2\epsilon}{dk^2},$$

from which, if we replace the summation by an equivalent integration over a continuum of wave vectors, we find that

$$\frac{dI_0}{dt} = \frac{1}{\hbar^2} e^2 E \frac{1}{\pi} \int \frac{d^2\epsilon}{dk^2} dk. \quad\quad\quad (F.3.17)$$

For N noninteracting *free electrons*, we obtain simply

$$\frac{dI_0}{dt} = \frac{e^2 E}{m_e} N, \quad\quad\quad (F.3.18)$$

---

[1]This is not the rate of change of the electric current under stationary conditions: to apply this to the steady-state, collisions must also be considered, as the flow of electric current is inherently a nonequilibrium phenomenon.

while for electrons in a crystal, we have instead

$$\frac{\mathrm{d}I_0}{\mathrm{d}t} = \frac{1}{\hbar^2}\, e^2 E\, \frac{1}{\pi} \int_{-k_a}^{k_a} \frac{\mathrm{d}^2\epsilon}{\mathrm{d}k^2}\, \mathrm{d}k$$

$$\equiv \frac{e^2 E}{m_e}\, N_{\text{eff}}. \tag{F.3.19}$$

We see from this expression that N_{eff} is given by

$$N_{\text{eff}} = \frac{1}{\pi}\, \frac{m_e}{\hbar^2} \left\{ \left(\frac{\mathrm{d}\epsilon}{\mathrm{d}k}\right)_{k=k_a} - \left(\frac{\mathrm{d}\epsilon}{\mathrm{d}k}\right)_{k=-k_a} \right\}$$

$$= \frac{2}{\pi}\, \frac{m_e}{\hbar^2} \left(\frac{\mathrm{d}\epsilon}{\mathrm{d}k}\right)_{k=k_a}. \tag{F.3.20}$$

This is a very important result, as it tells us that the effective number of electrons capable of contributing to electrical conduction depends upon the slope of the ϵ vs. k curve at the highest occupied energy level. As $\dfrac{\mathrm{d}\epsilon}{\mathrm{d}k} = 0$ at the highest level in the band, we can conclude that the effective number of electrons for a full band is zero, and hence for a filled energy band, there is no electrical conduction.

Appendix G
Hamiltonian Mechanics

G.1   Generalized Coordinates and Momenta

Generalized coordinates: any set of coordinates by means of which the position of the particles in a system may be specified.

While it is possible to set up Newton's equations of motion in terms of Cartesian components for each specific system with which we deal and then change over into generalized coordinates, it is both desirable and, ultimately, convenient to have a method for setting up equations of motion directly in terms of some convenient set of generalized coordinates, as was first accomplished by Lagrange in the eighteenth century.

Obviously, the number of coordinates in the new system must be the same as the number of Cartesian coordinates for all of the particles involved in the system of N particles under consideration, i.e., $3N$. Traditionally, each coordinate is designated by q_i, so that a set of $3N$ generalized coordinates (three coordinates for each particle) can be written as $q_1, q_2, \cdots, q_{3N}$.

Example A particle confined to move in a plane may be described by two generalized coordinates (q_1, q_2), which could be specifically Cartesian coordinates (x, y) or polar coordinates (r, θ) or, similarly, for a particle moving in the full 3-space, with generalized coordinates (q_1, q_2, q_3), or specifically, Cartesian coordinates (x, y, z), spherical polar coordinates (r, θ, ϕ), or cylindrical coordinates (r, ϕ, z).   □

For every configuration of an N-particle system, the generalized coordinates must have a definite set of values, each coordinate q_i will be a function of the set of Cartesian coordinates, plus time (for moving coordinate systems), so that we may write q_i as

$$q_i = q_i(x_1, y_1, z_1, \cdots, z_N; t), \qquad i = 1, \cdots, 3N. \qquad (G.1.1)$$

© Springer Nature Switzerland AG 2021

F. R. W. McCourt, *Statistical Thermodynamics for Pure and Applied Sciences*,
https://doi.org/10.1007/978-3-030-52006-9

Similarly, it must be possible to express the Cartesian coordinates in terms of the generalized coordinates as

$$x_i = x_i(q_1, q_2, \cdots, q_{3N}; t), y_i = y_i(q_1, q_2, \cdots, q_{3N}; t), z_i = z_i(q_1, q_2, \cdots, q_{3N}; t).$$
$$(G.1.2)$$

If one set of these equations can be constructed, it can be inverted (solved) to obtain the other set. As you will know from linear algebra and calculus courses (see also Sect. B.2 of Appendix B), such a solution is only possible if the Jacobian determinant

$$\frac{\partial(q_1, \cdots, q_{3N})}{\partial(x_1, y_1, z_1, \cdots, z_N)} \equiv \begin{vmatrix} \dfrac{\partial q_1}{\partial x_1} & \dfrac{\partial q_2}{\partial x_1} & \cdots & \dfrac{\partial q_{3N}}{\partial x_1} \\[2mm] \dfrac{\partial q_1}{\partial y_1} & \dfrac{\partial q_2}{\partial y_1} & \cdots & \dfrac{\partial q_{3N}}{\partial y_1} \\[1mm] \vdots & \vdots & \vdots & \vdots \\[1mm] \dfrac{\partial q_1}{\partial z_1} & \dfrac{\partial q_2}{\partial z_1} & \cdots & \dfrac{\partial q_{3N}}{\partial z_1} \end{vmatrix}$$

does not vanish at any point. We shall, however, treat this condition here only as a formal requirement.

Extended Example Two-dimensional motion in a plane expressed in terms of Cartesian and plane polar coordinates for comparison.

As you will recall, Cartesian and plane polar coordinates are related via

$$x = r\cos\theta, \quad y = r\sin\theta, \quad r = (x^2 + y^2)^{\frac{1}{2}}, \quad \theta = \tan^{-1}\frac{y}{x}.$$

We shall employ unit vectors $\mathbf{i}$ and $\mathbf{j}$ in the Cartesian system and unit vectors $\mathbf{n}$ and $\mathbf{l}$ in the polar coordinate system. Remember that the unit vectors $\mathbf{i}$ and $\mathbf{j}$, defined in the directions of increasing x and y, respectively, are constant vectors, and are thus independent of their location, while the unit vectors $\mathbf{n}$ and $\mathbf{l}$ associated with the directions of increasing r and θ, respectively, are functions of θ, though not of r.

It may be deduced from Fig. G.1 that

$$\mathbf{n} = \mathbf{i}\cos\theta + \mathbf{j}\sin\theta, \quad \mathbf{l} = -\mathbf{i}\sin\theta + \mathbf{j}\cos\theta,$$

from which we see that

$$\frac{d\mathbf{n}}{d\theta} = \mathbf{l}; \quad \frac{d\mathbf{l}}{d\theta} = -\mathbf{n}.$$

We shall now use the representation of $\mathbf{r}$ in the form $\mathbf{r} = r\mathbf{n}(\theta)$ to study the motion of our particle in plane polar coordinates. The velocity vector is given as

Fig. G.1 Polar coordinate representation of motion in a plane

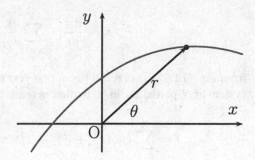

$$\mathbf{v} = \frac{d\mathbf{r}}{dt} = \frac{dr}{dt}\mathbf{n} + r\frac{d\mathbf{n}}{d\theta}\frac{d\theta}{dt} = \dot{r}\,\mathbf{n} + r\dot{\theta}\mathbf{l} \equiv \mathbf{v}(v_r, v_\theta)\,,$$

so that v_r and v_θ are given by $v_r = \dot{r}$ and $v_\theta = r\dot{\theta}$.[1]

Similarly, we may determine the acceleration of our particle as

$$\mathbf{a} = \frac{d\mathbf{v}}{dt} = \ddot{r}\mathbf{n} + \dot{r}\frac{d\mathbf{n}}{d\theta}\frac{d\theta}{dt} + \dot{r}\dot{\theta}\mathbf{l} + r\ddot{\theta}\mathbf{l} + \dot{\theta}\frac{d\mathbf{l}}{d\theta}\frac{d\theta}{dt}$$

$$= (\ddot{r} - r\dot{\theta}^2)\mathbf{n} + (r\ddot{\theta} + 2\dot{r}\dot{\theta})\mathbf{l}\,,$$

so that the polar coordinates of acceleration are

$$a_r = \ddot{r} - r\dot{\theta}^2\,, \qquad a_\theta = r\ddot{\theta} + 2\dot{r}\dot{\theta}\,.$$

We note in passing that $r\dot{\theta}^2 = v_\theta^2/r$ is called the *centripetal acceleration* arising from motion in the θ-direction, and the term $2\dot{r}\dot{\theta}$ is often called the *Coriolis acceleration*. The special case in which $\ddot{r} = \dot{r} = 0$ gives the path of a circle, with $a_r = -v_\theta^2/r$.

Exercise Show that the kinetic energy T has the structure

$$T = \tfrac{1}{2}m\dot{x}^2 + \tfrac{1}{2}m\dot{y}^2 = \tfrac{1}{2}m\dot{r}^2 + \tfrac{1}{2}m\dot{\theta}^2\,,$$

so that it is expressed as the sum of quadratic terms in both the Cartesian and polar coordinate systems. □

If a system of particles is described by a set of generalized coordinates $q_1, \cdots, q_{3N}$, we call the time derivative $\dot{q}_i$ the *generalized velocity* associated with this coordinate. For example, we may write

---

[1]Note that $\dot{r} \equiv \dfrac{dr}{dt}$ ($\dot{r}$ is the notation originally introduced by Newton and still used extensively in classical physics).

$$\dot{x}_i \equiv \frac{\mathrm{d}x_i}{\mathrm{d}t} = \sum_{j=1}^{3N} \frac{\partial x_i}{\partial q_j} \dot{q}_j + \frac{\partial x_i}{\partial t} \tag{G.1.3}$$

from Eq. (G.1.2). If we recall [from first year physics] that the kinetic energy T of a system of N particles can be written in terms of Cartesian coordinates as

$$T = \sum_{i=1}^{N} \tfrac{1}{2} m_i (\dot{x}_i^2 + \dot{y}_i^2 + \dot{z}_i^2) \equiv \sum_{i=1}^{N} \tfrac{1}{2} m_i \dot{\mathbf{r}}_i^2 , \tag{G.1.4}$$

then we can obtain the kinetic energy of the N particles in terms of generalized coordinates from Eqs. (G.1.3) and (G.1.4) as

$$T = \sum_{j=1}^{3N} \sum_{i=1}^{3N} \tfrac{1}{2} A_{ij} \dot{q}_i \dot{q}_j + \sum_{j=1}^{3N} B_i \dot{q}_i + T_0 , \tag{G.1.5}$$

with A_{ij}, B_i, and T_0 defined as

$$A_{ij} \equiv \sum_{k=1}^{N} m_k \left(\frac{\partial x_k}{\partial q_i} \frac{\partial x_k}{\partial q_j} + \frac{\partial y_k}{\partial q_i} \frac{\partial y_k}{\partial q_j} + \frac{\partial z_k}{\partial q_i} \frac{\partial z_k}{\partial q_j} \right) ,$$

$$B_i \equiv \sum_{k=1}^{N} m_k \left(\frac{\partial x_k}{\partial q_i} \frac{\partial x_k}{\partial t} + \frac{\partial y_k}{\partial q_i} \frac{\partial y_k}{\partial t} + \frac{\partial z_k}{\partial q_i} \frac{\partial z_k}{\partial t} \right) , \tag{G.1.6}$$

$$T_0 \equiv \sum_{k=1}^{N} \tfrac{1}{2} m_k \left[\left(\frac{\partial x_k}{\partial t} \right)^2 + \left(\frac{\partial y_k}{\partial t} \right)^2 + \left(\frac{\partial y_k}{\partial t} \right)^2 \right] ,$$

These quantities are functions of the coordinates $q_1, \cdots, q_{3N}$ and, for moving coordinate systems, also of t. If $A_{ij} = 0$ unless $i = j$, the coordinates are said to be *orthogonal coordinates*; the coefficients B_i and T_0 are zero when t does not occur explicitly in Eqs. (G.1.1).

The kinetic energy is thus in general the sum of three terms, one quadratic in, one linear in, and one independent of, the generalized velocities. Again, note that B_1 and T_0 appear only in moving coordinate systems; for fixed coordinate systems, which is all that we shall consider here, the kinetic energy is quadratic in the generalized velocities.

From Eq. (G.1.4) the Cartesian components of the linear momentum of particle i are

$$p_{ix} = m_i \dot{x}_i = \frac{\partial T}{\partial \dot{x}_i} , \quad p_{iy} = m_i \dot{y}_i = \frac{\partial T}{\partial \dot{y}} , \quad p_{iz} = m_i \dot{z}_i = \frac{\partial T}{\partial \dot{z}_i} , \tag{G.1.7}$$

which may be compared with the polar components of the momentum for a particle (labelled i) moving in a plane, viz.,

$$p_{ri} = \frac{\partial T}{\partial \dot{r}_i} = m_i \dot{r}_i , \quad p_{\theta i} = \frac{\partial T}{\partial \dot{\theta}_i} = mr^2 \dot{\theta}_i ,$$

in which p_{r_i} is the component of linear momentum in the direction of increasing r_i and p_{θ_i} is the angular momentum about the origin.

For any coordinate q_i that measures the linear displacement of any particle (or group of particles) in a given direction, the linear momentum of that particle (or group of particles) in the given direction is

$$p_i = \frac{\partial T}{\partial \dot{q}_i} , \tag{G.1.8}$$

while for any coordinate q_j that measures the angular displacement of a particle (or group of particles) about an axis, the corresponding angular momentum about that axis is given by $\partial T / \partial \dot{q}_j$.

Definition: the generalized momentum p_k associated with the coordinate q_k is defined as

$$p_k \equiv \frac{\partial T}{\partial \dot{q}_k} . \tag{G.1.9}$$

There are two special cases: q_i a distance gives p_i as linear momentum, and q_j an angle gives p_j as angular momentum.

G.2 Lagrangian Formulation

G.2.1 Generalized Forces

If forces $\mathbf{F}_1, \cdots, \mathbf{F}_N$ act on the particles in an N-particle system, the work done in moving the particles from $\mathbf{r}_i$ to $\mathbf{r}_i + \delta\mathbf{r}_i$ will be given by

$$\delta W = \sum_{i=1}^{N} \mathbf{F}_i \cdot \delta\mathbf{r}_i , \quad \text{with} \quad \delta\mathbf{r}_i = \sum_{j=1}^{N} \frac{\partial \mathbf{r}_i}{\partial q_j} \delta q_j . \tag{G.2.1}$$

We say that the set of $3N$ differences in the generalized coordinates represent a *virtual displacement* of the system: virtual because it may not be necessary that it represent any actual motion of the system. Equivalently, the work, δW, done in creating the virtual displacement may also be expressed in terms of the displacements δq_k as

$$\delta W = \sum_{k=1}^{3N} \mathcal{F}_k \, \delta q_k , \tag{G.2.2}$$

with $\mathcal{F}_k$ given by

$$\mathcal{F}_k = \sum_{i=1}^{N} \left(F_{ix} \frac{\partial x_i}{\partial q_k} + F_{iy} \frac{\partial y_i}{\partial q_k} + F_{iz} \frac{\partial z_i}{\partial q_k} \right)$$

$$= \sum_{i=1}^{N} \left(\mathbf{F}_i \cdot \frac{\partial \mathbf{r}_i}{\partial q_k} \right) . \tag{G.2.3}$$

The $\mathcal{F}_k$ thus defined are referred to as *generalized forces* associated with the coordinates q_k. Should the forces $\mathbf{F}_i$ be derivable from a potential energy function $V(\mathbf{r}_1, \mathbf{r}_2, \cdots, \mathbf{r}_N)$, then $\delta W = -\delta V$ and

$$\delta V = \sum_{i=1}^{N} \left(\delta \mathbf{r}_i \cdot \frac{\partial}{\partial \mathbf{r}_i} \right) , \qquad \frac{\partial}{\partial \mathbf{r}_i} \equiv \nabla_{r_i} . \tag{G.2.4}$$

This results in δW being related to $V(\mathbf{r}_1, \cdots, \mathbf{r}_N)$ via

$$\delta W = -\sum_{k=1}^{N} \frac{\partial V}{\partial q_k} \delta q_k , \tag{G.2.5}$$

so that

$$\mathcal{F}_k = -\frac{\partial V}{\partial q_k} . \tag{G.2.6}$$

Example For the motion of a particle in a plane and acted upon by a force $\mathbf{F} = \mathbf{i}F_x + \mathbf{j}F_y = \mathbf{n}F_r + \mathbf{l}F_\theta$, we obtain

$$\mathcal{F}_r = F_x \frac{\partial x}{\partial r} + F_y \frac{\partial y}{\partial r}$$

$$= F_x \cos\theta + F_y \sin\theta \equiv F_r ;$$

$$\mathcal{F}_\theta = F_x \frac{\partial x}{\partial \theta} + F_y \frac{\partial y}{\partial \theta}$$

$$= -r F_x \sin\theta + r F_y \cos\theta \equiv r F_\theta .$$

Note that $\mathcal{F}_r$ is the component of force in the r-direction, while $\mathcal{F}_\theta$ is the torque acting to increase θ.   □

G.2.2 The Lagrange Equations

Let us determine the time rate of change of the generalized momentum p_k as

$$\frac{dp_k}{dt} = \frac{d}{dt}\left(\frac{\partial T}{\partial \dot{q}_k}\right). \tag{G.2.7}$$

Newton's equations of motion in Cartesian form are

$$m_i \ddot{\mathbf{r}}_i = \mathbf{F}_i, \qquad (i = 1, \cdots, N). \tag{G.2.8}$$

We may utilize the form

$$T = \sum_{k=1}^{N} \tfrac{1}{2} m_i \dot{\mathbf{r}}_i^2$$

for the kinetic energy T to obtain the expression

$$\frac{\partial T}{\partial \dot{q}_k} = \sum_{i=1}^{N} m_i \left(\dot{\mathbf{r}}_i \cdot \frac{\partial \dot{\mathbf{r}}_i}{\partial \dot{q}_k}\right) \tag{G.2.9}$$

for the derivative of the kinetic energy with respect to the generalized velocity $\dot{q}_k$ that enters into the defining relation Eq. (G.2.7) for the time rate of change of the generalized momentum.

Let us examine briefly now an arbitrary component of the derivative $\dfrac{\partial \dot{\mathbf{r}}_i}{\partial \dot{q}_k}$, say $\dfrac{\partial \dot{x}_i}{\partial \dot{q}_k}$: from Eq. (G.1.3), we may write

$$\frac{\partial \dot{x}_i}{\partial \dot{q}_k} = \sum_{j=1}^{3N} \frac{\partial}{\partial \dot{q}_k}\left[\frac{\partial x_i}{\partial q_j}\,\dot{q}_j\right] + \frac{\partial}{\partial \dot{q}_k}\left(\frac{\partial x_i}{\partial t}\right) = \sum_{j=1}^{3N} \frac{\partial x_i}{\partial q_j}\,\delta_{jk} = \frac{\partial x_i}{\partial q_k}.$$

We obtain this result because the Cartesian coordinates x_i and their derivatives with respect to the generalized coordinates q_i are functions only of the q_i and t, while the time derivatives $\dot{x}_i$ [through Eqs. (G.1.3)] are also functions of the time derivatives $\dot{q}_j$ of the generalized coordinates. This result allows us to replace $\dfrac{\partial \dot{\mathbf{r}}_i}{\partial \dot{q}_k}$ in Eq. (G.2.9) with $\dfrac{\partial \mathbf{r}_i}{\partial q_k}$. Once we have carried out this step, we may differentiate Eq. (G.2.9) with respect to t to obtain

$$\frac{\mathrm{d}p_k}{\mathrm{d}t} = \sum_{i=1}^{N} m_i \left[\frac{\mathrm{d}\dot{\mathbf{r}}_i}{\mathrm{d}t} \cdot \frac{\partial \mathbf{r}_i}{\partial q_k} + \dot{\mathbf{r}}_i \cdot \frac{d}{dt}\left(\frac{\partial \mathbf{r}_i}{\partial q_k} \right) \right]$$

$$= \sum_{i=1}^{N} m_i \left[\ddot{\mathbf{r}}_i \cdot \frac{\partial \mathbf{r}_i}{\partial q_k} + \dot{\mathbf{r}}_i \cdot \frac{d}{dt}\left(\frac{\partial \mathbf{r}_i}{\partial q_k} \right) \right]. \tag{G.2.10}$$

The first term can now be simplified using Newton's equations in the form of Eq. (G.2.4) to obtain the kth component of the generalized force $\mathcal{F}$ as

$$\sum_{i=1}^{N} \left(m_i \ddot{\mathbf{r}}_i \cdot \frac{\partial \mathbf{r}_i}{\partial q_k} \right) = \sum_{i=1}^{N} \left(\mathbf{F}_i \cdot \frac{\partial \mathbf{r}_i}{\partial q_k} \right) = \mathcal{F}_k.$$

To simplify the second term of Eq. (G.2.10), let us note that

$$\frac{d}{dt}\frac{\partial}{\partial q_k} = \left[\sum_{j=1}^{3N} \dot{q}_j \frac{\partial}{\partial q_j} + \frac{\partial}{\partial t} \right] \frac{\partial}{\partial q_k} = \sum_{j=1}^{3N} \dot{q}_j \frac{\partial^2}{\partial q_j \partial q_k} + \frac{\partial^2}{\partial t \partial q_k}$$

$$= \frac{\partial}{\partial q_k}\left[\sum_{j=1}^{3N} \dot{q}_j \frac{\partial}{\partial q_j} + \frac{\partial}{\partial t} \right]$$

$$= \frac{\partial}{\partial q_k}\frac{d}{dt},$$

so that

$$\frac{d}{dt}\left(\frac{\partial \mathbf{r}_i}{\partial q_k} \right) = \frac{\partial}{\partial q_k}\left(\frac{\mathrm{d}\mathbf{r}_i}{\mathrm{d}t} \right) = \frac{\partial \dot{\mathbf{r}}_i}{\partial q_k}.$$

With this result, the second term in Eq. (G.2.10) becomes

$$\sum_{i=1}^{N} m_i \dot{\mathbf{r}}_i \cdot \frac{\partial \dot{\mathbf{r}}_i}{\partial q_k} = \frac{\partial}{\partial q_k}\left(\sum_{i=1}^{N} \tfrac{1}{2} m_i \dot{\mathbf{r}}_i^2 \right) = \frac{\partial T}{\partial q_k}.$$

Our desired expression for the rate of change of the generalized momentum thus becomes

$$\frac{\mathrm{d}p_k}{\mathrm{d}t} = \mathcal{F}_k + \frac{\partial T}{\partial q_k}, \qquad (k = 1, \cdots, 3N). \tag{G.2.11}$$

We note that this expression differs from the form taken by Newton's laws for rectilinear motion (due to the second term on the right-hand side of the expression).

Example Let us return briefly to our extended example involving nonrectilinear motion in a plane. We have seen that the kinetic energy can be expressed in plane polar coordinates as

$$T = \tfrac{1}{2}m(\dot{r}^2 + r^2\dot{\theta}^2),$$

which is a function of r, $\dot{r}$, and $\dot{\theta}$. Thus, we find that

$$\frac{\partial T}{\partial r} = \frac{\partial}{\partial r}\left[\tfrac{1}{2}m(\dot{r}^2 + r^2\dot{\theta}^2)\right] = m\dot{r}\,\frac{\partial \dot{r}}{\partial r} + mr\dot{\theta}^2 + mr^2\dot{\theta}\,\frac{\partial \dot{\theta}}{\partial r},$$

or

$$\frac{\partial T}{\partial r} = mr\dot{\theta}^2.$$

Now, we examine the equation of motion for p_r, which becomes

$$\frac{\mathrm{d}p_r}{\mathrm{d}t} = \frac{\mathrm{d}}{\mathrm{d}t}(m\dot{r}) = m\ddot{r}$$

$$= \mathcal{F}_r + \frac{\partial T}{\partial r},$$

or

$$m\ddot{r} = F_r + mr\dot{\theta}^2.$$

The second term in this result represents the so-called *centrifugal force*; components of this type always appear when the coordinate system employed in the description of the dynamics involves curvilinear coordinates (occurring when constant generalized velocities $\dot{q}_i$ result in curved motions in some parts of the mechanical system). □

Equations (G.2.11) are normally written down in the form

$$\frac{\mathrm{d}}{\mathrm{d}t}\left(\frac{\partial T}{\partial \dot{q}_k}\right) - \frac{\partial T}{\partial q_k} = \mathcal{F}_k, \qquad (k = 1, \cdots, 3N), \tag{G.2.12}$$

introduced by Lagrange in the 1770s. When the generalized forces $\mathcal{F}_k$ are derivable from a potential energy function $V(q_1, \cdots, q_{3N})$, a quantity referred to as the Lagrangian function, and designated by $\mathcal{L}(q_1, \cdots, q_{3N}; \dot{q}_1, \cdots, \dot{q}_{3N}; t)$ is introduced via

$$\mathcal{L}(q_1, \cdots, q_{3N}; \dot{q}_1, \cdots, \dot{q}_{3N}; t) \equiv T(q_1, \cdots, q_{3N}; \dot{q}_1, \cdots, \dot{q}_{3N}) - V(q_1, \cdots, q_{3N}), \tag{G.2.13}$$

so that

$$\frac{\mathrm{d}}{\mathrm{d}t}\left(\frac{\partial \mathcal{L}}{\partial \dot{q}_k}\right) = \frac{\mathrm{d}}{\mathrm{d}t}\left(\frac{\partial T}{\partial \dot{q}_k}\right) \tag{G.2.14}$$

and

$$\frac{\partial \mathcal{L}}{\partial q_k} = \frac{\partial T}{\partial q_k} - \frac{\partial V}{\partial q_k} = \frac{\partial T}{\partial q_k} + \mathcal{F}_k \qquad (G.2.15)$$

are the equations that govern the Lagrangian function $\mathcal{L}$. These equations allow us to write Eq. (G.2.12) in the form

$$\frac{\mathrm{d}}{\mathrm{d}t}\left(\frac{\partial \mathcal{L}}{\partial \dot{q}_k}\right) - \frac{\partial \mathcal{L}}{\partial q_k} = 0 , \qquad (k = 1, \cdots, 3N) . \qquad (G.2.16)$$

This is known as the Lagrangian form of the dynamical equations. Why do we consider these equations? Basically, we do so because they have the same form for any system of generalized coordinates, and hence they represent a means of obtaining equations of motion (for a system) that are independent of the coordinate system employed to describe the system and its components. As an aside, we note that this formulation is especially important for developing the general theory of relativity (in which Cartesian coordinates may not actually exist). We note also that if frictional forces are involved, it turns out not to be possible to write the equation of motion in the form of Eq. (G.2.16), so that the Lagrangian formulation is not as useful for some engineering applications as it is in physics. However, this formulation is still quite useful especially when electromagnetic forces and velocity-dependent potentials are involved.

G.2.3   Constants of the Motion and Ignorable Coordinates

Constants of the motion are functions of the coordinates and velocities that are constant in time. If the dynamical system can be characterized by a Lagrangian function in which a particular coordinate, say q_k, does not appear explicitly, then its Lagrange equation reduces to

$$\frac{\mathrm{d}}{\mathrm{d}t}\left(\frac{\partial \mathcal{L}}{\partial \dot{q}_k}\right) = 0 .$$

If we couple this result with $\dfrac{\partial L}{\partial \dot{q}_k} = \dfrac{\partial T}{\partial \dot{q}_k}$, as V does not depend upon $\dot{q}_k$, and the equation of motion, Eq. (G.2.7), we find that

$$\frac{\mathrm{d}p_k}{\mathrm{d}t} = \frac{\mathrm{d}}{\mathrm{d}t}\left(\frac{\partial T}{\partial \dot{q}_k}\right) = \frac{\mathrm{d}}{\mathrm{d}t}\left(\frac{\partial \mathcal{L}}{\partial \dot{q}_i}\right) = 0 ,$$

so that

$$p_k = \left(\frac{\partial \mathcal{L}}{\partial \dot{q}_k}\right) = \text{const.}$$

The absence of explicit dependence of the Lagrangian function on a coordinate q_k results in the conjugate momentum p_k being a constant of the motion. We then say that q_k is an *ignorable* coordinate.

G.3 Hamilton's Equations

For mechanical systems obeying the Lagrange equation, Eq. (G.2.16), with $\mathcal{L}$ a function of the generalized coordinates $\{q_k\}$, generalized velocities $\{\dot{q}_k\}$, and possibly time t, the state of the mechanical system at any time t is specified by specifying the generalized coordinates and velocities $\{q_k, \dot{q}_k\}$. The Lagrange equations connect the generalized accelerations $\{\ddot{q}_k\}$ to the generalized coordinates and velocities. We could, however, equally well specify the state of the system by giving the generalized coordinates $\{q_k\}$ and the corresponding momenta $\{p_k\}$ defined via

$$p_k = \frac{\partial \mathcal{L}}{\partial \dot{q}_k}, \qquad (k = 1, \cdots, 3N). \tag{G.3.1}$$

Equation (G.3.1) can be solved in principle for the $\{\dot{q}_k\}$ in terms of the $\{q_k\}$ and $\{p_k\}$.

Let us write the total differential $d\mathcal{L}$ of $\mathcal{L}$ as

$$d\mathcal{L} = \sum_{j=1}^{3N} \left(\frac{\partial \mathcal{L}}{\partial \dot{q}_j} d\dot{q}_j + \frac{\partial \mathcal{L}}{\partial q_j} dq_j \right) + \frac{\partial \mathcal{L}}{\partial t} dt. \tag{G.3.2}$$

The partial derivative in the first term can be replaced by p_j via Eq. (G.3.1), while a combination of Eqs. (G.2.16) and (G.3.1) leads to the partial derivative in the second term being $\dot{p}_j$, so that the total differential $d\mathcal{L}$ can be written in the form

$$d\mathcal{L} = \sum_{j=1}^{3N} (p_j d\dot{q}_j + \dot{p}_j dq_j) + \frac{\partial \mathcal{L}}{\partial t} dt. \tag{G.3.3}$$

If we now define a function $\mathcal{H}(q_1, \cdots, q_{3N}; p_1, \cdots, p_{3N}; t)$ via

$$\mathcal{H} \equiv \sum_{j=1}^{3N} p_j \dot{q}_j - \mathcal{L}, \tag{G.3.4}$$

in which the generalized velocities $\dot{q}_j$ are expressed in terms of coordinates and momenta, then the total differential $d\mathcal{H}$ can be expressed as

$$d\mathcal{H} = \sum_{j=1}^{3N}(\dot{q}_j\,dp_j + p_j\,d\dot{q}_j) - d\mathcal{L}$$

$$= \sum_{j=1}^{3N}(\dot{q}_j\,dp_j - \dot{p}_j\,dq_j) - \frac{\partial\mathcal{L}}{\partial t}\,dt\,. \qquad\qquad (G.3.5)$$

Notice that the function $\mathcal{H}(q_1,\cdots,q_{3N};\dot{q}_1,\cdots,\dot{q}_{3N};t)$ has been defined in Eq. (G.3.4) so that the total differential $d\mathcal{H}$ depends upon the differentials $\{dp_j\}$, $\{dq_j\}$, and dt explicitly: this structure for $d\mathcal{H}$ allows us to see that

$$\dot{q}_k = \frac{\partial\mathcal{H}}{\partial p_k}\,, \quad \dot{p}_k = -\frac{\partial\mathcal{H}}{\partial q_k}\,, \qquad (k = 1,\cdots,3N)\,, \qquad\qquad (G.3.6)$$

and

$$\frac{\partial\mathcal{H}}{\partial t} = -\frac{\partial\mathcal{L}}{\partial t}\,. \qquad\qquad (G.3.7)$$

Expression (G.3.4) may be considered as an example of a Legendre transformation (see Appendix B, Sect. B.2), similar to what is done in thermodynamics.

Equations (G.3.6) and (G.3.7) are known as Hamilton's equations of motion and $\mathcal{H}$ is called the classical Hamiltonian for a mechanical system, expressing $\dot{q}_k$ and $\dot{p}_k$ in terms of the generalized coordinates and momenta. When V is a function only of the coordinates, then for a stationary coordinate system, $\mathcal{H}$ is just the total energy expressed in terms of coordinates and momenta. If $\mathcal{L}$ does not contain time explicitly, then neither does $\mathcal{H}$, and $\mathcal{H}$ itself becomes a constant of the motion for the system (conservation of energy), as will be shown below.

As Hamilton's equations are simply a reformulation of Newton's equations of motion, they reduce for simple cases to the equations that could have been written immediately from Newton's laws. Why do we then concern ourselves with them? Firstly, they provide a useful starting point for setting up the laws of classical statistical mechanics (and of quantum mechanics too!). Secondly, many methods for solving these equations have been developed, such as the classical trajectory calculations for complex molecular systems that are employed in a number of modern molecular dynamics simulation packages. Hamilton's equations are valid not only for any set of generalized coordinates together with their corresponding momenta (referred to as the conjugate momenta) defined by Eq. (G.3.1), they are also valid for a much wider class of coordinate systems in which the coordinates and momenta can be expressed as functions of the original coordinates and momenta. This is discussed in greater detail in the standard textbook on classical mechanics by Goldstein.[2]

---

[2]H. Goldstein, *Classical Mechanics*, 2nd Edition (Addison-Wesley, Reading, MA, 1980).

G.3.1 The Hamiltonian and Energy Conservation

We may obtain an expression for the time derivative of a Lagrangian that does not depend upon time explicitly (i.e., a *conservative* system) from Eq. (G.3.2), namely,

$$\frac{d\mathcal{L}}{dt} = \sum_k \left(\frac{\partial \mathcal{L}}{\partial q_k} \dot{q}_k + \frac{\partial \mathcal{L}}{\partial \dot{q}_k} \frac{d\dot{q}_k}{dt} \right).$$

If we utilize Lagrange's equation to replace the factor $\dfrac{\partial \mathcal{L}}{\partial q_k}$ in the first term in this expression, we obtain

$$\frac{d\mathcal{L}}{dt} = \sum_k \left[\frac{d}{dt} \left(\frac{\partial \mathcal{L}}{\partial \dot{q}_k} \right) \dot{q}_k + \left(\frac{\partial \mathcal{L}}{\partial \dot{q}_k} \right) \frac{d\dot{q}_k}{dt} \right]$$

$$= \sum_k \frac{d}{dt} \left(\frac{\partial \mathcal{L}}{\partial \dot{q}_k} \dot{q}_k \right),$$

in which the final step is achieved by employing the product rule in reverse. We may now rearrange the terms in this result to obtain

$$\frac{d}{dt} \left[\sum_k \left(\frac{\partial \mathcal{L}}{\partial \dot{q}_k} \dot{q}_k \right) - \mathcal{L} \right] \equiv \frac{d\mathcal{H}}{dt} = 0$$

which, since the Hamiltonian is synonymous with the total energy in a conservative system, is an expression of the conservation of energy for a conservative system.

G.3.2 Phase Space

If a dynamical quantity, represented as a function of $(\mathbf{q}, \mathbf{p}, t)$, remains constant as the motion of a system unfolds in time, it is called a 'constant of the motion' or an 'integral of the system'. We have just seen that the Hamiltonian itself is a constant of the motion for a conservative system.

From Hamilton's equations, we obtain

$$\dot{p}_k = -\frac{\partial \mathcal{H}}{\partial q_k} \quad \Rightarrow \quad p_k(t) = p_k(0) - \int_0^t \frac{\partial \mathcal{H}}{\partial q_k} \, dt \,,$$

and

$$\dot{q}_k = \frac{\partial \mathcal{H}}{\partial p_k} \quad \Rightarrow \quad q_k(t) = q_k(0) + \int_0^t \frac{\partial \mathcal{H}}{\partial p_k} \, dt \,.$$

We can see from these two results that there will be $2N$ integration constants, corresponding to the set of values $\{q_k(0), p_k(0)\}$ at some time $t_0 = 0$.

Γ-space: for a system consisting of N molecules, each of which has s degrees of freedom, we have a total of $2sN$ generalized coordinates (q_k, p_k). We *define* a $2N$-dimensional Cartesian space whose axes are the $\{q_k, p_k\}$ variables, so that the state of an N-molecule system at any given instant of time, specified by the coordinates $(q_1, \cdots, q_N, p_1, \cdots, p_N)$ is represented by a single point in this $2N$-dimensional Cartesian space. This Cartesian space is called the *phase space* of the system or, for short, Γ-space. If we add time explicitly to our considerations, the augmented $(2N + 1)$-dimensional Cartesian space is called the $\widetilde{\Gamma}$-space.

G.3.3   Hamilton's Principle and Generating Functions

It is often useful to be able to transform from one set of coordinates $\{q_i\}$ to a new set of coordinates $\{Q_i\}$ in which the equations of motion can be simplified and a clean solution obtained. Of course, the new coordinates $\mathbf{Q}$ will be functions of the old coordinates $\mathbf{q}$ and t: i.e., $Q_i = Q_i(\mathbf{q}, t)$. When the new coordinates $\mathbf{Q}$ are functions only of the coordinates $\mathbf{q}$ and t, such a transformation is called a *point transformation*. However, in Hamiltonian mechanics the generalized momenta are also independent variables on the same level as the generalized coordinates, so that we must broaden our definition of the transformation to

$$Q_i \equiv Q_i(\mathbf{q}, \mathbf{p}, t) \,,$$

$$P_i \equiv P_i(\mathbf{q}, \mathbf{p}, t) \,,$$

in which both the old coordinates and the old momenta are involved. Only those transformations for which the new coordinates $\mathbf{Q}, \mathbf{P}$ are canonical coordinates will be of interest to us. The designation 'canonical' implies that we can find a function $\mathcal{K}(\mathbf{Q}, \mathbf{P}, t)$ that has the properties

$$\dot{Q}_k = \frac{\partial \mathcal{K}}{\partial P_k} \,, \quad \dot{P}_k = -\frac{\partial \mathcal{K}}{\partial Q_k} \,, \tag{G.3.8}$$

representing Hamilton's equations in the new coordinates, with Hamiltonian $\mathcal{K}$.

The real problem comes down to how do we find such transformations? It turns out that there are four basic functions that lead to the types of transformation that we seek. They are known as generating functions of type 1 through 4, and are designated by $F_1(\mathbf{q}, \mathbf{Q}, t)$, $F_2(\mathbf{q}, \mathbf{P}, t)$, $F_3(\mathbf{p}, \mathbf{Q}, t)$, $F_4(\mathbf{p}, \mathbf{P}, t)$. Which generating function is needed depends upon the details of the specific problem being considered. However, once F has been specified, the canonical transformation equations are themselves also completely specified: F will be a function of both the old and the new variables (and time), but only half of these variables will be independent because of the transformation equations (G.3.8). The origin of these generating functions for canonical transformations lies in Hamilton's Principle.

Hamilton's principle states that the motion of a system from time t_1 to time t_2 is such that the line integral $I = \int_{t_1}^{t_2} \mathcal{L}\, dt$, with $\mathcal{L} \equiv T - V$, is an extremum (normally a minimum) for the path of motion. In other words, any variation of the line integral I for fixed t_1 and t_2 must vanish: this requirement is expressed as

$$\delta I = \delta \int_{t_1}^{t_2} \mathcal{L}(\mathbf{q}, \dot{\mathbf{q}}, t)\, dt = 0\,.$$

The expression for δI is known as a variational principle, and comes out of a field of mathematics known as the calculus of variations, highlighted briefly in Appendix B. Hamilton's Principle must apply both to the old set of generalized coordinates $\{\mathbf{q}, \mathbf{p}, t\}$ with its corresponding Hamiltonian function $\mathcal{H}$, and to the new set of generalized coordinates $\{\mathbf{Q}, \mathbf{P}, t\}$ and its corresponding Hamiltonian $\mathcal{K}$. Thus, we have two conditions,

$$\delta \int_{t_1}^{t_2} \left[\sum_i P_i \dot{Q}_i - \mathcal{K}(\mathbf{Q}, \mathbf{P}, t) \right] dt = 0 \tag{G.3.9}$$

and

$$\delta \int_{t_1}^{t_2} \left[\sum_i p_i \dot{q}_i - \mathcal{H}(\mathbf{q}, \mathbf{p}, t) \right] dt = 0 \tag{G.3.10}$$

that must be satisfied simultaneously. This means that the two integrands can differ at most by a total time derivative of an arbitrary function $F(t)$ that is constrained so that

$$\int_{t_1}^{t_2} \frac{dF}{dt}\, dt = F(t_2) - F(t_1)\,,$$

which automatically vanishes because the variation vanishes by construction at the end-points of the integral.

The integrands (G.3.9) and (G.3.10) can be connected by the relation

$$\sum_k p_k \dot{q}_k - \mathcal{H} = \sum_k P_k \dot{Q}_k - \mathcal{K} + \frac{d}{dt} F_1(\mathbf{q}, \mathbf{Q}, t)\,, \tag{G.3.11}$$

with the total time derivative of F_1 given by

$$\frac{dF_1}{dt} = \sum_k \frac{\partial F_1}{\partial q_k} \dot{q}_k + \sum_k \frac{\partial F_1}{\partial Q_k} \dot{Q}_k + \frac{\partial F_1}{\partial t}\,.$$

This last result enables us to rewrite Eq. (G.3.11) as

$$\sum_k \left(p_k - \frac{\partial F_1}{\partial q_k} \right) \dot{q}_k - \sum_k \left(P_k + \frac{\partial F_1}{\partial Q_k} \right) \dot{Q}_k - \left(\mathcal{H} + \frac{\partial F_1}{\partial t} \right) = -\mathcal{K} . \quad \text{(G.3.12)}$$

Now, because both the q_k and Q_k coordinates are considered herein to be independent variables, Eq. (G.3.12) can hold identically only if the coefficients of $\dot{q}_k$ and $\dot{Q}_k$ vanish separately, which gives

$$p_k = \frac{\partial F_1}{\partial q_k} , \quad P_k = -\frac{\partial F_1}{\partial Q_k} , \quad \mathcal{K} = \mathcal{H} + \frac{\partial F_1}{\partial t} . \quad \text{(G.3.13)}$$

The equations for the p_k involve only p_k, q_k, Q_k, and t: therefore, we can solve for the Q_k in terms of the p_k, q_k, and t. Once this has been done, the results thereby obtained for the Q_k in terms of the q_k, p_k, and t may be employed to obtain the P_k in terms of the q_k, p_k and t. The second step is often considerably more difficult to achieve than is the first step in this process. We shall only illustrate the workings of this procedure via the following example.

Example The 1-D Simple Harmonic Oscillator.

To see how the above procedure works, let us examine how it can be applied to the description of the well-known one-dimensional simple harmonic oscillator (SHO) typically studied in first year Physics courses.

For the 1-D SHO, we may treat the linear momentum p and the position x as generalized coordinates p, q, with the Hamiltonian representing the total energy as

$$\mathcal{H} = \frac{p^2}{2m} + \tfrac{1}{2}kq^2 .$$

We shall identify the ratio of the force constant k and the SHO mass m as ω^2, the square of the radial frequency of the SHO, as is traditionally done in treatments of this problem. This allows us to rewrite the Hamiltonian as

$$\mathcal{H} = \frac{1}{2m}(p^2 + m^2\omega^2 q^2)$$

in terms of the generalized coordinates p, q. Because $\mathcal{H}$ consists of the sum of two squares, we shall try for a transformation in which $\mathcal{H}$ is cyclic in the new coordinates: to accomplish this, we shall seek a canonical transformation of the form

$$p = f(P)\cos Q , \quad q = \frac{f(P)}{m\omega}\sin Q .$$

The reason for introducing the sine and cosine functions is so that we can utilize the relation $\sin^2 Q + \cos^2 Q = 1$. This proposed transformation then gives rise to the transformed Hamiltonian

$$K = \mathcal{H} = \frac{f^2(P)}{2m}(\cos^2 Q + \sin^2 Q) = \frac{f^2(P)}{2m},$$

with the consequence that the new coordinate Q is cyclic. Now we need to determine $f(P)$ so that the transformation is indeed canonical.

We begin by taking the ratio p/q, which we may rewrite in the form

$$p = m\omega q \cot Q,\qquad\qquad (G.3.14)$$

which is independent of P. This result has the form of the first of Eqs. (G.3.13) for the F_1-type generating function, namely,

$$p = \frac{\partial F_1(q, Q)}{\partial q}.$$

The simplest function $F_1(q, Q)$ that will give the result (G.3.14) is

$$F_1(q, Q) = \frac{m\omega q^2}{2}\cot Q.$$

From the second of Eqs. (G.3.13), we can determine P as

$$P = -\frac{\partial F_1(q, Q)}{\partial Q} = \frac{m\omega q^2}{2\sin^2 Q}.\qquad\qquad (G.3.15)$$

We may solve Eq. (G.3.15) to obtain q as a function of P and Q:

$$q = \sqrt{\frac{2P}{m\omega}}\sin Q.\qquad\qquad (G.3.16)$$

If we substitute this result into Eq. (G.3.14) for p, we find that

$$p = m\omega\sqrt{\frac{2P}{m\omega}}\sin Q \cot Q = \sqrt{2Pm\omega}\cos Q \equiv f(P)\cos Q,$$

from which we obtain $f(P)$ as

$$f(P) = \sqrt{2m\omega P},\qquad\qquad (G.3.17)$$

while the Hamiltonian $\mathcal{K}$ takes the form

$$\mathcal{H} \equiv \mathcal{K} = \frac{f^2(P)}{2m} = \omega P.\qquad\qquad (G.3.18)$$

Now we note that because the Hamiltonian is cyclic in Q, its conjugate momentum P must be a constant which, when written in the form

$$P = \frac{E}{\omega},$$

allows us to determine the equation of motion for Q as

$$\dot{Q} = \frac{\partial H}{\partial P} = \omega.$$

We may now solve this equation for $Q(t)$ to obtain

$$Q(t) = \omega t + \delta.$$

The quantity δ (also called the phase factor) is a constant of integration determined by the initial conditions. We may now put all of this information together to obtain $q(t)$ as

$$q(t) = \sqrt{\frac{2E}{m\omega^2}} \, \sin(\omega t + \delta). \qquad \text{(G.3.19)}$$

$\square$

Note that if we set up the phase space representation of a 1-D SHO, we have

$$p(t) = \sqrt{2Pm\omega} \cos Q = \sqrt{2mE} \, \cos Q = \sqrt{2mE} \cos(\omega t + \delta)$$
$$\equiv b \cos(\omega t + \delta),$$

$$q(t) = \sqrt{\frac{2E}{m\omega^2}} \, \sin(\omega t + \delta) \equiv a \sin(\omega t + \delta),$$

from which we may write equivalently,

$$\left(\frac{q}{a}\right)^2 + \left(\frac{p}{b}\right)^2 = 1.$$

Thus, the state of a 1-D SHO at time t lies somewhere on an ellipse in Γ-space, so that the phase space trajectory for the 1-D SHO forms an ellipse in the two-dimensional phase space associated with its motion: in the three-dimensional time-augmented phase space $\widetilde{\Gamma}$, the trajectory for 1-D SHM is a spiral along the t-axis, with a projection onto the 2D phase space Γ which is the ellipse appearing in Fig. G.2.

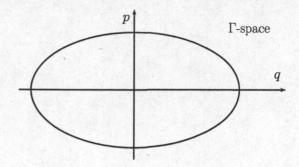

Fig. G.2 A typical phase space trajectory for the one-dimensional simple harmonic oscillator

G.3.4 Equation of Motion for a Dynamical Variable: Poisson Brackets

If $u(p, q)$ is a dynamical variable in Hamiltonian mechanics, its equation of motion is

$$\frac{du}{dt} = \sum_k \left(\frac{\partial u}{\partial q_k} \dot{q}_k + \frac{\partial u}{\partial p_k} \dot{p}_k \right) + \frac{\partial u}{\partial t} .$$

If we employ Hamilton's equations, we may convert this expression into

$$\frac{du}{dt} - \sum_k \left(\frac{\partial u}{\partial q_k} \frac{\partial \mathcal{H}}{\partial p_k} \frac{\partial \mathcal{H}}{\partial q_k} \frac{\partial u}{\partial p_k} \right) + \frac{\partial u}{\partial t} \tag{G.3.20}$$

Let us now define a quantity, referred to as the Poisson bracket between two dynamical quantities A and B as

$$[A, B] \equiv \sum_k \left(\frac{\partial A}{\partial q_k} \frac{\partial B}{\partial p_k} - \frac{\partial B}{\partial q_k} \frac{\partial A}{\partial p_k} \right) ,$$

in terms of which Eq. (G.3.18) can be obtained as

$$\frac{du}{dt} = [u, \mathcal{H}] + \frac{\partial u}{\partial t} . \tag{G.3.21}$$

The classical Poisson bracket is important in the sense that it corresponds to the commutator operator in quantum mechanics and, as such, it provides a means for making comparisons between quantum and classical mechanical expressions.

Index

© Springer Nature Switzerland AG 2021
F. R. W. McCourt, *Statistical Thermodynamics for Pure and Applied Sciences*,
https://doi.org/10.1007/978-3-030-52006-9

Printed in the United States
by Baker & Taylor Publisher Services